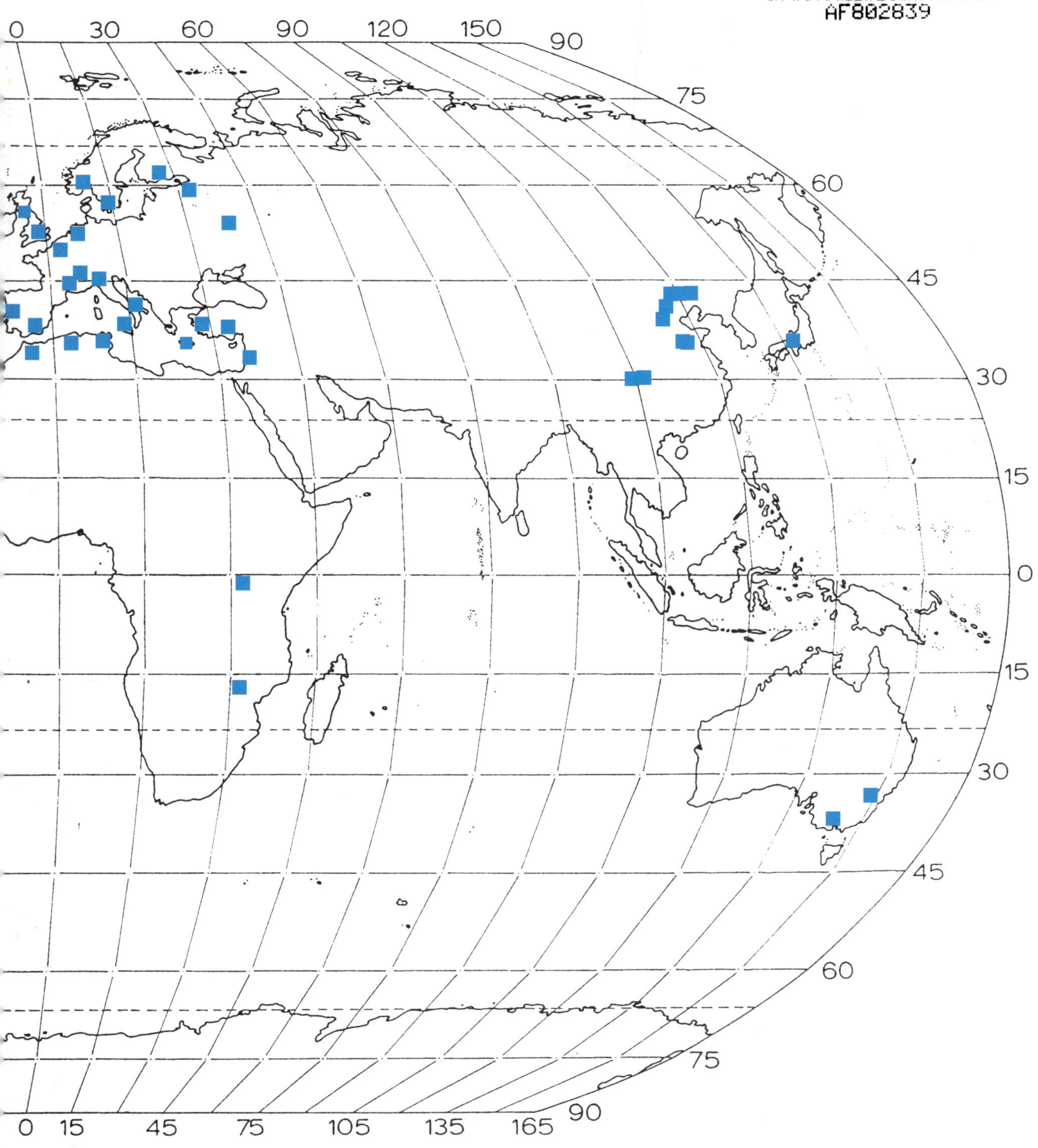

ɪAJOR CONCENTRATIONS OF
ɴAL AREAS OF GREENHOUSES

ːareas developed around large
ɔns with climates unfavorable
ɔre recently greenhouse culti-
ˈvorable transport facilities and
ɔor is available.

ECOSYSTEMS OF THE WORLD 20

GREENHOUSE ECOSYSTEMS

ECOSYSTEMS OF THE WORLD

Editor in Chief:

David W. Goodall

Centre for Ecosystem Studies, Edith Cowan University, Joondalup, W.A. (Australia)

ECOSYSTEMS OF THE WORLD 20

GREENHOUSE ECOSYSTEMS

Edited by

G. Stanhill

Institute of Soils, Water and Environmental Sciences,
Agricultural Research Station,
The Volcani Center,
P.O. Box 6,
Bet Dagan 50250,
Israel

H. Zvi Enoch

Institute of Soils, Water and Environmental Sciences,
Agricultural Research Station,
The Volcani Center,
P.O. Box 6,
Bet Dagan 50250,
Israel

ELSEVIER
Amsterdam – Boston – London – New York – Oxford – Paris – San Diego
San Francisco – Singapore – Sydney – Tokyo

ELSEVIER SCIENCE B.V.
Sara Burgerhartstraat 25
P.O. Box 211, 1000 AE Amsterdam, The Netherlands

First edition 1999
Second Impression 2002

Library of Congress Cataloging –in-Publication Data

Greenhouse ecosystems / edited by G. Stanhill, H. Zvi Enoch.
p. cm. -- (Ecosystems of the world ; 20)
includes bibliographical references and indexes.
ISBN 0-444-88267-7
1. Greenhouse plants-- Ecology. 2. Greenhouses. 3. Greenhouse management. I. Stanhill, G. II. Enoch, Herbert Z. III. Series.
SB415.G7 1998
635.9'823-- dc21 98-20251
CIP

ISBN: 0 444 88267 7

The paper used in this publication meets the requirements of ANSI/NISO Z39.48-1992 (Permanence of Paper).
Printed in The Netherlands.

PREFACE

The greenhouse agroecosysten is defined by its boundaries, as are all systems; as an ecosystem it is additionally defined by its constituent biota, and as an agroecosysten by the purposeful management required to repay its very high capital and management costs.

The structure of this volume follows the same order. The first section is devoted to the greenhouse itself with four chapters dealing with its history and current geography, the mechanical features of the rigid, perennial structures clad with light-transmitting materials, the effect of these structures on the physical processes which modify the above-ground environment, and the root-zone environment.

The second section consists of six chapters devoted to the biological components of the greenhouse ecosystem. These describe the most widely-grown edible, flower and foliage crops that are cultivated, the disease and insect pests which exploit them and the human inhabitants which manage both crops and pests.

The third section consists of four chapters each devoted to the fluxes of energy, carbon, water and nutrients within and across the greenhouse borders.

The last section comprises two chapters dealing with the management of the greenhouse and describe models used to estimate the potential yields and search for optimum management practices.

Finally, we wish to thank the many persons involved in the production of this volume. First and foremost, those timely authors who have accepted the numerous delays to the publication of their work, and also the many reviewers whose anonymous contributions are gratefully acknowledged.

G. Stanhill
H.Z. Enoch

LIST OF CONTRIBUTORS

A.M. ARMITAGE
Department of Horticulture
The University of Georgia
111 Plant Science Bldg
Athens, GA 30602-7273, USA

B.J. BAILEY
Process Engineering Division
Silsoe Research Institute
Wrest Park, Silsoe, Beds MK45 4HS, UK

A. BAILLE
INRA, Unité de Bioclimatologie
Domaine Saint Paul
84914 Avignon Cedex 9, France

M. BAKKER
Department of Horticulture
Wageningen Agricultural University
Haagsteeg 3, 6708 PM Wageningen, Netherlands

H. CHALLA
Department of Horticulture
Wageningen Agricultural University
Haagsteeg 3, 6708 PM Wageningen, Netherlands

C.A. CONOVER
Institute of Food and Agricultural Sciences
2807 Binion Road
Apopka, FL 32703, USA

S. DASBERG
Institute of Soils and Water
ARO, The Volcani Center
Bet Dagan 50250
Israel

W. DAY
Process Engineering Division
Silsoe Research Institute
Wrest Park, Silsoe, Beds MK45 4HS, UK

H.Z. ENOCH
Institute of Soils, Water and Environmental Sciences
Agricultural Research Station
The Volcani Center
P.O. Box 6
Bet Dagan 50250
Israel

Y. ENOCH
Department of Sociology and Political Science
The Open University of Israel
P.O. Box 39328
Ramat Aviv
Tel Aviv 61392
Israel

Y. ELAD
Department of Plant Pathology
Institute of Plant Protection
ARO, The Volcani Center
P.O. Box 6
Bet-Dagan 50250
Israel

C. GARY
INRA, Unité de Bioclimatologie
Domaine St.-Paul
Site Agroparc
84914 Avignon Cedex 9, France

J.H. GROENEWEGEN
Voorstraat 14/9
2685 EM Poeldijk, Netherlands

O. JOLLIET
IATE
Swiss Federal Institute of Technology
1015 Lausanne, Switzerland

P. LUNDQVIST
Division of Work Science and Technology
Department of Agricultural Biosystems & Technology
Swedish University of Agricultural Sciences
Box 88
230 53 Alnarp, Sweden

T.A. NELL
Institute of Food and Agricultural Sciences
2807 Binion Road
Apopka, FL 32703, USA

M.P. PARRELLA
Department of Entomology
University of California
Davis, CA 95616, USA

R.T. POOLE
Institute of Food and Agricultural Sciences
2807 Binion Road
Apopka, FL 32703, USA

I. SEGINER
Department of Agricultural Engineering
Technion, Haifa 32000
Israel

I. SOCOLOVSKI
Department of Horticulture
The University of Georgia
111 Plant Science Bldg
Athens, GA 30602-7273, USA

G. STANHILL
Institute of Soils, Water and Environmental Sciences
Agricultural Research Station
The Volcani Center
P.O. Box 6
Bet Dagan 50250
Israel

Chr. von ZABELTITZ
Institut für Technik in Gartenbau und Landwirtschaft
Universität Hannover
Herrenhauser Strasse 2
3000 Hannover, Germany

CONTENTS

Chapter 1

THE HISTORY AND GEOGRAPHY OF THE GREENHOUSE

H. Zvi ENOCH and Yael ENOCH

INTRODUCTION

This chapter briefly describes the development of the greenhouse from antiquity to the present. This includes a discussion of protected cropping in cold- and hot-frames, glass cloches, lean-to houses, winter gardens, orangeries, conservatories and glasshouses in their varied forms, and plastic-clad plant enclosures of all sizes ranging from low tunnels, walk-in tunnels, to large plastic houses.

For the purpose of this chapter, the greenhouse is defined as a transparent enclosure designed to grow or temporarily protect plants. Thus the definition does not include enclosures for growing mushrooms or tissue cultures.

In spite of the large literature on greenhouse cultivation (currently at least 28 books in English, 44 in German and 23 in French, are in print), there are relatively few works on the history of greenhouses, and these are generally on very selective aspects (Lenz, 1859; Fawkes, 1881; Andersen, 1891; Bouché and Bouché, 1895; Ravenscroft, 1901; Hasluck, 1907; Rowles, 1910; Feldhaus, 1931; Lawrence, 1948; Willmott, 1958; Spice, 1959; Lemmon, 1962; Paisley, 1962; Walls, 1973; Avery, 1974; Martincigh, 1980; Koppelkamm, 1982; Greene, 1983; Lipa, 1985; Kingman, 1986; Kohlmaier and von Sartory, 1986; Falland, 1987; Vickman, 1987; Woods and Warren, 1988; Brent, 1991; Geytenbeek, 1991; Marston, 1992; Saudan-Skira and Saudan, 1994; Jacobs, 1995; Meijgaard, 1995). A notable exception is the exhaustive book on the history of glasshouse architecture by Hix (1996), which includes a large number of illustrations.

A truly comprehensive manuscript by E.W.B. van den Muijzenberg was written in Dutch, and an English edition prepared for publication in 1980 in cooperation with Dr. Peter Newton, University of Manchester, U.K. The English manuscript "A history of greenhouses" (454 typed pages) with about 800 references, was distributed in a preliminary edition of 150 copies, by the Institute of Agricultural Engineering, Wageningen, the Netherlands. Unfortunately, this manuscript was never published. Much of the information in this chapter is based on van den Muijzenberg's manuscript.

Access to the earlier literature was kindly provided by the libraries of the Royal Botanical Garden, Kew, England, The Royal Horticultural Society, London, and The Royal Veterinary and Agricultural University, Copenhagen, Denmark. We appreciate the help offered by the above-mentioned institutions, and by Prof. H. Challa who provided a photocopy of van den Muijzenberg's manuscript.

FROM ANTIQUITY TO THE MIDDLE AGES

In antiquity, there were probably no greenhouses in the sense defined here, but there were various practices adopted to provide out-of-season flowers and vegetables. Thus one finds quoted in the Midrash (rabbinical comments and explanations to the Bible) to Ecclesiastes in the name of Rabbi Hanna Bar Hanina: "that there was never anything missing from the table of King Solomon, neither roses during the heat of summer nor squash [zucchini – *Cucurbita pepo*] during the rainy season [i.e. winter]". King Solomon lived in Jerusalem three millennia ago.

The cultivation of out-of-season plants was intensely studied in ancient Athens and Rome. This can be seen for instance in the work of Theophrastus [who lived from 372 to 287 BCE (before the common era)]. He mentioned moving plants into the peristyle (verandah in a closed courtyard of a Greek house) at night and heating the soil by mixing it with decaying manure or compost. Pedanius Dioscorides [first century, CE (Common Era)], described in an illustrated book about 600 different cultivated plants, some of which were grown out of season. Gaius Plinius Secundus ("Pliny the Elder") compiled from 2000 works, representing 146 Roman and 327 Greek authors, a large "Encyclopedia Historia Naturalis" of which Volume 19 treats horticulture (Steere, 1974). He mentions artificially forced asparagus.

Growers in China, Mesopotamia, Egypt, Israel, Greece and Rome grew valuable plants in pots and placed them indoors at night or during cold periods to protect them. Sometimes, plants were grown on wheelbarrows that could easily be brought into a grotto, cellar or room at night (Theophrastus, ~300 BCE; Singer, 1921; Sarton, 1953; Greene, 1983). Chinese plant culture may have included greenhouses that, according to tradition, go back to antiquity (Li, 1957). The traditional Chinese greenhouse consisted of a brick wall oriented East–West. On its southern side, a transparent surface consisting of bamboo sticks covered with oiled paper was placed at an angle of 30–40° to the horizontal. During the day, the brick wall became heated by the sun and at night it released its heat. In addition, at night the paper window was covered by rush mats which acted as thermal insulation. Similar simple greenhouses are in use today, and it is claimed that they increase the night temperature by up to 6°C (N. Zieslin, personal communication, 1995).

Enclosures clad with the transparent mineral mica were used in Roman times, according to Culumella (63 CE) in his book "De Re Rustica". According to archaeologists, a building found in Pompeii, Italy, was a heated greenhouse (Lemmon, 1962). The building was provided with a specularium (stove) and had tiers of masonry for displaying the pot plants. Hot-air flues were built into the walls and the building showed every indication of having possessed a frontage of talc or rough glass (van den Muijzenberg, 1980).

It has been suggested that out-of-season grapes, served at the tables of the rich Romans, were raised in such enclosures clad with transparent material (Gibault, 1898). Wright (1934) reported that "Beginning as a covered pit, plant enclosures later developed into a greenhouse. As sheet glass was not known till the 3rd century, the roofs were covered with thin slabs of talc or mica" (p. 80).

According to Lenz (1859), Gargidius Martialis wrote in the 3rd century about frost protection of tender vegetation, and advocated the use of transparent stones which resist storms and let the sunlight through undisturbed.

One of the driving forces behind the improvement of greenhouse technology was the desire of the rich for out-of-season fruit and flowers. This could be achieved either by protecting plants from cold weather, by placing pot plants indoors at night or during the entire cold season, by growing plants in greenhouses, or by transporting them from a warmer region, where they could be grown in the winter. The Romans used all these solutions. According to Lenz (1859), the Roman author Martialis reported that roses were exported to Rome from Egypt, but that they had to compete with the splendor of locally-grown flowers, which bloomed and gave perfume in profusion. Lenz (1859) also reported the forcing of roses as described by Palladius Rutilius Taurus Aemilianus, about 380 CE: "if one wishes to have roses before the ordinary time, then he digs around the plant at a distance of two palms [of the hand] a trench and pours two times a day warm water into it" (translated from Lenz, 1859, p. 645).

In the period from the end of the Roman Empire to the early Renaissance there are virtually no records of greenhouses in Europe. However, the rich wanted luxuries and there were reports that pomegranates and oranges were eaten at banquets in the Netherlands in the 15th century (van den Muijzenberg, 1980, p. 356). The first step towards greenhouses was the building and use of movable beds with plants. Wheelbarrows with plants were placed inside cellars, grottoes or buildings on cold nights and sometimes heated with charcoal pans. In northern Italy and southern Germany it became common to use wintering sheds, and there may have been a few hectares of such sheds in use for winter protection towards the end of the 15th century (van den Muijzenberg, 1980).

FROM THE MIDDLE AGES TO 1800

The fifteenth century was a period full of changes during which the European world expanded. Long journeys became common and, with them, the transport of seeds and plants. In 1536, the well-known author Rabelais sent seeds from Rome to his friend Geoffrey d'Estissas in France with instructions on how to manure the plants and protect them from frost by using reeds to cover the soil, and deep layers of compost to heat the root medium (van den Muijzenberg, 1980, pp. 37, 48). Wealthy people wanted rare produce and hence employed horticulturists to grow vegetables and flowers in protected cultivation outside their normal season.

Many books on horticulture appeared in the Mediterranean region and were later translated into the languages of northern Europe. A case in point is the book by Charles Estienne (Kaerle Stevens in Dutch) "De Re Hortensi Libellus", which was published in Paris in 1536, and then translated into Dutch 1566. Other books were written in northern Europe, such as that of de Crescentiis (1602).

Vegetables such as cucumbers and gherkins (*Cucumis sativus*) became common food for the rich, as did citrus fruits (Stevens, 1566). Since the plants had to be protected against frost, they were placed indoors to survive the winter. It is known that Cosimo de Medici (1519–1574), Grand Duke of Tuscany, had a temporary heated greenhouse for his orange trees in the mid-sixteenth century (de Bilderling, 1974). Lord Burghley built the first orangery in England and in 1562 imported orange trees from France, which survived the winter, presumably by being sheltered in the "Orange Court", a long room with many, large windows heated in winter by stoves (Lemmon, 1962).

Sir Hugh Platt wrote two books, "The Jewel House of Art and Nature" (1594) and "The Paradise of Flora" (1600). Platt (1594) advocated the use of waste heat when he wrote: "Why not utilize a kitchen fire planting them [i.e. apricots or vines] near a warm wall, or brewers, dyers, soap boilers or refineries of sugar, who have continued fire may easy convey the heat of steam of their fires (which are not utterly lost) into some private room, adjoining where to bestow their fruit trees" (Amherst, 1895, p. 193). The use of waste heat in greenhouses is still an active field of research (Hietala and Ashley, 1984) in the second half of the 20th century.

The interest in botanical gardens was revived in the Renaissance, in Italy and later elsewhere in Europe. Sylvester created the first botanical garden in Salerno, Italy, at the beginning of the 14th century and in 1333 a botanical garden is said to have existed in Venice (de Bilderling, 1974). Casper de Gabriel made a botanical garden in Padua in 1525, Enricus Cordus planted a private one in 1525 in Erfurt, Nordicus one in Cassel (Germany), and Conrad Gesno one in Zürich around 1540 (van den Muijzenberg, 1980).

In 1570, Graf Christof von Württemberg sent 120 species and varieties of plants to August of Saxony, who also received citrus plants from a gardener in Prague. During winter, these plants were covered by a structure (Koch, 1910).

The first university botanical garden was founded in Padua in 1545, and in 1550 it was equipped with a greenhouse (Viridarium) into which plants could be taken in the winter (Lemmon, 1962). The University in Pisa became in 1547 the second university to have a botanical garden, with an orangery with glass windows and possibly heating (Loudon, 1823; Noisette, 1829). Around 1600, stoves, soil warming and, inadvertently, enrichment with carbon dioxide (from decomposition of organic material) were becoming common. Oliver de Serres (also known as Seigneur du Pradel), who lived from 1539 to 1619, published his "Théatre d'Agriculture . . . " in 1600. The book was later banned, and only in 1802–1805 republished by A.M. Grisors. In this work de Serres described a removable wooden shed in Heidelberg (Germany) with great openings (which could be closed) and stoves for heating. He also described the growth of melons in hotbeds with a meter-deep layer of horse manure, thus providing enrichment with carbon dioxide, soil heating and plant nutrition. Some plants were grown in large boxes mounted on wheels so that they could be placed in caves or cellars during frost and placed in the sun on warm days. De Serres further described the use of cloches for growing plants. Cloches were glass bells about 30 cm in diameter (de Serres, 1600; Grisors, 1802–1805).

In the 17th century, orangeries became very fashionable in Europe, gradually replacing the wintering-sheds. The orangery had large windows on its south side in order to allow light to reach plants inside. Instead of many charcoal burners, a few larger stoves would be used to heat the orangeries in the winter. As the name suggests, the favoured plants grown were citrus species. Ferrarius (1646a,b) wrote textbooks on plant growth in greenhouses and elsewhere. He described the use of heating for orchards that were protected by temporary roofs in the winter. He also described various means for heating plants in rooms with large windows (the forerunner of orangeries). Some rooms were heated by the first truly pollution-free method, namely by hanging thick preheated iron plates on the inner walls of the orangery. The plates were heated in the morning and evening in a fireplace in the adjoining room.

Henry IV ordered the building of an orangery for the Tuileries in Paris and one at Fontainebleau. In 1626 Louis XIII ordered the "Jardin des Plantes" to be built in Paris, in which glasshouses were included. Louis XIV, who liked to have the biggest of everything, increased the orangery at Versailles to a length of 152 m and a width of 12 m. With a volume of 6000 m^3, it was the largest orangery in 1670 (Wright, 1934). However, Pieter van der Voort (de la Court), visiting Versailles, mentioned that the orangery had too little light (Driessen, 1928).

The fashion of producing out-of-season vegetables, for the rich, continued (Jacob, 1797). An expert horticulturist such as de la Quintinye, Director for King Louis XIV (of France) of the fruit and kitchen garden, was able to provide the king with fresh asparagus (*Asparagus officinalis*) in December; radishes (*Raphanus sativus*), lettuce (*Lactuca sativa*) and mushrooms (*Agaricus campestris*) in January; cauliflower (*Brassica oleracea* var. *botrytis*) in March; strawberries (*Fragaria* sp.) and peas (*Pisum sativum*) in April; and figs (*Ficus carica*) and melons (*Cucumis melo*) in June. De la Quintinye was a learned man who had studied in Italy, and had written a book, "Le parfait jardinier" (1695).

In the race to produce fruit even earlier, greenhouses with larger panes of glass came into use. The development of lean-to houses with flues may have been due to the need to obtain more light for the plants. Glasshouses began to be built with flues of greater surfaces so that the plants received a smaller heat-flux density than from ovens (van den Muijzenberg, 1980, p. 89).

Glass-covered bark pits became a common installation and were used to produce heat and carbon dioxide for a period of up to six months. Van der Voort (de la Court) (1737, 1750) explained how the bark pit gave off heat by the decomposition of organic material. Van der Voort used glass windows that were 2.44 m high and 0.9 m wide with 72 panes of glass, each measuring 18×13 cm^2. Greenhouses spread to the New World. The first orangery was built in New Amsterdam – now New York – in 1674 by Jacob Beekman; and the first greenhouses with opaque roofs were built in 1719 in Chicago and in Boston between 1710 and 1720 (Taft, 1898, reprinted 1920).

Another reason for building greenhouses with larger glass surfaces came from the desire to grow pineapples, at the time called Queen of Fruits. The fruit needed more light and higher temperatures than could be obtained in an orangery (Loudon, 1822).

Van der Voort (de la Court) (1737, 1750) gave the appropriate values of temperatures for greenhouse cultivation and recommended temperatures corresponding with 15–18°C for forcing vines in January in Northwestern Europe. His book on greenhouse management also made recommendations about building materials. According to van der Voort (1737, 1750) pine-wood, especially that obtained from old trees grown in Norway, was the best, followed by fir, then oak-wood. Van der Voort (1737, 1750) also mentioned that Swedish iron was more liable to rust than coarse-grained German iron and recommended dipping heated iron in oil to protect it. Diel (1798) recommended the cultivation of citrus fruit trees and other plants in clay pots.

The 18th century was a period of awakening experimental science. Boerhaave (1710, 1720) suggested that the glass surface should be perpendicular to the sun rays at noon during winter solstice, when the plants most needed warmth and light. This also appears to be the angle of the glass in the oldest drawing of a lean-to house with a chimney built in the botanical garden in Leiden, the Netherlands, in 1710 (Boerhaave, 1710: title page of "Index Plantarum"). Lean-to glasshouses became the fashion of the 18th century for growing plants needing high light levels. An additional advantage of the design was that the pot plants could be placed on shelves parallel to the glass, so that each pot plant would receive the maximum amount of sunlight without shading the plants on the shelves above or below (Agricola, 1719, 1721; Kennedy, 1777). To improve the light regime in the lean-to houses further, Le Berrugais (Berries) (1788) recommended to whitewash the inside wall with a suspension of lime several times a year.

Glass was fastened to the wooden frames either by lead or by putty made from chalk and oil (van der Voort, 1737, 1750).

However, plant injury caused by heavy metals, such as lead, was observed by Salzman in 1783 on melon plants. Salzman recommended the use of putty, even though it often cracked after a while. This appears to be the first report of heavy metal pollution in plant science.

There also was great interest in energy conservation in glasshouses. Van der Voort (1737, 1750) described several methods of heat conservation in hot-houses used for growing pineapples (*Ananas comosus*). He reported the use of: (1) curtains made of well-tanned Flemish sail-cloth; (2) pieces of dyed Osnabrück linen; (3) a double carpet, made of cows-hair and a reed mat five or six strings high; (4) a similar second reed mat if the weather was very cold; (5) a hair carpet or sometimes a third reed mat; and (6) more hair carpets to accelerate growth of vines.

Research on the direct influence of static electricity on plant growth and development was reported already in 1755 (van den Muijzenberg, 1980). A picture of a "machine for a perpetual electrified garden" from 1755 was reproduced by van den Muijzenberg (1980, fig. 36).

In the 18th century plants were grown in deep beds of horse manure and straw, which supplied the plants with extra carbon dioxide, heat and nutrients. As earlier mentioned, longer-lasting bark pits were sometimes used (shallow troughs in the ground filled with decomposing bark), particularly for the cultivation of pineapple (Miller, 1731; van der Voort, 1737, 1750).

In some cases, the soil was heated by a subterranean furnace and, as early as 1664, Evelyn wrote that at Chelsea, London "very ingenious was the subterranean heat conveyed by a stove under the conservatory which was all vaulted with brick". Flues under the floor were illustrated in the two-volume work by Volkamer (1708) and reproduced by van den Muijzenberg (1980, fig. 37).

The location of the flues changed with time. Initially, flues were built into the back wall, then partly in front of it; after that completely in front of the back wall (Chambers, 1763; Le Berrugais, 1788; Abercrombie, 1792; Robertson, 1798).

Much attention was paid to preventing flue gases from penetrating into the glasshouse. Agricola (1719) mentioned that it was important to smear all the joints of the flue gas system with loam so that no smoke could escape. Agricola further recommended a transverse pipe of iron to promote movement of air up the chimney.

Very early in the development of greenhouses, their observant users made recommendations for temperature regimes suitable for growing tropical plants. Thus, van der Voort (1750) indicated the equivalent of 10ºC as minimum, but 14ºC after January and 39ºC as the temperature threshold for ventilation. To appreciate the revolutionary nature of using such high temperatures, one may note that these temperatures were much higher than was usually maintained at the time in the fire-heated living rooms of wealthy homes, namely about 0ºC if it was −13ºC outdoors and 3ºC during light frost (van den Muijzenberg, 1980). For that reason it was common to wear numerous layers of clothing indoors, a habit that persisted in many regions until this century. Thus, van den Muijzenberg remembered that his mother told him that in 1912 their housemaid in winter wore six underskirts!

According to Pascottet and Dairat (1927), glasshouse heating by the use of circulating hot water appears to have been used by the Swede Martin Triewald in 1718 or 1724 in Newcastle-on-Tyne, England. Water or steam heating was also used in a greenhouse in the so-called Taurian Palace at St. Petersburg (Platt, 1594, 1600; Loudon, 1842; Lemmon, 1962). Moll (1829) wrote that water heating was invented by the French physician Bonnemain in 1777, but that he earned so little from his invention that the French government was asked in 1829, when Bonnemain was 80 years old, to save him from poverty.

THE NINETEENTH CENTURY

Industrialization in the nineteenth century also brought rapid developments to greenhouse technology.

From being a pastime of the rich, commercial horticulture began to develop around the large towns in Europe and the United States, largely based on the use of so-called "garden frames" or "cold frames" and "hot beds" for vegetable production (Fig. 1.1).

In the same period large glasshouses were built (Fig. 1.2), not only in the gardens of the rich, but also for public winter gardens, sometimes devoted to palm-tree cultivation.

Horse manure was a very important raw material for vegetable and grape production in hot beds (Fig. 1.3) and under cloches (Stanhill, 1977). According to Bixio (1847), one million cart-loads of horse manure, worth 14 million francs, were sold by horse owners to vegetable growers near Paris. The horse manure mixed with straw was a good source of heat, nutrients and

Fig. 1.1. Garden frames (2.5×1.9 m) often used for growing vegetables in a 30–50 cm layer of compost (hot frames) or in soil (cold frames) (Fawkes, 1881, fig. 49).

Fig. 1.2. Large greenhouse from the 19th century (Fawkes, 1881, p. 59).

carbon dioxide for plants grown under cloches or glass frames, and in glasshouses.

In the first part of the 19th century, most greenhouses were simple constructions (Fig. 1.4). Typically the greenhouse roof had a slope of 45° and the building was heated by the decomposition of organic materials or by stoves, or with steam circulating in metal pipes. Lean-to houses were often built of iron (according to the French encyclopedia, Petit Larousse, 1914).

Moll (1829) described the advantages of hot-water heating systems over greenhouses heated by flues because the latter nearly always produces a sultry, dry and unpleasant atmosphere. Often flues became leaky, cracked and smoked.

Loudon (1825) recommended building heating flues so that they are free of the wall, and as thin as possible. Moreover, the inside should be sealed with lime-putty, instead of plastering. The heating capacity should be about 100 cubic meters per flue.

According to Loudon (1825), Mackenzie in 1815 considered the optimal greenhouse shape to be a hemisphere. In his view this form of glass roof was best suited for the admission of the sun's rays. He also considered it to be the ultimate in regard to the principle and perfection of form.

Two-sided glasshouses and larger panes were the greatest new developments in the 19th century. The first two-sided glasshouse was designed in 1806 by Tod (Tod, 1823). Tod also used counterweighted vents, as well as overlapping glass with lead strips on panes of $40 \times 25\ cm^2$ or $22 \times 18\ cm^2$.

Loudon (1823, 1825) suggested building large, high glasshouses with a continuous north–south ridge-and-furrow design, and octagonal and hexagonal panes of glass resting on cast-iron supports, which could also serve as gutters for removal of rainwater. The supporting columns were to be 30–45 m high, so that the highest trees could be cultivated.

The greenhouses were often made to impress. Thus, Joseph Paxton built a gigantic glasshouse, in 1836–41, with a ground plan of $3444\ m^2$, a basilical cross-section with ridge-and-furrow glass roofs (Coats, 1968). Less wealthy people also wanted to make glasshouses look impressive (see Fig. 1.2). In 1851, Paxton built the "Crystal Palace" in London, as it was later dubbed. The building was designed in modular units; for example, only two different lengths of girders were used. Repetitive patterns throughout made it possible to use prefabricated components. The Crystal Palace was the showpiece of the Great Exhibition (van den

Fig. 1.3. Grape production in ground vinery (Fawkes, 1881, p. 76).

Fig. 1.4. Pit house or 'Bijou greenhouse' (Fawkes, 1881, p. 75).

Muijzenberg, 1980) requiring 83 610 m^2 glass. It was the first greenhouse assembled from standard parts. It was taken down in 1852 and rebuilt in 1854 at Sydenham.

Another large glasshouse of this period is the Palm House at Kew Gardens, designed and built in the 1840's by Burton and Turner, and used to house plants from tropical forests.

Automatic ventilation of greenhouses was introduced early in the 19th century. According to Loudon (1825) John Willeams of Pitmaston had invented a cylinder, in which a piston actuated by the expansion and contraction of the air operated a sliding sash.

In 1806, one Mr. Keely (Lemmon, 1962) made an "automation Gardener" to open and shut ventilators, shutting off a steam valve or damper and ringing an alarm bell in the gardener's booth, once a set temperature had been exceeded. Mr. Keely used his invention in his own greenhouse on the Isle of Man. "Its basic motivation was a thermometer placed horizontally and balanced like a beam of a weight scale". Later an electrical thermometer was recommended (Fawkes, 1881) "of which the dial portion or electric thermometer is placed in the green-house, the temperature of which is required to neither rise above nor fall below certain limits" (Fig. 1.5). "The bell may be placed in the gardeners bedroom ... and the bell continues to ring, until the temperature returns to its previous limits or one of the wires is disconnected at some point".

Thermal insulation by double glazing was mentioned by Moll in 1829. Double-roofed glasshouses made covering of the glass with mats during the night unnecessary. A double-roofed plant house with heating between the two layers of glass was already mentioned by Loudon, who liked to remove the upper sashes in spring.

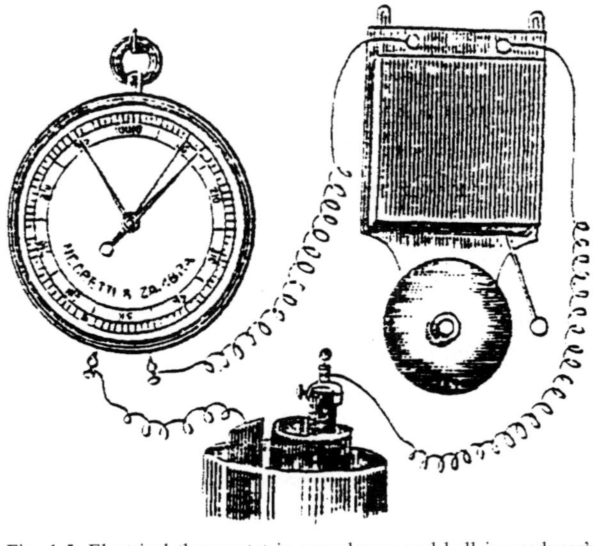

Fig. 1.5. Electrical thermostat in greenhouse and bell in gardener's bedroom (Fawkes, 1881, p. 244).

In 1834 Loddiges invented an overhead irrigation system that was used to water a glasshouse 12 m long (Loudon, 1878). The irrigation pipes were placed parallel to each other under the roof horizontally, 1.8 to 2.4 m apart. The lead pipes had a bore of 25 mm and were each perforated every 5 cm.

During the 19th century larger glass panes became available as a result of improved manufacturing methods and the demand for more light in the glasshouses. Loudon (1878) noted that, up to 1845, gardeners used very small panes. The small panes, 13×23 cm^2 according to Noisette (1829) with overlaps of 13 to 35 mm, were so thin that they often were destroyed during hailstorms. Thicker and larger sheet glass became available in London after 1833, and by the middle of the century the tax on glass was reduced in England. As a result, larger panes came into use which were 61 to 191 cm long, 30 cm wide and weighed 5.4 to 8.0 kg m^{-2}.

About 1880, the first Belgian glass 141 cm long and 73 cm wide was produced (Wiersma, 1918), though the smaller size of 56 cm long and 46 cm wide remained the most common up to 1898.

Loudon (1825) reported a very modern approach to greenhouse ecology, which was presented in 1803 by Dr. James Anderson, in his publication "A description of a patent hothouse". This greenhouse which was mainly heated by the sun's heat during the day and by collected heat during the night, without smoke flues, bark pit or steam. Anderson concluded that "Heat must not only be produced in hot-houses but its waste avoided ... The introduction and management of light is the most important point to attend to in the construction of hot-houses". Thus, Dr. Anderson initiated modern greenhouse-climate research.

According to Robinson (1974), the use of electricity for growing greenhouse plants, has a surprisingly long history. William Gilbert of England described frictional electricity already in 1600. Charles François de Cisternay Dufay, who was gardener to the King of France, is reported to have noticed the existence of two types of electrical forces, an attractive and a repulsive force (Robinson, 1974). Van den Muijzenberg (1980) reported that the first studies of the influence of static electricity on plants were made around the middle of the 18th century. In 1755 a publication, "Monthly Magazine", showed a drawing of "A machine for a perpetual electrified garden" in a greenhouse (van den Muijzenberg, 1980, fig. 36).

According to Rane (1894), Hervé Magnon (1825–1888) observed that chlorophyll was formed when young rye plants were exposed to electric light.

When Siemens in 1880 made his first experiments on the influence of light on vegetation, he found that the electric arc is also efficient in hastening the ripening process in strawberries and probably in nearly every fruit (Siemens, 1880). Arc light had the disadvantage that the intensity changed when the distance between the electrodes increased, as the electrodes were consumed. It was therefore a large improvement when Edison invented the first incandescent lamp in 1879, which had a power consumption of 4.9 W cd^{-1} and a lifetime of 300 hours. Improvements made over the next 50 years included the invention of the gas-filled lamp, with 0.7–0.8 W cd^{-1} and a lifetime of 1000 hours in 1929 (van den Muijzenberg, 1980).

Electrical heating of greenhouses was not used in the 19th century, although Hare already in 1831 had invented heating by electricity, letting a current pass through high-resistance wires. An electrical samovar was shown in 1883 in Vienna at the Great Exhibition. In 1891 platinum wire, encased in asbestos, and in 1894 a resistance wire within a ceramic block, were introduced (Schulz, 1931).

THE 20TH CENTURY

The extensive literature on the development of heating, irrigation and fertilization systems in greenhouses during this century lies beyond the scope of this chapter, but can be studied in the relevant chapters in this book.

In the 20th century, in both northwestern Europe and North America, there was a development from smaller to larger greenhouse units, usually even-span or ridge-and-furrow multi-span houses. The use of small electrical motors allowed automated climate control, including the opening of vents. Heating by thermosiphon hot water circulation was replaced by forced circulation. Electrical pumps were used for irrigation and to operate shading systems, etc.

Lean-to glasshouses slowly became obsolete and heating by smoke flues disappeared in the first decades of this century. However, these changes all occurred very gradually. An example of this development in the Netherlands is outlined below.

Simple over-wintering sheds were used in Aalsmeer, the Netherlands, up to the 1920's, and store-frames, placed against the walls to protect vines in the Westland region were photographed by van den Muijzenberg in 1928 (van den Muijzenberg, 1980).

At the beginning of the century, 90% of the area used for protected cultivation in the Netherlands consisted of cold frames or hotbeds (heated by decomposition of manure and compost). By 1950 the area was still about 30%, and in 1964 it was 10% (van den Muijzenberg, 1980).

At the beginning of this century the use of Dutch lights was introduced. Initially, only single-span frames were used; double-light frames date from about 1920 (van den Muijzenberg, 1980). Comparing the building practices in the Netherlands with those in the more advanced United States, one sees that in 1904, 90% of the glass-covered area in the Netherlands consisted of light frames, while 25% of the glass-covered area in the USA in the year 1900 was of proper glasshouses (Wright, 1917). The area under glass in the Netherlands increased from 1.6 km^2 in 1904 to 4.0 km^2 in 1912, to 32.9 km^2 in 1950, reaching 60.2 km^2 in 1964 (van den Muijzenberg, 1980).

Heating and carbon-dioxide fertilization of hotbeds through the decomposition of mixtures of horse manure and plant residues was widely used in the Netherlands at the beginning of the century, and remained the common way for growing cucumbers and melons until horses were replaced by tractors in the years after the Second World War.

Claassen and Hazeloop (1933) mentioned that at that time flue heaters were only used by a few growers in the Netherlands, whereas in Belgium, most of the 4.40 km^2 used for fruit-growing in 1939 was heated by flues (Danvig, 1939). Dutch greenhouses suitable for a variety of crops were developed gradually over half a century. In 1900, at Loosduinen, a steel-framed greenhouse was built which became the forerunner for the "Venlo Warenhuis"; a house that could be used for different crops (van den Muijzenberg, 1980). This glasshouse, first built in 1937, consists of large sheets of glass held within the grooves of the two sides of a separating narrow steel bar, permitting the maximum transmission of light. Venlo-type structures were becoming popular in Westland of the Netherlands in the 1950's (van den Muijzenberg, 1980). Burning natural gas for the supply of carbon dioxide to greenhouse crops began in the Netherlands in 1961, following work done by J. Stender from the Institute of Horticultural Engineering in Wageningen (van den Muijzenberg, 1980), and became globally applied in the next decade (Wittwer, 1986).

The introduction of glass substitutes

There has been an interest in glass substitutes for several centuries, glass having the disadvantage of fragility, especially under conditions of high wind, hail or snow load. Since the 17th century it was realized, in Europe and possibly even earlier in China and Japan, that oiled paper was a reasonably transparent material which could be used to cover frames and greenhouses (van der Groen, 1670; Commelyn, 1676; de la Quintinye, 1695; Miller, 1731).

According to the publication *PVC* (Anonymous, 1965), paper was widely used in Japan as a greenhouse cover, and in 1954 covered much greater surfaces than plastic and glass. Japan had 30.2 km^2 of crops under cover in 1954, of which 25.7 km^2 was covered with paper, the rest with polyvinyl chloride (PVC) film. In 1956, out of a total of 35.7 km^2 of protected crops, about half was covered with paper; by 1965 only about 12 km^2 (4% of the total 300 km^2) was covered by paper (Yanagisawa, 1970).

According to van den Muijzenberg (1980), cellulose products as cladding materials were described by Wright as early as 1917, and Claassen and Hazeloop (1933) mentioned celluloid film strengthened with copper gauze as not being very satisfactory. As a result of trials by van den Muijzenberg (1962) the replacement of glass by plastic was largely abandoned in the Netherlands. In fact, new plastic products were first used in Japan and the USA, and from there introduced into southern Europe and Israel.

Cotter (1968), describing the development of plastic greenhouses in the United States and southern Europe, mentioned that Emmert in 1948 grew crops in greenhouses and cold-frame structures that were covered with cellophane in Kentucky. By 1958 polyethylene film 0.03–0.05 mm thick was in wide use as cladding material, though the Dutch scientists Germing (1958) and Kostelijk and Spek (1967) did not consider that plastic would be a success for this purpose.

However, the use of plastic as a greenhouse cladding material soon gained in popularity, especially in Japan. Dalrymple (1973) estimated that there were 300 km^2 of greenhouses in the world, around 1970, covered with plastic film, of which at least 100 km^2 were in Japan. This compared with 240 km^2 of glasshouses worldwide, of which 72 km^2 were in the Netherlands (Dalrymple, 1973).

According to Abe (1964), the use of plastic films in Japan began in 1950, and by 1962 there were 60 km^2 under plastic cover, of which 93% were used for vegetables, 6% for flowers, and 1% for fruits of the area. Most (89%) was under PVC film, and the rest (11%) under polyethylene (PE) film. Greenhouses and high tunnels constituted 55%, frames 40%, and hotbeds and cloches 5% of the plastic-covered area (Abe, 1964). In Israel, the area of plastic-covered greenhouses in 1964 was about 1.0 km^2 and of plastic-covered tunnels about 10 km^2 (Ettinger, 1964).

In the United States many new types of greenhouses were built, including ones without any rigid support which consisted of a large plastic sheet of 0.13 mm polyester film with nylon mesh, supported by a small atmospheric over-pressure (Courter, 1962).

In addition to plastic films, rigid plastics came into use after the Second World War. Polyester fiberglass in corrugated or plain sheets became common in the United States during the 1950's (Courter, 1962) and in Europe (Nisen, 1963). In addition, Plexiglas (polymethyl methacrylate) was sometimes used, especially in Germany.

The use of rigid, plastic covers, as well as plastic films, is now common practice and will probably remain so in protected crop cultivation.

THE GEOGRAPHICAL DISTRIBUTION OF GREENHOUSES

It is not simple to find reliable information about the world-wide area distribution of greenhouses. The total world-wide area was 43 143 ha (431 km^2) in the seventies according to Hanan et al. (1978); 180 000 ha (1800 km^2) [45 500 ha greenhouses and 135 000 ha plastic tunnels] according to Meijgaard (1995); 265 800 ha (2658 km^2) greenhouses and high tunnels and 40 700 ha (407 km^2) glasshouses according to Wittwer and Castilla (1995) and 93 570 ha (936 km^2) greenhouses according to Boulard (1996). Wittwer (1993) estimated that there were 198 700 ha of plastic greenhouses and no less than 658 700 ha low tunnels and soil mulches in the world. The area of greenhouses in individual countries is frequently noted, but the figure is typically the estimate of (an) informed

Table 1.1
Geographical distribution of greenhouse area (ha)

Country	Year	Cladding: Glass	Cladding: Plastic	Cladding: Not specified	Total
Algeria[1]	1990			5000	5000
Australia[2]	1986	600	600		1200
Austria[2]	1986	385			385
Belgium[3,4]	1993	542	2535		3077
Bulgaria[2]	1986	400	400		800
Canada[5,6]	1994	350	600		950
China[7]	1994			600 000	600 000
Colombia[8]	1993			4500	4500
Cyprus[1]	1990			300	300
Czech Republic[19]	1990	50			50
Denmark[3,9]	1993			523	523
Egypt[1]	1990			600	600
Finland[3,5]	1994	180	120		300
France[3,5]	1994	3500	4300		7800
Germany[3,5]	1994	3700	1100		4800
Greece[5]	1989	100	3470		3570
(Crete)[1]	1990			(1900)	(1900)
Hungary[10]	1994	150	750		900
Iceland[2]	1986	15			15
Ireland[3,5]	1994	100	15		115
Israel[1]	1990			1400	1400
Italy[3,5]	1994	3750	25 000		28 750
(Sicily)[1]	1990			(8000)	(8000)
Japan[11]	1995	2000	40 000		42 000
Jordan[5]	1990		350		350
Kenya[12]	1994		200		200
Lebanon[1]	1990		1100		1100
Lithuania[14]	1994			26	26
Malta[1]	1990			15	15
Morocco[1]	1990			1600	1600
Mexico[15]	1991	200	200		400
Netherlands[16]	1994	10 000	100		10 100
New Zealand[5]	1990	120	80		200
Norway[2,3]	1994	170	10		180
Poland[17]	1992	468	200		668
Portugal[1]	1990	100	1400		1500
Romania[5]	1990	1450	4050		5500
Saudi Arabia[5]	1985	80	220		300
South Korea[13]	1990		5107		5107
Spain[2,3]	1994	180	25 000		25 180
Sweden[2,3]	1994	180	35		215
Switzerland[18]	1990	272			272
Tunisia[1]	1990			1150	1150
Turkey[1]	1990			9800	9800
UK[2,3]	1994	2400	380		2780
USA[20]	1993	4378	3182		7560
USSR[1]	1990			14 000	14 000
Zimbabwe[11]	1994		120		120
Total		35 820	125 124	634 414	795 358

References: 1, Gartenbau Report (1990); 2, Germing (1986); 3, AIPH (1995); 4, Rabobank (1992); 5, Anonymous (1990); 6, DFWB (1995a); 7, Dutch Embassy Beijing (1995); 8, PVS (1994); 9, DFWB (1995b); 10, DFWB (1994); 11, Anonymous (1995b); 12, ABN-AMRO (1995); 13, Larenstein (1995); 14, Anonymous (1995c); 15, Dutch Ministry of Agriculture (1992); 16, PVS (1995); 17, DFB (1993); 18, DFWB (1995c); 19, DFB (1995); 20, DFWB (1995d).

researcher(s), as for instance Amiram (1987); and Castilla and Hernandez (1995).

In the following, we have based our estimates on a survey (Tables 1.1, 1.2) made by a Dutch horticultural marketing company, HORTIMARC B.V. Amsterdam, in November, 1995 (Anonymous, 1995a). As can be seen in Table 1.1, the total world-wide area of greenhouses is currently about 800 000 ha (8000 km^2), or about 1.3 square meter per person in the world. Obviously, most of the world's population (6×10^9 persons) do not use greenhouse products. The largest area of protected cultivation can be found in China, which has 600 000 ha (6000 km^2) or about three-quarters of the total area. Nearly all the greenhouse area in China consists of simple plastic-clad houses, mostly used for vegetable production around large cities. Glasshouses in China are only used to a minor extent, mainly for research (personal communication from Dr. Yao Chou Ren, Republic of China, 1995).

The second-largest area of greenhouses, 42 000 ha (420 km^2), is found in Japan (5.3% of world area), only 2000 ha (20 km^2) of which consists of glasshouses. This

Table 1.2
Greenhouse area (excluding plastic tunnels) devoted to different crop categories in certain countries

Country	Vegetables		Flowers	
	Year	Area (ha)	Year	Area (ha)
Belgium [1,2]	1990	2535	1993	542
Canada [3,4]			1993	950
Colombia [5]			1993	4500
Denmark [1,6]	1990	140	1993	383
France [1,3]	1990	6450	1990	1889
Germany [1,3]	1990	2000	1988	2402
Hungary [7]			1993	900
Ireland [1,3]	1990	55		
Israel [8]			1990	1400
Italy [1,3]	1990	9000	1994	4402
Japan [9]			1994	18000
Kenia [10]			1993	200
South Korea [11]	1990	3807	1990	1300
Mexico [12]			1991	400
Netherlands [13]	1990	4500	1994	5500
Norway [1,14]			1994	45
Poland [15]			1992	668
Portugal [8]	1990	1250		
Spain [1,14]	1990	18500	1991	3012
Sweden [1,14]			1993	27
Switzerland [16]			1990	270
UK [1,14]	1990	2430	1993	859
USA [17]			1993	7560

References: 1, AIPH (1995); 2, Rabobank (1992); 3, Anonymous (1990); 4, DFWB (1995a); 5, PVS (1994); 6, DFWB (1995b); 7, DFWB (1994); 8, Gartenbau Report (1990); 9, Anonymous (1995b); 10, ABN-AMRO (1995); 11, Larenstein (1995); 12, Dutch Ministry of Agriculture (1992); 13, PVS (1995); 14, Germing (1986); 15, DFB (1993); 16, DFWB (1995c); 17, DFWB (1995d).

is followed by Italy with 28 750 ha (287 km^2) (3.6%) and Spain with 25 000 ha (250 km^2) (3.2%), and the Netherlands with 10 100 ha (101 km^2) (1.3%), most of which is glasshouses.

Thus, out of the total global greenhouse area, the bulk now consists of simple plastic houses, with the exception of a few European countries, such as the Netherlands, Italy, Germany, France and Denmark, in which a considerable proportion of protected cultivation still is in glasshouses.

The general world-wide trend of the greenhouse ecosystem has thus gone full circle, from simple plant shelters, used two millennia ago, through complex and expensive glasshouses back to simple inexpensive plastic shelters. The reason for this development can be found in the two alternatives to greenhouse cultivation by which the consumer can obtain out-of-season food and flower products, namely transport and storage.

The consumer requiring a year-round supply of fresh fruit, vegetables and flowers can be supplied either by:

- Changing the climate of a greenhouse covered by plastic or glass.
- Storing fruit, flowers and vegetable products after the normal harvest season until a later marketing date.
- Transporting the product across climate zones, from where it grows naturally to where it is marketed.

All three alternatives are in use at the present. The greenhouse ecosystems of northwestern Europe provide, at great cost, the climate changes needed to produce flowers (e.g. roses) and vegetables (e.g. tomatoes) in the winter. Consumers can also receive the same goods by importing them from warmer (Mediterranean) countries.

Some plant products, for instance carnation flowers, can be stored for prolonged periods. Then the stored carnations are marketed at a period when their price is high.

Some crops are grown outdoors and flown to their markets in other countries. A typical example is orchids; a large proportion of the world's orchid production is grown outdoors in Thailand, harvested, and flown to their markets in Japan, Australia, Europe and North America. There is no lack of knowledge about how to grow orchids in heated greenhouses in those countries where they are marketed, but apparently it does not pay to produce them locally. Roses are grown in winter in Israel and elsewhere along the Mediterranean, and sold in Europe, exactly as roses from Egypt competed with locally produced flowers from protected cultivation in Rome two millennia ago.

Enoch (1978a,b) analyzed these three options (climate change, storage, and transport over climate zones) and calculated the energy expenditure of these alternative options. Currently, economic considerations and not energy optimizations are paramount. If it is cheaper to produce field crops, or to store or transport products, then this will be done in preference to making the necessary climate changes within greenhouses, even if it is more expensive in energetic terms.

There is every reason to expect that in the future transport over climate zones and longer storage will become more efficient and thus cheaper, thereby decreasing the commercial importance of the greenhouse as an agricultural production system.

However, greenhouses will almost certainly survive for the same reasons that first favored their development, namely as a provider of luxuries for the rich, for their aesthetic and recreational importance, and their role in botanical and ecosystem research.

REFERENCES

Abe, S., 1964. L'utilisation des plastiques dans l'horticulture au Japon [The use of plastic in horticulture in Japan]. 7e *Colloq. Agr. Comité Plastiques en Agriculture, Paris*: 16–24.

Abercrombie, J., 1792. *Der Treibhausgärtner, Über die Kultur der Ananasse, Weintrauben u.s.w. Wien.* [The Greenhouse gardener, Pineapple cultivation, grape cultivation, etc]. Vienna. Translated from English into German. [Quoted by van den Muijzenberg (1980)].

ABN-AMRO, 1995. *Competition-profile: African-Dutch Rose Growing.* ABN-AMRO, pp. 6–17. In Dutch.

Agricola, G.A., 1719. *Nieuwe en ongehoorde, dog in de natuur welgegronde queek-konst. Uit Hoogduits vert. door J. de Long* [New and unheard of cultivation yet well documented in nature.] Translated from German into Dutch by J. de Long, Amsterdam. 208 pp. [Quoted by van den Muizenberg (1980).]

Agricola, G.A., 1721. *A philosophical treatise of husbandry and gardening,* the whole revised and compared with the original, together with a preface by Richard Bradley, London, 304 pp. [Quoted by van den Muizenberg (1980).]

AIPH, 1995. *AIPH 1994,* Vol. 43. The Hague, pp. 19–22.

Amherst, Alicia, 1895. *A History of Gardening in England.* London, 398 pp.

Amiram, D.H.K., 1987. Land transformation in Israel. In: M.G. Wolman and F.G.A. Fournier (Editors), *Land Transformation in Agriculture.* Wiley, Chichester, pp. 291–317.

Andersen, M.P., 1891. *Frukthuset med dertill hörande Kulturer* [The Fruit House and its Crops]. C.J. Lundgren, Jönköping, Sweden. 97 pp. In Swedish.

Anonymous, 1965. Japanese Plastic Industry Publication.

Anonymous, 1990. *The 1992 Euromarket for Dutch Greenhouse Builders.* Annex 15. In Dutch.

Anonymous, 1995a. *Global Distribution of Greenhouses: A Survey of the Greenhouse Area in Selected Countries and Regions.* Compiled by Hortimarc B.V., Amsterdam, the Netherlands. November 1995, 7 pp, with 20 references.

Anonymous, 1995b. *Horticulture in Japan.* 45, pp. 1–13.

Anonymous, 1995c. Gärtnerbörse 1995.

Avery, G.S., 1974. Botanical gardens, In: *Encyclopedia Britannica, Macropedia 3.* Encyclopedia Britannica Inc., Chicago, IL, pp. 63–65.

Bixio, A., 1847. *Die Gärtnerei.* Translated from French by F. Mandry, Stuttgart [quoted by van den Muizenberg (1980)].

Boerhaave, H., 1710. *Index plantarum quae in Horto academico Lugduno Batavorum repetiunter.* Lugduni-Batavorum, 290 pp. [quoted by van den Muijzenberg (1980)].

Boerhaave, H., 1720. *Index alter plantarum quae in Horto academico Lugduno Batavorum colunter.* Lugduni-Batavorum, 270 pp. [Quoted by van den Muijzenberg (1980)].

Bouché, C.D. and Bouché, J., 1895. *Das Gewächshaus, Ein Handbuch für Gärtner* [The Greenhouse, A Handbook for Gardeners], 2nd edition. Emil Strauss, Bonn, 352 pp.

Boulard, T., 1996. *Charactérisation et modelisation du climat des serres: Application à la climatisation estivale,* Ph.D. Thesis. Preparée à l'Unité de Bioclimatologie, I.N.R.A. d'Avignon. France.

Brent, E., 1991. Conservatory catalogues of yesteryear. Treasures of the Lindley Library. *The Garden. Journal of the Royal Horticultural Society* (London), 116: 172–176.

Castilla, N. and Hernandez, J., 1995. Protected cultivation in the Mediterranean area. *Plasticulture,* 107(3): 13–20.

Chambers, W., 1763. *Plans, Elevations, Sections and Perspective Views of the Gardens and Buildings at Kew in Surrey, London.*

Claassen, C.H. and Hazeloop, J.G., 1933. *Leerboek voor de groenteteelt. 2. De teelt onder glas* [Textbook on greenhouses. 2. The Glasshouse]. Tjeenk Willink, Zwolle, 274 pp.

Coats, P., 1968. *Great Gardens of the Western World.* Hamlyn Publ. Group, Weidenfeld and Nicolson, London, 288 pp.

Commelyn, J., 1676. *Nederlantze Hesperides, dat is, oeffening en gebruik van de Limoen- en orangeboomen.* [Dutch plant enclosures, that is, cultivation and use of Lemmon and orange trees]. Amsterdam, 49 pp. [Quoted by van den Muijzenberg (1980)].

Cotter, D.J., 1968. A comparison of plastic greenhouse development in the United States and Southern Europe. *Acta Horti.,* 9: 43–46.

Courter, J.W., 1962. *Plastic Greenhouses.* Univ. of Illinois, Coll. Agr. Coop. Ext. Serv. Circ., 857, 32 pp. [Quoted by van den Muijzenberg (1980)].

Dalrymple, D.G., 1973. *Controlled Environment Agriculture: A Global Review of Greenhouse Food Production, Foreign Agri. Econ. Rep., no. 89.* U.S. Department of Agriculture, Washington, D.C., 150 pp.

Danvig, A.M., 1939. *Plantedrivning* [Plant production by forcing]. Almindelig Dansk Gartnerforenings Bogforlag, Copenhagen, 216 pp.

de Bilderling, N., 1974. The evolution of horticulture. *Chron. Horti.,* 14(2): 17–24.

de Crescentiis, P., 1602. *Neu Feldt und Acherbau.* [New range field (crop) cultivation]. Lazari Zetzner, Strassburg, 670 pp. [Quoted by van den Muijzenberg (1980)].

de la Quintinye, J., 1695. *Le Parfait Jardinier ou Instruction pour les Jardins Fruitiers et Potages. Avec un traite des orangeries.* Paris. [The Perfect Gardener, or, Instructions for the Fruit Grower; with a chapter on citrus cultivation]. [Quoted by van den Muizenberg (1980).]

de Serres, O., 1600. *Traité d'Agriculture et Menage des Champs. Remis en francais par A.M. Grisors 1802–1805* [Agricultural writings and field management], 614 pp. [Quoted by van den Muizenberg (1980).]

DFB, 1993. *Country-report Poland.* Dutch Flower Board (DFB), 7 pp. In Dutch.

DFB, 1995. *Country-report Czech Republic.* Dutch Flower Board (DFB), 7 pp. In Dutch.

DFWB, 1994. *Country-report Hungary.* Dutch Floricultural Wholesale Board (DFWB), 7 pp. In Dutch.

DFWB, 1995a. *Country-report Canada.* Dutch Floricultural Wholesale Board (DFWB), 7 pp. In Dutch.

DFWB, 1995b. *Country-report Denmark.* Dutch Floricultural Wholesale Board (DFWB), 7 pp. In Dutch.

DFWB, 1995c. *Country-report Switzerland.* Dutch Floricultural Wholesale Board (DFWB), 7 pp. In Dutch.

DFWB, 1995d. *Country-report USA.* Dutch Floricultural Wholesale Board (DFWB), 7 pp. In Dutch.

Diel, D.A.F.A., 1798. *Ueber die Anlegung einer Obst Orangerie in Scherben und die Vegetation der Gewächse.* [About the construction of a fruit orangery in clay pots and the cultivation of plants]. The Andrean Bookshop. Frankfurt am Main, Germany.[Quoted by van den Muijzenberg (1980)].

Driessen, F., 1928. *De reizen der de la Courts in 1641, 1700 & 1710* [The Journeys of de la Courts in 1641, 1700 & 1710]. Leiden, Netherlands, 121 pp. [Quoted by van den Muijzenberg (1980)].

Dutch Embassy Beijing, 1995. *The Chinese Agribusiness.* Dutch Embassy, Beijing, 18 pp.

Dutch Ministry of Agriculture, 1992. *Horticulture in Mexico.* Dutch Ministry of Agriculture, annexure 3.13. In Dutch.

Enoch, H.Z., 1978a. A theory for optimalization of primary production in protected cultivation. I. Influence of aerial environment upon primary plant production. *Acta Horti.*, 76: 31–43.

Enoch, H.Z., 1978b. A theory for optimalization of primary production in protected cultivation. II. Primary plant production under different outdoor light regimes. *Acta Horti.*, 76: 45–57.

Ettinger, J., 1964. Semi-forçage et serres plastiques [(Partly) Forcing a crop and plastic greenhouses] *7e Coll. Plast. Agri. 33–34* (Proc. 7th Int. Agri. Plastics Conf.), Paris.

Evelyn, J., 1664. *Kalendarium Hortense* (Agricultural calender) (Quoted by Lemmon, 1962).

Falland, J., 1987. Greenhouses in plant propagation: An historical perspective. *Proc. Int. Soc. Plant Propagators.* Seattle, WA, pp. 158–164.

Fawkes, F.A., 1881. *Horticultural Buildings.* B.T. Batsford, London, 255 pp.

Feldhaus, F.H., 1931. *Die Technik der Antiken und des Mittelalters* [Technology in the Antique Times and the Middle Ages]. Potsdam, 442 pp. [Quoted by van den Muijzenberg (1980)].

Ferrarius, J.B., 1646a. *Hesperides sive de malorum aureorum cultura usu. Romae.* 480 pp. [Quoted by van den Muijzenberg (1980)].

Ferrarius, J.B., 1646b. *Flora, seu de florum cultura.* Amstellodami. [Quoted by van den Muijzenberg (1980)].

Gartenbau Report, 1990. *Gartenbau Report* 492: 29.

Germing, G.H., 1958. Kunststof folie in de tuinbouw. *Meded. Dir. Tuinbouw Neth., Inst. Tuinb. Techn. Meded.*, 37: 2–13.

Germing, G.H., 1986. *Chronica Horticulturae*, vol. 26, no. 1, April.

Geytenbeek, E., 1991. *Oranjerieën in Nederland.* [Orangeries in the Netherlands]. Carnaletto, Alphen aan den Rijn.

Gibault, G., 1898. Les origines de la culture forcée [The origin of forcing in crop culture]. *J. Soc. Hort. De France:* 1109–1117.

Greene, E.L., 1983. *Landmarks of Botanical History.* Stanford University Press, Stanford, CA, pp. 128–211.

Grisors, A.M., 1802–1805. see de Serres, 1600.

Hanan, J.J., Holley, W.D. and Goldsberry, K.L., 1978. *Greenhouse Management.* Springer, New York, 530 pp.

Hasluck, P.N., 1907. *Greenhouse and Conservatory Construction and Heating.* Cassel and Company Ltd, London.

Hietala, J.S. and Ashley, G.C., 1984. History and status of selected waste heat greenhouse projects. *Acta Horti.*, 148: 233–237.

Hix, J., 1996. *The Glasshouse.* Phaidon Press, London, 240 pp.

Jacob, R., 1797. *Economical Handbook for the Promotion of the Early Growth of Vegetables and Fruit Varieties in Hotbeds and Greenhouses with the Necessary Aeration Regime and Supplied with Continuous Weather Tables.* F. Esslinger, Frankfurt Am Main. In German.

Jacobs, J.M., 1995. The history of greenhouse production. In: J.C. Bakker, G.P.A. Bot, H. Challa and N.J. Van de Braak (Editors), *Greenhouse Climate Control – an integrated approach.* Wageningen Pers, Wageningen, pp. 3–7.

Kennedy, J., 1777. *Treatise upon Planting, Gardening and the Management of the Hothouse,* Vols. I, II. S. Hooper, London. [Quoted by van den Muijzenberg (1980).]

Kingman, D., 1986. *A History – Colorado flower growers and its people.* Colorado Greenhouse Growers Association, Denver, CO, 234 pp.

Koch, H., 1910. *Sachsische Gartenkunst.* Berlin, 407 pp. [Quoted by van den Muijzenberg (1980).]

Kohlmaier, G. and von Sartory, B., 1986. *Houses of Glass.* MIT Press, Cambridge, MA.

Koppelkamm, S., 1982. *Glass Houses and Wintergardens of the Nineteenth Century.* Granada Publishing, London.

Kostelijk, S. and Spek, J.C., 1967. Kassen met Kunststof-omhulling. [Plastic-clad greenhouses]. *Inst. Tuinb. Techn., Wageningen,* Publ. 13: 124–137.

Larenstein, I.A.H., 1995. *Cut Flowers in South Korea.* pp. 12–13.

Lawrence, W.J.C., 1948. *Science and the Glasshouse.* Oliver and Boyd, Edinburgh, 183 pp.

Le Berrugais, M.L.B., 1788. *Traité des Jardins, Tome IV. Traité de L'Orangerie, des Serres-chaudes et Châssis.* Caen, 523 pp. [Quoted by van der Muijzenberg (1980)].

Lemmon, K., 1962. *The Covered Garden.* Museum Press, London, 284 pp.

Lenz, H.O., 1859. *Botanik der alten Griechen und Römer, Deutchen Auszügen aus deren Schriften.* [Botany of the old Greeks and Romans, summaries of their writings in German]. Thienemann, Gothaus, 766 pp.

Li, H.L., 1957. *The Garden Flowers of China.* The Ronald Press, New York, Ch. 2., pp. 12–21.

Lipa, J.J., 1985. History of biological control in protected culture: Eastern Europe. In: N.W. Hussey and N. Scopes (Editors), *Biological Pest Control: The Glasshouse Experience.* Poole, Dorset, England, pp. 23–29.

Loudon, J.C., 1822. *Pine-Apple. The different modes of cultivating the pine-apple from its first introduction into Europe to the late improvements of T.A. Knight.* Esq. London, 184 pp. [Quoted by van den Muijzenberg (1980).]

Loudon, J.C., 1823. *Eine Encyclopädie des Gartenwesens* [An Encyclopedia of Horticulture], Vol. 1, Weimar. [Quoted by van den Muijzenberg (1980).]

Loudon, J.C., 1825. *Eine Encyclopädie des Gartenwesens* [An Encyclopedia of Horticulture], Vol. 2, Weimar. [Quoted by van den Muijzenberg (1980).]

Loudon, J.C., 1842. *The Suburban Horticulturist.* London, 732 pp. [Quoted by van den Muijzenberg (1980).]

Loudon, J.C., 1878. *Encyclopedia of Gardening.* London. [Quoted by van den Muijzenberg (1980).]

Mackenzie, 1815. *Hort. Trans.,* 11: 149 pp. [Quoted by van den Muizenberg (1980).]

Marston, P., 1992. *The Book of the Conservatory.* Wiedenfeld and Nicolson, London, p. 176.

Martincigh, L., 1980. *Greenhouse Architecture. History, current technology, future prospects*. Genio-Rurale, Bologna. Vol. 43(9), pp. 33–42.

Meijgaard, D., 1995. The greenhouse industry in The Netherlands, in: J.C. Bakker, G.P.A. Bot, H. Challa and N.J. Van de Braak (Editors), *Greenhouse Climate Control – an integrated approach.* Wageningen Pers, Wageningen, pp. 7–13.

Miller, P., 1731. *The Gardeners Dictionary,* London. [Quoted by van den Muizenberg (1980).]

Moll, G., 1829. *Over het Verwarmen van Stookkassen met Heet Water* [About heating greenhouses with hot water], J. Muller and Comp., Amsterdam. [Quoted by van den Muizenberg (1980).]

Nisen, A., 1963. Tendances actuelles dans la construction d'abris horticoles. *Ann. Gembloux,* 69(2): 533–555.

Noisette, L., 1829. *Manuel du Jardinier.* Bruxelles. 612 pp.

Paisley, K., 1962. *Protected Cultivation, modern management of greenhouse and frame.* C. Arthur Pearson Ltd., London.

Pascottet, P. and Dairat, J., 1927. *Cultures de Serres, Construction et Chauffage des Serres Forçage des Raisins et des Fruits des Primeurs.* Introduction par R. Regnard. J.B. Bailliere et fils, Paris, 540 pp. [Quoted by van den Muizenberg (1980).]

Platt, H., 1594. *The Jewel House of Art and Nature.* [Quoted by van den Muizenberg (1980), p. 52].

PVS, 1994. *Colombia on the European Cut Flower Market.* PVS, 1994, 16 pp. In Dutch.

PVS, 1995. *Statistics of the Dutch Horticultural Industry 1994.* PVS, 10 pp.

Rabobank, 1992. *The International Vegetable Industry.* Rabobank, Utrecht.

Rane, T.W., 1894. *Electro-horticulture.* West Virginia Agricultural Experimental Station, Morgantown, West Virginia, Bul. 37, 37 pp.

Ravenscroft, B.C., 1901. *Greenhouse Construction and Heating.* L. Upcott Gill, London, 224 pp. [Quoted by van den Muizenberg (1980).]

Robertson, W., 1798. *A Collection of Various Forms of Stoves.* London. [Quoted by van den Muizenberg (1980).]

Robinson, F.N.H., 1974. Electricity. *Encyclopedia Britannica*, Vol. 6. Encyclopedia Britannica Inc., Chicago, IL, p. 540.

Rowles, W.F., 1910. *Greenhouses, How to Make and Manage Them.* C. Arthur Pearson Ltd., London, 108 pp.

Sarton, G., 1953. *A History of Science: Ancient science through the Golden Age of Greece.* Oxford University Press, Oxford, pp. 547–561.

Saudan-Skira, S. and Saudan, M., 1994. *Orangeries. Splendeurs et Metamorphoses* [The Beauty of Orangeries and their changes]. Le septième Fou, La Bibliotheque des Arts, Geneva.

Siemens, C.W., 1880. On the influence of electric light upon vegetation and on certain physical principles involved. *Proc. Roy. Soc. London,* 30: 210–219.

Singer, C., 1921. Greek biology and its relation to the rise of modern biology. In: C. Singer (Editor), *Studies in the History and Methods of Science.* Clarendon Press, Oxford, pp. 1–101.

Spice, H.R., 1959. *Polythene Film in Horticulture.* Faber and Faber, London, 176 pp.

Stanhill, G., 1977. An urban agro-ecosystem: The example of Nineteenth Century Paris. *Agro Ecosystem,* 3: 259–284.

Steere, W.C., 1974. Botany. In: *Encyclopedia Britannica, Macropodia 3.* Encyclopedia Britannica Inc., Chicago, IL, pp. 65–71.

Taft, L.R., 1898. *Greenhouse Construction.* New York, 210 pp. [Quoted by van den Muizenberg (1980)]. Reprinted 1920.

Theophrastus, ~300 BCE. *De Causis Plantarum.* English translation: B. Einarson and G.K.K. Link 1976, William Heineman, London. *Enquiry into Plants.* English translation: Sir Arthur Hort, 1968, 1980, 2 volumes. Harvard University Press, Cambridge, MA.

Tod, G., 1823. *Plans, Elevations of Hothouses–Greenhouses.* London, 23 pp. [Quoted by van den Muizenberg (1980).]

van den Muijzenberg, E.W.B., 1962. Toepassingen van kunststoffen in den tuinbouw [Use of plastics in the greenhouse]. *Plastica,* 15(10): 504–512.

van den Muijzenberg, E.W.B., 1980. *A History of Greenhouses,* Preliminary edition, Translated and edited by Peter Newton, Manchester University. 150 Copies distributed by The Institute of Agricultural Engineering, Wageningen, 435 pp. with 800 references.

van der Groen, J., 1670. *Den Nederlandtsen Hovenier* [The Dutch Gardener]. Amsterdam, 106 pp. [Quoted by van den Muijzenberg (1980).]

van der Voort (de la Court), P., 1737. *Byzondere Aenmerkingen over het Aenleggen van Pragtige en Gemeene Landhuizen, Lusthoven, Plantagien en Aenklevende Cieraden.* Leiden, 412 pp. [Quoted by van den Muijzenberg (1980).]

van der Voort (de la Court), P., 1750. *Les agréments de la campagne.* Leyde, 412 pp. [Quoted by Muijzenberg, 1980.]

Vickman, L., 1987. *A Brief History of Solar Greenhouses,* New Alchemy Quarterly. New Alchemy Institute, East Falmouth, MA, 36 pp.

Volkamer, J.C., 1708. *Nürnbergische Hesperides oder gründliche Beschreibung der edlen Citronat, Citronen- und Pomerantzen Früchte* [The Orangery in Nuremberg, or the Description of the Noble Citrus, Lemon- or Pomerans (grapefruit) Fruits]. Nürnberg, 2 volumes. [Quoted by van den Muizenberg (1980).] Second edition: 1714.

Walls, I.G., 1973. *The Complete Book of the Greenhouse.* Ward Lock Ltd, London, 298 pp. 4th Edition: 1988.

Wiersma, K., 1918. *Kassen en Kassenbouw* [Greenhouses and Greenhouse Construction]. Bootsma, 's Gravenhage, 96 pp. [Quoted by van den Muizenberg (1980).]

Willmott, P.K., 1958. *Dutch Lights and Frames; A Guide for the Commercial Grower.* Ernest Benn Ltd, London, 136 pp.

Wittwer, S.H., 1986. Worldwide status and history of CO_2 enrichment – an overview. In: H.Z. Enoch and B.A. Kimball (Editors), *Carbon Dioxide Enrichment of Greenhouse Crops.* CRC Press, Boca Raton, FL, pp. 3–15.

Wittwer, S.H., 1993. Worldwide use of plastics in production. *Hort. Technology,* 3(1): 6–19.

Wittwer, S.H. and Castilla, N., 1995. Protected cultivation of horticultural crops worldwide. *Hort. Technology,* 5(1): 6–23.

Woods, M. and Warren, A.S., 1988. *Glasshouses: A History of Greenhouses, Orangeries and Conservatories.* Rissoli, New York, 216 pp.

Wright, R., 1934. *The Story of Gardening from the Hanging Gardens of Babylon to the Hanging Gardens of New York*. George

Routledge, London, 457 pp. [Quoted by van den Muizenberg (1980).]
Wright, W.J., 1917. *Greenhouses, Their Construction and Equipment.* Orange Judd Comp., New York, 269 pp. [Quoted by van den Muizenberg (1980).]
Yanagisawa, M., 1970. Development of agricultural plastics application in Japan. *C.R. 4e Congr. Intern. Plast. Agr.,* Paris: 61–65.

Chapter 2

GREENHOUSE STRUCTURES

Christian VON ZABELTITZ

INTRODUCTION

Greenhouses are buildings which should provide the conditions for satisfactory plant growth throughout the year. Plastic tunnels and walk-in tunnels should improve the climatic conditions for plants compared to the outside climate. The growth factors light, temperature, humidity and air composition should be delivered and maintained at optimum levels for plant growth.

Daylight, the visible part of the global radiation, is one of the essential growth factors in the greenhouse; therefore greenhouses and tunnels have to be planned and built with a light-transmissive cladding so that enough light reaches the plants, especially in seasons when the sun is low in the sky. Some greenhouses containing special crops may need shading during periods of high irradiation. Heat must be supplied when the outside temperature falls below the desired inside temperature. The heat loss through the covering material should be as low as possible. The greenhouse cladding should provide maximum light transmission at a minimum heat requirement during the cold season.

When the inside temperature exceeds the desired value the surplus heat has to be dissipated by ventilation or, in case of a very high outside temperature, by cooling with other than outside air.

The purpose of good greenhouse construction is to create the necessary climatic conditions for plant production throughout the year, and to control the climate factors as far as possible to the specified optimum.

Greenhouses are required to have high light transmission, low heat consumption, sufficient ventilation efficiency, adequate structural strength, low cost of construction, and low operating cost (Bailey and Richardson, 1990).

Light transmission depends on the transmission of the cladding material and on the number and thickness of structural components. To maximize the transmission of solar radiation it is necessary to employ a minimum of opaque structural members. A greenhouse has to withstand loads generated by permanent structures, imposed crop loads, and wind and snow loads. The structural components have to have structural strength not less than the minimum given by load standards. The wind load, for example, depends on the height of the greenhouse and the roof angle.

Heat consumption depends on the cladding material (single or double cladding), on the tightness of the greenhouse (air exchange through leakages) and the surface area (side-wall height and roof angle). Increasing thermal resistance can reduce light transmission. Heat consumption is directly proportional to the surface area. For structural strength and energy conservation greenhouses should be of low height, but from the point of view of climate control the volume of the greenhouse should be high. The best design represents a compromise between these conflicting requirements. One should use optimization methods to find a good compromise (Kurata and Tachibana, 1978; Bailey and Richardson, 1990).

Construction may be optimized by combining standards of design, construction and loading with simulation models for light transmission, heat consumption and climate control.

CLIMATE REQUIREMENTS AND CLIMATIC SUITABILITY

Greenhouses should be built less in accordance with national traditions and more with regard to general

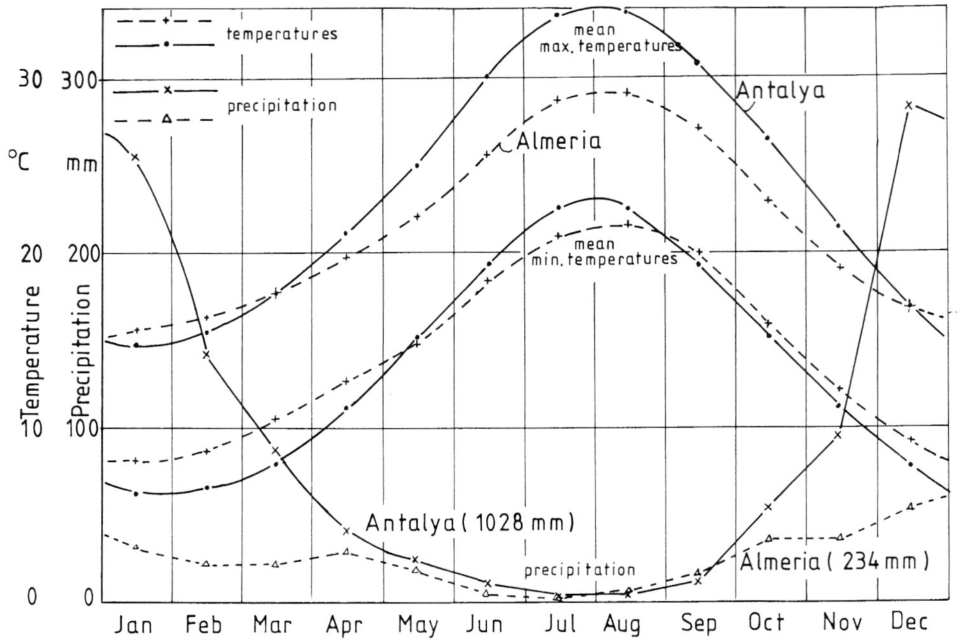

Fig. 2.1. Mean daily minimum and maximum temperatures and mean monthly precipitation for two subtropical climates.

cropping needs and conditions of the different climate zones.

Greenhouse types have been developed for the following different climates:

(1) temperate climates;
(2) subtropical climates and highlands;
(3) tropical climates;
(4) dry and arid climates.

To check the suitability of a region for protected cultivation, the climate data should be compared with those for other regions and with the main requirements for plants.

Most of the plants grown in greenhouses are warm-season species. The climatic requirements of plant growth can be defined and summarized as follows (Sirjacobs, 1989; Baudoin et al., 1990; Verlodt, 1990; Krug, 1991):

(1) Plants can be killed by frost. The absolute minimum temperature in the greenhouse has to be above 0°C. The risks of subzero temperatures can be neglected when the daily minimum outside temperature exceeds 7°C.
(2) Plants grown under protected cultivation are mainly adapted to average temperatures ranging from 17° to 27°C. Taking into account the warming-up effect of solar radiation in greenhouses one can define the climatic limits of suitability to be from 12° to 22°C average daily outside temperature if the greenhouses are not heated.
(3) If the average daily outside temperature is below 12°C greenhouses have to be heated, particularly at night. With an average daily temperature above 22°C artificial cooling may be necessary, or cultivation in greenhouses may have to be abandoned. With average temperatures between 12° and 22°C natural ventilation is sufficient.
(4) The absolute maximum temperature for plants should not be higher than 35 to 40°C.
(5) A minimum of 500 to 550 h of sunshine for the three winter months (November, December and January in the Northern Hemisphere) is desirable. This corresponds to a daily insolation of about 2300 $W\,h\,m^{-2}\,d^{-1}$. The limit for effective production is 1000 $W\,h\,m^{-2}\,d^{-1}$ (Krug, 1991). Artificial lighting may be used for intensive production.
(6) The minimum soil temperature is 15°C.
(7) Verlodt (1990) suggests a threshold for the average night temperature of 15 to 18.5°C for heat-requiring plants such as tomato (*Lycopersicon esculentum*), pepper (*Capsicum* spp.), cucumber (*Cucumis sativus*), melon (*Cucumis melo*) and beans (*Phaseolus* spp.).
(8) Relative humidities of 70–90% can be regarded as being within a safe range.

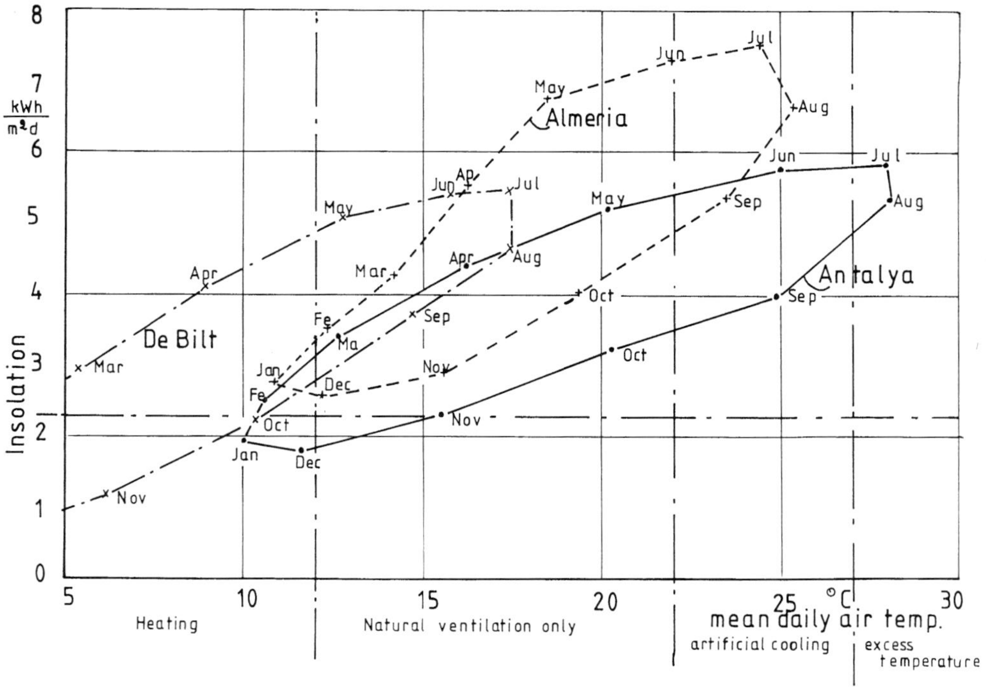

Fig. 2.2. Mean daily insolation versus mean daily temperature for subtropical and temperate climates.

Figure 2.1 shows the mean daily maximum and minimum temperatures and the mean monthly precipitation for two Mediterranean regions, Antalya (Turkey) and Almeria (Spain). Almeria is the most important region for protected cultivation in Spain, with a greenhouse area of about 15 000 ha; 6000 ha of greenhouses are concentrated in the Antalya region.

Almeria is a very dry area with low precipitation in winter and summer. The total yearly precipitation is 234 mm. Antalya has high precipitation mainly in winter. The total yearly precipitation is 1028 mm.

This means that gutters are necessary for greenhouses in Antalya to lead off the rainwater and to protect the crops from water penetrating through the side walls. Collecting the rainwater for irrigation is recommendable and profitable.

The mean daily maximum temperatures are higher in Antalya in summer, and the mean daily minimum temperature is a little lower in Antalya than in Almeria. This means that greenhouses in Antalya need more ventilation efficiency in spring, summer and autumn. The typical Almeria greenhouse type is not transferable to Antalya or other regions.

The conditions for climate control are shown in Fig. 2.2, in which the mean daily insolation is plotted against the mean daily temperature for two subtropical climates, Almeria and Antalya, and for a temperate climate, De Bilt (the Netherlands). If the minimum global radiation is considered to be 2.3 kW h m^{-2} d^{-1}, the Antalya region is below this figure in December and January. If the threshold for heating greenhouses is 12°C, they must be heated from December to the end of February in Antalya. Heating is necessary in January in Almeria. During summer there is lower insolation but higher mean temperatures in Antalya than in Almeria. In both centers crop production without artificial cooling is not possible from June to September. Insolation is comparable in Antalya and De Bilt during summer, but heating is necessary from September to May in the Netherlands.

Figure 2.3 shows the average daily insolation for Antalya (Turkey), and De Bilt (the Netherlands). The total global radiation is higher in Antalya than in the Netherlands. But one has to take into consideration the seasons of production. In the Mediterranean regions transplanting starts in September/October and the plants are developing with decreasing light intensity up to the middle of the cropping season when light intensity is at a minimum. In northern countries like the Netherlands, transplanting starts in February and crops grow with increasing light intensity, with a

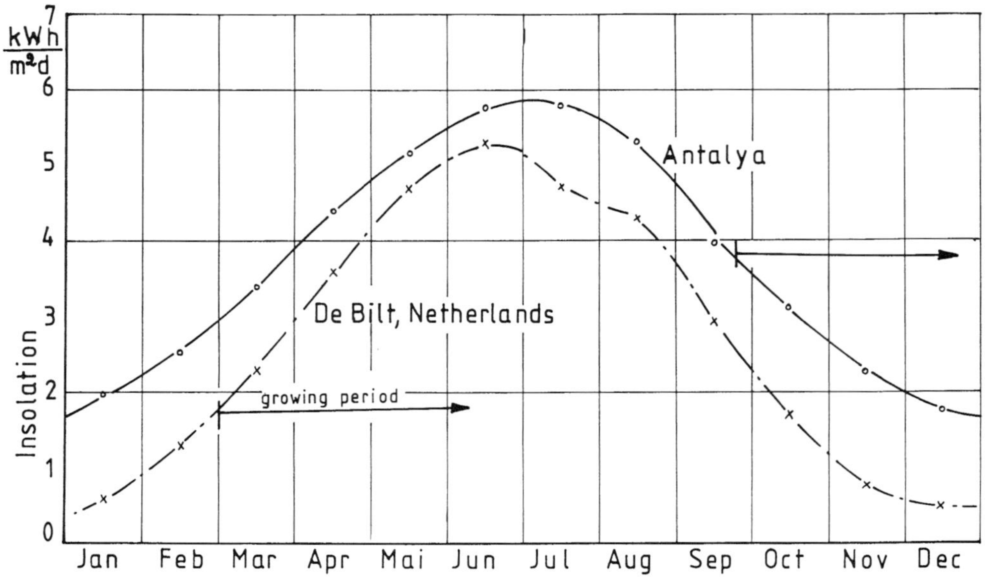

Fig. 2.3. Mean daily insolation for subtropical and temperate climates.

maximum of light intensity in the middle of the cropping season.

The total insolation is nearly the same for a growing period of about 9 months, from September to May in the South and from February to October in the North (von Zabeltitz, 1988).

As light is a limiting factor during the growing season in the South one has to increase light transmissivity of greenhouses as much as possible.

GENERAL DESIGN CRITERIA, GREENHOUSE MODELLING AND STANDARDIZATION

General design criteria

Structures for protected cultivation can be categorized as low plastic tunnels, walk-in tunnels or greenhouses.

Greenhouses can also be classified according to the following scheme.

(1) Construction characteristics: width, single-span or multi-span, side-wall height, roof shape, and roof slope.
(2) Cladding materials: glass, rigid plastics, plastic film or combinations of them.
(3) Construction materials: steel, aluminum, wood, or combinations of them.

In Central and Northern Europe most greenhouses are covered with glass, whereas most of the greenhouses in warmer climates, Japan and the United States are covered with plastic film. Globally, there are more plastic-film greenhouses than glass greenhouses.

The main problems for plant production in greenhouses in *temperate climates* are:

(1) low outside temperature in winter;
(2) wind and snow;
(3) insufficient light in winter;
(4) occasional hot summers.

Greenhouses for temperate climates should have the following characteristics:

(1) sufficient stability against wind and snow loads, taking account of national standards;
(2) an efficient heating system;
(3) high light transmissivity;
(4) efficient insulation against heat losses at low outside temperatures;
(5) sufficient ventilation and shading in summer.

The main problems for plant production in greenhouses in *subtropical (Mediterranean) regions* are:

(1) temperatures below the biological optimum on winter nights, making heating necessary for one to three months;
(2) high temperature during the day, even in spring;
(3) high humidity at night;
(4) wind;
(5) poor water quality and water shortage;
(6) deficiency of carbon dioxide in closed greenhouses during the day.

Greenhouses for subtropical climates should have the following characteristics:
(1) high light transmissivity of both construction and cladding material, and low heat losses even in unheated greenhouses;
(2) heating equipment to increase the temperature during the night;
(3) efficient ventilation;
(4) high stability to wind;
(5) gutters, which are recommended to collect rainwater for irrigation.

In *arid zones* the following conditions have to be considered:
(1) temperatures during the day may be extremely high;
(2) strong winds with dust and sand are likely;
(3) the outside humidity is low;
(4) nights are often cold;
(5) the water supply is frequently brackish or saline, and must be desalinated.

Greenhouses for arid zones should have the following characteristics:
(1) effective ventilation;
(2) sufficient tightness to sand, dust and losses of moisture;
(3) wind-proof construction;
(4) evaporative cooling if necessary;
(5) avoidance of heat losses at night;
(6) simple but efficient solar desalination systems.

Special closed-system greenhouses have been developed for *desert regions* which are subject to the following demands:
(1) water-saving operation and sufficient humidity level (closed atmosphere);
(2) tightness to dust and sand;
(3) solar water desalination;
(4) low requirement of water and energy for cooling.

In some *tropical regions* plant production in greenhouses is possible and advantageous. The production can take place throughout the year, whereas production in the open field is affected by heavy rainfall and wind. The main demands on greenhouses and shelters in tropical regions are:
(1) protection from extremely heavy rainfall, high irradiation and wind;
(2) high ventilation efficiency; no closeable ventilators are necessary;
(3) durability of plastic film (one year at least);
(4) rainwater collection for irrigation in dry seasons.

Energy fluxes, energy and mass balances

The radiant energy in all spectral ranges of the global radiation can be converted into heat energy. In a greenhouse the irradiated energy is partly used for the photosynthesis of plants and the evaporation of water, and partly converted into heat energy. The different heat fluxes in the greenhouse result from heat transfer by conduction, convection, radiation and evaporation. When making an energy balance one has to consider particularly the heat flux in the soil, the evaporation of plants and soil, the radiation from plant and soil, the radiation from construction parts and cladding material, convection exchanges at the surface of soil, plants and greenhouse, heat storage and heat exchange during ventilation (Fig. 2.4).

The energy balance of a greenhouse can be expressed as follows (ASHRAE, 1985):

$$Q_h + Q_{rs} + Q_e + Q_r = Q_c + Q_{rt} + Q_g + Q_p + Q_l + Q_s, \tag{2.1}$$

where Q_h = supplement of heating energy,
Q_{rs} = solar radiation gain,
Q_e = equipment heat gains, artificial lighting and CO_2-burner,
Q_r = plant respiration gains,
Q_c = conductive and convective heat loss,
Q_{rt} = thermal radiation heat losses,
Q_g = ground conduction heat loss,
Q_p = photosynthesis heat losses,
Q_l = latent heat loss,
Q_s = sensible heat loss.

Often it is necessary to determine different climate factors or climate-control measures, for example the temperature of air and plants, or to predict the rate of energy supply. This can be done by establishing energy balances for all components in the greenhouse which interact thermally, usually including the air, plants, soil and roof. Each energy balance consists of equations which describe the gains and losses of energy, and also the rate of change in temperature of the component if a time-dependent analysis is undertaken.

The steady-state energy balances for the components air, plants, floor and roof can be written as follows, using the subscripts to denote the interacting components (Bailey, 1988):

$$\text{Air:}\quad Q_{c,p} + Q_{c,f} + Q_{c,r} + Q_{c,h} + Q_{i,a} = 0, \tag{2.2}$$

$$\text{Plants:}\quad Q_{c,a} + Q_{l,a} + Q_{r,f} + Q_{r,r} + Q_{r,h} + Q_{sr,p} + Q_{r,sk} = 0, \tag{2.3}$$

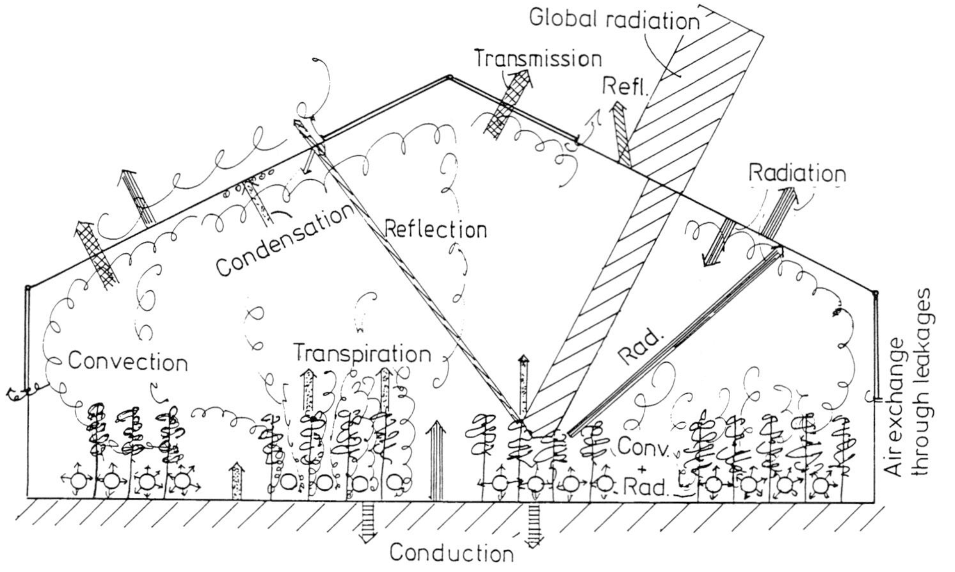

Fig. 2.4. Heat fluxes in the greenhouse.

$$\text{Floor:}\quad Q_{c,a} + Q_{l,a} + Q_{r,p} + Q_{co,f} + Q_{r,r} + Q_{r,h} + Q_{sr,f} + Q_{r,sk} = 0, \tag{2.4}$$

$$\text{Roof:}\quad Q_{c,a} + Q_{l,a} + Q_{r,p} + Q_{r,f} + Q_{r,h} + Q_{sr,f} + Q_{c,ea} + Q_{r,sk} = 0, \tag{2.5}$$

where the left- and right-hand subscripts characterize the heat transfer and the greenhouse component, respectively: c, convection; co, conduction; i, infiltration, leakage; l, latent heat; r, thermal radiation; sr, solar radiation; and a, air; ea, external air; f, floor; h, heater; p, plant; r, roof; sk, sky.

Appropriate values have to be selected for the heat transfer coefficients, radiation shape factors and the properties of the materials such as emissivity and transmissivity of the greenhouse cladding, the thermal conductivity of the floor, etc. It is also necessary to choose values for the boundary conditions, which influence the energy fluxes, but which are not affected by them. Depending on the problem, these can include the soil temperature at a sufficient depth to remain constant, outside air temperature, sky temperature, wind speed, solar radiation and outside air humidity. With the energy-balance equations the different component conditions can be calculated.

Depending on the problem which has to be solved many simulation models have been developed and investigated.

Tables 2.1 and 2.2 give a survey of the simulation models developed for different purposes. The tables were published by Lacroix and Zanghi (1990) and de Halleux (1989) and finally listed by Hölscher (1992), the author of the tables presented here. Table 2.1 shows the balances of single-component models and Table 2.2 shows the balances of multi-component models, distinguishing between static and dynamic models.

In summer the temperature in an unventilated greenhouse can increase considerably. This is the "greenhouse effect", a phrase now widely used by analogy on the macroclimate scale. Part of the irradiated energy is absorbed by soil, plants, interior equipment and construction parts, and raises their temperature. The heat energy is partly stored and partly emitted. In the field, the radiant energy of the sun is also converted into heat by soil and plants. The temperature rise in the greenhouse has two causes:

(1) In comparison to the conditions in the field the air movement in the greenhouse is reduced due to the cover. Convection – that is, the heat release from soil and plants to the atmosphere by air movement – is largely prevented in an enclosed greenhouse. The temperature rises in comparison to the field if the greenhouse is not ventilated thoroughly.
(2) By the laws of radiation, the heat release by radiation is proportional to the fourth power of the absolute temperature T:

$$Q \sim T^4. \tag{2.6}$$

The wavelength of the energy emitted by a heated

Table 2.1
Single-component simulation models

1	2	3	4	5	6	7	8	9	10	11	12	13	14	15			
•	•	•	•	•	•	•	•	•	•	•	•	•	•	•	temperature		climatic boundary conditions
•	•	•	•		•	•	•	•	•	•	•	•	•	•	global radiation		
				•			•								direct – diffuse		
															PAR – Non-PAR ‡		
•							•								thermal radiation sky		
•													•		cloud cover		
•	•		•		•		•		•		•				velocity	wind	
															direction		
	•				•	•				•	•		•		humidity		
							•			•					soil temperature		
				•	•										Bowen ratio		
															CO_2–concentration		
•	•	•	•	•	•	•	•	•	•	•	•	•	•	•	energy balance		balances
					•						•				humidity balance		
															CO_2–balance		
•	•	•	•	•	•	•	•	•	•	•	•	•	•	•	solar heat gain		fluxes considered
•	•	•	•	•	•	•	•	•	•	•	•	•	•	•	heat transfer cover		
•				•			•			•		•	•		thermal radiation		
						•	•			•					conduction soil		
•							•								losses to the surroundings		
•			•					•		•					heat storage		
•	•			•			•		•		•				infiltration	air exchange	
•	•	•													ventilation natural		
•	•				•	•	•	•		•	•	•	•	•	ventilation forced		
										•		•			respiration		
						•				•		•			photosynthesis		
															transpiration (crop)	latent fluxes	
															evaporation (soil)		
•							•	•	•	•	•	•	•		evapotranspiration		
•	•	•	•	•			•	•	•	•	•	•	•		heating system		
			•									•	•		equipment heat		
	•				•						•				cooling system		
	•			•							•				solar energy system		
										•		•			reject heat utilization		
										•					losses by gutter		
															artificial light		
							•		•		•				insulation		
•	•			•				•		•	•				thermal screen		

‡ PAR = photosynthetic active radiation

References

1, Jolliet (1988, edited by de Halleux, 1989);
2, Kimball (1986, Type 15);
3, van de Braak et al. (1984);
4, Wass and Barrie (1984);
5, Milburn (1982);
6, Landsberg et al. (1979);
7, Garzoli and Blackwell (1973);
8, Arinze et al. (1982);
9, Duncan et al. (1981);
10, Damrath (1980, edited by Hölscher, 1992);
11, Shen (1977);
12, Rotz (1977);
13, Price and Peart (1973);
14, Walker (1965);
15, Morris et al. (1958).

Table 2.2
Multi-component simulation models

		1	2	3	4	5	6	7	8	9	10	11	12	13	14	15	16	17	18	19	20	21	22	23	24	25	26
climatic boundary conditions	temperature	•	•	•	•	•	•	•	•	•	•	•	•	•	•	•	•	•	•	•	•	•	•	•	•	•	•
	global radiation	•	•				•		•	•		•	•	•	•	•			•			•	•		•	•	•
	direct – diffuse	•		•	•	•		•			•						•	•						•			
	PAR – Non-PAR ‡						•							•		•			•					•			
	thermal radiation sky	•		•	•	•	•	•	•	•	•	•	•	•	•	•	•	•	•	•	•	•	•	•	•	•	•
	cloud cover	•		•	•			•		•	•				•								•		•		
	wind: velocity	•	•	•	•	•	•	•	•	•	•	•	•	•	•	•	•		•		•	•	•	•	•	•	•
	wind: direction																										
	humidity	•	•	•	•	•	•	•	•	•	•	•	•	•		•	•	•	•	•	•	•	•	•	•	•	•
	soil temperature	•	•	•	•		•			•	•	•		•		•	•	•									
	soil water potential													•								•					
	CO_2–concentration						•		•					•		•			•								
analysed "nodes"	air	•	•	•	•	•	•	•	•	•	•	•	•	•	•	•	•	•	•	•	•	•	•	•	•	•	•
	cover	•		•	•	•	•	•	•	•	•	•	•	•	•	•	•	•	•	•	•	•	•	•	•	•	•
	soil	•	•	•	•	•	•	•	•	•	•	•	•	•	•	•	•	•		•			•	•		•	•
	crop	•		•	•	•	•	•	•	•		•	•	•	•	•	•	•	•			•	•	•	•	•	•
balances	energy balance	•	•	•	•	•	•	•	•	•	•	•	•	•	•	•	•	•	•	•	•	•	•	•	•	•	•
	humidity balance	•	•	•	•	•	•	•	•	•	•	•	•	•	•	•	•	•	•	•	•	•	•	•	•	•	•
	CO_2–balance													•					•								•
climate control equipment	heating system	•	•	•	•	•	•	•	•	•	•	•	•	•	•	•	•		•	•			•	•			•
	solar energy system		•		•		•							•									•				
	heat storage system		•		•	•	•							•								•	•				
	cooling system		•				•		•					•		•	•		•	•			•	•	•		
	humidification system												•							•							
	dehumidification system						•														•						
	artificial light	•																	•								
	CO_2–application													•					•								
generalities	insulation				•										•								•				
	multi layer cover	•	•	•			•		•	•					•								•	•			
	crop development			•		•	•			•				•		•											
	quasi one–dimensional	•		•	•	•	•	•	•	•	•	•	•	•	•	•		•	•	•		•		•	•		•
	multi–dimensional		•														•						•			•	
	thermal inertia: air	•		•	•	•		•		•	•			•	•	•	•										
	thermal inertia: cover	•			•	•		•		•	•	•		•	•	•											
	thermal inertia: soil	•	•	•	•	•	•	•	•	•	•	•	•	•	•	•	•	•									
	thermal inertia: crop	•			•	•		•		•					•	•	•										
	humidity inertia	•			•	•		•		•	•			•	•	•	•										
	CO_2–inertia													•													

Models 1–17: dynamic; models 18–26: static

‡ PAR = photosynthetic active radiation

References

1, de Halleux (1989, edited by Hölscher, 1992);
2, Kimball (1986, Type 5);
3, Cormary and Nicolas (1985);
4, Arinze et al. (1984);
5, Bot (1983, edited by de Halleux, 1989);
6, Cooper and Fuller (1983);
7, Ahmadi and Glockner (1982);
8, Avissar and Mahrer (1982);
9, von Elsner (1982, edited by Hölscher, 1992);
10, Glaub and Trezek (1981);
11, Kitas, 1980 (edited by de Halleux, 1989);
12, Kindelan (1980);
13, van Bavel and Sadler (1979);
14, Tantau (1975, edited by Hölscher, 1992);
15, Soribe and Curry (1973);
16, Takakura et al. (1971);
17, Selcuk (1970);
18, Seginer et al. (1986);
19, Seginer and Kantz (1986);
20, Silveston et al. (1980);
21, Takami and Uchijima (1977);
22, Kimball (1986, Type 14);
23, Kimball (1973);
24, Maher and O'Flaherty (1973);
25, Selcuk (1971);
26, Businger (1966, edited by de Halleux, 1989).

body depends on the temperature of the body. This dependence between the wavelength at which radiation reaches a maximum and the temperature is expressed by the following law:

$$\lambda_{max} = \frac{2\,885\,000}{T} \text{ (nm).} \quad (2.7)$$

The sun has a surface temperature of some 5700 K. Thus the energy maximum in the spectral wavelength range lies at

$$\lambda_{max} = \frac{2\,885\,000}{5700} = 506 \text{ nm.} \quad (2.8)$$

The wavelength range of the global radiation is from 290 to 3000 nm. The spectral transmissivity of glass lies in the wavelength range from 320 to 2900 nm. Thus glass is transmissive to global radiation except the ultra-violet and extreme infrared parts of the spectrum.

If an object in a greenhouse has reached a temperature of 25 to 30°C (i.e. 298 to 303 K) it emits the heat in a wavelength range with a maximum of $\lambda_{Bmax} \approx 9800$ nm. Figure 2.5 shows the wavelength ranges of global radiation and long-wave radiation (IR).

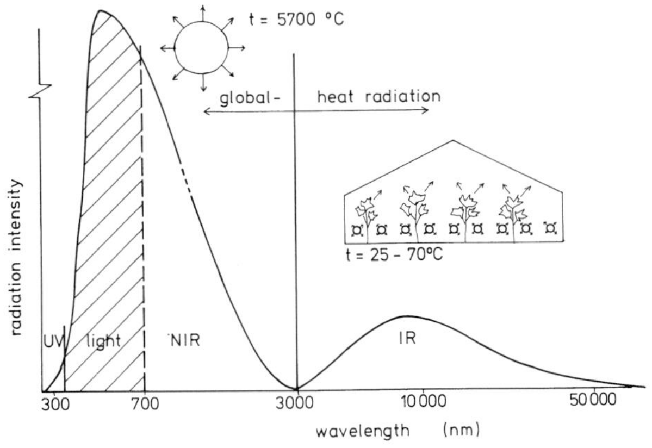

Fig. 2.5. Wavelength of global radiation and long-wave radiation.

Long-wave radiation is not transmitted by glass; thus, the heat energy released by soil, plants and interior equipment remains in the greenhouse. The greenhouse acts as a radiation trap.

Comparing the spectral transmissivity of glass and, as an example, polyethylene film, the latter is to a large extent transmissive to the wavelength range around 10 000 nm, which means that the emitted energy can pass the film. Nevertheless, the temperature rise in a greenhouse covered with plastic film is the same as in a greenhouse covered with glass. This is because the temperature rise in a greenhouse results mainly from the prevention of air exchange with the higher air layers of the atmosphere and to a lesser extent by the radiation trap effect. It is therefore wrong to explain the heating "greenhouse effect" on the basis of radiation trapping alone, or to treat it as equivalent to the radiation trap.

A considerable proportion, 30–70%, of the radiation coming into the greenhouse is used for the evaporation of water.

Loads and standardization

Greenhouses are lightweight structures, which have to have high light transmission for optimum plant growth. The structures also have to withstand the outside weather stresses such as wind, snow, rain and hail. Damage to greenhouses, and especially to plastic-film greenhouses, is caused very often by storms and heavy snow-fall, because the main construction components are too weak. To avoid severe damage, the greenhouse structure should be calculated and designed in accordance with standards which give figures for the different minimum loads which the greenhouse should withstand.

Standards for glass greenhouses have been prepared and are used in several countries with temperate climates, for example France (Afnor U 57-060, 1989), Germany (DIN 11535, 1982), Italy (UNI 6781-71, 1971), the Netherlands (NEN 3859, 1990), the United Kingdom (BS 5502: Part 22, BSI, 1987) and the United States (ASAE EP288.4, 1990).

Similar standards for plastic-film greenhouses are needed for all climate regions. Some standards or Codes of Practice have been prepared, for example for France (Afnor U 57-063, 1989) and the United Kingdom (BSI Document No. 91, BSI, 1991).

A Working Group of the International Society for Horticultural Science (ISHS, 1991) has prepared a draft Code of Practice for the design and construction of, and loadings for, greenhouses. The European Committee for standardization worked out a Greenhouse standard for all countries of the European Community (CEN, 1997).

The requirements will vary with location and topography, but some basic figures for the design and loads can be given.

Some standards of European countries classify greenhouses in different classes according to the intended minimum life for which they are designed. This classification represents the traditional and necessary

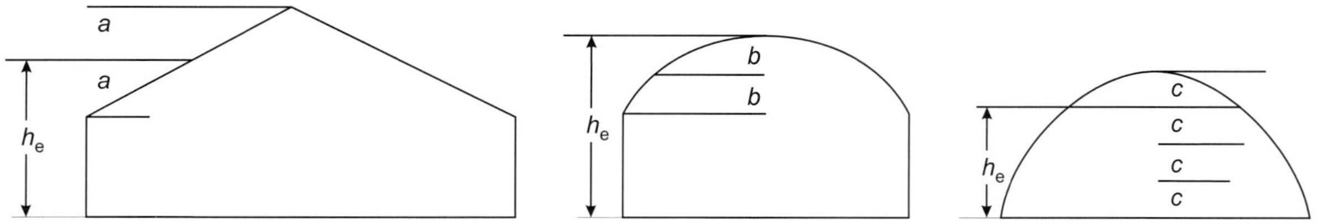

Fig. 2.6. Effective height of greenhouses.

shorter economic and technical lifes of greenhouses compared with those of conventional buildings. The main loads which have to be taken into account are:

(1) dead load or permanent load;
(2) imposed loads (crop loads);
(3) snow load;
(4) wind load.

Special load combinations, given in the standards, should be considered in the design.

Dead load

Dead load should be calculated from the unit weights of structure and cladding material. In some cases the weights of fixed equipment such as heating, lighting, irrigation systems, shading and thermal screens are included in the dead load (UK: BSI, 1987, 1991; France: Afnor U 57-060, 1989; Germany: DIN 11535, 1982). Other standards exclude the weight of installation even if permanent (the Netherlands: NEN 3859, 1990).

Crop loads

Where structures support crops, special crop loads have to be considered. Examples for minimum vertical loads are: tomatoes, cucumbers, $0.15\,\mathrm{kN\,m^{-2}}$; pot plants suspended from the structure, $1.0\,\mathrm{kN\,m^{-2}}$.

Where crops are suspended on separate horizontal wires, the horizontal tensile forces transmitted to the structure have to be taken into account. For example, the horizontal load for trained tomatoes and cucumbers is typically 1.2 kN per wire.

Snow load

Snow loads are calculated on the horizontal projection of the roof by using a basic snow load given in snow maps adjusted by a multiplicative coefficient which normally reduces the snow loads compared to conventional buildings. A minimum snow load of $0.25\,\mathrm{kN\,m^{-2}}$ has to be applied for heated greenhouses in the Netherlands and Germany. The snow loads are higher in other areas and in mountain regions.

Wind load

The wind load normally will be calculated by the following equation:

$$F_{\mathrm{w}} = ACp_{\mathrm{w}}, \qquad (2.9)$$

where F_{w} is the wind force, A is the greenhouse surface affected by wind pressure or wind suction, C is the wind-pressure coefficient, and p_{w} is the dynamic wind pressure. Wind forces appear as pressure forces and suction forces on the surface of the greenhouse.

The dynamic wind pressure depends on the effective height of the greenhouse, which is defined as the distance between ground level and the average height

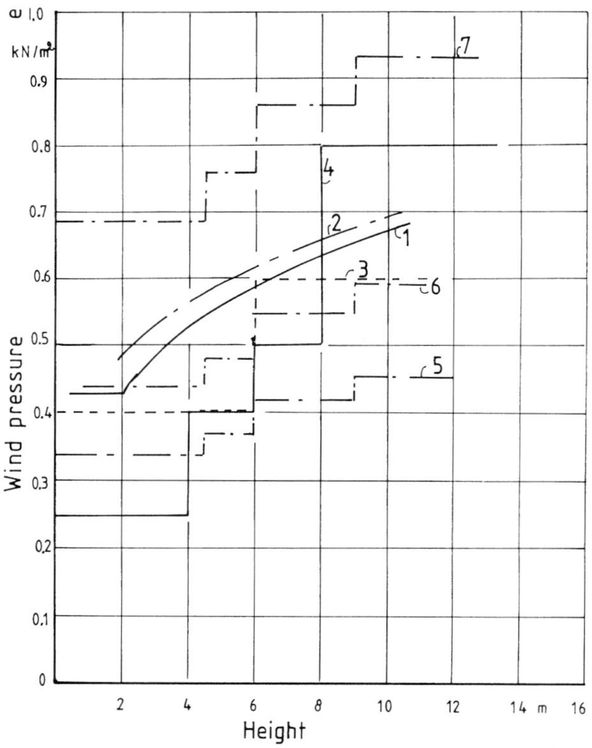

Fig. 2.7. Dynamic wind pressures depending on greenhouse height. 1, NEN 3859 (1990), Netherlands; 2, Draft 01 for European Greenhouse Standard (ISHS, 1991); 3, UNI 6781-71 (1971), Italy; 4, DIN 11535 (1982) Germany; 5–7, ASAE EP288.4 (1990) for wind speeds of $70\,\mathrm{km\,h^{-1}}$, $80\,\mathrm{km\,h^{-1}}$ and $100\,\mathrm{km\,h^{-1}}$.

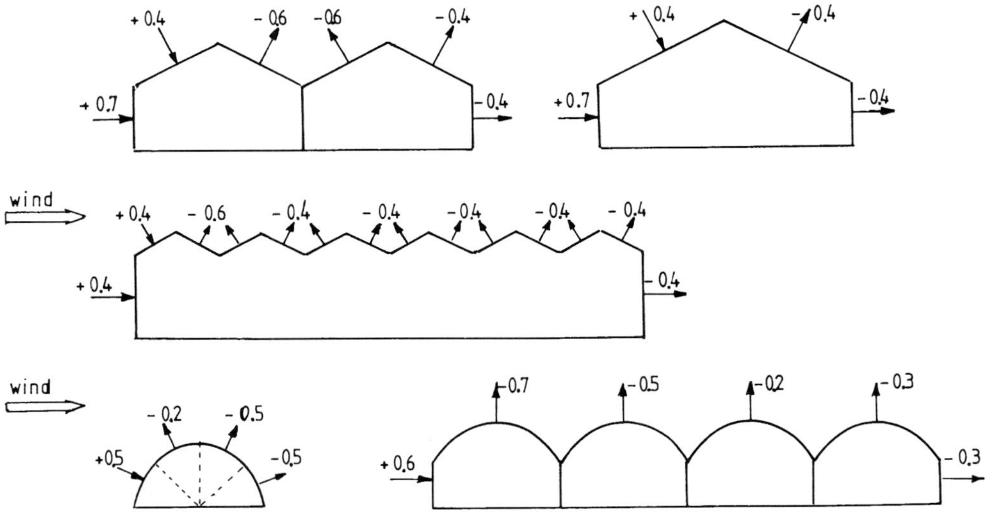

Fig. 2.8. Wind pressure coefficients c: +, pressure forces; –, suction forces.

between gutter or eaves and the highest point of the roof (NEN 3859, 1990) (Fig. 2.6). Some figures of dynamic wind pressures depending on the height are given in Fig. 2.7 in accordance with different standards.

The wind pressure coefficient takes into consideration the pressure and suction forces, and depends on the surface area for which the wind forces are calculated. Some examples are given in Fig. 2.8 (NEN 3859, 1990). Higher local pressure coefficients due to higher wind forces have to be applied at the gable corners, and the gable and side-wall roof sides (Fig. 2.9). The width of glass panes in those areas should be smaller than in the other surfaces. The continuous ventilators of glass and plastic-film greenhouses should end 2 m before the gable.

Wind speeds for design can be taken from wind maps, but local wind speeds can vary considerably from those in area maps. Special reduction factors may be applied for different classes of greenhouses.

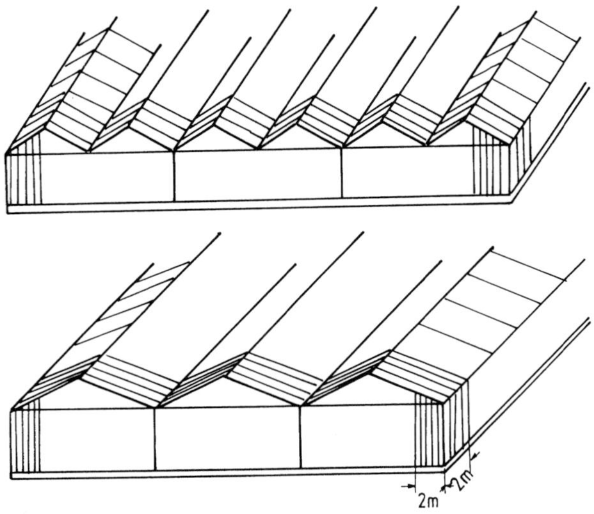

Fig. 2.9. Local pressure forces at gables and sidewalls.

Foundation

The foundation of a greenhouse structure should safely sustain and transmit to the ground the worst combination of the different loads. The allowable bearing pressures of the soil have to be taken into consideration, as for instance, in BSI Document No. 91 (BSI, 1991) and DIN 1054 (1979). In some temperate-climate regions the outer foundations for glass greenhouses have to go down to a frost-free depth (80 cm in Germany).

Because of the light weight of greenhouses, special attention should be paid to the ability of the foundations to resist uplift under wind action. For concrete foundation blocks the maximum upward vertical load should not exceed the uplift force F_u (BSI, 1991; Afnor U 57-060, 1989):

$$F_u = \rho_f V_f + \rho_S V_S + F_F, \qquad (2.10)$$

$$F_F = S\left[\rho_S\left(D + \tfrac{1}{2}H\right) K_0 \tan \Phi + c\right], \qquad (2.11)$$

where ρ_f is the density force of foundation material (kN m^{-3}), V_f is the volume of foundation block (m^3), ρ_S is the density force of soil (kN m^{-3}), V_S is the volume of soil above the foundation block (m^3), S is the area of friction faces (m^2) – i.e. lateral area of buried foundation block, D is

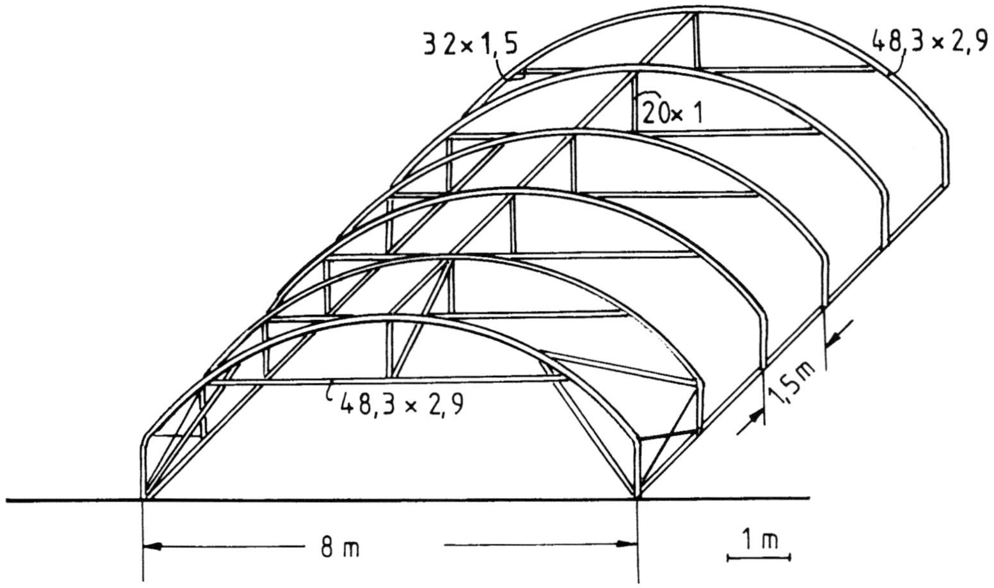

Fig. 2.10. Plastic-film greenhouse with dimensions calculated in accordance with NEN 3859 (Waaijenberg and Denkov, 1992).

Table 2.3
Some average soil properties

Soil	ρ_S ($kN\,m^{-3}$)	Φ (°)	c ($kN\,m^{-2}$)
Sandy	20	30	0
Sandy clay loam	18	20	0
Clay	18	20	4

the depth of soil over foundation block (m), H is the depth of foundation block from top surface to bottom surface (m), K_0 is the coefficient of horizontal earth pressure at rest ($=1-\sin\Phi$), Φ is the angle of shearing resistance (deg), and c is the cohesion ($kN\,m^{-2}$). Some average soil properties are given in Table 2.3.

Calculation example for a plastic-film greenhouse

The results of an analysis of storm damage after a severe storm in the Netherlands in 1990 can be summarized for plastic-film greenhouses as follows (Waaijenberg and Denkov, 1992):

(1) profiles used in the bent trusses for plastic-film greenhouse tunnels were too light and the arches were deformed by wind;
(2) plastic film sometimes appeared to be stronger than the structure;
(3) plastic-film tearing started at sharp edges of construction components which were in contact with the film;
(4) the foundations often were not sufficiently secured against uplift;
(5) damage to a plastic-film greenhouse often started from the ventilation openings;
(6) the connections of arch, purlin and ridge tubes often were not sufficient, especially clamp connections;
(7) too few bracings were fixed to the structure to give sufficient stability.

Experience from other countries shows that these weak points occur frequently.

The requirements for a plastic-film greenhouse in the Netherlands have been calculated in accordance with the Netherlands Standard NEN 3859 as shown in Fig. 2.10 (Waaijenberg 1990, Waaijenberg and Denkov, 1992). These calculations allow for a combination of wind pressure and wind suction, with a snow load of $0.25\,kN\,m^{-2}$, but without crop load. Table 2.4 lists the dimensions of the arch tubes for single-span greenhouse tunnels with and without a straight side wall of 1 m, for different widths and different arch distances. The dimensions of the arch tubes can be reduced if the arch distance is 1.50 m instead of 2.0 m.

Waaijenberg and Denkov (1992) made similar calculations in accordance with climate loads of Tunisia and with a crop load of $0.15\,kN\,m^{-2}$. Figure 2.11 shows a structure with horizontal trellis girders and another

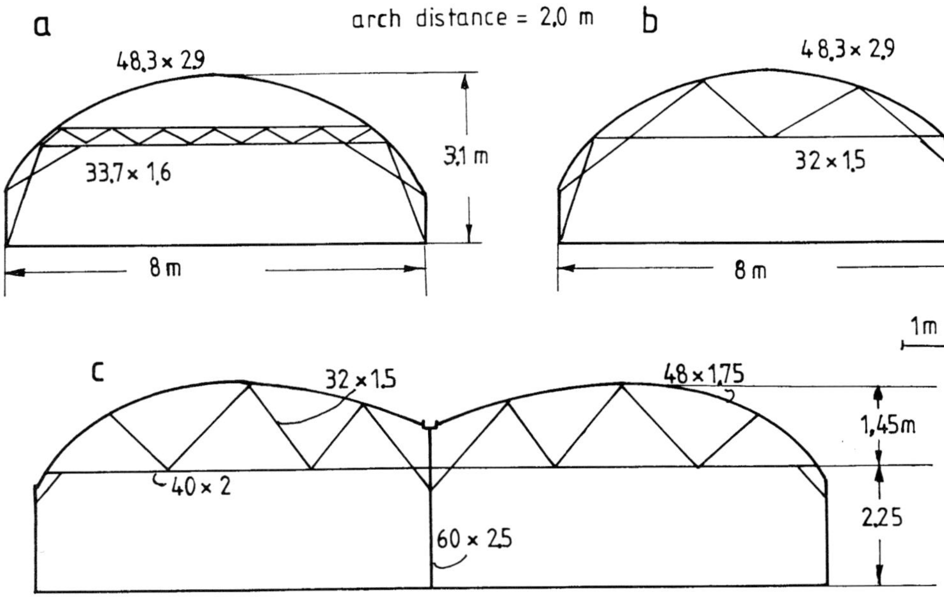

Fig. 2.11. Plastic-film greenhouses, (a) with trellis girder, (b) with extra braces and (c) a bitunnel. Dimensions calculated with crop loads for subtropical regions (Waaijenberg and Denkov, 1992).

Table 2.4
Minimum dimensions of arch tubes for plastic-film tunnel greenhouses [1]

Arch distance (m)	Width [2]				
	4.4 m, no side wall	6 m		8 m	
		no side wall	with side wall	no side wall	with side wall
2	48.3×2.9	60.0×2.0	60.3×3.0	60.0×3.0	60.0×4.0
1.5	42.4×2.6	48.3×2.9	48.3×2.9	60.0×2.0	48.3×2.9

[1] Waaijenberg and Denkov (1992).
[2] Values given as (tube diameter in mm)×(wall thickness of tube in mm).

one with extra bracings. The arch tubes combined with trellis girders and extra bracings can be made of tubes with dimensions 48.3 × 2.9 mm even if the arch distance is 2 m. The same figure shows the dimensions of a bitunnel calculated also with crop load.

Generally cold-formed circular steel tubing is used for the framework of arched plastic-film greenhouses. The dimensions of the tubes should be

$$\frac{D}{t} < 50 \quad \text{or} \quad t > \frac{D}{50},$$

where D is the outside diameter and t is the wall thickness in mm. The wall thickness must not be less than 1.2 mm (BSI, 1991) or 1.5 mm (NEN 3859, 1990).

GREENHOUSE STRUCTURES

Shapes of greenhouse structures

Greenhouses have to provide optimal climate conditions for the plants growing inside. The shape of the greenhouse structure influences the internal climate temperature, humidity and light transmission. The shapes that appear most frequently are (Fig. 2.12):

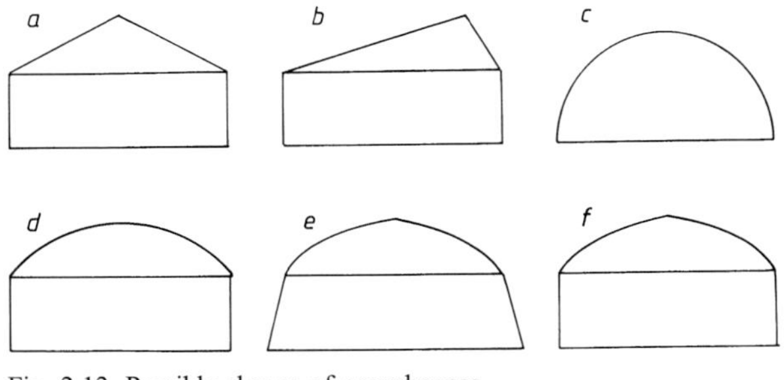

Fig. 2.12. Possible shapes of greenhouses.

(a) saddle roof; (b) saw-tooth or shed roof; (c) round-arched tunnel; (d) round arch with vertical side wall; (e) pointed arch with sloping side wall; and (f) pointed arch with vertical side wall.

Height of the greenhouse

Ventilation efficiency by natural ventilation depends on the height of the greenhouse, if ventilation openings are positioned at the ridge and on the side wall. The higher the ridge and the greater the distance between

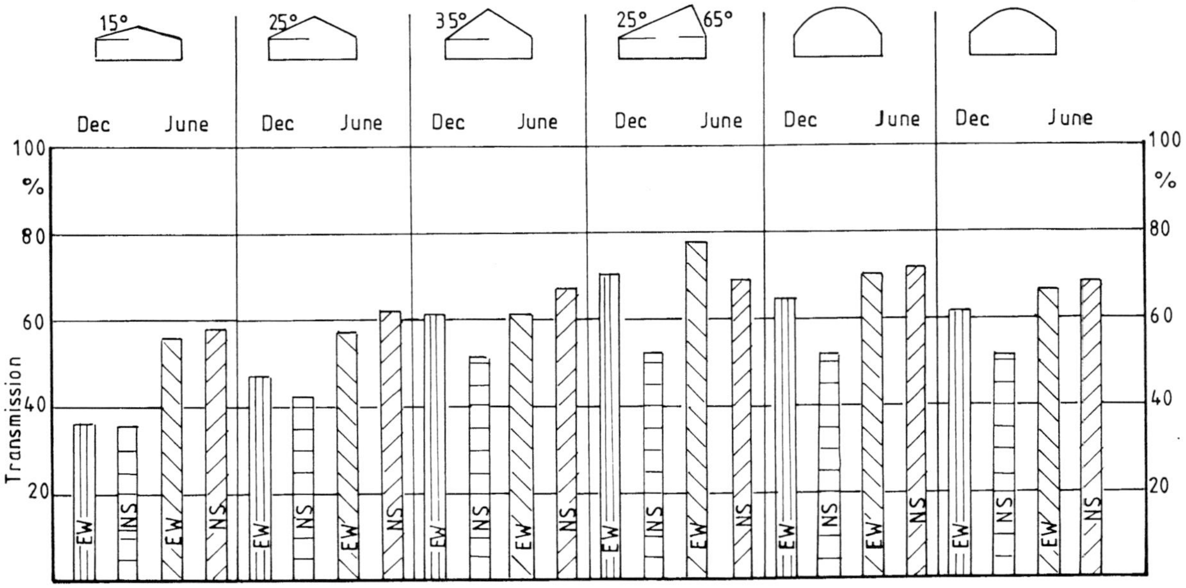

Fig. 2.13. Average light transmission for different greenhouse types.

ventilators at the ridge and side wall, the higher are the pressure differences. The ventilation efficiency is proportional to the pressure differences (Bot, 1983). On the other hand, wind loads and the strength of the structural components depend on the greenhouse height, just as the heat requirement depends on the surface area. High greenhouses with large volumes provide better climatic conditions but also increase the heat requirement if heating is needed.

Light transmission of greenhouses

Only a certain percentage of global radiation penetrates through the cladding of the greenhouse. The global radiation is reduced by:

(1) absorption and reflection at the cladding material,
(2) shading by greenhouse structural components,
(3) dirt on the cladding material, and
(4) condensation on the cladding material.

The light transmission is also influenced by the orientation of the greenhouse and the altitude of the sun.

The transmission of a greenhouse is defined by the relation of solar radiation inside the greenhouse (q_{rg}) and outside (q_{ro}):

$$T_r = \frac{q_{rg}}{q_{ro}}. \tag{2.12}$$

Different researchers have measured and calculated the transmission of greenhouses (Schulze, 1954; Stoffers, 1967; Nisen, 1969; Kirsten, 1973; Kozai et al., 1978; Bot, 1983; Critten, 1983, 1989; van den Kieboom, 1984)).

Figure 2.13 shows the average transmission for different greenhouse types with East–West and North–South orientation in December and June (Nisen, 1969). The following conclusions can be drawn from this figure:

(1) light transmission increases with roof slope in saddle-roof constructions;
(2) Light transmission in East–West-oriented greenhouses is higher in winter and lower in summer than in North–South oriented greenhouses;
(3) A saw-tooth or shed roof with the steeper and shorter roof side to the South in the Northern Hemisphere has a better transmission than a saddle roof, but the surface area is larger;
(4) Greenhouses with curved roofs have better transmission than greenhouses with saddle roof and 25° roof slope.

Figure 2.14 shows some calculations for transmission at different times of the day (Kirsten, 1973). A reduction of 10% by dirt and reduction by structural components is taken into consideration. Best transmission in winter is given by the curved roof of single-span greenhouses with an East–West orientation, followed by the saw-tooth and saddle roof. The transmission is a little better under a saddle roof in summer than under a saw-tooth roof. The transmission through multi-span

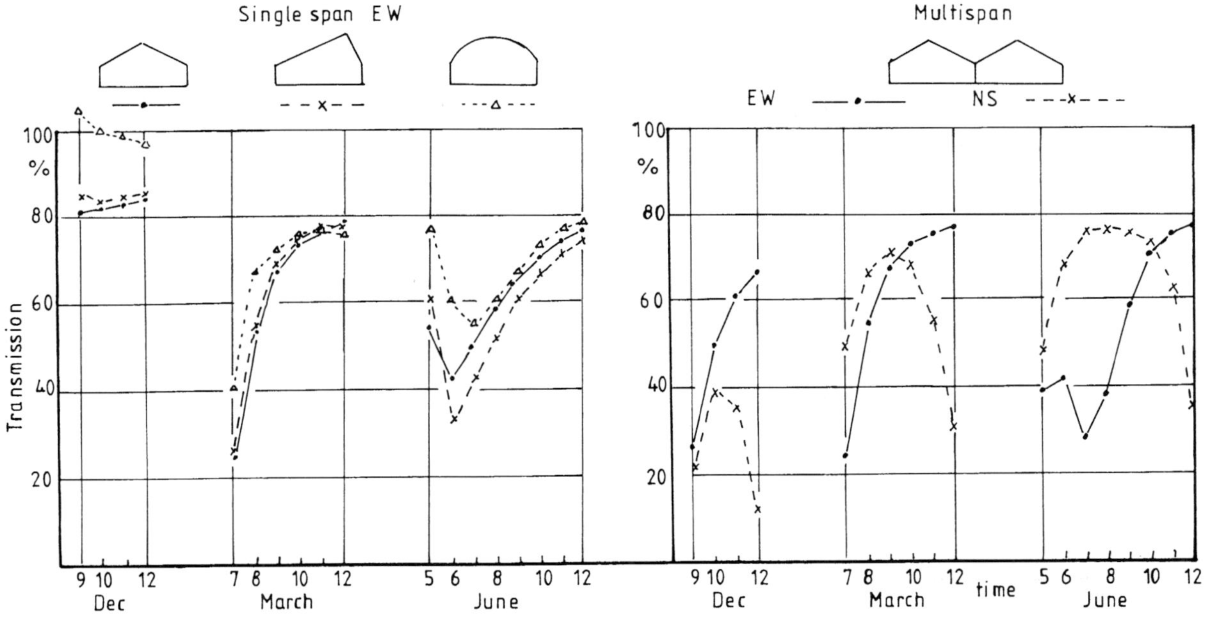

Fig. 2.14. Hourly light transmission for different greenhouse types.

Table 2.5
Light transmission (%) of a Venlo-type greenhouse

Date	Orientation	
	East–West	North–South
21 December	45	35
21 February	58	53
22 March	59	62
21 April	60	67
21 June	65	70

saddle-roof constructions is higher in winter with East–West orientation than North–South. There is more light in the greenhouse with North–South orientation in the morning in spring and summer. Table 2.5 gives average daily figures for a Venlo-type glass greenhouse, calculated by Bot (1983).

Condensation on the greenhouse cladding

Water condensation on the inner surface of the cladding influences the light transmission and the heat transfer through the cladding material. One has to distinguish different kinds of condensation. On untreated plastic materials condensation appears in drop form, which reduces light transmission considerably as incoming light will be totally reflected inside the drops (Hsieh and Rajvanshi, 1977). Treated, so-called "Anti-Drop" materials have film condensation, which can improve the transmission. On glass, water condenses as a water film or in form of long drops which run off. Semicircular drops on plastic material can drop down and drench the plants, if the roof angle is too low. With increasing water condensation the heat transfer through the cladding material increases (Walker and Walton, 1971; Delwiche and Willits, 1984; Garzoli, 1984; Weimann, 1986). In heated greenhouses the heat loss increases by condensation; in unheated greenhouses condensation can have a cooling effect both by day and by night. Eggers (1975) measured a reduction of 15–18% in light transmission under a plastic-film tunnel with drop condensation.

The light transmission of different cladding materials with condensation has been measured in Hannover (Fähnrich et al., 1989). Figures 2.15–2.17 demonstrate the form of condensation on glass, untreated polyethylene film, and treated polyethylene "Anti-Drop" film (at roof slopes of 15–30°). A water film and flat drops, which run off but do not fall down, are formed on glass at 15° and 30°. Semicircular drops of different sizes are formed on untreated polyethylene film. The drops are larger with a roof slope of 15°. Bigger drops fall down from the film on to the plants. A new "Anti-Drop" film shows only film condensation. Table 2.6 shows the light transmission of the materials in dry condition and with condensation. The results can be summarized as follows:

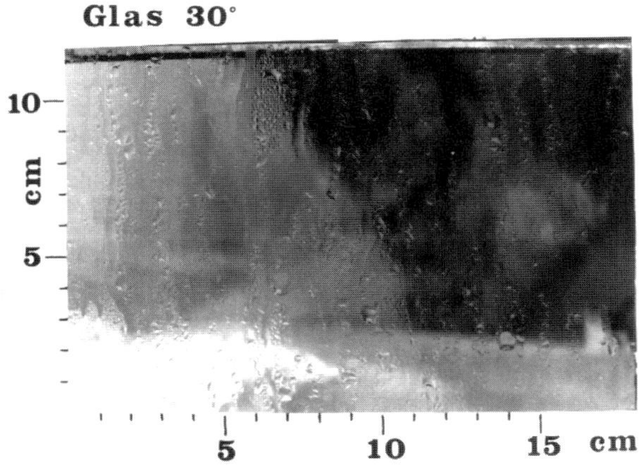

Fig. 2.15. Water condensation on glass, 30° roof slope.

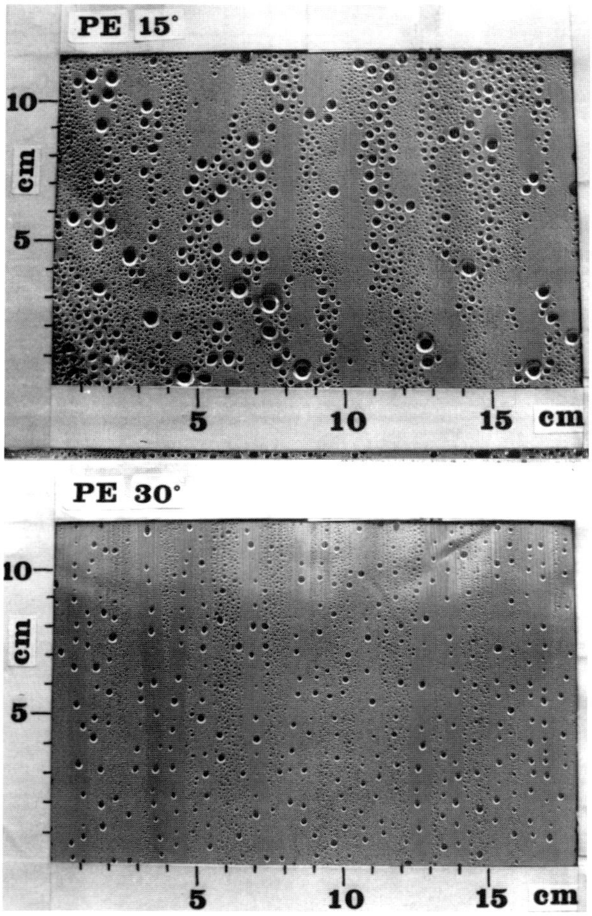

Fig. 2.16. Water condensation on untreated polyethylene film, 15° and 30° roof slope.

(1) with glass, heavy condensation reduces light transmission due to flat drops, but drops do not fall off;

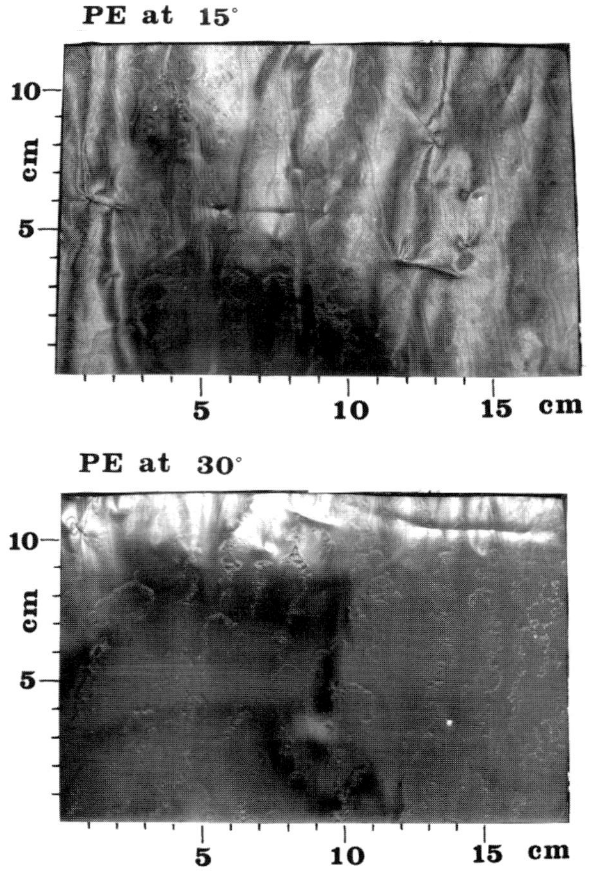

Fig. 2.17. Water condensation on treated polyethylene "Anti-Drop" film, 15° and 30° roof slope.

Table 2.6
Light transmission of glass and polyethylene film in dry and wet conditions

Material	Roof slope	Light transmission (%)	
		dry	with condensation
Glass	15°	90.8	82.7
	30°	89.9	81.9
Polyethylene film	15°	90.8	82.7
	30°	89.9	81.9
Polyethylene film,	15°	87.1	88.1
"Anti-Drop"	30°	86.4	88.3

(2) with polyethylene film, drop condensation reduces light transmission; many drops fall off with low roof slopes, and a minimum roof slope of 20–25° should be used to avoid heavy drop fall from untreated plastic film onto the plants;

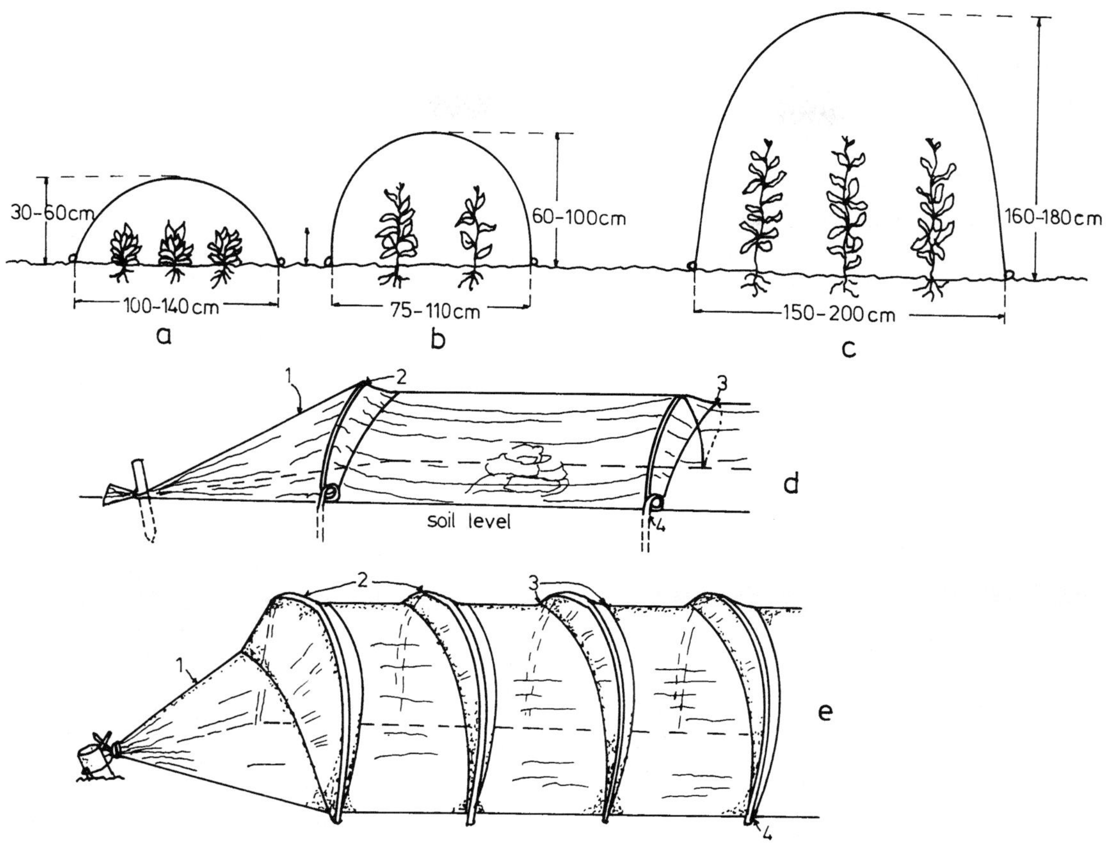

Fig. 2.18. Low tunnel and walk-in tunnel. (a) Low tunnel for low-growing plants. (b) Low tunnel for high-growing plants. (c) Walk-in tunnel. (d) Construction of low tunnel. (e) Construction of walk-in tunnel. (1) Plastic film. (2) Supporting hoops. (3) Retaining cord or wire. (4) Soil anchor.

(3) with treated polyethylene or "Anti-Drop" film, light transmission is not reduced.

Low plastic tunnels and walk-in tunnels

Low tunnels and walk-in tunnels could be called miniature greenhouses. Typical dimensions are shown in Fig. 2.18. Many types of tunnels have been developed consisting of a semi-circular supporting framework covered with plastic film. The supporting framework can be made of wood, wire, plastic-coated wire or bamboo cane. The hoops are stuck in the ground at a spacing of 2 to 3 m. Low tunnels are normally put up only for one cropping period and taken down for tillage. Figure 2.18 shows a low plastic tunnel. In addition to the supporting hoop a string or wire fixes the film. The film can be lifted between the supporting hoop and the retaining wire on both sides for ventilation.

A simple method is to stretch a film over the hoops and dig it into the ground at the sides.

Low tunnels protect plants against low temperature, wind, rain, hail, birds and insects. They are normally unheated. Low tunnels are applied to quicken plant growth, to increase production and to improve the quality of the products.

The main advantages of low tunnels are low cost and simple construction. The disadvantages are difficulties in heating, ventilation and plant husbandry.

Walk-in tunnels are miniature greenhouses 1.8–2.0 m high covered with plastic film. They can remain in place for longer than one cropping period. The grower can walk into the tunnel for planting, sowing, husbandry and harvest. If ventilation is necessary the gables are opened, or, in case of outside temperature rise, holes are cut into the film.

Plastic-film greenhouses

In many countries, especially in those with a warmer climate, plastic-film greenhouses are used to a much greater extent than glass greenhouses for cultivation. Plastic-film greenhouses are generally cheaper than those covered with glass. Plants can be cultivated easily if the greenhouse is built to certain specifications.

Several of the existing greenhouse designs have disadvantages. For example:

(1) the work required for assembly and replacement of the film,
(2) reduction of film stability by global radiation and friction against the structure,
(3) fluttering of the film against the structure, sensitivity to storm,
(4) falling of drops condensed on the film,
(5) inadequate ventilation in multi-span greenhouses, resulting on poorer quality.

The following design criteria can be defined for the construction of plastic-film greenhouses:

(1) Film destruction by fluttering in the wind has to be avoided. Films must be stretched tightly with simple devices.
(2) Simple methods should be available to replace the film with minimum time and effort – for example detachable fastening devices on the structure.
(3) Wind loads have to be considered.
(4) Insulation may be needed for those parts of the structure which are in contact with the film and are heated by solar radiation (mainly metal structures). An insulating white plastic strip can be put between film and metal parts, or the film can be painted white on the outside over the metal parts. This is not necessary when wooden structures are used.
(5) Efficient ventilation is required.
(6) The greenhouse must be tight to air exchange when closed. In case of inversion conditions, the ventilators have to be opened slightly so that indoor temperatures do not sink below outdoor temperatures and condensation on the plants does not occur.
(7) Greenhouse volume should be as large as possible especially in warm climates.
(8) Greenhouses with gutters are recommended for collecting rainwater for irrigation.
(9) The film should have a long life.
(10) Falling of drops of condensed water should be avoided by appropriate inclination of the roof, film treatment or "Anti-Drop" film.
(11) The height of the side walls should be sufficient; construction parts in the house should be few, so as to permit use of machines.
(12) The width of plastic film available limits the width of individual spans in multi-span greenhouses, if the film is stretched longitudinally.

In addition to these general demands, there are influences which depend on the climatic conditions of the different regions. Two different forms of greenhouses have to be distinguished to meet the requirements of different climatic zones:

(1) Constructions with permanently open ventilators and large internal volume for ventilation efficiency for tropical wet and tropical dry climates.
(2) Constructions with closeable ventilators, mainly for temperate subtropical climates and for tropical highlands when temperatures drop below the prescribed internal temperature at night.

A large number of plastic-film greenhouses have been developed world-wide. They range from simple structures of wood with nailed-on plastic film and round-arched tunnels to complex constructions. In the following some types are described (von Zabeltitz, 1990a,b).

The simplest form of plastic-film greenhouse is the single-span round-arched greenhouse. Figure 2.19 shows the dimensions of tunnels made of galvanized steel pipes. The width of the tunnels is 7–9 m, the height 3–3.3 m, the length 40–60 m. The space between the stanchions is 2.5 m in the middle, but it is 2 m at the gables because of higher wind loads. Longitudinally, five pipes are installed and also 26–29 wires attached. Wires in plastic-film greenhouses have the disadvantage that the film lying on the wires will be damaged by friction. A space of 2 m is left between two tunnels. This causes a loss of about 20% of area which cannot be cultivated.

Round-arched greenhouse tunnels have some disadvantages (von Zabeltitz, 1990a,b). The ventilation efficiency is insufficient when the overlapping plastic films are drawn to the side. A roll-up system for ventilation openings in the side walls is much better (Fig. 2.20).

The advantage of the single-span tunnels is the relatively simple construction system and the low wind resistance.

Different types of saddle-roof greenhouses with plastic-film cladding are being built around the world (Fig. 2.21). The stanchions are made of wood, steel or concrete. The roof structure is made of wood. The plastic film is often nailed on wooden laths, which is a disadvantage for changing the films and for film durability. Roll-up ventilators are preferable.

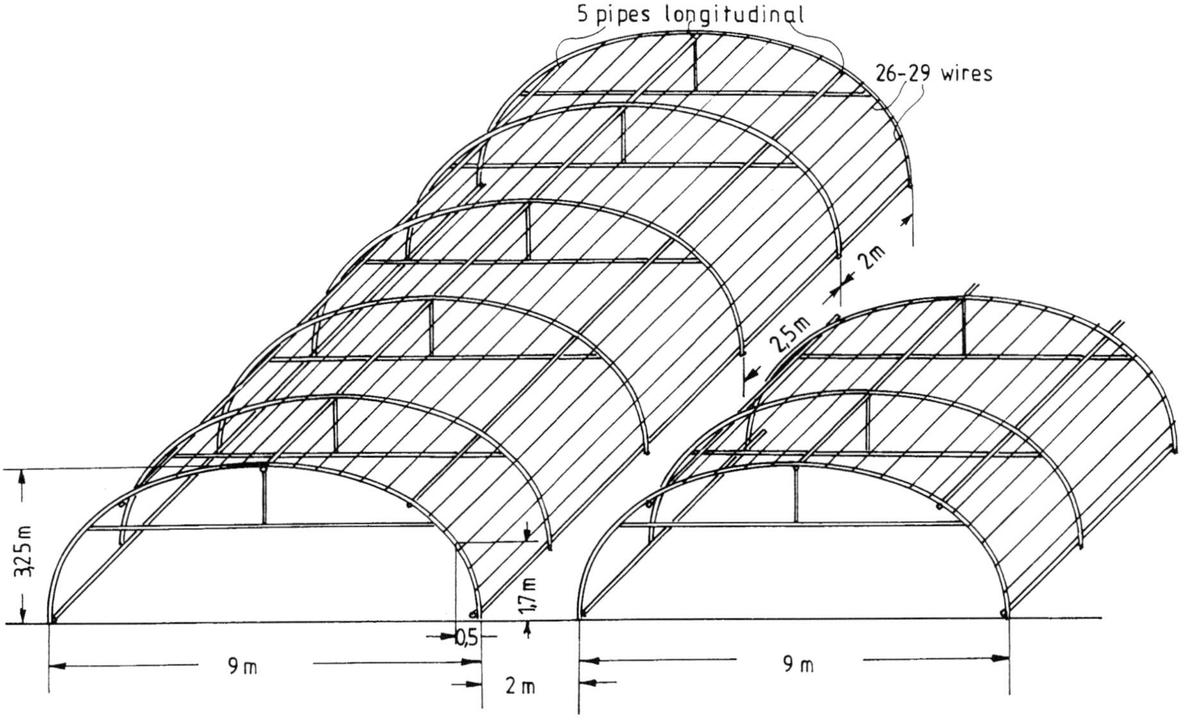

Fig. 2.19. Dimensions of round-arched tunnel greenhouses.

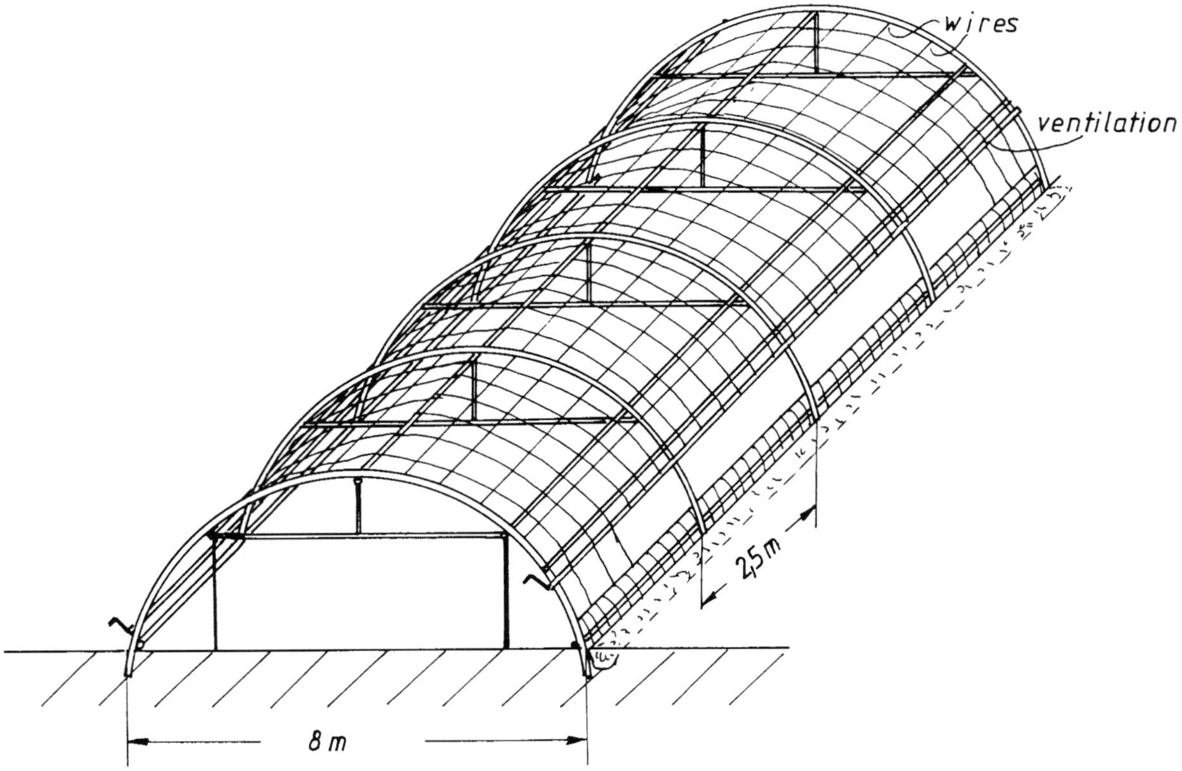

Fig. 2.20. Round-arched tunnel greenhouse with continuous side-wall ventilation.

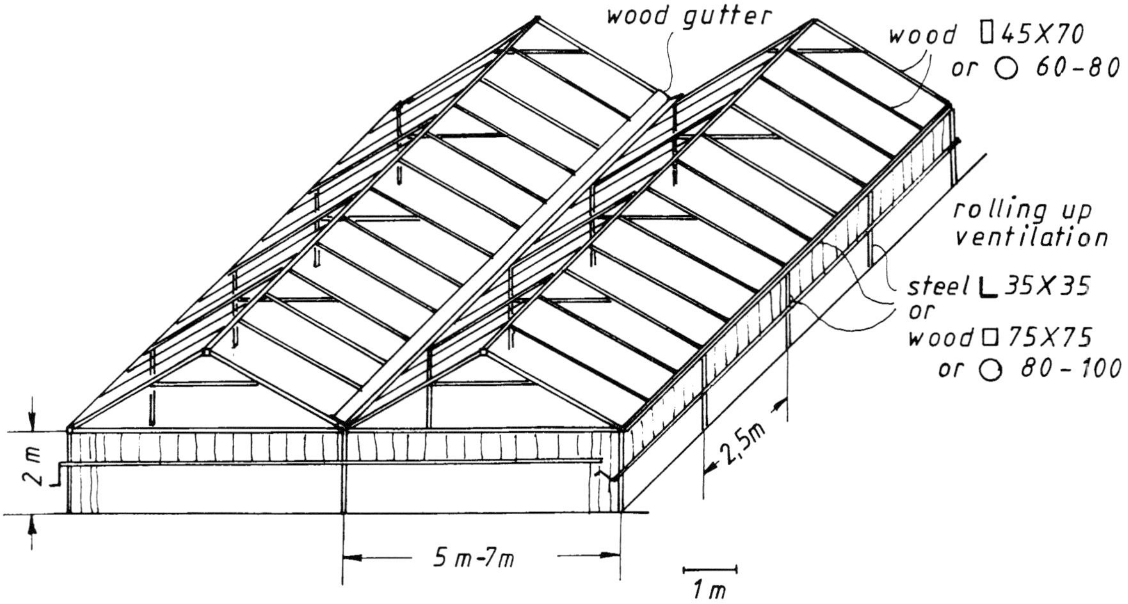

Fig. 2.21. Saddle-roof construction.

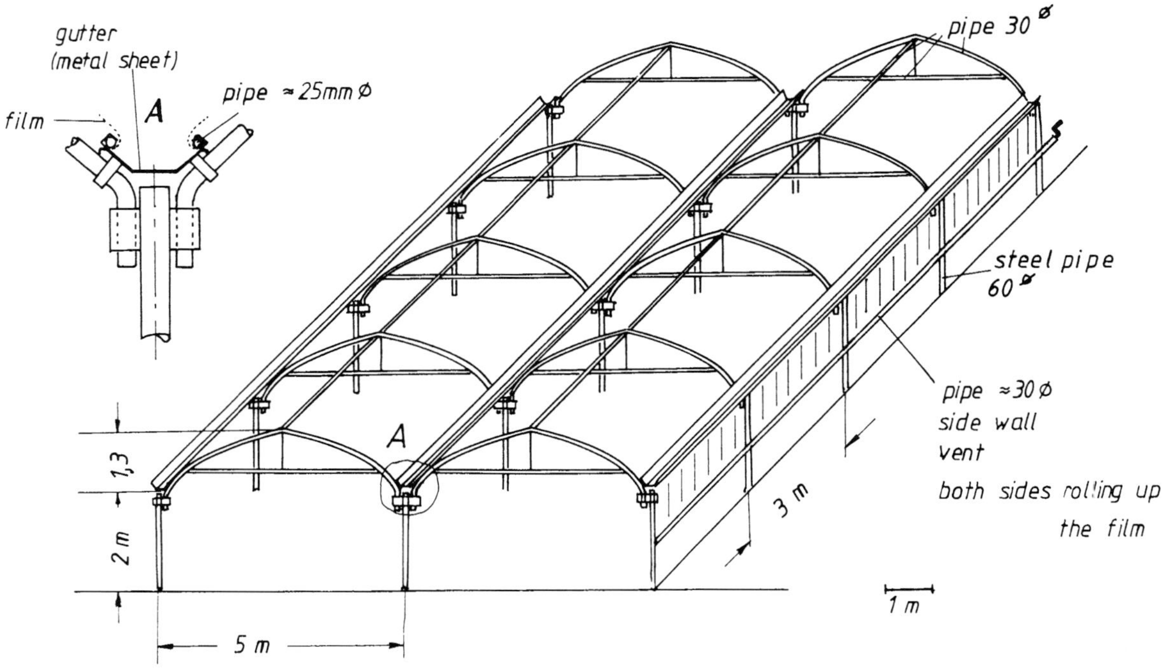

Fig. 2.22. Pointed-arched steel-pipe construction.

Steel-tube greenhouses are built in round-arched and pointed-arched shape (Fig. 2.22). Pointed-arched or Gothic-arched structures have advantages over round arches because condensation water can flow off better on the inner side of the film and, therefore, little water drops down on the plants. The film is fixed and stretched by rolling it on a pipe in the gutter. In many regions greenhouses should have gutters to collect rainwater for irrigation.

The greenhouse shown in Fig. 2.23 has sloping guy

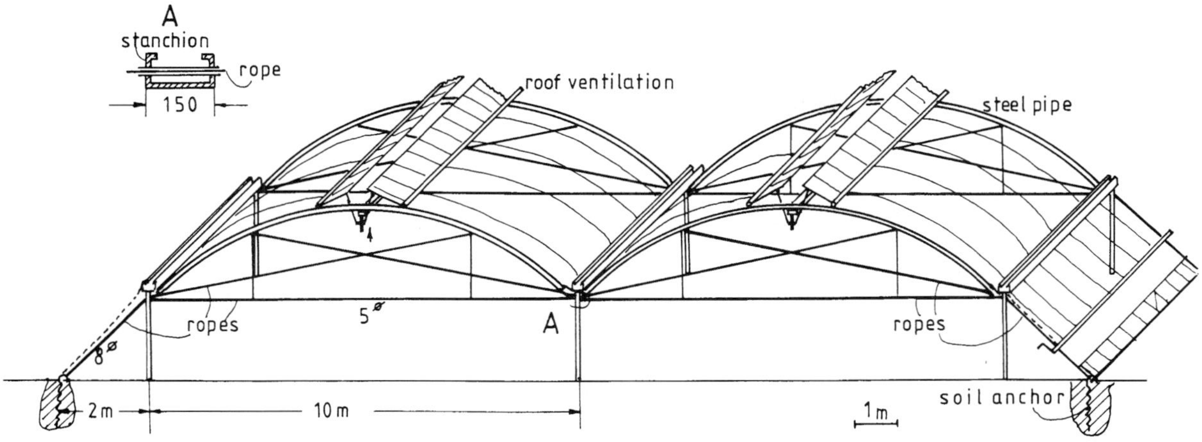

Fig. 2.23. Greenhouse with sloped side wall.

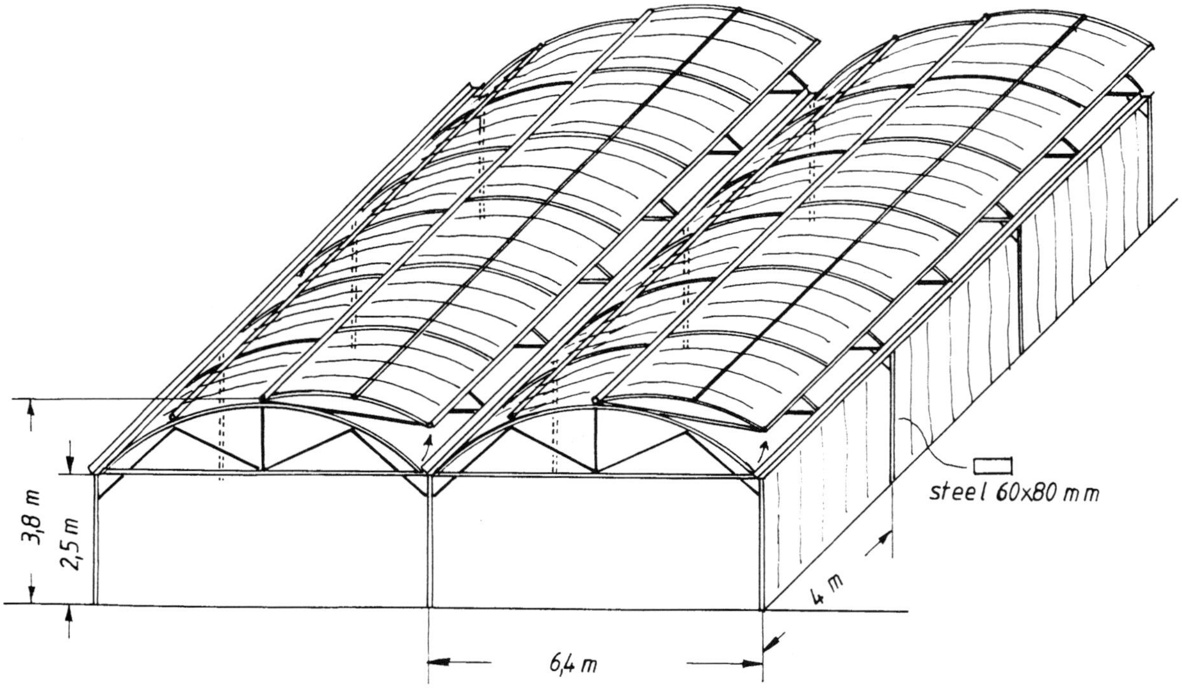

Fig. 2.24. Double plastic-film construction with continuous roof ventilation.

wires to stabilize the construction. The film is rolled up on the sloping wires for ventilation purposes. The construction is designed with tension ropes, which have small dimensions and result in good transmission of light.

Double plastic-film structures have been developed, in which one-half of the roof can be opened for ventilation (Fig. 2.24). Many plastic-film greenhouses are covered with double film in order to conserve energy or to maintain higher temperature differences during the night in unheated greenhouses. The space between the films is inflated to a pressure of 40–60 Pa, which increases the ability to withstand wind or snow load.

Greenhouses with ridge ventilation can also be equipped with rollable ventilators that can be closed overnight (Fig. 2.25). This construction may be a little more expensive than that described earlier, but ridge ventilators give the best ventilation efficiency in combination with side-wall ventilation.

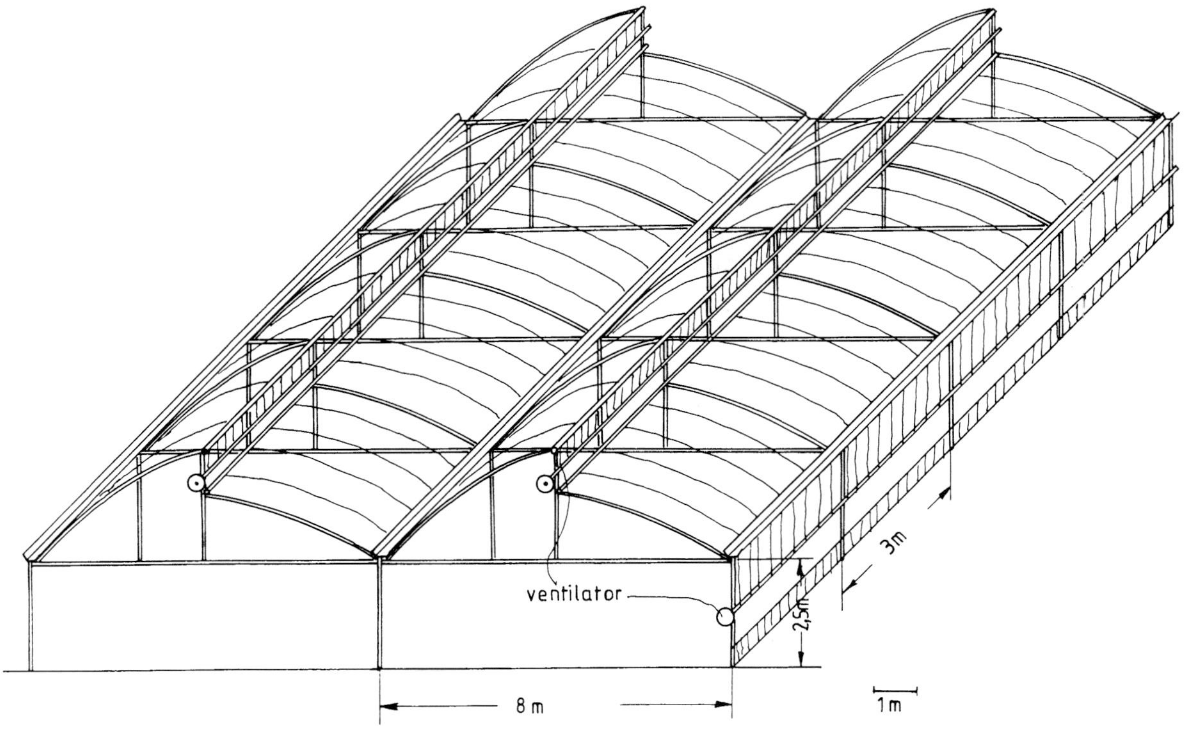

Fig. 2.25. Greenhouse with closeable ridge ventilation.

For tropical regions the measures most important to improve the ventilation efficiency are:

(1) Volume/floor ratio as large as possible, if local wind speed is not too high.
(2) Single-span and multi-span constructions with open side walls, gables and ridge ventilators. All ventilators should remain open.
(3) Nets on ventilators, if protection from insects and birds is necessary.

Figure 2.26 shows a simple single-span structure that can be used, for example, in the first stage of vegetable production. The growers can build them from wood or steel parts. The construction must be so simple that the assembly is relatively easy to learn. A single-span mode of construction is easy to ventilate and is suitable for smaller holdings.

As an alternative to the simple wooden construction, steel pipes can also be used for the stanchions and other connecting pieces. In this case, steel pipes would be driven approximately 1–1.5 m into the earth, and the tubular stanchions would then be stuck into and pinned to these steel pipes.

The choice of the construction material depends on the local prices for timber and steel. At present, wood construction is cheaper in many countries. However, with increasing scarcity of wood, it may be necessary, in the long term, to change over to steel construction.

In Colombia, single-span and multi-span greenhouses are built with a longitudinal angle of inclination (slope) of the ridge of 2 to 3° (Fig. 2.27). The construction is highest in the middle and slopes down towards the gables to improve the drainage of rainwater during heavy rainfall, especially in the case of multi-span greenhouses. In the middle, the greenhouse ridges are up to 6 m high. The ventilators at the ridge are 0.7 m wide. Due to the height and the wide ventilator openings, good ventilation efficiency can be achieved. The roof overhangs at the side walls so that rainwater can flow off without problems. Figure 2.28 shows a saw-tooth construction with gutter. The roof consists of vertical and sloped surfaces. The vertical surfaces consist of ventilation openings. For a maximum incidence of light, the vertical surfaces must face the sun in winter.

If the vertical surface with the ventilation openings is placed away from the wind, the wind current produces a negative pressure which sucks the warm air from the greenhouse, if the air inlet is in the side wall. This can improve the ventilation efficiency. The sloped

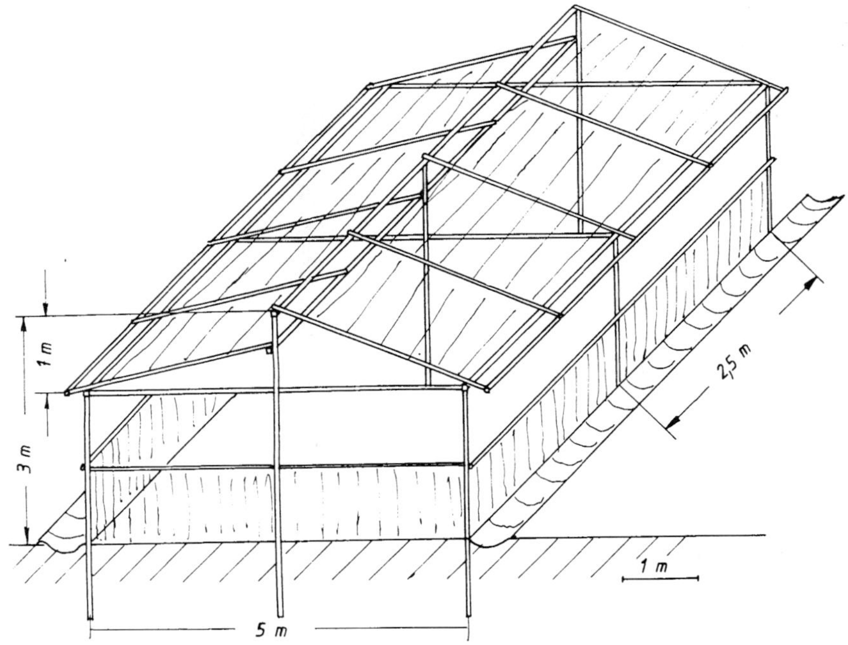

Fig. 2.26. Simple single-span construction for tropical regions.

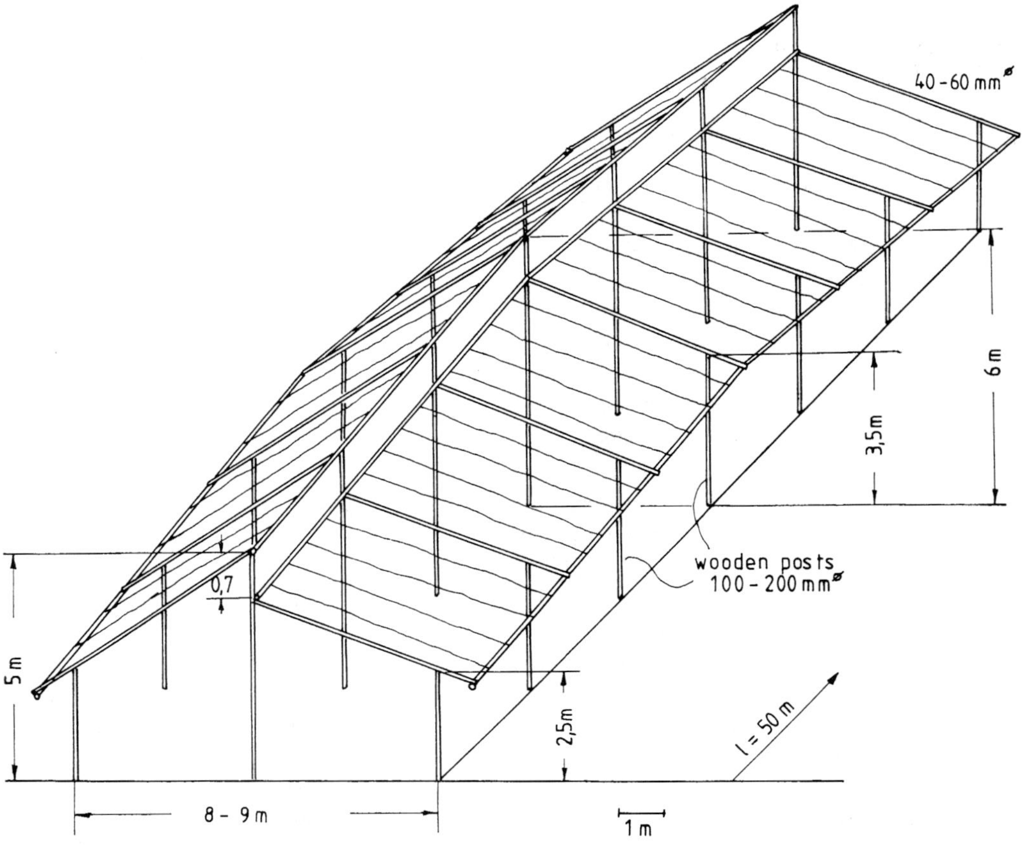

Fig. 2.27. Greenhouse construction with sloping ridge (Colombia).

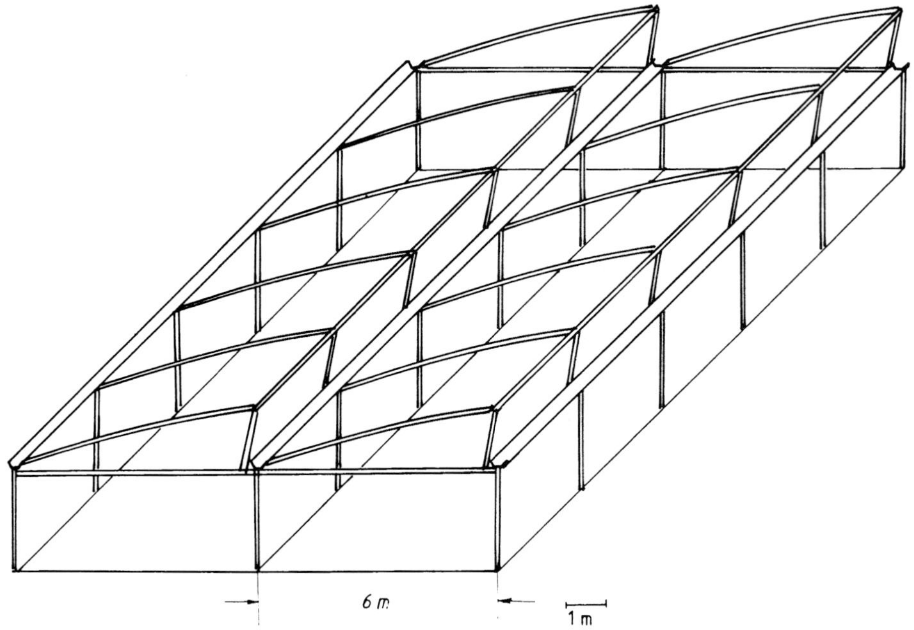

Fig. 2.28. Saw-tooth construction.

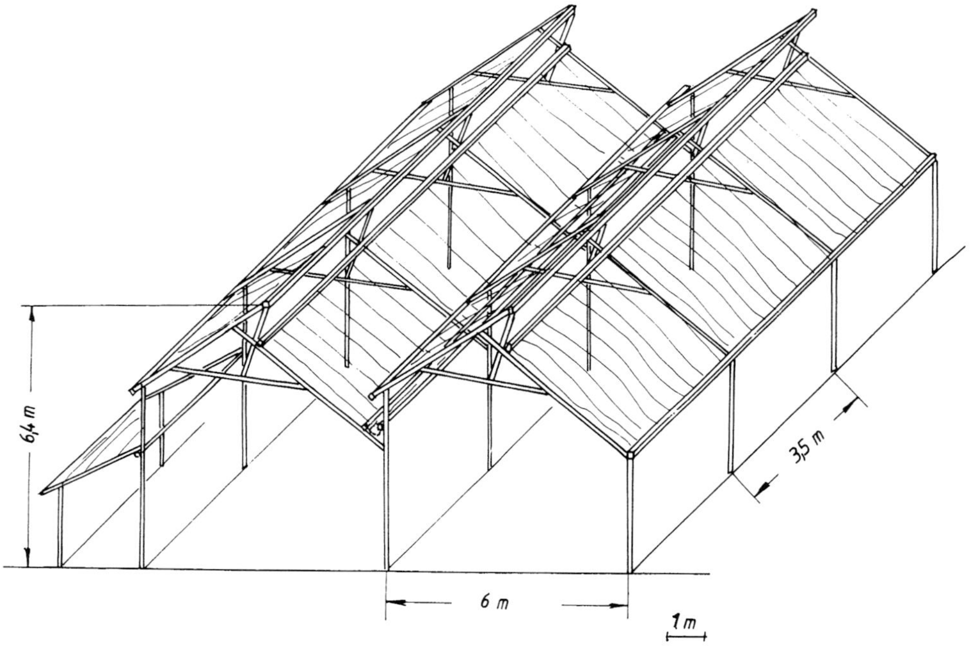

Fig. 2.29. Multi-span wooden construction (Kenya).

construction parts can also be built of flexible wood or bamboo sticks.

In Kenya, very high wooden greenhouses with ventilation openings at ridge and gutter are built for cut flowers (Fig. 2.29). The gable and side-wall surfaces are open. The height of the ridge is 6.4 m. The gutter is made of plastic film. Generally, plastic-film gutters have the disadvantage of forming water bags if the plastic film is not stretched very tightly. On the sides facing the wind, a kind of porch or

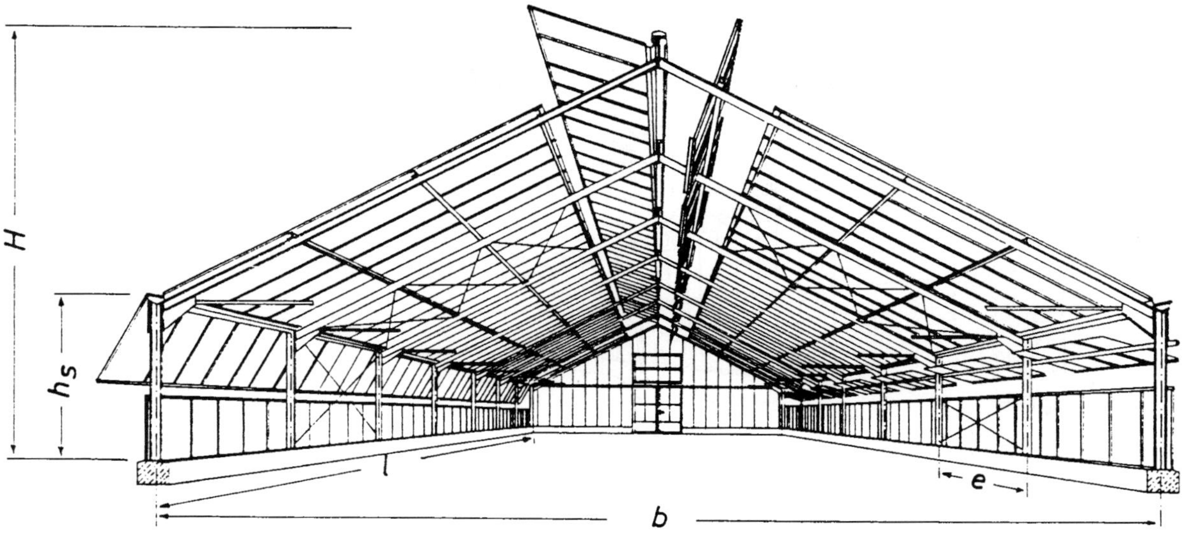

Fig. 2.30. Wide-span glass greenhouse.

canopy is built as a means of protection from wind and rain.

Glass and rigid plastic greenhouses

Greenhouses covered with glass or rigid plastics can be classified as follows:
(1) wide-span greenhouses,
(2) Venlo-type greenhouses.

A characteristic feature of the *wide-span greenhouse* is the greater number of glass panes on the roof between gutter and ridge. The *Venlo-type greenhouse* on the other hand has only one glass pane placed in glazing bars between gutter and ridge.

The gutters have several functions. The first is the drainage of rainwater. Collecting rainwater is becoming more and more important because of the shortage of irrigation water. The second function is drainage of melt water to prevent snow from accumulating near the gutters. Finally, the gutter has a structural function, with stanchions and rafters.

Figure 2.30 shows a wide-span greenhouse, with its main dimensions: length (l), width (b), height of ridge (H), height of side wall (h_s) and distance between stanchions (e). These greenhouse types can be built single-span or multi-span connected at the gutter; they have ventilators at the ridge and, if needed, in the side wall. Figure 2.31 shows a Venlo-type construction with a standard width of 3.20 or 4 m. There are two possibilities: a stanchion under each gutter, or a stanchion every second or third gutter, the gutter in between being supported by a lattice girder.

Materials for greenhouse construction

The material for greenhouse construction should be a domestic product, or at least should be available on the local market. And the cheapest material should be chosen.

Wood, bamboo and steel are usually suitable for the construction of plastic-film greenhouses. Aluminum and steel are used for glass and rigid plastic greenhouses. Concrete is used for foundations.

Due to the humidity, wood can swell and shrink. Encouraged by humidity, fungi and insects can infest the wood, causing rotting and decay. Therefore, wood has to be protected for longer usage. In addition, it has to be protected from termites. Wood is protected by impregnation with substances such as synthetic or natural resins, or chemicals such as zinc chloride, copper sulphate or borax. Toxic substances should be avoided. Before the treatment the wood has to be dry; it is then put into the chemical solution for a certain time, ranging from hours up to days. A more efficient process is vacuum treatment, where the air is sucked from the pores. While this is done, the wood is put into the solution. Afterwards under normal pressure, the protective substance penetrates the timber deeper, providing for longer life. Another simple method is also to burn or char the surfaces of the posts.

Steel has to be protected from corrosion. Nowadays,

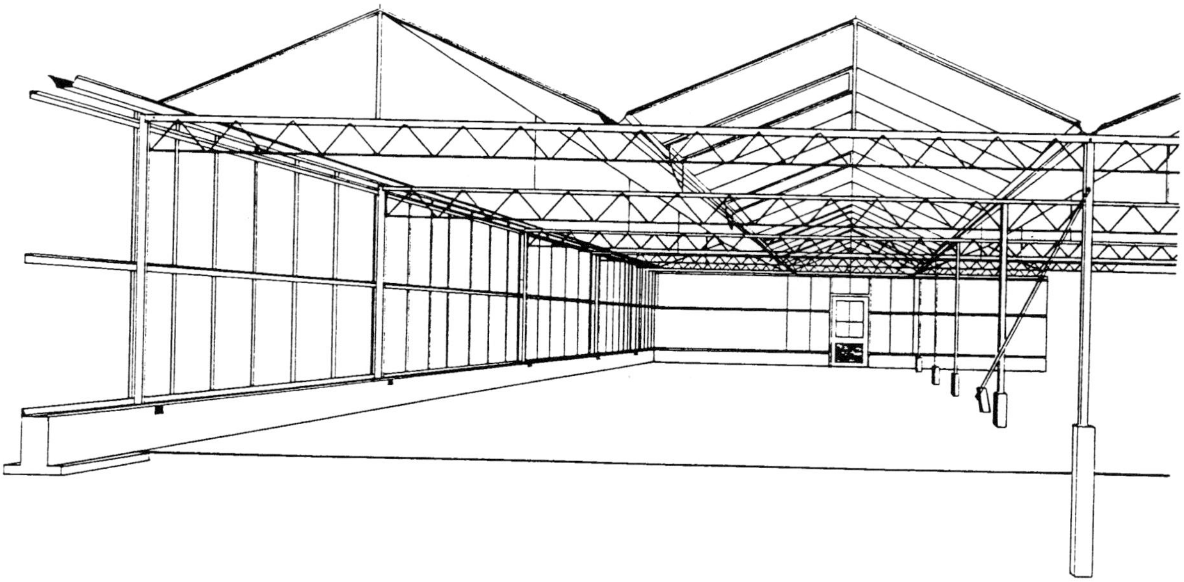

Fig. 2.31. Venlo-type greenhouse with lattice girder.

steel is usually galvanized with zinc, though galvanized steel parts are often twice as expensive as non-galvanized parts. Steel with factory-applied paint is also used. Aluminum does not need to be protected from corrosion.

CLADDING MATERIALS

The materials used for greenhouse cladding are: glass, rigid plastic, and plastic film. The following characteristics of the cladding material play important roles:

(1) transmissivity of global radiation in the range of ultra-violet radiation (UV), visible light (L) and near-infrared radiation (NIR) – that is, the wavelength range from 280 to 3000 nm;
(2) transmissivity of long-wave radiation (IR) with a wavelength of more than 3000 nm;
(3) dependence of radiation transmissivity on age;
(4) dropwise condensation and the fall of condensation drops from the inside of the roof;
(5) water vapor permeability;
(6) durability to wind, hail and snow;
(7) fabrication width and cost of production.

Only a certain percentage of the global radiation in the field penetrates into the greenhouse through the cladding material. The irradiated energy is reduced by absorption and reflection by the cladding material, shading by parts of the greenhouse structure, and dirtiness of the cladding.

The radiation transmissivity T_r of a greenhouse is defined as the ratio of radiant energy on a horizontal area in the greenhouse (q_{rg}) to the radiant energy on a corresponding area in the field (q_{ro}):

$$T_r = \frac{q_{rg}}{q_{ro}}. \tag{2.13}$$

Thus the irradiation in the greenhouse becomes

$$q_{rg} = T_r q_{ro}. \tag{2.14}$$

T_r depends on the position of the sun, orientation, cladding material, grade of dirtiness of the cladding, quantity and arrangement of structural members in the roof, and condensation on the inside of the cladding (see pp. 30–33).

Glass

The advantage of glass is its durability if it is not destroyed by storm or hail. The global radiation has no influence on the durability or transmissivity of glass.

Single glass used as cladding material in the Netherlands has a thickness of 3–5 mm and a width between 45 and 120 cm. The length of the panes is 50–200 cm.

The spectral transmissivity of glass lies in the range of 320–2900 nm. Glass is mainly opaque to ultraviolet radiation and long-wave infrared radiation of more than 3000 nm. For visible light, glass has a transmissivity of 89–92% with vertical light incidence and in a clean state. The light transmissivity can decrease considerably when the glass is dirty.

Glass has an emission coefficient of about 0.9 – that is, with regard to heat-radiation characteristics glass acts as a black body.

Coated glass

For greenhouse cladding a special glass was developed in Belgium, coated with a thin layer of stannic oxide on the outside, which reduces the emission coefficient for long-wave heat radiation to 0.25–0.3. If the greenhouse is heated and the glass cladding warms up, less heat is emitted outward by radiation and heating energy is conserved. This effect only occurs if the glass is dry, as a wet pane has a high emission coefficient. In Central Europe, in comparison to normal glass, the energy requirement with this cladding is 20–23% less. The light transmissivity, however, is lower than that of normal glass.

Rigid plastic materials

Plastic materials used for greenhouse cladding include acrylic, polycarbonate (PC), polyvinyl chloride (PVC), and polyester reinforced with glass fibers.

Acrylic is used mainly for double cladding (*Stegdoppelplatte*), Fig. 2.32). It is available both in double and three-layer sheets. The thickness of the double sheets is 16 mm and that of the three-layer sheets is 16 or 32 mm. The length of the sheets is 2–6 m. Today the sheets have additives against drop condensation – "Anti-Drop". Water condenses as a water film so that a better light transmissivity is achieved. The energy requirement, in the case of double web sheets, is 40–45% less than with double glass.

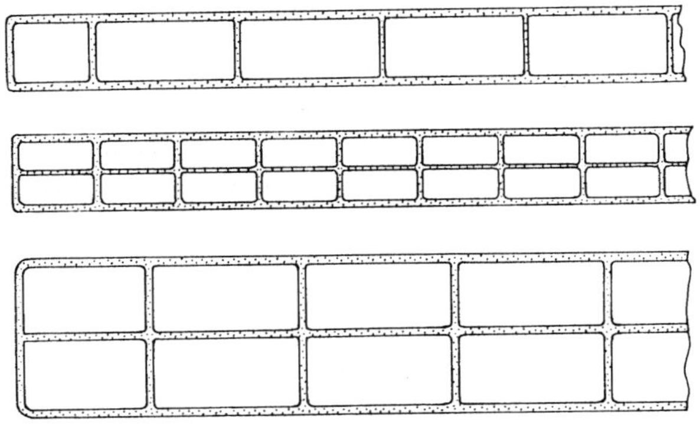

Fig. 2.32. Multi-layer acrylic material for greenhouse covering.

The light transmissivity with vertical light incidence is 86%. It decreases about 1% within 10 years. Acrylic is opaque to long-wave infrared radiation. Sheets are available with and without ultraviolet radiation transmissivity. Acrylic is permeable to water vapor. Water vapor diffuses through the material, so that water can condense within the sheets, and must be able to run out to avoid problems with the growth of algae. Acrylic expands considerably when exposed to heat and dampness. This has to be taken into account during construction.

Like acrylic, **polycarbonate** is used mainly for double cladding. The sheets have a thickness of 4–16 mm and a length of up to 6 m. Polycarbonate sheets are remarkably hail-proof. The light transmissivity is less durable than that of acrylic. In order to improve the durability of light transmissivity, polycarbonate sheets are coated with acrylic. The water-vapor permeability is similar to that of acrylic material. At the same thickness polycarbonate sheets have a lower light transmissivity than acrylic sheets.

Fibre-glass is mainly used for single cladding. The material is formed into corrugated plastic sheets which are highly resistant to hail. The light transmissivity decreases rapidly in regions with high insolation. The conditions can be improved by coating with polyvinyl fluoride.

Plastic films

Various types of plastic film are used as cladding materials for greenhouses.

Polyethylene (PE) film, unstabilized, has a maximum durability of one year in Central Europe. New polyethylene film should be put into place in autumn so that the light transmissivity is the highest in winter, and the film is not destroyed by storms during that season. This film is not recommended for warm climates.

Ultraviolet-stabilized polyethylene film is stabilized against ultraviolet radiation, and has a durability of two to three years in Central Europe. Only the stabilized film should be used for greenhouse cladding. In the dry state the film is largely transmissive to long-wave infrared radiation.

Polyethylene-infrared film has a reduced transmissivity in the long-wave radiation range (IR).

On **polyethylene "Anti-Drop" film** the water does not condense in the form of micro-drops but as a film and therefore runs off better. Drop condensation on plastic films is a problem, as the water drops reduce the light transmissivity and are harmful to the plants when they drip from the roof (see above, p. 31).

Whether a greenhouse is covered every year with a new, thin film (0.12 mm) or every other year with a thicker film (0.2 mm) is a question of the durability of polyethylene films and cost. Under high insolation plastic film becomes brittle earlier, so that its light transmissivity is reduced and the film is more easily torn by wind. Under these circumstances it is advisable to change the plastic film more often, provided that the local prices for the crop are high enough, and those for labor and plastic film low enough.

Transmissivity of plastic films

The light transmissivity of different films to direct light, in comparison to glass are shown in Table 2.7. The ranges for light transmissivity result from different film thicknesses and qualities.

Besides the mechanical resistance, the decrease in light transmissivity with time is very important for the use of films for greenhouses. The following tendencies were found for the decrease of light transmissivity:

polyethylene film, ultraviolet stabilized: 5–7% within three years;

ethylvinyl acetate film: about 6% in the first year, only a small decrease afterwards;

Hostaflon: 4% within five years.

Light transmissivity is influenced by dirtiness. Long-lasting films (especially polyvinyl chloride) get dirty as a result of electrostatic charging. If water condenses in drop form on the film the light transmissivity can be reduced by about 15% because of total reflection of the light by the drops.

Table 2.7
Transmissivity for visible light (direct light) of different films, with vertical light incidence

Material [1]	Thickness (mm)	Light transmissivity (%)
Glass	3.8	89–92
PE, UV stabilized	0.1–0.2	89–92
PE-IR	0.1–0.2	89–92
PE "Anti-Drop"	0.2	90
EVA	0.18	91.5
PVC	0.1–0.2	87–91
Polyester (Melinex)	0.12	94
Hostaflon ETFE	0.1	93
Tedlar (PVF)	0.05–0.2	93–94
Teflon (FEP)	0.05	96

[1] Abbreviations: ETFE, ethylene tetrafluoroethylene; EVA, ethylvinyl acetate; FEP, fluorinated ethylene propylene; PE, polyethylene; PVC, polyvinyl chloride; PVF, polyvinyl fluoride.

Transmissivity to ultraviolet radiation : Stabilized polyethylene film, polyvinyl chloride film and ethylvinyl acetate film are not transmissive to ultraviolet radiation, which is said to have an influence on the flower coloring of some ornamental plants.

Polyvinyl chloride film is more expensive than polyethylene film but lasts longer. It can, however, only be delivered in smaller widths. Polyvinyl chloride film is used in Japan much more than polyethylene film because of its lower transmissivity to long-wave radiation.

Polyvinyl chloride and polyethylene film can be differentiated by their density. Polyethylene film floats on water, whereas polyvinyl chloride film sinks. When burnt, polyvinyl chloride film smells of chlorine gas.

Ethylvinyl acetate film (EVA) combines the qualities of polyethylene and polyvinyl chloride film, especially with regard to transmissivity of visible light and infrared radiation, and fabrication width. The transmissivity for IR radiation depends on the vinyl acetate (VA) content which lies between 4% and 18%. Ethylvinyl acetate film with a high vinyl acetate content is softer and extensible. Therefore it can flap on the structure at high temperatures. In order to compensate for this disadvantage **coextruded films** are produced made of two or three layers of polyethylene film, ethylvinyl acetate film and polyethylene "Anti-Drop" film.

On **polyester film** (Melinex), water condenses in a film as on glass, and not in drops. This film is very durable.

Hostaflon (ethylene tetrafluoroethylene – ETFE) has a relatively high light transmissivity and a long-term durability.

Tedlar film (polyvinylfluoride – PVF) is used for the covering of solar collectors and greenhouses. The film has a longer durability than polyvinyl chloride or polyethylene film. The light transmissivity is relatively high.

Transmissivity to infrared radiation: The transmissivity for long-wave infrared radiation, in the range above 3000 nm, has an influence on heat transfer through the cladding material and on the plant temperature. The highest radiation intensity of the energies

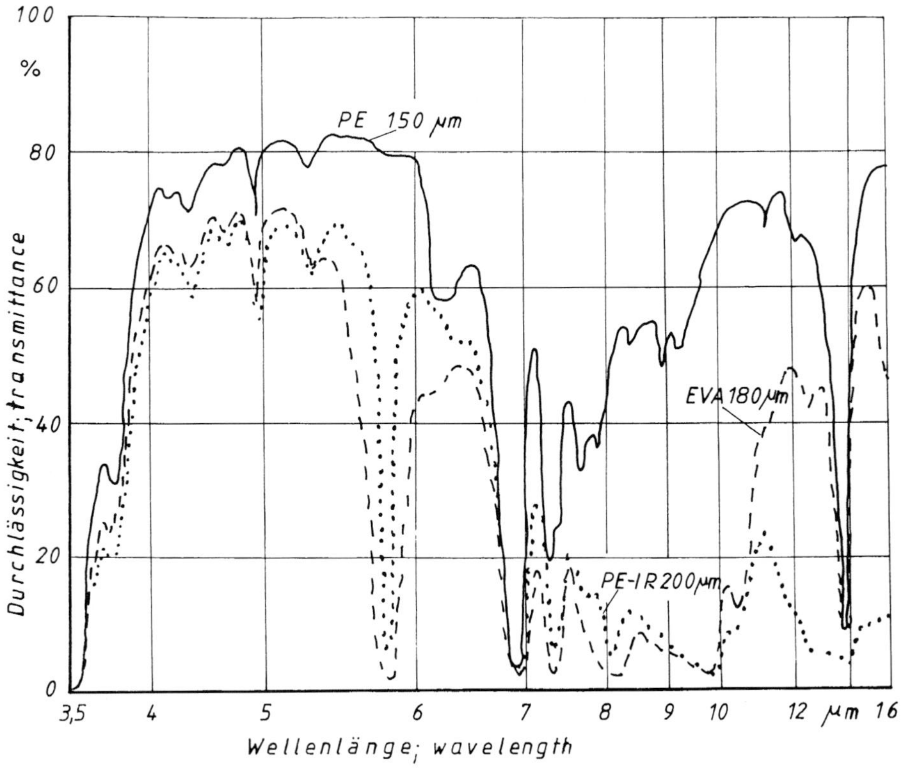

Fig. 2.33. Spectral transmissivity of different films.

emitted by plants, soil and interior equipment lies in the wavelength range between 6000 and 14 000 nm.

If films have a high infrared transmissivity, the heat loss from the greenhouse is higher in this range, and the plant temperature can fall below that of plants from greenhouses covered with films of lower infrared transmissivity. In the case of unheated greenhouses and transmissive film, damage can occur on cold nights with a clear sky. Figure 2.33 shows the spectral transmissivities of different films for long-wave radiation (infrared radiation, IR). Table 2.8 shows the mean values for transmissivity of infrared radiation in the wavelength range between 5000 and 14 000 nm (5–14 μm). These values are only valid for dry films.

Wet films are not transmissive to infrared radiation. Dry ultraviolet stabilized polyethylene film has a relatively high transmissivity for infrared radiation. This transmissivity is reduced by absorbents, in the case of both infrared film and "Anti-Drop" film. The transmissivity for infrared radiation depends also on the thickness of the film. The thicker the film, the lower the infrared transmissivity (Fig. 2.34).

Table 2.8
Mean transmissivity for long-wave radiation in the range 5000 to 14 000 nm

Material [1]	Thickness (mm)	IR transmissivity (%)
PE	0.1	72
PE	0.2	55
PE-IR	0.2	29
PE "Anti-Drop"	0.18	37
PE-IR "Anti-Drop"	0.18	24
PE with water film		0
EVA 13% VA	0.1	42
EVA 8% VA	0.1	53
EVA 4% VA	0.12	51
EVA	0.18	22
EVA "Anti-Drop"	0.18	22
PVC	0.1	30
PVC	0.2	17
Polyester (Melinex)	0.125	5
Hostaflon ETFE	0.1	22

[1] See Table 2.7 for abbreviations.

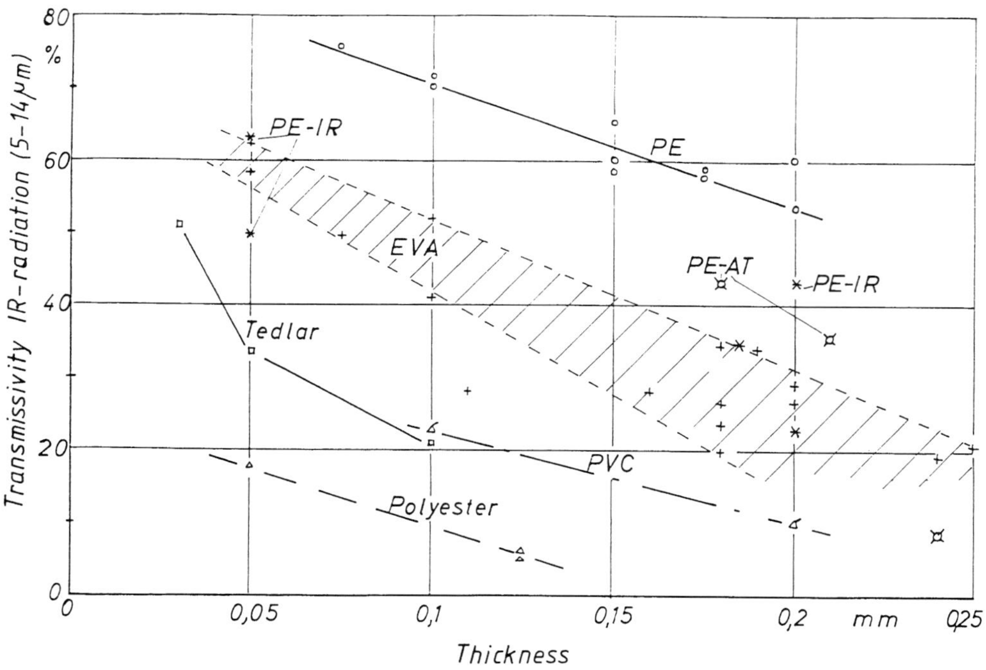

Fig. 2.34. Dependence of long-wave radiation on thickness.

Fastening and stretching of plastic films: Plastic films must be fastened and stretched on the greenhouse structure. They must be stretched tightly so that they cannot flutter in the wind and be damaged. Fastening of the film to the structure, and its replacement when necessary, must be done quickly, and with limited requirements for labor.

A simple method is to nail the plastic film to wooden slats. If possible this method should be avoided, because it takes too much time. It is better to screw the film between wooden slats, or between steel members of the structure and wooden slats, so that the film is easier to remove. Some designs have wires between which the film is clamped.

Plastic films can be fastened with plastic clips to steel pipes. A very good solution for simultaneous fastening and stretching of films on the structure is to roll up the film on a steel pipe in the gutter and to fix the steel pipe in the gutter profile (see p. 35–37).

Fastening devices made of plastic, aluminum and steel have also been developed (Fig. 2.35). By use of these devices the film can quickly be fastened to and removed from the structure. Unfortunately these fastening devices are relatively expensive. Some other systems for fastening plastic films which are part of the greenhouse structure have been described above (p. 34). After being fastened to the structure the films have to be stretched tightly. This can be done by the following methods:

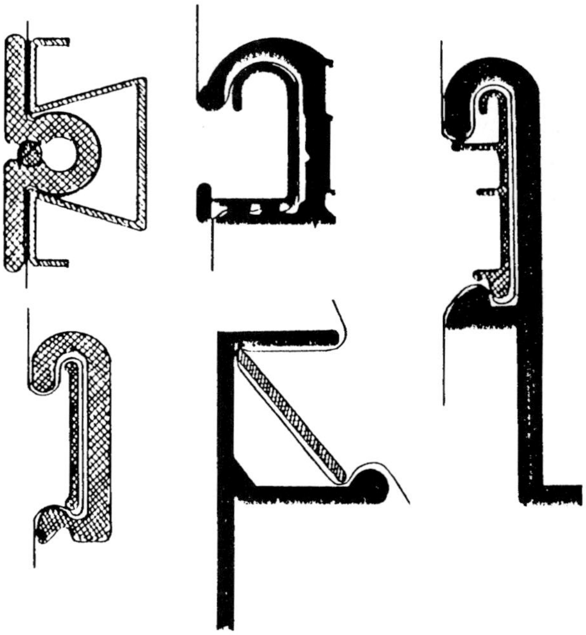

Fig. 2.35. Fastening devices for plastic films.

(1) fastening devices;
(2) stretching plastic ropes over the construction;
(3) rolling up the film on a steel pipe in the gutter;

(4) mechanical stretching devices;
(5) inflated double film.

Double films inflated at an internal pressure of 50–60 Pa offer a good resistance to wind and snow forces and have a good thermal-insulating effect.

Fastening and stretching of the film is very important for the durability of film and construction. When building new greenhouses in the future one should invest more in the construction and thus have more security for plant production and less labor cost.

VENTILATION AND COOLING

Ventilation

The ventilation of a greenhouse is the exchange of air between the inside and outside atmosphere with the purpose of:

(1) exchange of carbon dioxide and oxygen;
(2) dissipation of surplus heat, and temperature control;
(3) humidity control.

Sufficient ventilation is very important for optimal plant growth, especially in the case of high outside temperatures, high solar radiation and high humidity. A distinction has to be made between natural ventilation by openings in the side wall, gable and roof on the one hand, and forced ventilation by fans on the other.

Characteristics for greenhouse ventilation are:

(1) The volume of air (V) exchanged in relation to the greenhouse volume (V_G),

$$\text{air-exchange figure: } Z = \frac{V}{V_G}[\mathrm{h^{-1}}]. \tag{2.15}$$

(2) The volume of air exchanged in relation to the greenhouse ground area (A_G)

$$\text{ventilation rate} \qquad V_A = \frac{V}{A_G}[\mathrm{m^3\,m^{-2}\,h^{-1}}]. \tag{2.16}$$

With the mean greenhouse height h given, the relation between the two characteristics can be calculated:

$$V_A = hZ. \tag{2.17}$$

The ventilation rate necessary to keep a fixed temperature difference between inside and outside can be calculated by an energy balance (Bailey, 1988; von Zabeltitz, 1988). The energy balance of a ventilated greenhouse can be written as

$$q_{sr} = (q_c + q_l) + q_v + q_f + q_p, \tag{2.18}$$

where q_{sr} is the solar radiation absorbed within the greenhouse, q_c is the heat conducted through the cladding, q_l is the heat loss by air leakage, q_v is the heat removed by ventilation, q_f is the heat flow into the soil, q_p is the energy used in photosynthesis, and all these quantities are expressed in units of $\mathrm{W\,m^{-2}}$.

The energy used in photosynthesis is about 2–3% of the total solar radiation and can be ignored. The soil heat flux is very important but can be ignored in calculation of the ventilation rate. The solar radiation absorbed inside the greenhouse, q_{sr}, depends on the external solar radiation:

$$q_{sr} = T_r(1 - r_e)q_o, \tag{2.19}$$

with T_r the transmission of the greenhouse, r_e the reflection inside the greenhouse, and q_o the external solar radiation. The conduction and air-leakage heat losses (q_c and q_l) can be calculated by

$$(q_c + q_l) = \frac{A_c}{A_g}u(t_i - t_o), \tag{2.20}$$

where A_c is the area of greenhouse cladding, A_g is the ground area of the greenhouse, u is the overall heat-loss coefficient (in units of $\mathrm{W\,m^{-2}\,K^{-1}}$), t_i is the inside temperature, and t_o is the outside temperature.

The energy removed by ventilation consists of sensible and latent heat:

$$q_v = q_s + q_{la}, \tag{2.21}$$

each quantity being expressed in $\mathrm{W\,m^{-2}}$. The sensible heat transfer, q_s, is given by

$$q_s = V_A \rho c_p(t_i - t_o), \tag{2.22}$$

with V_A the ventilation rate and c_p the specific heat of air (in units of $\mathrm{W\,s\,kg^{-1}\,K^{-1}}$). The latent heat transfer, q_{la}, is given by

$$q_{la} = \varepsilon T_r(1 - r_e)q_o, \tag{2.23}$$

where ε is the fraction of the absorbed solar radiation which is used in transpiration.

Combining the above equations one can calculate the ventilation rate depending on the temperature difference. Figure 2.36 shows some calculated figures for a greenhouse with a single cladding layer at a global radiation rate of $700\,\mathrm{W\,m^{-2}}$. The results correspond very well with experimental results (Baytorun, 1986; von Zabeltitz, 1986).

For the greenhouse ventilation to be sufficient to maintain an interior temperature no more than 5 K

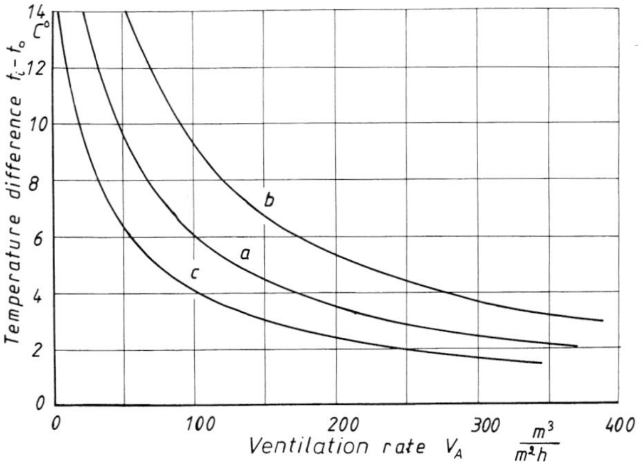

Fig. 2.36. Temperature difference depending on ventilation rate (in $m^3\,m^{-2}\,h^{-1}$). (a) $\varepsilon = 0.35$; (b) $\varepsilon = 0$; (c) $\varepsilon = 0.5$.

above that outside, the quantities must have approximately the following values:

$$Z > 55\,h^{-1}; \qquad V_A > 170\,m^3\,m^{-2}\,h^{-1}.$$

Natural ventilation

For effective ventilation, ventilators should, if possible, be located at the ridge, on the side walls and the gable. Continuous roof and side-wall ventilation is more effective than trap-window ventilation.

In tropical climates with high temperatures the ventilators should be opened as wide as possible during the day. They need not to be closed at night, because the temperatures are high enough. In subtropical temperate climates and highlands the ventilators must be closed during the night if the temperature falls below the biological optimum.

To provide sufficient ventilation, the total ventilator opening in relation to the greenhouse floor area should be 15–25%:

$$\frac{\text{Ventilator area } A_V}{\text{Floor area } A_G} = 0.15\text{–}0.25.$$

Forced ventilation

Forced ventilation by fans is the most effective way to ventilate a greenhouse, but consumes electricity. The principle of forced ventilation is to create an air flow through the house. Fans suck air out on the one side, and openings on the other side let air in. Some key elements are:

(1) The fans must exhaust air from the greenhouse. Exhaust fans improve the temperature distribution in the house.
(2) The distance between two fans should not exceed 8–10 m.
(3) Ventilation fans should develop a capacity of about 30 Pa static pressure (3 mm on a water gauge).
(4) If possible, the fans should be located on the lee side or the lee end of the greenhouse.
(5) A space of at least 1.5 times the fan diameter should be left between the fan discharge and the nearest obstruction.
(6) The inlet opening on the opposite side of a fan should be at least 1.25 times the fan area.
(7) The velocity of the incoming air must not be too high. In the plant area, the air speed should not exceed 0.5 m s^{-1}.
(8) The openings must close automatically when the fans are not in operation.

The air rate of all fans together (in $m^3\,h^{-1}$) can be calculated directly from the ventilation rate and greenhouse ground area:

$$V_f = V_A A_g. \tag{2.24}$$

Cooling

In case of high irradiance in summer, the plant temperature in greenhouses can exceed the air temperature by 5 to 15°C, even with good ventilation. Efficient ventilation provides insufficient cooling for many plants. In many climates also the air temperature is too high. The air temperature in the greenhouse cannot be decreased below the outside temperature by using ventilation only. For additional cooling of greenhouses, evaporative cooling with the fan-and-pad system or a spray cooling system may be used.

The precondition for evaporative cooling is a low outside air humidity. The rates of air and water flow necessary for air cooling inside the greenhouse depend on:

(1) the radiation inside the greenhouse, influenced by cladding material and shading;
(2) heat transfer through the cladding material;
(3) the evaporation of the cooling system,
(4) the efficiency of the equipment.

Before installing a system, the air and water flow rates required must be calculated.

The fan-and-pad cooling system is most commonly used in horticulture (Fig. 2.37). Air from outside is blown through pads with as large a surface as possible. The pads are kept permanently wet by sprinkling. The water from the pads evaporates and cools the air. For this reason, the outside air humidity must be low.

There are basically two systems of fan-and-pad cooling (Fig. 2.37): the negative-pressure system, and

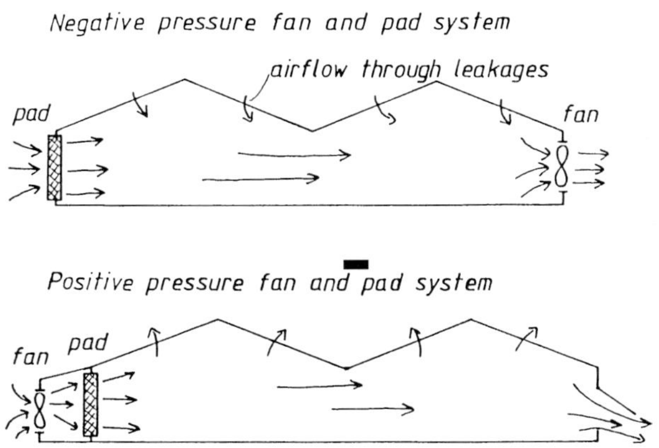

Fig. 2.37. Negative and positive pressure fan-and-pad cooling systems.

the positive-pressure system: The **negative-pressure system** consists of a pad on one side of the greenhouse and a fan on the other side. The fans suck the air through the pad and through the greenhouse. The pressure inside the greenhouse is lower than the pressure outside. Thus hot air and dust can get into the greenhouse. There is a temperature gradient from pad to fan. The **positive-pressure system** consists of fans and pads on one side of the greenhouse and vents on the other side. The fans blow the air through the pads into the greenhouse. The pressure inside the greenhouse is higher than outside. Dust cannot get into the greenhouse. In order to achieve an optimal cooling the greenhouse should be shaded.

The water flow rate, water distribution system, pump capacity, recirculation rate and output rate of the fan-and-pad cooling system must be carefully calculated and designed to provide a sufficient wetting of the pad and to avoid deposition of material in it. The manufacturers' guidelines for pad selection and installation have to be observed. The following rules should be considered when designing a fan-and-pad cooling system:

1. **Pad efficiency**. The cooling efficiency should provide inside air humidity of about 85% at the outlet. A higher air humidity slows down the transpiration rate of the plants. The plant temperature can then increase above air temperature.

2. **Pad material and thickness**. Different materials are available. The pad material should have a high surface, good wetting properties and high cooling efficiency. It should cause little pressure loss, and should be durable.

The average thickness of the pad is about 100 to 200 mm. It is very important that there are no leaks in the pad where the air can pass through without making contact with the pad. Different pad materials are available, such as wood wool ("wood excelsior"), swelling clay minerals, and specially impregnated cellulose paper.

3. **Pad area and fan-to-pad distance**. The pad area depends on the air flow rate necessary for the cooling system and the permissible surface velocity over the pad. Average face velocities are 0.75–1.5 m s^{-1}. Excessive velocities may cause problems with entrainment of drops into the greenhouse. The pad area should be about 1 m^2 per 20–30 m^2 greenhouse area. The maximum fan-to-pad distance should be 30–40 m.

4. **Pad location.** Pads may be oriented horizontally of vertically, the latter being more common. Vertical pads are supplied with water from a perforated pipe along the top edge (Fig. 2.38). In the case of horizontal pads, the water is sprayed over the upper surface. Air flows upwards through the pad (Fig. 2.38). Where the

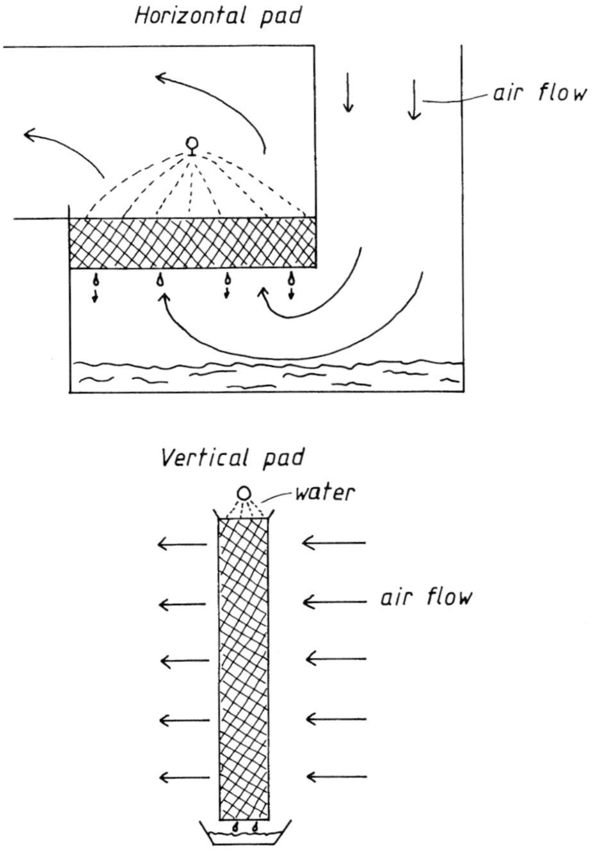

Fig. 2.38. Vertical and horizontal pads.

air is dusty, horizontal pads remain efficient for a longer time; these are, thus, preferable for desert conditions. The water distribution must be such as to ensure an even wetting of the pad. Pads have to be protected from direct sunlight to prevent local drying-out. Salt and sand might clog them, if they become locally dry. In areas with frequent sandstorms it is recommended to protect the wet pad with a thin dry pad serving as a sand filter. The pads have to be located and mounted in a way which permits easy maintenance and cleaning. They should be located on the side facing the prevailing wind.

5. **Fans**. Belt-driven or direct-driven propeller fans are used. Direct-driven fans are easier to maintain. Fans should be placed on the lee side of the greenhouse. If they are on the windward side, an increase of 10% in the ventilation rate will be needed. The distance between the fans should not exceed 7.5–10 m, and the fans should not discharge towards the pads of an adjacent greenhouse less than 15 m away. All exhaust fans should be equipped with automatic shutters to prevent air exchange when the fans are not operating, and also to prevent back-draughts when some of the fans are not being used.

6. **Operation of fan-and-pad system**. When starting the cooling system, the water flow through the pad should be turned on first to prevent the pads from clogging. The fans should be started only after the whole pad has been completely wetted. When stopping the cooling system in the evening, the fan should be turned off before the water flow through the pad. It is recommended to operate the cooling system by a simple control system depending on the inside temperature.

7. **Air flow rate**. The air flow rate depends on the solar radiation inside the greenhouse – that is, on the cladding material and shading – and on the evapotranspiration rate from the plants and soil. The air flow rate can be calculated by an energy balance. Generally, a basic air flow rate of 120–150 m^3 per m^2 greenhouse area per hour will permit satisfactory operation of an evaporative cooling system.

8. **Water flow rate**. The following water flow rates can be assumed for vertical pads (Hellickson and Walker 1983; ASAE EP406, 1988; Munters, 1992):

(1) 250–360 ℓ per m horizontal length of pad per hour, or

(2) 7 to 8 times the water loss due to evapotranspiration.

The manufacturer of "Celdec" pads (impregnated cellulose paper) prescribes 360 $\ell\, h^{-1}\, m^{-1}$ for a pad of 100 mm thickness. If the height of the pad is 2 m, the water flow rate is 125–180 ℓ per m^2 pad area per hour.

The pipe system should include control valves and a gauge for measuring the water flow.

Salty and brackish water may be used for fan-and-pad cooling. The high amount of water flow needs a recirculation system. To avoid a concentration of salt, some of the recirculating water needs to be discharged and replaced by fresh water. The relation of discharged water (bleed-off) to evaporated water depends on the water quality and has to be calculated according to the manufacturers' guidelines.

HEATING AND ENERGY CONSERVATION

When the outside temperature falls below the temperature required by the plants in the greenhouse, heating becomes necessary. In order to determine the maximum capacity of the heating system it is necessary to calculate the heat requirement. The actual fuel consumption is determined by a heat consumption calculation.

Calculation of the heat requirement

The maximum heat requirement occurs during the night when the outside temperature reaches its minimum. Its calculation requires the overall heat consumption coefficient u[1]. Table 2.9 shows mean values for the overall heat consumption coefficient at a wind speed of 4 $m\, s^{-1}$. The value of u depends on the cladding material of the greenhouse, its air tightness, the heating system, the irrigation system, wind speed, cloudiness and occurrence of rain; it can vary more than 10% for different heating systems; for example, pipe heating at the height of the eaves has $u = 8.8\ W\, m^{-2}\, K^{-1}$; pipe heating under the benches has $u = 7.4\ W\, m^{-2}\, K^{-1}$.

The mean minimum outside temperature, t_o, for the design of the heating system depends on the

[1] The overall heat consumption coefficient is the heat energy required per square meter of roof surface to maintain a temperature difference between inside and outside of 1 K.

Table 2.9
Overall heat consumption coefficient u for different covering materials and a wind speed of $4\,m\,s^{-1}$

Material	u ($W\,m^{-2}\,K^{-1}$)
Single glass	6.0–8.8
Double glass	4.2–5.2
Double acrylic SDP 16 (16 mm)	4.2–5.0
Triple acrylic SDP (32 mm)	3.0–3.5
Single film	6.0–7.8
Double film	4.2–5.5
Thermal screens under single cover	
Single film, woven materials	4.1–4.8
Double film, double woven materials	3.4–3.9
Aluminized materials, single layer	3.4–3.9

region. The following values may be used for selected countries:

Bosnia, Croatia, Serbia: −5 to −21°C,
Czechoslovakia: −12°C,
England: −5°C,
France: −5 to −7°C,
Germany (Hannover): −14°C,
Hungary: −15°C,
Israel: 0 to +5°C,
Tunisia, coast: +5°C,
Tunisia, inland: 0 to +3°C.

The heat requirement for the greenhouse, Q (in W), is then:

$$Q = A_c u(t_i - t_o), \tag{2.25}$$

where A_c is the roof area, and t_i the desired inside temperature.

The heat requirement per square meter of greenhouse area is:

$$q = (A_c/A_g)u(t_i - t_o) \text{ (W)}, \tag{2.26}$$

where A_g is the ground area of the greenhouse.

As an example, the heat requirement for a single-film greenhouse with an inside temperature of 12°C can be calculated as follows:

$t_o = -14°C$: $q = 1.5 \cdot 7 \cdot [12 - (-14)] = 273\ W\,m^{-2}$,

$t_o = 0°C$: $q = 1.5 \cdot 7 \cdot (12 - 0) = 126\ W\,m^{-2}$.

Actual heat consumption

For the calculation of the heat consumption, both the actual outside temperature and the global radiation have to be considered. During the day the heating energy is supplied by solar energy plus additional heating. The effective heat consumption q_H (in $W\,m^{-2}$) can be estimated as follows:

$$q_H = (A_c/A_g)u(t_i - t_{oe}) - q_{Gl}D\eta. \tag{2.27}$$

In this equation, t_{oe} is the actual outside temperature (°C), q_{Gl} is the intensity of global radiation ($W\,m^{-2}$), D is the transmissivity for radiation, η is the conversion factor of global radiation energy to thermal energy in the greenhouse, and other symbols have the same meanings as before. The transmissivity for radiation of single film ranges from 0.6 to 0.7, for double-film greenhouses from 0.5 to 0.6. The conversion factor η is 0.4 to 0.7, depending on the crop.

The heat consumption q_H has to be calculated for every hour of the day with corresponding values of t_{oe} and q_{Gl}. Figure 2.39 shows the values of heat consumption for Germany, Cyprus and Crete.

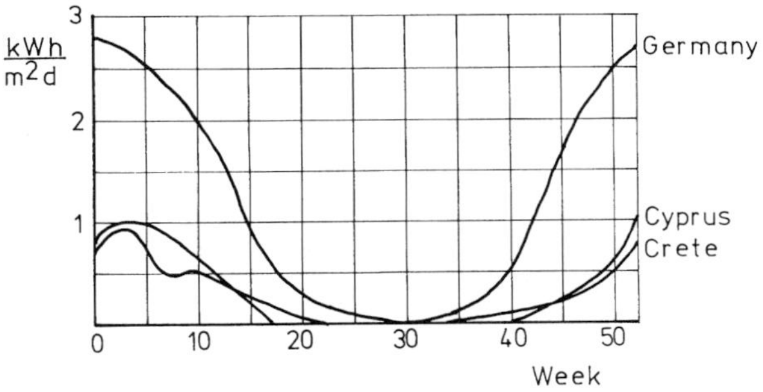

Fig. 2.39. Daily heat consumption (q_H), for an interior temperature of 16°C.

Heating systems

Heating systems supply heat energy to the greenhouse so that air and plant temperature remain rather constant. In the assessment of heating systems, the following desiderata should be taken into account:

(1) fuel consumption as low as possible;
(2) temperature distribution in the greenhouse as uniform as possible in the vertical and horizontal direction;
(3) easy control.

The heating systems themselves influence the heat consumption of the greenhouse. The following conditions have to be fulfilled to lower heat consumption:

(a) Regular heat distribution. The heating system should be installed as low as possible, preferably within the plant canopy.

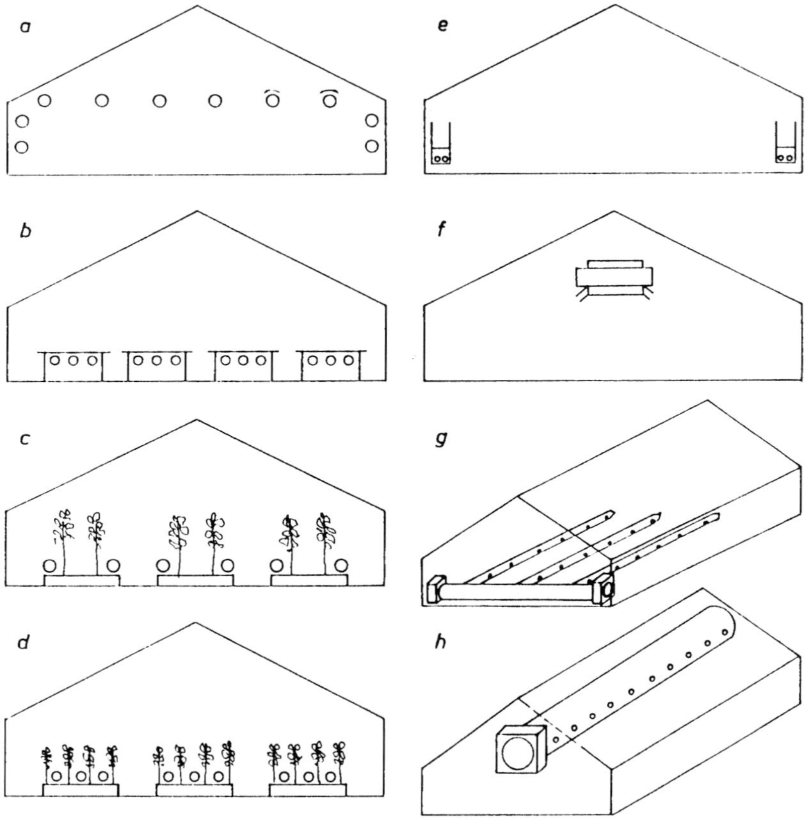

Fig. 2.40. Heating systems for greenhouse heating. (a) Pipe heating at the height of the eaves and side-wall heating. (b) Pipe heating under benches. (c,d) Ground-level pipes. (e) Side-wall convector. (f) Free-discharge air heater. (g) Ducted air heater on the ground. (h) Ducted air heater at eaves level.

(b) Prevention of heat radiation to the roof. The heat radiation should be directed to the plants.
(c) Screening of the side walls to prevent the penetration of cold air to the plants.

Figure 2.40 shows schemes of different heating systems.

Energy conservation

In order to minimize the use of fossil energy sources for greenhouse heating one may look to the development of new greenhouse concepts, or alternative energy sources. However, there are also considerable possibilities for energy conservation in conventional greenhouses and boiler houses (von Zabeltitz, 1982, 1986, 1988).

It is important that energy-conservation measures should not reduce light intensity too much, and that they should not influence other climatic factors such as carbon dioxide concentration and air humidity too strongly. The following methods for energy conservation can be distinguished:

(1) reducing heat losses through the cladding material and the greenhouse structure by double cladding, thermal screens, avoiding thermal bridges and leaks, and a favorable ratio of surface to ground area;
(2) heat transfer in the greenhouse by heating systems;
(3) better use of the cultivation area;
(4) control of the climatic factor temperature;
(5) wind breaks.

One of the most important measures is the use of thermal screens.

Thermal screens

Movable thermal screens are closed at night for energy conservation. When the screen is opened during the day the light comes nearly undiminished into the greenhouse. In northern regions with temperate climate (such as Germany) 70–75% of the heating

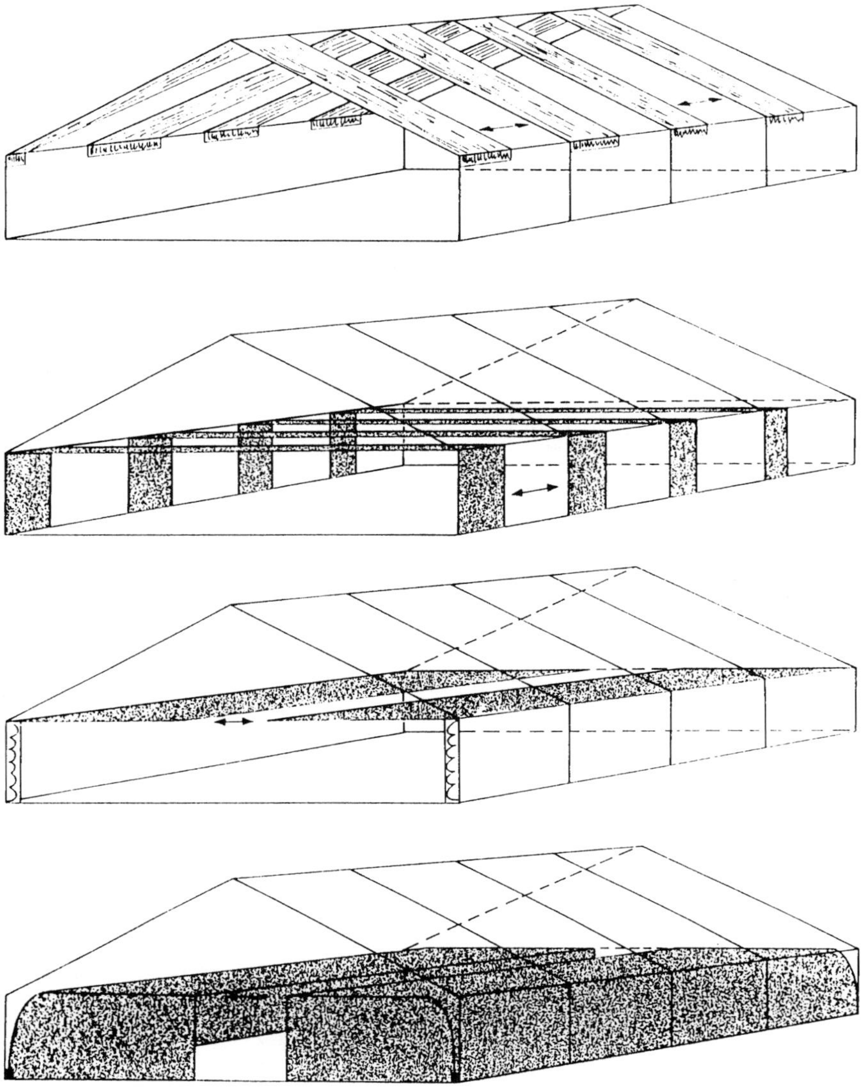

Fig. 2.41. Different systems of movable thermal screens.

energy is required during the night. By closing the thermal screens only at night a relatively high energy conservation can be achieved.

Some of the requirements for the screen material and the mechanical system are:

(1) low transmissivity for long-wave (infrared) radiation;
(2) low radiation efficiency and high reflection for long-wave radiation at a wavelength of more than 3000 nm;
(3) the material should be permeable to air and water vapor if transpiration is high, impermeable if it is low.

Different materials like plastic film, woven and non-woven materials are used for thermal screens. Aluminized materials have very favorable properties.

Requirements for the mechanical systems of thermal screens are:

(1) superior tightness at gable, side wall and foundation to prevent the penetration of cold air;
(2) low light reduction during the day;
(3) small surface of the screen;
(4) applicability to different materials;
(5) no tension stress on the material.

Thermal screens may be installed in the following ways (Fig. 2.41):

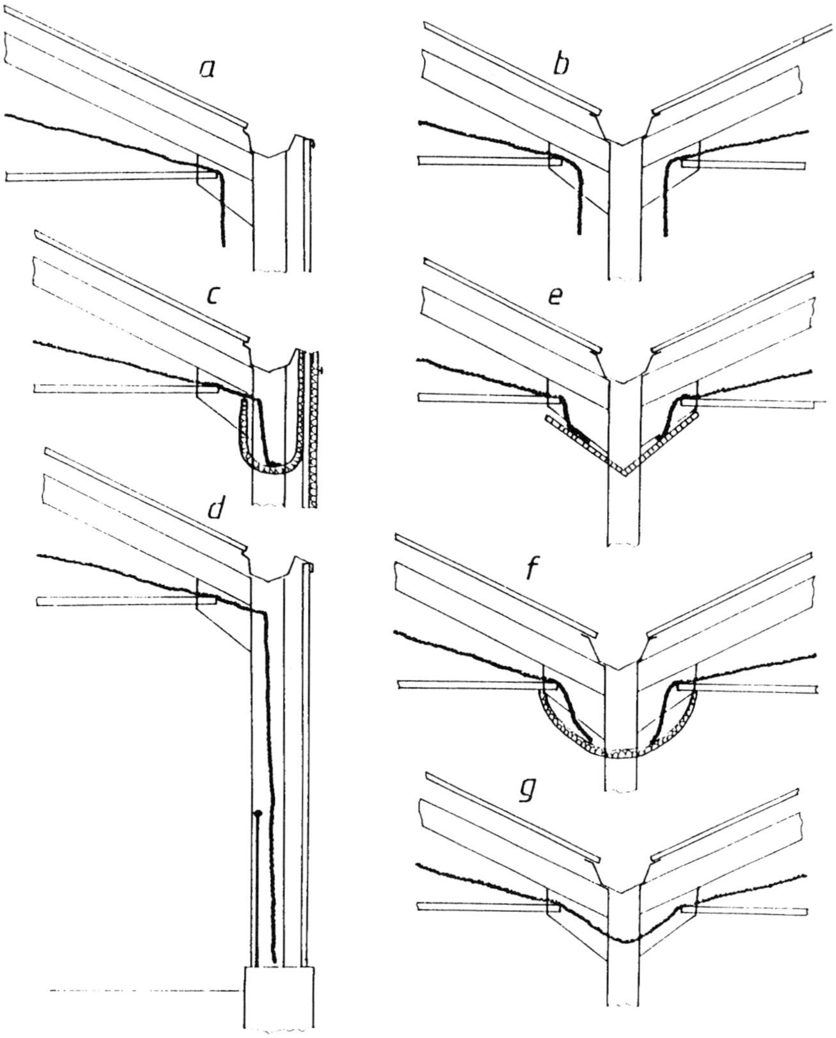

Fig. 2.42. Airtightness of thermal screens. (a,b) Not tight. (c)–(g) Well-sealed screens.

(1) parallel to the roof from stanchion to stanchion;
(2) horizontal at eave height, movable between the stanchions;
(3) horizontal at eave height, movable between the gutters;
(4) horizontal at eave height, movable down to the foundation.

Thermal screens are ineffective in conserving energy if they are loose, even if the material is effective. The edges of the screen materials must not hang down (Fig. 2.42a,b), as cold air might descend from the roof. They have to be led into special gutters (Fig. 2.42c,e,f) or should hang down at the side wall and be tightened at the foundation (Fig. 2.42d). When closed the screens have to overlap each other.

Energy conservation with a movable thermal screen depends on the type of material and the effectiveness of the barrier between the heated and the non-heated space in the greenhouse.

Meyer (1981, 1982) presented results of a thermal-screen system with optimum characteristics. Control procedures were described for checking the fuel conservation by thermal screens in commercial greenhouses.

The heat loss from a greenhouse at night has been given above (Eq. 2.26, p. 51), and can be calculated by using the overall heat loss coefficient u. This coefficient enables one to compare the energy consumption of greenhouses with different technical equipment. As this coefficient depends also on the

Table 2.10
Percentage energy saving with thermal screens ($t_i = 18°C$, wind speed $4\,m\,s^{-1}$)

Material	Energy conservation	
	At night	Total
Polyethylene film, clear	38	27
Polyethylene film, black	45	32
Polyester film, aluminized	54	38
Acrylic shading cloth	38	27
Polyester, felted material	40	28
Polyester, felted material, aluminized	43	31
"Blackout"[1], felted material	47	33
"Aircap"[2] film, polyethylene clear	50	35
Polyethylene film, black, double	54	38
"Aircap" film, aluminized	57	40
Polyester, felted material, double	59	42

[1] "Blackout": black plastic sheeting.
[2] "Aircap": an insulating film with air bubbles.

Table 2.11
Factor F_a at a wind speed of $4\,m\,s^{-1}$

Material	Factor F_a
Polyethylene film, clear	0.46
Polyethylene film, black	0.41
Polyester film, aluminized	0.27
Acrylic shading cloth	0.70
Polyester, felted material	0.39
Polyester, felted material, aluminized	0.52
"Blackout", felted material	0.40
"Aircap" film, polyethylene clear	0.44
Polyethylene film, black, double	0.33
"Aircap" film, aluminized	0.36
Polyester, felted material, double	0.31

external climatic conditions (principally the wind speed, but also rain and snow), it is given in relation to the wind speed. In order to calculate the percentage of energy conservation, one takes the value of u at a wind speed of $4\,m\,s^{-1}$ and relates it to a standard coefficient of energy consumption ($7.6\,W\,m^{-2}\,K^{-1}$ at $4\,m\,s^{-1}$, Meyer, 1981). Table 2.10 gives the savings obtained at night using different screen materials, with an internal temperature of 18°C, and also the average daily conservation. The results can be summarized as follows:

(1) For single-layer screens, both film and woven materials give maximum savings in the range of 38–45%. The differences between the materials result from the different permeabilities to long-wave (infrared) radiation. The differences in air permeability of the screen material become less significant when the tightness between the screen and the greenhouse structure is good.
(2) Double-layer systems allow a conservation of 50–60%. Their permeability to long-wave radiation and to air is similar to that of the single-layer materials.
(3) It is possible to get a 50% conservation with materials having good reflective properties (used as a single-layer system).
(4) The dependence of the coefficient of thermal transmission on the wind speed is very much reduced by the use of thermal screens.

Control of the airtightness of screen installations in practice

The standard of airtightness between the heated space and the unheated section of the greenhouse has a considerable influence on the possible conservation of energy. If the screen system does not close tightly the energy conservation can be reduced to half or one-third of the amount possible (Müller, 1987).

The airtightness can be estimated in practice by use of the control factor F_a, which is easy to determine from the expression

$$F_a = \frac{t_a - t_o}{t_i - t_o}, \tag{2.28}$$

where t_a is the temperature between screen and roof (glass), t_i is the temperature below the screen (= the internal temperature), and t_o is the external temperature. F_a expresses the degree of diminution of the temperature above the screen in relation to the internal temperature. It also depends on the wind speed. Table 2.11 gives F_a values for a wind speed of $4\,m\,s^{-1}$ and for the materials tested in the studies on heat consumption. Commercial installations can be checked by comparing the measured values with the performance of a thermally similar material shown in Table 2.9. If the differences are considerable, the screen system has leakages.

Energy conservation in commercial installations: A heating system must be maintained at a higher temperature than the internal air in order to obtain a specified temperature difference with the outside temperature. This temperature increase depends on the type and location of the heating equipment and on the thermal insulation of the greenhouse. A thermal screen makes an additional barrier; the lowering of the required excess temperature consequently provides a measure of the energy saving thus obtained.

The excess temperature is expressed by the factor

$$F_H = \frac{t_{Hm} - t_i}{t_i - t_o},$$

where t_{Hm} is the average temperature of the heating system, t_i is the inside temperature, and t_o is the outside temperature.

The relation between the two figures measured with open and closed screens gives an approximate value of the percentage of energy conservation.

The calculation of factors F_a and F_H enable one to assess a thermal-screen installation with regard to the standard of airtightness and the percentage of energy conservation. The most suitable period for recording the measured temperatures is from about 20.00 to 03.00 to ensure stable climatic conditions. The calculation of the factors F_a and F_H is based on the averages of the temperature readings during this period. The measurement of F_H requires readings with the screen closed and open during several nights.

HUMIDITY AND CARBON DIOXIDE CONTROL

Humidity control

The air humidity and the carbon dioxide concentration are, with light and temperature, important growth factors for the plants. The air humidity in the greenhouse is influenced by the plants themselves, by evaporation, by condensation on the roof materials and by exchange with the outside atmosphere. The air humidity in greenhouses is often expressed as relative humidity, which depends on the temperature. Problems with humidity are often due to temperature influences. Therefore, the temperatures of the external air, the internal air, the plants and the soil, and the temperature of the inner surface of the roof, are relevant. Water in plants acts as a growth element, effects cooling by evaporation, controls pressure in the cells, and acts as a medium for nutrient transport. Some of these functions take place only if the plants can transpire.

In summary, one can say the following:

(1) If the saturation deficit (on which the evaporation depends) is too high or too low, it has an adverse influence on plant physiology and plant growth.
(2) The air humidity inside the plant canopy influences plant diseases.
(3) If the plant temperature sinks below the dew point of the air, water condenses on the plants.
(4) The development of plant diseases depends on the duration of wetness of the plant surfaces.

The air in a greenhouse may be too dry and must be moistened, or too humid and must be dehumidified.

Increasing humidity

The humidity of the air in greenhouses can be increased by evaporation of water. All evaporative cooling systems also raise the air humidity, especially fan-and-pad cooling (see above, pp. 48–50), spray-cooling, and cooling by fog nozzles. The latter system is good for the humidification of the air.

Cooling by fog-nozzles: If water can be atomized so that the fog drops evaporate in the air, then the air in the greenhouse can be cooled without wetting the plants. For this to occur, the fog drops must have a diameter of 5–50 μm. There are three principal methods of atomizing water which give such small drops:

(1) revolving discs or brushes;
(2) high-pressure fog systems;
(3) air–water nozzle system with low pressure.

In the system with **revolving discs or brushes**, water is applied to revolving discs or brushes, which rotate at high speed, and therefore spray the water in fine drops (Fig. 2.43). The durability of the brushes is limited.

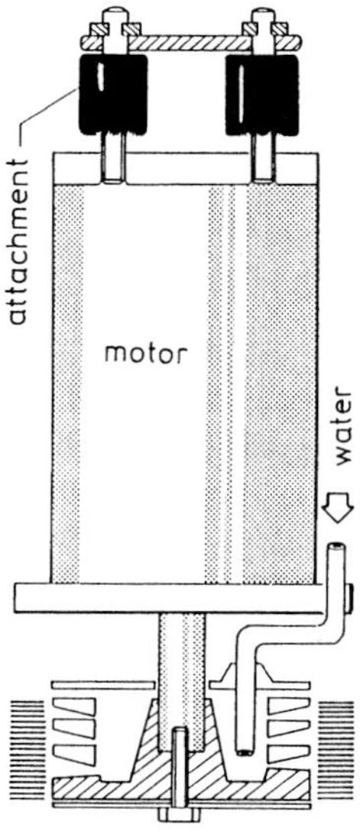

Fig. 2.43. Revolving brushes.

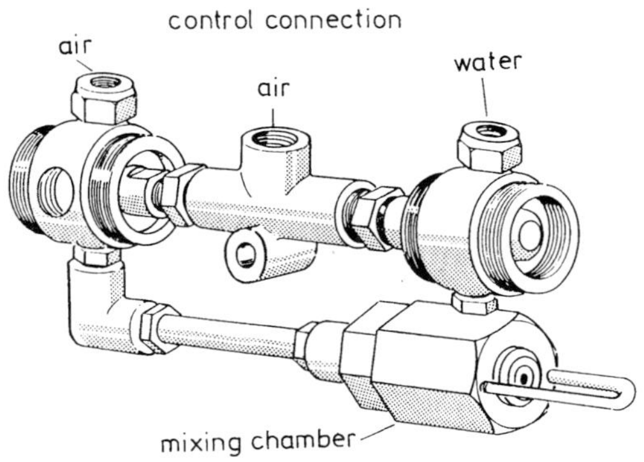

Fig. 2.44. Air–water nozzle system.

An advantage of the system is that the water quality does not need to be high. One unit is enough for a greenhouse area of 30–50 m^2.

In the **high-pressure fog systems**, the water is forced at high pressure through fine nozzles. A pressure of 3–6 MPa (30–60 bar) is needed to atomize the water. Very fine filters are needed to clean the water, and water with a high salt concentration is not suitable.

The **air–water nozzle system** uses air to atomize water in a mixing chamber with a nozzle (Fig. 2.44). The air and water pressure are much lower than in the high-pressure system. The air and water pipelines need a filter. One unit is enough for a greenhouse area of 20–30 m^2.

The fog-nozzle systems are used for cooling and air moistening. An exchange of internal and external air by ventilation is necessary if the air is to be cooled. The cooling efficiency depends on the air humidity and can be calculated as for the fan-and-pad cooling system.

Dehumidification

Air humidity and condensation are influenced by the following factors:

(1) The temperature of the inner surface of the roof. This depends on the roof material, heating system, outside temperature, clouds and wind speed. The temperature of the inner surface of the roof determines the condensation point on the roof and therefore the humidity level.
(2) Air exchange by leakages in the structure.
(3) Thermal screens. Tightness of the system and permeability for water vapor.
(4) External temperature and external humidity as well as internal greenhouse and plant temperature.
(5) Heating system.
(6) Accuracy of the temperature control.
(7) The irrigation system and the evaporation from the soil and substrate.

Methods for influencing the air humidity in greenhouses and avoiding condensation on the leaves of the plants (von Zabeltitz, 1984, 1986) may be listed as follows:

(1) In single-cladded greenhouses, dehumidification may occur by condensation on the roof material and by air leakages.
(2) Ventilation in unheated greenhouses.
(3) Heating and ventilating at the same time, with a good regulation of the ventilation, heating and air humidity. Much heat is lost by this method.
(4) Dehumidification of the plants by appropriate heating systems inside the plant canopy – "vegetation heating systems", air heating systems with air distribution through perforated plastic tubes, or pipe heating under benches.
(5) Air circulation between the plants.
(6) Water-saving irrigation systems with small evaporation areas and short evaporation times.
(7) Reduction of evaporation areas by mulching.
(8) Exact temperature control with small fluctuations.
(9) Timing of the irrigation. Irrigation in the morning means that the air can take up more moisture with rising temperatures and then can be expelled through ventilators.
(10) Controlled forced ventilation with heat recovery by a heat exchanger. A reduction of the carbon dioxide concentration below the value outside is then avoided simultaneously.
(11) The use of dehumidification devices such as a dehumidification heat pump or latent-heat converter.

In system (10), the cold air from outside is heated in a heat exchanger by the moist warm air from inside the greenhouse (Fig. 2.45) (Rüther, 1989). Compared with system (3), energy conservation is relatively high. Still, the application of heat recovery is an economic problem. Low energy costs for natural ventilation and high investment costs for the heat exchanger, as well as electricity costs, limit a positive result under Central European climatic conditions. In principle, however, heat exchangers for heat recovery are suitable for controlling and limiting greenhouse humidity. At the same time, it may be possible to avoid a reduction of carbon dioxide concentration below the value outside. If the control of humidity and carbon dioxide concentration leads to higher quality and yield, the system can become economic.

In Israel, a so-called latent-heat converter for the

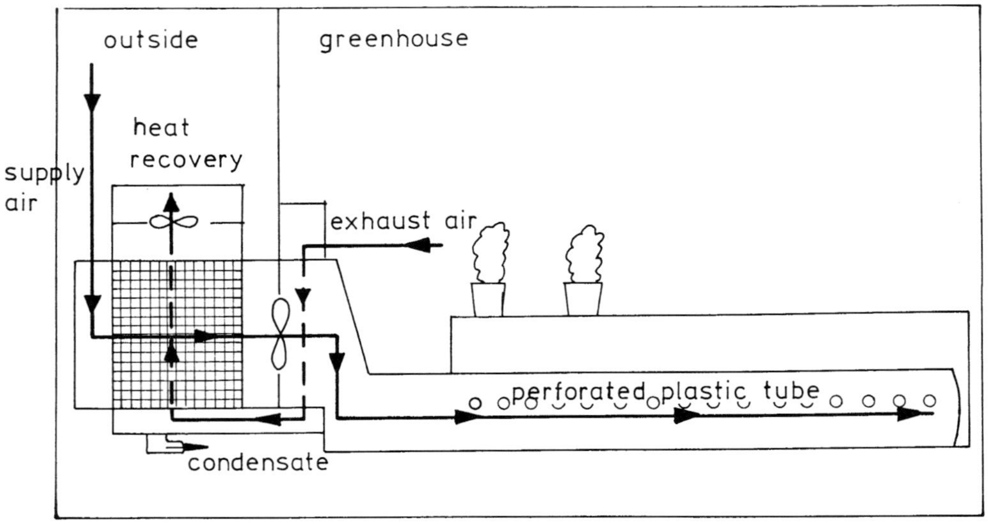

Fig. 2.45. Controlled forced ventilation with heat recovery.

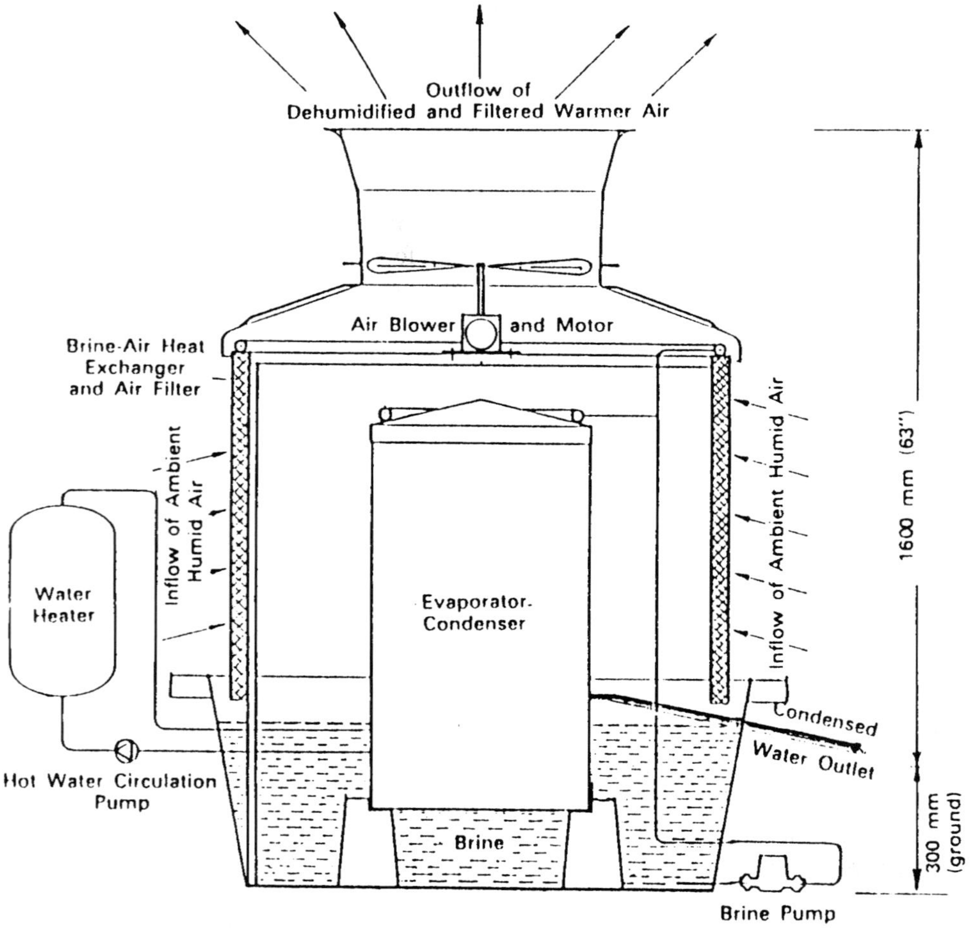

Fig. 2.46. Latent-heat converter.

dehumidification of greenhouses has been developed (Fig. 2.46) (Assaf, 1986). The moist greenhouse air is sucked through a heat exchanger by a ventilator. Brine flowing through the heat exchanger has direct contact with the air. The brine is hygroscopic, so that the moisture from the humid air condenses. The air is dehumidified and then flows, dried and filtered, back into the greenhouse. Heat is released by the condensation which then raises the temperature of the brine, which in turn raises the temperature of the air flowing through the heat exchanger.

The concentration of the brine must be raised again, as it was reduced by the condensed water. This happens in the brine evaporator–condenser system. The brine is heated by a warm-water heat exchanger. The warm water comes from an integrated water heater. Water evaporates from the brine and the vapor condenses on the inner side of the cylindrical brine–vapor heat exchanger. The condensate is then cooled by the same brine, which is pumped to the brine–vapor heat exchanger. The condensed water then flows out of the system.

The economy of a system and method of greenhouse air dehumidification depends on the efficiency of the installation, on the costs of electric power used, and on the investment costs, which have to be compared with the increase in quality and yield.

If the plants and the soil have abundant water available for evaporation, it can happen that the air humidity is not reduced very much, although the dehumidifier takes much water out of the air. In this case the rate of evaporation from the soil and plants has been raised.

Carbon dioxide enrichment

Enrichment of greenhouse air with carbon dioxide leads to better plant growth, shorter cropping times and higher quality. The carbon dioxide concentration can be raised as follows:

(1) Straw between the plant rows enriched with fertilizer and wetted. Carbon dioxide is released during decomposition. The amount of carbon dioxide cannot be controlled.
(2) Technical carbon dioxide from bottles or tanks.
(3) Exhaust gases from gas burners.

Exhaust gases from oil and coal heaters must not be used, because of the content of sulphur dioxide. Exhaust gases from gas burners may be used in two ways:

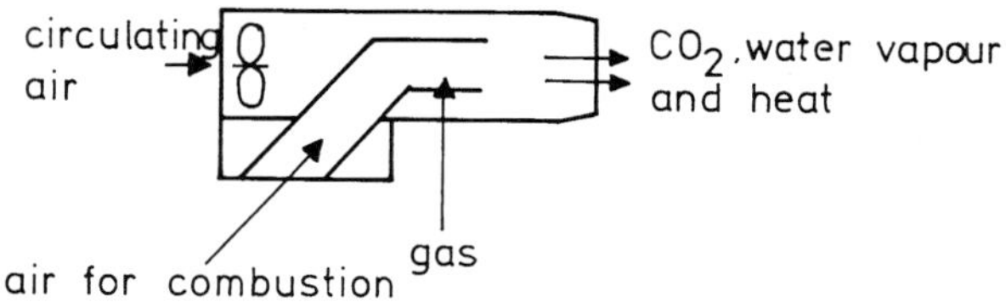

Fig. 2.47. Burner for carbon dioxide enrichment.

(1) piping of exhaust gases from a central heating system;
(2) direct use of exhaust gases from direct-fired gas burners inside the greenhouse.

Exhaust gases from gas burners can be led directly into the greenhouse. Special control systems are necessary, and care must be taken that no carbon monoxide is formed. It is practical to mix the exhaust gases with fresh air.

Carbon dioxide from a central heating system is only produced while it is in operation and when heat is needed. If carbon dioxide is needed, the system must operate during the day even if no heat energy is required.

The direct introduction of combustion gases from direct-fired gas heaters installed in the greenhouse is very beneficial, because energy is saved. Figure 2.47 shows a so-called "CO_2-gun". Gas is burned and the carbon dioxide is blown with the circulating air into the greenhouse. No carbon monoxide develops, as there is enough surplus air. Burners producing low levels of nitrogen oxides are to be preferred.

The maximum concentration of carbon dioxide permitted in a workplace for human beings should not be exceeded. In Germany, this maximum is 5000 ppm by volume. Forced air exchange may perhaps become necessary.

Figure 2.48 shows a system with a direct-fired gas burner with heat exchanger for greenhouse heating.

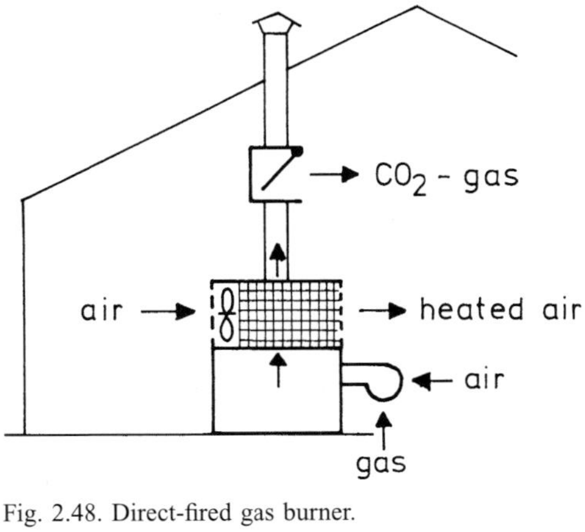

Fig. 2.48. Direct-fired gas burner.

Some of the combustion gases are tapped off and used for carbon dioxide enrichment of the greenhouse air. In this system measurement of the carbon dioxide concentration is necessary, because the entry of surplus air at the burner is usually set at a low level to ensure a high efficiency. The use of so-called "CO_2-guns" (Fig. 2.47) is preferable.

ARTIFICIAL LIGHTING

Light is one of the vital climate factors for plant growth. Besides natural daylight which comes into the greenhouse through the cladding material, artificial light is increasingly used, especially in temperate climates. Between the 40th and the 80th parallels light is definitely a limiting factor for plant growth in winter.

Three light-requiring processes rule plant growth.

Photosynthesis is the most important one, being essential for the growth of the plant. Radiant energy is converted into chemical energy which is necessary for the synthesis of the organic components in the plant.

Photomorphogenesis is the formative effect of light on plants.

Photoperiodism is the phenomenon through which plants react in different ways to the length of the daily light period to which they are exposed.

In horticulture, artificial light can be applied in various ways:

(1) to supplement daylight in greenhouses, in order to increase the irradiance level for photosynthesis;
(2) to prolong the effective day length;
(3) as a substitute for daylight in growing rooms, where plants are grown for commercial purposes;
(4) as a substitute for daylight in laboratories, where plants are grown for research purposes.

Terms and units in plant irradiation and lighting

When plants are lit artificially, it has to be borne in mind that the sensitivities of plants and the human eye to light are very different. The sensitivity curve of the human eye has its peak in the yellow-green spectral range, at a wavelength of 555 nm. The sensitivity curve of plants for photosynthesis has a peak in the red range of the spectrum, and has its minimum in the green range.

For the design and evaluation of artificial lighting systems one has to know and to compare the terms and units of radiant and luminous energy.

Radiant energy (watt × second, or joule: W s): energy which is emitted, transferred or absorbed in the form of radiation. In plant irradiation, only the spectral range between 400 and 700 nm is taken into account.

Luminous energy (lumen × second: lm s): light quantity, measured with the sensitivity of the human eye, radiated or received during a certain period.

Radiant flux (watt: W): the flow rate of the total energy from a light source.

Luminous flux (lumen: lm): the rate of emission of light, derived from the radiant flux by measuring the radiation with the sensitivity of the human eye.

Irradiance (mW per square meter: $mW\,m^{-2}$): the radiant flux density of photosynthetically active radiation (PAR) received by a surface. The PAR is the irradiation in the spectral range between 400 and 700 nm.

Illuminance ($lux = lm\,m^{-2}$: lx): the luminous flux density received by a surface.

Radiant efficiency ($mW\,W^{-1}$): the relation between the radiant flux emitted by a source of radiation in a spectral range between 400 and 700 nm and the energy consumed. The conversion of electric energy into radiant energy by a lamp.

Luminous efficiency ($lm\,W^{-1}$): the ratio of the total luminous flux emitted by a lamp to the energy consumed. There are large differences between lamp types.

When designing installations for artificial lighting the irradiance ($mW\,m^{-2}$) is the best basis for calculation. When measuring the level which is achieved by an existing installation the illuminance (lx) is often the more practical unit. Each lamp type has its special conversion factor for converting luminous flux into radiant energy (Table 2.12).

Lamps and lamp housings

The choice of a lamp depends on the purpose of application. Influencing variables are irradiation level, time of irradiation, spectral distribution, photoperiodic characteristics, switching cycles, available space and cost breakdown. Table 2.12 shows different types of lamps and their characteristics.

Incandescent lamps are easily installed and maintained. They need no ballast or starter. Their radiant efficiency, however, is rather poor. Figure 2.49 shows

Table 2.12
Characteristics of different lamps for artificial lighting in horticulture

Lamp type	Rated wattage (W)[1]	Luminous flux (lm)	Conversion factor ($mW\ lm^{-1}$)	Radiant flux (mW)	Radiant efficiency ($mW\ W^{-1}$)	Life (h)
Incandescent 150 W	150	2220	4.2	9320	62	1000
Fluorescent 58 W	70	5400	2.9	15 600	216	7500
High-pressure mercury 400 W	423	23 000	2.9	66 700	158	12 000
Metal halide 400 W	413	31 500	2.8	88 200	214	8000
High-pressure sodium 400 W	436	47 000	2.3	108 100	250	12 000

[1] Ballast included.

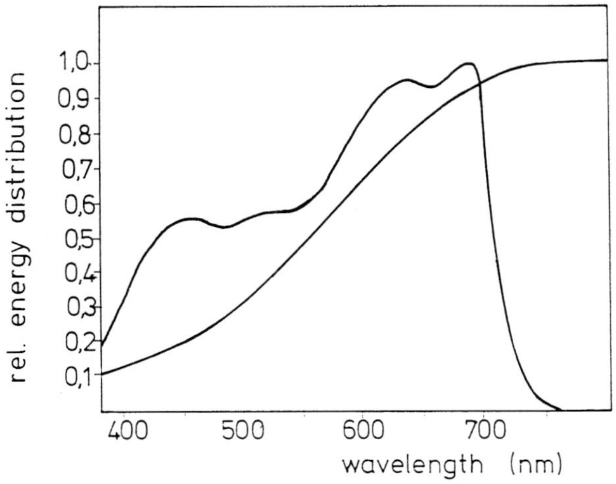

Fig. 2.49. Spectral energy distribution of an incandescent lamp in comparison with the sensitivity curve of photosynthesis.

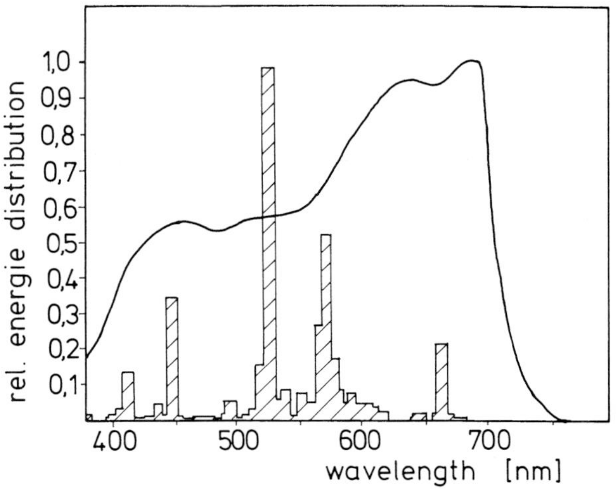

Fig. 2.50. Spectral energy distribution of a high-pressure metal halide lamp (the histogram) in comparison with the sensitivity curve of photosynthesis.

the spectral distribution compared to a sensitivity curve for photosynthesis. Incandescent lamps are mainly used for photoperiodic effects.

Tubular fluorescent lamps are suitable for phytotrons and multi-layer crops. They are mainly used to control photoperiodism in greenhouses.

High-pressure mercury lamps have frequently been used in horticulture. This lamp type is being replaced more and more by high-pressure sodium lamps.

High-pressure metal halide lamps have a high radiant efficiency. Their spectral distribution is shifted slightly into the blue range (Fig. 2.50).

High-pressure sodium lamps combine a high radiant efficiency with a spectral energy distribution which is suited to various crops (Fig. 2.51). High-pressure sodium lamps have largely replaced the high-pressure metal halide lamps.

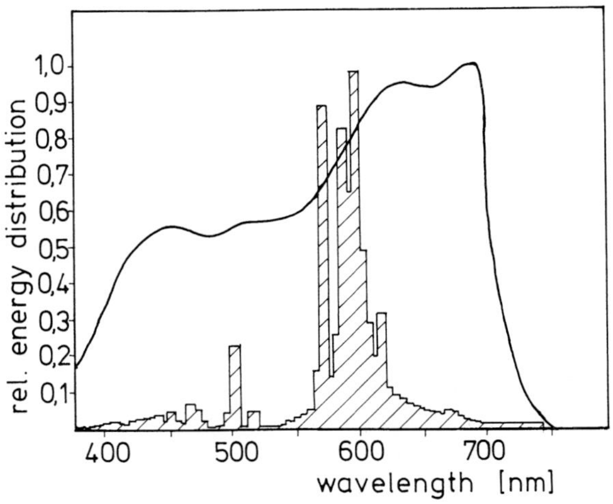

Fig. 2.51. Spectral energy distribution of a high-pressure sodium lamp (the histogram) as compared with the sensitivity curve of photosynthesis.

If the level of daylight is very low during winter, it is advisable to use high-pressure sodium lamps in combination with high-pressure metal halide lamps so that the blue content of the spectrum can be increased.

For a uniform light distribution at plant height it is necessary to install the lamps as high as possible. As the light from lamps in greenhouses is reduced by the greenhouse structure and interior fittings (thermal screens) it is necessary to have lamp housings giving a wide-angle light distribution. Nowadays producers have lamp housings for a restricted mounting height of 1.6–1.9 m. Different types of light distribution (circular, rectangular, elliptical) are available.

The energy demand for artificial lighting depends on the light intensity or illuminance and on the lamp type. For photosynthetic plant irradiation an electric power of 40 to 50 W m^{-2} is commonly used. Only a relatively low percentage of the electric energy consumed by a lamp will be converted into effective radiation for the plants. This can be seen in the data about radiant efficiency (Table 2.12). The rest of the electric energy is dissipated by heat radiation and convection. The radiant efficiency of a high-pressure sodium lamp is 250 mW W^{-1}. This means that 25% of electrical energy is converted into light. The energy converted into radiation and convection is used directly to heat the greenhouse. But the radiation in the visible range will also be absorbed and partly used for the supply of heat energy to the greenhouse.

In a greenhouse with single cladding an electric installation for artificial lighting of 50 W m^{-2} increases the inside temperature by approximately 5 to 6°C. In a double-covered greenhouse the temperature can be increased by up to 8°C.

The energetic evaluation of an artificial lighting system must include three considerations (Meyer, 1989):

(1) Light conditions; depending on installed electric power and duration of lighting.
(2) Electric energy demand; depending on installed electric power and duration of lighting.
(3) Heat energy demand; depending on inside temperature, greenhouse insulation, installed electric power, duration and time of artificial lighting.

These three parts influence each other and are connected by mode of operation and climate control. The control strategy of artificial lighting is therefore an important influence on light conditions, total heat and energy consumption, and cost. The different factors should not be considered separately.

BENCHES AND MOVABLE PALLETS

Plants in greenhouses have to be grown so that they are within reach for work, husbandry and harvest. Up to now in most cases a worker has gone to the plant for all husbandry work. The greenhouse area has then to be divided up so that the cultivation areas are within reach. New developments for the production of ornamental plants tend to use transport systems to move plants from their place in the greenhouse to a central workplace for the necessary operations.

In dividing the greenhouse into paths, benches and beds, the cultivation area should be made as large as possible in relation to the ground area, and the transport time required, especially for manual transport, should be made as short as possible. The ratio of the cultivation area to the total floor area of the greenhouse depends on the widths of paths and benches. Up to now, ground beds without permanent paths have been used for production of vegetables and cut flowers. If soilless culture is increasingly applied in future, the greenhouse floors will be concreted to avoid losses of fertilizer and plant-protection agents into soil and ground water.

Another new trend in pot-plant production is towards concreted ground beds (Fig. 2.52). The pot plants are placed on the ground and watered by stowing irrigation.

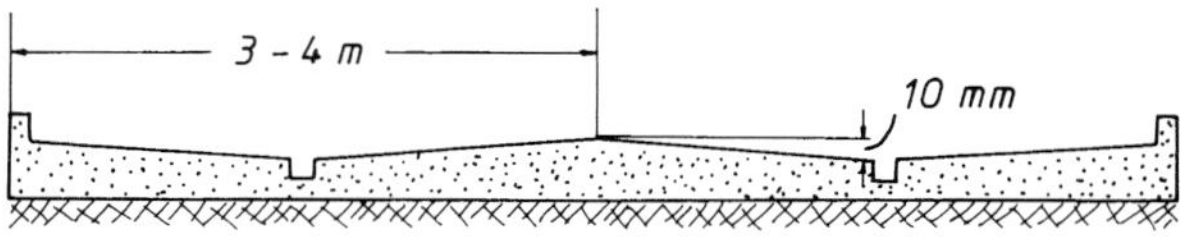

Fig. 2.52. Ground beds made of concrete.

Benches with permanent paths are made of wood, aluminum or concrete. Figure 2.53 shows an example of benches made of prefabricated concrete elements. The paths between the benches are 45–50 cm wide. The net cultivation area with fixed benches amounts to 65–75% of the total greenhouse area.

In the case of movable benches, the bench top can be moved 50–60 cm sideways (see Fig. 2.54). Thus the path area can be reduced considerably. The net cultivation area can be increased to 83–88% by use of movable benches. Movable benches consist of a substructure, roller tubes on which the benches are moved, and the bench top. The substructure is made of concrete or steel, the bench tops are made of aluminum bars with different tops. The benches are moved on two roller tubes installed in the longitudinal direction,

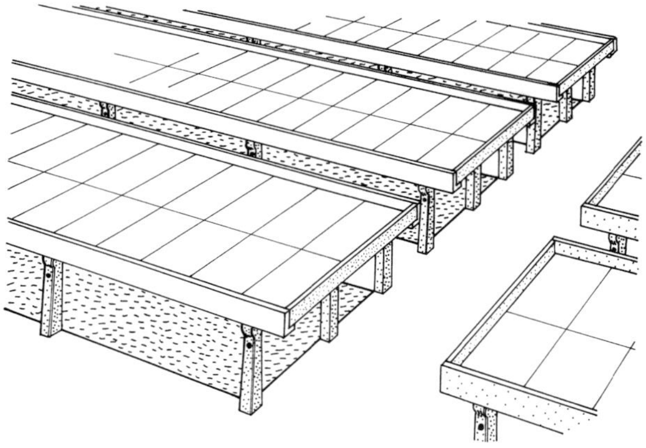

Fig. 2.53. Benches with permanent paths.

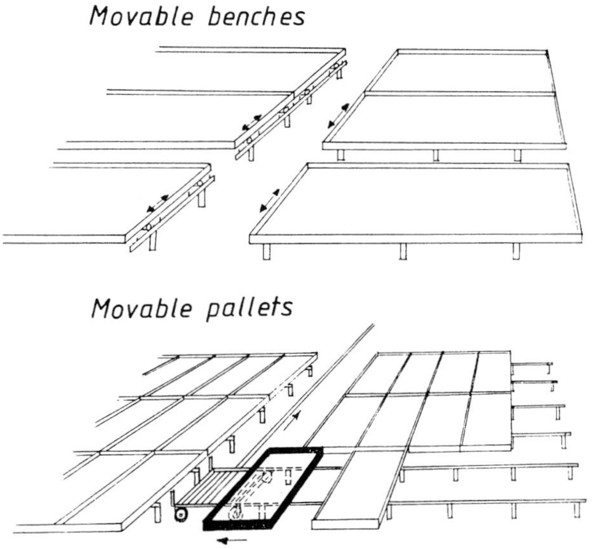

Fig. 2.54. Movable benches and movable pallets.

which roll between bench top and substructure. A space of about 10 cm should be left between benches which are moved together, so that warm air from the heating system can rise between the benches. When planning greenhouses with movable benches, the approximate labor requirement of the different crops should be considered. If several paths are necessary in a house, one fixed bench should be installed to every five or ten movable benches.

On movable pallets the pot plants are transported to a central working hall (Fig. 2.54). The net cultivation area achieved with the movable pallet system is slightly larger than that with movable benches. Theoretically, the whole greenhouse area except the main path can be put to use. When movable pallet systems are installed, however, paths between the pallet rows are necessary to tend the plants. Movable pallet systems will be appropriate if as many work operations as possible are transferred to the workroom, and if the transport problems are solved. A suitable dimension for movable pallets is 3 m × 1.6 m. The pallets have rollers and are moved to the beds on tubes. They must be as light as possible, and therefore are made of aluminum bars. They often have polystyrene tops.

The pallets are moved to the workroom on transport trolleys (Fig. 2.55) on which they are rolled from the beds. The transport trolleys can be automatically directed to the workroom by built-in control systems along concrete paths. A permanently installed transport system to move pallets along the paths is illustrated in Fig. 2.56. Roller tubes (2) which lie at right angles to the transport direction are lifted by means of pneumatic cylinders (1). The pallets are rolled on these tubes and lowered until they lie on the transport rollers (3). These rollers driven by an electric motor (5) using a belt system (4) transport the pallets to the workroom. Another possibility is to install the transport system over the benches.

The useable cultivation area in greenhouses can be further increased by multi-layer production. Here the production takes place on two levels. One practicable method is the horizontal rotor conveyer, Fig. 2.57, where pallets hang on a horizontal transport chain. The pallets loaded with pot plants are placed in the upper or lower level in turn. The lower level has to be lighted artificially. At one end of the rotor is a workplace. The plants are automatically irrigated in turn. This system is applicable only for relatively small plants. The cultivation area is increased by a factor of 1.6 to 1.8.

Another variation of the multi-layer system is the installation of movable pallets on two levels in the greenhouse. The lower pallets are lighted artificially. In order to move the pallets from one level to the other and

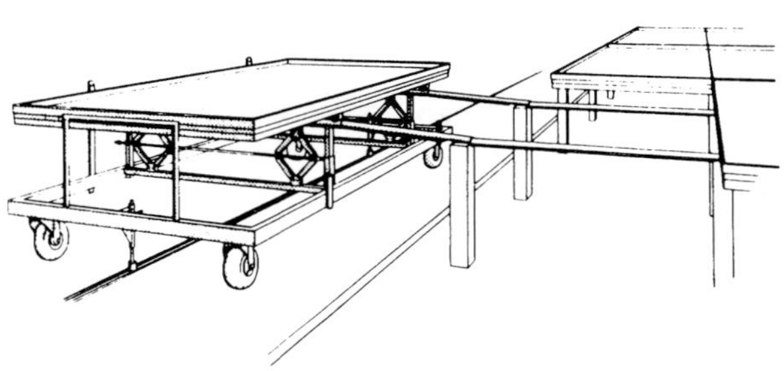

Fig. 2.55. Transport trolley for movable benches.

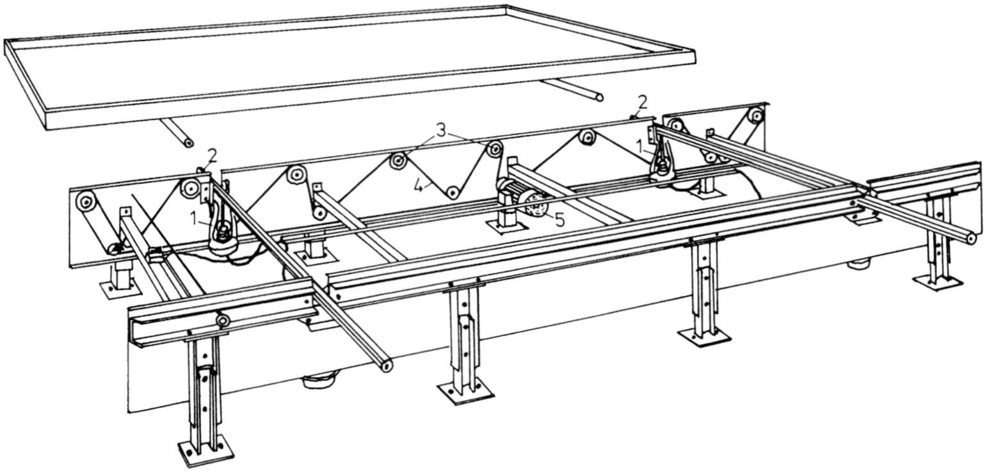

Fig. 2.56. Transport system for movable pallets.

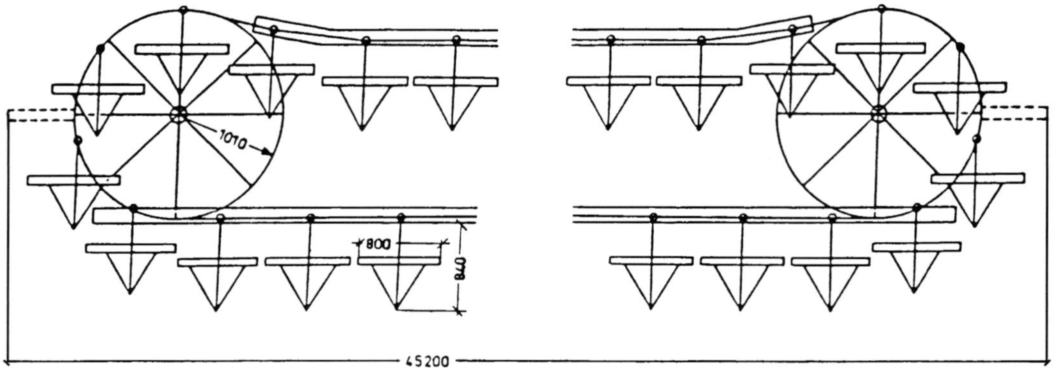

Fig. 2.57. Horizontal rotor for multi-layer production.

to transport them to the workroom special lifting and transporting equipment is required. Figure 2.58 shows a machine for automatically moving the pallet from one level to the other. There is a machine on each side of the pallet row.

In Denmark, Germany and the Netherlands, for example, pot plants are usually grown on movable benches and movable pallets. There are some 15 double-layer systems in Germany.

TECHNICAL EQUIPMENT FOR SOILLESS CULTURE

The application of soilless culture is constantly increasing in horticulture (see also Chapter 4). The reasons are:

(1) increased yield and improved quality by regular supply of mineral nutrients and better hygiene;
(2) saving of peat and protection of natural resources used as substrates;
(3) protection of the environment: no contact of fertilizer and plant-protection agents with the soil and the ground water.

Fertilizer and plant-protection measures are required for intensive plant production in greenhouses. When plants are grown in soil, surplus fertilizer and plant-protection agents inevitably are leached to deeper soil horizons and to the ground water. Many countries may therefore pass laws for the protection of the ground water, and thus will permit only soilless cultivation and closed-loop irrigation systems for plant production in greenhouses.

Technical equipment for a soilless culture system (Fig. 2.59) is required for processing of water and nutrients and nutrient control, and to provide space for plant growth. Figure 2.59 shows a complete system for

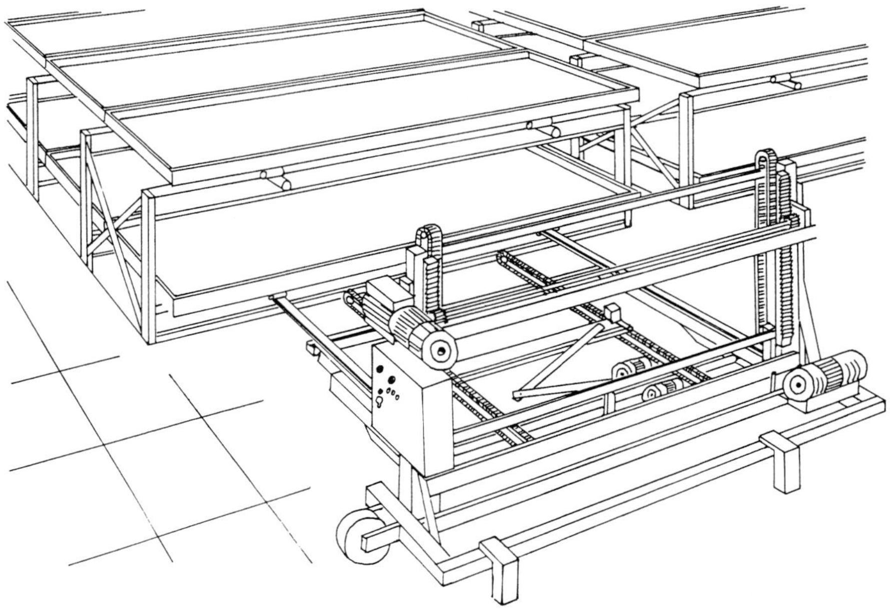

Fig. 2.58. Multi-layer production on movable pallets. Machine for automatically moving the pallets.

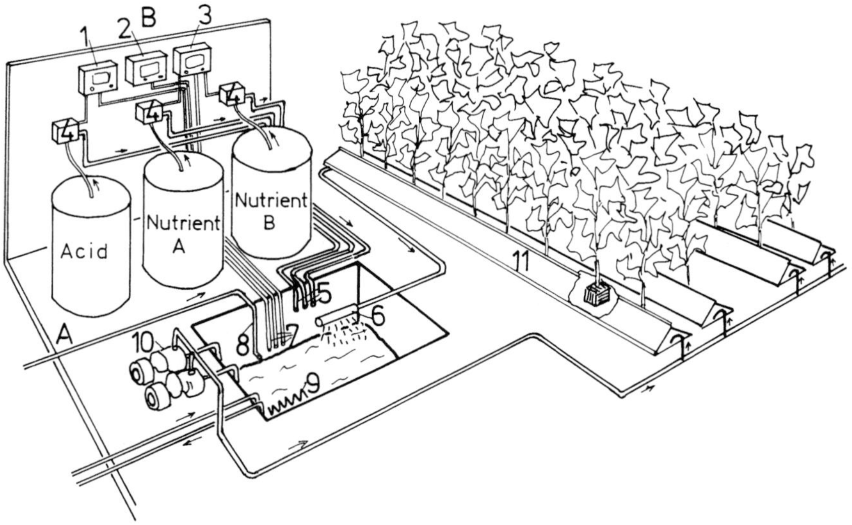

Fig. 2.59. Scheme of a nutrient-film system. 1, pH control; 2, temperature control; 3, salt control; 4, pumps for nutrients; 5, nutrient supply; 6, aeration; 7, measurement of pH, temperature and salt concentration; 8, water supply; 9, heating; 10, circulation pump; 11, NFT channel.

the nutrient-film technique (NFT). This illustrates the requirements for any soilless culture system, though simplifications may be possible.

By means of a control system, fertilizers are pumped into a catchment tank to maintain a given concentration. The irrigation and fertilizer solution is warmed up in the tank. The solution is then pumped through supply tubes to trays and canals in which the plants are standing. The surplus water is collected at the end of the canals and returned to the catchment tank. Here it is mixed with air, as a sufficient oxygen concentration in the root space is essential for all soilless cultivation.

In different systems root spaces may be found, with or without artificial substrates, which are either drenched with nutrient solution or through which the water flows. Figure 2.60 gives a schematic overview of different systems.

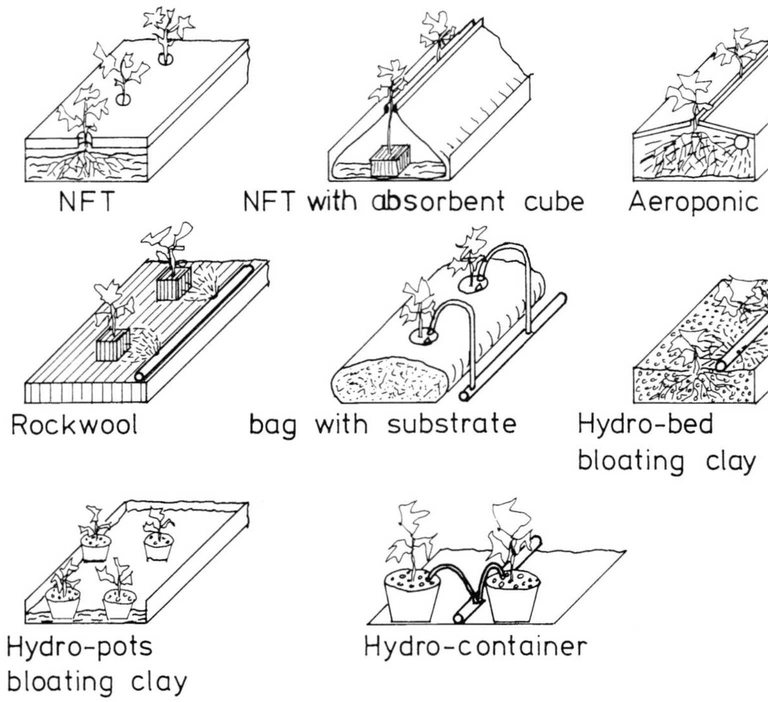

Fig. 2.60. Different systems for soilless culture.

When using nutrient-film culture the nutrient solution runs through the trays which have holes on the top and are tightened to avoid algal growth. The plants are held in mounts to that the roots grow in the nutrient solution without substrate. In case of other NFT methods seedlings are placed into the nutrient solution in absorbent cubes. In the aeroponic system, the nutrient solution is sprayed into the root space from pipes with nozzles. The plant roots hang in an air space with tiny drops or fog.

Plant cultivation on rockwool is widely applied. Seedlings in cubes are placed on rockwool pads into which the roots grow. The rockwool has to be provided with nutrient solution continuously by a drip irrigation system. Surplus solution runs off into trays. The rockwool pads are sometimes covered with plastic film to avoid evaporation. The disadvantage of rockwool is its limited suitability for frequent re-use. Moreover it is often difficult to dispose of the fertilizer-drenched rockwool pads after use, in a way harmless to the environment.

In some countries sand and other artificial substrates are filled into plastic bags. Seedlings in cubes are placed on the substrates through holes in the plastic and are irrigated by drip irrigation. In this case water and nutrient consumption have to be quite accurately adapted to plant consumption.

The use of bloating clay[2] for hydroponic culture has been known for a long time in floriculture. During the last few years bloating clay has also been applied in vegetable cultivation (Fölster, 1988). The advantage of bloating clay is that it can be recycled without limit. The grain size of bloating clay should be between 8 and 16 mm.

Culture in bloating clay can take place in containers with drip irrigation, in "hydro pots" with stowing watering, in which case the pots have to be 2–4 cm deep in the water, in "hydro trays" or in "hydro beds" with irrigation pipes. The plants in the bloating clay are continuously provided with nutrient solution. Normally the nutrient solution is not recirculated. Therefore the concentration and the nutrient ratio have to be adjusted to the plants' nutrient demand as accurately as possible. In the case of a long growing period, salts can accumulate at the surface due to evaporation and capillary action. It is best to supply the nutrient solution to the substrate from above. If the trays or beds have a slight slope, surplus nutrient solution can run off and

[2] "Bloating clay": the clay particles are aggregated into small balls, inflated with air bubbles, and then baked.

be recirculated. Bloating clay is also used in mixtures with other substrates.

As there is no or little buffer action for nutrients or water in the substrate for soilless cultures, exact measurement and stable concentration of the nutrients is necessary. Moreover a sufficient oxygen supply to the roots is important. For these reasons technical equipment with automatic control is necessary. These methods will gain more and more acceptance in order to protect the environment.

REFERENCES

Afnor U 57-060, 1989. Serres et équipements. Règles de stabilité pour la construction des serres à vitrages plats, Dec. 1989.

Afnor U 57-063, 1989. Serres et équipements. Règles de stabilité pour la construction des serres-tunnels couvertes de materiaux plastiques souples ou cintrables à froid, Dec. 1989.

Ahmadi, G. and Glockner, P.G., 1982. Dynamic simulation of the performance of an inflatable greenhouse in the southern part of Alberta. I. Analysis and average winter conditions. *Agric. Meteorol.* 27: 155–180.

Arinze, E.A., Schoenau, G.J. and Besant, R.W., 1982. *Simulation study on the effectiveness of external thermal insulations on energy conservation in greenhouses,* Paper No. 82-4539. ASAE, St. Joseph, MI.

Arinze, E.A., Schoenau, G.J. and Besant, R.W., 1984. A dynamic thermal performance simulation model of an energy conserving greenhouse with thermal storage. *Trans. ASAE* 27(2): 508–519.

ASAE EP288.4, 1990. Agricultural building snow and wind loads. ASAE standards, Engineering Practice: EP288.4

ASAE EP406, 1988. Heating, Ventilation and cooling greenhouses. ASAE standards, Engineering Practice: EP406.

ASHRAE, 1985. *ASHRAE Handbook of Fundamentals.* ASHRAE, Atlanta, GA.

Assaf, G., 1986. LHC – The latent heat converter. *Heat Recovery systems* 5(5), 369–379.

Avissar, R. and Mahrer, Y., 1982. Verification study of a numerical greenhouse microclimate model. *Trans. ASAE* 25(6): 1711–1720.

Bailey, B.J., 1988. In: Chr. von Zabeltitz (Editor), *Energy Conservation and Renewable Energies for Greenhouse Heating,* REUR Technical Series, No. 3. FAO Regional Office for Europe, pp. 17–39.

Bailey, B.J. and Richardson, G.M., 1990. A rational approach to greenhouse design. In: P. Feuilloley (Editor) *Second Workshop on Greenhouse Construction and Design, Acta Horti.* 281: 111–117.

Baudoin, W., Grafiadellis, M., Jiminez, R., La Malfa, G., Martinez Garcia, P.F., Garnaud, J.C., Monteiro, A.A., Nisen, A., Verlodt, H., de Villelle, O. and von Zabeltitz, C., 1990. *Protected Cultivation in the Mediterranean Climate. FAO Plant Production and Protection Paper,* No. 90.

Baytorun, A.N., 1986. Bestimmung des Luftwechsels bei gelüfteten Gewächshäusern. *Gartenbautechnische Information* (ITG Hannover), Heft 27, 155 pp.

Bot, G.P.A., 1983. *Greenhouse climate: from physical processes to a dynamic model,* Ph.D. Thesis. Agricultural University, Wageningen, Netherlands, 240 pp.

BSI, 1987. British Standard BS 5502, *Buildings and Structures for Agriculture.* Part 22: Code of Practice for design, construction and loading.

BSI, 1991. Document No. 91: *Film-plastic and Rigid Plastic Clad Tunnel Buildings.* Code of practice for design, construction and loading. (Draft, Dec. 1991).

Businger, J.A., 1966. The glasshouse (greenhouse) climate. In: W.R. van Wijk (Editor), *Physics of Plant Environment,* 2nd edition. North-Holland, Amsterdam, pp. 277–318.

CEN, 1997. European Committee for Standardization, Brussels; Draft prEN 13031-1. Greenhouses: Design and construction, Part 1: Commercial production greenhouses.

Cooper, P.I. and Fuller, R.J., 1983. A transient model of the interaction between crop, environment, and greenhouse structure for predicting crop yield and energy consumption. *J. Agric. Eng. Res.* 28:401-417.

Cormary, Y. and Nicolas, C., 1985. *La Thermique des Serres,* Collection de la Direction des Etudes et Recherches d'Electricité de France, Vol. 58. Eyrolles, Paris, 357 pp.

Critten, D.L., 1983. A computer model to calculate the daylight integral and transmissivity of a greenhouse. *J. Agric. Eng. Res.* 28: 61–76.

Critten, D.L., 1989. Light transmission into greenhouses. In: *Engineering and Economic Aspects of Energy Saving in Protected Cultivation. Acta Horti.* 245,: 377–383.

Damrath, J., 1980. *Tabellen zur Heizenergieermittlung von Gewächshäusern, Gartenbautechnische Informationen,* Vol. 8. Institute for Horticultural Engineering, University Hannover.

de Halleux, D., 1989. *Modèle dynamique des échanges énergétiques des serres: étude théorique et expérimentale,* Ph.D. Thesis. Faculté des Sciences Agronomiques de Gembloux, Belgium, 278 pp.

Delwiche, S.R. and Willits, D.H., 1984. The effect of condensation on heat transfer through polyethylene film. *Trans. ASAE,* pp. 1476–1482.

DIN 1054, 1979. *Zulässige Belastung des Baugrundes.* Deutsches Institut für Normung.

DIN 11535, 1982. *Gewächshäuser. Grundsätze für Ausführungen und Berechnung,* Teil I (March).

Duncan, G.A., Loewer, O.J. and Colliver, D.G., 1981. Simulation of energy flows in a greenhouse: magnitudes and conservation potential. *Trans. ASAE* 24(6): 1014–1021.

Eggers, H., 1975. Untersuchungen zum Klima unter verschiedenen Folienabdeckungen. *Gartenbautechnische Information* (ITG Hannover), Heft 2.

Fähnrich, I., Meyer, J. and von Zabeltitz, Chr., 1989. Influence of condensation on light transmission and heat transfer through greenhouse covering materials. *Plasticulture* 84(4): 13–18.

Fölster, E., 1988. Blähton im Gemüsebau – eine Alternative zur Steinwolle. *Dtsch. Gartenbau* 30:1844–1847.

Garzoli, K., 1984. *Solar greenhouse design and performance,* Doctoral Thesis, Agricultural Engineering Section, University of Melbourne.

Garzoli, K.V. and Blackwell, J., 1973. The response of a glasshouse to high solar radiation and ambient temperature. *J. Agric. Eng. Res.* 36:75–85.

Glaub, J.C. and Trezek, G.J., 1981. *Heat and mass transfer analysis of greenhouses,* Paper No. 814030. ASAE, St. Joseph, MI.

Hellickson, M.A. and Walker, I.N., 1983. Ventilation of Agricultural

Structures. ASAE Monograph no. 6. ASAE Am. Soc. Agric. Eng., St. Joseph, MI.

Hölscher, Th., 1992. *Ein Simulationsmodell für das Gewächshausklima,* Doctor Thesis. University of Hannover.

Hsieh, C.K. and Rajvanshi, K., 1977. The effect of dropwise condensation on glass: solar properties. *Solar Energy* 19: 389–393.

ISHS, 1991. *Greenhouses Design, Construction and Loading.* European Greenhouse Standard, Draft 01.

Jolliet, O., 1988. *Modelisation du comportement thermique d'une serre horticole,* Ph.D. Thesis No. 713. Ecole Polytechnique Fédérale de Lausanne, Switzerland, 246 pp.

Kimball, B.A., 1973. Simulation of the energy balance of a greenhouse. *Agric. Meteorol.* 11:243–260.

Kimball, B.A., 1986. *A modular energy balance program including subroutines for greenhouses and other latent heat devices.* Agricultural Research Service, United States Department of Agriculture (ARS 33), 356 pp.

Kindelan, M., 1980. Dynamic modeling of greenhouse environment. *Trans. ASAE* 23(6):1232–1239.

Kirsten, W., 1973. *Die natürliche Einstrahlung in Gewächshäuser in Abhängigkeit von deren Form und Aufstellungsrichtung sowie der Beschaffenheit des Füllstoffes,* Doctoral Thesis. Universität Hannover.

Kozai, T., Goudrian, I. and Kimura, M., 1978. *Light Transmission and Photosynthesis in Greenhouses,* Simulation Monograph. Dudoc, Wageningen.

Krug, H., 1991. *Gemüseproduktion* [Vegetable production]. Paul Parey Verlag, Berlin/Hamburg, 541 pp.

Kurata, K. and Tachibana, K., 1978. Greenhouse design with the optimization technique. *Acta Horti.,* 37: 21–30.

Lacroix, R. and Zanghi, J.C., 1990. *Étude comparative de la structure des modèles de transfert d'énergie et de masse dans les serres.* Can. Agric. Eng. 32:269–284.

Landsberg, J.J., White, B. and Thorpe, M.R., 1979. Computer analysis of the efficacy of evaporative cooling for glasshouses in high energy environments. *J. Agric. Eng. Res.* 24:29–39.

Maher, M.J. and O'Flaherty, T., 1973. An analysis of greenhouse climate. *J. Agric. Eng. Res.* 18:197–203.

Meyer, I., 1981. *Bewegliche Energieschirme* [Movable thermal screens]. *Gartenbautechnische Information* (ITG Hannover), Heft 10, 108 pp.

Meyer, I., 1982. *Bewertung von beweglichen Energieschirmen im Hinblick auf den Energieverbrauch. Gartenbautechnische Information* (ITG Hannover), Heft 11, 143 pp.

Meyer, I., 1989. *Analyse von Kunstlichtsystemen mit einem modularem Simulationsprogramm zur Datengewinnung. Gartenbautechnische Information* (ITG Hannover), Heft 33, 155 pp.

Milburn, W.F., 1982. *Analysis of thermal energy systems in greenhouses,* Ph.D. Thesis. Pennsylvania State University, University Park, PA, 167 pp.

Morris, L.G., Trickett, E.S., Vanstone, F.H. and Wells, D.A., 1958. Limitation of maximum temperature in a glasshouse by the use of a water film on the roof. *J. Agric. Eng. Res.* 3:121–130.

Müller, G.I., 1987. *Energieschirme unter Praxisbedingungen – Bewertung und Optimierung im Hinblick auf Energieverbrauch und Klimaführung. Gartenbautechnische Information* (ITG Hannover), Heft 28, 175 pp.

Munters, 1992. *The importance of bleed-off,* Leaflet E.25.4 184.10.Munters Company, Sollentuna, Sweden.

NEN 3859, 1990. *Greenhouses Structural Requirements.* Unofficial translation, March.

Nisen, A., 1969. *L'éclairement Naturel des Serres,* Edition I. Duclot S.A., Gembloux.

Price, D.R. and Peart, R.M., 1973. Simulation model to study the utilization of waste heat using a combination multiple reservoir and greenhouse complex. *J. Environ. Qual.* 2:216–224.

Rotz, C.A., 1977. *Computer simulation to predict energy use and system costs for greenhouse environmental control,* Ph.D. Thesis. Pennsylvania State University, University Park, PA, 132 pp.

Rüther, M., 1989. *Energieeinsparung durch Wärmerückgewinnung beim Zwangsluftwechsel in abgedichteten Gewächshäusern* [Energy conservation by forced ventilation with heat recovery in tightened greenhouses]. *Gartenbautechnische Information* (ITG Hannover), Heft 31, 160 pp.

Schulze, L., 1954. *Die Einstrahlung natürlichen Lichts in Gewächshäuser, gemessen an Modellen und Folgerungen für den Gewächshausbau,* Doctoral Thesis. Universität Hannover.

Seginer, I. and Kantz, D., 1986. In-situ determination of transfer coefficients for heat and water vapour in a small greenhouse. *J. Agric. Eng. Res.,* 35: 39–54.

Seginer, I., Angel, A., Gal, S. and Kantz, D., 1986. Optimal CO_2 enrichment strategy for greenhouses; a simulation study. *J. Agric. Eng. Res.* 34:285–304.

Selcuk, M.K., 1970. Use of digital computers for the heat and mass transfer analysis of controlled environment greenhouses. In: *Proc. 1st Symp. on the Use of Computers for Engineering related to Buildings.* National Bureau of Standards, Gaithersburg, MD, pp. 557–577.

Selcuk, M.K., 1971. Analysis design and performance evaluation of controlled-environment greenhouses. *ASHRAE Trans.* 23(6):1240–1245.

Shen, H., 1977. *Dynamic simulation of the utilization of industrial waste heat for greenhouses,* Ph.D. Thesis. Rutgers University, New Brunswick, NJ, 154 pp.

Silveston, P.L., Costigane, W.D., Tiessen, H. and Hudgins, R.R., 1980. Energy conservation through control of greenhouse humidity. I. Condensation heat losses. *Can. Agric. Eng.* 22(2):125–132.

Sirjacobs, M., 1989. Greenhouses in Egypt. In: *Protected Cultivation in the Mediterranean Climate.* FAO, Rome.

Soribe, F.I. and Curry, R.B., 1973. Simulation of lettuce growth in an air-supported plastic greenhouse. *J. Agric. Eng. Res.* 18:133–140.

Stoffers, I.A., 1967. *Lichtdoorlatendheid van met vlakke materialen bedekte warenhuizen.* IMAG Wageningen, Publikatie, Nr. 14.

Takakura, T., Jordan, K.A. and Boyd, L.L., 1971. Dynamic simulation of plant growth and environment in the greenhouse. *Trans. ASAE* 14(5):964–971.

Takami, S. and Uchijima, Z., 1977. A model for the greenhouse environment as effected by the mass and energy exchange of a crop. *J. Agric. Meteorol.* (Japan) 33(3):155–166.

Tantau, H.-J., 1975. *Der Einfluß von Einfach- und Doppelbedachungen auf das Klima und den Wärmehaushalt von Gewächshäusern,* Ph.D. Thesis. *Gartenbautechnische Informationen,* Vol. 4. Institute for Horticultural Engineering, University Hannover, 238 pp.

UNI 6781-71, 1971. Serre a struttura metallica. Istruzioni per l'importazione generale del progretto strutturale e per calculo.

van Bavel, C.H.M. and Sadler, E.J., 1979. *SG79 – a computer simulation program for analyzing energy transformations in a solar greenhouse.* Texas A&M University, College Station, TX, 75 pp.

van de Braak, N.J., Breuer, J.J.G. and Heyna, B.J., 1984. Cumulative frequency distribution curves of greenhouses heat requirements. *Acta Horti.* 148: 337–344.

van den Kieboom, A., 1984. *Stralingseigenschappen van Kasbedekkingsmaterialen en constructies.* IMAG Wageningen.

Verlodt, H., 1990. Greenhouses in Cyprus, In: *Protected Cultivation in the Mediterranean Climate.* FAO, Rome.

von Elsner, B., 1982. *Das Kleinklima und der Wärmeverbrauch von geschlossenen Gewächshäusern,* Ph.D. Thesis. *Gartenbautechnische Informationen,* Vol. 12. Institute for Horticultural Engineering, University Hannover, 197 pp.

von Zabeltitz, Chr. (Editor), 1982. Energieeinsparung und Alternative Energiequellen im Gartenbau [Energy conservation and alternative energies in horticulture]. Ulmer Verlag, Stuttgart, 115 pp.

von Zabeltitz, Chr., 1984. Luftfeuchte im Gewächshaus – ein sehr komplexes Thema. *Gärtnerbörse und Gartenwelt, Gb+Gw,* 47: 1114–1119.

von Zabeltitz, Chr., 1986. *Gewächshäuser,* 2nd edition. Ulmer Verlag, Stuttgart, 284 pp.

von Zabeltitz, Chr. (Editor), 1988. *Energy Conservation and Renewable Energies for Greenhouse Heating,* REUR Technical Series, No. 3. FAO Regional Office for Europe, 167 pp.

von Zabeltitz, Chr., 1990a. Greenhouse constructions in function of better climate control. In: *Symp. on Simple Ventilation and Heating Methods for Greenhouse in Mild Winter Climates, Acta Horti.,* 263: 357–374.

von Zabeltitz, Chr., 1990b. Greenhouses and walk in tunnels. In: *Protected Cultivation in the Mediterranean climate,* FAO Plant Production and Protection Paper, No. 90. FAO, Rome.

Waaijenberg, D., 1990. Standard for film-covered greenhouses. In: *Second Workshop on Greenhouse Construction and Design, Acta Horti.* 281: 129–137.

Waaijenberg, D. and Denkov, D., 1992. Reduction of wind damage to tunnel greenhouses by structural improvements. *Proc. Workshop on European Greenhouse Standards,* Barcelona. Technical Series 25, 48–56. FAO, Rome.

Walker, J.N., 1965. Predicting temperatures in ventilated greenhouses. *Trans. ASAE* 8(3):445–448.

Walker, J.N. and Walton, L.R., 1971. Effect of condensation on greenhouse heat requirement. *Trans. ASAE,* pp. 282–284.

Wass, S.N. and Barrie, I.A., 1984. Application of a model for calculating glasshouse energy requirements. *Energy Agric.* 3:99–108.

Weimann, G., 1986. Untersuchungen zu neuen Foliengewächshauskonzeptionen. *Gartenbautechnische Information* (ITG Hannover), Heft 24, 152 pp.

Hannover, 1995

Chapter 3

PHYSICAL PRINCIPLES OF MICROCLIMATE MODIFICATION

W. DAY and B.J. BAILEY

INTRODUCTION

Concepts of microclimate physics

In order to understand the constraints on biological productivity in any ecosystem, it is necessary to define the physical environment within that ecosystem. The rates of underlying biochemical processes depend on temperature, moisture status, and the supply of various substrates. The development of the whole plant or animal is also strongly dependent on the environment which existed throughout its growth, and this influences both the structure of the organism and its response to various aspects of its surroundings. In the natural environment, there is an extremely wide range of conditions that will support the growth of living organisms. Some microorganisms can grow at temperatures from −6 to 100°C, whilst higher plants can survive at temperatures below zero or above 40°C. The greenhouse ecosystem will in general involve a narrower range of environments, as the purpose of human intervention is to provide conditions most appropriate to the growth of particular species.

Radiant energy in those wavelengths that can excite photosynthetic pigments in plant tissue is the driving force for carbon fixation in higher plants. This carbon fixation sets up gradients of carbon dioxide concentration and a flux of carbon dioxide into the plant from the surrounding air. The radiation environment over a wider range of wavelengths can influence other biological processes, such as those which control the development of the whole organism. Perception of day length and its changes leads to switches in patterns of development, for instance from vegetative to floral in plants, and directional stimuli from light can influence the way in which the plant grows. Of course, the total amount of radiant energy has indirect significance because of its effect on the temperature equilibrium of the system.

The flux of carbon dioxide has already been mentioned. Of comparable importance is the flux of water vapor. In higher plants, these exchanges predominantly occur through pores in the plants outer surface (stomata), and the resistance to gas exchange through these pores is to a degree under the control of the plant, thus giving the plant some ability to influence its own environment. The plant must maintain a high degree of hydration in its tissues if they are to continue to function well. This is achieved by the presence of waxes and similar materials over much of the plant's surface, so that closure of stomata can reduce water loss to a minimal level. Minimizing water loss in this way is not an ideal solution, as it also minimizes the exchange of carbon dioxide through the same pores. It is also important for plants to have a significant flux of water through their transpiration stream in order to ensure a sufficient supply of some mineral nutrients into various parts of the plant (Ho, 1988).

Moisture is also important to other organisms that occur in the greenhouse ecosystem. Many fungal organisms that can live parasitically on the leaves, stems and fruits of greenhouse plants require the presence of free water on the plant surface; this enables the fungal spores to germinate, penetrate the leaf epidermis and then colonize the leaf tissue. Such conditions are also required for some fungi used as biological control agents, through their ability to grow in the bodies of insect pests and hence to kill them.

Microclimate physics is predominantly concerned with transport processes, where physical laws can be defined that govern the exchange of heat, radiation and mass between the various parts of the system in question through its microclimate (Fig. 3.1). The underlying laws are fundamental in nature –

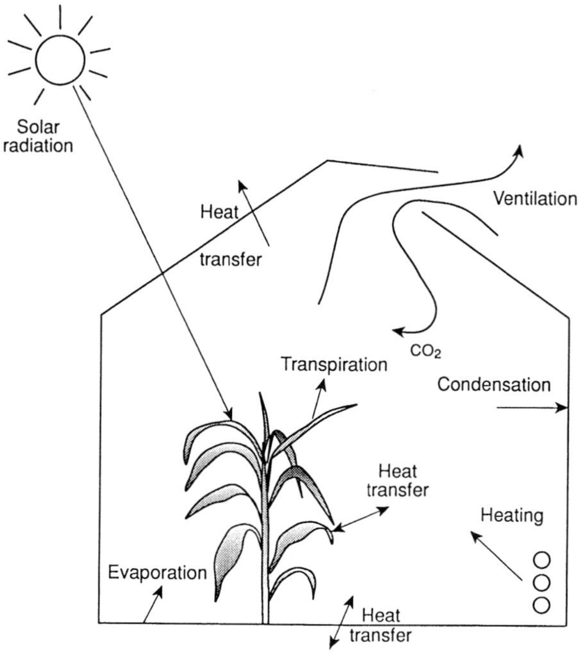

Fig. 3.1. Schematic diagram of a greenhouse showing the major processes which influence the microclimate.

conservation of energy and conservation of mass. The rates of transfer of mass and energy between the entities in the greenhouse ecosystem depend on complex processes such as turbulent heat transfer in the greenhouse air, and carbon dioxide assimilation within leaf tissue. Therefore some empiricism is necessary to give a useful description of all the key processes in the greenhouse environment. The first requirement is to define the components of the greenhouse that influence the transport processes. In the next step of describing the interactions, some approximations are necessary in order to make it possible to solve the equations and establish how the greenhouse climate will respond to the driving forces of the outside weather conditions, or of imposed heating or shading, or of design changes to the structure or fabric of the greenhouse. Current developments in mathematical modelling, computer hardware and software, and the technology of greenhouse production systems, are making it increasingly important to understand the physics of the greenhouse ecosystem if full advantage is to be gained from its improved environment.

The greenhouse microclimate

The predominant role for the greenhouse is to provide a cost-effective environment for economic plant production (Hanan et al., 1978). Because of the high capital and operational cost of the structure, greenhouses are mostly used for growing high-value crops, whether they be vegetables or flowers, because of the extra value that can be derived from greater production, higher quality or harvesting when prices are high because outdoor crops are not available locally, or are expensive to import. In addition, it should be remembered that the increasing use of plastic coverings, loosely applied over field crops, involves modifications to the crop environment comparable with those achieved in more formal greenhouse systems. A smaller use of greenhouses is for experimentation and testing of agrochemicals and new plant lines. For these uses the main requirement is for an environment that mimics as closely as possible the one in which it is intended to use the test material. A further minor use with different requirements is as a working building – for example as part of a store selling garden materials, or as an amenity area within an office complex. In these situations the plant material is likely to play a less important role in setting the environmental conditions, though this may make it all the more difficult to provide an appropriate environment for plant growth. The laws that govern the physical environment in any of these situations will of course be essentially the same.

With these general characteristics of greenhouse systems in mind, one can outline the principal features of the physical environment (Mastalerz, 1977). The prime aim of the greenhouse is to grow plants, and therefore high transmission of solar radiation in the waveband from 400 to 700 nm is essential to maximize photosynthesis rates. The amount of structural material and the properties of the cladding will influence the proportion of incident radiation transmitted to the plants. The photosynthetically active radiation will be accompanied by radiation at other, mostly longer, wavelengths. All the radiation entering the greenhouse will contribute to the potential elevation of the greenhouse temperature above that of the external air. The greater the insulating properties of the house the greater will be the elevation, though as a general rule those cladding materials that might be chosen for good thermal resistance will also tend to be less good at admitting radiation for plant growth (Bailey and Hunter, 1988). Elevated temperatures will only be desirable when outside conditions are below the optimum for plant growth. To make full use of an expensive structure through as much of the year as possible generally requires methods of cooling the house to be available. The most common is by natural ventilation, exchanging hot and humid air inside the house with cooler, drier air from outside. Forced ventilation using fans is also used, and sometimes increased cooling is obtained by passing very dry external air through a wet pad before it enters the house (Bailey, 1968; see also Chapter 2, pp. 48–50). Obtaining the right combination of radiation environment, and temperature and humidity regime inside the house is an important challenge. The use of shading during the day to exclude some part of the solar radiation in order to prevent excessive heating, or the use of screens at night to reflect back far-infrared thermal radiation in order to reduce heat loss at night (see Chapter 2, pp. 52–56), are two other practical means to manipulate the physical environment in order to optimize conditions (Bailey, 1978). Of course, in cool climates greenhouses require additional heating for a part of the year, and the heat must be provided efficiently and distributed to create a uniform temperature through the house.

Enclosing the crop within a greenhouse structure obviously affects the transfer of gases between the crop and the atmosphere. The need for air exchange by ventilation has already been mentioned. It is also necessary for there to be a continuing supply of carbon dioxide for assimilation by the crop. Under low light conditions, the supply by leakage can be sufficient to keep the internal concentration near the ambient value, but higher concentrations can give useful enhancement of crop growth, justifying the use of supplementary carbon dioxide (Enoch and Kimball, 1986). The economics of adding extra carbon dioxide depends on the balance between the amount of gas that needs to be added and the resulting benefit in terms of increased production. The degree of air exchange will also affect water vapor. Control of plant water balance is an important feature of greenhouse management, and the physical properties of the greenhouse system will have a major influence on this balance. It will also influence the humidity of the greenhouse air, thus affecting the environment for all organisms within the greenhouse ecosystem.

The other climate modification achieved by the greenhouse is to exclude some of the major climatic events that can be very damaging to plants. Wind, rain, snow and hail can all cause significant damage to crops – the greenhouse structure has to be designed to cope with the loads that these events introduce (Hoxey and Richardson, 1989), but in terms of the physical environment these conditions are effectively excluded.

Principles of energy and mass transfer

In this chapter we will describe how the principal physical interactions in the greenhouse ecosystem can be described in terms of physical laws. The basis of these laws can be spelled out for the key areas – exchange of electromagnetic radiation, heat transfer by conduction and convection, latent heat exchange by evaporation and condensation, and mass transfer between various parts of the microclimate and the outside air.

Electromagnetic radiation is a form of energy that comprises coupled oscillating electrical and magnetic fields and propagates at high speed. All matter can emit and absorb electromagnetic radiation, and these exchanges are associated with changes in the energy status of the atoms and molecules in the material, whether it be gas, liquid or solid. In general, the hotter a body is, the more radiation it emits and the higher the frequencies (shorter wavelengths) that are present in the spectrum of emitted radiation. The absolute amount of energy emitted at any wavelength depends on the absorptance/emissivity of the body. A completely absorbing and emitting body radiates a characteristic spectrum that is determined by its temperature – it is called a black body, and the black-body radiation spectrum is shown in Fig. 3.2. Real objects will emit less energy as radiation at some or all wavelengths; their absorptance/emissivity is a function of wavelength, and is equal to the ratio of their

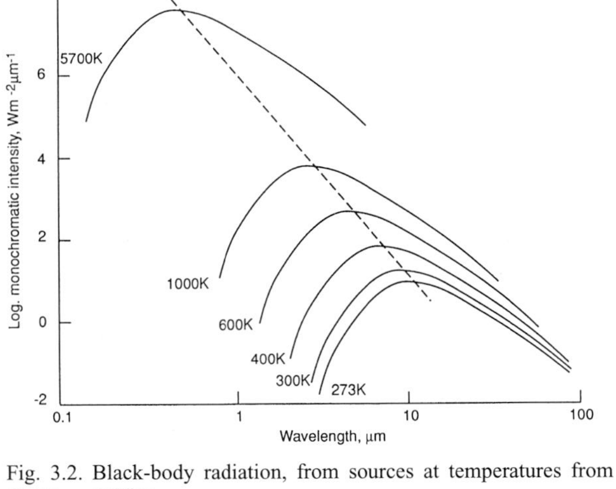

Fig. 3.2. Black-body radiation, from sources at temperatures from 273 K to 5700 K.

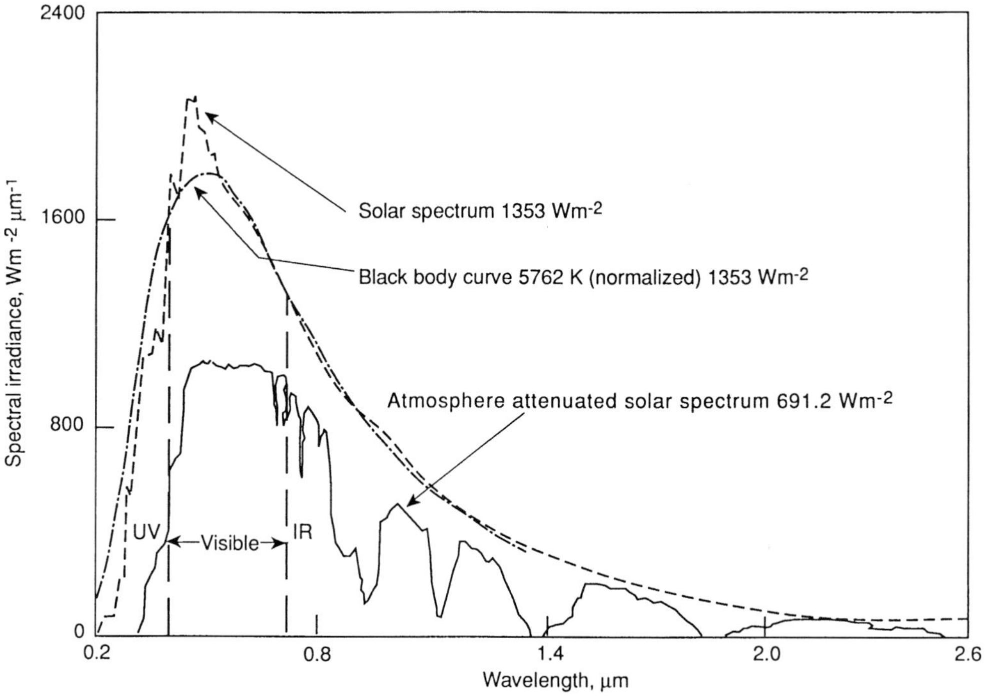

Fig. 3.3. Solar irradiance above the atmosphere and at the Earth's surface.

absorption/emission at that wavelength divided by the emission of a black body at the same temperature.

The sun behaves as a near black-body radiator at a temperature of almost 6000°C. Its radiation spectrum (Fig. 3.3) shows that the peak energy occurs at about 500 nm, in the middle of the visible spectrum for human sight. This spectrum is modified to some extent by passage through the earth's atmosphere. In particular, the amount of ultraviolet radiation at wavelengths below 300 nm is reduced to negligible levels, and there are several gaps in the infrared spectrum due to absorption by water molecules. The other bodies that are involved in the greenhouse ecosystem are at much lower temperatures, in the range from about −20°C for a clear sky on a cold day to perhaps 120°C for the heating pipes. The emission spectrum for these surfaces peaks at around 10 μm. At each surface, or for each body, the First Law of Thermodynamics demands that energy is conserved. Thus, the net balance of the radiation components at different wavelengths must be set against other energy-exchange processes. This can be summarized in terms of the following equation:

$$R + M = C + \lambda E + J, \tag{3.1}$$

where R is the net gain of energy from radiation, M the gain from chemical and biochemical transformations, C the net loss of sensible heat to the environment by conduction and convection, λE the net loss of latent heat by evaporation (less any gain by condensation) and J the rate of increase of stored heat.

The term M is primarily concerned with plant respiration and photosynthesis. It is not very important in the heat balance of the ecosystem; for example the photosynthetic fixation of carbon dioxide usually accounts for around 3–4% of the energy in photosynthetically active wavelengths incident on the leaf surfaces, even for leaves photosynthesizing close to their maximum efficiency.

The term C for sensible heat loss covers a range of different processes depending on the part of the system being considered. Convective heat exchange is of predominant importance – for instance, between leaves and air in the house, and between the greenhouse structure and both internal and external air. Convection is the transfer of heat by the movement of a fluid, as compared to the transfer from molecule to molecule in a solid or a still fluid which is termed conduction. The pattern of heat transfer by convection in a bulk of fluid such as air can be very complex as it depends directly upon the pattern of fluid flow, which can be influenced by ventilation of the building or the temperature differences between bodies in the space. In forced convection, rates of air movement are relatively rapid and the rate of heat exchange larger than for free convection, where heat transfer depends on the movement of the fluid which is created and maintained by the temperature differences and the changes in air density that they cause. The term C will also include heat conduction for the heat balance in some parts of the greenhouse ecosystem. For example, there is sensible heat transfer between the outside air and that inside the house by conduction through the cladding, and heat conduction in the soil is important for the energy balance at the greenhouse air/soil interface.

The term λE is predominantly associated with transpiration, the loss of water from the plants by evaporation within the stomatal cavities, followed by diffusion to the air outside the leaf. The air inside the cavities is close to saturation, so the diffusion gradient is largely determined by the leaf temperature, the humidity of the external air and the diffusion resistance of the stomata.

The final term J in Eq. (3.1) represents the rate of change of stored heat as the greenhouse air, the crop or whatever part of the system is under consideration increases in temperature. This buffering by the heat capacity of the various parts of the greenhouse leads to shifts in time in the pattern of temperature change for the different parts, and makes the dynamic changes in the greenhouse climate in response to outside weather quite complex.

RADIATION

Radiant energy

The major component processes in radiation exchange are illustrated in Fig. 3.4. The radiant energy important in determining the greenhouse environment can be considered in three wavebands. The most significant band stretches from about 400 to 700 nm and contains the photosynthetically active radiation (PAR), which contributes directly to photosynthesis and hence to crop growth. It corresponds closely with the human visible waveband. Another waveband in the solar spectrum is the near-infrared band from 700 to about 3000 nm. Because it cannot contribute to photosynthesis, it is not so important that this radiation enters the greenhouse, but where the outside climate is not warm enough for optimal crop growth the energy in this waveband will

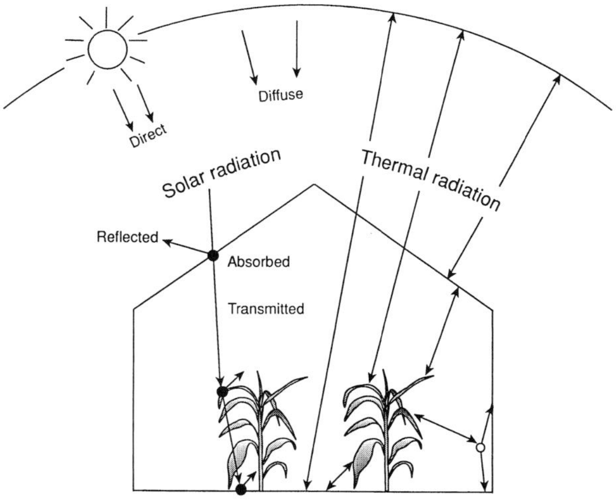

Fig. 3.4. Interaction of solar and thermal radiation in the greenhouse.

be a major contributor to the heating of the house. In round terms, there is a similar amount of energy in the PAR and near-infrared wavebands. The third waveband is the far-infrared or thermal radiation, which for objects close to ambient temperatures covers wavelengths from 5 to 30 μm. There is another waveband present in incident solar radiation, the ultraviolet at wavelengths from 300 to 400 nm, but after atmospheric absorption this waveband only represents some 3% of total solar energy at the earth's surface and so has little importance in greenhouse energy balance. It can have important side effects by leading to the breakdown of plastic films.

The primary physical properties of greenhouse materials that influence the relationship between radiation and the conditions inside the greenhouse are their transmittance, absorptance and reflectance in the different wavebands. Because the wavebands are wide and there are many specific absorption bands within each range, it is not possible to give single values for each band; but Table 3.1 gives some general figures that are indicative. In the far infrared, there is a further parameter of importance – the emissivity. This is the ratio of the thermal radiation emitted by the body to that which would be emitted by a black body at the same temperature. Kirchhoff's Law states that this is equal to the absorptance which is given in the table. The higher the value of transmittance in the PAR band, the greater will be the potential photosynthesis and crop production. Materials like double polycarbonate, which are of interest because of their increased thermal resistance and hence better energy conservation, can be

Table 3.1
Radiation exchange properties, where τ is the transmittance and α the absorptance

Material	PAR (direct)		Solar (direct)		Thermal	
	τ	α	τ	α	τ	α
Leaves	0.09 [1]	0.85 [1]	0.3 [1]	0.5 [1]		0.98 [2]
Glass	0.89	0.03	0.91 [3]		0	0.94 [4]
Polyethylene	0.87 [5]	0.04	0.86 [3]		0.4	0.55
Polycarbonate, twin-walled			0.72 [3]			
Water						0.95 [4]

References
[1] Jones (1983);
[2] Idso et al. (1969);
[3] Landgren (1985);
[4] McAdams (1953);
[5] Roberts et al. (1985).

seen to have less favourable properties for PAR. There is a major contrast between glass and polyethylene in the transmittance in the thermal region. This results in much faster loss of crop heat by back radiation through polyethylene on clear nights, though this difference is ameliorated by condensation on the film surface, as water has an absorptance close to that of glass.

Solar radiation

The primary purpose in using a greenhouse is to create an environment which improves the productivity and quality of plants compared to growing them in the open. This is strongly dependent, particularly at high latitudes, on the transmission of solar radiation by the structure, to maximize photosynthesis. The proportion of the external solar radiation which enters the greenhouse and is absorbed by plants depends on the shape and structure of the house, the cladding materials, and on the amount, distribution and orientation of the leaves of the crop.

Solar radiation entering the greenhouse

The transmittance of the cladding is maximal for radiation incident at right angles to the surface; thus, for maximum transmission of all sky radiation, the optimal cover for a plant would be a hemisphere with the plant at its centre. Hemispherical structures have been used, but are generally not cost-effective as an environment for crop production, since they are of limited plan area, and cannot be packed together without interfering with each other's radiation environment. For structures covering a large area, the key design feature influencing radiation transmission is the multispan roof, which is generally made up of repeated units of a common cross-section. The form chosen for the cross-section depends upon a number of considerations, including the angular distribution of incident solar radiation (Kozai et al., 1978).

The incident solar radiation can be considered as made up of direct sunshine, circumsolar radiation and diffuse sky radiation. The direct radiation comes from angles that can be determined from knowledge of the latitude, time of year and time of day. The circumsolar radiation represents the radiation coming from the region close to the sun, and is greatest when atmospheric scattering of radiation is large. Littlefair (1983) has suggested from United Kingdom data that, when averaged over a long period, the circumsolar radiation can be expressed as

$$S_c = a \exp(-\theta/40), \tag{3.2}$$

where θ is the angle between the sky element and the sun in deg. Thus there is a wide region of the sky over which the radiation input is dependent on the relationship to the position of the sun. The diffuse or sky radiation component is not dependent on the position of the sun and is commonly expressed as

$$S_d = b + c \sin \zeta, \tag{3.3}$$

where b and c are functions of the solar elevation and ζ is the elevation of the sky element (Littlefair, 1983).

This is an approximate representation as it is based on measuring average properties of the solar radiation distribution over many days. However, it is sufficiently detailed to indicate how the crop environment in the greenhouse will be affected. In practice, a further simplification is usually made, and the circumsolar component is not considered separately. This is partly because the available data on the distribution of solar radiation is predominantly in terms of a direct and a diffuse component, obtained from the readings on two radiometers: one measures total irradiance on a horizontal surface, and the other has a shade ring to exclude the direct sun component by masking the daily route of the sun across the sky. The shade ring is wider than the angle subtended by the sun's disc, and therefore excludes a significant component of the circumsolar radiation, particularly along the direction of travel of the sun across the sky. The readings from the shaded radiometer are taken to give the diffuse component, and the direct component is taken as the difference between the two instruments. This approach

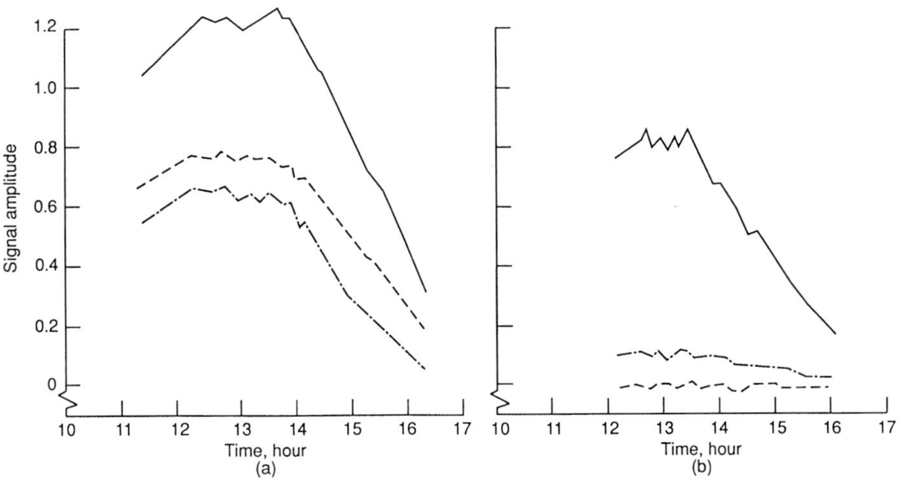

Fig. 3.5. Amplitudes of the major component of skylight distribution for (a) a stable blue sky and (b) a stable overcast sky. The coefficients are: D_{10} (solid curves), coefficient of first symmetric component, which varies as $\sin\zeta$; D_{11} (dashed curves), coefficient of first antisymmetric component, which varies as $\sin\zeta \sin\chi$; D_{30} (dot-dashed curves), coefficient of second symmetric component, which varies as $(\sin^2\zeta - 0.6\sin\zeta)$. (Critten and Stammers, 1986.)

is satisfactory for most purposes, but a significant part of the diffuse radiation will in fact be from parts of the sky closer to the sun's position than is assumed.

There is also a significant variation in the sky radiation with azimuth and elevation. Critten (1988a) has shown that this can be described by the first few terms in an expansion of the distribution as harmonic components of the form $D_{nm}P_n(\Phi)F_m(\chi)$, where D_{nm} is a constant, P_n is the nth Legendre polynomial, F_m is the mth term in a Fourier series, $\Phi = \sin\zeta$, ζ is the elevation of the sky element, and χ is its azimuth. The variations that occur in skylight distribution described in this way are exemplified by the results from Critten and Stammers (1986) in Fig. 3.5; they show, for instance, that there is considerable enhancement of the antisymmetric and second symmetric terms of the sky radiation in direct sun conditions, beyond that contained in the circumsolar term. Each harmonic component comprises a complete hemispherical distribution of known variation in amplitude. When it passes through a transmitting cladding surface, the amplitude will be modified, and the resulting angular distribution can be described in harmonic terms in the same way. In general, each incident harmonic term can contribute to all the harmonic terms in the transmitted distribution. When the answer of interest is the irradiance on a horizontal surface within the greenhouse, then the problem reduces to determining just the zero-order harmonic in azimuth and the first-order harmonic in elevation, as all other harmonics sum to zero. The gain/loss coefficients that define the contribution of the significant harmonics in the incident distribution of solar energy to the internal irradiance can be expressed as analytical functions of roof angle, for example, and thus the consequence of changing roof angle for particular distributions of sky radiance can be determined.

Light transmission into the greenhouse can be estimated by computer simulation, tracing the path of a ray of light from a sky element through an element of the greenhouse surface, and then following the path of the resultant reflected and transmitted rays until they either reach the floor of the house, or leave the house in an upwards direction, or fall outside the house. Repeating this for all sky elements, ascribing to each a radiance corresponding to the skylight distribution of interest, and for all elements of the greenhouse surface, taking account of structural members, allows transmission to be calculated (Critten, 1983). This procedure is workable for reasonably straightforward structures, but simpler methods are attractive and particularly necessary for more complex systems such as multispan houses.

A simpler method for estimating light losses due to reflection from the cladding is based on two hypotheses: first, that transmittance gain at any point inside or outside an assembly of roofs is equal to the sum of the gains that would occur for each roof in isolation; and second, that the total reflected energy per unit area of reflecting surface is independent of the angle between the surface and the incident light.

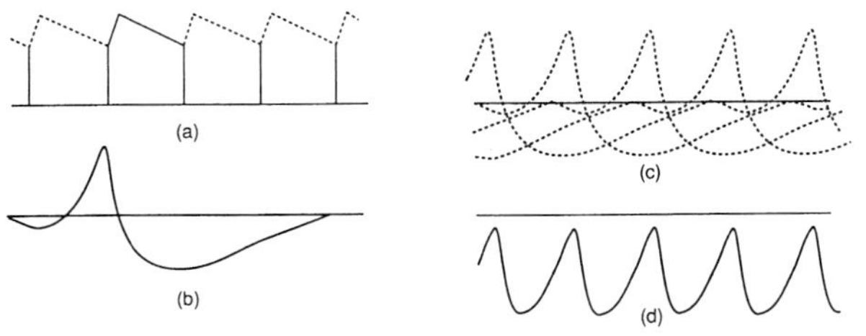

Fig. 3.6. The irradiance pattern for a multispan greenhouse (a) is synthesized from the contributions from individual spans, all having an identical form (b), by summing the contributions (c); the result is shown in (d).

The first hypothesis is illustrated in Fig. 3.6, where the individual patterns of transmission for roof sections are summed together to give the net transmittance gain for the complete structure. Critten and Legg (1987) have shown that errors are of the order of the square of the reflection coefficient of the cladding material, and less than this for low-angle roofs. This summation hypothesis is general, applying to any light-absorbing three-dimensional structure; the infinite multispan greenhouse forms a specially simple case as the transmittance gain need only be estimated for one roof span. The mean transmittance is equal to the extended integral, normal to the axis of the span, of the transmission pattern of a single span, divided by the span width. This means that a calculation, often in the form of a computer simulation, of the fate of radiation from all sky directions that passes through a single roof section can be used to give the transmission of more extensive structures in a straightforward way.

The second hypothesis allows the transmittance losses from individual planes (such as roof planes) to be estimated analytically. The major component in the loss of radiation due to the cladding is by upward reflection. The reflection coefficient of transmitting materials varies with angle of incidence on to the surface. These coefficients are determined by the differences in refractive index between air and the cladding material, and are called the Fresnel coefficients. The reflectance $r(\alpha)$ of cladding materials generally increases with the angle of incidence, α. However, when considering the effect of unit area of material, account needs also to be taken of the decrease in light intercepted by unit material as the angle of incidence increases. Combining the two effects as $r(\alpha)\cos(\alpha)$ gives a term that is remarkably constant to angles of incidence as large as 85°. Using this essentially first-order approach, Critten (1987) integrated the total energy reflected over all angles of incidence that gave upward reflections, to

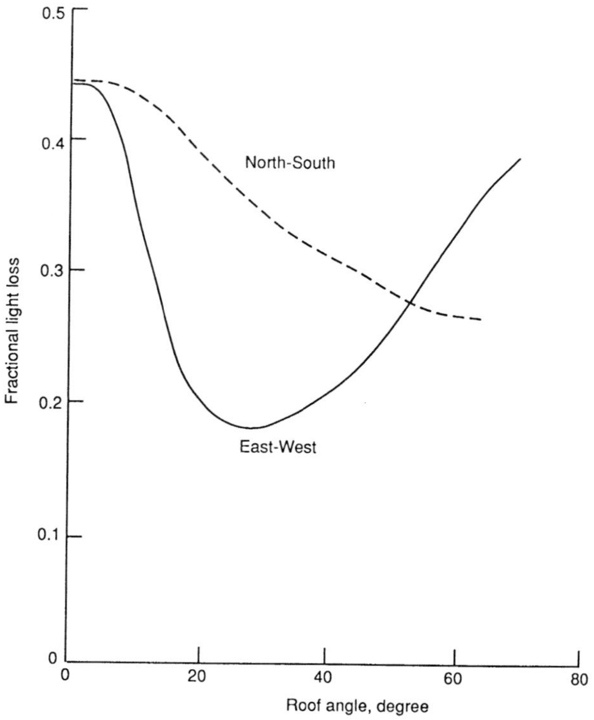

Fig. 3.7. Reflective light losses averaged over a day for a multispan, double-glazed greenhouse roof with a solar elevation at noon of 20° (Critten, 1987, 1988c).

obtain an analytical expression for total energy lost, for a skylight distribution described analytically. The loss due to upward reflection under direct light conditions depends strongly on house orientation and latitude. Minimizing the angle to the vertical at which most of the light strikes the cladding will minimize the reflection. The simplest example is for the (unlikely) case of a greenhouse on the equator at the equinox. Here for the pitched-roof house it is obvious that a North–South orientation will minimize the reflective loss. At northern (or southern) latitudes and in the winter, the direct light comes from low in the sky for much of the day, and at the middle of the day may still be from quite low elevations. In general there will be an upward reflection, and thus a loss, when $2\alpha < \pi - \phi$ where ϕ is the solar elevation. Thus, when the roof angle is less than $\phi/2$ for the midday value of the solar elevation, there will be an upward reflection at all times of day. For houses aligned East–West this results in an optimal roof angle for minimum transmission loss (Fig. 3.7). The compromise between different roof angles thus needs to take into account the period of the year when maximal light transmission is most important to the profitability of the enterprise.

For houses oriented North–South at high latitudes, transmission loss decreases progressively as the roof angle is increased and can be significantly less than for East–West houses in the summer months, though much greater in the winter. The choice of roof angle and orientation therefore requires careful assessment of the radiation environment to which the house is exposed, and in addition the requirements of the crops being grown. For example, midsummer production may not be possible because of the higher ambient temperatures, or the cheapness of crop production in the open field, and maximal radiation penetration in those months will increase the amount of radiative heating at a time when increased temperatures are likely to be detrimental to productivity and to quality.

These examples have been taken from calculations for symmetric-roofed houses. The option of an asymmetric roof can be used as a way of increasing the penetration of low-angle winter light at higher latitudes. The potential benefit of this approach, and particularly the interaction between its effect on radiation and the potential benefit in decreased heating costs if insulating cladding is used, can only be realistically assessed by using accurate descriptions of the physics of radiation transmission (and other factors affecting profitability) to evaluate concepts and optimize design (Bowman, 1970; Kurata and Tachibana, 1978; Thomas, 1978; Bailey and Critten, 1985; Bellamy and Ward, 1985).

Solar radiation within the greenhouse

Once radiation is in the house, there is still scope for increasing its use by the crop. For single-span houses and at the edges of multispan houses, there can be significant loss of light that entered through the roof but leaves through the side wall. It can be worth while to attempt to return this light to the crop. A reflecting surface on the north wall (in the Northern Hemisphere) can increase the average irradiance on the floor of a single-span house by as much as 12.5% on a sunny winter day, though it will reduce the diffuse radiation received (Andrews et al., 1982). An alternative of using reflecting slats hanging in the ridge of the house can achieve even greater benefits – up to 45% on a sunny winter day. The design of such reflectors and their potential benefit in a particular location needs careful calculation, as the angular direction of the radiation received, and particularly the proportion of direct sun in the total radiation, affect their performance (Critten, 1988b). In all houses some benefit can be achieved by redirecting light that is not intercepted by the crop and reaches the greenhouse floor. Gains in light intercepted by the foliage in response to reflective floor coverings have been found to be as large as 18% for a crop with a Leaf Area Index (LAI) of 2, decreasing as crop cover increases to a gain of 8% at a LAI of 4. When crops are grown in substrates, it is common practice to cover the greenhouse floor with white plastic film to reflect light. This increases the light available to plants by at least 2% (Andrews et al., 1982).

To be effective in crop production, the radiation in the photosynthetically active waveband (PAR) must be intercepted by the crop foliage and used to fix carbon dioxide. A large proportion of radiation striking a leaf is absorbed – close to 80%, with 10% reflected and 10% transmitted – and so the proportion of light entering the greenhouse that is intercepted by foliage is a close measure of the effectiveness of the crop at capturing PAR. An individual horizontal leaf of area A, in a flux of solar radiation coming from an elevation ϕ and with irradiance on a horizontal surface S, will intercept a flux of $SA\sin\phi$. If a canopy of leaves are all at the same angle ϕ to a direct beam of radiation, then the amount of radiation $\mathrm{d}S$ that is intercepted by a small element of Leaf Area Index dLAI is given by

$$\mathrm{d}S = -\Phi\,\mathrm{dLAI}, \qquad (3.4)$$

where $\Phi = \sin\phi$ and the minus sign indicates that the radiation flux has been decreased. If this equation is integrated from the top of the crop downwards,

$$S(L) = S(0)\exp(-\Phi\,\mathrm{LAI}) \qquad (3.5)$$

represents the direct solar irradiance on a horizontal surface below a canopy of leaf area index LAI. This assumes the leaves are randomly distributed in space in such a way as to form a homogeneous canopy.

In a real crop, leaves are distributed over a range of angles. A common approximation used to estimate light interception is that the distribution is uniform over all elevations and azimuths. The leaves could then be arranged uniformly over the surface of a sphere, and this is termed a spherical leaf distribution. The shadow cast on a horizontal surface by a sphere of radius r in radiation from an elevation ϕ has area $\pi r^2/\sin\phi$, and as both sides of the leaves are equally effective at intercepting incident radiation, the Φ value for this leaf distribution is

$$\Phi = 0.5/\sin\phi. \qquad (3.6)$$

In practice, interception by a crop canopy can differ significantly from this. The leaves may not be randomly distributed in space; any clumping of leaves will tend to decrease the interception. The leaf–angle distribution may not be spherical. A more vertical leaf canopy can be advantageous at high leaf area indices, as it means that radiation will penetrate to greater depths when solar radiation is at its most intense – that is, at large solar elevations; a more horizontal leaf canopy will maximize interception at low leaf areas. The distribution of plants in greenhouses may also have a major influence on light interception, as the crop canopy is rarely uniform. The need for access to the crop and other crop-management concerns mean that crops are often grown in rows with quite wide gangways between. The "hedgerow" effect that this creates is a complicated one for the calculation of radiation interception. The most effective approach is to consider the crop as having near-uniform distribution of leaf area in the hedges with clear gangways between, and then calculating the fate of the distribution of incident radiation by a computation over a representative range of radiation paths (e.g. Mutsaers, 1980).

The final stage in consideration of the physics of photosynthetically active radiation in the crop canopy is to look at the efficiency of utilization in photosynthesis. Leaf photosynthesis is not a linear function of the incident or absorbed PAR (Fig. 3.8). At low irradiance, the response to increasing irradiance is almost linear;

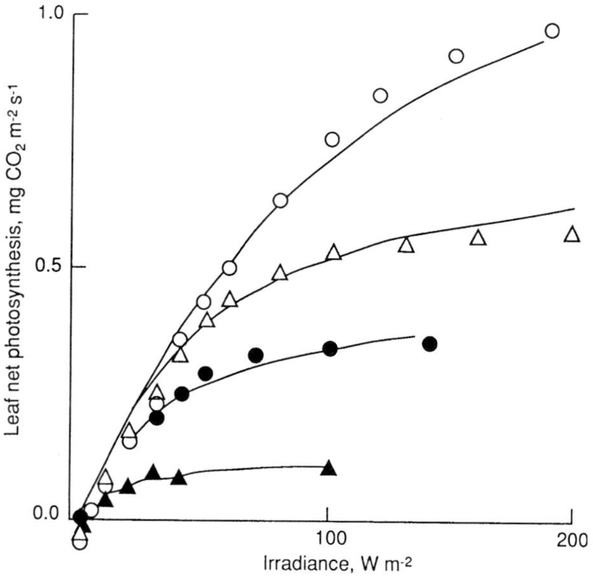

Fig. 3.8. The measured response of leaf net photosynthesis to irradiance for unshaded upper leaf (open circles), shaded upper leaf (solid circles), lower leaf adjacent to shade netting (open triangles), and severely shaded lower leaf (solid triangles) (Acock et al., 1978).

but at high irradiance photosynthesis tends to be saturated, as the assimilation reactions become limited by the availability of substrate rather than of energy. Saturation begins at quite low irradiance, but because of the distribution of leaf angles and mutual shading of leaves it applies predominantly to the leaves at the top of the crop. During periods of direct sun, saturation is an important phenomenon. The consequence for the physical environment in the greenhouse can be considered in terms of radiation distribution, the effect of shading and the potential benefit from carbon dioxide supplementation.

When a crop is exposed to direct sun from reasonably large elevations, there will be a part of the canopy whose photosynthesis is saturated, and other parts where the irradiance is low and photosynthesis is limited by energy supply. Aikman (1989) has suggested that a 20% gain may be possible by using semi-reflecting strips in the inter-rows of row crops such as tomatoes to redistribute the direct beam over a greater proportion of the leaf canopy. Alternatively the diffuse component of the radiation within the greenhouse could be increased with a diffusing cladding material. The problem with diffusing materials is that they tend to have a decreased transmittance, offsetting any gain from reduction in photosynthetic saturation. Shading the greenhouse may be necessary to reduce the energy load, so that the greenhouse temperature does not increase excessively. It is often considered that crop productivity is directly proportional to the integral of radiation interception by the crop, and this is probably an accurate representation of the situation when the probability distribution of different irradiances is not greatly modified. However, if partial shading is used in a dynamic way, covering the crop when the irradiance is above a certain threshold value and removing the partial shade when irradiance has fallen, the decrease in productivity may be considerably smaller than would be expected on the basis of direct proportionality to the total radiation integral, because of the nonlinear photosynthetic response to light intensity (de Koning, 1988).

The absolute response of the photosynthetic rate of leaves to an increase in the concentration of carbon dioxide is greater when the leaves are light saturated. In principle it should be possible to calculate the optimal concentration of carbon dioxide, given information on the responsiveness of photosynthesis, and the relationship between the added value of crop and the cost of supplemental carbon dioxide. One of the key problems in resolving this calculation is likely to be the need to know what proportion of the crop canopy is saturated at any particular time, given reasonable knowledge of crop size and structure, and of the amount and distribution of radiation in the greenhouse. This problem brings together all the issues of understanding the radiation environment in the greenhouse in a complex of considerable importance to the industry.

The topics that have been presented in this sub-section have been mostly expressed in terms of PAR and the potential for photosynthesis. The near-infrared component in the incident solar radiation is governed by the same laws, and though, of course, it does not contribute to photosynthetic productivity, some wavebands are important for plant development. The greater transmittance of leaf tissue to the near-infrared implies that the spectral distribution changes with level in the crop canopy (Szeicz, 1974). At the bottom of a dense canopy there can be three or four times as much infrared radiation as PAR, whereas at the top of the canopy the amounts are typically about equal.

Thermal radiation

The net transfer of energy between two black bodies by exchange of thermal radiation (5–30 μm wavelength) is given by

$$L_{12} = A_1 F_{12} \sigma (T_1^4 - T_2^4), \qquad (3.7)$$

where T_1 and T_2 are their respective absolute temperatures, σ is the Stefan–Boltzmann constant, A_1 is the area of body 1 and F_{12} is a factor which represents the angular view of body 2 from body 1. Thermal radiation is assumed to be diffuse and can be calculated from the geometrical interrelationship between the two bodies. For non-black bodies, some thermal radiation is reflected and a proportion of this will return to the emitter. For the restricted example of a pair of bodies that exchange radiation with each other but with no other body, it can be shown that

$$L_{12} = \frac{A_1 \sigma (T_1^4 - T_2^4)}{\frac{1}{\varepsilon_1} + \frac{A_1}{A_2}\left(\frac{1}{\varepsilon_2} - 1\right)}, \qquad (3.8)$$

when ε_1 and ε_2 are the emissivities of the two surfaces.

In a real greenhouse there are many bodies with a complicated exposure one to another, and some with emissivities significantly less than one. In some circumstances it is possible to reduce the problem to a tractable form by a set of assumptions about the main bodies within the greenhouse. In an extensive greenhouse with a fully developed crop, the main bodies are the roof, the crop, the floor and the heating pipes, and it is usually possible to consider each as having a single temperature value for the purpose of thermal radiation exchange. The geometrical factors governing the exchange between each pair can be estimated and the effect of emissivity taken into account.

Though thermal radiation emission is proportional to the fourth power of the absolute temperature of a body, for most objects in the greenhouse the temperature differences are sufficiently small for the net exchange to be proportional to temperature difference, that is $T_2^4 - T_1^4 = 4\bar{T}^3(T_2 - T_1)$ where $\bar{T}$ is the mean temperature. This can be further simplified to a linear proportionality to the temperature difference alone by replacing $\bar{T}$ by the greenhouse air temperature. The only exception is likely to be the exchange with the heating pipes, where the term proportional to the square of the temperature difference may approach 25% of the linear term.

The sky temperature is an important component in the thermal radiation balance. With a clear sky, one can see the stars; the thermal background temperature in outer space is around 4 K (–269°C). The earth's surface is protected from this extreme cold by the absorptance of water vapor and carbon dioxide in the lower atmosphere. This absorptance is sufficiently great that thermal radiation exchange from the earth's surface is with the atmosphere only up to 5000 m, and about half the radiation exchange is with the lowest 100 m. It has been shown from measurements at a number of locations (Monteith, 1973) that the sky radiation is in fact quite well expressed as if it were from a black body at a temperature, t_b, related to that of the air at screen height, t_a, by

$$t_b = (t_a - 21) + 0.2 t_a, \qquad (3.9)$$

where the temperatures are expressed in °C. Thus the net loss of thermal radiation to a clear sky from a black body at air temperature ranges from 107 to 87 W m^{-2} between 0 and 20°C.

Clouds emit thermal radiation as black bodies at the temperature of their constituent water droplets or ice crystals. This is a much higher temperature than the effective temperature of clear-sky radiation, because the gaseous absorbers are nearly transparent in the wavelength range 8 to 13 μm. Again, the temperature of the clouds can be related on average to the air temperature at screen height. There is some variation with cloud type, but a reasonable average value can be used; observations at Kew in England (Monteith, 1973) suggest an average temperature difference of 2°C, corresponding to a net flux of 9 W m^{-2}.

Glass has a negligible transmittance in the far infrared, and so this radiation exchange with the sky directly involves just the cladding and not the crop itself. With a polyethylene covering, the crop itself will lose radiation directly to the sky, and so crop surface temperatures on clear nights will be colder under polyethylene covers than under glass. The cooling of the cladding and of the crop can lead to surface temperatures lower than the dewpoint of the air, and this can result in condensation on these surfaces. Water has a much lower transmittance than polyethylene, and so a layer of moisture on the cover can lead to less direct long-wave exchange between sky and crop directly, and therefore higher crop temperatures, than with a dry polyethylene cladding. Condensation on the crop is generally detrimental, often leading to the development of fungal diseases, and is therefore best avoided.

AIR MOVEMENT

One of the major differences between the conditions inside a greenhouse and those occurring outside is the reduced influence of wind. Average wind speeds in the open can be in the region of 4 m s^{-1}, whereas inside a greenhouse air speeds are lower by more than an order of magnitude. The provision of shelter by the greenhouse is important, because the relative growth rate of leaves is reduced if they are subjected to air speeds much in excess of 0.5 m s^{-1} (Morse and Evans, 1962). However, some movement of the air is necessary in order to facilitate the exchange of energy, water vapor and carbon dioxide between plants and their surroundings. The air in a greenhouse is unlikely to be completely stationary because differences in temperature usually exist, which create air movement by convection. In addition, greenhouses are not completely airtight, and air movement can be induced by the wind. Air movement is also created when heating and ventilation are used to control the climate in a greenhouse, particularly if the air is circulated inside the greenhouse by fans.

Boundary layers

Transfer of heat between a surface and the air adjacent to it requires a difference in temperature between the surface and the air. Similarly, the evaporation of water requires there to be a difference between the vapor pressure at the surface and in the air. The region in which these gradients occur is known as the boundary layer. The concept of a boundary layer is also applied to the motion of air in the vicinity of a surface. In this case the air speed decreases towards the surface as a consequence of friction between the surface and the air, and of the viscous forces in the air. Thus in general a boundary layer represents the region close to a surface in which properties such as air speed, temperature and water vapor or carbon dioxide concentration differ from their values in the surrounding air.

The creation of a boundary layer by air flowing over a flat plate is illustrated in Fig. 3.9. At the edge of the plate, there is a laminar region in which the air streamlines are continuous and almost parallel to the plate surface. The laminar boundary layer increases in thickness with increasing distance from the upstream edge of the plate, but eventually the flow becomes unstable and breaks down into a chaotic pattern of swirling eddies which constitute the turbulent boundary layer. A thin layer of laminar flow may still exist close to the plate surface beneath the turbulent layer. In the laminar boundary layer, viscous forces, which tend to oppose changes in velocity, dominate over inertial forces, which produce changes in velocity. The ratio of the inertial to the viscous force is characterized by the Reynolds number

$$\mathrm{Re} = \upsilon l/\nu, \tag{3.10}$$

where υ is the free-stream air speed, l the characteristic length of the system, commonly the distance downstream from the edge of the body, and ν is the kinematic viscosity of the air. Above a critical value of Reynolds number, inertial forces become dominant and

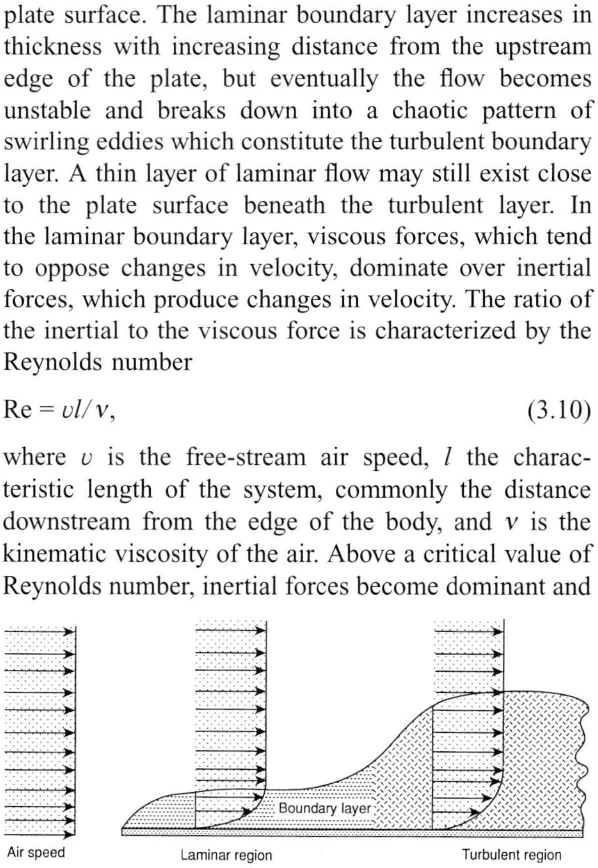

Fig. 3.9. Diagrammatic representation of a boundary layer over a smooth flat plate in a laminar air stream (Monteith, 1973).

the flow becomes turbulent. In practice the transition between the laminar and turbulent boundary layers also depends on the smoothness of the plate. For laminar flow in the boundary layer, the boundary layer thickness δ (Monteith, 1973) is given by

$$\delta = 1.72\sqrt{\nu l/\upsilon}. \tag{3.11}$$

Convection

When the temperature of a surface differs from that of the surrounding air, heat may be transferred to or from it by convection. This is an important process in the exchange of energy between leaves and air in a greenhouse, and between the house structure and both internal and external air. Convection is the transfer of heat by the movement of a fluid, as compared to the transfer from molecule to molecule in a solid or a still fluid which is termed conduction. The pattern of heat transfer by convection in a bulk of fluid can be very complex, as it depends directly upon the pattern of fluid flow, which can be influenced by ventilation of the building or the temperature differences between bodies in the space. In many cases it is deemed sufficient to simplify these aspects of heat transfer by considering the greenhouse air as a single "lumped" component. Convective heat transfer to the air then depends on the temperature difference between the body and the air, and the transfer resistance:

$$C = \rho c_p(t_s - t_a)/r_h, \tag{3.12}$$

where ρ is the density and c_p the specific heat of air, t_s and t_a the temperature of the body surface and the air respectively, and r_h is the heat transfer resistance. It is useful to express the resistance as a function of other variables in order to establish the values it will take in different circumstances. The heat transfer equation can also be expressed as

$$C = k(t_s - t_a)/\delta, \tag{3.13}$$

where k is the thermal conductivity of still air, and δ is the effective thickness of the boundary layer at the surface of the object. The value of δ cannot be determined directly, but it appears in the dimensionless group Nu, the Nusselt number which is the ratio l/δ, where l is a characteristic dimension of the object, logically taken to be its length in the direction of air movement. Comparing the two equations gives

$$r_h = \rho c_p l/k\,\mathrm{Nu} = l/K\,\mathrm{Nu}, \tag{3.14}$$

where K is the thermal diffusivity of air ($k/\rho c_p$).

The value taken by Nu depends upon the movement of the fluid. In forced convection, where rates of air movement are relatively rapid, Nu is a function of the Reynolds number Re and can be expressed as

$$\mathrm{Nu} = a\,\mathrm{Re}^n, \tag{3.15}$$

where a and n depend on the geometry of the body. In free convection, heat transfer depends on the movement of the fluid which is created and maintained by temperature differences, and the changes in air density that they cause. In this case Nu is a function of a different dimensionless group, the Grashof number Gr, which is the product of a buoyancy force and an inertial force divided by the square of a viscous force; Nu is given by

$$\mathrm{Nu} = b\,\mathrm{Gr}^m, \tag{3.16}$$

where b and m depend on geometry, and

$$\mathrm{Gr} = \beta g l^3(t_s - t_a)/\nu^2, \tag{3.17}$$

with β the coefficient of thermal expansion of the fluid and g the gravitational acceleration.

To decide whether the heat transfer is likely to be by free or forced convection, Gr is compared with the square of Re. The ratio of these two represents the ratio of buoyancy to inertial forces, so when the ratio is much greater than unity, heat transfer is dominated by free convection, and when much less than unity, it is dominated by forced convection (Kreith, 1958).

Examples of heat transfer resistances, boundary layer thicknesses and typical modes of convection are given in Table 3.2.

The heat transfer term C will also include heat conduction for the heat balance for some parts of the

Table 3.2
Convective heat transfer and boundary layer resistances

Surface	Type of convection	C ($m^2\,K\,W^{-1}$)	r_h ($s\,m^{-1}$)	δ (mm)[3]
Roof exterior	forced	0.07[1]	80	4
Roof interior	free	0.3[1]	350	50
Leaf				
$l = 5$ cm, $\upsilon = 5\ cm\,s^{-1}$	free	0.25[2]	300	7
$l = 5$ cm, $\upsilon = 50\ cm\,s^{-1}$	forced	0.08[2]	100	2
$l = 20$ cm, $\upsilon = 5\ cm\,s^{-1}$	mixed	0.45[2]	550	13

[1] Papadakis et al. (1992).
[2] Parkhurst et al. (1968).
[3] Equation (3.11).

greenhouse ecosystem. Sensible heat transfer between the inside air and that outside the house is by conduction through the cladding, whose thermal conductivity can vary enormously between single and multiple-layer materials. At the soil surface a significant part of the energy balance is represented by heat conduction into the soil.

Convective air movement

In addition to the convection created when leaves are heated by solar radiation, air movement is also created by greenhouse heaters. Heating implies the existence of temperature differences and consequently convective motion of the air. One common form of heating involves the passage of water at 70–120°C through steel pipes. Heat is transferred from the pipes as natural convection and thermal radiation in similar proportions. The air speeds created by this source of natural convection are low, typically $0.05\,m\,s^{-1}$ but these can create persistent large-scale air circulation in the greenhouse.

Placing the pipes uniformly over the floor and along the walls of a long, narrow greenhouse will result in a series of circulation cells along the house in which the air rotates in a vertical plane. Adjacent cells rotate in opposite directions and the ascending and descending air produces alternate regions of high and low temperature (Fig. 3.10a). Placing all the pipes on the side walls creates two counter-rotating cells with their axes parallel to the walls (Fig. 3.10b). As the heat sources are concentrated in a small region, this installation generates more rapid air movement, providing good uniformity of temperature over the greenhouse. In large greenhouses it is not easy to arrange the heating pipes to provide local heating and still generate the circulation of air required to give uniform temperature. An additional requirement is that large relatively unobstructed spaces are required for the air circulation to become established, and the presence of crops, especially tall plants grown in rows, limits the volume available for air circulation and presents a resistance to flow. The trend towards insulating greenhouses using thermal screens (see Chapter 2, pp. 52–56), which extend horizontally across the house just below the gutters, considerably reduces the available height, and this tends to suppress the air circulation.

Placing heating pipes overhead, which has the desirable effect of heating the plant canopy by thermal radiation and removes potential obstructions from the floor, does create air circulation. Unfortunately this occurs predominantly in the roof spaces of the greenhouse and does not penetrate into tall crops; consequently there can be gradients in temperature near the ground.

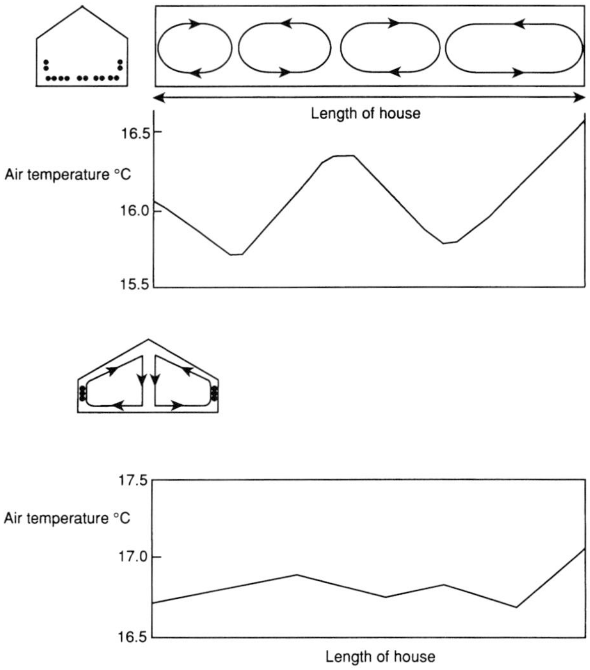

Fig. 3.10. Convective air movement created by piped heating systems: (a) pipes over the floor and on side walls; (b) similar heating capacity but all pipes on side walls (Morris, 1963; Carpenter and Bark, 1967).

Some greenhouses are also heated by the distribution of warm air which has passed through an air heater. In this situation the heat is transferred by forced convection. The air can leave the heater as jets directed into the space between the top of the crop and the greenhouse roof, or it can be distributed through the greenhouse by perforated tubes of plastic film. The former method can produce air speeds of $10\,m\,s^{-1}$ and large gradients in temperature occur. The latter gives a more uniform distribution of air, although the perforations in the tubes should be positioned to give uniform heating rather than the uniform supply of air. The air tubes are frequently placed on the greenhouse floor and thus provide air movement within the plant canopy.

Mass transfer

The energy gained by leaves from solar radiation or

by convection provides the latent heat required for transpiration, in which water evaporates from the leaf. The vapor diffuses through the stomata of the leaf and then passes through the boundary layer. As with the convective transfer of heat, the rate of vapor transport (E) is proportional to a driving force and inversely proportional to a resistance. The driving force is the difference between the water vapor concentration in the stomatal cavity (c_l) and in the free air (c_a).

$$E = (c_l - c_a)/r_v, \tag{3.18}$$

where r_v is the total resistance to water vapor transfer.

As the concentration of water vapor is already linked with the partial pressure of the vapor (e), an alternative version of this equation can be formulated which is often used in practice:

$$E = \frac{\rho}{P}\frac{m_w}{m_a}\frac{(e_l - e_a)}{r_v}, \tag{3.19}$$

where m_w and m_a are the molecular weights of water and air, ρ is the density of the air, and P is the atmospheric pressure.

Inside the leaf, the air is close to saturation, and as a consequence the vapor pressure is only dependent on leaf temperature. The resistance is the sum of two resistances, one for the diffusion resistance of the boundary layer (r_b) and another (stomatal resistance, r_s) for the resistance to the diffusion of vapor through the stomata. In the boundary layer the diffusion of vapor is analogous to the convective transfer of heat, as both occur by eddy diffusion.

The total resistance to the flow of vapor must take into account the distribution of stomata over the two leaf surfaces. If the stomata are uniformly distributed over the upper and lower surfaces, and have the same resistance, the total resistance for each surface is the sum of stomatal and boundary layer resistances in series, and the total resistance for the whole leaf is the sum of the two surface resistances in parallel. Thus for leaves of sugar-beet (*Beta vulgaris)* it is $(r_b + r_s)/2$. If the majority of stomata occur on only one surface, as with leaves of tomato (*Lycopersicon esculentum*) or barley (*Hordeum vulgare*), the total leaf resistance is $(r_b + r_s)$. The same reasoning can be applied to the transfer of carbon dioxide between the air and the stomata. In this case there is a flow of carbon dioxide towards the leaf when photosynthesis is occurring, but away from it during the night because of respiration.

Reducing the boundary layer resistance, by providing air movement, will generally increase the rates of transpiration and carbon dioxide exchange. However, because the stomatal resistance is in series with the boundary layer resistance, the net reduction in total resistance for mass transfer will not be as marked as for convective heat transfer (Stanghellini, 1987).

Air movement caused by mass transfer

In summer, ventilation is used to cool the greenhouse by replacing the air in the house, which has been heated by absorbing solar energy, with cooler external air. The nature and speed of the air movement created depends on how the ventilation is provided. There are two basic methods: forced and natural ventilation (see Chapter 2, pp. 47–50). The former uses fans to extract air from the greenhouse, which is replaced by outside air entering an inlet on the opposite side of the house. Natural ventilation relies on air movement created by differences in density and by the influence of wind. The air exchange occurs through openings created in the roof, and, in small greenhouses, in the side walls.

It is becoming increasingly common for fans to be used in greenhouses to create air circulation in a horizontal plane (Koths and Bartok, 1985). As in the case of vertical circulation, the energy required to maintain a slowly circulating mass of air is not high, and small fans can induce motion on a large scale. The narrow jet of air issuing from a fan entrains the surrounding air, and an expanding cone of moving air is created. As the momentum of the air leaving the fan is transferred to the entrained air the average speed decreases. If unobstructed, the cone of moving air expands with an angle of approximately 22° and the effective throw of the air is proportional to the product of the initial air speed and the fan diameter. The fans are positioned to reinforce the motion and to control the direction of flow (Fig. 3.11). This technique is commonly used in the United States, the aim being to provide an air speed of 0.2 m s^{-1}. The motion of the air not only improves the uniformity of air temperature but also increases the transfer of carbon dioxide and water vapor to and from leaves. The reduced boundary layer resistance obtained with an air speed of 0.5 m s^{-1} at a carbon dioxide concentration of 200 ppm by volume has been found sufficient to give the same rate of photosynthesis as a concentration of 300 ppm when the air was not moving (Waggoner et al., 1963). The relative humidity at plant height in a house with air movement has been increased by 11% compared to when the air was not moving (Cotter and Seay, 1961).

Using fans for ventilation and air cooling provides

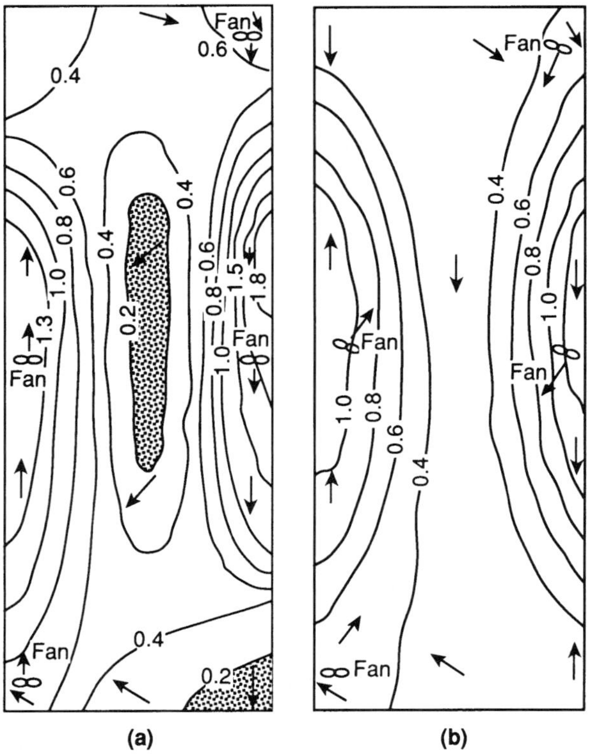

Fig. 3.11. Air velocities with horizontal air circulation fans; air speed in m s^{-1} (Walker and Duncan, 1974): (a) fan axes parallel to walls; (b) axes at 15° away from walls.

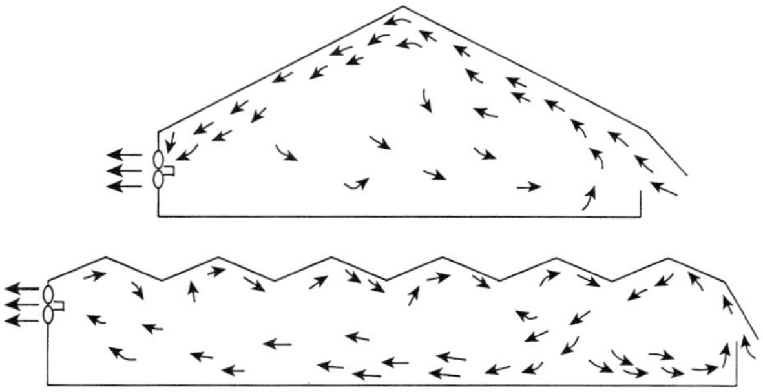

Fig. 3.12. Air-flow pattern in glasshouses as affected by roof configuration when inlet directs the incoming air upwards (Wolfe and Cotton, 1973).

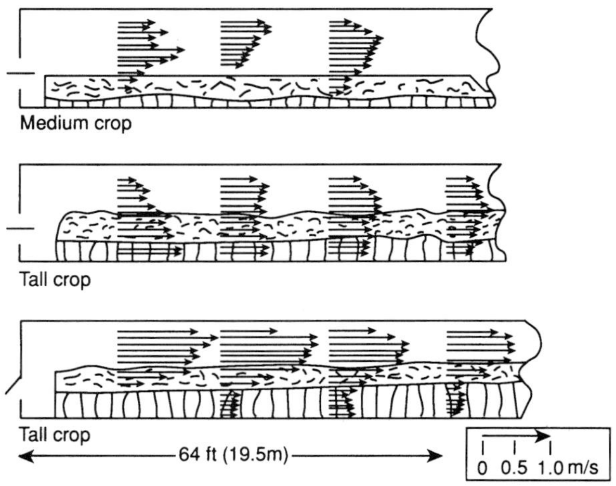

Fig. 3.13. Influence of air inlet and crop height on the distribution and speed of ventilation air through the greenhouse (Wolfe and Cotton, 1973).

well-defined patterns of air flow and high air speeds. The flow is generally directed horizontally from inlet to fan, though the greenhouse and the nature of the air inlet exert a strong influence over the flow pattern (Fig. 3.12). If the inlet is angled to direct the incoming air onto the roof, the air flow will continue along that surface. This attachment to the roof is referred to as the Coanda effect (Jones, 1980). In a single-span greenhouse this can result in the air bypassing the crop zone. In multispan houses the flow separates from the roof at the first gutter and then passes more or less uniformly towards the fans. The motion of air within the greenhouse is also influenced by the size and density of the plants (Fig. 3.13). At low rates of ventilation, air directed into the crop, even if it is dense, remains there as it travels towards the fans. At high ventilation rates the resistance of the crop causes the air to move preferentially into the unobstructed space above the crop, even if initially directed into the crop zone. In order to limit the temperature rise along the path of air flow, and also to avoid high air speeds, the distance from inlet to the fans is usually limited to 40 m, which gives a mean air speed, in a greenhouse 3 m high, of 0.5 m s^{-1}. In geographic regions with low ambient humidity, it is possible to improve the effectiveness of fan ventilation by evaporating water to precool the ventilation air (Mastalerz, 1977). Generally this is achieved by passing the air horizontally through a wetted pad, where the temperature is reduced by up to 80% of the wet-bulb depression. The density of the air is increased by the cooling and it tends to pass through the greenhouse at a low level. Thus, care has to be taken in designing fan-and-pad cooling systems to ensure that the top of the air stream does not fall below the plant canopy. The air flows produced by natural ventilation tend to be in the vertical direction. If both side wall and roof ventilators are open, the air enters through the opening in the side wall and leaves through the roof opening. In narrow houses this provides very effective ventilation, driven by a combination of the chimney effect, due to the density difference between the external and internal air, and the pressure differences between different ventilator openings caused by the wind. In large greenhouses, side wall ventilators are of limited value and ventilation is by roof ventilators. This form of ventilation is generally not so effective as the air enters and leaves through the same opening – although the ventilation flows can be influenced by wind speed and direction (Bot, 1983). In the absence of wind, ventilation occurs because of variations in density arising from differences between the inside and outside temperatures. The force which generates the air movement through a single opening is proportional to the temperature difference ($t_i - t_o$) and to the vertical height of the opening (h_v) (Bot, 1983). An energy balance shows the ventilation rate (V) is

$$V = \sqrt{\beta g(t_i - t_o)/F_v} h_v^{3/2} l_v/3, \tag{3.20}$$

where β is the expansion coefficient of air, g is the acceleration of gravity, F_v is related to the aspect ratio of the ventilator and l_v is the length of the ventilator opening. If two ventilators at different heights are open, h_v involves the vertical distance between them.

Wind produces ventilation by virtue of the pressures it creates at open ventilators. These can be positive, as at a windward wall, or negative, as at a leeward wall and over most of the roof. The wind pressure is related to wind speed (υ) by the relation

$$p = K_p \upsilon^2/2, \tag{3.21}$$

where K_p is the dimensionless wind pressure coefficient which characterizes the pressure distribution over the greenhouse (Kozai et al., 1980).

The inflow of air, which equals the outflow, can be obtained from an energy balance

$$V = 0.5 A_v \sqrt{K_f f_w / F_v}, \tag{3.22}$$

where A is the area of the opening, K_f is a pressure coefficient which takes account of the fluctuations in wind speed and f_w is a factor, $0 < f_w < 1$, which depends on the ventilator geometry (Bot, 1983).

The turbulence inherent in wind ensures that there will always be spatial variations in the wind pressure over a greenhouse roof. Consequently, ventilation will always occur even if only the leeward ventilators are open.

Bot (1983) showed that ventilation due to wind effects will dominate over that caused by temperature difference if $3\upsilon > \sqrt{t_i - t_o}$, where the factor 3 has dimensions $K^{1/2}\,s\,m^{-1}$. Thus, even for an extreme temperature difference of 25°C, wind-induced ventilation will dominate at wind speeds greater than $1.7\,m\,s^{-1}$. In addition, as the ventilation rate increases, the temperature difference diminishes, which further reduces the rate of ventilation due to the temperature effect.

The movement of air inside a greenhouse during natural ventilation is complex. The air currents differ from opening to opening, and they are influenced by wind speed. Bot (1983) showed that fluctuations in the speed of air passing through ventilators had a characteristic frequency that was dependent on the size of the opening. The frequency decreased as the ventilator was closed, so that the ventilator can be seen to act as a tunable high-pass filter. Fluctuations in internal air movement can be considered as being normal during natural ventilation.

On a summer day the rate of photosynthesis is such that the uptake of carbon dioxide by a crop can reach $3\,g\,h^{-1}\,m^{-2}$ (of ground area), and this can rapidly reduce the carbon dioxide concentration in the greenhouse atmosphere. When the internal concentration is lower than outside, carbon dioxide will be added during ventilation, but, even with the ventilators fully open, the concentration within the canopy can be less than 200 ppm. Plant transpiration increases the humidity of the greenhouse, and so ventilation provides a means for removing water vapor from the greenhouse and lowering the humidity. Temperatures in greenhouses are generally higher than outside, so ventilation can almost always be used to reduce humidity. However, in cool climates it may be necessary to heat at the same time, to maintain the required greenhouse temperature.

Excessive air speeds

Air movement affects the energy and mass exchanges of leaves by its influence on the boundary layer resistance for instance, increasing air speed increases convective heat transfer, and hence reduces the leaf–air temperature differential. The effect on transpiration is not so obvious, because the exchange rate depends on the stomatal and boundary layer resistances and on the gradient of the saturated vapor pressure curve at the temperature of the leaf. Under low irradiance, increasing air speed will increase the temperature if the stomatal resistance is low, while the converse is true at high irradiance for any stomatal resistance. Increasing air speed might be expected to increase photosynthesis because of the reduced boundary layer resistance to the transport of carbon dioxide. This has been observed (Waggoner et al., 1963), but if the increased air speed raises transpiration to such an extent that water stress and hence stomatal closure occur, then photosynthesis will be reduced as a consequence.

Information on the influence of wind speed on the growth of plants is not consistent. The evidence suggests that, for plants grown in the soil or solid media, the optimum wind speed is 0.3–0.7 $m\,s^{-1}$ and that speeds of 3–4 $m\,s^{-1}$ reduce growth appreciably (Morse and Evans, 1962). However, this is not found where plants are grown using hydroponics. Grace (1977) considered that this difference might be because wind increases transpiration to the point where the supply of water from the growing media to the roots did not match the rate of loss, whereas with hydroponic culture water was always available to the roots.

It is known that shading, stroking or gently rubbing leaves reduces growth, possibly due to the effects of plant growth substances (Biddington, 1986). Consequently air speeds greater than 1.5–2 $m\,s^{-1}$ which induce excessive movement of leaves, are undesirable. Higher air speeds which result in physical damage to leaves should obviously be avoided, although some species appear able to compensate for some loss or damage to leaves.

Disease transmission

The spread of plant pathogens in the field is influenced by environmental factors such as wind, air turbulence and rain, which cause shaking of leaves and liberation of fungal spores. In the greenhouse these factors are either absent or are reduced in magnitude. However, because the greenhouse environment is relatively constant and more favourable than the field for the development of fungal pathogens, for instance through the maintenance of high humidities, the build-up of inoculum can be substantial. The pathogens can be introduced by the passage of air through open ventilators and through gaps around closed ventilators, doors and glazing. Once established in the greenhouse, the spores can be disseminated by moving air, the movement of diseased plant material, and the movement of workers. Some spores are liberated by positive actions, such as the "squirt gun" of certain ascomycetes, whilst others, such as those of powdery mildews, are liberated passively. Once in the air their flight distance depends on the rate of sedimentation and the speed and turbulence of the air. The sedimentation rate is defined for smooth spheres by Stokes Law, in which important properties are spore diameter and density. However, spores are seldom spherical, and their morphology influences the sedimentation velocity.

Frinking and Scholte (1983) found that the dissemination of fungal spores in a glasshouse was greatest when the source was near a door through which there were frequent air currents caused by the entrance of workers. Dissemination was far less in areas where there was little air movement, such as at the end of the house remote from the door, where the conditions were not as frequently disturbed by workers.

TEMPERATURE

Temperature, together with visible radiation, water and nutrition, is one of the major environmental factors affecting plant growth. Plant metabolism and the rates of biochemical reactions are strongly dependent on temperature, and the growth rate of many plants growing at relatively cool temperatures can be doubled by an increase of 10°C in temperature. Both low and high extremes of temperature are detrimental to good plant growth (Hanan et al., 1978). Chilling or freezing temperatures at critical times during growth can cause flower abortion and reduce yield. Denaturation of enzymes and other proteins occurs at temperatures above or below certain critical values. Because of the influence of temperature on the saturated vapor pressure of water and on the vapor pressure deficit, temperature also has a major effect on the rate of water loss from plants.

Temperature can be thought of as the property of a body which determines the direction and the rate of heat exchange with other bodies. Heat is transferred across the boundary between objects as a consequence of and in proportion to the difference in temperature. The temperature of an object is independent of its size, mass or extent, and is known as an *intensive* property of matter. Conversely, the energy stored in a system does depend on its mass or extent; hence thermal energy is an *extensive* property of matter.

Energy balances and greenhouse simulation models

The temperature of an object is a consequence of its energy gains and losses. If the gain of energy exceeds the loss, the temperature will rise, and vice versa. When the temperature remains constant, the rate of energy loss equals the rate of supply, and the object is in thermal equilibrium with its surroundings. The fundamental physical law which governs energy exchange is the First Law of Thermodynamics (see p. 74). For a system which does not involve mechanical work, this law states that any increase in the intrinsic energy of a system is equal to the algebraic sum of the energy flows to and from it. This balance between the rates of energy gain, loss and storage of an object is expressed mathematically by an energy balance equation such as Eq. (3.1). As has been shown elsewhere, with the exception of solar radiation, the energy fluxes created by the individual heat transfer processes can be formulated in terms of differences in temperature. Thus, in principle, if the temperatures of all objects which interact with an object of interest are known, it is possible to determine the temperature of that object by solving its energy balance equation.

The flow of energy into and out of a greenhouse is illustrated in Fig. 3.14. During the day, solar radiation is the major mode of heat input. Most is absorbed by the plants, but some is absorbed by the soil, and a small amount by the greenhouse roof. The plants transfer energy to the greenhouse air by convection, to the roof and possibly the soil by thermal radiation, and energy is also removed by transpiration. Water lost by transpiration, together with any evaporated from the soil, will condense on the roof or be removed from the greenhouse by ventilation. Consequently, in order to calculate, for example, the plant temperature, the temperatures of the air, roof and soil must be determined. Thus it is necessary to create energy balance equations for all greenhouse components which interact thermally, and generally

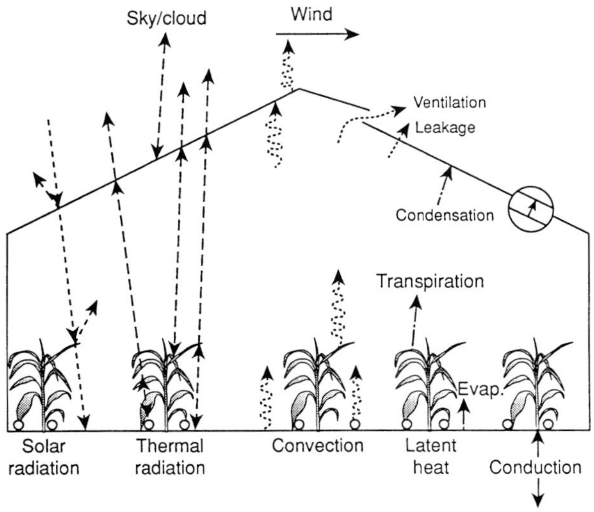

Fig. 3.14. Modes of heat transfer in a greenhouse.

these consist of the air, plants, floor and greenhouse cladding.

The temperature of the greenhouse air can only be changed by the convective transfer of heat, and therefore the energy balance for air will include transfer to or from the plants, floor, cladding, and the heater by convection, and to or from the outside air by leakage and ventilation.

The energy exchange of plants includes convection with the air, the loss or gain of latent heat through transpiration or condensation, the gain from solar radiation, and the exchange of long-wave thermal radiation with the floor, the greenhouse cladding, and possibly the greenhouse heater and the sky.

The inner surface of the greenhouse cladding interacts thermally with the internal air through convection, which can be influenced by the convection generated by the greenhouse heater. It gains latent heat if water condenses on it and exchanges thermal radiation with the plants, the floor, and possibly the greenhouse heater. The outer surface of the cladding exchanges thermal radiation with the sky, where the effective temperature is influenced by the amount of cloud cover. The cladding also exchanges heat with the outside air through forced convection. Usually the thermal resistance within the greenhouse cladding is omitted because, for single layers of plastic film and glass, it contributes less than 5% to the total thermal resistance of the cladding. The inner and outer surfaces of the cladding are then assumed to be at the same temperature. When the cladding has a high thermal resistance, this assumption cannot be made. Individual energy balances must be made for the inner and the outer surfaces, and each will include the heat transfer between the two surfaces by conduction, if the cladding is homogeneous, or by convection, thermal radiation and conduction if it contains air spaces.

The heat fluxes in these equations will be similar to those described earlier for heat transfer by convection, radiation, and conduction. The flux of latent heat is given by the product of the rate of water vapor transfer and the latent heat of evaporation. For a leaf the latent heat energy flux, λE, is

$$\lambda E = \frac{c_p \rho}{\gamma} \frac{(e_l - e_a)}{r_v}, \tag{3.23}$$

where γ is called the psychrometer constant and has the value $c_p P m_a / \gamma m_w$. The rates of latent heat transfer by transpiration, evaporation and condensation which contribute to the energy fluxes are themselves dependent on the humidity of the greenhouse air. This can be found by invoking another fundamental physical law which states that mass must be conserved. Thus, the humidity of the greenhouse air is the value at which the rates of gain from transpiration and evaporation are balanced by the losses due to condensation and ventilation.

To use the energy and mass balance equations, appropriate values are required for the parameters and coefficients in the equations for the heat fluxes. Examples include the heat transfer coefficients, radiation shape factors, and properties of materials such as the transmittance and absorptance of the greenhouse cladding to solar radiation, its emissivity and transmittance for thermal radiation, and the thermal conductivity of the soil beneath the greenhouse. The energy fluxes also depend on conditions outside the greenhouse such as solar irradiance, the temperature and humidity of the external air, wind speed, sky temperature or cloud cover, and the temperature of the soil at a sufficient depth beneath the greenhouse to be treated as constant. Representative values can be selected for these "boundary conditions" or they can be taken from records of meteorological data. Then the set of energy balance equations and the water-vapor mass-balance equation can be solved simultaneously to give the temperatures of all the greenhouse components.

These equations, which provide a mathematical description of the physical conditions in the greenhouse, constitute a greenhouse simulation model. Using such a model, with a suitable set of boundary conditions, it is possible to predict, or simulate, the conditions which occur inside a greenhouse. The simulation model provides a very powerful way of determining how changes to individual parameters, such as the thermal resistance or emissivity of the cladding, influences the greenhouse climate or rate of heat loss.

Energy balance of leaves

The usual objective of using a greenhouse is to provide conditions more conducive to plant growth than those outside. Thus, control of leaf temperature and transpiration are the major requirements. During the day, solar radiation is usually the major mode of energy supply to the greenhouse. It is absorbed by leaves, which then transfer energy to the air and other objects in their vicinity. As was indicated earlier, the energy involved in photosynthesis and respiration is small and unimportant when considering plant temperature. Also, the thermal capacity of leaves is small, so that the change in energy storage can be omitted in comparison to the other energy fluxes involved in the overall energy balance. Consequently, for leaves, the energy balance equation (3.1) reduces to

$$R = C + \lambda E. \tag{3.24}$$

This means that the leaves achieve a temperature at which there is a balance between the fluxes of net radiation, sensible heat (predominantly due to convection), and latent heat (mainly transpiration, but can include condensation).

The net radiation flux is equal to the radiant energy absorbed by the leaf minus the energy radiated from it. During the day there are two components to the absorbed energy: solar radiation transmitted by the greenhouse cladding, and the long-wave thermal radiation emitted by all objects in the plant's environment (see p. 74). The absorptance of leaves is high for photosynthetically active radiation (400–700 nm) and low for the near-infrared portion of solar energy (700–3500 nm). Thus, leaves are efficient absorbers of the radiation required for growth but relatively poor absorbers of the approximately 50% of solar radiation in the near-infrared which produces heating. Plants exchange energy by long-wave thermal radiation with surrounding surfaces such as the roof, walls, floor and the heat distribution system in a greenhouse. The absorptance and emissivity of leaves for this radiation is high, typically 0.97 (Idso et al., 1969). The greenhouse cladding is the most important of these surfaces and, with the exception of polyethylene film, they are made from materials which have high absorptance for long-wave thermal radiation. Because plants are usually warmer than the cladding, they lose heat by thermal radiation exchange with the cladding. However, compared to plants growing in the open, the loss of energy by thermal radiation is much reduced. The temperature of a clear sky can be 20°C below the ambient value, but the greenhouse cladding will only be slightly below ambient at night if the house is unheated. If the greenhouse is heated, the cladding temperature will be intermediate between the ambient and the internal temperature. As a consequence, the radiative loss is reduced from typically 100 W m^{-2} for plants in the open to 20–40 W m^{-2} for plants in a heated greenhouse in northern Europe (Bailey and Cotton, 1977).

The transfer of sensible heat to a leaf occurs predominantly by convection. The amount transferred by conduction along the plant stem is very small in comparison. As has been described above (p. 83), convective heat transfer occurs when energy is transported by motion of a fluid, in this case air. Inside a closed greenhouse, wind is virtually absent and, even if fans are used to induce air circulation, the mean air speed is unlikely to exceed 0.5–1 m s^{-1} except close to and in line with a fan. Thus free convection, in which the air movement occurs because of differences in buoyancy arising from temperature differences, is generally more important than forced convection. The latter only becomes important when both windward and leeward ventilators are open.

The normal situation is for a plant leaf to be transpiring, in which case the evaporating water will remove latent heat. The rate of transpiration depends on the vapor pressure difference between the leaf and the air, and also on the leaf stomatal resistance. Because the plant can change its stomatal resistance in response to its environment, it can control the rate of transpiration. Thus the loss of energy by latent heat depends not only on the other energy fluxes but also on the condition of the leaf. Latent heat is supplied to a leaf when water condenses on it. However, in the greenhouse this is to be avoided whenever possible, in order to reduce the risk of fungal diseases.

The environmental influences on leaf temperature

The energy balance equation and the equations for the individual heat transfer processes can be used to show how leaf temperature is influenced by its environment. The leaf is not a passive partner. It responds to its environment, and this feedback process modifies the influences of the environment. The feedback is provided by variations in the stomatal resistance.

If the leaf resistance is infinite, no transpiration or evaporation occurs and Eq. (3.24) then becomes

$$R + C = 0. \tag{3.25}$$

The leaf gains energy by radiation and, under steady-state conditions, it will reach a situation where that rate of energy gain will be equalled by the rate of heat loss by convection. During the day, this implies the non-transpiring leaf must be hotter than the surrounding air (Fig. 3.15). Increasing solar irradiance will raise leaf temperature, and increasing the rate of convective heat transfer, by creating air movement, will lower leaf temperature. At night, there is no solar radiation and the net radiative flux will be directed away from the leaf. The leaf will then be cooler than the air, maintaining

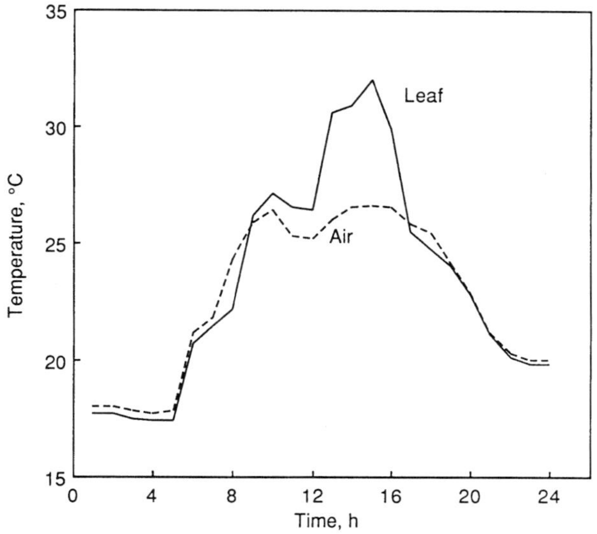

Fig. 3.15. Tomato leaf and air temperatures measured in a greenhouse in the United Kingdom during May.

a temperature where the supply of heat by convection matches the radiative loss (Fig. 3.15).

When the leaf is completely wet, as might occur when it is covered in dew, the leaf resistance is zero. The latent-heat cooling is then a maximum for any value of the boundary layer resistance, and the leaf temperature is the minimum which can occur for given values of the other heat fluxes.

In practice, the leaf resistance lies between these two extremes, and it is influenced by environmental factors (Jones, 1983). Stomata are most sensitive to visible radiation, opening rapidly as radiation levels increase to around 100 W m^{-2} (Stanghellini, 1987). Stomata are also sensitive to leaf water status, tending to close to conserve moisture when the leaf becomes water-stressed (Ludlow, 1980). They also close with decreasing ambient humidity, but open with increasing temperature. With a finite leaf resistance, the leaf temperature tends towards the air temperature as the boundary layer resistance approaches zero. As the leaf resistance decreases, the amount of transpirational cooling increases for normal values of boundary layer resistance. Whether this is sufficient to cool the leaf below air temperature depends on other factors, particularly the net radiation flux and the vapor pressure difference. Increasing the vapor pressure deficit will always cool the leaf, especially if both leaf and boundary layer resistances are small. Leaf temperature will always tend to rise as the leaf resistance increases. Increasing the carbon dioxide concentration in the air can also lead to an increase in stomatal resistance, and, in consequence, a small rise in leaf temperature (Stanghellini, 1987).

The influence of the boundary layer resistance on leaf temperature is complex, since increasing it can raise or lower leaf temperature depending on the environmental conditions and on the leaf resistance. When the leaf is hotter than the air, increasing the boundary layer resistance will always increase leaf temperature.

Greenhouse heating

In the winter the greenhouse temperature can fall below the optimum value, and heating may be required. The rate of heat supply required to achieve a particular temperature can be determined using a detailed greenhouse simulation model, but in practice a simplified energy balance method has been found to be satisfactory. The transfer of heat from the greenhouse is assumed to take place in two ways: by conduction through the cladding, and by air infiltration. There will also be a flow of heat into and out of the soil beneath the greenhouse, as well as losses through the soil around the perimeter of the house. Except for small greenhouses, the soil heat fluxes can usually be neglected. The loss of heat from a greenhouse can be expressed per unit ground area as

$$H = (h + h')A_r(t_i - t_o) = UA_r(t_i - t_o), \tag{3.26}$$

where U is the overall heat transfer coefficient of the greenhouse and A_r is the ratio of cladding to floor areas. Typical values of U are given in Table 3.3.

The coefficient h is the conductance of the cladding, which includes the resistances at the inner and outer surfaces as well as the thermal resistance of the cladding material. The resistance at the inner surface depends on the air speed in the vicinity of the cladding, which is influenced by the type of heater used. The external heat transfer resistance is affected markedly by wind, which has the effect of making the overall heat-transfer coefficient dependent on wind speed. A clear sky has a lower effective radiative temperature than a cloudy sky, so that the heat loss by thermal radiation is higher under a clear sky. This means that the effective heat transfer resistance at the external surface is lower under a clear sky. In the case of a greenhouse covered with a single layer of glass or plastic film, the two boundary layer resistances constitute more than 95% of the total resistance of the cladding. Coefficient h' refers to the heat removed by air infiltration, and, for a well-maintained greenhouse, it should only be 10 or 20% of h.

Table 3.3
Total heat loss coefficient U at wind speed of 4 m s^{-1} (von Zabeltitz, 1986)

Material	U (W m^{-2} K^{-1})
Single glass	6.0–8.8
Double glass, 9 mm air space	4.2–5.2
Double acrylic 16 mm	4.2–5.0
Triple acrylic 32 mm	3.0–3.5
Double polycarbonate, 10 mm air space	4.7–4.8
16 mm air space	4.2–5.0
Single plastic film (not aluminized)	6.0–8.0
Double plastic film (not aluminized)	4.2–6.0
Single glass plus energy screen of	
single film, single fabric, non-woven film	4.1–4.8
double film, double fabric, air-bubble film	3.4–3.9
aluminized single film	3.4–3.9
aluminized air-bubble film	3.2

Greenhouses can be heated by a variety of systems which supply energy by a combination of radiation, convection and conduction. Predominantly radiative heating is provided from pipes heated to 200–400°C using flue gases from combustion of natural or petroleum gas. Typically, these supply 70% of the total heat as radiation and 30% as natural convection (Businger, 1963). In the piped hot-water heating systems more commonly used (containing water at temperatures up to 120°C), heat is transferred by radiation and natural convection in approximately equal amounts. Heating by forced convection is provided by using air heaters, although, if inflatable tubes of film plastic are used to distribute the air, some radiation and natural convection also occur. Heating predominantly by conduction is obtained when plants are grown on heated floors or benches, although some energy is released by radiation and convection from areas not covered by plants. All the modes of heat transfer occur when water at a temperature of less than 50°C is circulated through large plastic tubes which rest on the greenhouse floor. Heat is then transferred into the soil by conduction, to the air by natural convection, and to the plants by natural convection and thermal radiation.

Greenhouse cooling

During the day, particularly in summer, solar energy can increase the greenhouse temperature to values

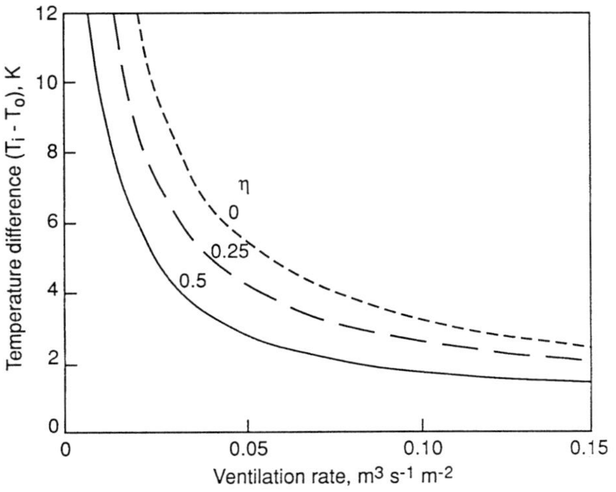

Fig. 3.16. Calculated temperature rise in a ventilated greenhouse, for three different values of the fraction of solar energy that is used for transpiration.

which are detrimental to plants. It is then necessary to cool the greenhouse by ventilation. The ventilation requirement of a greenhouse can be established (Bailey, 1988) using an energy balance equation of the form

Solar input = cladding conduction + ventilation.

The rate of removal of sensible heat by ventilation is

$$C_{\text{vent}} = V\rho c_{\text{p}}(t_{\text{i}} - t_{\text{o}}). \qquad (3.27)$$

Assuming the solar energy used in photosynthesis and absorbed by the soil is small, and that the proportion of the solar energy used for transpiration is η, the ventilation rate to give an internal temperature of t_i is

$$V = \frac{\tau(1-r)(1-\eta)S}{\rho c_{\text{p}}(t_{\text{i}} - t_{\text{o}})} - \frac{UA_{\text{r}}}{\rho c_{\text{p}}}. \qquad (3.28)$$

The influence of ventilation rate on the elevation of greenhouse temperature above the ambient value is shown in Fig. 3.16. The effectiveness of ventilation in reducing temperature is very dependent on the value of η. Thus a much higher ventilation rate is required to cool a greenhouse containing a few plants than in one with a full crop canopy.

Temperature control

Changes are induced in the conditions inside the greenhouse as the external climate varies. The rate at which these changes occur depends on the rate of change of the energy fluxes and on the thermal capacity of the greenhouse components. The ratios of the thermal capacities of the air, greenhouse cladding, plants and the soil to a depth of 200 mm are approximately 1:3:10:100. Therefore the greenhouse air responds most quickly to changes and the soil most slowly. The response time is defined as the time to reach 63% of the final value following a step change. The response of the greenhouse air to heating and cooling depends on the type of heater and ventilation system. In the case of piped hot-water heating the response time is 40–60 min, compared to the time of 10 min typical with an air heater.

The differences in thermal response can be both advantageous and detrimental. The slow response of the soil means that energy absorbed during the day from solar radiation is retained near the surface. The depth to which heat is conducted depends on the period of the heating cycle and on the thermal conductivity and volumetric heat capacity of the soil. At night, when the air temperature falls below the temperature of the soil surface, this energy is returned to the greenhouse by convection and thermal radiation. Heat transfer to and from the soil therefore limits the maximum temperature during the day and raises the minimum temperature at night. A disadvantage follows from the relatively slow response of fruit on plants, as their temperature increases more slowly than air temperature following sunrise. It is then possible for the fruit temperature in this period to be lower than the dew point of the air, thereby causing water to condense on the fruit, which in turn affects quality and favours the development of fungal diseases.

WATER VAPOR

Water is a vital commodity in the greenhouse, and understanding the physical laws that govern its exchange is important in the proper management of the crop environment. The main contributor to water vapor exchange is transpiration from the leaves of the crop, but there can also be significant contributions from evaporation from moist soil surfaces that may be present, and also from condensation on cool surfaces such as the inside of the cladding (Fig. 3.17). The net loss of water from the greenhouse depends on the rate of air exchange between internal air and the exterior. The greenhouse provides a high-humidity environment, and it might be thought that this would lead to much lower evaporation rates than in the open. This is not necessarily the case, and to understand the

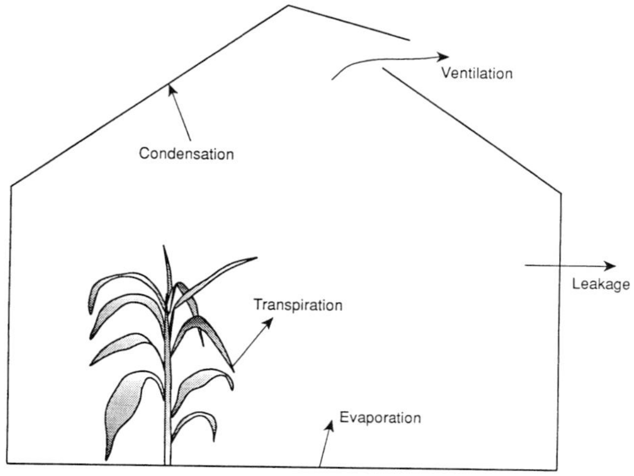

Fig. 3.17. Sources and sinks of water vapor in a greenhouse.

reason requires consideration of the driving forces for evaporation and transpiration. Before looking at the equations, it is necessary to clarify the terminology involved.

Water vapor concentration and humidity

The water vapor content of the air can be expressed in a number of interrelated ways. The concentration, in mass per unit volume, and the partial pressure or vapor pressure, e, are directly interchangeable definitions because water behaves like an ideal gas. These terms describe the absolute humidity of the air, but because both liquid and vapor phases of water can coexist at ambient temperatures, there is an effective upper limit to the humidity which represents the concentration at equilibrium with the liquid phase. This is termed the saturation vapor pressure, e_s – the vapor pressure in air whose composition is unchanging in the presence of a plane surface of the condensed phase (usually liquid water, but sometimes ice) at the same temperature and pressure. The relative humidity is the ratio of the actual vapor pressure to the saturation vapor pressure of air at that temperature. The dewpoint temperature is the temperature to which moist air would have to be cooled, at constant pressure and specific humidity, to become saturated with water vapor. When surfaces in the greenhouse reach temperatures that are below the dewpoint, water will begin to condense on them. The final term that needs to be defined is the saturation vapor pressure deficit, or just saturation deficit, which is the difference between saturation vapor pressure and the actual vapor pressure of the air. This difference is the driving force for diffusion of water vapor from a source that is at equilibrium with liquid water at air temperature, and air at a lower humidity. It is therefore an important term governing evaporative fluxes in the system.

The change of phase from liquid to vapor is associated with a large latent heat of vaporization – in other words, a lot of energy needs to be supplied to molecules of water to transfer them from the liquid to the vapor phase. When heat is absorbed causing changes in both the air temperature and the humidity, the change in the total energy content of the air (enthalpy, H) can be expressed as the increase in sensible plus the increase in latent heat,

$$H = \rho c_p(t_2 - t_1) + \rho c_p(e_2 - e_1)/\gamma. \quad (3.29)$$

The saturation vapor pressure increases rapidly with increasing temperature, because the thermal energy of the water molecules in the liquid phase becomes closer to the energy required to break the bonds with adjacent molecules and escape through the water–air interface. Thus, the same relative humidity represents a much larger saturation vapor pressure deficit at higher temperatures, and corresponds to the potential for much larger evaporation rates. This gives an indication of the reason why evaporation rates in the greenhouse can, in some circumstances, be close to rates in the open field. Although greenhouses are generally environments with high relative humidity, they operate at higher temperatures than the ambient value, thus restoring some of the driving force for water vapor diffusion.

Evaporation and transpiration

The evaporation rate from a wet surface can be analyzed by considering the sensible and latent heat fluxes given by Eqs. (3.12) and (3.23), and Eq. (3.24) for the overall energy balance. Assuming that differences in temperature and humidity are small, so that they can be linearly related by the slope of the saturated vapor pressure curve de_s/dt or Δ, Eq. (3.23) can be written as

$$\lambda E = \frac{C_p\rho}{\gamma}\,\frac{\Delta(t_s - t_a) + [e_s(t_a) - e_a]}{r_v}, \quad (3.30)$$

Substituting for $(t_s - t_a)$ from Eq. (3.12) and for C from Eq. (3.24) this equation can be rearranged to

$$\lambda E = \frac{\Delta R + C_p\rho[e_s(t_a) - e_a]/r_h}{\Delta + \gamma^*}, \quad (3.31)$$

where $\gamma^* = \gamma r_v / r_h$.

This equation is commonly referred to as the Penman–Monteith equation, or the combination equation, as it is the combination of two terms driven, respectively, by energy and the humidity gradient (Monteith, 1973).

The evaporation rate has two driving forces: the net radiation and the vapor pressure deficit of the air. Under most climatic conditions, it is the radiation term that dominates. Because net radiation is largely determined by solar energy flux, the rate of transpiration is often nearly proportional to solar irradiance (Stanhill and Albers, 1974).

These equations deal with energy balances and vapor exchanges for individual components, and they assume that the key variables such as temperature and vapor pressure, and the key parameters such as the resistances, can be given appropriate average or lumped values to describe the environmental conditions associated with the main components in the greenhouse ecosystem. Legg (1989) has used this approach to compare environments and transpiration rates in the greenhouse with those in the open field. The consequence of covering a crop with a well-ventilated greenhouse is to reduce transpiration rates by perhaps 20% compared to the field. Decreasing the ventilation rate somewhat can result in an increase in transpiration (Legg, 1989). The resulting increase in leaf temperature has a more marked effect on vapor pressure deficit than on leaf-to-air temperature difference, because of the strong non-linearity in the relation between saturation vapor pressure and temperature.

As ventilation is further decreased, the situation becomes more complex than this simplified approach can illustrate, because other modes of heat loss from the glasshouse come into play – for instance the elevated temperature of the house increases the importance of heat loss by thermal radiation; and, as relative humidity approaches 100%, condensation on the inside of the cladding provides further heat transfer without vapor exchange.

Transpiration from the leaf has an important role in maintaining the temperature of the leaf at acceptable levels under high levels of insolation. Using Eqs. (3.12), (3.23) and (3.24) enables the leaf-to-air temperature difference to be expressed as

$$t_l - t_a = \frac{Rr_h}{\rho c_p\left[1 + \left(\frac{r_h \Delta}{r_v \gamma}\right)\right]} - \frac{r_h}{\gamma r_v} e_s(t_a). \qquad (3.32)$$

From this it can be clearly seen that there are two terms – a temperature rise proportional to the net radiation balance of the leaf, offset by a cooling that is proportional to the transpiration rate.

The potential cooling that can be achieved by evaporation can also be utilized for control of the greenhouse environment. If the external air is very dry and air temperatures are above the optimal temperature for the greenhouse, then cooling can be achieved by passing the ventilation air through evaporation pads that are kept permanently wet. However, because this approach increases the humidity in the greenhouse and decreases the potential for transpiration from the leaves of the crop, the differential between leaf and air temperature within the house will fall.

Low humidities in the greenhouse can also be a problem, leading to high evaporative demands on the plant. This problem can arise particularly when plants are small, so that a relatively small proportion of the incident energy is used to drive transpiration. In these circumstances, an alternative source of water may be provided to absorb some of this energy, for instance by using fogging nozzles to inject a spray of fine water droplets. Evaporation of these droplets leads to an increase in air humidity and a decrease in air temperature, as required.

Condensation

The physical laws that govern condensation are a natural extension of those for evaporation. Evaporation occurs when the surface temperature is greater than the dewpoint temperature of the air, and condensation occurs when the opposite is true.

The major source of water vapor in a greenhouse is transpiration from plants, although some may be produced by evaporation from the soil. Condensation is therefore likely to occur when there are surfaces in the greenhouse at temperatures lower than parts of the crop, particularly during periods when there is a low rate of water loss by ventilation. In practice, when the ventilators are closed and heat is being received either from the heating system or from the sun or both, it is condensation of water on the roof which controls the moisture content of the greenhouse atmosphere. If there is no transpiration or evaporation taking place in the greenhouse, the dewpoint cannot exceed the temperature of the inner surface of the greenhouse roof, although it can be less. However, transpiration takes place from plants by day and also at night, although at a reduced rate, and under these conditions the dewpoint must exceed the roof temperature in order to create the gradient in vapor pressure necessary to promote transport of water to the roof when condensation takes place. The energy supplied to the greenhouse roof by the latent heat of condensation can contribute up to 15% of the total heat transferred by convection, radiation and latent heat.

Condensation can modify a number of the other physical processes important in the greenhouse. When condensation forms on glass, the water film spreads over the surface because the angle of contact between water and glass is small. However, the angle of contact between water and most plastics materials is large, and, as a consequence, the water condenses as drops which do not coalesce. When these drops appear on the inside of the cladding material they will modify the transmission of rays of light passing through the cladding. Total internal reflection in the water drop can result in a ray that would previously have entered the greenhouse being reflected away, thus decreasing the internal light level. To overcome these problems, plastic films are now available with additives that decrease the surface tension and thus result in more uniform wetting of the film. With polyethylene films there is an advantage of condensation because it provides a barrier to loss of long-wave radiation. These films are generally quite transparent to long wavelengths, whereas water is not. Therefore the formation of a water film will intercept part of the long-wave radiation exchange and result in less heat loss from the greenhouse by this mechanism.

Dehumidification can also be used as a technique for positive environmental control, with potential benefits for energy consumption. In cold conditions where no greenhouse ventilation is required to maintain the correct greenhouse temperature, high humidities can build up, posing risks to the crop both through suppressed transpiration and the possibility of the growth of fungal disease organisms. Using some energy to run a refrigerator unit can remove the high humidity problem without ventilating and thus losing heat from the greenhouse, and also make use of the latent heat of the water vapor to heat the house. This is, of course, only of major value when the requirements for dehumidification and for heating the greenhouse occur at the same time. An alternative to electrical energy in refrigerative dehumidification is to use brine solutions as the sink for excess humidity. Saturated solutions of salts have a well-defined equilibrium humidity, so that passing air of a higher humidity over the solution leads to water vapor condensing out, thus reducing the air humidity (Assaf, 1986). The latent heat released heats the air passing through the unit. The brine is reconcentrated by heating, and the resulting vapor is condensed so the latent heat can be recovered and supplied to the greenhouse. The only energy losses are the sensible heat of the condensate and that used to recirculate the greenhouse air.

Notation

A	Area (m^2)
A_r	Ratio of greenhouse surface to ground area
C	Sensible heat flux density ($W\,m^{-2}$)
c	Concentration of water vapor ($kg\,m^{-3}$)
c_p	Specific heat of air at constant pressure ($J\,kg^{-1}\,K^{-1}$)
D_{nm}	Amplitude of n, m harmonic of incident light
E	Water vapor flux density ($kg\,m^{-2}\,s^{-1}$)
e	Partial pressure of water vapor (subscripts: l, leaf; a, air; s, saturated) (Pa)
F_{12}	Shape factor for thermal radiation exchange
F_m	Fourier polynomial of order m
F_v	Function of ventilator aspect ratio
f_w	Factor dependent on geometry of greenhouse ventilator $0 < f_w < 1$
Gr	Grashof number
g	Acceleration of gravity ($m^2\,s^{-1}$)
H	Total flux density of sensible and latent heats ($W\,m^{-2}$)
h	Heat transfer coefficient for greenhouse cladding
h_v	Vertical height of ventilator opening (m)
h'	Equivalent heat transfer coefficient for greenhouse air exchange
J	Rate of change of stored energy per unit area ($W\,m^{-2}$)
K	Thermal diffusivity of air ($m^2\,s^{-1}$)
K_f	Dynamic wind pressure coefficient
K_p	Steady state wind pressure coefficient
k	Thermal conductivity of air ($W\,m^{-1}\,K^{-1}$)
LAI	Leaf area index

L	Thermal radiation flux density (subscripts: u, up; d, down) ($W\,m^{-2}$)
l	Characteristic dimension of a body (m)
l_v	Length of ventilator opening (m)
M	Rate of heat production or absorption by chemical or biochemical process per unit surface area ($W\,m^{-2}$)
m	Molecular weight (subscripts: w, water; a, air)
Nu	Nusselt number
P	Atmospheric pressure (Pa)
P_n	Legendre polynomial of order n
p	Dynamic wind pressure (Pa)
R	Net radiation flux density ($W\,m^{-2}$)
Re	Reynolds number
r	Reflectance of crop
r_b	Resistance of boundary layer for water vapor transfer ($s\,m^{-1}$)
r_h	Resistance of boundary layer for convective heat transfer ($s\,m^{-1}$)
r_s	Resistance of leaf stomata to diffusion of water vapor ($s\,m^{-1}$)
r_v	Total resistance for water vapor transfer ($s\,m^{-1}$)
S	Total solar irradiance on a horizontal surface ($W\,m^{-2}$)
S_c	Circumsolar irradiance ($W\,m^{-2}$)
S_d	Diffuse solar irradiance on a horizontal surface ($W\,m^{-2}$)
T	Absolute temperature (K)
t	Temperature (subscripts: a, air; b, black-body temperature of sky; i, inside greenhouse; l, leaf; o, outside greenhouse; s surface) (°C)
U	Overall heat transfer coefficient of greenhouse ($W\,m^{-2}$)
V	Ventilation rate per unit greenhouse plan area ($m^3\,m^{-2}\,s^{-1}$)
v	Air/wind speed ($m\,s^{-1}$)
α	Angle of incidence/absorptance
β	Coefficient of thermal expansion of air (K^{-1})
γ	Psychrometer constant ($Pa\,K^{-1}$)
γ^*	Apparent value of psychrometer constant ($\gamma r_v/r_h$) ($Pa\,K^{-1}$)
Δ	Rate of change of saturation vapor pressure with temperature ($Pa\,K^{-1}$)
δ	Thickness of equivalent boundary layer (m)
ε	Emissivity
ζ	Elevation of sky element
η	Proportion of solar radiation absorbed in a greenhouse used for transpiration
θ	Angle relative to sun (°)
λ	Latent heat of vaporization of water ($J\,kg^{-1}$)
υ	Kinematic viscosity of air ($m^2 s^{-1}$)
ρ	Density of air ($kg\,m^{-3}$)
σ	Stefan–Boltzmann constant ($W\,m^{-2}\,K^{-4}$)
τ	Transmittance
ϕ	Solar elevation
Φ	$\sin\phi$ or $\sin\zeta$
χ	Angle of azimuth

REFERENCES

Acock, B., Charles-Edwards, D.A., Fitter, D.J., Hand, D.W., Ludwig, L.J., Warren Wilson, J. and Withers, A.C., 1978. The contribution of leaves from different levels within a tomato crop to canopy net photosynthesis: experimental examination of two canopy methods. *J. Exp. Bot.*, 29(11): 815–827.

Aikman, D.P., 1989. Potential increase in photosynthetic efficiency from the redistribution of solar radiation in a crop. *J. Exp. Bot.*, 40: 855–864.

Andrews, R.A., Bailey, B.J. and Cotton, R.F., 1982. *The effect on glasshouse light transmission of covering the walls and floor with plastics materials,* Div. Note DN/1122. Nat. Inst. Agri. Eng. Silsoe, 12 pp.

Assaf, G., 1986. LHC – The latent heat converter. *Heat recovery systems*, 6: 369–379.

Bailey, B.J., 1978. Applications of plastics screens in glasshouse environmental control. In: *Proc. Natl. Conf. on Advances in Use of Plastics in Agriculture and Horticulture and Produce Distribution.* British Agric. and Hort. Plastic Assoc. National College of Agricultural Engineering, Silsoe.

Bailey, B.J., 1988. Principles of environmental control. In: Chr. von Zabeltitz (Editor), *Energy Conservation and Renewable Energies for Greenhouse Heating,* CNRE Guideline No. 2. FAO.

Bailey, B.J. and Cotton, R.F., 1977. *Reducing glasshouse heat loss with reflective thermal screens,* Div. Note DN/G/797/04013. Nat. Inst. Agri. Eng., Silsoe, 23 pp.

Bailey, B.J. and Critten, D.L., 1985. The vertical south roof multispan greenhouse. *Acta Horti.,* 170: 381-386.

Bailey, B.J. and Hunter, A., 1988. Plant response and energy use in five high thermal resistance greenhouses. *Acta Horti.,* 229: 165–171.

Bailey, W.A., 1968. Fan and pad cooling of greenhouses. *Acta Horti.,* 6: 109–121.

Bellamy, L.A. and Ward, G.T., 1985. Optimisation of the design of a solar heated greenhouse. *Acta Horti.,* 170: 47–57.

Biddington, N.L., 1986. The effects of mechanically-induced stress in plants – a review. *Plant Growth Regulation,* 4: 33–41.

Bot, G.P.A., 1983. *Greenhouse climate: from physical processes to a dynamic model,* Ph.D. Dissertation. Landbouwhogeschool, Wageningen, Netherlands, 240 pp.

Bowman, G.E., 1970. The transmission of diffuse light by a sloping roof. *J. Agric. Eng. Res.,* 15: 100–105.

Businger, J.A., 1963. The greenhouse climate. In: W.R. Van Wijk (Editor), *Physics of the Plant Environment.* North-Holland, Amsterdam, pp. 277–318.

Carpenter, W.J. and Bark, L.D., 1967. Temperature patterns in greenhouse heating. *Florists' Review,* 139(3610): 43–45.

Cotter, D.J. and Seay, P.T., 1961. The effect of circulating air on the environment and tomato growth response in a plastic greenhouse. *Proc. Am. Soc. Horti. Sci.,* 77: 643-646.

Critten, D.L., 1983. A computer model to calculate the daily light integral and transmissivity of a greenhouse. *J. Agric. Eng. Res.,* 28: 61–76.

Critten, D.L., 1987. An improved theory for reflective losses from infinitely long greenhouses. *J. Agric. Eng. Res.,* 38: 301–311.

Critten, D.L., 1988a. A theoretical assessment of conventional symmetric roofed multispan E–W greenhouses compared with vertical south roofed greenhouses under natural irradiance conditions. *J. Agric. Eng. Res.,* 32: 173–183.

Critten, D.L., 1988b. Light enhancement using E–W aligned long prismatic arrays at high latitude. *Sol. Energy,* 41: 583–591.

Critten, D.L., 1988c. Direct sunlight losses in north–south aligned multispan greenhouses with symmetric roofs at UK latitudes. *J. Agric. Eng. Res.,* 40: 71–79.

Critten, D.L. and Legg, B.J., 1987. A general theory of the light transmittance of complex structures. *J. Agric. Eng. Res.,* 36: 125–140.

Critten, D.L. and Stammers, V.A., 1986. *The use of natural daylight variations to measure the light transmittance of harmonic components of daylight in greenhouses,* Div. Note DN 1350. Nat. Inst. Agri. Eng., Silsoe, 19 pp.

de Koning, A.N.M., 1988. Schermen in de Zomer. "Een procent minder licht kost één procent produktie" gaat niet altijd op. *Groenten en Fruit,* 8 April: 32–33.

Enoch, H.Z. and Kimball, B.A. (Editors), 1986. *Carbon Dioxide Enrichment of Greenhouse Crops.* CRC Press, Boca Raton, FL. Vol. I, 181 pp; Vol. II, 230 pp.

Frinking, H.D. and Scholte, B., 1983. Dissemination of mildew spores in a glasshouse. *Philos. Trans. R. Soc. London Ser. B,* 302: 575–582.

Grace, J., 1977. *Plant Response to Wind,* Experimental Botany Series. Academic Press, London, 204 pp.

Hanan, J.J., Holley, W.D. and Goldsberry, K.L., 1978. *Greenhouse Management.* Springer, New York, 530 pp.

Ho, L.C., 1988. Effects of high humidity and high salinity on calcium distribution in tomato plants. In: K.E. Cockshull (Editor), *The Effects of high Humidity on Plants Grown in Energy-saving Greenhouses.* European Communities, EUR 11261, Luxembourg, pp. 71–79.

Hoxey, R.P. and Richardson, G.M., 1989. Design of agricultural buildings to resist wind. In: V.A. Dodd and P.M. Grace (Editors) *Proc. 11th Int. Congress on Agricultural Engineering,* CIGR, Dublin, Ireland. A.A. Balkema, pp. 1165–1172.

Idso, S.B., Jackson, R.D., Ehrler, W.L. and Mitchell, S.T., 1969. A method for determination of infra red emittance of leaves. *Ecology,* 50: 899–902.

Jones, H.G., 1983. *Plants and Microclimate.* Cambridge University Press, Cambridge, 323 pp.

Jones, W.P., 1980. *Air Conditioning Applications and Design.* Edward Arnold, London, 494 pp.

Koths, J.S. and Bartok, J.W., 1985. Horizontal air flow. *The Connecticut Greenhouse Newsletter,* No. 125, February. Conn. Cooperative Extension Service, Storrs, CT.

Kozai, T., Goudriaan, J. and Kimmur, M., 1978. *Light Transmission and Photosynthesis in Greenhouses.* Pudoc, Wageningen, 99 pp.

Kozai, T., Sase, S. and Nara, M., 1980. A modelling approach to greenhouse ventilation control. *Acta Horti.,* 106: 125–136.

Kreith, F., 1958. *Principles of Heat Transfer.* International Text Book Company, Scranton, PA.

Kurata, K. and Tachibana, K., 1978. Greenhouse structure design with the optimisation technique. *Acta Horti.,* 87: 21–30.

Landgren, B., 1985. Aging test with covering materials for greenhouses. *Acta Horti.,* 270: 119–125.

Legg, B.J., 1989. Micrometeorological conditions in plant canopies under glasshouse and field conditions. *Aspects Appl. Biol.,* 21: 13–23.

Littlefair, P.J., 1983. Modelling real sky daylight availability with the BRE average sky. *Interior Environment and Lighting Design,* CVA 20th Session, D302/1–3.

Ludlow, M.M., 1980. Adaptive significance of stomatal responses to water stress. In: N.C. Turner and P.J. Kromer (Editors), *Adaptation of Plants to Water and High Temperature Stress.* Wiley, New York, pp. 123–128.

Mastalerz, J.W., 1977. *The Greenhouse Environment.* Wiley, New York, 629 pp.

McAdams, W.H., 1953. *Heat Transmission,* 3rd edition. McGraw-Hill, New York, 459 pp.

Monteith, J.L., 1973. *Principles of Environmental Physics,* Contemporary Biology Series. Edward Arnold, London, 241 pp.

Morris, L.G., 1963. Some recent advances in the control of plant environment in glasshouses. In: *Proc. Symp. on Engineering Aspects of Environment Control for Plant Growth.* CSIRO, Melbourne, Australia.

Morse, R.N. and Evans, L.T., 1962. Design and development of CERES – an Australian Phytotron. *J. Agric. Eng. Res.,* 7: 128–140.

Mutsaers, H.J.W., 1980. The effect of row orientation, date and latitude on light absorption by row crops. *J. Agri. Sci., Cambridge,* 95: 381–386.

Papadakis, G., Frangoudakis, A. and Kyritsis, S., 1992. Mixed, forced and free convection heat transfer at the greenhouse cover. *J. Agric. Eng. Res.,* 51: 191–205.

Parkhurst, D.F., Duncan, P.R., Gates, D.M. and Kreith, F., 1968. Wind tunnel modelling of convection from leaves. *Agric. Meteorol.,* 5: 33–47.

Roberts, W., Simpkins, J.C. and Mears, D., 1985. Performance of i r polythene over two years' trials. *Acta Horti.,* 170: 81–90.

Stanghellini, C., 1987. *Transpiration of greenhouse crops, an aid to climate management,* Ph.D. Dissertation. Agricultural University, Wageningen, Netherlands, 150 pp.

Stanhill, G. and Albers, J.S., 1974. Solar radiation and water loss from glasshouse roses. *J. Am. Horti. Soc.,* 99: 107–110.

Szeicz, G., 1974. Solar radiation in plant canopies. *J. Appl. Ecol.,* 11, 1117–56.

Thomas, R.B., 1978. The use of specularly reflecting back walls in greenhouses. *J. Agric. Eng. Res.,* 23: 85–97.
von Zabeltitz, Chr., 1986. *Gewächshäuser: Planung und Bau.* Eugen Ulmer Verlag, Stuttgart.
Waggoner, P.E., Moss, D.N. and Hesketh, J.D., 1963. Radiation in the plant environment and photosynthesis. *Agron. J.,* 55: 36–39.
Walker, J.N. and Duncan, G.A., 1974. Effectiveness of recommended greenhouse air circulation systems. *Trans. Am. Soc. Agri. Eng.,* 17: 371–374, 378.
Wolfe, J.S. and Cotton, R.F., 1973. Fan ventilation of greenhouses, Paper No. 5. In: *Proc. of the Subject Day on Greenhouse Engineering,* Report No. 9. Nat. Inst. Agri. Eng., Silsoe, UK.

Chapter 4

THE ROOT MEDIUM

S. DASBERG

INTRODUCTION

The main considerations in starting a greenhouse operation are marketing, climate and the requirements for climate control for the chosen crop under the local conditions. Traditionally, greenhouses were erected using the local soil, with little consideration of soil quality. For example, rose culture developed in the area of Aalsmeer, because of its proximity to the markets in the cities of the Netherlands. The local soil, a heavy sea clay, served as growth medium, in spite of the fact that it had very undesirable characteristics for intensive use. Gradually, methods were developed to improve the local soil by deep tillage, the introduction of drainage and by adding large amounts of organic matter. These practices have been accepted in greenhouse culture all over the world, since the high-value crops make it possible to invest not only in structures and climate control, but also in the improvement of the local soil. Parallel with this development, methods have been developed to grow plants in artificially prepared soil mixes, with addition of organic or mineral materials. A further development is the use of hydroponics with inert media which, besides the possibility of obtaining improved growth conditions, separates the root system from the local soil, and thus results in a greatly reduced infestation by soil-borne diseases. The use of rock-wool, a recently developed technique of soilless culture, has spread very fast during the 1980s. In the Netherlands more than 90% of the vegetables under glass are grown in rock-wool while the same trend occurs in flower culture. In this chapter, the important properties of the root media for greenhouse culture will be described.

GREENHOUSE SOILS

The traditional attitude of horticulturists has been that there is no substitute for a well-drained, fertile soil. Crops can be grown with less risk, and less technical problems with the supply of water, electrical power or fertilizers. With artificial media or soilless culture, there is much less margin for errors, since the water and nutrient reserves in these systems are much smaller than in the natural soil. They are characterized by a relatively low buffer capacity, and are more vulnerable to technical failures. On the other hand, the small root volume provides an opportunity for quick adjustments and a good crop management. The major soil properties for successful horticultural production are as follows.

Soil chemical properties

The colloidal soil particles [i.e. the clay fraction ($<2\,\mu m$) and the humus (decayed organic matter)] determine the cation exchange capacity. This is the storage for nutrients and other ions, determining the nutrient supply or soil fertility. The greater the cation exchange capacity, the greater the buffer capacity of the soil and the greater the margin for error in fertilization. Sandy soils are considered poor because of the small nutrient storage. The organic matter of the soil, which is a large pool for potential nitrogen supply to the soil solution, is largely determined by the climate. In hot and dry climates, the organic matter added to the soil decomposes fast, and the equilibrium organic-matter content is very low (0.1–1%). However, in temperate, humid climates the organic-matter content of the soil tends to be high (5–10%). In some areas greenhouses are built on soils with up to 40% organic matter.

Soils provide plants with nutrient elements (N, P, K, Ca, Mg, S and trace elements). Nitrogen is taken up by the plant roots mainly in the form of NO_3^-, and to some extent as NH_4^+, which are formed from the soil organic matter by microbial transformations or added as fertilizers. Large amounts of potassium are adsorbed on the soil depending on the type of clay mineral and the pH. Potassium interactions with other ions, such as Ca^{2+} and Mg^{2+} must be taken into account. Only a small part of the total phosphorus in most soils is in soluble or available form. Therefore, continuous growth of the root system into the soil provides the main mechanism of phosphorus uptake in natural soils. Yet, as soil can seldom provide intensively grown crops with an ample supply of nutrients, fertilization is essential, and large amounts of fertilizers are usually applied to greenhouse-grown crops. A more detailed treatment of plant nutrition can be found elsewhere in this volume (Chapters 5, 6, 7, 14).

Soil pH is another important chemical property, determining the availability of nutrients. The pH recommended for most greenhouse crops is 6.0–7.0; lower pH (more acid) causes aluminum and manganese toxicity, higher pH (more alkaline) causes precipitation of phosphorus and problems with micronutrient uptake. On soils high in organic matter, pH values of 5–6 are advised. Soil pH can be increased by liming, and can be decreased by proper composition of the nutrient solution. Changes in soil pH are very slow because of the large buffer capacity of natural soils.

Soil salinity affects plant growth by increasing the osmotic potential of the soil solution, causing plant water stress and decreasing water uptake. Specific ions may cause toxic effects. Soils containing soluble salts, giving an electrical conductivity above 3–4 $dS\,m^{-1}$ ($= mS\,cm^{-1}$) in the saturation extract are not suitable for greenhouse culture. For salt-sensitive plants, even lower threshold values apply. Excess salts should be leached from the soil prior to greenhouse construction. Problems of salinity are associated in many cases with the quality of irrigation water (see page 107).

Soil physical properties

The physical properties of the soil are of greatest importance for success in greenhouse culture. Soil texture – that is, the size distribution of the ultimate soil particles – is an important criterion. Soils with high clay content (containing more than 40% particles $< 2\,\mu m$ diameter) are more difficult to handle, tend to hold excess water and drain slowly, and are therefore less suitable for greenhouse culture. On the other hand, sandy soils (containing more than 90% particles > 0.2 mm) are less fertile, have a low water capacity and need better fertilization and frequent irrigation. The most important soil physical property is soil structure, this being the nature and stability of the bonds between soil particles, causing aggregation and large soil pores. Good and stable soil structure is enhanced by organic matter, which should be added to the soil if not present originally. Soil texture and structure determine the rate of water movement in the soil, the water holding capacity and the drainage. Soils which tend to drain slowly show aeration problems, since soil pores filled with water prevent oxygen diffusion and the uptake of oxygen by the plant roots, necessary for respiration and ion uptake. If a soil has severe drainage problems, artificial drainage (requiring the installation of drainage pipes at a depth of 60–100 cm) is necessary. One should bear in mind, however, that artificial drainage will remove excess water from the soil only when the water table is close to the soil surface, or when the pipes are installed in an impermeable soil horizon which causes a perched water table.

Comparing different soil types for greenhouse culture

Since crop yield is affected by different soil properties coupled with environmental conditions and management, it is not easy to compare crop response to different soil situations without taking the necessary measures to ensure a reliable trial. This may explain the scarcity of relevant data on the relationship between different soil conditions and crop yield. A review of several publications dealing with rose culture will be given in order to illustrate the importance of certain soil properties to the horticultural results. A survey, providing information on the relations between root development and soil profile properties, was carried out in 15 rose glasshouses in reclaimed land at Aalsmeer, the Netherlands (Arnold Bik et al., 1962). The soil consisted of a sandy clay subsoil, overlain by a 30–50 cm clay layer with a high organic-matter content (10–15%), mixed with additional organic material and covered with organic mulch. In most cases roots were very shallow and did not penetrate into the subsoil. This situation stemmed from a lack of a good subsoil structure, which resulted in excess water occurring due to over-irrigation and the absence of natural drainage. In cases of thorough mixing of part of the subsoil with the topsoil, and where drains were installed (including

Table 4.1
Flower production of Baccara roses on four soils[a]

Location	Type	Clay content (%)	Cation exchange capacity (mmol kg^{-1})	Porosity (%)	Soil air content (%)	Calcium carbonate content (%)	Annual dry matter production (kg m^{-2})	
							with peat	without peat
Netanya	Rhodoxeralfs	8	70	44	26	0.1	9.2	9.1
Gilat	Xerochrepts	16	130	46	9	18	10.3	9.8
Negba	Haploxeralfs	39	310	53	20	15	8.9	9.1
Yifat	Vertisols	69	560	55	15	5	9.0	8.3

[a] Based on Dasberg and Feigin (1978).

the removal of drainage water by pumping), deep rooting patterns were observed and better crop growth resulted.

A more detailed survey, including yield and quality of Baccara roses from 35–40 farms on several soil types (ranging from peat soils to heavy clay) was carried out during three years in the flower-growing region of Aalsmeer (van der Knaap, 1976). In spite of the large differences in soil properties (available soil water, drainage depth) no significant differences in the total yield and quality of flowers were observed. The peat soils yielded more flowers during the summer, but lower quality during the winter period. The mean groundwater level in the peat soils was much higher than in the clay soils (53 and 108 cm below soil level respectively). The annual income on all soil types was similar. A main outcome of this study was that proper management can result in good rose yields despite differences in soil properties.

Rose culture in Israel has undergone rapid development since 1970. In 1969 an experiment was initiated in order to compare the growth and yield of roses on four different soils under the same growing conditions during four years. Organic matter (10% by volume of *Sphagnum* peat) was added to half of all the soil beds. A summary of the main soil properties and of the flower yields is given in Table 4.1. The results of this experiment indicate that the soil had no overriding effect on rose production. High yields of flowers were obtained in all soils tested, despite large variations in clay and lime content and soil physical properties (Dasberg and Feigin, 1978).

It can be concluded that flower production can be successful under widely different soil conditions. Proper management seems to be more important than the soil per se. The advantages of natural soils over artificial media are the large volume of available water and the large reserves of nutrients, allowing some errors in fertilization and irrigation. Yet natural soils tend to be very variable, which makes it sometimes necessary to have management programs differing, even in different parts of the greenhouse. The degree of control by the grower over the soil conditions is limited, because of long-term processes which may bring about changes in soil properties as a result of intensive use, causing impaired water movement and gas exchange. Therefore, there has always been a tendency in horticulture to improve the soil by mulching in order to prevent soil crusting, and by mixing the soil with all kinds of materials in order to improve and stabilize its properties.

SOIL AMENDMENTS

Many types of amendment have been used to improve the physical properties of the local soil and in some cases also to improve its fertility. The choice of the proper amendment depends on the original soil properties, but the overriding factor is the local availability. Large volumes of amendments (at least 10–20% of the soil volume) have to be added to the soil in order to change its properties appreciably, and the cost of transport is usually a large part of the total expenses.

Manure, compost

Farmyard manure has been the traditional soil amendment in horticulture. This can be explained by its role as a source of plant nutrients, as well as its positive effects on soil structure. Increase in soil organic-matter content, improved aggregation and water infiltration, a decrease in bulk density and soil crusting following the addition of large amounts of manure to the soil have been reported (Meek et al., 1982; Sommerfeldt and Chang, 1985). The chemical composition of the manure varies, according to the type of livestock, its management and the degree of composting. The effect of manure on soil is not permanent, its decomposition is rapid and the organic-matter content of the soil tends to reach an equilibrium based on the climatic conditions. The short-term effects of manure addition are not always beneficial – the rapid increase of microbial populations may cause temporary immobilization of plant nutrients, especially nitrogen; however, the short-term effect on aggregation and soil structure may be beneficial (Tate, 1987). Manure should be properly prepared by adding straw, sometimes lime, and proper mixing and ageing. If not properly prepared, it may contain excess salts, weed seeds and even toxic materials. After proper preparation, manure loses its original properties and becomes compost. The material is humidified, friable, and without offensive odors, and is easy to mix with soil. Compost can be prepared not only from manure but also from plant residues, and chemical fertilizers are sometimes added in order to increase its effects on plant growth. Sometimes composts are used not only as soil amendments, but also as a medium for plant growth. Soil may be part of the compost, as in the case of John Innes compost, which consists of loamy soil, *Sphagnum* peat and coarse sand (Lawrence and Newell, 1952).

Peat

Peat has been used as a soil amendment in horticulture, as an important constituent of commercial composts, and as a medium for plant growth without admixture. The use of peat as a soil amendment developed in those areas where manure was less available. The advantages of peat over manure or compost were that the material was relatively stable (causing more permanent soil improvement), sterile, free from weed seeds, highly uniform and easy to mix with the soil. The recommended doses were 10–20% by volume, in order to affect the soil properties appreciably. The use of peat as a soil amendment has declined because of the high costs of transportation. Only high-quality *Sphagnum* peat should be used.

Sand, gravel

Sand and gravel, if locally available, are sometimes used to improve the physical properties of local soil. This practice should be used with caution. Under certain conditions, small amounts of sand added to a heavy soil may actually decrease the total soil porosity.

SUBSTRATES BASED ON MIXTURES OF SOIL AND ORGANIC MATERIALS

Natural soils are variable. Therefore the need was recognized for a uniform soil medium with consistent and stable properties, allowing the development of standard methods of fertilization and irrigation in intensive horticulture. Several representative mixtures based on soil, sand, peat and other materials are listed in Table 4.2. All of these mixtures (sometimes erroneously called composts) are based on peat as the organic constituent. *Sphagnum* high-moor peat has a very high porosity (> 90%) and water holding capacity (> 60%). It is a uniform, sterile material, as opposed to other organic amendments which have variable properties.

The main constituent of the John Innes composts was loamy soil of high fertility with high organic matter content (Lawrence and Newell, 1952). The University of California mixtures (Baker, 1957) contain fine, noncalcareous sand. The Cornell mixtures (see Hanan et al., 1978) are based on vermiculite and perlite as inert materials. Most of these mixtures were marketed with added fertilizer, notably superphosphate and lime. Many more mixtures are in use – "sphagnum", for instance, the Dutch substrates based on the local peat material (*Sphagnum* peat, frozen black high-moor peat). The advantages of these mixtures are that they have consistent physical and chemical properties, they are sterile and they can be stored indefinitely.

SOILLESS CULTURE

The elimination of soil-borne diseases was a major incentive for the development of soilless culture techniques (Baker, 1957; see also Chapter 2) and it is still an important reason for the use of such growing systems.

Table 4.2
Formulations of commercial "compost" mixtures (%)[a]

Ingredients	Loam (soil)	Sand	Peat	Vermiculite	Perlite
John Innes Seed	50	25	25		
John Innes Potting	60	25	15		
University of California, UCA		100	0		
University of California, UCB		75	25		
University of California, UCC		50	50		
University of California, UCD		25	75		
University of California, UCE			100		
Peatlite Cornell A			50	50	
Peatlite Cornell B			50		50
Peatlite Cornell C			50	25	25

[1] Based on Hanan et al. (1978).

Materials

The possibility to grow plants in nutrient solutions without soil has been known for many years and has been in use for research purposes, in order to determine the nutritional requirements of plants (Hewitt, 1966). Commercial development of soilless culture, sometimes called hydroponics, has been more recent. Pure solution culture on a commercial basis is not very widespread. Mostly some inert material is used as a substrate for the plant root system. This means that the substrate is chemically inert, has no base-exchange capacity and no organic matter to support microbial activity. Not all substrates used are completely chemically inert. Most materials are not physically inert; they have distinct moisture-retention curves depending on pore size distribution, which determine their suitability as growth substrates, as will be shown later. Some of the most common materials are as follows.

Gravel. In order to be chemically inert, the gravel should be noncalcareous. The water holding capacity is very low; the gravel serves mainly as physical support for plant roots, while the nutrient solution has to be circulated continuously, in order to provide ample amounts of water and nutrients to the plants. This is the classical hydroponic system, requiring pumps and periodic flooding (Steiner, 1977).

Sand should also be noncalcareous. Coarse sand of uniform particle diameter has a low porosity, but water is released from the pores very slowly when suction is increased. This means that appreciable depths of substrate must be used, otherwise the material will remain saturated and aeration will be poor. The irrigation and drainage of sandy material may be problematic and depends very much on particle size distribution.

Vermiculite is a secondary soil mineral, which in its natural condition tends to release magnesium and to absorb large amounts of potassium and ammonium. After high-temperature treatment it becomes chemically inert, but is not very stable physically. Its porosity and water holding capacity tend to decrease with time.

Perlite is a silica–alumina mineral treated at high temperatures to form a very light and porous material with very favorable water-retention properties, making it very suitable as a substrate for plant growth. Growing systems using perlite have been developed in Scotland (Hall et al., 1988) and in the United States (Hanan et al., 1978).

Rock-wool is a fibrous material manufactured from diabase and limestone by a special procedure at very high temperatures, resulting in a product with very high porosity (3–5% dry matter). The rock-wool slabs have a stable structure, are homogeneous, sterile and chemically inert. The use of rock-wool as a growth substrate has grown at a very fast rate during the last decade. More than 90% of the Dutch vegetable crop (2500 ha) is grown on rock-wool (Sonneveld, 1989) and the area of flower culture on rock-wool is expanding rapidly. Also in England 85% of the hydroponic area for tomatoes (*Lycopersicon esculentum*) and cucumbers (*Cucumis sativus*) is on rock-wool (Vaughan, 1989). Rock-wool slabs of 7.5–

10 cm thickness, 15–30 cm wide are used for most crops, resulting in 10–15 ℓ material per m^2. The young plants are propagated in rock-wool plugs and placed on top of the slabs. Water and nutrition is given by means of trickle irrigation, mostly one dripper per plant. Excess water is drained or can be reused (see below). Reuse of rock-wool slabs after steam sterilization is possible. Attention has been paid recently to developing suitable ways for the disposal of used rock-wool material.

Straw bales and peat bags. These substrates, which have a high biological activity and tend to decompose during one growing season, have been in use for cucumber and similar crops in Western Europe. This system was used mainly in glasshouses where the local soil had deteriorated, and before the advent of the inert materials described previously. These organic materials pose problems in fertilization and in disposal at the end of the growing season, and are no longer much used.

Tuff or scoria. This is porous volcanic material, formed during rapid cooling of lava. It has a high porosity and a low bulk density. Many variations in physical and chemical properties occur, depending on the location and conditions during formation. Generally, tuff has favorable physical properties, and is not completely inert chemically. Its cation exchange capacity varies according to the proportion of the clay fraction. Sometimes tuff material shows properties of variable charge, resulting in phosphate fixation. Tuff is used successfully where local deposits are available (Hanan et al., 1978; Feigin et al., 1980).

Nutrient-film technique. The ultimate soilless culture is without substrate, the root system grows directly in the nutrient solution. This so-called nutrient-film technique was developed on a commercial basis by Cooper (1976, 1979) and Graves (1983). With this system a very shallow stream of nutrient solution (nutrient film) is continuously recirculated past the roots of plants grown in long, narrow gullies made of polythene installed at a small slope (1–2%). The solution is pumped from a drainage trench and continuously recirculated. This continuous movement and the thin film of water covering the root system provides enough aeration. This system is very vulnerable to equipment failure, since it has virtually no buffer capacity, and to spread of disease. The irrigation water has to be of high quality.

Culture technique

Drainage

Adequate drainage is necessary in greenhouse crops grown in soil or in other substrates. Drainage can be natural or artificial. With natural drainage, excess water and nutrient percolate into the subsoil below the root zone without the need for interference. This happens in deep, permeable, light soils with no restricting layers. Eventually, the excess salts, nutrients and pesticides will reach the aquifer or surface water. Drainage water from soilless substrates (such as rock-wool) can also penetrate the greenhouse border soil and reach the deep groundwater or the surface water. Artificial drainage, or the installation of drains below the root zone in order to remove excess water, to facilitate leaching of excess salts and pesticides and to provide sufficient aeration in the root zone, is necessary under most greenhouse conditions. One has to bear in mind, however, that drains will only remove excess water when the soil is saturated – that is, when they are placed in a permeable layer. If the water table of the area is high (less than 1 m from the soil surface) excess water has to be removed from the drain outlets by pumping. With all drainage systems, whether natural or artificial, the subsoil or surface waters are eventually contaminated. Moreover, the use of nutrients and water by the plants is not very efficient, and part of the supplies are wasted.

A comparison of several growth substrates is given in Table 4.3, based on data from Chen et al. (1980). Terra rossa soil has a well-developed structure and relatively high porosity. At a suction of 1 kPa (equivalent to a depth of substrate of 10 cm), only 5% of the pores release water and contain air. The situation is even worse with sandy material – only 2% of the pores contain air at a suction of 1 kPa. This means that, with a sandy material, a depth of 50 cm is needed in order to provide proper aeration. Rock-wool, perlite and tuff (especially the coarser fractions) release appreciable amounts of water at a depth of 5 cm. These materials, with high total porosity, rapid drainage and high water retention in the 1–5 kPa suction range, are very suitable as growth substrates. It is possible (and necessary) to irrigate these materials frequently (several times daily) without problems of waterlogging and aeration. With soil or sandy materials high-frequency irrigation may pose aeration problems.

Recirculation

In many substrate cultures (e.g. the nutrient-film

Table 4.3
Physical properties of growth substrates[1]

Substrate	Specific surface area ($m^2 g^{-1}$)	Porosity (%)	Air content (%)		Available water (%)	
			0–0.5 kPa	0–1 kPa	1–5 kPa	5–10 kPa
Terra Rossa soil	284	56	3	5	7	22
Red Sandy soil	29	42	1	2	22	7
Coarse sand	5	32	1	2	26	3
Coarse Perlite (0.2–2 mm)	–	97	42	55	18	4
Fine Perlite (0–0.2 mm)	65	94	5	14	35	15
Red Tuff (0–4 mm)	–	–	5	22	15	2
Red Tuff (2–4 mm)	57	67	12	32	3	1
Rock-wool	–	98	7	58	36	2

[1] Based on Chen et al. (1980) and Sonneveld (1989).

technique), the nutrient solution is continuously recirculated, allowing for less waste. No system can apply the exact amounts of water and nutrients taken up by the plants. There is always some variability in water distribution and water use, since not all plants are of equal size and the greenhouse environment is never completely uniform. The necessary excess nutrient solution (about 20–30% with a rock-wool substrate) can be either recirculated or drained away. Recirculation causes less pollution and less waste of water and fertilizer, but adds the hazard of spread of plant diseases, since viruses and fungi are recirculated too. Methods are being developed to disinfect the recirculated solution (Runia et al., 1988). This subject is treated in greater detail in Chapter 14.

A variation of the recirculation system by pumping is the ebb-and-flood system, where the water table in the growth medium is periodically raised and lowered. This system is widely used in the growth of pot plants and in growing seedlings.

The velocity of circulation affects the level of nutrients necessary for obtaining good plant growth; the necessary nutrient influx is obtained at lower nutrient concentrations where circulation velocity increases (Asher and Edwards, 1983).

Water quality

The total salt content of the irrigation water, expressed as the electrical conductivity is an important criterion determining water quality. The main effect of salinity is osmotic – that is, the higher the concentration of salts, the higher the osmotic pressure of the solution, according to the approximate relationship

$$\text{osmotic pressure (MPa)} \approx 0.033 \times \text{electrical conductivity } (\text{dS m}^{-1}).$$

Distilled water has the highest quality, since only added nutrient (beneficial) salts are present and the osmotic pressure is minimal. The higher the osmotic pressure, the higher the water stress, and the water is usually less available to the plants. Not only is the total salinity important, but also the concentration of specific ions causing toxic effects. Moreover, the interactions among nutrient and other ions sometimes cause problems. The effects of salinity on crop yield and quality are not simple and straightforward, and interactions with other growing conditions occur (Sonneveld, 1988). There are differences in sensitivity of different crops, and even between cultivars. Flower crops are usually more sensitive than vegetables. Sensitivity of vegetables decreases in the order: beans (*Phaseolus vulgaris*), pepper (*Capsicum* spp.), cucumber, tomato, lettuce (*Lactuca sativa*). Crops are more sensitive when grown under dry conditions than under wet climatic conditions. Positive effects of salinity on fruit quality (tomatoes) have been observed (Sonneveld and Welles, 1988; Greenberger et al., 1996).

Nutrition

Ideally, nutrients should be supplied to the crop at the rate they are taken up, in order that no storage or excess should occur. This is only possible under two conditions: the exact rate of uptake is known; and no

Table 4.4
Composition of some representative nutrient solutions used in horticulture (mg kg^{-1})

Solution	Hoagland [1]	Long Ashton [1]	Nutrient-film [2]	Rock-wool [3]
Nitrate nitrogen	212	113	150–200	190
Ammonium nitrogen	11	57	0–20	18
Potassium	235	156	300–500	360
Phosphorus	62	41	50	48
Sulphur	32	48	50	60
Calcium	160	160	150–300	170
Magnesium	24	36	50	42
Iron	1.1	5.6	3	1
Manganese	0.1	0.55	1	0.5
Copper	0.03	0.06	0.1	0.05
Zinc	0.1	0.07	0.1	0.1
Boron	0.3	0.5	0.3–0.5	0.3
Molybdenum	0.05	0.05	0.05	0.05

[1] Hewitt (1966); [2] Graves (1983); [3] Sonneveld (1989).

interactions with the growth medium occur. The second condition is only met if the plants are grown in nutrient solution, or if the growth substrate is chemically inert. In the case of soil or organic material, cation and/or anion exchange may occur, resulting in ion interactions. The actual soil solution in contact with the plant roots will be different from the nutrient solution supplied with the irrigation water.

Unfortunately not enough data are available on the rates at which different crops take up different ions. A review on ion uptake rates of several horticultural crops has been published by van Goor et al. (1988). The rate of uptake changes according to the growth stage, the growing conditions (radiation, carbon dioxide concentration, irrigation) and the ion source. Theoretically, the composition of the nutrient solution should be changed continuously, in order to be adjusted to the plant uptake. In practice, this is not possible, especially with a chemically active growth medium. The nutrient concentrations can be adjusted approximately by monitoring the composition of the drainage solution. Approximate recipes are available for crops grown under different conditions. Details on plant nutrition for the rock-wool substrate are given by Sonneveld (1989).

Typical, traditional and recently suggested nutrient solutions are presented in Table 4.4.

CONCLUSIONS

The root medium traditionally should give support to the root system and to the whole plant, should supply nutrients and should store water for the transpirational demand. These three main tasks have become less important in modern greenhouse production with advances in technology.

Plants are no longer supported by their root system, but by trellises and wires shaping their development to the convenience of the grower. The uncontrolled supply of nutrients by soil or soil-mixes interferes more than helps in meeting the precise nutritional demands of the growing crop. The more chemically active the growth medium, the less control the grower has on plant nutrition. Modern irrigation systems (nutrient-film, drip, low-volume sprayers) make it possible to supply water according to the need of the plants without relying on the storage capacity of the soil. The trend in greenhouse culture is towards more inert and more uniform growth media, providing better control of growth conditions. Not only nutrition and irrigation are controlled better, but also climate (the possibility of heating or cooling the root system) and sanitary conditions (it is much easier to disinfect or sterilize an artificial growth medium than a large volume of soil). Also, reuse of water and fertilizer reduces pollution of the environment.

Computer-based control systems will make it possible to plan the production of greenhouse crops in great detail. For this purpose, the variability of the root medium has to be minimized, in order to obtain a uniform product of high quality.

REFERENCES

Arnold Bik, R., van der Kloes, L.J.J., van Marsbergen, W. and Schellekens, A.F.C.M., 1962. Rooting of roses under glass. *Meded. Dir. Tuinb.,* 25: 716–725.

Asher, C.J. and Edwards, D.G., 1983. Modern solution culture techniques. In: A. Lauchli and R.L. Bicliski (Editors), *Inorganic Plant Nutrition, Enclyclopedia of Plant Physiology.* Springer, Berlin, pp. 94–119.

Baker, K.F. (Editor), 1957. *The UC system for producing healthy container-grown plants,* Manual 23. University of California, Division of Agricultural Science, 32 pp.

Chen, Y., Banin, A. and Ataman, Y., 1980. Characterization of particles and pores, hydraulic properties and water–air ratios of artificial growth media and soil. In: *Proc. 5th Int. Congr. on Soilless Culture,* International Society for Soilless Culture ISOSC, Wageningen, Netherlands, pp. 63–82.

Cooper, A.J., 1976. Crop production with nutrient film technique. In: *Proc. 4th Int. Congr. on Soilless Culture,* pp. 121–136.

Cooper, A.J., 1979. *The ABC of NFT.* Grower Books, London, 181 pp.

Dasberg, S. and Feigin, A., 1978. The effects of irrigation, fertilization and organic matter on roses grown in four soils in the greenhouse. *Scientia Hort.,* 9: 181–188.

Feigin, A., Dasberg, S. and Singer, Z., 1980. Rose culture in Scoria. *Acta Hort.,* 99: 131–139.

Graves, C.J., 1983. The nutrient film technique. *Hort. Rev.,* 5: 1–44.

Greenberger, A., Silverman, D., Itshak, A. and Deco, Z., 1996. Improving the taste of tomatoes by salination. *Proc. 7th Int. Conf. on Water and Irrigation,* Agritech, Tel Aviv, Israel, pp. 146–152.

Hall, D.A., Hitchan, G.M. and Schmidt, R.A.K., 1988. Perlite culture: a new development in hydroponics. In: *Proc. 7th Int. Congr. on Soilless Culture,* International Society for Soilless Culture ISOSC, Wageningen, Netherlands, pp. 177–183.

Hanan, J.J., Holley, W.D. and Goldsberry, K.L., 1978. *Greenhouse Management.* Springer, Berlin, 530 pp.

Hewitt, E.J., 1966. *Sand and water culture methods used in the study of plant nutrition.* Commonwealth Agricultural Bureau, Fernham Royal, Bucks, UK.

Lawrence, W.J.C. and Newell, J., 1952. *Seed and Potting Composts.* Allen & Unwin, London, 176 pp.

Meek, B., Graham, L. and Donovan, J., 1982. Long-term effects of manure on soil nitrogen, phosphorus, potassium, sodium, organic matter and water infiltration. *Soil Sci. Soc. Am. J.,* 46: 1014–1019.

Runia, W.Th., van Os, E.A. and Bollen, G.J., 1988. Disinfection of drainwater form soilless cultures by heat treatment. *Neth. J. Agric. Sci.,* 36: 231–238.

Sommerfeldt, T.G. and Chang, C., 1985. Changes in soil properties under annual applications of feedlot manure and different tillage practices. *Soil Sci. Soc. Am. J.,* 49: 983–987.

Sonneveld, C., 1988. The salt tolerance of greenhouse crops. *Neth. J. Agric. Sci.,* 36: 63–73.

Sonneveld, C., 1989. Rock-wool as a substate in protected cultivation. *Chronica Hort.,* 29: 33–37.

Sonneveld, C. and Welles, G.W.H., 1988. Yield and quality of rock-wool-grown tomatoes as affected by variations in EC value and climatic conditions. *Plant Soil,* 111: 37–42.

Steiner, A.A., 1977. The development of soilless culture. In: *Proc. 4th Int. Congr. on Soilless Culture,* Las Palmas, pp. 21–37.

Tate III, R.L., 1987. *Soil Organic Matter, Biological and Ecological Effects.* Wiley, New York, 291 pp.

van der Knaap, W.C.A., 1976. *The suitability of peat and clay soils for the growth of Baccara roses in glasshouses.* STIBOKA, Wageningen, Report 1297. In Dutch.

van Goor, B.J., de Jager, A. and Voogt, W., 1988. Nutrient uptake by some horticultural crops during the growing period. In: *Proc. 7th Int. Congr. on Soilless Culture,* International Society for Soilless Culture ISOSC, Wageningen, Netherlands, pp. 163–175.

Vaughan, J., 1989. Comparing ways of growing hydroponics. *The Grower,* 112: 45–63.

Chapter 5

EDIBLE FOLIAGE AND FRUIT CROPS

J.H. GROENEWEGEN

INTRODUCTION

Greenhouses are often used to grow edible foliage and fruit crops during seasons when these crops cannot be grown locally under outdoor conditions, and also when high quality crops are wanted. The range of crops grown and the cultivation practices adopted is very large, the choice being determined mainly by the local climate and by the market requirements and returns.

In this chapter, the commercial practices recommended for growing edible crops in greenhouses in the Netherlands, one of the largest and most advanced production centers in the world, will be outlined for the following crops: lettuce (*Lactuca sativa*), endive (*Cichorium endive*), radish (*Raphanus sativus*), tomato (*Lycopersicon esculentum*), cucumber (*Cucumis sativus*), sweet pepper (*Capsicum frutescens*), melon (*Cucumis melo*) and eggplant (*Solanum melongena*).

These recommendations are largely based on the author's experience gained over nearly forty years as a research worker and extension officer concerned with vegetable production in greenhouses in the Netherlands. Use has also been made of a wide range of written material appearing in the annual reports of Naaldwijk Research Station for Horticulture under Glass [Proefstation voor de Tuinbouw onder Glas (BTG), but now Proefstation voor Bloemisterij en Glasgroente (PBG)], in the weekly journals for growers "Groenten en Fruit" and "De Tuinderij" and in specialized publications from Naaldwijk and other research institutes in the Netherlands.

EDIBLE FOLIAGE

Lettuce

Lettuces are grown during the winter in many countries in the temperate and subtropical zones. In the summer, lettuce is grown in greenhouses as well as in the open field. Lettuce consumption is increasing over the last few decades, stimulated by the ever-increasing variety of shapes and colors and by the general opinion that it is healthy to eat uncooked vegetables.

In temperate-zone greenhouses, lettuce crops are mainly grown during the winter in rotation with fruit crops grown during the summer. Although many cultivars and varieties of lettuce are grown, including crinkly (iceberg) lettuce, Roman (cos) lettuce, oakleaf lettuce, curled lettuce, lollo rosso, etc., the methods described in this section will be restricted to butterhead lettuce, probably the most widely grown variety.

Butterhead cultivars

Within the butterhead lettuce group there are many cultivars but each cultivar is only grown commercially during a certain time of the year – for instance, only in autumn or only in spring. Cultivars adapted to the day length in autumn, winter or early spring are not used in summertime, because premature bolting will occur under the influence of the longer days (shorter nights).

During the last 30 years, much work has been done to breed lettuce cultivars adapted to greenhouse conditions in the winter, and other cultivars which are suitable for the summer conditions. The breeding of cultivars that are able to develop a head under poor winter light conditions has made it possible to grow and harvest lettuce in northwestern Europe during the whole winter. A few decades ago "tipburn" (e.g., tissue damage at the leaf edge) was an important problem in midsummer. However, now cultivars have been developed which are highly resistant to this physiological disorder.

In northwestern Europe (e.g., the Netherlands) the shape and size of a lettuce head are important for its market value. From the growers' point of view, cultivars are preferred which have a slightly pointed bottom. The advantage of this form is that fungus infection on the lower older leaves is diminished. The cross-section of the butt which turns brown after harvest should be as small as possible. When fully grown, the head should have a given size and weight. This is possible but not easy to achieve in the middle of the winter. Planting distance is also of influence. Bigger heads obtain a higher price. Under limited light conditions lettuce can develop an open base. The color is also important, although the color variation of the marketed produce is limited, because growers do not like to market heads with a light color, thinking that this indicates an excessive sensitivity to handling, while consumers do not like a dark green color, thinking that this is associated with thick and less tender leaves.

A large amount of plant breeding work has been done and is still being done to introduce cultivars which are resistant to *Bremia lactucae*, a downy mildew of which a new race, to which current varieties are no longer resistant, appears nearly every year. Nevertheless, the possibilities of growing lettuce free of this disease have increased greatly. Greenhouse lettuces have also been bred with diminished sensitivity to manganese toxicity, which often occurs on clay and calcareous soils after they have been steam-sterilized. Currently, it is considered important to breed cultivars with a low nitrate content, as the existing cultivars tend to accumulate relatively high amounts of nitrate under poor light conditions. This is regarded by the health authorities to be a danger to the consumer.

Growing systems

The time needed for plant propagation (from sowing to planting-out) in heated greenhouses varies considerably in the Netherlands (52° northern latitude), and as shown in Fig. 5.1 extends from less than a fortnight in early autumn to two months in mid-winter.

The time from planting to harvest (Fig. 5.2) varies

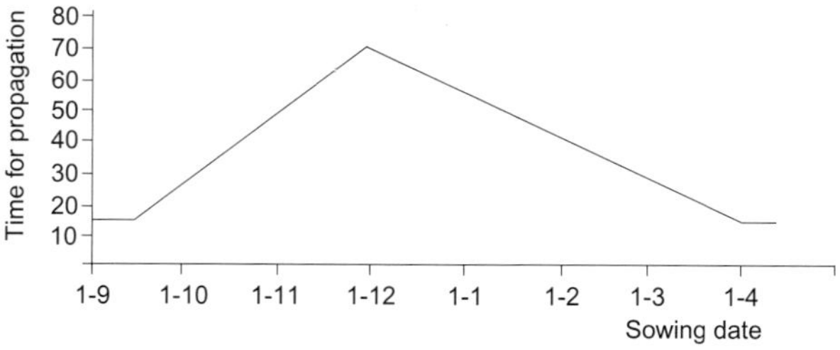

Fig. 5.1. Time needed for propagation of lettuce plants.

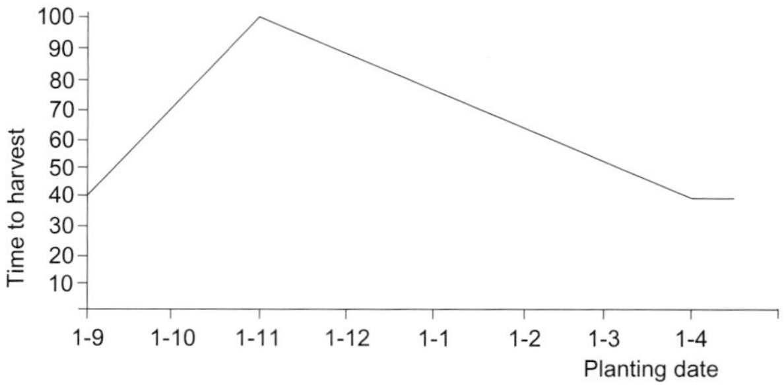

Fig. 5.2. Time needed between planting-out and harvesting.

from about five weeks to more than three months. The length of the growth season depends mainly on the light conditions. In practice, there is also a variation amongst growers because greenhouses are different, in particular with respect to light transmission. The grower can determine the temperature regime and decide to harvest the lettuces relatively early, when the heads are not fully developed or later when they have reached a larger weight. It is also possible to plant out bigger seedlings. This is often done in mid-winter (2nd or 3rd crop). In this case the time for plant propagation is longer and the time between planting out and harvesting shorter. In a heated greenhouse, under Dutch conditions, three or even four lettuce crops can be grown during the winter half-year:

Cropping calendar for 3 crops:
1st planting 15/9 – harvest 5/11;
2nd planting 8/11 – harvest 8/2;
3rd planting 11/2 – harvest 11/4.

Cropping calendar for 4 crops:
1st planting 1/9 – harvest 7/10;
2nd planting 10/10 – harvest 5/1;
3rd planting 7/1 – harvest 15/3;
4th planting 17/3 – harvest 27/4.

In mid-summer, it is possible, in a greenhouse, to raise a suitably-sized seedling in 11–12 days and to harvest one month after planting out.

The most difficult periods for growing lettuce are midwinter and mid-summer. Planting out just after the middle of October gives the longest growing time. However, when the growth period is long the plants are highly susceptible to fungal diseases such as *Botrytis, Pythium* etc. on the older leaves. In mid-summer, high temperatures often lead to a high proportion of the lettuce showing signs of tipburn and becoming infected with the fungus *Rhizoctonia.* Thus, year-round growing of lettuce is difficult and only takes place on a limited scale. Most growers use a certain period for disinfecting the soil, etc. so that seven crops a year are often the maximum. In soilless culture, growth and development is faster than in soil, and less time is needed for preparation between crops. Hence up to nine crops a year can be grown. Summer crops of greenhouse lettuce always compete with outdoor-grown lettuce crops and therefore usually fetch a lower price than winter-season crops. Many growers grow lettuce only once or twice in the winter season. With this schedule, the main crop can be grown during longer periods.

The time needed for the propagation of plants as well as the time needed between planting out and harvest is longer in northwestern Europe. In the south of Europe it is possible to grow a heavier head in the same period. This makes it difficult for northern greenhouse lettuce to compete, in the winter period, with lettuce from southern Europe. However, when the weather is unfavorable for lettuce production in southern regions, the northern greenhouse industry has an advantage.

Fig. 5.3. Pelleted seeds in soil blocks.

Seedling production

Seedling production mostly takes place in specialized nurseries which are highly mechanized and can deliver better and cheaper plants. Characteristically, a great number of lettuce seedlings are needed by each grower. Plants are nearly always raised in soil blocks.

The art of propagating lettuce plants has developed rapidly in recent years. Previously, seeds were sown on a seedbed of soil and the rooted plants later planted out. A first development was that the plants, after being sown on a seedbed, were pricked out in soil blocks, as soon as the cotyledons developed. Later, the seeds were directly sown in the soil blocks, a method which became the rule after the introduction of pelleted seeds, as a result of which sowing by machine became much easier (Fig. 5.3). At present, the soil blocks containing the seeds are directly placed in the same plastic boxes in which the seedlings are delivered to the growers.

Plants should be uniform and of the right size, free of diseases and sufficiently hardy. A first condition for uniform plants is good seed, which means seed with a high germination rate, a short germination period and minimal variation in size. New seed may still be in a state of dormancy, with consequent insufficient and irregular germination. This is likely to happen when the temperature is too high. A suitable temperature for germination is 12–15°C. After sowing, the surface of the soil blocks must be kept moist, and when there are extended periods with sunshine, the soil blocks should be protected from solar radiation by thermal insolation materials, for instance by covering them by plates of expanded polystyrene. The soil blocks are usually cones with a surface area of between 6 to 25 cm^2, the size depending on the desired plant size. The time needed to raise plants and the size of plants desired has been discussed previously. During propagation, disease control is important and where possible, steps are taken to exclude harmful insects and fungi. In foliage crops, which have a large surface area per unit weight, there is a risk of excessive chemical residues at harvest, and therefore chemicals should be used only when the plants are still small, preferably during propagation. It is also economical to apply such materials at this early stage when the plants are still close together. Hardening of the plants is necessary when planted out in the open field. Hardening can be achieved by increasing the spacing of the plants (size of soil blocks) and by lowering the temperature before transporting and planting out.

Preparations before planting

These depend on the growth medium. Soil is still by far the most common growth medium for lettuce production, soilless culture being presently practised on only a limited scale. Here, I shall concentrate on a lettuce crop grown in soil. The example concerns a typical growing system for a lettuce crop in the autumn and/or winter following a fruit-bearing summer crop.

The preparations for the lettuce crop consist of:

(1) cleaning and disinfecting the greenhouse;

(2) soil disinfection;
(3) leaching; and
(4) soil cultivation.

Cleaning and disinfecting the greenhouse usually is by spraying with formaldehyde or quaternary ammonium solutions. The spraying should be done directly after the fruit-bearing crop has been removed. In particular, the transparent greenhouse roof must be cleaned of dust, dirt and algae etc. to secure maximum light transmission.

Soil disinfection takes place primarily in preparation for growing fruit-bearing crops, but it can also be important for lettuce. The fungus *Olpidium brassicae* may be the vector of lettuce big-vein and ring necrosis virus. The soil must be dry before using steam sterilization. A disadvantage of this method is that, in particular on heavier and calcareous soils, manganese is made soluble. This results in plants suffering from manganese toxicity, in particular when the crop has been planted shortly after steaming and when sensitive varieties are used. Before a chemical treatment the soil must be cultivated to let the chemical penetrate into the soil. A certain period of time must be allocated for steaming as well as chemical treatment.

Soil with a high salt content requires leaching before seedlings are planted. Nutrients must be added to and mixed with the soil by tilling or rotary cultivation. The soil should have a good structure, and be neither too loose nor too compact. If the soil is too loose a small amount of water should be applied after the last tilling. The greater the water absorption capacity the looser the topsoil can be. Finally, the surface is smoothed by raking.

Planting-out and plant density

Nowadays only seedlings raised in soil blocks are used. These are often planted high in a shallow planting hole to prevent the lower leaves from touching the soil during the early stage of plant development. The workers who plant the seedlings often use a low platform on wheels (planting carriage) which marks the planting holes for 5 to 7 rows, and carries the boxes with seedlings and two people, who place the plants in the holes. The velocity of the platforms should be adjusted to the rate at which the people are planting.

Plants touching each other in the boxes may also stick to each other when taken out to be planted, thus damaging the leaves. This can be prevented by spraying a thin film of water over the plants. Directly after planting out, the plot should be irrigated in order to ensure a good contact between the soil blocks and the greenhouse soil. In case of intensive sunshine irrigation should be repeated frequently, in particular when the water absorption capacity of the greenhouse soil is limited. If the soil blocks dry out, this increases the incidence of *Pythium*. In order to keep the plants from touching the soil surface, plastic film with planting holes has been used experimentally. The experiments indicate that white plastic, which acts as a light reflector, results in yield increase. It also prevents the incidence of fungi diseases but practical application has been difficult.

Plant density (i.e., distance between plants) is important to production and quality. In the Netherlands, the usual practice is as follows:

Harvest date	Plant density (m^{-2})
September to mid-January	16–17
Mid-January to end February	18–20
March, April	20–24
Summer	20–22

Larger plant distances result in larger and heavier heads and in better-shaped lower leaves. With the more compact varieties, a smaller plant distance can be used.

At harvest all heads should have about the same weight. This may be achieved by using uniform plants, and by providing uniform growing conditions throughout the greenhouse.

Nutrition and water supply

The recommended nutrient content in the root zone of lettuce is given in Table 5.1. As can be seen, the recommended nutrient content in the root zone depends on the light conditions. For growing in water also, when light conditions are bad, higher concentrations are needed and will be recommended. Lettuce is very sensitive to a high content of Na^+ and Cl^- and to a high ionic concentration or electrical conductivity (EC). In case the root medium is too saline, it may be necessary to leach the soil. A high salt content promotes the incidence of tipburn, one of the most important physiological disorders in lettuce.

Highly concentrated chemical fertilizers are preferred as nutrients. They can either be spread over

Table 5.1
Lettuce: recommended nutrient content in the root zone

Nutrient	Units	Growing in soil [1]		Growing in water
		Good light conditions	Bad light conditions	
NH_4^+	mmol ℓ^{-1}	<0.2	0.2	<0.5
K	mmol ℓ^{-1}	2.0	2.5	6.0
Ca	mmol ℓ^{-1}	2.5	3.25	7.0
Mg	mmol ℓ^{-1}	1.0	1.0	1.5
NO_3^-	mmol ℓ^{-1}	2.0	3.0	19.0
SO_4^{2-}	mmol ℓ^{-1}	2.5	3.5	2.0
P	mmol ℓ^{-1}	0.15	0.15	1.0
Si	mmol ℓ^{-1}			0.5
Fe	μmmol ℓ^{-1}			40.0
Mn	μmmol ℓ^{-1}			1.0
Zn	μmmol ℓ^{-1}			5.0
B	μmmol ℓ^{-1}			50.0
Cu	μmmol ℓ^{-1}			1.0
EC	mS cm^{-1} at 25°C	1.2		2.5

[1] These concentrations are measured in an extract of 1 part of soil and 2 parts of water.

the soil and worked in, or applied in the water. After watering, the soil always requires tilling and leveling with a rake.

A good irrigation system provides a uniform water application. The amount of water that should be applied during the growing season depends on the potential evapotranspiration and on the amount of available water the soil can store. Sufficient water in the rooting zone and a quickly drying surface layer is ideal for lettuce crops. A wet surface layer creates a microclimate between the soil and the lower leaves that is favorable for fungal diseases. Therefore, irrigation should cease towards the final stage of crop development when the soil is completely covered by the crop canopy and the surface layer is drying slowly.

On less absorbent soils with low water-holding capacities, frequent watering may cause the nutrient concentration in the topsoil to become too low. In that case top dressings are needed – that is, nutrients applied to the soil surface, in particular nitrogen. During the darker period of the year relatively high nitrate concentrations occur in lettuce. Nitrate levels should be kept to a minimum, as high nitrate content of foliage crops is considered a health risk (see p. 112).

Climatological conditions

The light, temperature and carbon dioxide levels in the greenhouse have an important influence on the development of well-shaped lettuce heads. If the rate of transpiration is too low some physiological disorders such as glassiness (water in the intercellular spaces, causing dying-off of cells) occur, and if transpiration is too high, some types of tip-burn are frequent. In comparison with fruit-bearing crops, lettuce can be grown at a relatively low temperature, and therefore is often grown in partly heated greenhouses. During the day, the temperature may rise when there is more light, which in general means when the conditions for photosynthesis are better. Too high a temperature results in narrow leaves and looser heads. Temperatures that are too low result in slow growth and a longer growing period. The night temperature during the darker winter months should be kept at 5–6°C, and the day temperature during that time of the year at 10–12°C or higher when there is more solar radiation. Too high a temperature should be prevented primarily by ventilation. In summer it may be necessary to shade the greenhouse and/or to spray small amounts of water over the crop. In this, rather risky, treatment only the leaves exposed to sunshine should be moistened. Under winter conditions, heating and ventilating should be regulated to give the right rate of transpiration. When the air humidity is too high and/or when the crop is wet, glassiness appears. Glassiness often occurs when a sunny day is followed by a clear and cold night and by dull and dark weather the next day. The sunny day increases the soil temperature and root pressure, whereas the cold night lowers the plant temperature below that of the air, because of heat radiation from the lettuce to the cold greenhouse cover. The low plant temperature and the dull weather during daytime reduce the potential transpiration. Heating is the only measure at the grower's disposal to prevent this unfavorable situation. Occasionally the temperature is brought up to 16–17°C, a few hours after which the ventilators are opened. This causes a rapid temperature decrease within a few hours – a so-called "temperature shock". The use of alternating heating and ventilating in order to increase transpiration also diminishes the incidence of a number of fungal diseases. The use of energy screens, which reduce the light on the crop and promote a high humidity, is harmful, in particular when those screens cannot be opened at day time. It is important to mention that lettuce react very positively to carbon dioxide enrichment. In the Netherlands, lettuces are sometimes grown in unheated greenhouses with very limited possibilities for climate control. The nearer to midwinter this is done, the riskier this practice.

Quality, harvest, grading, packing and cooling

The quality of a head lettuce is good when it has reached the marketable weight and is sufficiently compact, when the base is closed, the butt surface small, the color medium green-yellow, and the nitrate content low, when the head is free inside and outside from chemical residues, free from crumbs of soil, fresh-looking and free from damage, disease and physiological disorders. Finally, in any one lot, the heads should not have large weight differences. The year-round average weight of a lettuce head when harvested is about 250 g, less in winter and more in spring.

There are several harvesting systems. The traditional but still widespread system is for the workers to wear waterproof trousers, bend down on their knees and cut the heads with a short-bladed knife. The older, yellowing leaves are removed, after which the head is placed in a plastic bag and put into a cardboard box which should be protected from dirt. As soon as a box is full it is closed, and when a number of boxes are ready they are collected and taken to a cool place, protected from sun. During harvesting the smaller heads are collected separately; grading is done by estimating size and hence weight. This, of course, requires some experience. Small differences in weight are unavoidable and allowed. The principle is that, during harvesting of lettuce, grading and packing are done simultaneously. This reduces labor and increases quality because each time the lettuce is handled a number of leaves will be damaged, in particular in winter-time, when the leaves are very fragile. For this reason systems using mechanical handling and a turning packing table have not been widely adopted, because each head has to be touched twice.

Where large quantities of lettuces have to be stored, for instance, in the Netherlands in the auctions hall, the lettuces are often vacuum-cooled to reduce post-harvest respiration and transpiration. In such coolers, air pressure is reduced from atmospheric pressure to 0.5% of atmospheric pressure. Consequently, the water in and on the lettuce evaporates, the energy needed for this being drawn from the lettuce. If 1% of the water in the crop evaporates the temperature decrease is 6°C. The time between harvesting and cooling should be as short as possible, and this is especially important when the temperature in the greenhouse is high. During the cooling process the temperature of the lettuce is brought down to 1°C or 2°C. This cooling is most important for the post-harvest quality of the lettuce when the temperature is high; even at temperatures of 7–8°C it still has an advantageous influence on post-harvest quality. The capacity of vacuum coolers is typically 5000 heads (Fig. 5.4). The total cooling process takes about 20 to 30 minutes. The advantage of using cooling are: good quality is maintained longer, visual appearance is improved, and it permits transport over longer distances, and harvest in the greenhouse during all hours of the day. The costs of cooling can be low when carried out on a large scale. Cooling is often done in the Netherlands by a growers' cooperative. Ideally, cooling should be followed by refrigerated transport.

Fig. 5.4. Vacuum cooling.

New developments in lettuce growing

A new development in lettuce growing is soilless culture including hydroponics – growth in nutrient solutions. In comparison to soil-cultured crops, a large investment is involved. Especially expensive are the many sloping gutters, either made of concrete or of plastic, and pumps and pipes for distributing the nutrient solutions.

In combination with hydroponics or soilless culture, some mechanization of the harvest is possible.

Though technically possible, soilless culture has not

become common. The main reason is that the many expensive installations only give a sufficient financial return when used intensively all the year round. However, year-round lettuce growing seems to be too risky for most lettuce-growers in the Netherlands.

Endive

Endive is grown on a limited scale in northwestern Europe, mainly in greenhouses and in the open field during the summer, and in the Mediterranean region mainly in the open field during the winter. In many respects, endive cultivation is similar to that for lettuce. However, the crop is less fragile, less susceptible to fungal diseases, more resistant to frost, more open-headed and in general easier to grow. A specific property of endive is its sensitivity to vernalization (low temperature influencing the early stages of development) resulting in early bolting – that is, development of reproductive organs. Even ripening seeds still on the mother plants can be vernalized. Germination of seeds and propagation of plants should always take place at high temperatures. Early bolting is also promoted by long days. Endive can be grown in greenhouses throughout the winter season. An autumn crop always grows fast, and there is no risk of early bolting. The growth can be so fast that the crop becomes weak and very sensitive to tip-burn. The recommended crop density is 16 plants per square meter. After planting in the autumn it is recommended to ventilate the greenhouse most of the time. For mid-winter harvest, a higher plant density of a few more plants per square meter is possible. Though endive has a moderate resistance to frost, it pays to protect the crop during the winter, primarily by heating and, when heating is not possible, by using a thermal screen for heat retention. The spring crop is the most sensitive to early bolting. The spring crops are sown from the beginning of December until the beginning of February. The optimum temperature regime involves a germination temperature of 25°C followed by 20–22°C by day and 16–17°C by night. Temperatures below 16°C should be avoided. Because of the high growth temperatures required, the use of heat retaining screens at night can be economical. The later the harvest takes place in spring, the heavier the heads that may be expected; but also the risk of bolting increases. Later in spring, irrigation should be increased since endive develops a bitter taste in response to a large water stress. This may well be the reason why endive is not grown under warm and dry climatic conditions. Endive is sold by weight. In winter endives yield 3–4 $kg\,m^{-2}$, but later in spring the yields increase to 6–8 $kg\,m^{-2}$.

Radish

Twenty-five years ago radish was only grown in greenhouses in the spring in the Netherlands and on a very limited scale. Since about 1970 radish growing has increased, because of the better cultivars available and increased mechanization for sowing, harvesting and washing. No doubt an increase in consumption also contributed to the more widespread greenhouse production of radish. Over recent decades, the growing season has extended gradually, and nowadays year-round production is possible. Radish growing has developed from a catch crop to a main crop on many holdings.

The light and temperature regime influences radish production and quality. Days with high light intensities result in a more rapid development of the edible roots. In the Netherlands it takes 10–12 weeks in the winter, and just over three weeks in the summer, to grow roots of a marketable size. The temperature in the greenhouse is usually higher when there is more light. A higher temperature at a certain light level results in relatively greater foliage development. Irrespective of the microclimate, a balance is needed between the development of roots and foliage.

Cultivars used in spring differ from those used in the autumn, because a spring cultivar grown in the autumn will develop too much foliage. A low tendency to sponginess, a physiological disorder in radish which often occurs in summer, is an important feature for successful radish cultivars. A red color is wanted, but not dark red. Red radish with a white tip to the root is grown on a limited scale. A high-round shape is ideal, while the small root at the base of the edible portion should be thin.

In greenhouses it is possible to grow about eight crops within one year. A year-round radish grower divides his house into several compartments and draws up a schedule of sowing and harvesting dates, taking into account that harvesting is labor-intensive. The period needed for soil disinfection is always planned for the summer. In the summer it is difficult for the greenhouse-grower to compete with outdoor-grown radishes. Therefore, some radish growers grow a crop of melons or sweet pepper in their greenhouses during the summer. In the Netherlands all combinations exist between year-round growing and growing only one radish crop a year.

Radishes are mainly grown on sandy soils. The more clay or organic matter a soil contains, the greater the risk of thick main roots and the longer the growing period. Harvesting is also easier on sandy soils. Also, these soils maintain their structure better despite pressure from the machinery used for sowing and/or harvesting. Before sowing the nitrogen content should be brought to 2–3 mmol ℓ^{-1} and the K content to 1.5–2 mmol ℓ^{-1}. Too low a nutrient concentration leads to lighter colored leaves which wilt soon after harvest, and shortens the shelf life. A low nutrient concentration also promotes sponginess.

Sowing is mostly done by a machine covering a strip 1 m wide, and sowing 12 rows (see Fig. 5.5). The distance between the seeds within the rows varies with the time of year from 4 to 5.5 cm. Thus, the number of seeds per square meter varies from 220 in the autumn (when the light/temperature ratio is the smallest) to 300 seeds in early spring. The germination temperature should be at least 10°C and preferably 14°C, in order to obtain high germination rate and percentage, both being important in order to get a uniform crop. This means that in winter heating of the greenhouse is necessary. It is also important to use large seeds. Smaller seeds below 2 mm in length take 10 days longer to develop from sowing to harvest than bigger ones of 3 mm. Nowadays growers improve the amount of

Fig. 5.5. Radish sowing by machine.

reflected light in winter by scattering granulated white polystyrene on the ground after sowing. This demands extra attention when harvesting, because the marketable radish should be free of this material.

Temperature control is the most important tool for regulating growth between sowing and harvesting. Growth will be too slow at winter temperatures lower than 6–7°C at night and 12°C by day. More light allows higher temperatures, but foliage development may be excessive; this may lead to greater incidence of sponginess, because water is withdrawn from the edible root by the transpiring leaves. This condition is often a problem in late spring and summer. In late autumn and winter it may be necessary to promote transpiration by heating and ventilating at the same time. This keeps the crop more or less dry, thus preventing downy mildew.

Radishes are usually harvested by specialized part-time workers and marketed in bunches of 20 roots. In one year, 75 bunches can be produced per square meter. After bunching, the radishes are washed and packed in one layer in wooden boxes with 36 bunches per box; the combination of fresh green and red looks very attractive. Radish is a perishable product, particularly because of rapid yellowing and rotting of the leaves; drying-out of the leaves should be prevented by covering the produce. Cooling shortly after harvest lengthens the shelf life.

Harvesting can also be by machine, followed by mechanical treatment to remove the leaves, washing the roots, grading and packing them in bags. This saves labor and the final product is also attractive.

FRUIT CROPS

Tomatoes

Tomatoes are grown in many parts of the world and in countries with very different climatological conditions. In countries with a relatively cool summer they are mostly grown in heated greenhouses. In countries with a Mediterranean climate tomatoes are grown in greenhouses in the cooler seasons and in the open field in summer-time. In tropical countries tomatoes are sometimes grown in pad-and-fan cooled greenhouses, but mostly they are grown in the open field. In the context of this chapter, I restrict myself to tomato crops grown in the temperate zone in greenhouses and plastic-covered tunnels, and shall generally refer to the agrotechnical practices adopted in the Netherlands.

Cultivars

The cultivars used in the Netherlands in recent decades can be subdivided in the following ways:

The stem: Self-topping (bush-type) and indeterminate growth (i.e., non-selftopping).

Shape of fruits: Round and large, round and small (cherry), cylindrical, chunky (so-called beef tomatoes which are mainly multilocular)

Color of the unripe fruits: Green, half-green, or pale around the stalk.

Color of the ripe fruits: Red, pink or yellow.

Resistance against diseases: To define a cultivar the following characteristics are used:

Tm, resistance against tomato mosaic virus;

C_5, resistance against 5 races of *Cladosporium*;

V, resistance against *Verticillium*;

F_2, resistance against 2 races of *Fusarium*;

Wi, resistance against white heads (a physiological disorder);

N, resistance against nematodes;

P, resistance against the fungus *Pyrenochaeta* (corky roots).

For example, with the above-mentioned characteristics, the variety "Marathon" will be called: Marathon TmC_5VF_2Wi.

All new cultivars for protected cultivation are first-generation hybrids (F_1) and therefore the price of the seeds is very high.

Crop growing systems

Under the moderate climatic conditions of north-western Europe a tomato crop is typically planted out in the heated greenhouses in the middle of the winter, and grows there until October. In unheated greenhouses planting before April is risky. At more northerly latitudes the winters are too dark and often too cold to grow tomatoes profitably in winter-time. Hence, at these latitudes, tomatoes can only be grown without heating during a short growing season in the summer. In the south of Europe, summer temperatures are too high for growing tomatoes in greenhouses unless evaporative cooling is used. In winter growing tomatoes is unheated greenhouses is possible as research carried out in Israel showed but in some Mediterranean areas night temperatures can be so low that stagnancy of fruitsetting appaers.

Tomatoes are sometimes grown in rotation with a foliage winter crop (mainly lettuces) or as an autumn crop after a first crop of cucumbers or tomatoes. The time between planting and removing a tomato crop varies from 4 to 11 months, depending mainly on the outside conditions and the degree of climate control in the greenhouse. When a crop is grown for 11 months the greenhouse should be higher and have better ventilation (larger openings) than those used for a shorter and smaller four–month crop. Keeping one crop of tomatoes nearly all the year round can only succeed in a high greenhouse, partly because of the space needed for the long-lasting crop and also because of the better ventilation which reduces high summer temperature. In lower greenhouses the best solution for year-round tomato cropping is by interplanting. In this case a first crop is maintained until 17–18 inflorescences (trusses) have been formed, and a second crop is planted about 5–6 weeks before the first crop is finished.

Seedling production (plant propagation)

Seedlings for planting out should be uniform in size, free of diseases, grown in pots or plugs in order to prevent stunted growth, and have a certain minimal size and firmness. In the last decade, the production of tomato seedlings has developed in the direction of less transplanting, less root damage, less stunting of root growth, more mechanization, better controlled environmental conditions and a transfer of plant-propagation activities from crop growers to specialist seedling nurseries. There are several systems for propagating tomato plants. For any system, the seeds should be uniform and of a good quality, and should be covered with 5 mm of soil and sufficiently moistened. This, in combination with a germination temperature of 25°C, will promote quick germination and uniform plants.

The traditional practice was to sow in boxes or in a seedbed in a mixture of potting compost. As soon as the cotyledons were developed, pricking-out in pots took place. Space on the seedbed should be wide enough to allow the seedlings to be kept until plant selection was possible. In practice this meant less than 1 g of seed per square meter of seedbed. Nowadays, it is more common to use trays with pyramid-shaped holes, filled with compost (when growing in soil) or granulated rock-wool (when growing in rock-wool). Sowing is done mechanically. After two or three weeks the seedlings and soil are lifted out of the holes and placed in pots. Pots can be made of soil (blocks), clay, polythene, or rock-wool. When growing in soil, *soil-blocks* are to be preferred, in particular if the potting compost is of high quality. The roots develop mainly on the outside of

Fig. 5.6. Cultivar and rootstock before grafting.

the soil block because of better aeration (see Fig. 5.6). There is a risk of too much rooting in the greenhouse soil, resulting in cracking of the roots when the plants are taken up. Plants in soil blocks can therefore be kept only during a limited time. Advantages of soil blocks are that they are easy to handle (less labor) and that they are cheap. They are widely accepted. In *clay pots* roots form against the inside of the pot. In older clay pots the plants sometimes grow less well, because of a high salt content in the pots. The porosity of the pot material causes lateral transport of moisture and salts. On the outside the water evaporates and the salts remain. To reduce the salt content older pots should be soaked in water before use.

Besides soil blocks and clay pots, *polythene pots* are widely used. They have several advantages. As a result of regular root development plants can be kept in the pot for a long period. This enables the growers to reach the desired balance between vegetative and generative growth. To obtain a well-developed plant, a pot containing 0.75 l of pot mixture is most suitable.

Rock-wool-plugs can, as far as the root development is concerned, be as good as soil blocks. However, the moisture-holding capacity of rock-wool is very limited so the water supply needs constant attention. Plant-propagation in rock-wool plugs mostly takes place on a concrete floor. The irrigation water always contains a certain level of chemical nutrition. Experienced growers are of the opinion that plants to be planted out in rock-wool should be propagated in rock-wool pots. The proper *size of the plants* depends on the season. The earlier in the season or, in general, the more expensive a tomato culture, the bigger the plants before they are placed in the final position and also the bigger the plug. The later in spring or early summer, the smaller the plants can be. Plants must not become too big to handle, and the bigger the plants the greater the risk of non-optimal growth. Plant size corresponds with plant weight. A plant of 20–25 g is a well-developed plant for early season. Only 16–18 of these plants per square meter can be kept during the last week of propagation. Spacing the plants should keep pace with leaf development.

Temperature after germination influences the position of the first truss and the sturdiness of the plants. A 24-hour mean temperature of 24°C during the first three weeks after sowing results in a first truss above the 10th or 11th leaf. This high position of the first truss is desirable in an early crop under limited light conditions. The greater number of leaves promote a good development of that truss. After three weeks the 24 h mean temperature may be lowered to 17–19°C. For plants intended for planting later in the season, this temperature may be used directly after germination. In this case, the first truss will appear above the 6th or 7th leaf. Day and night temperatures influence the sturdiness of the plants (see Fig. 5.8, p. 124).

Varieties or related species with resistance to soil-borne diseases can be used as a rootstock when a grower wants to make use of grafted plants. Because many growers sterilize their soils or grow in a sterile substrate, the use of grafted plants is exceptional. A good method of grafting is pricking out the rootstock and the scion plant together in one pot (Fig. 5.6). When they are ready for grafting an upward cut is made in the scion stem, while the top of the rootstock is removed by a diagonal cut; the rootstock is then placed into the scion stem. Then a little strip of lead is wrapped around to keep both together and to prevent the wounds from drying out. After ten days the two stems have grown together. There are then two possibilities: keeping both root systems, or cutting the scion stem just below the graft. When grafting to prevent *Fusarium* or *Verticillium* attack, the latter method is necessary; but when grafting against *Pyrenochaeta* or nematodes, diseases which do not grow into the stem, both root systems can be maintained (Fig. 5.7).

Fig. 5.7. Cultivar and rootstock after grafting.

Preparations before planting

In general it is necessary to disinfect and replace or clean a greenhouse roof thoroughly once a year to improve light transmission. Another treatment may be soil sterilization when tomatoes are grown almost year-round in soil. The type of soil-borne disease determines the sterilization needed, whether steaming or chemical disinfection, and the choice of chemical.

Sterilization requires time. When steam sterilizing is used, only a small area can be treated within a given time. After steaming, it is necessary to wait a few weeks until the normal microbial soil population is re-established. The steam is mostly produced in the heating boilers. Before chemical disinfection, the soil must be wetted and tilled. It takes some time before the roots which are left in the soil are sufficiently decomposed. Afterwards, a certain time is needed before the chemicals are decomposed. In many countries, only specially trained people may apply the chemicals. Most repairs and maintenance of the building should also be done when the greenhouse is empty. Finally, soil preparation, including nutrient application (base dressing) and tilling, is carried out. A few days before the plants are brought into the greenhouse, heating is begun, in order to reach the desired soil temperature and also to get rid of possible poisonous gases which may be present after painting of the heating pipes or the substructure with oil or paint.

When growing in rock-wool, the time between clearing an old crop and starting a new one is shorter than when growing in soil, as replacing the rock-wool slabs (new or steam-sterilized old ones) takes less time than soil preparations. This means that the growing season can be a few weeks longer. Before placing rock-wool slabs on the ground it may be necessary to improve the slope of the soil surface. This can be done by machine. It is often advantageous to cover the ground with white plastic in order to increase light reflection. Heating need only start a few hours before the plants are brought into the greenhouse.

Planting-out and plant density

At the beginning of the year the soil temperature may be too low, and in that case it should be brought up to 15°C before planting out. Early in the year, and in general under poor light conditions, plants should be well developed before they are planted out; ideally, the first truss should be visible or even flowering. There is a risk that less-developed plants may have a wrong top-to-root ratio. In order to acclimatize the potted plants they are sometimes placed for one or two weeks in the greenhouse before planting. Plants stand during that period in their pots beside their final planting position. Later in the year, and in general under brighter light conditions, smaller or less-developed plants may be planted out without this having a large effect on subsequent development and yield.

It is important to prevent root damage when planting out, and it is for this reason that plants should be raised in pots. Often it is recommended to add a limited amount of water just around the root system directly after planting. When growing in rock-wool, the plants are placed on top of the rock-wool slabs, which are wrapped in a white plastic sleeve. When the plants have developed sufficiently they are placed on the rock-wool in holes in the plastic sleeves and rooting into the slabs will follow.

Recommended plant density varies with the time of the year. The usual distance between the rows is 80 cm, and the distance within the row is 55–60 cm when planting in soil and 50 cm in rock-wool. This corresponds to a plant density of between 2 and 2.5 plants per square meter of ground area. When planting density is lower the early yield is a little less, but the total yield may be the same because of bigger fruit size. There is also a relation between plant density and post-harvest quality. In spring, a lower plant density results in fruits which keep somewhat longer after picking whereas in summer the reverse is true. The improved greenhouse climate in summer, due to a greater crop volume, is probably one of the causes. The incidence of russetting (hairline cracks in the fruit) increases when the planting density is lower. For early crops in northerly latitudes, white plastic is often used as a soil cover in order to improve light reflection. This results in slightly earlier and higher production. When growing in rock-wool the white plastic must be placed before bringing in the rock-wool slabs; when growing in soil it is placed after planting out. A disadvantage can be a slower rise of the soil temperature.

Plant nutrition: Base dressing

A certain level of nutrients in the root environment is required to keep a good balance between vegetative and reproductive growth, resulting in a good crop production and quality. The composition of the nutrients (base dressing) depends on the root medium – soil or artificial substrate (e.g., rock-wool). The recommended concentrations for the root environment are as shown in Table 5.2.

Table 5.2
Tomato: recommended nutrient content in the root zone

Nutrient	Units	Growing in rock-wool		Growing in soil [1]
		Good light conditions	Bad light conditions	
NH_4^+	mmol ℓ^{-1}	<0.5	<0.5	
K	mmol ℓ^{-1}	7.0	8.0	2.2
Ca	mmol ℓ^{-1}	7.0	10.0	2.5
Mg	mmol ℓ^{-1}	3.5	4.5	1.7
NO_3^-	mmol ℓ^{-1}	17.0	23.0	5.0
SO_4^{2-}	mmol ℓ^{-1}	5.0	6.75	2.5
P	mmol ℓ^{-1}	0.5	1.0	
pH		5.5	5.5	5–6.5
Fe	μmmol ℓ^{-1}	25.0	25.0	
Mn	μmmol ℓ^{-1}	5.0	5.0	
Zn	μmmol ℓ^{-1}	7.0	7.0	
B	μmmol ℓ^{-1}	50.0	50.0	
Cu	μmmol ℓ^{-1}	0.7	0.7	
EC	mS cm^{-1} at 25°C	2.7	3.7	1.4

[1] These concentrations are measured in an extract of 1 part of soil and 2 parts of water.

The electrical conductivity (EC) maintained indicates the nutrient concentration. In most situations the EC will be higher because the water used is seldom completely pure, but often contains a certain amount of Na^+, Cl^- and other ions. When growing tomatoes early in the year under poor light conditions the concentrations should be higher, as can be seen. It should be clear that these figures are not absolute, they are merely a guideline.

When soil analysis shows that the nutrient content is too low, the quantities applied as a base dressing are approximately as follows (assuming a root depth of 25 cm):

30 g m^{-2} N (10 g m^{-2} increases concentration by 1.6);
60–70 g m^{-2} K [600–700 kg ha^{-1}] (10 g m^{-2} increases concentration by 0.4);
10 g m^{-2} Mg [100 kg ha^{-1}] (10 g m^{-2} increases concentration by 1.2).

For phosphate it is not possible to give a figure. Soils in older greenhouses often have a certain reserve of phosphate, but in soils recently covered with greenhouses it may be necessary to supplement phosphate. It is important that the nutrients are well mixed in the entire topsoil.

When growing in an artificial substrate (soilless culture) the necessary concentration can be reached by irrigating the substrate with the recommended solution of fertilizer, which should include trace elements.

Additional nutrient supply: top dressing

Some weeks after planting the concentration of nutrients will have changed because of leaching (in particular N), adsorption to the soil complex, and uptake by the plant during growth.

Because of this, additional nutrient supply as top dressing is desirable. Factors of influence are:

The base dressing. If the base dressing appears insufficient the top dressing should start very soon after planting.

Leaf and fruit development. When the plants are young with mainly leaf development, mostly nitrogen is taken up, whereas during fruit development mostly potassium is taken up.

The quantity of water. Irrigation should provide a surplus of water in excess of the rate of evapotranspiration. Consequently, leaching, in particular of nitrogen, takes place during the growing season.

Soil characteristics. Each soil has its own properties, for instance, clay often binds potassium, whereas a high calcium content can make phosphate insoluble. The pH of the soil can be a criterion in choosing particular fertilizers which are adapted to this property.

Production level. Top dressing is more important when the growing season is long and production high.

The other growing conditions. Bad growing conditions, for instance as a consequence of soil-borne diseases, results in a limited uptake of nutrients and less need of top dressing.

The use of organic manure. The nutrients in organic manure are released gradually and become available for the plants during the decomposition process. The more organic manure used before planting, the less important the top dressing will be.

The end of the growing season. By the end of the growing season the nutrient levels may be lower than the values specified. One month before the end of a crop the application of top dressings can be stopped.

Occasionally nutrients and in particular trace elements, are applied as foliar feeds. Research has shown that in protected cultivation this method only affects production and quality when there is a deficit of certain nutrients in the root-zone. An exception must be made for magnesium, which may be available in the root-zone but sometimes is taken up insufficiently by the plants. In particular, this often occurs in autumn tomatoes. In that crop, spraying with magnesium sulphate is normal practice.

When growing in rock-wool one cannot solely depend on experience, and regular analysis of nutrient solution is essential because the level of nutrients and the pH can change very rapidly. The composition of the fertilizers applied should be monitored frequently.

Training, deshooting and deleafing

Soon after planting, side shoots (laterals) begin to develop in the leaf axils. It is common to remove them and only leave one main stem per plant. The main stem is twisted around a string. This is called "training", and is done at least once a week. Shoots developing at the fruit trusses are also removed. The supporting strings are fastened between the plants and steel wires stretched over the rows. Their length depends on the height of the wires, the system of training and the duration of the crop. In a long-lasting crop the stem length at the end of a season can be 10 m, so a string of at least that length is needed. In this case a winch can be used from which the string is unwound during the growing season. The winches are hooked onto the wire and moved horizontally when unwinding. The result is that the older deleafed part of the stem is gradually lowered into a horizontal position near soil level (layering system). The top of the plant is always kept vertical. This layering system is only possible in high greenhouses, otherwise the trusses of unripe fruits would touch the ground after lowering.

Another possibility is to allow the tops of the plants to hang downwards. When the tops have almost reached ground level a new string can be fixed and the stem can be trained upwards again.

In this system wide paths are needed to prevent the plants from closing the gap between rows. Another option is to slope the wires sideways.

A third possibility is to allow the tops of the plants to grow in a horizontal direction over the wires. An example of the latter is the so-called "Guernsey arch system". This system is very labor-intensive. The last two systems mentioned are mainly used in lower greenhouses.

It will be clear that the above-mentioned methods concern long-season crops. Short-period crops are always trained to grow vertically. In this case, the plants are usually stopped by removing the apex of the plants after the 8th or 10th cluster.

Deleafing is necessary because a tomato plant forms about three new leaves every week and this would result in too much foliage after a given time. Deleafing promotes air movement and the ripening of the fruits through heat radiation from the heating pipes onto the fruits. When deleafing takes place under low-light growing conditions, *Botrytis* will often appear on the wounds on the stem. Deleafing should be done by breaking the leaves upwards and more than three leaves should not be removed at a time. Sufficient leaves should be left on the plants to prevent the fruits from being exposed to direct sunlight, and to maintain sufficient transpiration to ensure an optimum humidity in the greenhouse.

The principle behind the choice of plant distance and system of training is to fill the greenhouse space with enough leaves in a relatively short time so that most of the solar radiation is intercepted. This results in a high level of photosynthesis, high productivity and high quality.

Climate control

Temperature. The grower needs to control the mean daily temperature, the night temperature and the day temperature. A relatively high mean temperature causes fast growth and development and early picking, but a lower total yield because of smaller fruits and sometimes a shorter season of production. The optimum mean temperature depends on many factors. In general, the better the conditions for photosynthesis are – light, carbon dioxide concentration – the higher the mean temperature should be maintained. When economics are considered the market price of the tomatoes and the fuel price also play a part.

A given mean temperature, such as 20°C, may be reached in several ways; for example:

(A) 24°C at day and 16°C at night;
(B) 20°C at day and 20°C at night; and
(C) 19°C at day and 21°C at night.

Plants exposed at the temperature regime as mentioned under A show spindly growth, whereas C results in more sturdy growth (Fig. 5.8). Temperature regime C is, however, in general neither practical nor economical because of the naturally higher temperature in the daytime. This indicates that when a plant shows a tendency towards too spindly growth the direction in which the day and night temperatures should be changed is known.

Large differences between day and night temperatures are not desirable because of the risk of condensation on plants and fruits in the early morning, when the air temperature is higher than that of the plants.

The greenhouse temperature can be controlled in various ways by a hot water pipe system, a hot air heating system, ventilation, evaporative cooling inside

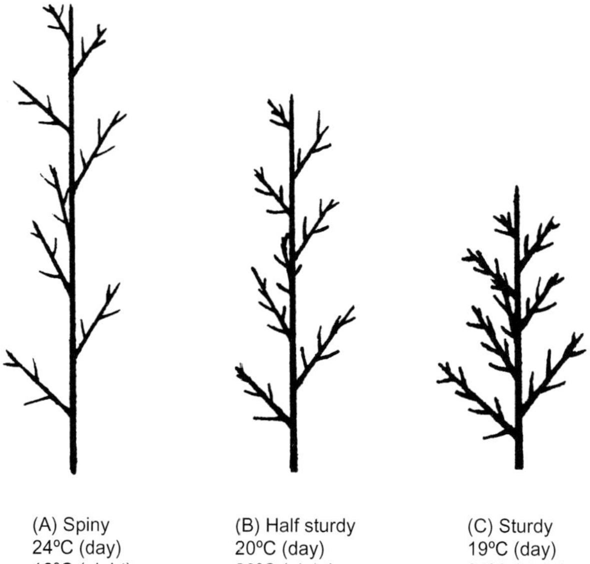

Fig. 5.8. Influence of day and night temperature.

the greenhouse, spraying water on the outside of the greenhouse, movable thermal screens, or whitewashing the glass.

A hot water pipe system is preferable to a hot air system. Advantages of a pipe system are safety, the absence of flue gases and the lesser horizontal difference in temperature.

When heating pipes are placed just above ground level there are the following additional advantages:

(1) a positive effect on the soil temperature early in the season;
(2) increased vertical air movement resulting in an even temperature;
(3) prevention of excessively high relative humidity and condensation in the lower part of the canopy; and
(4) the possibility of using the heating pipes as a rail system.

A certain minimum pipe temperature of 45–50°C is maintained when weather is dull. Depending on the light level this can be reduced to 30°C on a bright day. Not all of the heating pipes should be installed just above ground level. One or two pipes in the top of the house can, during cold outside conditions, prevent a current of cold air from the roof reaching the tops of the plants. Part of the energy from the heating pipes is emitted by radiation, which heats the plants, keeping them dry, promoting transpiration and advancing fruit ripening.

Ventilation is used to control temperature and humidity. The ventilating capacity (air exchange rate) of lower and older greenhouses is often insufficient. Simultaneously heating and ventilating is a very effective method of reducing humidity in greenhouse. The capacity of the ventilators, light intensity, the inside and the outside air temperature, the inside humidity, outside wind speed and direction, and rainfall are all factors which influence the ventilation rate. It is thus a very complex process best controlled by a computerized climate-regulation apparatus.

Evaporative cooling can be achieved with a pad-and-fan-cooling system or by fogging. It is often used in areas with hot summers where the outside humidity is sufficiently low. A disadvantage of a pad-and-fan cooling system is the difference in temperature and humidity between the pads and the fans. A fog installation has the disadvantage of wetting the plants near the mist nozzles. Nevertheless, evaporative cooling is essenial when growing greenhouse tomatoes in a hot climate.

Spraying water on the roof of a greenhouse is also effective when the relative humidity is low and evaporation high, and when the water is slightly acid, preventing calcium deposits forming on the greenhouse roof cover. Part of the water spray will penetrate the greenhouse via the ventilators and it is therefore not a method recommended for regular use, but it is useful for the hottest days or the hottest hours of a day.

Movable thermal screens (see Chapter 2, pp. 52–56) are used to avoid the effect of cold nights and to save fuel. These energy screens are not the most suitable method of improving the summer climate but in practice they are often used for that purpose, that is, to reduce the temperature during the hottest hours of a day. They can lower the air temperature by 2–6°C depending on the type of screen. More important, they reduce direct sun radiation on fruits and leaves, resulting in a lower surface temperature and less discoloration of the fruits. Because they also reduce the light level their effect on production will be slightly negative.

Whitewashing the greenhouses can prevent excessive solar radiation on the plants, but this method always reduces the light intensity, even during hours when the intensity is low. Nevertheless under certain conditions whitewashing can be a "necessary evil".

Carbon dioxide supply. The carbon dioxide content of the air has an important influence on the production and earliness. Carbon dioxide enrichment can be applied: to prevent too low a concentration (i.e. lower than 340 ppm) or to increase it to a higher value, such as 700–800 ppm, this being about the optimal level for photosynthesis in a greenhouse.

In a closed greenhouse with a high leaf area index and high solar radiation flux density, the carbon dioxide content can drop within a short time to values below 200 ppm. This diminishes the rate of photosynthesis by 50% compared to the normal rate at 340 ppm. Enrichment to 700–800 ppm can result in a increase in photosynthesis and productivity by more than 20%.

The question when to enrich, with closed or partly open ventilators, and to what level, is an economic one. The proper decision depends on the price of the available carbon dioxide, and the extra yield. A computer program can help to decide from moment to moment if enrichment is justified and if so, to what level.

In parts of Western Europe, for instance the Netherlands, growers can burn natural gas for heating, and part of the flue gas can be used for carbon dioxide enrichment. As soon as the heating plant operates carbon dioxide enrichment is possible, even when the vents are open and the carbon dioxide therefore has a short residence time in the greenhouse. When there is no need for heating but carbon dioxide is required, the heat is stored in a special tank to be used during the next night. Pure carbon dioxide in cylinders or tanks is much more expensive, and in general not economic.

Fruit setting

Proper fruit setting is an essential requirement for good fruit quantity and quality. In particular, the fruits of bilocular cultivars can be deformed as a result of insufficient setting. Non-viable flowers are formed under conditions when soil temperature is below 15°C, or when air temperature is above 30°C, or under low light and/or low carbon dioxide conditions or when the growth is unbalanced (e.g., too much vegetative development).

Well-developed flowers always produce fertile pollen but the pollen is not always easily released. This occurs, for example, when the relative humidity is too high, and in such cases heating and/or ventilation may be necessary to make the pollen drier. Release of the pollen and promotion of fruit-setting can be enhanced or promoted by vibrating the trusses or stems electrically or by hand. This procedure should be carried out 2 or 3 times a week around midday when the pollen is dry and therefore easily released.

A new method of promoting fruit setting is the introduction of bees into the greenhouse. Bumblebees (*Bombus terrestris*) can in some countries be bought commercially, and beehives placed in the greenhouses. The results are good, even in summertime when the temperature is high and the ventilators (windows) wide open. After a period of 5–8 weeks the bee population must be replaced. Both methods, vibrating and bees, are very costly, but in intensive greenhouse tomato culture one of the two methods is needed.

Sometimes hormones such as chlorophenoxyacetic acid are used to promote fruit setting, but when the conditions for fruit setting are good there is no need for their use. In cold greenhouses where the temperature can be too low early in the season for sufficient pollen production hormones can be used successfully. However, they can have a negative influence on quality, and can reduce production when plant growth is slow.

For beef tomatoes, pruning is done in such a way that only four flowers or fruits are left on each truss. Fruits from the tips of big trusses are always smaller and slower to develop than fruits close to the main stem. This treatment does not influence the total production because, if numbers are reduced, the remaining fruits tend to become bigger.

Harvesting, internal transport, grading, packing and cooling

Harvesting should preferably be done during the morning hours when the turgor pressure in the fruits is highest. Tomatoes should be harvested at least three times a week, and more often in hot weather. Usually the fruits are harvested when the red coloring has just started, but sometimes they are allowed to ripen on the plants during periods with low outside temperatures.

In some countries the fruits are picked with the calyx attached and in other countries without. The calyx can indicate how fresh the fruits are when sold, but unfortunately they also cause damage to adjacent fruits. New are the so-called "truss tomatoes" whereby the whole truss is harvested after all the fruits (5 or 6 per truss) have ripened. This is only possible with cultivars whose fruits have the property to stay firm a long time after coloring (ripening).

Ethrel [(2-chloroethyl) phosphonic acid] can be used to promote the ripening process of the fruits. For this treatment 1.5 ℓ of ethrel per hectare is needed. The easiest method is by applying a mixture of ethrel and yoghurt (1:3) or ethrel and starch on the stems; this admixture improves the spreading and adherence of the ethrel. When the temperature is sufficiently high the plants will turn yellow, and the fruits ripen after 3–4 days. Ethrel does not have a negative influence on quality when the fruits were fully grown at the time of treatment. Of course, ethrel treatment is only possible at the end of the season.

Quality and shelf life are influenced by damage caused by rough treatment during harvesting, transport and grading, for instance, by dropping the fruits into baskets or boxes, by overloading the containers for transport, by transport over bumpy roads and by letting the fruit drop during grading. Damage can also be caused when the grading machines are not cleaned, so that the surface is very rough. To minimize the damage, the picking baskets or boxes should be lined with soft material, such as foam plastic.

In order to prevent damage during internal transport in the greenhouses, in particular to the heavier beef tomatoes, the fruit may be transported in a gutter 40 cm wide through which water runs with a high specific gravity (indicated by an electrical conductivity of at least 5 mS cm^{-1}) to make the fruits float to the grading machine (Fig. 5.9). There they are taken out of the water and dried. The gutter should be covered to prevent dirt falling in and the development of algae. The desired density of the solution can be obtained by dissolving potassium fertilizer (not nitrogen). A pH higher than 6 promotes the growth of algae, but at a pH below 5 more fruits will crack. The water must be kept free from organic material by filtering and/or by replacement in order to prevent the development of fungi on the fruits a few days later. Also in relation to quality, it is recommended that the fruits be not kept in the water longer than one hour. Bilocular cultivars or fruits with russetting or dry cracks should not be transported this way, because then their shelf life will be reduced dramatically.

Fig. 5.9. Gutter for transport.

Modern grading machines grade according to color and size. After grading, high-quality tomatoes should be packed in attractive boxes, usually made of wood or cardboard and nonreturnable. An attractive option is to place one layer of beef tomatoes in a box lined with paper.

Cold storage of tomatoes can slow down the ripening process. Unripe (green or half-green) tomatoes, however, should never be stored below 9°C, because at that temperature the ripening process stops completely and will not begin again even if the tomato is stored at higher temperatures later on.

Quality of the fruits

The quality of tomato fruits is determined by their flavor, shape, coloring, firmness, shelf life, and, of course, also by the absence of damage by viruses, fungi and insects. Fruit with visible disease damage is of lower quality. Good growing conditions are important for good quality. The influence of plant distance, nutrition, deleafing, fruit setting and truss pruning on quality have been described. Additional reasons for low quality are:

Hollow and angular fruits appear mostly when the growth is too strongly vegetative, often as a consequence of the nutrient concentration or the temperature regime being too low.

Cracked fruits and russetting can occur when the skin is inelastic. This occurs when the fruits are exposed to direct sunshine, when the fruits are wet in the morning, and when the growth of the fruits is slow. This last often happens to the youngest fruits at the end of the truss.

Blossom-end rot is caused by a shortage of calcium in the fruits or, indirectly, by the electrical conductivity in the root zone being too high, and also under extreme high transpiration conditions when the movement of water and calcium is mainly to the leaves.

Blotchy ripening is caused by a low nutrient concentration in the root zone.

Yellow/green shoulders are the result of too much direct sun radiation on the fruits, causing fruit temperatures of over 30°C.

Gold specks consist of calcium oxalate crystals. Fruits with specks have a shorter shelf life. These crystals appear mainly in unheated greenhouses, where transpiration is limited during the night and root pressure is high.

The flavor of tomato fruits is influenced by the cultivar, by the conductivity in the root zone, and the content in the fruit of acid and soluble solids. High conductivity in the root zone, and ripening of the fruits on the plants, seem to have a positive influence on fruit flavor.

Cucumber

Cucumbers probably originated in northern India; they have been grown in Egypt for 5000 years, were introduced to China around 86–40 BC, and were known in France in the 9th century. Nowadays cucumbers are grown throughout the world, either in the open field or as a protected crop in heated or unheated plant enclosures.

Types and cultivars

Many types and cultivars of cucumbers are grown. The cucumber cultivars used for pickling are mainly grown in the open, whereas salad cucumbers are mostly grown as a protected crop. In this chapter only salad cucumbers will be discussed.

Fruits may be distinguished according to their color: green, yellow and white (salad) cucumbers. The last two types can tolerate lower temperatures than the green types, and their fruits are somewhat drier. Many decades ago these were grown on a limited scale in cold frames, but they are not marketed nowadays. Green salad cucumbers are generally grown in greenhouses.

According to length, the green types can be divided into the "mini" type (short), the "midi" type (middle) and the "long" type. The mini types, about 15–18 cm long, are mainly grown in the subtropical zone, and are common in all of the Arab countries. These fruits are sold by weight. The midi type is grown on a very limited scale for special markets. The long type are about 25–40 cm long and are usually sold by the piece, and are the most commonly cultivated cucumbers in Northwestern Europe. It is this type whose cultivation will be described in the following section.

Most of the long green cultivars are hybrids, have only female flowers, and therefore bear parthenocarpic fruits. These female varieties are very productive. When male flowers are needed for seed production, plants are treated with the plant hormone gibberellic acid. The varieties (hybrids) used nowadays are resistant to the fungus *Cladosporium cucumerinum,* and do not produce fruits with a bitter taste. A few newly-bred cultivars are also resistant to powdery mildew (*Sphaerotheca fuliginea*). A good-quality fruit has a dark green color, which it keeps several days after picking, and is straight with a well-proportioned shape.

Seedling production

Cucumbers need a relatively high temperature and good aeration around their roots, not only after being sown but also during germination. This is achieved in the following way: Sowing takes place in special galvanized trays (Fig. 5.10). A set of trays consists of a galvanized iron outer tray in which water stands 2 cm deep, and an inner tray, also galvanized, with many small holes in the bottom and partly filled with wet coarse sand or perlite. When in place, there is a 3 cm

Fig. 5.10. Set of trays with cucumber seedlings.

space between the bottoms of the two trays. The seeds are covered with 1 cm of the same material. The water in the bottom tray is kept at 27°C by an electric heating plate controlled by a thermostat. The top tray is covered with glass in such a way that a little ventilation occurs. The purpose of this sheet of glass is to prevent the sand or perlite from drying out.

The water in the bottom tray rises by capillary forces into the top tray and evaporates slowly thereby keeping the sand or perlite in the top tray moist. A thermal screen above the trays is required to prevent direct sunshine from heating the trays, and the trays from losing heat to the greenhouse roof when the outside temperature is low. Sudden temperature changes promote the incidence of *Pythium*, a fungus to which cucumber seedlings are very susceptible. As soon as the seedlings appear, ventilation is increased by lifting the sheet of glass a little more. After three days this glass can be removed, and four days after sowing pricking out can begin. This is done into pots or on a bed with potting compost maintained at a temperature of at least 21°C. When the seedlings are pricked out, the soil should not be pressed around the roots. In soil blocks the central hole is not filled. It is sufficient to press the roots against the inside wall of the hole, as aeration is important.

In the Netherlands the use of artificial light in the winter produces a sturdier plant. Because only a minority of growers wish to specialize in the production of young cucumber plants, because of their need for special equipment and frequent supervision of the propagation process; most growers leave the raising of young plants to specialized plant propagators.

Fig. 5.11. Cucumber grafting on *Cucurbita ficifolia.*

In order to prevent certain soil-borne diseases, plants may be grafted onto a resistant rootstock.

As many growers now sterilize their soils and/or grow in sterile substrates, soil-borne diseases are no longer an important problem. For this reason, the use of grafted plants is limited to smaller unheated greenhouses where soil disinfection is difficult. In such cases cucumbers are grafted on the pumpkin *Cucurbita ficifolia* used as a rootstock. *Cucurbita ficifolia* is resistant against *Fusarium oxysporum* and less susceptible to *Phomopsis,* and is able to grow at a relatively low soil temperature.

Grafting can be done by several methods. A common method is to use a 15-day-old cucumber plant with one normal leaf 3–5 cm in diameter and a rootstock of the same size, about 9 days old. These are fixed together (see Fig. 5.11), wrapped with a strip of thin lead, planted in pots, and kept for a few days at a high relative humidity by covering the plants with a thin plastic sheet. The rootstock, having a faster growth rate than the cucumber scion, is allowed to retain only one or two leaves.

Twelve to fourteen days after the grafting, when the graft and rootstock have merged, the stem of the cucumber can be cut off. A few days later the plants are ready for planting out (see Fig. 5.12).

Fig. 5.12. Graft and rootstock grown together.

Grafted plants as well as ungrafted ones must be supported by a stick and a ring to keep the main stem straight. Cucumber plants have prickly hairs, which can damage adjacent plants when they are being moved. The larger the plants the greater this risk. When cucumber plants are transported, one should also keep in mind that the plants are very sensitive to low temperatures and wind. Cucumbers need a relatively high temperature, in particular during propagation. The time needed to produce a fairly well-developed plant varies between 3 and 5 weeks, depending on the time of the year and amount of artificial illumination. The propagation time for grafted plants is at least one week longer.

Preparations before planting out

Proper management of greenhouse cultivation aims at the longest possible crop growing season and the shortest possible time for repairs, maintenance and preparations for the next crop. The water-supply system should be disinfected, and the greenhouse roof and walls cleaned. Disinfection of the soil is needed before the new crop is planted in it. Most growers in the Netherlands who grow in soil choose steam sterilization. This must be done thoroughly, and deep enough to prevent re-infection from the subsoil. Disinfection also kills the natural enemies of parasites. Neglect may result in an explosive growth of the parasites in question.

To promote root activity, the temperature in the root zone should be maintained above 21°C and preferably around 24–26°C. This means that in winter a soil heating system is needed. The traditional way is to make hotbeds with organic matter, for instance, 7–8 kg of straw per square meter in furrows, wetted and supplied with nitrogen or stable manure in order to reach at C:N ratio of about 35:1. This promotes bacterial activity and starts the fermenting process. The organic material should be covered with a thin layer of 10–15 cm of soil. A greater depth prevents sufficient aeration, which is needed for the proper aerobic decomposition of the organic matter by bacteria.

The rate of decomposition also depends on the greenhouse temperature. It may be necessary to bring the heating system into operation. After one week the soil temperature is usually high enough to plant out. Disadvantages of using hot beds are the limited control of the soil temperature and the amount of labor and time involved. Advantages are the positive effect on the soil structure and the production of carbon dioxide. Nowadays soil heating often takes place by a special system consisting of subsoil tubes filled with circulating water at 30–35°C under each row. This soil heating system is often supplied with water from the flue-gas condenser of the heating plant. Nowadays cucumbers are often grown in rock-wool. In such cases the heating tubes are situated just underneath the slabs. When rock-wool is used, the white plastic reflectant sheets are put in place before the rock-wool slabs.

Advantages of growing in an artificial substrate are that the time needed for preparation is shorter than in soil and that production is higher. This way of growing is now common practice.

Planting out

There is a choice between "spot sowing", and using propagated seedlings as described previously, the decision depending on the specific situation. Where growth is slow and the cost of fuel and equipment is high, large seedlings are used in order to make the initial period till first harvest as short as possible. This is usually done in hothouses in moderate climates. In Mediterranean climates spot sowing is common practice, particularly as cucumber is a fast-growing

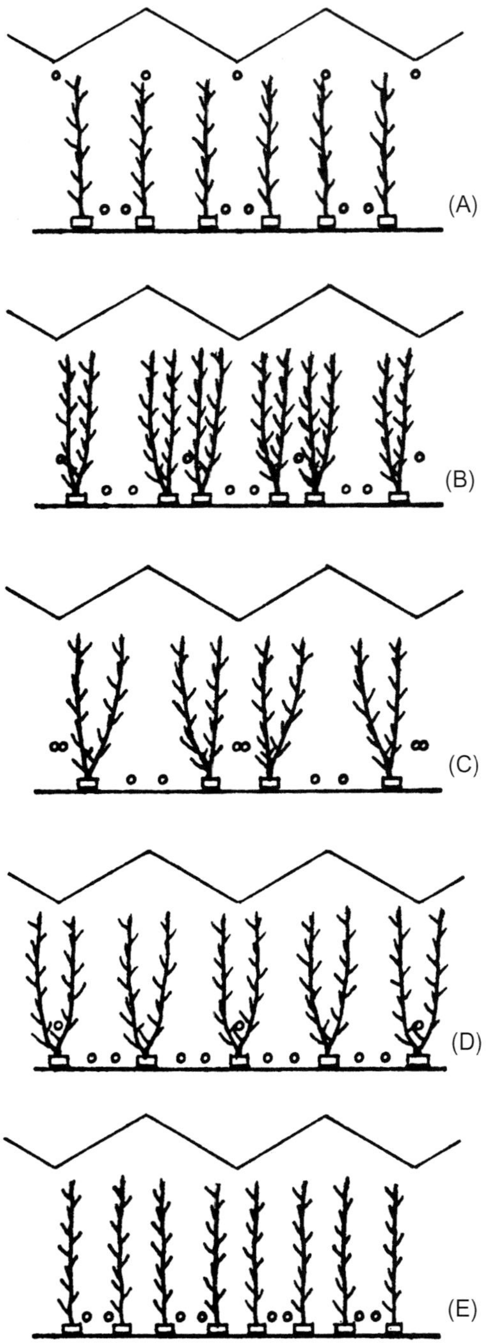

Fig. 5.13. Planting systems for cucumbers: (A) 6 rows per 6.40 m; (B) 6 rows in V system per 6.40 m; (C) 2 rows in V system per 3.20 m; (D) 4 rows in V system per 6.40 m; (E) 4 rows per 3.20 m.

crop. This saves labor in transplanting, and also the subsequent check to plant growth is prevented. The root system tends to develop deeper from spot-sown plants, resulting in faster growth later on.

The varieties used now are very fertile, and the fruits develop parthenocarpically easily and without problems of flower quality, pollination, unbalanced growth etc., which are typical for tomatoes. The cucumber never grows too fast in relation to fruit setting, which is why it is possible to plant out relatively young plants and to allow the plants to grow quickly directly after planting out.

Plant density considerations

The number of plants per unit area – the plant density – determines the production time and yield. It is also important to provide each plant with an equal area, and to provide enough room for the greenhouse workers to avoid damaging the plants.

Planting systems are chosen to fit the greenhouse dimensions, taking into account the width of the bays, the position of the heating pipes and the substrate used. Five examples of planting systems are presented in Fig. 5.13. These are:

(A) Plants are not equally spaced in available space. Crop can be treated very easily. 1–1.1 plants per m^2.
(B) Plants are very well spaced. Crop can be easily treated. Highest leaf area index. 1.5 plants per m^2. High production capacity.
(C) An old system used in older greenhouse types. 1.3 plants per m^2. Suitable when growing in hot beds.
(D) Position of heating pipes is such that changing to tomato or pepper growing is easily possible. 1.2 plants per m^2.
(E) In the Netherlands the Venlo type of greenhouse is commonly used, which has a width of 3.2 m. To accommodate four rows per 3.20 m the distance within the rows is relatively wide, usually 80–85 cm. The crop density is 1.5 plants per m^2. This system is used when a second (autumn) cucumber crop is grown after a first crop of tomatoes, in particular when tomatoes are grown on four rows of rock-wool per 3.20 m.

Base dressing and top dressing

For cucumbers the desired nutrient concentrations in the root zone are as shown in Table 5.3. The nutrient level for growing in soil is lower than for tomatoes (Table 5.2), resulting in a lower conductivity. The ion concentrations for growing in rock-wool differ little from those for tomatoes. An important difference is that, under bad light conditions, high concentrations should be avoided. High conductivity values for a few

Table 5.3
Cucumber: recommended nutrient content in the root zone

Nutrient	Units	Growing in rock-wool	Growing in soil[1]
NH_4^+	mmol ℓ^{-1}	0.5	
K	mmol ℓ^{-1}	8.0	1.8
Ca	mmol ℓ^{-1}	6.5	2.2
Mg	mmol ℓ^{-1}	3.0	1.2
NO_3^-	mmol ℓ^{-1}	18.0	4.0
SO_4^{2-}	mmol ℓ^{-1}	3.5	1.5
P	mmol ℓ^{-1}	0.9	0.15
Si	mmol ℓ^{-1}	0.6	
Fe	μmmol ℓ^{-1}	15.0	
Mn	μmmol ℓ^{-1}	7.0	
Zn	μmmol ℓ^{-1}	7.0	
B	μmmol ℓ^{-1}	50.0	
Cu	μmmol ℓ^{-1}	1.5	
EC	mS cm^{-1} at 25°C	3.0	1.0
pH		5.5	5–7

[1] These concentrations are measured in an extract of 1 part of soil and 2 parts of water.

weeks do not have much effect on production, but when maintained for a longer time an important decrease of the total yield occurs. Higher conductivity leads to an increased incidence of calcium deficiency and chlorosis in the leaves. Cucumbers can be grown in a wide range of cation ratios and there is a positive relationship between the nutrient concentration and shelf life, in particular with respect to the color during storage.

In connection with top dressing, one must keep in mind that, shortly after planting out, and before the period in which there is rapid fruit development, the increase of leaf area can be very fast resulting in a rapid drop in the nitrogen content. Furthermore, in periods with heavy fruit load, a very rapid decrease of potassium can occur. Because of the different uptake rates at the different stages of plant development, top dressing requires much attention. Control is more difficult in crops growing on hot beds which release a certain amount of nutrients during the decomposition process.

Water supply

Cucumbers are more sensitive than other crops to a regular and sufficient water supply. An irregular water supply has negative effects on growth, production and quality, and is likely to increase incidence of *Pythium*, not only during the germination and propagation of plants but also in the period after planting out.

A cucumber crop can grow very fast, and reaches its maximum leaf area index very quickly. Cucumbers are generally grown at a relatively high temperature, and both the large leaf area and the high temperature lead to high transpiration. The irrigation regime and influencing factors are similar to those described for tomatoes. Therefore, cucumbers are, when growing in soil, mostly irrigated by minisprinklers, whereas, for crops grown in rock-wool, trickle irrigation is the standard practice.

A high growth rate demands a high root activity and furthermore a good aeration of the root environment. In soil it is often difficult to irrigate without diminishing soil aeration. For plants growing in rock-wool, this problem does not arise.

Training and pruning

To train the plants, one end of each string is attached to a plant and the other to the wire on top of each row at a height of 2 to 2.5 m. The strings are twisted around these plants twice a week. Although not always done, the top of the main stem of each plant is often pinched off 10–15 cm above the wire. When the grower is pruning, he has to decide what to do with the lateral shoots? what to do with the (older) leaves? and which fruits to keep on the main stem? Pruning of the lateral shoots depends on whether the plants develop spindly or sturdy shoots. Spindly growth can often be observed under low light conditions, when the temperature is high, and when fruit set occurs relatively early. In such a case, the lateral shoots are maintained above the height of 1 m and pruned from the first or the second leaf. On a spindly plant less laterals will develop and therefore growers sometimes try to promote a more or less spindly growth. This minimizes the amount of pruning to be done.

When the growth is more sturdy, which is always the case later in the season when there is more light, all laterals are removed except for a few at the top of the plant near the wire. These laterals are not pruned at all, but placed above the wire. This is also the case for laterals of the next order. These too are made to grow downwards.

Leaves should be pruned when the growth is sturdy and young laterals would become shaded by the older leaves. In particular, leaves on the main stem can intercept the light for the younger leaves at the top of the laterals, which always need maximum light.

The amount of fruit that should be kept on the main stem depends on the leaf area and the light intensity. Dependent on these factors, a plant may be able to bear only a few fruit at a height of 1 to 1.2 m, or may carry fruit at nearly every second axil. When many fruits are developing at the same time a number of young fruits will abort, so fruit pruning occurs naturally. Nevertheless, there may be too many fruits at a given time; in this case some fruits will be partly misshapen, and it will take longer before other fruits are fully grown. This is undesirable, as it results in a poorer quality and a shorter shelf life.

Climate control

At 52°C latitude and further north the possibilities for successfully growing a cucumber crop in the winter is restricted because there is too little light for crop photosynthesis. However, if the greenhouse is heated sufficiently it can be profitable to start a new crop in December.

Directly after planting out a mean diurnal temperature above 20°C, and preferably about 22–23°C, should be maintained. The high temperature promotes quick growth, and prevents the main stem from growing too thick, which always leads to a shorter crop duration. High temperature also restricts the development of the lower lateral shoots, thus saving labor. The influence of different combinations of day and night temperature is similar to the effect on tomatoes described earlier.

More fruit growth always means less growth of the roots and higher risk for the reserves in the plants to be exhausted within too short a time. Therefore plants in full production need a lower diurnal temperature than during vegetative growth. A night temperature even as low as 14–15°C is permissible; but in this case it is recommended to heat the greenhouse and plants before sunrise. This prevents the plants becoming wet in the early morning, which could result in more fungus diseases and a poorer fruit quality. Growing cucumbers in unheated greenhouses in Northwestern European countries is hardly possible. In particular, downy mildew is a problem in unheated greenhouses. In Israel and many Arab countries cucumber growing in unheated greenhouses is common practice.

Heating is carried out not only to achieve a given temperature, but is also to promote air movement, to prevent a high relative air humidity, and in particular to prevent the plants from becoming wet by condensation in the early morning. Cucumbers are very sensitive to extreme conditions. Therefore ventilation of the greenhouse should be done very carefully, avoiding conditions of excessive humidity. Cucumbers are also very sensitive to poor transpiration conditions, caused by high humidity or lack of sunshine. Under these circumstances the leaves senesce earlier, turn yellow and die off sooner than normal. This physiological disorder is common, and can be observed early in the year and in autumn. Extra heating and ventilation can partly prevent it. A high carbon dioxide content also promotes this disorder, in particular in the early morning when it is responsible for the stomata remaining partly closed. In the Netherlands it is recommended to maintain the carbon dioxide content at 700–800 ppm but enrichment should only start after the stomata have opened. Older cucumber crops suffer badly under high summer temperatures. Growth is impaired, the fruit-bearing capacity is reduced, and the quality of the fruits is unsatisfactory. Whitewashing the greenhouse reduces the temperature, but the reduction of light is bad for the crop. Because of this dilemma, more and more cucumber growers finish their crop in June and start with young plants. Even three plantings in one year are sometimes made. Interplanting, as with tomatoes, is exceptional. The risk of re-infection by spider mites, thrips and other insects is too great. Another reason is the high growth rate of cucumbers; which makes it unnecessary to seek the extra production which might result from interplanting; in the three weeks after planting out, the first fruits can be picked from the new crop. The fact that long cucumber cultivars are sensitive to high summer temperatures may be the reason that in warmer countries shorter cucumber, the mini, varieties are grown.

Harvesting, grading, packing and storing

If cucumbers are grown nearly the year round the total yield per year can approach 60 kg m^{-2}. To reduce the growers' workload many improvements were made in recent years to establish mechanical transport systems in cucumber greenhouses. To obtain high quality, the fruit is harvested every second day in the early morning hours when the turgor pressure in the fruits is at its highest. The fruits are cut from the plants with a sharp knife and carefully placed in plastic boxes, transported out of the greenhouse, graded the same day according to weight, mostly on specially developed grading machines, packed in attractive cardboard boxes and transported to the auction or market where they can be sold the following morning. Sometimes, it is necessary to store the fruits a couple of days, for instance during the weekends. The best storage is in a special cool room at 12ºC. Some of the fruits are repacked by the wholesalers and wrapped in shrink foil. This is done on fruits of high quality to lengthen their shelf life by a few days.

Cucumbers are very sensitive to ethylene. Therefore, it is undesirable to store cucumber together with ripening apples, tomato or other fruits. Even the small amount of ethylene produced by a tractor used to bring the cucumbers into a storeroom can influence the ripening. When cucumbers are exposed to ethylene they turn yellow after a few days and become unsaleable. Transport of cucumbers in the same lorry with other fruits can have the same effect.

Quality of the fruits

The quality of cucumber fruits is determined by color, shape, and the absence of damage, physiological disorders and various diseases. Dark green fruits are favored by the public.

Yellowing appears when the fruits are too old. The longer the fruit has been growing from the time of flowering to harvest, the sooner the picked fruits yellow. Rapid growth is therefore important for obtaining a good shelf life. High temperatures and poor light penetration in the canopy to the fruits during the growing process also promote yellowing, and finally the fruits yellow rapidly if they are exposed to ethylene.

The *shape* must be cylindrical. Irregular growth, for instance as a result of variation in light conditions early in the year, can often result in a bad fruit shape. Cracked fruits will appear when the canopy is dense. Fruits on the main stem are always shorter than fruits on the laterals. Relatively short and thick fruits are less desirable (Fig. 5.14).

Damage is caused by rough treatment at harvesting, grading and transport, particularly when the fruits are very spindly.

An important *physiological disorder* is corky skin (corked hair-line cracks), which is believed to be caused by wetting of young fruits. This may happen in the early morning, when the fruits can be colder than the surrounding air. Another physiological disorder occurs when the fruits are exposed to temperatures below 12ºC; this results in sunken spots, "low temperature deterioration", some days later. Therefore fruits must also be kept above 12ºC also during transport. The fungal disease *Didymella bryoniae* is very harmful to quality. This fungus can attack the fruits through the skin when there has been contact between the fruits and infected plants. This often happens in a dense canopy. Also, an internal infection of the fruit through its blossom-end is possible.

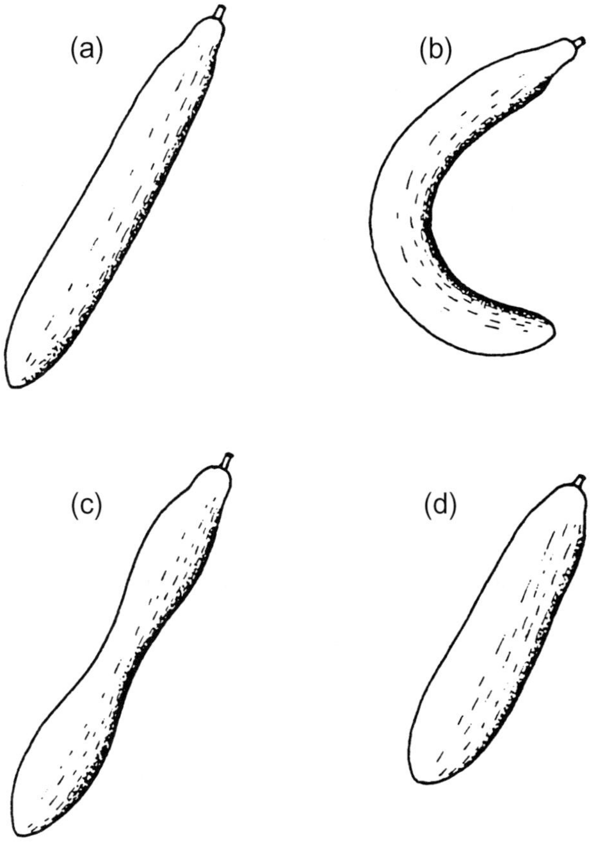

Fig. 5.14. Shapes of cucumber fruits: (a) nicely shaped: cylindrical with neck; (b) crooked (canopy too tight); (c) badly shaped (unbalance between number of fruits and photosynthesis); (d) less wanted (thick and short, without neck).

Fungal development on the fruit skin is not always visible at harvest, but appears later on. The fruit starts rotting, and this can spread quickly. In general, it holds true that the older a crop, the more infection may be expected. Other problems which can affect the quality seriously are thrips and several virus diseases.

It will be clear that the quality of the fruits is affected by many factors, and that much knowledge and attention are required to produce and market good-quality cucumbers.

Sweet pepper

Introduction

According to most historians, sweet peppers were originally grown in Central America and Brazil. Later, the crop spread to North America and Mediterranean countries. Nowadays, they are grown outdoors in the temperate zone during the summer. During the last few decades, sweet pepper production in greenhouses has increased. In the Mediterranean countries, sweet pepper is mainly grown in plastic greenhouses and tunnels, and in western Europe mainly in glasshouses. These latter developments are expected to continue in the years to come because it is possible to produce high quantity and quality in protected cultivation.

In comparison with other fruit-bearing crops, sweet peppers have the advantage that the labor requirement is less and is relatively constant during a season. The reason is that sweet peppers grow slowly, harvesting is possible at different stages (from green to red), and ripening is slow.

Cultivars

A large number of local cultivars are grown. The blocky (cubical shaped) type with green immature fruits and red mature fruits is mostly used. The fruits of these cultivars are picked either when green or red. People like variation in color. The cultivars being grown today are listed in Table 5.4 according to type (shape) and color.

Most of the blocky varieties have a thick pericarp and are firmer than the pointed types with their thin pericarp (Fig. 5.15). Firmness is important in relation to quality, as is a lower susceptibility for spots and swelling cracks. Most cultivars are susceptible to the pepper strains of Tomato Mosaic virus.

In greenhouses, cultivars with a sturdy growth are favored. Breeding work is in progress in order to get

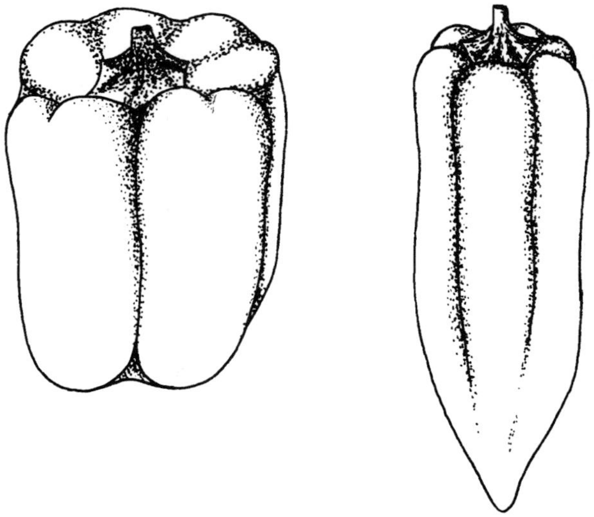

Fig. 5.15. Sweet pepper fruit: (left) block type and (right) pointed type.

higher production, better quality, longer growth season and resistance against diseases, especially Tomato Mosaic virus. The cultivars used nowadays in protected cultivation are all hybrids.

Table 5.4
Sweet pepper cultivars

Type	Color when immature	Color when mature
Block	green	red
	green	yellow
	green	orange
	green	brown
	purple	red
	white	orange-red
Pointed	violet	orange-red
	white	orange-red
	white	yellow-red
	yellow	red

Growing systems

Peppers need a relatively high temperature and good light conditions. At lower temperatures, deformed fruits occur more frequently; and when the crop grows at low light levels the productivity is low. In comparison with cucumbers, and even with tomatoes, peppers grow slowly and a long growth period is needed, an important consideration in the "blueprints" for growing pepper crops. In the following section, crop growth in heated and in unheated greenhouses, and productivity in northern as well as more southerly latitudes, will be described. In the Netherlands in winter, light is limited, but, if the greenhouse is heated, planting out in that season is possible, because development in the vegetative stage is less light-dependent. The crop is often planted at the end of November or beginning of December. In the Netherlands, clearing of the old crops often takes place in October in order to begin a new crop. Later planting also occurs, for instance in January–February. This saves fuel but shortens the growing season. The decision when to plant a new crop depends mainly on the price of fuel on the one hand, and the price and size of the expected production on the other. Peppers are also grown as a second crop after lettuces or strawberries, or as an autumn crop. Occasionally, peppers are interplanted between tomatoes. Because of the relatively slow development of peppers, planting out should be done by July 10 at the latest.

In unheated greenhouses in northern latitudes the growing season is much shorter, which makes the profitability of the crop doubtful. In more southerly latitudes like the Mediterranean countries, peppers are mostly grown in unheated plastic-covered greenhouses. During the winter the light conditions are better than in northern latitudes but night temperature can be too low, resulting in a fruit of poorer quality. Other disadvantages of growing pepper in protected cultivation (in greenhouses) in the sub-tropical zone are that summer temperatures during the daytime can be too high. In these regions, outdoor production is possible in the summer. For these reasons planting out often takes place at the earliest in August.

Seedling production

The commercial pepper-grower generally obtains his seedlings from a specialized seedling propagator. Because the pepper crop occupies the greenhouse for such a long period, there is not enough time for the grower to grow seedlings himself. In northern latitudes, the time needed to raise a well-developed seedling varies from 50–70 days in autumn and winter to 30–40 days in spring and summer. For the earliest plantings, at the end of November to the beginning of December, the seeds are sown during late September. Artificial illumination combined with a high temperature can reduce the propagation time.

The quality of the seeds is important. Seeds from mature fruit with a high weight are preferred. In order to reduce the chance of virus infecting the crop the seeds are disinfected for two hours in a 10% solution of trisodium phosphate.

If the mature plants are to be grown in rock-wool, the seedlings should be raised in rock-wool pots too. Sowing can best be done in "multitrays", in which the small holes are filled with rock-wool granulate. Sowing in soil and then pricking out plantlets in rock-wool would increase the incidence of fungi on the root collar. The best germinating temperature is between 24°C and 28°C. Below 15°C germination is very poor. Covering the seeds with insulating material may be necessary to protect them from direct solar radiation. In winter, when sun irradiance is weak, a plastic sheet is often sufficient to prevent dehydration. In well-equipped seedling nurseries, germination often takes place in special germination rooms where the temperature and air humidity is controlled. Before the plants begin to shade each other, they are pricked out in pots. The day before pricking out, the seedlings can be sprayed with skimmed milk, which immobilizes the virus particles. After pricking out, a night temperature of 21°C and a day temperature of 23°C or higher is recommended, depending on the light intensity. When the plants begin to touch each other, they should be spaced out again.

Using relatively big plants for planting out saves fuel for the crop, as a relationship between plant size at planting and both earliness and total yield has been established. It is advantageous to use seedlings up to 50 g, higher weights not being practical. The positive influence of plant weight on production continues to increase during the season. The same rule applies for the autumn crop. In practice, 40-day-old plants are recommended.

Preparations before planting

The time available for crop preparation is often limited, because the old pepper crop can be kept till late in the summer. On the other hand, growers wish to plant out the new crop early because of the slow growth of pepper crops. The preparation to be done in the period between the two crops when they are grown on rock-wool will be described below. Preparations for the new crop include, first of all, clearing away the old crop. A pepper crop nearly one year old has a heavy mass of plant material. The thick stems should be cut off from the rock-wool slabs first to prevent damaging the rock-wool. This is important when the rock-wool is to be re-used. The old crop should be moved far from the growing sites to reduce the chances of re-infection by diseases and pests. Before re-using rock-wool it must be steam sterilized, and this is often done in rented steam containers. The easiest way to do this is to pile the slabs on pallets and place them in a steam container. The drier the slabs are before steaming the better. Re-use of rock-wool is more time-consuming than replacing it. The re-use of rock-wool is, however, more environment-friendly, and diminishes the problems of disposing of the old material.

Before a new planting the trickle irrigation system must be cleaned to avoid variation between the emitters. Sometimes cleaning with water at a pressure of 5 bar is sufficient. When the deposits consist of calcium carbonate, iron compounds, or particles of fertilizers, they can be dissolved with a solution of 3% nitric acid. When the clogging consists of algae or other organic material, the irrigation equipment should be soaked in

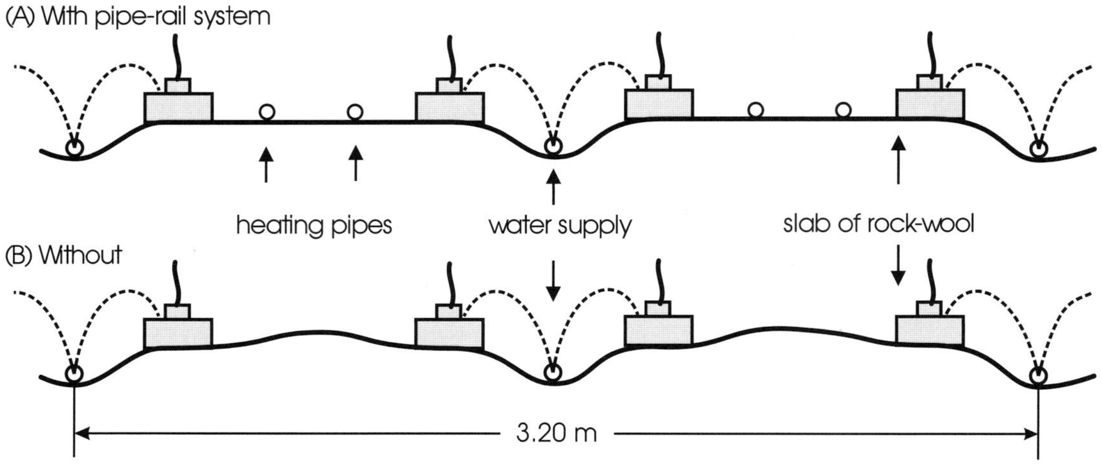

Fig. 5.16. Examples of soil profiles and installations.

a 10% hypochlorite solution for 24 hours. When rock-wool is used in combination with gutters for water circulation, these gutters should be cleaned, particularly if the old crop was infected with virus.

Next, the greenhouse roof should be cleaned. The energy screen needs attention too, in particular when it is non-movable. In such cases, new polythene foil must be installed for each new crop.

If rock-wool is used, it is placed on a plastic film on the ground; the slope of the soil surface is important. Often, new levelling may be necessary. The slabs are placed on a ridge where the surplus water from the slabs can be drained off, and the trickle lines are positioned at the lowest levels (see Fig. 5.16). After levelling the soil, reflecting foil is placed on the ground. The trickle lines are then placed in the troughs. In order to secure the maximal amount of light to the crop, the rock-wool slabs are placed in new white sleeves, and the young plants placed in the holes. The trickle installation is put into operation until the slabs are saturated with water and nutrients.

If the crop is grown in soil, it will often be necessary to sterilize the soil before the new seedlings are planted. This can only be done after the old crop has been removed. The cleaning and disinfection of greenhouses for fruit-bearing crops is time-consuming and cumbersome, and this is one of the reasons why artificial root media have replaced soils.

Planting out and plant distance

On rock-wool, a hole is made in the rock-wool sleeves for each plant. The slabs are then soaked with water and nutrients, the plants are placed on the holes and the emitters of the trickle irrigation are placed on the rock-wool pots.

Before planting out, the soil or rock-wool slabs are heated to at least 20°C. Just before or just after planting out, the slabs should be drained by making small slits in the plastic cover just above ground level.

In soil, one should take care to plant out no deeper than to the cotyledons. The stem of young plants is very sensitive to infection by *Pythium*.

Historically, peppers were grown in hot beds with a large distance between the rows. In modern greenhouses, such as the Venlo type in the Netherlands, it is recommended to use four rows per 3.20 m (0.8 m between rows). The plants should be trained according to the so-called "stem system". The distance within the rows depends on the number of stems retained per plant. A distance of just over 40 cm and two stems per plant (6 stems per square meter) has proved to be the most profitable in the Netherlands with respect to yield, the price of the plants and the amount of labor involved.

In autumn crops it is common to leave more stems per plant, and keep a plant distance within the row of 50 cm. If peppers are interplanted between tomatoes, the same distance has to be used as those of the tomatoes.

Fertilization and water supply

In soil as well as in rock-wool the desired nutrient content is very similar to that used in cucumber crops, but under low light conditions the recommended concentration of fertilizers for peppers are approximately the same as for tomato. The electrical conductivity (EC) influences the production. An experiment showed that increasing the EC value of a nutrient solution from 2.2 to 3.7 mS cm^{-1} reduced the marketable yield by 4%, while an increase to 5.3 mS cm^{-1} reduced the yield by 16%. The EC is also important in relation to quality. The shelf life of many fruit-bearing plants is better when they are grown at a relatively high EC. Peppers, however, react differently, as a high EC promotes softness during shelf-life. A low EC promotes russetting. The uptake of Ca++ can be diminished by a high EC, and will result in blossom-end rot. A high uptake of calcium leads, on the other hand, to brown spots on the outside of mature, colored fruits, in cultivars which are susceptible to this disorder. These spots are the result of the formation of calcium-oxalate crystals in the cells. Hence, the recommended EC early in the season is 3.4 to 3.5 mS cm^{-1}.

The uptake of chloride ions by peppers is very low, and consequently accumulation of this element in the root environment will take place if its concentration in the irrigation water is more than 3 mmol ℓ^{-1}. A high content of chloride ions promotes the incidence of blossom-end rot and reduces the marketable yield. Leaching may be necessary during the growing season because of high chloride levels.

The instructions previously presented concerning top dressing and water supply of other crops, also apply to the pepper crop.

Training

In long-season crops such as pepper, it is common practice in the Netherlands to maintain two stems per plant and about six stems per square meter. More than this leads to an excessively dense crop. Compared with the old multi-stem system, the yield increases by 1.5 kg m^{-2}, and the fruits are bigger and of better quality. The two-stem system demands a little more labor for pruning but this is compensated by a lesser labor requirement at harvest. Three stems per plant is acceptable, but then a wider plant distance is necessary to obtain the same number of stems per square meter. This saves on the cost of plants on the one hand, but on the other hand the crop production is a little lower.

To compare the merits of these training systems, one must understand how pepper plants grow and develop. After germination a main stem is formed with a number of sessile leaves (see Fig. 5.17A). The growing point of the main stem ends in a flower bud (see B). The

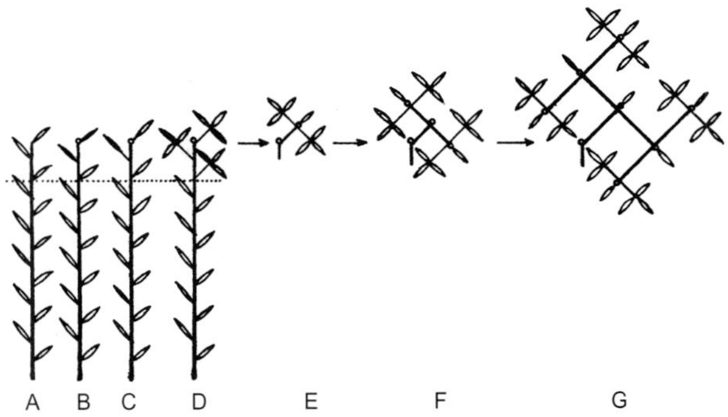

Fig. 5.17. Development of peppers.

leaves just below this bud develop petioles (C). Next, two more leaves are formed in the axils (D), again a bud is initiated and after that all the leaves have petiolles. In Fig. 5.17, it can be seen how this branching and flower initiation is repeated. In the 2–3 stem system the main stem is in fact very short, and what is done in practice is to grow stems consisting of branches of the 1st, 2nd, 3rd, 4th and higher orders (Fig. 5.18).

Pruning causes each plant to be sufficiently covered with leaves and flower buds to form a dense canopy. The stems are twisted around a string fixed to a wire on top of each row. Because peppers grow slowly, it takes a long time before the stems reach the wire. Nevertheless, when the greenhouse is low, it is difficult to maintain a pepper crop for a very long time. Layering of the stems, as in tomatoes, is not possible since the stems are too thick and not sufficiently pliable. In a high

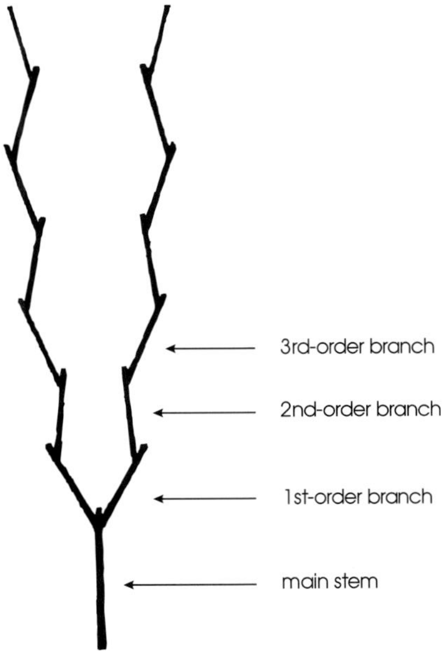

Fig. 5.18. The stem of a pepper plant

Fig. 5.19. Pipe-rail system.

greenhouse a long-lasting crop of peppers can reach such a height that a wagon travelling on a pipe-rail system is recommended to facilitate access for workers to the upper part of the crop (Fig. 5.19).

The branches should be pruned at least once a fortnight when the plants are young. Later, when growth slows down during the fruit-development phase, intervals of three to four weeks between prunings are possible. In order to prevent the transfer of virus from one plant to the next the grower should dip his fingers in skimmed milk. When the crop is planted out later in the year and in particular for an autumn crop, the old multi-stem system is mostly applied. In this case, five or six stems per plant may develop. This system is possible in autumn because the short growing season limits the height of the plants. In this system, no branches are removed, but are all kept and supported by a number of strings.

Climate control

For the growth and development of pepper crops, the temperature in the root zone, the diurnal mean temperature, the daytime temperature, the temperature in the early night and the temperature in the late night are all important.

The root-zone temperature should be at least 20°C. If it is less, growth will be too slow and incidence of *Pythium* increases. Early in the year, when the outside temperature and the temperature of the greenhouse roofing material are low, heat will be lost from the soil to the colder greenhouse cover. During this period, an installation to heat the root environment is essential. Heating pipes close to the soil level, as in the pipe-rail system, have a positive effect, but are insufficient for soil heating early in the year. For summer plantings, soil heating can be dispensed with. The root zone is heated as for cucumbers, by using polythene tubes with circulating water. Temperatures above 25°C should be avoided to prevent swelling cracks in the fruit. The diurnal mean temperature can be measured exactly with a thermocouple thermometer linked to a data-logger or computer. The diurnal mean temperature is a very important factor influencing the rate of growth, development and earliness. It should be 21°C in the period after planting out, and 20°C later on when the fruits are developing. A higher temperature results in a poorer shelf life, and a lower temperature in more misshapen fruits. Dependent on the light intensity, a limited variation in the diurnal mean temperature, 1°C higher or lower, is beneficial, as mentioned previously.

As with tomatoes, a given diurnal mean temperature can be maintained in several ways, for instance, by a larger or smaller temperature amplitude. The larger the day–night amplitude is, the more spindly and elongated the crop. A similar growth response to tomatoes can be found in peppers but because peppers grow more sturdily a greater amplitude is permissible. For this reason, some growers maintain 15°C at night and 24°C in the daytime. A night temperature of 15°C is favorable if it is compensated by the day temperature. If not, fruit setting will be worse and many misshapen fruits can be expected. In an experiment where night temperatures of 15°C and 18°C with the same 24-h mean temperature were compared, the 18°C treatment resulted in a slightly earlier, but lower, total production. A daytime temperature of 24°C is very suitable; higher as well as lower temperatures reduce yield. Around 1980, when fuel prices were very high, growers started using "energy screens", movable or fixed. In an early crop they can be used for ten weeks. Peppers can withstand the higher humidity that is created under the screens very well. With a movable screen, it can be more profitable to maintain a higher night and a lower day temperature, 20°C at night and 23°C by day. The night temperature depends also on the development stage of the crop. When flower development is weak or when the fruit setting is insufficient, the night temperature should be lowered a few degrees in order to promote the transport of carbohydrates to the reproductive organs of the plant.

Maintaining a relatively low night temperature throughout the night means a sudden change in temperature in the morning. This is undesirable because there is a higher risk of water condensing on the fruits, which may cause russetting. It is better to maintain a low night temperature in the early night and gradually to bring the temperature up at the end of the night to the daytime level. This can be easily achieved with a modern climate-control system. In the early night-time, a temperature as low as 13°C is permissible if it is increased later in the night. Crops in heated greenhouses transpire not only during the daytime when the stomata are wide open, but also during the night. The lower the night temperature and the smaller the vapor pressure deficit of the greenhouse air, the smaller the night-time transpiration rate. Peppers can tolerate a high relative humidity (low vapor pressure deficit) better than cucumbers and tomatoes, in particular when the plants are young and the leaf area is small. In the later stage, a very high humidity should be prevented because lack of transpiration limits the uptake of nutrients and can cause deficiencies, in particular of calcium and magnesium. The diminished uptake of nutrients can be overcome by applying a higher nutrient concentration during that time of the year. In greenhouses with fixed polythene screens (thermal curtains) there is considerable light reduction even during periods with low light levels. For a new polythene screen the light transmittance is 83% when dry; but when moistened by condensed water, which is often the case, the transmittance is only 68%. This is why anticondensation foil is used, but after 6–7 weeks the anticondensation treatment often becomes ineffective. In spite of the disadvantages, fixed screens are nevertheless used because, on a yearly basis, the reduction of light in the winter months is only a minimal factor. Fixed screens are often removed 7–8 weeks after planting.

As a closed screen makes a greenhouse nearly airtight, its carbon dioxide concentration may become insufficient for growth. Carbon dioxide levels can drop to below 200 ppm within one hour, which results in a very low rate of photosynthesis. A level of 700–800 ppm of carbon dioxide should be maintained. Like other crops, peppers respond favorably to carbon dioxide enrichment.

Harvesting, grading, packing and storing

Fruits can be picked at the green or at the mature stage when the fruit is either red, brown, yellow or orange. Allowing fruits to reach the mature stage on the plant reduces the number of fruits by 15–20%, but dry weight is higher in comparison with crops picked at the green stage. Picking at the red stage also increases the risk of fruit spots and russetting. Fruits at the immature green stage should not be picked before they are hard and shining, otherwise the shelf life deteriorates rapidly, and the fruits become soft sooner. The time between fruit setting and the green picking stage is often 30–35 days in spring and summer. Harvesting at the red stage requires a further 20–30 days. In autumn the duration from flower setting to red stage can be 90 days. The picking frequency varies according to the cultivar and the thickness of the pericarp. The thicker the pericarp, the less the picking frequency can be. When picking at the mature stage, the frequency has to be higher because this stage has a duration about 60% of the green stage. A low picking frequency results in a higher percentage of soft fruits with a reduced shelf life. It is neccessary to pick once a week in the green stage and every five days at the mature stage. Partly colored fruits, between red and green, are not wanted.

In harvesting the fruits it is better to cut the stalks than to break them, as a smooth wound leads to less rotting. Harvesting when the temperature is very high should be avoided. Rough treatment damages the fruits; in particular, the stalks may damage other fruits. Sometimes the remains of the style of the former flower is still visible as a projection (tail) at the end of the fruit. This can break, creating an entrance for fungi. After being picked, the fruits should be moved from the greenhouse as soon as possible to a cool and shaded location. Grading according to weight is done on specially-developed grading machines. Grading according to quality can only be done by hand. Packing should be done in attactive boxes each containing 5 kg. Ripe fruits can be stored for two weeks at a temperature of 7–8°C; green fruits need 10–12°C. Lower temperatures increase the occurrence of "low temperature deterioration" (sunken spots). Ideally, the atmospheric humidity should be between 90–95%. When green peppers are stored together with other ripening fruits, the ripening process can be enhanced by ethylene production. During transport the fruits should be protected against sunshine, wind, cold and precipitation. In a long-lasting pepper crop, the yearly yield can be over 25 kg m^{-2}.

Quality of the fruits

As stated above, fruit quality depends on the shape, size, firmness, color and freedom from any damage caused by the growing conditions, viruses, fungi and pests. The influence of nutrition, the electrical conductivity, the water supply the air and soil temperature, the treatment during harvest, transport, grading and packing, sunshine, humidity, storage and cooling on quality have already been referred to. Well-known physiological disorders include blossom-end rot (see also under tomatoes, p. 127), "tails", softness, flat fruits, russetting, spots, sunscald, low temperature deterioration and internal growth of the pericarp. A number of viruses can have an extremely bad effect on quality. Fungi such as *Pythium, Botrytis* and *Sclerotinia,* and, last but not least, insects including thrips and aphids, and Begonia and spider mites (*Polyphagotarsonemus latus* and *Tetranychus urticae*), can cause severe damage. The conclusion is that to achieve a good quality the environmental conditions, during the growing season and after harvest, must be controlled, as well as pests and diseases. Protected cultivation offers the possibility to control these many factors, resulting in a high-quality product which meets market requirements.

Melons

In Dutch greenhouses only sugar melons (*Cucumis melo*) are grown, though elsewhere the watermelon (quite a different plant – *Citrullus lanatus*) is also cultivated. There is a similarity between the way cucumbers and melons are grown, but there are also important differences between the two crops.

One important difference is that melons need to grow at higher temperatures and a higher level of solar radiation. For this reason, melons are grown on a large scale in the Mediterranean countries. In the Netherlands, the first plantings begin in the second half of February; it is only possible to grow melons in unheated greenhouses there in midsummer.

Fruit setting is fundamentally different from cucumbers because fruit development of melons only occurs after pollination. In practice, pollination is promoted by introducing bees into the greenhouse. A third difference is the alternation in vegetative and reproductive growth in melons as a consequence of a temporarily heavy fruit load on the plants. The fruit load of melons can be so heavy that the plant reserves are exhausted, and do not recover sufficiently to produce a complete second set. For this reason the duration of a melon greenhouse crop is often relatively short.

Propagation of seedlings for planting is practically the same as for cucumbers.

When plants are to be grafted in order to prevent *Fusarium oxysporum* infection, *Benincasa cerifera* is recommended as a rootstock in the Netherlands. This species grows slowly and should therefore be sown a few days earlier than the melons. In principle, melons can be grown horizontally on the ground or trained vertically on strings. The latter method leads to a higher leaf area index, and is therefore preferable. The only reason to use the former system is when a cultivar produces very big fruits, which could fall from a stem trained above the ground.

There are many planting and training systems, which differ little from those used for cucumbers. In heated greenhouses melons should be planted at a density of 1.5 plant per square meter, in particular when grown in soil. In unheated greenhouses in northern countries 1.2 plants per square meter is sufficient; but 1.6 plant per square meter is possible in Mediterranean countries. Closer spacing will result in a few more fruits, but these fruits will be smaller and total yield is unchanged. Too dense a crop will affect the quality, and the surface of the fruits does not develop the netlike pattern wanted by buyers. The later-developed laterals are trimmed after the first leaf, and when the fruits in this leaf axil develop, the lateral of the next order in that axil is removed. The first fruits develop mostly at about 1 m height. A second setting appears higher in the canopy.

The average weight per melon depends on the cultivar used and, as stated above, also on the plant density and the number of fruits developing at the same time. Fruits from the second setting are nearly always smaller. The average fruit weight is often about 1 kg, and a yield of $10\,kg\,m^{-2}$ is considered a very good result under Dutch climatic conditions.

In respect of flavor, the sugar content is important. This is positively affected by the amount of solar radiation during fruit development and by plant health, and negatively by a high weight of fruits in relation to leaf area and by strong vegetative growth. A moderate growth rate results in a higher sugar content than a high growth rate. Investigations have shown that, in general, the public prefer a sugar content (as measured by the refractive index of the juice) of 11–12%.

Because of the difficulty of controlling soil-borne diseases, including *Fusarium* and nematodes, the option of growing melons in soilless culture can provide a solution, especially for long-season cropping.

Eggplant

The eggplant originated in India. It is adapted to

warm countries, where it is grown on a large scale, mainly outdoors. Today, in Mediterranean countries it is grown as a protected crop in plastic greenhouses and high tunnels, to lengthen the cropping season and improve quality. In northwestern Europe the eggplant was grown on a very limited scale in the mid-seventies, and today the greenhouse area is still small, about 80 ha in the Netherlands. The most important developments in eggplant cultivation in greenhouses were the introduction of new cultivars and the change from growing in soil to growing in rock-wool.

In order to improve eggplant cultivation in greenhouses, over 200 cultivars from different countries were tested and some of them were used as sources for hybrids. The properties of the cultivars are very different. The shape of the fruits can be like eggs, more cylindrical, or pear-shaped. The color can be white, cream, yellow, flame or purple. The weight of the fruits can vary from under 100 g to over 700 g. The growth rate and the production capacity also vary, as does the compactness of the plants and their sensitivity to diseases and physiological disorders. Breeding work has been carried out on a considerable scale and new hybrids, better adapted to growing in greenhouses, have been developed and introduced. In general, cultivars with purple colored and well-developed fruits are required, the demand for cultivars with small fruits of other colors being small. For greenhouse-growing in the Netherlands, a rather compact growth habit is wanted, because many greenhouses are not high enough to accommodate very elongated plants.

The growth and development pattern of the eggplant is similar to that of the tomato. The main stem is relatively thick and stiff which prevents layering. Though the vertical growth rate is about half that of tomatoes, high greenhouses are needed for eggplants if the growing season is long. In northern countries eggplants are mostly grown as a main crop planted in December, for instance, and kept until October. Eggplants grown as an autumn crop are the exception.

Eggplants need high temperatures, and for this reason growing in unheated greenhouses cannot be successful in northern latitudes. The optimum germination temperature is 23–25°C. After pricking out, the plants are rooted at a night temperature of 19°C and a day temperature of 21°C in dull weather; some degrees higher are required when the weather is bright. Production is lower when lower temperatures occur shortly after planting out. The roots should have a temperature of 22°C. This means that soil heating or slab-heating (when growing on rock-wool) is necessary. A minimum pipe temperature of 35–40°C is needed to prevent *Botrytis,* and this will activate the plants when sunshine is lacking. The effect of carbon dioxide enrichment on yield is positive, but concentrations of over 800 ppm may cause, more than in other crops, closing of the stomata, lower transpiration, a higher leaf temperature and earlier yellowing of the leaves. A high relative humidity results in a higher production and less incidence of browning of the calyx. This may be related to a lower uptake of calcium when the relative humidity is high. In particular during the night, a higher relative humidity is desirable.

Fertilization can be similar to that for sweet pepper and tomato but, because of the sensitivity of eggplant to magnesium deficiency, the magnesium content should be higher and the content of potassium and calcium a little lower. An electrical conductivity above 4, when maintained for a long time, decreases production.

Propagation of eggplants is similar to that of pepper or tomato. The growth rate is intermediate between these crops. When *Verticillium* may be a problem, it is possible to graft eggplant on *Verticillium*-resistant tomato cultivars, using the method applied in grafting cucumbers.

The recommended planting distance between the rows in the Netherlands is 80 cm and that within the row 55 or 70 cm, depending on whether two or three stems per plant are required. The numbers of plants per square meter are, respectively, 2.25 and 1.8, resulting in 5.5 stems per m^2 in both cases. Previously, three stems per plant was the normal practice, but recent experiments at Naaldwijk have suggested that two stems per plant may be economical in spite of the large number of plants per unit area.

As with tomatoes, laterals should be removed from eggplants, as should the yellow leaves. The flower bunches have one big flower and a few smaller ones, developing later into one larger and a few smaller fruits. For this reason, the smaller flowers are mostly removed. Parthenocarpic fruit-setting occurs but fruits with seeds develop quicker and become larger. Spraying hormones once a week has the same effect on fruit development as seed-setting and the application of hormones has become normal practice. This treatment also prevents the plants from allocating too much assimilate to vegetative growth. If hormones are sprayed, they should be concentrated on the flowers as much as possible and the leaves avoided. If the flowers are large, they can be sprayed from two days before opening till four days after. Fruits develop quickly and can be harvested 5 to 6 weeks after flowering using a pair of scissors; in a long-lasting crop the yield may be 30 kg m^{-2}, with the weight per fruit varying between 250 and 500 g.

Chapter 6

CUT FLOWERS IN THE GREENHOUSE

Allan M. ARMITAGE and Ishai SOCOLOVSKI

INTRODUCTION

Cut flowers have long been an important part of human society, used to celebrate rites of passage such as marriage, birthdays and funerals as well as expressing love and caring, and asking forgiveness. Flowers were placed in the tombs of pharaohs and continue to be part of man's social evolution. They have been grown under protection for hundreds of years and have spawned commercial industries in many countries. In the commercial sense, cut-flower species may be grown for the long stem and flower such as roses (*Rosa* × *hybrida*), and baby's breath (*Gypsophila paniculata*), but also consist of species used for their foliage (Eucalyptus, *Eucalyptus* spp.; leatherleaf fern, *Rumohra adiantiformis*) or colorful, unique stems (willow, *Salix* spp.; dogwood, *Cornus* spp.). The most widely cultivated cut-flower crops are roses (*Rosa* × *hybrida*), carnations (*Dianthus caryophyllus*) and chrysanthemums (*Dendranthema* × *grandiflorum*), the majority of which are produced in the greenhouse. These are the *major cut-flower crops* worldwide.

The large number of lesser-known species of cut flowers sold throughout the markets of the world today are referred to as *specialty cut-flowers* (Armitage, 1993). Some species such as alstroemeria (*Alstroemeria* spp.), lilies (*Lilium* spp.), gladioli (*Gladiolus* spp.), baby's breath (*Gypsophila paniculata*) and statice (*Limonium* spp.) are very important and grown in the greenhouse as well as in the field. Other specialty flowers such as gayfeather (*Liatris* spp.), aster (*Aster* spp.), lisianthus (*Eustoma* spp.), goldenaster (× *Solidaster* spp. and snapdragons (*Antirrhinum majus*) continue to increase in popularity. Many specialty cut flowers have historically been field-grown but, as demand increases, so does their value. The result has been a steady rise in greenhouse production of these specialty species. As some flowers increase in importance and volume, the distinction between *major* and *specialty* cut-flower crops diminishes.

Consumption of cut flowers increased by 11% between 1985 and 1990, with sales concentrated in three regions: western Europe, North America and Japan. Some trends in cut-flower cultivation in these regions are shown in Table 6.1.

Approximately 28% of the area used for cut flowers is under protection of glass or plastic, although percentages differ among countries (Haak et al., 1992).

Table 6.1
Flower usage in Western Europe, North America and Japan [1]

Country	Consumption (million guilders) [2]	Per capita (units of cut flowers) [3]	Total crop area (ha) [4]	Production value (million guilders) [5]
France	2800	26	– [6]	– [6]
Germany	5700	54	2500	550
Italy	7300	– [6]	7000	2600
Japan	11 000	40	16000	2800
Netherlands	1000	150	5500	3300
Spain	500	26	2300	550
U.K.	2100	50	– [6]	– [6]
U.S.A.	11 000	14	6500	1000

[1] Haak et al. (1992).
[2] Total consumption in 1990.
[3] Includes *Alstroemeria*, carnation, chrysanthemum, *Freesia*, *Gladiolus*, *Narcissus*, rose, tulip and bouquet bunches.
[4] Israel, Colombia and Kenya had approximately 1800, 3700 and 1000 ha in 1989.
[5] Total production value at grower level as of 1989. Israel, Colombia and Kenya had production levels of 200, 400 and 100 million guilders.
[6] Data not available.

In some countries of western Europe (e.g. Germany), and in the United States, the open field area is decreasing due to land costs, increased demand for off-season flowers and demand for specialty flowers. The protected area is increasing in Italy, Spain and Colombia. The area of both open and protected cultivation has also increased in other countries (e.g. Kenya, the Netherlands, and the United States).

Although there is less production of cut flowers in the greenhouse than in the field, the value of the products is far greater. This is because prices for flowers during the winter and important holidays in consuming countries are significantly higher than prices of "summer flowers". The greenhouse ecosystem is unique as a result of the ability to modify the environment to produce flowering crops throughout the year or for major holidays. Temperature, carbon dioxide enrichment, light intensity and photoperiod may be artificially manipulated and maintained, while media, fertility and irrigation systems can also be developed to produce crops on demand with high yields and excellent quality.

Important cut flowers

There are literally dozens of cut-flower species on the market place, and diversity of production appears to be a major trend in many countries. Chrysanthemum, rose, lily, carnation, Peruvian lily (*Alstroemeria* hybrids), baby's breath and gladioli are important throughout world markets. Many other specialty flowers are relatively minor in comparison, but are increasing in importance in some countries. Countries such as Colombia, Equador and Israel are directing much of their area expansion into specialty flowers. In Israel, the greatest area is in waxflower (*Chaemelaucium* spp.) and baby's breath (Haak et al., 1992). Some of the more important flowers are listed in Table 6.2.

Table 6.2
Important cut flowers in world floriculture

Scientific name	English name
Alstroemeria hybrids	Peruvian lily
Dendranthema × *grandiflorum*	chrysanthemum
Dianthus caryophyllus	carnation
Eustoma grandiflorum	Eustoma, lisianthus
Freesia spp.	freesia
Gerbera jamesonii	gerbera
Gladiolus spp.	gladiolus
Gypsophila paniculata	baby's breath
Liatris spicata	gayfeather
Lilium spp.	lily
Limonium spp.	statice
Narcissus spp.	daffodil
Rosa × *hybrida*	rose
Tulipa hybrids	tulip

Cut flowers are exported from many countries. Export of cut flowers has become somewhat of a stepping stone to economic development, providing developing countries with a means of obtaining foreign currency. European countries account for 76% of world exports, with the Netherlands accounting for approximately 59% of world exports in 1990. Colombia is responsible for about 10% of world exports, but ships mainly into the United States, thus accounting for a considerable share of American imports (Haak et al., 1992).

The popularity of cut-flower species varies with season. In the United States, the purchase of major species (roses, carnations, gladioli, and chrysanthemums) coincides closely with holidays, particularly St. Valentine's Day, Mother's Day and Easter, although sales are very stable from year to year (Behe, 1991; Behe and Wolnick, 1991). Sales of specialty flowers also fluctuate with holidays; however, baby's breath, statice and other species are not associated with any particular event or holiday. Many specialty flowers are often associated with summer production, long thought to be the "off-season". The mix of specialty species changes from year to year as new species are tested, then produced as cut flowers. As new crops become less of a novelty and more mainstream, growers turn to the greenhouse for winter production.

THE GREENHOUSE AS AN ECOSYSTEM FOR PROPAGATION AND PRODUCTION

Propagation systems

The method of propagation depends on the species and the facilities available. The major species are all propagated vegetatively (often through tissue culture) whereas many specialty species are propagated by seed.

Propagation of major species is accomplished by specialist propagators, and seldom carried out by the commercial flower-grower. Facilities for maintaining disease-free, productive stock plants and warm, humid rooting areas are necessary. Such facilities include highly accurate temperature control, a misting or fogging system designed to maintain turgidity of cuttings, and equipment for disease and insect control at all times. Different species may require different environments for optimum propagation. For example, stock plants of chrysanthemums are grown under long days, but should be kept for less than 15 weeks, otherwise premature flower buds may occur (Kofranek, 1992). Potential new cultivars of carnations and chrysanthemums are subjected to an extensive heat-treatment program for eradicating virus and bacteria, then propagated from tip cuttings. Once introduced to the grower, carnation stock plants are discarded after one season (Besemer, 1992). Rose cultivars are generally grafted or budded to a vigorous rootstock (Hasek, 1992). Gerbera (*Gerbera jamesonii*) may be propagated from seed, by division of older plants, or from tissue culture (Rogers and Tjia, 1990). All new cultivars of roses, carnations and chrysanthemums are patented; therefore only the originator and licensed propagators may propagate these cultivars. The cuttings, however, are rooted by the flower grower or by the specialist propagator.

Seed-propagated species are often propagated by specialist "plug growers" or by the growers themselves. Most seed-propagated species are annuals (or treated as annuals), although perennial species are also multiplied from seed. Plug growers produce seedlings in trays with many small (approximately 1 cm^3) "plugs" which are then shipped to the grower for transplanting to the greenhouse. Seed propagation requires facilities to maintain high humidity (90–95%) and sensitive temperature control. Plug growers require highly technical facilities and equipment such as automatic seeders, mist systems and sophisticated benching and handling systems (Koranski and Laffe, 1985). Although manual controls in which the workers open and close the vents, water by hand and control the thermostats are still used in many areas, sensitive control of the greenhouse ecosystem becomes more important as the value of the crop increases. Beginning growers may propagate their own seedlings with nothing more than a heated area and a basic mist system. However, in modern greenhouses the propagation and production areas, as well as many of the cultural procedures, are under sophisticated computer control.

Production systems

Production practices vary widely among producers of cut flowers, but all systems must be organized around the basic greenhouse environment under which the plants are grown. Irrigation and the growing medium are important parameters for growing any crop, whether in the greenhouse or field, but will not be dealt with here as they are considered in other chapters. However, in view of the effects on aeration and drainage of pots and greenhouse beds, attention to irrigation and media is extremely important.

An understanding of the basic fundamentals of light intensity, photoperiod and temperature, and their relationship to production is essential for optimum production and profitability. The principal environmental parameters affecting cut-flower production which can be controlled through greenhouse manipulation are light flux, photoperiod, temperature and carbon dioxide. The selection of media, fertility and irrigation practices directly relates to the control of the environment under which the plants are grown.

Light flux

The yield of most cut flowers is directly related to light flux. Light is the source of energy for plants and, together with water and carbon dioxide, results in the production of carbohydrates and oxygen through photosynthesis. The light energy is fixed in the carbohydrate molecule, which is then translocated throughout the plant. When optimal levels of carbon dioxide, water and temperature are provided, an optimal light intensity results in greatest dry-weight accumulation and in the case of non-photoperiodic species, optimal flowering. Reducing light flux reduces growth, and supraoptimal light may destroy chloroplasts, thus also reducing growth.

In Pennsylvania, U.S.A. (lat. 41° N), crops in the greenhouse are subjected to a fifty-fold range in light intensities as cloudy, winter days change to bright, sunny days during the year (Nelson, 1991). While the same percentage seasonal change may be less or greater in other areas, a significant seasonal change occurs regardless of where the greenhouse is located. The minimum intensity above which a net gain of carbohydrates in the plant is produced is referred to as the light compensation point. When additional light no longer results in additional carbohydrate gain, the plant is said to be light-saturated (Fig. 6.1).

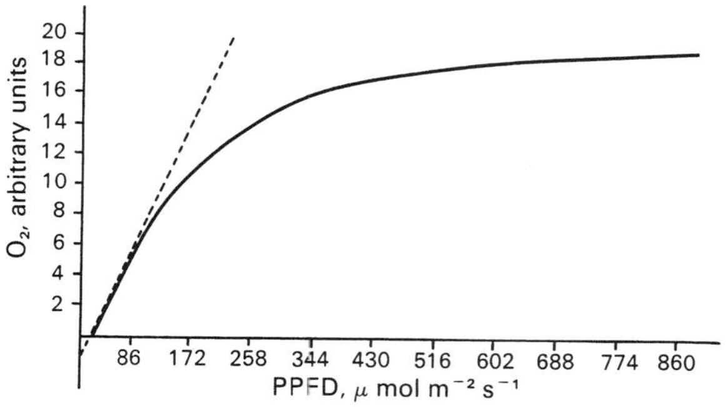

Fig. 6.1. Rate of photosynthesis (measured as O_2 evolution) as a function of light intensity (PPFD). At a PPFD of approximately 400 μmol m^{-2} s^{-1}, the rate of photosynthesis is close to its maximum (light saturation). After Walker (1992).

Light saturation levels vary from crop to crop. Roses and carnations may grow well under full light intensities in the summer. As intensities fluctuate over seasons, yield and quality of carnations and roses (Table 6.3) are significantly affected (Post and Howland, 1946; Holley, 1959).

Light intensity should be maximized during the winter, because yield declines and stem strength of most cut-flower species is weakened under low light. Maximizing light is accomplished through greenhouse design, construction materials and orientation. All-metal open-span houses allow up to 78% of the incident light into the greenhouse. Below 40° N latitude in the Northern Hemisphere, a north–south orientation results in fewer shadows than east–west orientation of the same greenhouse. Further north than 40° N, greenhouses benefit from an east–west orientation (Manbeck and Aldrich, 1967). The covering material (glass transmits more light than double polyethylene), size, position and persistence of shadows (Mattson and Maxwell, 1971) must be considered, and timely maintenance is also important to keep the covering clean. Maximizing light intensity to the plants themselves is accomplished through optimal spacing of the plants. Plants should be more widely spaced in the winter than in the summer, and additional spacing should also be provided for plants in the middle of the bed. (Fig. 6.2). This permits light to reach the potentially darkest area of the production bed, resulting in increases in yield and quality.

Light is the most important environmental influence in flowering of many greenhouse crops. In carnations, flower initiation is strongly influenced by the amount of solar radiation received, and delays in flower initiation have been reported under low daily light integrals (Harris and Harris, 1962; Bunt, 1973). The response to light integral is exerted through a direct effect on photosynthesis and, therefore, on the availability of assimilates for growth. Most crops that are produced for major holidays in the Northern Hemisphere are produced under suboptimal light levels, and light intensity can be increased by artificial means to increase yield and flower quality. Although most cut-flower species tolerate and often thrive under high light conditions, summer intensities are usually reduced because of the supraoptimal heat associated with the additional light. Shading the greenhouse reduces temperature and light intensity significantly. Shade may be provided

Table 6.3
Number of bunches of spray-type chrysanthemums produced during the year [1]

Planting date	Bunches (m^{-2})	Mean number of flowers per stem
July 16	9.7	14.0
Augustus 15	8.9	11.6
September 14	6.2	5.9
October 15	5.5	5.2
November 15	7.5	7.9
December 16	9.4	10.0
January 14	10.2	10.3
February 15	9.9	11.3
March 15	11.4	13.1
April 15	15.5	18.0
May 14	12.2	13.0
June 16	10.43	11.3

[1] After Post (1948).

Fig. 6.2. A planting arrangement for chrysanthemums.

through application of greenhouse shading compounds, shade cloth placed over the greenhouse roof, or with automated sunscreen material which is activated when light levels rise above a predetermined level. Some areas of the world are more suitable for production of cut flowers than others because greenhouses may be situated where high light can be combined with cool temperatures, as in locales of little cloud cover at high elevations. In the United States, carnations were almost exclusively produced in regions of California, Colorado and the midwest where light levels were high in the winter and shading was not required in the summer. One of the reasons why Colombia has become a leader in carnation production is because of its optimum environmental conditions. The excellent environment, combined with low production cost, helped elevate Columbia into a leading carnation producer.

Table 6.4
Influence of supplemental light on gerbera and rose[1]

Light	Flowering stems per plant	Stem length (cm)	Fresh weight (g)
Gerbera[2]			
Natural light	8.2	40.4	14.2
Supplemental light[3]	10.6	45.6	16.2
Rose "Forever Yours"			
Natural light	19.8	43.5	15.9
Supplemental light[4]	40.1	39.3	16.5

[1] Adapted from Tsujita (1983) and Carpenter (1974).
[2] Cultivars were "Labiro", "Frederello" and "Gallant".
[3] 24 h HPS light (3800 lux).
[4] 21 h fluorescent (lamps: 320 W m^{-2}).

Supplemental light

Greenhouse crops are provided with additional light by using high-intensity discharge (HID) lamps in months of low light. Metal halide and sodium lamps are the principal light sources used for supplemental light. Cut-flower crops are excellent candidates for supplemental lights because of their relative permanence in the greenhouse. Supplemental light increases yield and quality of all cut crops. The yield of gerbera increased 30–50% with sodium light levels of 54 μmol m^{-2} s^{-1} (Tsujita, 1983) and similar increases in yields occurred with cut roses (Moe, 1972; Armitage and Tsujita, 1979), chrysanthemums (Hughes and Cockshull, 1971; Cockshull, 1972) and snapdragons (Petersen, 1955; Hedley, 1974). Supplemental lights are generally burned for 18 hours per day (assuming non-photoperiodic effects) with a minimum light intensity of 50 μmol m^{-2} s^{-1}. Higher intensities may be used; the decision must be based on the return on investment and subsequent quality of the stems. For short-day plants such as chrysanthemums, supplemental light is provided during the long-day portion of growth. The length of time chrysanthemums remain under long days is directly related to the photosynthetically active radiation; in winter as many as 6 weeks may be necessary prior to application of short days, whereas in summer only 3 weeks of natural long days may be necessary. Therefore the use of HID lamps during the long-day period can save significant crop time in the greenhouse. If conditions are continuously cloudy, supplemental light may also be used during the short-day portion of the life cycle. Roses and gerberas, which are less affected by photoperiod than chrysanthemums, also benefit from supplemental light (Table 6.4).

In order to reduce cost per plant, supplemental lighting is generally provided when plants are young and densely spaced. The effectiveness of supplemental light is reduced as plants mature (Mastalerz, 1977). During the early stages of growth, plants may be placed close together, thus reducing the cost per unit. Also, a large proportion of the supplemental light is intercepted by the plant foliage. All cut crops respond positively to supplemental light, an important improvement in the greenhouse environment. While not all crops respond sufficiently to justify the additional expense of lamps and electrical power, the use of supplemental light for propagation, stock-plant maintenance or enhancement of yield and/or quality of flowers is common in many flower-producing regions of the world.

Photoperiod

Flowering of many cut crops responds significantly to photoperiod (Vince-Prue, 1975). Of the major crops, carnations, gypsophila and snapdragons are long-day plants, whereas the chrysanthemum is a short-day plant. Other crops such as *Cosmos bipinnatus, Cattleya* sp. and *Zinnia elegans* are classified as short-day plants while *Callistephus, Campanula* and *Phlox* are long-day plants. Some species (or cultivars) have an absolute requirement for the proper day/night period and are called qualitative or obligate plants for short days (*Dendranthema* × *grandiflorum*) or long days (*Campanula carpatica*). With other species, the response is simply accelerated, but the response (e.g., flowering) will occur regardless of photoperiod. Such plants are quantitative or non-obligate. Many interactions with photoperiod and temperature may be found in cut-flower species. The understanding and control of photoperiod and its attendant interactions are important for control of flowering under commercial conditions. The greenhouse environment may be manipulated to induce vegetative growth, flower initiation and development, or dormancy. As physiological processes are studied in more detail, photoperiodism appears to be involved in some aspect or other of most plant species. The flowering response of a few cut crops to photoperiod are briefly discussed below. The influence of photoperiod on annual statice (*Limonium sinuatum*), perennial aster (*Aster novae-angliae, A. novi-belgii*), snapdragon and baby's breath are discussed in detail later in the chapter.

Influence of photoperiod on specific crops

***Alstroemeria* hybrids (Peruvian lily):** Long days result in faster flowering than short days. The optimum photoperiod appears to be 12 to 14 hours, and more than 14 hours decreases total shoot production. The long-day treatment is not effective unless rhizomes have been subjected to cool temperatures (Healy and Wilkins, 1981).

***Callistephus chinensis* (annual aster):** Flowers develop most rapidly when a period of long days is followed by short days (Cockshull, 1985a). That is, plants are "primed" or induced to flower under long days and then develop more rapidly under short days after induction. If plants are given a continuous period of long days, they flower more slowly but on longer stems than those which are exposed only to short days, a treatment used for cut-flower production. Flowering of laterals is also delayed when plants do not receive short-day treatment (Goldsberry et al., 1989). Approximately 4–5 weeks of long days are necessary to "prime" the plant for short-day treatment. Less than 4 weeks of long days followed by short days do not accelerate flowering compared with plants which receive only short days. The critical long-day photoperiod is approximately 14 hours, and satisfactory long-day effects can be obtained by extending the daylength to 16 hours with incandescent lights, or by continuous light (Cockshull and Hughes, 1969).

***Freesia* spp.:** Flower initiation is enhanced by short days whereas flower development (the stage after initiation) is enhanced by long days (Gilbertson-Ferriss and Wilkins, 1978). Short days increase the number of flowers per raceme, the length of stem and the number of lateral flower stems (DeLint, 1969; Gilbertson-Ferriss and Wilkins, 1978).

***Gerbera jamesonii* (gerbera):** Short days promote flowering in some, but not all, cultivars of gerberas. Short winter days and low night temperatures (<13°C) increase flowering by causing increased development of basal branching, which gives rise to additional flowers as light levels increase (Lin and French, 1985).

***Gladiolus* spp.:** Long days, either natural or by applying night-break lighting in the greenhouse, reduce the incidence of blindness (shoots with no flowers) and improve the length and quality of the spikes (Shillo and Halevy, 1976; McKay et al., 1981). Long days, however, may delay flowering in some cultivars. Corm development continues during short days but is arrested when long days are applied and does not resume until after flowering (Shillo and Halevy, 1981). During winter production, day extension of 4–5 hours or a night break of 2–4 hours are used (McKay et al., 1981; Shillo et al., 1981).

***Liatris spicata* (gayfeather):** Flower initiation occurs regardless of photoperiod; however, plants respond to photoperiod through accelerated flowering and longer stems when given long days. If corms are provided with a sufficient cooling period, long days result in accelerated flowering. However, if plants receive a short cooling period, then short days enhance flowering (Berland, 1983). Similar interactions with forcing temperature were found if corms were forced in the greenhouse. At cool forcing temperatures (13°C), long days hastened flowering, but at warmer forcing temperatures photoperiod had little effect (Espinosa et al., 1987). For greatest flower acceleration, long days (14 hours) should be applied in the first 5 weeks after emergence of the foliage. Long days also result in greater stem elongation regardless of cooling treatments or forcing temperatures (Geller, 1981; Espinosa et al., 1987).

***Lilium* spp.:** In Easter lilies (*L. longiflorum*), application of long days can substitute for the cold requirement (Rees, 1985).

***Matthiola incana* (stock):** Long days provided during or before the cold treatment usually result in faster flowering at a lower leaf number compared to short days (Post, 1942; Biswas and Rogers, 1963; Heide,

1963). Long days can partially substitute for cold in some cultivars, and result in earlier flowering if temperature requirements are not fully satisfied (Roberts and Struckmeyer, 1939; Biswas and Rogers, 1963; Heide, 1963). Long-day treatments may be provided by day-length extension or night-break lighting. The response to long days is cultivar-dependent.

***Phlox paniculata* (summer phlox):** Phlox is a long-day plant; the critical photoperiod is between 8 and 16 hours. All plants of "Fairy's Petticoat" flowered when exposed to 16- and 24-hour days, but no flowering occurred under 8 hours, regardless of cold treatment (Iversen, 1989).

***Rosa* × *hybrida*:** No photoperiodic response is evident.

***Tulipa* hybrids:** No photoperiodic response is evident.

One of the reasons why the greenhouse is essential for profitable production of cut crops is because of the ability to control flowering time. Control of flowering time results in flowers being available for major holidays. The photoperiod may be manipulated in the greenhouse through the use of lights for long-day plants and shading for short-day plants. Flowering is inhibited in many short-day plants by night-break lighting with incandescent lamps. The time and duration of the night break differs with different species. Flowering in the short-day plant *Xanthium strumarium* (cocklebur) is inhibited with a few minutes of low-intensity lighting, but a single 12 minute interruption at 6400 lux did not inhibit flowering in chrysanthemum. However, when a total of 12 minutes was given intermittently at 600 lux, flowering was completely inhibited in chrysanthemum (Bothwick and Cathey, 1962; Cathey and Bothwick, 1964). Because of the response difference between species, growers maintain long days in the greenhouse by lighting for 2–4 hours in the middle of the night (night-break lighting), or by extending the day length, or lights may be cycled on and off for short periods of time (cyclic lighting). Cyclic lighting (for both photoperiodic and photosynthetic light treatment) is most cost-efficient, and is used in many countries to reduce lighting costs. Red and far-red wavelengths are essential for triggering the photoperiodic response by activation of phytochrome, a light-sensitive pigment in the foliage. Phytochrome exists in two interconvertible states with absorption maxima in the red and far-red regions of the spectrum. The red absorbing form (P_{660}) is converted into the far-red absorbing form (P_{730}) by red light or sunlight. Similarly, far-red light causes P_{730} to revert to P_{660}. During the day or through the use of lights, phytochrome mainly exists in the P_{730} form. When darkness begins, P_{730} is slowly converted to P_{660}. The longer the dark period, the more P_{660} is formed. When P_{730} predominates (when lights are used to extend the day length), flowering is inhibited in short-day plants and promoted in long-day plants. The form P_{730} also predominates when lights break up the dark period (night-break or cyclic lighting). Therefore, any light source rich in red light which extends the day or breaks up the night causes short-day plants to remain vegetative or long-day plants to flower. The most effective light sources for photoperiodic manipulation are those rich in red wavelengths, such as incandescent lamps. Fluorescent lamps and high-intensity discharge lamps are not as rich in red wavelengths, although growers have been successful with almost all light sources. The onset of darkness causes P_{730} to revert to P_{660}, and if a long night occurs, levels of P_{660} increase. When P_{660} becomes predominant, flowering is inhibited in long-day plants and promoted in short-day plants. Therefore, a long uninterrupted period of darkness (caused by long natural winter nights or by shading with cloth) results in buildup of P_{660}, and short-day plants initiate flowers. When cloth is used during summer months, the amount of light under the cloth should not exceed 10 lux. The previous discussion demonstrates that short-day plants flower, not because of the length of the day, but because of the length of the night. The length of the dark period required varies with the species or cultivar.

Photoperiodic response can, in many crops, be significantly influenced by temperature. For example, the critical photoperiod for flower-bud initiation in the chrysanthemum cultivar Encore is 13.75 hours at 10°C, 14.5 h at 15°C and 15.25 h at 25°C (Cathey, 1957). Chrysanthemums have been classified into three groups, depending on their photoperiod–temperature interaction (Cathey, 1955). Those in which temperature has little effect on photoperiod-controlled flowering are thermozero; cultivars in which temperatures below 15°C inhibit flower development (although initiation may not be affected) are thermopositive; and cultivars in which temperatures above 15°C delay flower development are thermonegative. Selection of thermopositive cultivars should be avoided for winter production, thermonegative cultivars should not be selected for summer production. Much of Cathey's work was done with cultivars which have since disappeared from commercial culture. Most cultivars used today in year-round flowering programs would be classified as thermozero, as they are capable of flowering in a wide range of temperatures (de Jong, 1978) and show no carry-over effect of spending the long-day period at 10ºC (de Jong and Smeets, 1982) or 25ºC (Runger, 1977). Other cut-flower species demonstrate similar temperature–photoperiod interactions. For example, *Delphinium* is a long-day plant at low temperature, but day-neutral at high temperatures; stock (*Matthiola*) is a long-day plant at high temperature, but day-neutral at low temperatures (Spector, 1956).

Temperature

The greenhouse provides an ecosystem in which day and night temperatures may be manipulated to control yields and quality of cut-flower crops. Temperature affects the growth and flowering of all cut-flower species by influencing basic physiological processes such as photosynthesis, respiration and nutrient uptake. Effects of poor temperature control are manifested in reduction of cutting production and quality on stock plants, abnormal flowers and foliage under temperature extremes, excessive wilting, chlorosis, internode elongation and delay of flowering. Plants respond to temperature at all stages of growth, and an optimum temperature at one stage may not be so at another stage. Although the influence of temperature is often described as an independent variable, study after study has shown that temperature closely interacts with light, carbon dioxide, water and other environmental inputs.

Temperature has a greater influence on respiration than on photosynthesis between 15ºC and 30ºC. In general, a 10ºC rise in temperature approximately doubles the respiration rate of most greenhouse species. As temperatures increase, the resulting reduction of carbohydrates derived from photosynthesis has the effect of greatly reducing the net gain of carbohydrates, perhaps to zero. Plant growth is determined by the difference in the quantity of carbohydrates manufactured in photosynthesis and the quantity consumed by respiration. To add dry weight, the rate of photosynthesis must be appreciably greater than the rate of respiration. Under sunny conditions, photosynthesis is 10–30 times greater than respiration; however, during dark, cloudy winter days, respiration rates may be similar to or even exceed photosynthesis. The interaction of day temperature and solar radiation was studied extensively in carnations. The optimum day temperature decreased as the solar radiation decreased over the season. When intensities were high ($>400\ \mathrm{cal\ cm^{-2}\ day^{-1}}$), optimum day temperatures for maximum yield were 18.3ºC; if light was less than $400\ \mathrm{cal\ cm^{-2}\ day^{-1}}$, the optimum temperature was 15.6ºC. Similarly, night temperatures were also adjusted, the highest optimum night temperatures occurring during the summer months, the lowest during the winter (Manring and Holley, 1960). The major effect of temperature on carnations is on flower-bud development, which is significantly delayed at low temperatures. The interval from initiation to visible bud is 40 days at 10ºC and 17 days at 20ºC; the time from visible bud to anthesis is 100 days at 10ºC and 33 days at 20ºC (Bunt, 1973; Bunt and Powell, 1980). A good deal of work to determine the influence of day and night temperature was conducted with chrysanthemum, a short-day plant. Essentially, the data suggest that the mean temperature over the entire daily cycle determines the time of flowering (Cockshull et al., 1982). To conserve energy, temperatures may be set lower than optimum. Work with temperature reduction showed that there was an improvement in quality; however this was often offset by a delay in maturity, reduction in yield, and an increase in the proportion of malformed flowers (Boodley and Seeley, 1960; Munoz and Holley, 1972).

Warm temperatures are necessary for propagation of many cut-flower species. Rooting of chrysanthemums and roses occurs most rapidly at media temperatures of 18–21ºC (Kofranek, 1992; Hasek, 1992) whereas carnations are best rooted at a medium temperature of 15ºC (Besemer, 1992). Germination of most species propagated from seed benefits from temperatures of 20–28ºC. High temperatures often delay flowering in greenhouse crops by delaying flower initiation. Temperatures greater than 26ºC can prevent floral initiation, and delay or distort floral development in chrysanthemums and kalanchoes (Whealy et al., 1987; Gehrke, 1988). Flowering plants are most sensitive to high temperature during the early stages of flower initiation, and this effect may be cultivar-dependent (Whealy et al., 1987).

Carbon dioxide enrichment

The greenhouse lends itself well to carbon dioxide injection to enhance yield and quality of greenhouse crops (Kessler and Armitage, 1993; Hicklenton, 1988). Carbon dioxide may be limiting in greenhouses during bright, cool winter days when air infiltration is restricted (Wittwer and Robb, 1964; Hicklenton, 1988). Levels of carbon dioxide may be reduced to 100–150 ppm if ventilation is terminated and plants are actively photosynthesizing. When this point is reached, carbohydrate reserves are depleted and plant growth is reduced (Mastalerz, 1977). Gaastra (1959) showed that an increase in temperature was necessary to increase the rate of photosynthesis at saturating levels of carbon dioxide or radiant energy. Attempts have been made to apply this theory in the greenhouse, but results have not always been consistent. For roses, with carbon dioxide levels above 1500 ppm, recommendations called for the delay of ventilation until the greenhouse reached 29°C. Although stem length and fresh weight were reduced, the increase in yield more than compensated for the decrease in quality (Hanan, 1973). However, an increase in carbon dioxide did not compensate for reduction in stem and flower quality of greenhouse roses when temperatures were raised above 30°C (Shaw and Rogers, 1964). With carnations, both yield and quality are reduced when day temperatures are set too high. Elevating day temperatures from approximately 15°C to approximately 20°C resulted in restricted plant growth, reduced yield and quality, even though carbon dioxide was elevated to 700 ppm (Holley et al., 1964). Additional research with carbon dioxide showed that significant increases in temperature were not necessary to increase yield and quality in carnations (Goldsberry and Holley, 1962; Holley and Juengling, 1963), roses (Goldsberry and Holley, 1962; Lindstrom, 1965; Mastalerz and Langhans, 1969) or chrysanthemums (Rogers, 1962; Koths and Adzima, 1964). Snapdragons vary in their response to elevated carbon dioxide. Some research showed dry-weight increases of 67–90% when carbon dioxide was raised to 1200–2000 ppm (Lindstrom, 1965). Others (Nelson and Larson, 1969) attributed the fact that benefits of elevated carbon dioxide were insignificant to the necessity of having to keep ventilators open to maintain optimal temperatures. Flower quality, but not yield, of cut gerberas increased with 600–800 ppm carbon dioxide (Tsujita, 1983). However, response to carbon dioxide treatment appears to be cultivar-dependent in gerberas. Work in the Netherlands demonstrated that the yield of "Marleen" decreased above 800 ppm, but the yield of "Appleblossom' and "Gosta" continued to increase until 1200–1600 ppm was reached (van Berkel, 1986). Without doubt, cultivar–environmental interaction occurs with all cut-flower species. Most crops benefit if the ventilators can remain closed while carbon dioxide is supplied; but the poorer the response to carbon dioxide enrichment, the more need for ventilation. In areas where ventilators remain closed, the only limiting factor restricting the use of carbon dioxide is the amount of solar radiant energy. Supplemental lighting is used in the greenhouse to enhance the response to carbon dioxide. The yield of roses increased 40% with 1000 ppm carbon dioxide fertilization, 140% with high-intensity discharge lamps and 226% when the two were supplied together (Langhans et al., 1973). The uptake of carbon dioxide by whole carnation plants increased with increasing irradiance (Enoch and Hurd, 1977; Enoch and Sacks, 1978).

PRODUCTION STUDIES OF FOUR GREENHOUSE CUT-FLOWER CROPS. SOME SPECIFICS

To illustrate some of the points made in the discussion above, specific information is presented for four greenhouse cut-flower crops. They have been chosen to illustrate the importance of environmental control on widely differing species, and to provide specific information on lesser-known but important commercial crops. Although the major crops such as roses and carnations have more economic value than the species chosen, vast amounts of literature are presently available on those crops. This is not the case for the four crops which follow.

As a result of studies on these crops, control of flowering in the greenhouse is possible, and economic production is feasible. The various headings are used to demonstrate the interaction between the crop and the greenhouse environment under which it is produced.

Limonium sinuatum (Plumbaginaceae)

Limonium sinuatum is known as annual statice, or "sinuata statice" in the trade. Plants grow approximately 60–90 cm tall and about 60 cm wide. The species is native to the Mediterranean region, and many cultivars have been bred and selected for cut-flower production. One of the most popular cut flowers in the world, annual statice continues to be produced in vast quantities. Breeding of annual statice has provided numerous hybrids of various strains and colors suitable both for drying and for fresh use. Flowers are produced in all flower-producing areas, and flourish in areas of warm days and cool nights. Flower stems are produced for use both fresh and dry; however, dry-flower usage is increasing most rapidly.

Propagation

Control of greenhouse microclimate is essential for optimum germination of statice seed. There are approximately 400 clean seeds per gram (350 raw seeds g^{-1}). Clean seed is more expensive than unprocessed seed but is easier to handle and germinates more rapidly. The seeds should be sown and covered lightly (approx. 0.5 mm deep) or left exposed to insure that light reaches them. At a temperature of 20–22°C seeds germinate in 5–14 days, depending on cultivar, and the seedlings may be transplanted in 3 weeks. Seed flats or beds benefit from 25% shade, applied by drawing shade cloth over them during the brightest hours of the day. Large plugs or transplants may be placed on the greenhouse bench or in the field 6–8 weeks after sowing. Propagation by tissue culture has become more important because it leads to high productivity, uniformity of habit and flower color, and a relatively greater resistance to disease.

Growth

Temperatures of 15°C are beneficial if provided for 1–2 weeks after the cotyledons emerge to encourage rooting. Plants flower more rapidly and are of better quality if vernalized when young, and should initially be grown at cool temperatures (10–12°C). Warmer temperatures result in non-uniform, poor-quality plants with a pale green color. The normal rosetting behavior is inhibited, and the stem begins to elongate the and leaves to turn up. The seedling is most responsive to vernalization, but cold treatment of the seeds is ineffective. Cooling may be started when plants reach the 2–4 leaf stage, and continued for 4–8 weeks (Socolovski, 1993). In many regions natural cool temperatures do not occur, and some growers germinate seed in the greenhouse and then place flats of seedlings or plugs in cold frames for cool-temperature treatment. As day length increases, the length of the cold treatment can be decreased. If seedlings are vernalized under long days, however, flowering is also enhanced.

Environmental effects

Photoperiod: Long days promote flowering in two ways. The optimum photoperiod is above 13 hours (Shillo and Zamski, 1985). If long days are provided during the seedling stage, the length of the vernalization treatment may be reduced. Long days without vernalization are relatively ineffective. Long days applied to actively growing plants induce a greater percentage of plants to flower, compared with short days, regardless of temperature. Long days applied to seedlings have little effect if not accompanied by cool temperatures. Long days are most effective after plants have been cooled and are actively growing, and they promote earlier flowering, a greater percentage of flowering plants, and higher yields (Table 6.5).

Table 6.5
Effect of photoperiod on flowering in annual statice [1]

Cultivar	Plants flowering (%)	
	long-day [2]	short-day [2]
"American Beauty"	100	48
"Midnight Blue"	100	55
"Iceberg"	100	83
"Twilight Lavender"	100	52

[1] After Semeniuk and Krizek (1973).
[2] Long days were 16 hours, short days were 8 hours. Plants were treated 8 weeks from sowing. Temperatures not reported.

Temperature: A cold treatment hastens flower initiation. High temperatures promote leaf initiation, leaf growth and stem elongation, but inhibit flowering. This helps to explain why non-vernalized plants placed under warm conditions in field or greenhouse grow well but have poor yield. Krizek and Semeniuk (1972) placed seedlings in controlled temperature chambers for 6 weeks. They showed that plants grown at 26/22°C day/night temperature formed a vegetative rosette and remained so for approximately 4 months. Day/night temperatures of 21/18°C caused flowering in only 20% of the plants but temperatures less than 18°C caused flower initiation in all plants. An optimum temperature regime was 15/13°C. In general, blue-flowered cultivars have the greatest cold requirement (lower temperature and longer duration) followed by lavender, pink, white and yellow (Semeniuk and Krizek, 1973).

In summary, temperature during the seedling stage is the most important constraint to flowering. Temperatures of 10–13°C should be applied at the seedling stage for 3–8 weeks depending on season and cultivar. This treatment may occur naturally in the field with spring-planted crops, or may be applied artificially in coolers for summer-planted crops.

The length of precooling varies with the season. This was shown in Israel, where plants are cropped throughout the winter; in fall plantings, a maximum flowering response was obtained with 8 weeks of precooling, but in early spring plantings only 3 weeks precooling was necessary (Shillo, 1977). The reduction in precooling time in the spring plantings was likely due to cooler night temperatures when the plants were young, and longer photoperiods as the plants matured.

After vernalization is completed, high temperatures speed up the flower development process, but should not be applied immediately after planting. In practical terms, this means that precooled plants should not immediately be transplanted to warm greenhouses or fields; spring transplanting in greenhouse or field is recommended. An optimum temperature regime for early and profuse flowering is 22/16°C day/night temperatures (Semeniuk and Krizek, 1973).

Watering and fertilization: In general, watering and fertilizing depends on the greenhouse environment; fertilizer is generally applied with the irrigation water, referred to as "constant liquid feed" (CLF). After every fourth or fifth CLF, plain water is applied to reduce the build-up of soluble salts. During the spring and summer, 3 to 4 irrigations per week are generally satisfactory. The ratio by weight of nutrients in the fertilizer is recommended to be 6:2:6:6:3 (N:P:K:Ca:Mg), and treatments should start with approximately 50 ppm N about 3 weeks after sowing. The concentration may be raised to 150 ppm nitrogen 6 weeks after sowing. The nitrate form of nitrogen and an application of boron is also recommended.

Gibberellic acid: Gibberellic acid (GA_3) has been used to reduce the need for precooling in annual statice. Work in Florida and Israel has shown that application of GA_3 partially overcomes the precooling requirement. Gibberellic acid is relatively ineffective on statice if the precooling requirements have been satisfied (Wilfret and Raulston, 1975; Shillo, 1977). As more precooling was applied, GA_3 became less effective, but with 0–12 days of precooling the application of GA_3 was effective in promoting flowering (Table 6.6).

The optimum application time and concentration are approximately 12 weeks after seeding (or when the rosette is 15–20 cm wide) and 500 ppm, respectively. Gibberellic acid should not be applied after the plants have already formed flower-buds. Ten ml of solution applied per plant is effective. Work by Wilfret and Raulston (1975) showed that "Iceberg" and "Midnight Blue" flowered in 134 and 160 days from seed, respectively, with a gibberellic acid application but required 238 and 229 days without GA_3.

Table 6.6
Influence of gibberellic acid (GA_3) on flowering of "Midnight Blue" statice

Number of days of precooling	Plants flowering (%)	
	with GA_3	without GA_3
0	83	40
6	94	42
12	100	72
28	100	95

Greenhouse culture

Seedlings are spaced at six plants m^{-2} and grow in a cool greenhouse (10–13°C) to initiate flowers. Cool temperatures must be maintained for at least 6 weeks. After the cool treatment, temperatures may be raised to 15–21°C night/day. In the absence of cool temperatures, a spray of 500 ppm gibberellic acid may be applied when plants are 15–20 cm across. Plants benefit by being maintained under photoperiods of approximately 16 hours. The photoperiod during the summer may be sufficiently long without the expense of additional lights; however, photoperiodic lighting is beneficial in the winter.

A nitrate–nitrogen fertilizer at a concentration of 100–150 ppm nitrogen at each irrigation may be applied when plants are grown cool, and the concentration is increased when temperatures are raised. After transplanting, Socolovski (1993) suggested three stages for plant fertilization: the vegetative stage (transplant to visible peduncles), 120–150 ppm N, 30–40 ppm P, 70–80 ppm K; flower development stage (visible peduncles to flower-buds coloring), 120–150 ppm N, 30–40 ppm P, 120–150 ppm K, 50–100 ppm Ca; production stage, 80–90 ppm N, 30–40 ppm P, 70–80 ppm K, 50–100 ppm Ca.

Productivity of 20–25 bunches (10–12 stems per bunch) m^{-2} yr^{-1} is possible. In areas with cool nights and bright days, plants can be in greenhouse production 52 weeks of the year. Plants sown in November and January will flower in March and May, respectively, depending on cultivar and location.

Stage of harvest

Harvest should take place when the individual flowers are mostly open and showing color. The white of the flowers (petals) should be visible and the rest of the flower color (sepals) well developed.

Antirrhinum majus (Scrophulariaceae)

Antirrhinum majus is grown almost everywhere in the world. Native to the Mediterranean area, it grows 1–2 m tall and 1 m wide. Cultivars have been bred to take advantage of seasonal variations in climate. The market for snapdragons fluctuates a great deal, primarily as a result of over- and under-production. When prices are low, production is decreased and, in time, prices become sufficiently strong to warrant additional production volume, resulting in another cycle. The length of time the high prices persist is directly related to the number of growers producing snapdragons. While this is true of all crops, snapdragons appear to be affected more than most. In the 1990's, the world market for snapdragons is strong and getting stronger. Crops are usually greenhouse-produced, but may also be field-produced in the spring.

Propagation

All plants are raised from seed, generally in plug trays with 200 to 400 cells. For optimum germination, seeds should be sown, watered in and cooled at 5°C for two weeks (Nau, 1989). Although cooling is beneficial, it is not a requirement. After cooling, the seeds may be germinated at 22–24°C under a fine mist. Seedlings emerge in 7–14 days. Seedlings may be propagated using a complete liquid fertilizer in the proportions by weight of 2N:2K:2Ca:1Mg and containing 50 ppm nitrogen at this stage. Nitrogen levels may be increased by 25 ppm per week until 100 ppm nitrogen is delivered.

Growth

After plants have three to five true leaves, night temperatures are lowered to 10–13°C and day temperatures are maintained as close to 15°C as possible. Temperatures above 17°C cause stem elongation, resulting in poor-quality transplants. Approximately 150–200 ppm of nitrogen is applied when the plants are grown in warm temperatures, but reduced by half when growing at 10–13°C. Plants are transplanted to the greenhouse bench as soon as they can be handled without damage (usually five to six weeks). In the greenhouse, transplanting may take place year-round.

Environmental effects

Photoperiod: The snapdragon is a quantitative long-day plant (Laurie and Poesch, 1932; Maginnes and Langhans, 1961). Although this is a typical response, some cultivars are virtually unaffected by day length, particularly the late-flowering cultivars. Plants grown under short days produce more vegetative growth; long days mainly affect the initiation of flowers. Once flowers have been initiated, the subsequent development is relatively unaffected (Maginnes and Langhans, 1961). Photoperiod also affects the juvenile stage, that is, the number of leaves which must form before flowers are initiated. The minimum number of leaves is higher under short days than long days (Cockshull, 1985b). Greenhouse (forcing) cultivars have been bred to flower virtually at any time of year under natural day lengths, and photoperiod is of less importance today than when older cultivars were used.

Light intensity: Snapdragons flower with stronger stems, more rapidly and with higher yields under high light. Low light results in more blind shoots and reduction in yield and stem quality. Light intensity is a major concern for winter greenhouse production in northern latitudes. Some cultivars have been bred for low light response and may be used for this purpose.

Temperature: Snapdragons have no vernalization requirement, and those grown cool (10°C or lower) initiate and open flowers later than those grown warmer (Maginnes and Langhans, 1960; Maginnes and Langhans, 1961). Although warm temperatures hasten flowering, they reduce spike length and stem strength (Post, 1955).

Greenhouse culture

Cultivar selection is important in greenhouse production. Winter cultivars are selected for winter forcing, summer varieties for summer. The density of plants may vary from 90 to 25 plants m^{-2}, depending on whether plants are pinched (with 3 to 4 nodes of terminal growth removed) or unpinched. After transplanting, as much light as possible must be provided. If cultivars are properly selected, manipulation of photoperiod is not particularly important nor generally practiced. However, providing long days (12–16 h) ensures flowering. Temperatures are set to 10–13°C throughout the crop cycle. Higher temperatures during the day are unavoidable in many crop-producing areas, and occasional temperatures of 20–22°C are not detrimental, if not too persistent. A relatively high nitrogen concentration (150–200 ppm) is maintained until plants are 40–45 cm tall, then nitrogen levels are reduced to 40–50 ppm. Potassium and calcium may be increased at this stage. Nitrogen should be approximately 85% in nitrate form and 15% in ammoniacal form (Socolovski, 1993).

Stage of harvest

For local production, the inflorescences are harvested when 1/2 to 2/3 of the flowers are open. This generally consists of 10 or more open flowers on the spike. For long-distance shipping, stems may be harvested when 1/3 of the flowers are open (Sacalis, 1993). For long-term storage, or if preservatives are to be used, flowers can be cut when only 2–3 buds are showing color (Nowak and Rudnicki, 1990).

Aster novae-angliae and *Aster novi-belgii* (Asteraceae)

Aster novae-angliae (New England Aster) and *Aster novi-belgii* (New York Aster, Michaelmas Daisy) are native to North America and have been extensively hybridized and bred for cut-flower production. Asters have evolved a complex temperature/photoperiod interaction, and optimum growth and flower production is highly dependent on manipulation of the greenhouse environment.

Propagation

All of the named cultivars are propagated by basal cuttings, often referred to as vegetative ground shoots. The stock plants are often produced through tissue culture and grown under cool temperatures (approximately 10°C). At these temperatures, stock plants may remain under short days to produce a rosette. Cuttings from stock plants held under cool temperatures and short days produce many lateral branches once transplanted. Unrooted cuttings are placed in peat/perlite or a comparably well-drained mix at 21–24°C under a humidity tent or intermittent mist. Cuttings root in 14–21 days. Rooting hormone is useful but not necessary. Cuttings should be maintained under a minimum daylength of 16 hours (McGrew, 1990) with incandescent lights, to keep plants vegetative.

Environmental effects

Aster novi-belgii has evolved into a highly regulated species in its flowering behavior. The species requires a specific sequence of vernalization to break dormancy, long days for stem elongation, shorter days for flower initiation, and finally very short days to induce dormancy. In nature, winter, long days of spring and summer, shortening days of late summer and finally the short days of late fall satisfy these requirements.

Vernalization: Plants should be chilled for 4–8 weeks at 5°C. Vernalization causes the plants to be more responsive to subsequent photoperiod (Schwabe, 1986). Vernalization also results in greater stem lengths of flowering shoots.

Photoperiod: Long days result in shoot elongation, and are necessary for flowering. Even without chilling, long days promote flowering, although the number of flowers is significantly less than with chilling. Approximately 4–5 weeks of long days of at least 18 hours are promotive, but continuous lighting (24 hours) is even more effective. The number of long-day cycles differs with cultivar; some may require up to 8 weeks of long days (Schwabe, 1986).

"Short days" are not particularly short for New England and New York asters. A short day of 14 hours is normal for many cultivars although some cultivars will even flower with 16 hour photoperiods. Day lengths below 12 hours result in flower abortion and onset of dormancy and should be avoided (Schwabe, 1986).

Temperature: During the short-day period, temperatures of 30°C and 22°C resulted in more flowers than 15°C. Increasing temperatures and longer photoperiods (although still effectively short-days) resulted in the greatest flower number (Schwabe, 1986). Low temperatures during short days caused a tendency to flower abortion.

Gibberellic acid: Gibberellic acid substitutes for vernalization and long days, provided the appropriate short days are given. Treatments are seldom applied commercially (see also below).

Greenhouse culture

For greenhouse forcing, cuttings or clumps should be vernalized. Rooted cuttings are cooled to 3–5°C for 4 to 6 weeks. Approximately 20 plants m^{-2} are placed in ground beds by the end of July. Photoperiods longer than 17 hours and temperatures of 15–18°C should be maintained. Approximately 8–10 stems are allowed to remain on each plant. If too many stems are allowed to mature, flower quality declines. Long days are applied until the stems are 30–45 cm long, followed by short days of 13–14 hours. Day lengths of less than 11 hours should be avoided. Photoperiods can be manipulated with incandescent lamps (100 lux) and shade cloth. Temperatures during the short days should average approximately 18°C for longest stem lengths. Flowering time is 6–10 weeks after the beginning of short-day treatment, depending on cultivar and temperature.

Table 6.7
Guideline for foliar analyses of *Aster* from field trials in Athens, Georgia and Watsonville, California[1]

Cultivar	Location	Element									
		N (%)	P (%)	K (%)	Ca (%)	Mg (%)	Fe (ppm)	Mn (ppm)	B (ppm)	Al (ppm)	Zn (ppm)
"Rosa Seiger"	Athens	3.0	0.45	3.29	1.68	0.20	168	65	37	83	72
"Climax"	Athens	3.1	0.65	3.64	0.98	0.18	162	88	39	48	121
	Watsonville	2.2	0.24	3.67	1.39	0.35	180	273	46	52	26

[1] Foliage was sampled from vigorously growing healthy plants when flower buds were visible, but prior to flower opening. These are guidelines only and should not be considered as absolute standards. Based on dry weight.

If noncooled cuttings are used, stems may be shorter and flowering delayed compared with cooled stems (Schwabe, 1986). Cool temperatures during long days should be maintained in the greenhouse for vernalization. Unheated greenhouses or plastic frames can be used for late spring and early summer crops. Cuttings planted in November–December will be vernalized naturally after approximately 6 weeks of temperature below 5°C. At that time, the long-day–short-day sequence described above may begin.

Gibberellic acid: Gibberellic acid is occasionally used to obtain better uniformity of the crop and to stimulate growth of plants after pruning. It may be applied at the concentration of 35–100 ppm during the early vegetative stage; late application results in leaf deformities.

Watering and fertilization: Asters require a good deal of water. It has been estimated that 6–7.5 $\ell\,m^{-2}$ of water is necessary under normal greenhouse conditions and 12–13.4 $\ell\,m^{-2}$ during summer days (Socolovski, 1993). Fertilization is generally provided at approximately 100–150 ppm N, 20–40 ppm P and 80–110 ppm K. A guideline for foliar analyses for two cultivars of *Aster* is given in Table 6.7.

Stage of harvest

Stems may be harvested when 2–4 flowers per inflorescence are open.

Gypsophila paniculata (Caryophyllaceae)

Gypsophila paniculata grows 1–1.5 m tall and up to a meter wide. The species originated from Europe and northern Asia. Although botanically a perennial, greenhouse-grown crops are generally treated as annuals. One of the mainstays of the cut-flower industry, *Gypsophila* is grown for the airy panicle of single and double flowers. The flowers are produced in the field and year-round in greenhouses.

Propagation

Seed is used for a few cultivars (e.g. "Snowflake") and the wild species, but most cultivars are propagated vegetatively because of the variability of the seedlings and the quality of resultant plants. In the 1940's and 1950's, double-flowered cultivars were grafted on the more vigorous single *G. paniculata* as a rootstock. Although still occasionally used, the practice is very labor-intensive and expensive, and has all but disappeared. Today, vegetative cuttings are taken from mother plants produced from meristem culture. Cuttings root in 3–5 weeks under mist and warm temperatures. Cuttings are more uniform and healthy, and flower production is equal to that of grafted plants.

Growth

Plants remain vegetative, developing a large rosette, under short days. Therefore, plants are grown under short days, high light and cool temperatures (10–13°C) until they are large enough to be transplanted to the greenhouse bench. Cool temperatures result in better branching than warmer (14–17°C) temperatures.

Environmental effects

Photoperiod: *Gypsophila* species are quantitative long-day plants, and flowering is inhibited by short days. In general, the critical photoperiod for flowering is 12–14 hours for most cultivars (van der Krogt, 1982). Plants must have at least 12 nodes before long days are perceived (Kusey et al., 1981). Under normal growing conditions, this occurs 5–9 weeks after transplanting. The more long-day cycles have been received, the more flowers are produced. Also, the longer the photoperiod the earlier the flowering (Shillo, 1985). Stem length and quality are greatest with 16–18 h photoperiods (Kusey et al., 1981).

Temperature: Night temperatures of approximately 13°C are necessary for plants to respond to subsequent long days. If temperatures are below 13°C, long-day induction will not be perceived (Shillo, 1985) and plants remain vegetative. At approximately 13°C all plants flower, but flowering is significantly delayed compared with plants grown under warmer temperatures (Shillo and Halevy, 1982; Moe, 1988). Day temperatures may be maintained by setting venting temperature to 25°C.

Rooted cuttings cooled at 1–3°C in the dark for 7 weeks become vernalized (Shillo, 1985). The cooling treatment results in plants flowering regardless of day length (Table 6.8).

Table 6.8
Effect of cooling rooted cuttings of *Gypsophila* at 1°C in the dark for 7 weeks[1]

Cooling	Flowering (%)		Number of flowering stems per plant	
	short-day	long-day	short-day	long-day
Noncooled	0	71	0	15.3
Cooled	80	97	15.3	28.5

[1] From Shillo (1985).

This use of precooled clumps permits flower production in cool greenhouses under natural winter day lengths and promotes winter flowering of field-grown plants in warm climates (e.g., Florida). Plants may also be cooled at 10 to 12°C for seven weeks if sufficient light is provided to maintain green leaf color (Shillo, 1985). Plants which had been cooled had more flowers under long days compared with those which had no vernalization (Shillo, 1985).

Carbon dioxide: Addition of carbon dioxide is effective in promoting growth and development (Shillo, 1985).

Light intensity: The number of flowers increases with an increase in light intensity. Lack of light is often a limiting factor in the northern greenhouse during the winter.

Gibberellic acid: Gibberellic acid causes elongation of the flower stems, but does not cause flower initiation. It may substitute for the warm temperatures necessary for plants to respond to subsequent long days but does not substitute for the long days themselves (Shillo, 1985). That is, plants respond to long days even below 13°C if GA_3 (250–500 ppm) is applied. Gibberellic acid is more useful in the winter than in the summer.

Greenhouse culture

Gypsophila can be produced year-round, but the main time for greenhouse culture is for winter and spring production.

Cuttings may be planted in August and September, and grown at 15–21°C until the plants have at least 12 nodes (3–5 weeks). Long days (18–20 hours) should be maintained until the stems are 30–40 cm long. Double-row spacing may be used, in which plants are spaced 45–50 cm apart and rows are about 50 cm apart. According to Socolovski (1993), pots may be used at a very dense spacing of 10×15 cm at the early stage, to reduce costs and improve handling. The use of potted plants provides more uniform plants for transplanting, and less time in the greenhouse between transplanting and flowering. Long days of at least 16 hours should be provided, at least until flower stalks are visible. Continuous lighting (24 hours) is better than lighting for 16 hours but night-break lighting and cyclic lighting may also be used. Incandescent bulbs are best for photoperiodic lighting. Night/day temperatures are set at approximately 15/21°C. Average day/night temperatures should not fall below 13°C. If precooled crowns or cuttings are used, long-day treatments are not necessary for flowering; however, weaker stems may result.

In northern latitudes, the use of sodium high-intensity discharge supplemental lights during early development enhances subsequent flower yield and quality (Hicklenton, 1987). Approximately 6 weeks of high-pressure sodium lighting (93 μmoles $m^{-2} s^{-1}$) applied after pinching (removal of terminal apex) resulted in flowers approximately 10 weeks later (Hicklenton, 1986). The number of flowering stems and the quality of stems and flowers are highly correlated with light intensity in northern climates. In southern regions, high intensity lighting is not as useful as in the North.

If gibberellic acid (250–500 ppm) is applied, the first application occurs at the beginning of the long day treatment. Additional applications may be made at 7 to 10 day intervals.

Recommendations for fertilization in the greenhouse have been developed by Socolovski (1993). Plants are initially fertilized with 60 ppm nitrogen (60% in nitrate form), 10 ppm phosphorus and 70 ppm potassium until new shoots form. During the growing-on stage and until shoots elongate, the concentrations should be

Table 6.9
Guideline for foliar analyses of *Gypsophila* from field trials in Athens, Georgia, USA [1]

N (%)	P (%)	K (%)	Ca (%)	Mg (%)	Fe (ppm)	Mn (ppm)	B (ppm)	Al (ppm)	Zn (ppm)
3.72	0.68	2.15	5.32	1.30	287	70	29	159	76

[1] Foliage was sampled from vigorously growing healthy plants when flower buds were visible, but prior to flower opening. These are guidelines only and should not be considered as absolute standards. Based on dry weight analysis.

130–180 ppm N (60% nitrate), 25–30 ppm P, and 100–140 ppm K. During floral development, Socolovski recommended 150–200 ppm N (70–80% nitrate), 30–40 ppm P, 150–200 ppm K and 100 ppm Ca. For final flowering and harvest time, his recommendations were 120–130 ppm N, 20–40 ppm P, 160–200 ppm K, and 100 ppm Ca.

Plants utilize high amounts of magnesium and calcium, as reflected in foliar analyses (Table 6.9).

Stage of harvest

For the fresh-flower market, stems should be cut with 60–70% open flowers. For drying, 80–90% of the flowers should be open. For immediate local sale of fresh flowers, 80–90% open flowers is also appropriate. For the export market, plants may also be harvested with 25–30% open flowers, then placed in a bud-opening solution (a solution of sucrose and biocides) until 60–70% are open.

GENERAL CONCLUSIONS

The greenhouse is a controllable ecosystem, without which the cut-flower industry would not exist. The ability to control light flux, photoperiod, carbon dioxide, irrigation and temperature is essential, as they determine flowering time, yield and quality.

The future of the greenhouse industry and that of the global cut-flower industry are intrinsically linked. The success of both depends on understanding the needs of the cut-flower crops and on the ability to provide those needs on demand through control of the greenhouse ecosystem.

REFERENCES

Armitage, A.M., 1993. *Specialty cut flowers, the production of annuals, perennials, bulbs and woody plants for fresh and dried cut flowers.* Timber Press, Portland, OR.

Armitage, A.M. and Tsujita, M.J., 1979. Supplemental lighting and nitrogen nutrition effects on yield and quality of Forever Yours roses. *Can. J. Plant. Sci.,* 59: 343–350.

Behe, B.H., 1991. 1992 holiday marketing ideas. *Society of American Florists Mag.,* (December) 8(9): 38–40.

Behe, B.H. and Wolnick, D.J., 1991. Type of floral product purchased and demographic characteristics and floral knowledge of consumers. *HortScience,* 26(4): 414–416.

Berland, M., 1983. *Growth and flowering in Liatris spicata,* M.S. Thesis. Agricultural University of Norway, Aas, Norway.

Besemer, S.T., 1992. Carnations. In: R.A. Larson (Editor), *Introduction to Floriculture*, 2nd edition. Academic Press, New York.

Biswas, P.K. and Rogers, M.N., 1963. The effects of different light intensities applied during the night on the growth and development of column stocks (*Matthiola incana*). *Proc. Am. Soc. Hortic. Sci.,* 82: 586–588.

Boodley, J.W. and Seeley, J.G., 1960. Some effects of adjusting night temperatures on growth of greenhouse roses. *Pa. Agr. Exp. Sta. Prog. Rept.,* 218.

Bothwick, H.A. and Cathey, H.M., 1962. Role of phytochrome in control of flowering of chrysanthemum. *Bot. Gaz.,* 125: 227–231.

Bunt, A.C., 1973. Effect of season on the carnation (*Dianthus caryophyllus* L.). II. Flower production. *J. Hortic. Sci.,* 48: 315–325.

Bunt, A.C. and Powell, M.C., 1980. *Ann. Rep. Glasshouse Crops Res. Inst., 1979*, pp. 92–93.

Carpenter, W.J., 1974. High intensity lighting in the greenhouse. *Michigan Agr. Exp. Sta. Farm Sci. Res. Rep.,* 255.

Cathey, H.M., 1955. Temperature guide to chrysanthemum varieties. *N.Y. State Flower Grower Bull.,* 119: 1–4.

Cathey, H.M., 1957. Chrysanthemum temperature study. F. The effect of temperature upon the critical photoperiod necessary for the initiation and development of flowers of *Chrysanthemum morifolium. Proc. Am. Soc. Hortic. Sci.,* 69: 485–491.

Cathey, H.M. and Bothwick, H.A., 1964. Significance of dark reversion of phytochrome in flowering of chrysanthemum. *Bot. Gaz.,* 125: 232–236.

Cockshull, K.E., 1972. Photoperiodic control of flowering in chrysanthemum, In: A.R. Rees, K.E. Cockshull, D.W. Hand and R.G. Hurd (Editors), *Crop Processes in Controlled Environments.* Academic Press, London.

Cockshull, K.E., 1985a. *Callistephus chinensis.* In: A.H. Halevy (Editor), *Handbook of Flowering*, Vol. 2. CRC Press, Boca Raton, FL, pp. 112–114.

Cockshull, K.E., 1985b. *Antirrhinum majus.* In: A.H. Halevy (Editor), *Handbook of Flowering*, Vol. 1. CRC Press, Boca Raton, FL, pp. 476–481.

Cockshull, K.E. and Hughes, A.P., 1969. Growth and dry-weight distribution in *Callistephus chinensis* as influenced by lighting treatment. *Ann. Bot.,* 33: 367–379.

Cockshull, K.E., Hand, D.W. and Langton, F.A., 1982. The effects of day and night temperature on flower initiation and development in chrysanthemum. *Acta Hortic.,* 125: 101–110.

de Jong, J., 1978. Selection for wide temperature adaption in *Chrysanthemum morifolium* (Ramat.) Hemsl. *Neth. J. Agric. Sci.,* 26: 110–118.

de Jong, J. and Smeets, L., 1982. The effect of day and night temperatures during long photoperiods on the vegetative growth and flowering of *Chrysanthemum morifolium* (Ramat.). *Sci. Hortic.,* 17: 271–275.

DeLint, P.J., 1969. Flowering in *Freesia*: Temperature and corms. *Acta Hortic.,* 14: 125–131.

Enoch, H.Z. and Hurd, R.G., 1977. Effect of light intensity, carbon dioxide concentration, and leaf temperature on gas exchange of spray carnation plants. *J. Exp. Bot.,* 28: 84–95.

Enoch, H.Z. and Sacks, J.M., 1978. An empirical model of CO_2 exchange of a C_3 plant in relation to light, CO_2 concentration and temperature. *Photosynthetica,* 12: 150–157.

Espinosa, I., Healy, W. and Roh, M., 1987. Interactions of photoperiod and temperature on flowering of *Liatris. Acta Hortic.,* 205: 113–119.

Gaastra, P., 1959. Photosynthesis of crop plants as influenced by light, carbon dioxide, temperature and stomatal diffusion resistance. *Meded. Landbouwhogesch. Wageningen,* 59(13): 1–68.

Gehrke, P.W., 1988. *The effects of supraoptimal temperatures on the growth and flowering of Kalanchoe blossfeldiana (Poelln.),* M.S. Thesis. University of Florida, Gainsville, FL.

Geller, Z., 1981. *Horticultural and physiological aspects in growing of Liatris spicata,* M.S. Agr. Thesis. Hebrew University of Jerusalem, Rehovat.

Gilbertson-Ferriss, T.L. and Wilkins, H.F., 1978. Flower production of *Freesia hybrida* seedlings under night interruption lighting and short day influence. *J. Am. Soc. Hortic. Sci.,* 103: 587–591.

Goldsberry, K.L. and Holley, W.D., 1962. Carbon dioxide research on roses at Colorado State University. *Colo. Flower Growers Assoc. Bull.,* 151: 1–5.

Goldsberry, K.L., Kell-Gunderson, L. and Silver, R., 1989. Scheduling single stem Japanese cut asters. Part 2. Winter and spring responses of Japanese cut asters. *Colo. State Univ. Res. Bull.,* 463.

Haak, M., Tap, H. and Haybroek, A.M.A., 1992. *A view of international competitiveness in the floristry business.* Rabobank Nederland, Eindhoven, the Netherlands.

Hanan, J.J., 1973. Ethylene pollution from combustion in greenhouses. *HortScience,* 8: 23–24.

Harris, G.P. and Harris, J.E., 1962. Effects of environment on flower initiation in carnation. *J. Hortic. Sci.,* 37: 219–234.

Hasek, R.F., 1992. Roses. In: R.A. Larson (Editor), *Introduction to Floriculture*, 2nd edition. Academic Press, New York.

Healy, W.E. and Wilkins, H.F., 1981. Interaction of soil temperature, air temperature and photoperiod on growth and flowering of *Alstroemeria* 'Regina'. *HortScience,* 16: 459.

Hedley, C.L., 1974. Response to light intensity and day-length of two contrasting flower varieties of *Antirrhinum majus* L. *J. Hortic. Sci.,* 49: 105–112.

Heide, O.M., 1963. Juvenile stage and flower initiation in juvenile stocks (*Matthiola incana* R. Br.). *J. Hortic. Sci.,* 38: 4–14.

Hicklenton, P.R., 1986. The effect of supplemental lighting on winter flowering of transplanted *Gypsophila paniculata. Can. J. Plant Sci.,* 66: 653–658.

Hicklenton, P.R., 1987. Flowering of *Gypsophila paniculata* cv. Bristol Fairy in relation to irradiance. *Acta Hortic.,* 205: 103–111.

Hicklenton, P.R., 1988. *Carbon Dioxide Enrichment in the Greenhouse,* Timber Press Growers Handbook Series, Vol. 2. Timber Press, Portland, OR.

Holley, W.D., 1959. Some major factors affecting quality and grade of chrysanthemums. *Colo. Flower Growers Assoc. Bull.,* 89: 1–3.

Holley, W.D. and Juengling, C., 1963. Effects of additional CO_2 and automated temperatures on carnations. *Colo. Flower Growers Assoc. Bull.,* 149: 1–3.

Holley, W.D., Goldsberry, K.L. and Juengling, C., 1964. Effects of CO_2 concentration and temperature on carnations. *Colo. Flower Growers Assoc. Bull.,* 192: 2–3.

Hughes, A.P. and Cockshull, K.E., 1971. A comparison of the effects of diurnal variation in light intensity with constant light intensity on growth of *Chrysanthemum morifolium* cv. Bright Golden Anne. *Ann. Bot.,* 35: 927–932.

Iversen, R.R., 1989. *Greenhouse forcing of herbaceous garden perennials,* Ph.D. Dissertation. Cornell University, Ithaca, NY.

Kessler, J.R. and Armitage, A.M., 1993. Effects of carbon dioxide, light and temperature on seedling growth of *Begonia* × *semperflorens-cultorum. J. Hortic. Sci.,* 68: 281–287.

Kofranek, A.M., 1992. Cut chrysanthemums: In: R.A. Larson (Editor), *Introduction to Floriculture,* 2nd edition. Academic Press, New York.

Koranski, D.S. and Laffe, S.R., 1985. Plug production. In: J.W. Mastalerz and E.J. Holcomb (Editors), *Bedding Plants III.* Penn. Flower Growers, University Park, PA.

Koths, J.S. and Adzima, R., 1964. The effect of CO_2 enriched atmosphere on snapdragons. III. *Conn. Flower Growers Supplement Bull.,* 356: 1–2.

Krizek, D.T. and Semeniuk, P., 1972. Influence of day/night temperature under controlled environments on the growth and flowering of *Limonium* 'Midnight Blue'. *J. Am. Soc. Hortic. Sci.,* 97: 597–599.

Kusey Jr., W.E., Weiler, T.C., Hammer, P.A., Harbaugh, B.K. and Wilfret, G.J., 1981. Seasonal and chemical influences on the flowering control of *Gypsophila paniculata. J. Am. Soc. Hortic. Sci.,* 106: 84–88.

Langhans, R., Seeley, J. and Hammer, P.A., 1973. HID lights on roses. *Florists Rev.,* 152(3952): 22.

Laurie, A. and Poesch, G.H., 1932. Photoperiodism – the value of supplementary illumination and reduction of light on flowering plants in the greenhouse. *Ohio Agric. Exp. Stn. Res. Bull.,* 512: 1–42.

Lin, W.C. and French, C.J., 1985. Effect of supplemental lighting and soil warming on flowering of three *Gerbera* cultivars. *HortScience,* 20: 271–273.

Lindstrom, R.S., 1965. Carbon dioxide and its effect on the growth of roses. *Proc. Am. Soc. Hortic. Sci.,* 87: 521–524.

Maginnes, E.A. and Langhans, R.W., 1960. Daylength and temperature affect flower initiation and flowering of snapdragons. *N.Y. State Flower Grow. Bull.,* 171(1).

Maginnes, E.A. and Langhans, R.W., 1961. The effect of photoperiod and temperature on initiation and flowering of snapdragon (*Antirrhinum majus* – variety Jackpot). *Proc. Am. Soc. Hortic. Sci.,* 77: 600–607.

Manbeck, H.B. and Aldrich, R.A., 1967. Analytical determination of direct visible solar energy transmitted by rigid plastic greenhouses. *Trans. Am. Soc. Agr. Eng.,* 10: 564–567, 572.

Manring, J. and Holley, D.W., 1960. Optimum temperatures for carnations in Colorado. *Colo. Flower Growers Assoc. Bull.,* 128: 1–3.

Mastalerz, J.W., 1977. *The Greenhouse Environment.* Wiley, New York.
Mastalerz, J.W. and Langhans, R.W., 1969. Roses. *Pennsylvania Flower Growers*, Penn. State Univ., State College, PA.
Mattson, R.H. and Maxwell, T.J., 1971. Studies of greenhouse orientation in a heliodon. *HortScience,* 6: 209–210.
McGrew, J., 1990. Personal communication.
McKay, M.E., Blyth, D.E. and Tommerup, J.A., 1981. The influence of photoperiod and plant density on yield of winter-grown gladioli in Queensland. *Sci. Hortic.,* 14: 171–179.
Moe, R., 1972. Effect of daylength, light intensity and temperature on growth and flowering of roses. *J. Am. Soc. Hortic. Sci.,* 97: 796–800.
Moe, R., 1988. Flowering physiology of gypsophila. *Acta Hortic.,* 218: 153–158.
Munoz, F. and Holley, W.D., 1972. Another look at temperature effects on carnations. *Colo. Flower Growers Assoc. Bull.,* 267: 1–7.
Nau, J., 1989. *Ball Culture Guide, The Encyclopedia of Seed Germination.* Ball Seed Co., West Chicago, IL.
Nelson, P.V., 1991. *Greenhouse Operation and Management.* Prentice Hall, New York.
Nelson, P.V. and Larson, R.A., 1969. The effects of increased CO_2 concentrations on chrysanthemum (*Chrysanthemum morifolium*) and snapdragon (*Antirrhinum majus*). *N.C. Agric. Exp. Stn. Tech. Bull.,* 194: 1–7.
Nowak, J. and Rudnicki, R.M., 1990. *Postharvest Handling and Storage of cut Flowers, Florist Greens and Potted Plants.* Timber Press, Portland, OR.
Petersen, H., 1955. Artificial light for seedlings and cuttings. *N.Y. State Flower Grow. Bull.,* 122: 2–3.
Post, K., 1942. Effects of daylength and temperature on growth and flowering of some florist crops. *Cornell Univ. Agr. Exp. Sta. Bull.,* 787: 58–61.
Post, K., 1948. Daylength and flower bud development in chrysanthemums. *Proc. Am. Soc. Hortic. Sci.,* 51: 590–592.
Post, K., 1955. *Florist Crop Production and Marketing.* Orange Judd Publ., New York.
Post, K. and Howland, J.E., 1946. Influence of nitrate level and light intensity on the growth and production of greenhouse roses. *Proc. Am. Soc. Hortic. Sci.,* 47: 446–450.
Rees, A.R., 1985. *Lilium.* In: A.H. Halevy (Editor), *Handbook of Flowering*, Vol. 1. CRC Press, Boca Raton, FL, pp. 288–293.
Roberts, R.H. and Struckmeyer, B.E., 1939. Further studies on the effects of temperature and other environmental factors upon the photoperiodic responses of plants. *J. Agric. Res.,* 59: 699–709.
Rogers, M.N., 1962. Carbon dioxide found to accelerate mum growth. *Florists Rev.,* 130(3376): 21–22.
Rogers, M.N. and Tjia, B.O., 1990. *Gerbera Production,* Timber Press Growers Handbook Series, Vol. 4. Timber Press, Portland, OR.
Runger, W., 1977. Temperatur und Nachtunterbrechung bei chrysanthemum. *Gartenbauwissenschaft,* 42: 246–248.
Sacalis, J.N., 1993. *Fresh (cut) Flowers for Designs. Postproduction Care and Handling.* Ball Pub., Batavia, IL.
Schwabe, W.W., 1986. *Aster novi-belgii.* In: A.H. Halevy (Editor), *Handbook of Flowering*, Vol. 5. CRC Press, Boca Raton, FL, pp. 29–41.
Semeniuk, P. and Krizek, D.T., 1973. Influence of germination and growing temperature on flowering of six cultivars of annual statice (*Limonium* cv.). *J. Am. Soc. Hortic. Sci.,* 98: 140–142.
Shaw, R. and Rogers, M.N., 1964. Interactions between elevated carbon dioxide and levels and greenhouse temperatures on the growth of roses, chrysanthemums, carnations, geraniums, snapdragons and african violets. *Florists Rev.,* 135(3489): 21, 56–60; 135(3491): 19, 37–39.
Shillo, R., 1977. Influence of GA and the number of cold days on flowering of statice 'Midnight Blue'. *Ann. Rept. Dep. Ornamental Hortic.,* Hebrew Univ. Rehovot: pp. 26–40. In Hebrew.
Shillo, R., 1985. *Gypsophila paniculata.* In: A.H. Halevy (Editor), *Handbook of Flowering*, Vol. 3. CRC Press, Boca Raton, FL, pp. 83–87.
Shillo, R. and Halevy, A.H., 1976. The effect of various environmental factors on flowering of gladiolus. II. Length of the day. *Sci. Hortic.,* 4: 139–146.
Shillo, R. and Halevy, A.H., 1981. Flower and corm development in gladiolus as affected by photoperiod. *Sci. Hortic.,* 15: 187–196.
Shillo, R. and Halevy, A.H., 1982. Interaction of photoperiod and temperature in flowering control of *Gypsophila paniculata* L. *Sci. Hortic.,* 16(4): 385–393.
Shillo, R. and Zamski, E., 1985. *Limonium sinuatum.* In: A.H. Halevy (Editor), *Handbook of flowering,* Vol. 3. CRC Press, Boca Raton, FL, pp. 292–301.
Shillo, R., Valis, G. and Halevy, A.H., 1981. Promotion of flowering by photoperiodic lighting in winter-grown gladiolus planted at high densities. *Sci. Hortic.,* 14: 367–375.
Socolovski, I., 1993. Personal communication.
Spector, W.S., 1956. *Handbook of Biological Data.* W.B. Saunders Co., New York.
Tsujita, M.J., 1983. Greenhouse gerbera production. *Ont. Min. Agr. Food Factsheet,* 83066: 1–4.
van Berkel, N., 1986. CO_2 enrichment in the Netherlands. In: H.Z. Enoch and B.A. Kimball (Editors), *Carbon Dioxide Enrichment in Greenhouse Crops,* Vol. 1. CRC Press, Boca Raton, FL.
van der Krogt, T.M., 1982. Betere bloeispreiding in *Gypsophila* door teeltplanning. *Bloemisterij,* 6: 34–35.
Vince-Prue, D., 1975. *Photoperiodism in Plants.* McGraw Hill, London.
Walker, D.A., 1992. Excited leaves. *New Phytologist,* 121: 325–345.
Whealy, C.A., Nell, T.A., Barrett, J.E. and Larson, R.A., 1987. High temperature effects on growth and floral development of chrysanthemum. *J. Am. Soc. Hortic. Sci.,* 112: 464–468.
Wilfret, G.J. and Raulston, J.C., 1975. Acceleration of flowering of statice. *HortScience,* 10: 37–38.
Wittwer, S.H. and Robb, W.M., 1964. Carbon dioxide enrichment of greenhouse atmospheres for food crop production. *Econ. Bot.,* 18: 34–56.

Chapter 7

INDOOR ORNAMENTAL FOLIAGE AND FLOWERING POT PLANTS

C.A. CONOVER, T.A. NELL and R.T. POOLE

INTRODUCTION

Ornamental plants are those crops that are selected for their aesthetic value rather than for their nutritional value in human or livestock consumption or for their use for other human purposes, such as medicine, clothing or rubber. They are used by the final consumer in interior locations in pots, larger containers or beds. One commonly used definition of a foliage plant is "any plant grown primarily for its foliage and utilized for interior decoration or interior landscape purposes" (Conover et al., 1971), although some areas in the tropics and subtropics can utilize foliage plants in outdoor locations also. Flowering plants are marketed for the aesthetic value of flowers in homes, offices and commercial buildings. Some plants, such as *Kalanchoe blossfeldiana* (kalanchoe), are commonly maintained as foliage plants after the flowers have died. In contrast, some plants produced by foliage growers, such as *Spathiphyllum* (spathiphyllum) and *Aphelandra squarrosa* (aphelandra), are not generally marketable without the presence of a flower.

The ecosystem of the greenhouse, as a means for the production of ornamental foliage and flowering crops, varies widely in location and type of crop because of the capability of humans to engineer a controlled-environment atmosphere in almost any global location. These controlled-environment operations are modified to simulate the natural conditions in which the original ornamental foliage and flowering crops were native. In the case of flowering potted plants, the greenhouse environment is modified to induce flowering year-round, as with *Dendranthema grandiflora* (chrysanthemum), or for holiday periods with *Euphorbia pulcherrima* (poinsettia) and *Lilium longiflorum* (Easter lily).

The ability to grow crops in the greenhouse in cold climates depends largely on the reliability of the heating system. Coal, oil and gas are the primary fuels used to heat greenhouses. Wood, waste heat and geothermal energy can also be utilized. To conserve energy, curtains designed for energy saving can serve as blankets over crops to reduce losses by radiation and convection. Curtains or even plastic coverings can reduce energy costs by as much as 30% to 75%. Summer cooling is also important, and is accomplished by natural ventilation, fan-and-pad systems and, in some greenhouses, fog systems. Plant growth in greenhouses can also be improved by carbon dioxide enrichment, photoperiodic control and supplementary greenhouse lighting.

The foliage and flowering crop "ecosystem" is unique since its components are artificially created and maintained specifically to benefit one dominant life form, which is best described as an alien to the natural system from which the components are derived.

BACKGROUND INFORMATION

Taxonomy

Foliage and flowering potted plants are a diverse group and include pteridophytes, gymnosperms, and both monocotyledons and dicotyledons of the angiosperms. Thousands of different types of plants are sold as ornamentals; these are from many genera, many species within the genera, and many cultivars within some of the species. A cultivar is a subgroup within a species that is sufficiently different from the remaining members of the species to be easily distinguished. Cultivars may vary in growth rate, as in *Ficus* (ficus) (Ottosen, 1988), flower color, leaf shape and color, number of lateral shoots and flowers produced and response to the production and post-production environment (Nell,

1989), as with the response of chrysanthemum cultivars to high production temperatures (Whealy et al., 1987a,b), *Cyclamen persicum* (cyclamen) cultivars to growing and interior conditions (Molinar and Williams, 1977) and poinsettia to epinasty following sleeving (wrapping in paper or plastic for protection) and transport conditions (Nell, 1990). Ornamental plants are often bought and sold by cultivar name only, but the binomial should be used as it specifies the exact plant in question.

Native habitat

Foliage potted plants are mostly natives or derivatives of natives of the tropics where climatic conditions are conducive to rapid growth, and variations in climatic tolerance levels are developed because of the changes in environment, ranging from rainforests to dry, which may be due to differences in land elevation, cloud coverage and rainfall in various regions. Many of these climatic variations are also seen in the subtropics, and some foliage plants are natives of this zone. Only a few foliage plants are native to temperate regions. Flowering potted plants, however, are native to a wide range of climatic regions, varying from tropical to alpine regions. Plants chosen for commercial production fit into the high-volume production systems which have been developed to grow crops for the mass market.

Commercial products

Popularity of any given ornamental plant may vary; some plants have a yearly supply/demand cycle, especially some flowering potted plants used for holidays and special events, while others show a steady, almost constant, year-round desirability. Prominent groups of foliage plants, some of which remain in constant demand because of the introduction of attractive new cultivars, include: *Aglaonema* (aglaonema), *Dieffenbachia* (dieffenbachia), *Dracaena* (dracaena), *Epipremnum aureum* (epipremnum), ferns, *Ficus*, *Maranta leuconeura* (maranta), palms, *Peperomia* (peperomia), *Philodendron scandens* subsp. *oxycardium* (philodendron), *Spathiphyllum* and *Syngonium podophyllum* (syngonium) (Table 7.1).

Sales of flowering pot-plants, on the other hand, are dominated by holiday usage, as poinsettias during the Christmas season and Easter lilies at Easter. Chrysanthemums, potted *Rosa* spp. (roses), kalanchoes,

Table 7.1
Most important species of pot plants for foliage and flowers in the U.S.A., listed by percent of wholesale value of sales

Foliage plants	% [1]	Flowering pot-plants	% [2]
Dracaena	9.4	*Euphorbia pulcherrima* (poinsettia)	28.9
Epipremnum	9.1	*Dendranthema* (chrysanthemum)	13.9
Ficus	6.5	*Rhododendron* (azalea)	8.6
Dieffenbachia	5.9	*Lilium* (Easter lily)	4.8
Palms (several genera)	5.8	*Saintpaulia* (African violet)	3.6
Philodendron	4.7		
Aglaonema	4.0		
Brassaia/*Schefflera*	3.5		
Spathiphyllum	3.3		
Hedera	3.1		
All other combined	44.7	All other combined	40.2

[1] McConnell et al. (1989). [2] U.S.D.A. (1994).

Sinningia speciosa (gloxinia) and other flowering potted plants are grown year-round for daily use in homes and offices, or for special occasions, such as weddings and birthdays. Availability of a larger selection of flowering plants year-round has been possible by altering the environmental conditions under which the plants are produced, thus inducing flowering outside of the natural season. Market demand varies with country, time of year and climatic conditions. Sales of flowering pot-plants and fresh cut flowers are dominated by holidays and special occasions in the United States, whereas daily and weekly use of flowering pot-plants and fresh cut flowers is common in Europe. Per capita consumption of flowering plants in Holland ($66.02) and Denmark ($101.40) is considerably greater than in the United States ($43.31) (Table 7.2). Danish buyers purchase more plants than cut flowers, whereas Dutch customers purchase a larger number of fresh flowers (Lembke, 1992). Consumers in the United States are nearly equally divided in the use of plants and flowers. Plant selection varies among countries also. For instance, potted roses are the most popular flowering plants in Denmark, while poinsettia is the most economically important flowering plant in the United States (U.S.D.A., 1994). Kalanchoe is the leading flowering potted plant in Holland (Flower Council of Holland, 1992).

Table 7.2
Per capita consumption[1] in Netherlands currency (Dfl) of cut flowers and flowering pot-plants in various countries (1989–1990)[2]

Country	Cut flowers	Flowering pot-plants	Total
Norway	110	130	240
Switzerland	105	120	225
Sweden	77	119	196
Denmark	63	129	192
Italy	127	51	178
Germany	87	83	170
Austria	67	64	131
Netherlands	77	48	125
Belgium	59	56	115
France	51	46	97
Japan	90	–	–
United States	45	37	82
Greece	34	34	68
United Kingdom	39	14	53
Spain	23	17	40

[1] Currency fluctuations will have influenced the calculation of these amounts.
[2] From Flower Council of Holland (1992).

One method of ensuring the demand for a given plant is by providing a number of different interior-design usages through creative "packaging". Use of small pots or large specimen pots, totem pole designs, hanging baskets, trees and any number of unique items are marketing tools to which consumers respond.

PROPAGATION

Propagation methods

Most flowering and foliage potted plants now grown commercially propagate easily, but good cultural management practices are essential to ensure quality, percentage and speed of rooting. Foliage plants can be propagated by cuttings, seed, air layers, spores, division and tissue culture (Conover and Poole, 1970). Flowering plants are propagated primarily by cuttings, seeds or tissue-culture procedures.

Until the advent of tissue-culture propagation, the most common method of propagation had been the use of vegetative cuttings – either stem, tip, single or double eye, leaf or cane. For many plants, this method continues to be preferred. Selection of the specific propagation method depends on plant form and availability of propagative material, and may also depend on the position of the cuttings on the stock plant (Hansen and Mose, 1987) and the time frame desired to bring a crop to completion, since some vegetative-cutting methods take longer for the crop to reach saleability than other methods. Limited research has been conducted on the effects of storage duration and storage temperatures on foliage-plant cuttings (Poole and Conover, 1988a,b), and generally show that, for optimal results, cuttings should be planted immediately. Cuttings of some flowering plants, such as chrysanthemum and *Dianthus caryophyllus* (carnation), may be stored for up to two weeks provided temperatures are maintained at 2°C to 4°C with 90% humidity. Storage at cool temperatures reduces transpiration, respiration and injury due to ethylene. Cuttings of plants which are sensitive to chilling, such as poinsettia and *Hibiscus rosa-sinensis* (hibiscus), should be immediately inserted into the rooting medium or stored overnight at 10°C to 13°C. Extended storage (more than 24 hours) reduces rooting of these cuttings. Low-pressure storage extends the storage life of poinsettia and *Pelargonium hortorum* (geranium) by two weeks compared to conventional storage procedures (Eisenberg et al., 1978). Successful low-pressure storage of cuttings is dependent on plant species, since recent studies with hibiscus demonstrated that rooting was adversely affected by storage (Andersen and Kirk, 1986). Currently, low-pressure storage is not used extensively due to lack of reliable commercial equipment at reasonable cost.

Seed propagation is an important method for some ornamental plants. For many foliage-plant species, seeds are not available, plant type is not consistent from seeds, or seed storage procedures have not been developed, but *Araucaria*, *Brassaia*, *Coffea*, *Dizygotheca*, *Podocarpus* and palms are propagated almost exclusively from seeds. Many tropical seeds decline in viability immediately after harvest; therefore, they should be planted when harvested. If storage is necessary, most tropical seeds store best at warm temperatures and high humidity. Although there is some information about time of planting and percentage of propagation (Mikorski and White, 1977; Perry and Boodley, 1980), little information is available concerning the best methods for handling, storing and germinating most tropical ornamental seeds. Limited data on palm-seed germination requirements are available (Poole and Conover, 1974; Broschat and Donselman, 1986; Carpenter, 1988a,b; Lyhne and Hegnelt, 1988) and generally show that palm seeds should be stored at warm temperatures.

Propagation from seeds has a significant role in a number of the minor flowering potted-plant crops – cyclamen, *Senecio* × *hybridus* (cineraria), *Gerbera* sp. (gerbera daisy) and gloxinia. Breeders have been successful in developing dwarf gloxinia and fast-crop dwarf cyclamen, which have resulted in renewed customer interests in these traditional crops. Seed suppliers keep these seeds under precise storage conditions to assure good germination, and specialized propagators germinate and grow prefinished plants for commercial finishing growers.

Four propagation methods used nearly exclusively in the production of potted foliage plants are: air layering, division, tissue culture, and use of spores. Air layering is still used for some foliage plants such as *Codiaeum*, *Ficus* and *Monstera*, but the associated high labor costs and need for very large stock areas make this method less favorable than in the past and one reason why growers are starting to use tissue cultures for these genera.

Sansevieria is the major foliage plant still being propagated by division, although this method is labor-intensive and transmission of pathogens, insects and/or diseases can occur easily, making this method a less favorable option. Some foliage plants previously propagated by division, such as *Calathea*, are now commonly propagated through tissue culture.

Spores are still used to propagate some ferns, although saleable ferns can take 1 to 2 years of production time when propagated by spores. Some ferns are also propagated by offsets or division; but tissue culture has also become the primary method used with ferns.

As indicated, tissue culture is currently very important in propagation of many foliage plants. There are several reasons for the increased popularity of tissue-cultured plants. First, new cultivars can be multiplied rapidly and their genetic consistency maintained with the use of tissue culture. It has also been shown that micropropagated plants may be more vigorous (Kristiansen, 1991). Additionally, tissue-culture methods allow rapid reproduction of plants in an environment free of diseases and pests, which has made it the method of choice for many cultivars. In fact, several popular foliage genera including *Dieffenbachia*, *Nephrolepis*, *Spathiphyllum* and *Syngonium* are now more commonly propagated through tissue culture than by any other means. Table 7.3 provides a listing of other plants propagated by tissue culture.

Table 7.3
Plants produced in tissue culture

Adiantum tenerum	*Episcia cupreata*
Aechmea fasciata	*Ficus benjamina*
Anthurium andraeanum	*Mammillaria woodsii*
Begonia semperflorens	*Nephrolepis exaltata*
Caladium × *Hortulanum*	*Peperomia obtusifolia*
Coffea arabica	*Platycerium stemaria*
Cordyline terminalis	*Saintpaulia ionantha*
Cryptanthus bivittatus cultivar Minor	*Spathiphyllum floribundum*
Dracaena deremensis cultivar Warneckii	*Syngonium podophyllum*
Dieffenbachia maculata	*Tupidanthus calyptratus*
Epipremnum aureum	*Yucca* sp.

Removal of tissue-cultured plants from the test tube and placement of them into the greenhouse was formerly a major obstacle to the use of tissue culture, since the change in environment was so extreme for the plantlets. With the development of procedures that facilitate the adaptation of plantlets from the high-humidity, low-light, and controlled-temperature environment of the tissue-culture laboratory environment to the more changeable, yet still controllable, conditions of the greenhouse environment (Poole and Conover, 1983; Henley, 1986), tissue-culture propagation has become more successful and, therefore, increasingly popular.

The species of plant grown and method of propagation will determine the facilities and the type of propagative materials required for successful production. For instance, commercial growers must either provide the space and environmental conditions required for stock plants, or purchase seeds or unrooted, callused, or rooted cuttings from specialist growers/propagators. Foliage-plant growers in tropical regions usually produce their own stock plants for cuttings since environmental conditions are conducive to their production. Flowering-plant producers, on the other hand, usually purchase cuttings, rooted or unrooted, from specialist propagators. Growers choosing to purchase seeds or unrooted cuttings from specialist growers must be able to provide optimum germination/rooting conditions. Otherwise, they must rely on specialist propagators who can provide an optimum propagation environment.

Stock plants

Stock plants, plants produced for the exclusive production of cuttings, must be grown in a manner which provides: proper light, optimum temperature, nutrition, and water; a well-drained growing substrate; freedom from insects and diseases (Knauss et al., 1981); and availability of an ample supply of electricity if photoperiodically sensitive plants are being grown. Failure to provide lighting to stock plants of chrysanthemum or kalanchoe will result in floral initiation in the plants prior to harvest of the cuttings (Horridge and Cockshull, 1989). Stock-plant producers must constantly monitor stock to assure the absence of insects and diseases, and use only clean plants in establishing stock beds. For instance, dieffenbachia cuttings used for stock plants must be free of *Erwinia,* and stock plants of chrysanthemum and geranium must be subjected to a series of laboratory and greenhouse tests, and to rigorous selection and sanitation procedures to eliminate vascular pathogens in the cuttings ("culture-indexing"). Culture-indexing insures that cuttings from the stock plants are virus-free (Oglevee-O'Donovan, 1993). Depending upon the crop, stock plants can be field-grown, or grown in raised beds in the greenhouse or in hanging baskets.

The age of stock plants cannot be overlooked. Cutting production declines as stock plants age (Eng et al., 1985), except on large foliage plants maintained outdoors, regardless of production techniques or environmental conditions. The optimum production time for stock plants depends on the plant being grown. For instance, chrysanthemum stock should be replaced every 4 to 5 months to obtain the maximum number of high-quality cuttings, whereas some *Dracaena* species produce highest yields at 10 to 15 years of age.

Growers must evaluate the actual costs of stock-plant production relative to the costs of purchasing unrooted or rooted cuttings from specialist growers and propagators. For instance, growers in temperate areas may find the costs of heating greenhouses for stock-plant production exceed the value of the cuttings produced, thus justifying purchase of cuttings from propagators in tropical or subtropical regions. In either case, facilities should be available for storage of cuttings unless cuttings will be used on-site and inserted into the rooting medium quickly after harvest.

Site selection

Stock plants should be produced in locations providing environmental conditions which favor rapid plant growth with good branching and strong stems, while maximizing the production of vegetative cuttings.

Outdoor field stock: When choosing a location for outdoor/shadehouse stock-plant production, factors such as wind speed and direction, minimum and maximum temperatures, slope of land, substrate characteristics and rainfall should be considered before selecting a location. All of these factors can have a significant effect on yield as well as the cost of production of the crop. For instance, high and consistent winds can cause mechanical damage, tipburn and/or wilt to foliage as well as structural failure of the entire plant, while temperature extremes may cause entire crop failure or decreased production during important sales periods. Temperature tolerances of different crops must be considered in order to locate an operation with optimum cost–benefit balance for any given plant type, since heating and/or cooling requirements may make the location unprofitable. In tropical areas, stock plants can be grown under natural shade trees or full sun, but commonly they are placed under polypropylene shade structures so that some control over the environment can be achieved (Fig. 7.1). Overhead irrigation, utilizing the latent heat of fusion of ice, is often used if cold protection is necessary in non-heated structures (Stamps and Haman, 1991), but shadehouses in the areas subject to occasional freezing are usually covered with plastic in winter and are generally equipped with some type of temporary heating system. At very extreme temperatures, cold

Fig. 7.1. Shadehouses are extensions of greenhouse ecosystems where light, temperature and irrigation can be manipulated.

protection by overhead irrigation is only effective for subtropical and tropical crops that can withstand temperatures of 0°C to –1°C, since it may be necessary to freeze ice on the plants (Stamps, 1992, 1993). In colder areas such as Europe or the northern United States and Canada, stock is grown in controlled-environment greenhouses to protect the plants from frost.

Most foliage and some other types of stock plants, such as chrysanthemums and azaleas (*Rhododendron* spp.), grow well in ground beds although saturated substrate can reduce cutting yield, therefore, slope of the land, substrate characteristics and need for amendments, and amount of rainfall should be carefully considered before initiating ground-bed production. Additionally, disease and insect problems are usually more difficult to control in ground beds.

Greenhouse stock: When selecting a production site, producers of stock in the greenhouse must, in addition to considering the factors discussed under outdoor field stock, emphasize light – both intensity and quality, temperatures – especially minima and durations, and total cost of energy – electricity, gas, oil, coal, etc. Although most stock plants may be grown in greenhouses, increased costs of production can limit this as an alternative. Greenhouses are expensive to build and maintain, and these costs will directly affect the profitability of stock production and should be considered before finalizing decisions on whether propagation materials should be produced on-site, or purchased.

Cultural systems

Substrate selection for stock plants will influence other cultural activities and is, therefore, of utmost importance. A well-drained substrate is necessary for good growth. Most stock plants will grow in ground beds, but native soils usually need to be amended to maximize production. Addition of peat will improve water- and nutrient-holding capacity of sandy soils and aeration of heavy clay soils. Bark, fired clay, coarse sawdust and rice hulls are other materials that will improve aeration and, therefore, yields in most substrates. Fertilizer supplements will be necessary for most native soils.

Bench growing of stock plants is preferable to ground-bed production, especially for many foliage and flowering plants, for several reasons: substrate and air temperatures are higher in benches; insect, nematode and disease problems are reduced for bench-grown plants; and working conditions are more comfortable, which improves employee productivity. The selection of the substrate to be used will depend upon ingredient characteristics, availability and cost. Some commonly used mixes that provide a good growing environment for stock-plant production in raised benches and containers include peat moss, peat-moss–pine-bark mixtures (1:1 by volume), peat-moss–builders'-sand mixtures (4:1 by volume) and peat-moss–perlite mixtures (3:1 by volume). Components should not decompose rapidly because of the extended period of time that stock plants remain on benches.

Raised benches should provide sufficient volume for adequate root growth, drainage and aeration. Some stock plants, such as dieffenbachia, are grown in individual containers to avoid rapid spread of diseases, such as *Erwinia*, from plant to plant. Wire bottoms and sides 15 cm high provide good protection for most containerized plants, but 10-cm sides are used for some smaller pot sizes. Wood, cement, aluminum and other types of solid materials are sometimes used for benches, especially if stock plants are not in pots; but spaces for aeration and drainage should be provided.

Irrigation for stock plants must be provided as needed, frequency and amount will depend on light intensity, humidity, temperature and plant use. Irrigation of plants grown in full sun or under shadecloth is usually accomplished with impulse or spinning sprinklers, while those in a greenhouse may be under an overhead system, on individual drip tubes or in an ebb-and-flow system. Since fertilizer is often applied with irrigation (i.e. "fertigation"), the importance of a properly engineered system which will provide optimum and uniform coverage cannot be overemphasized. In bed production with benches, irrigation tubing which allows seepage through the entire surface wall or which emits water at specific intervals is preferred. Spacing at 30-cm intervals across the bench will provide excellent coverage when used with most substrates and will prevent an increase in soluble salts in the substrate. Use of non-overhead irrigation systems minimizes problems of foliar disease by keeping the foliage dry (Knauss, 1971), and again emphasizes the importance of careful consideration in the selection of watering systems. Also, when foliage is kept dry, use of pesticides is decreased, thus reducing foliar residues. If problems of foliar disease are minimal for the plants grown, overhead sprinklers or spray stakes can be utilized.

The light intensity under which the stock plants

are grown affects the size, shape, color and density of foliage, carbohydrate level, insect and disease susceptibility and yield and, therefore, cutting quality and ability to root. The optimum light level for stock plants varies with the plant species but, in most cases, is similar to light levels used for production of the crop. Production of stock plants under low light levels results in thin cuttings, extended internodes and small leaves. Foliage plants grown under high light intensities usually are not attractive, but they yield a high number of cuttings. Cutting yield of *Epipremnum aureum* grown under 40% shade was almost double the yield of cuttings taken from plants grown under 80% shade (Conover and Poole, 1974). Yield of *Peperomia obtusifolia* increased 77% when shade level was decreased from 80% to 60%. Since cuttings taken for propagation, and to be grown on for several months subsequently, do not need to be attractive, stock production is usually done under high light conditions.

Spacing of cuttings in stock-plant beds often affects the number of cuttings obtained on flowering plants since these plants require high light for optimum growth and branching. Light must reach a large part of the leaf canopy to maximize photosynthesis. In some cases, a row of cuttings is omitted from the center of the stock-plant bed in order to allow additional light into the center of the bed. For the grower, spacing of stock plants in the bed is a balance between quality and cost of cutting production. Too many plants per unit area results in low-quality cuttings, proper spacing will produce good quality at a marketable price, and too few plants per unit area provide excellent cuttings at an unreasonably high cost.

The optimum temperature for stock-plant production varies between 14°C and 20°C. Some foliage stock plants may be produced at temperatures up to 35°C without impairing cutting quality or yield. Stock plants are commonly grown at warmer temperatures to hasten growth. In flowering-plant production, these higher temperatures are acceptable since environmental conditions necessary for floral initiation are not important in successful production of stock plants.

One final cultural procedure which must be considered to optimize a sustained yield from foliage stock plants is harvesting technique. Taking too many cuttings from a plant at any one time will reduce the photosynthetic capacity of the plant and have a negative impact on future yields.

Fig. 7.2. Mist propagation of *Philodendron scandens* ssp. *oxycardium*.

Propagation systems

The most important requirements for a propagation system are to minimize physiological stress (desiccation and temperature) on the cuttings, and to root them as quickly as possible. This is accomplished by placing the cuttings in an environment which will provide optimum moisture, temperature and light conditions.

The most common propagation system is a mist or high-pressure fog area (Fig. 7.2) where unrooted cuttings are inserted into the medium in a cell-pak or tray system on a bench, and then pulled and potted when they are rooted. The frequency of misting depends on light intensity, temperature, humidity and plant type, and should be timed to keep some moisture on the foliage at most times for most plant types. In warmer climates where temperature and humidity present an extremely stressful environment, a fine-spray mist or fog is set for very frequent intervals. The fine mist or fog keeps the foliage moist and lowers the temperature, but does not saturate the substrate; therefore, it can be used frequently without serious disease problems. White polyethylene tents or covers over propagation units can be used in the cool climates of Europe or North America to provide 100% humidity, thus eliminating the need for misting. However, the use of propagation tents in high-light areas can result in high temperatures which will restrict root initiation and development. Additionally, for foliage plants, tents should not reduce light levels below those used for production so that plants will be acclimatized to production conditions.

In addition to moisture, the physical characteristics and temperature of the substrate are the major factors which control the efficiency of rooting. The major requirements for a propagation substrate are good aeration, good drainage, proper pH and cation exchange capacity; satisfying these requirements will provide an optimum rooting environment and readily available nutrients for developing roots. Ornamental plants have been propagated in a large variety of mixes. The most common propagation mixes contain sphagnum peat moss – either as the only ingredient or in combination with perlite, polystyrene foam, vermiculite or pine bark – amended with dolomite for proper pH adjustment. A number of synthetic materials, such as rockwool and open-celled polyphenolic foams, are being used successfully in propagation of flowering plants. These materials provide a consistent reproducible propagation substrate with good physical characteristics which are conducive to good root initiation and development. Incorporated fertilizer may be a helpful addition for the propagation of some plants. Nutrient mist is sometimes used on cuttings, but this is not recommended until roots are initiated because of algal problems.

Substrate temperature greatly influences the rate of root initiation; propagation beds should be maintained at 21ºC or higher, but temperatures above 30ºC will inhibit root growth of some species (Poole and Waters, 1971; Conover and Poole, 1973; Hartmann and Kester, 1983; Carpenter, 1989; Vogelezang, 1991). Since heating the substrate is much more efficient than heating the air with space heaters, providing heat to the substrate is the more cost-effective choice. Methods of supplying heat to the substrate include heating pads, substrate cables, hot-water pipes and under-bench hot-air tubes. Germination chambers with substrate temperatures maintained at 27ºC to 32ºC have been beneficial for many seed propagated foliage plants, especially palms. A high temperature (40ºC) for 2 to 3 weeks is necessary prior to germination in *Butia capitata* to terminate thermal dormancy (Carpenter, 1988a,b). Scarification to remove the embryo cap may substitute for high pre-germination temperature (Carpenter et al., 1993). Most seeds of flowering-plant crops require substrate temperatures of 20–22ºC for optimum germination (Cathey, 1969).

The time of transplanting rooted cuttings and their treatment during transplanting greatly influence the growth and quality of plants (Murakami and Rauch, 1984). Transplanting should be done as early as possible after roots are developed to decrease the potential for root damage, and methods employed should create the minimum stress possible on the cuttings. In recent years some producers have shifted to a direct-stick method in which cuttings are placed directly into the pot in which they will be sold. This method decreases handling of the cuttings, and is especially adapted to plants which root rapidly and grow quickly, such as *Hedera*, *Kalanchoe* or *Pelargonium*. Direct-stick propagation is dependent upon a high rooting percentage and a large area equipped with misting and irrigation capability which will be available for the whole period from stick to sale; while this is common in southern production areas where growth rates are rapid, in cooler regions where plant growth may be slower it is not always practical unless the plants are sold at a small size.

PRODUCTION

Practices vary greatly among producers of ornamental plants, but an understanding of the interrelationships among the factors of fertilization, light and temperature is important to produce good-quality plants. Knowledge of the interaction of the various cultural and environmental factors is critical to making logical decisions leading to production of high-quality plants at a profit without application of excessive amounts of fertilizer or water, or the use of unnecessarily high temperatures.

Substrates

Important characteristics to consider when selecting a production substrate are aeration (depending on both capillary and non-capillary pore space), moisture retention or water-holding capacity, nutrient retention or cation exchange capacity, and pH. Other factors such as uniformity of the substrate, consistent availability, specific gravity or weight, and the cost of the substrate or its components and mixing costs, are also important.

Potting substrates used to grow ornamental plants range from entirely organic materials to a mixture of organic and inorganic materials, although organics are used considerably more often (Worrall, 1981; Tosi and Tesi, 1987; Hansen, 1988; Pivot, 1988; Rishani and Rice, 1988; Welleman and Smulders, 1988; Wang, 1989).

Lime, to adjust pH, is usually added at the time the potting substrate is prepared; the best pH range appears to be between 5.5 and 6.5, where nutrient availability

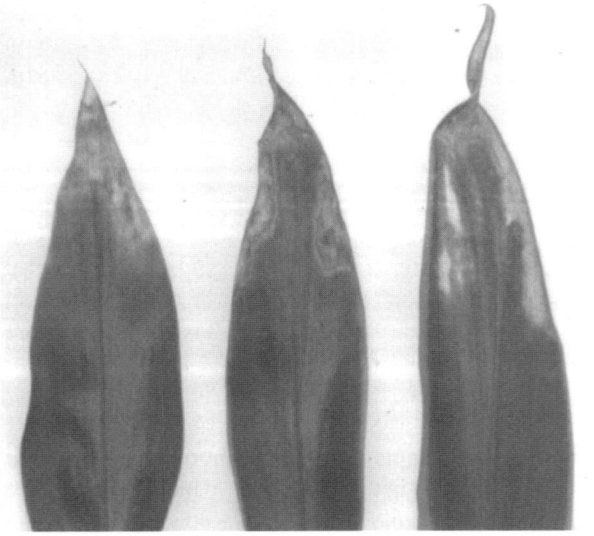

Fig. 7.3. Effects of fluoride toxicity on *Cordyline terminalis* cultivar "Baby Doll".

is optimal, although many excellent plants have been produced outside this range. Ferns and marantas grow satisfactorily at 5.0–6.0, whereas *Cordyline terminalis* (cordyline, ti) and *Dracaena* are usually more attractive when grown at a pH of 6.0–7.0. Similarly, azaleas develop good growth at 5.5 while chrysanthemums are best grown between 5.8 and 6.5. Although calcitic limestone is sometimes used, dolomitic limestone is more common because it provides magnesium, as well as calcium to the plants. Superphosphate, which contains about 2% fluoride, is sometimes incorporated into the substrate to supply phosphorus and some microelements, but its use should be restricted to plants that are not sensitive to fluoride (Fig. 7.3) (Conover and Poole, 1971; Marousky, 1981; Poole and Conover, 1981b; Ben-Jaacov et al., 1984). Production of *Dracaena* and *Lilium* is affected severely when either a substrate or irrigation water containing fluoride is used. Micronutrients are also frequently added to the potting substrate if these nutrients will not be supplied by fertilizer application. In fact, *Euphorbia* has such high micronutrient requirements that micronutrient incorporation into the growth substrate, in addition to fertilization with minor elements throughout production, is recommended (Hammer, 1988).

Some common ingredients of potting substrates and an indication of their properties are shown in Table 7.4, which also includes mixes used in the commercial trade. Aeration and water- and nutrient-holding capacities are good for all mixes shown.

Table 7.4
Substrates for production of potted plants and their physical and chemical characteristics

Substrate	Aeration	Water-holding capacity	Cation exchange capacity	Weight
Individual components				
Composted pine bark	H[1]	M	M	M
Perlite	H	L	L	L
Polystyrene foam	H	L	L	L
Sand	M	L	L	H
Sphagnum peat moss	M(V)	H	H	L
Vermiculite	M	M	M	L
Combinations				
Peat:bark (1:1)	H	H	H	L
Peat:bark:polystyrene foam (2:1:1)	H	M	H	L
Peat:bark:vermiculite (2:1:1)	H	H	H	L
Peat:perlite (2:1)	H	M	M	L
Peat:sand (4:1)	M	H	H	H

[1] Abbreviations used: H, high; M, average; L, low; V, variable.

Many growers have resorted to commercially prepared growing substrates in order to have a uniform and consistent substrate year-round. To optimize plant quality, irrigation and fertilizer practices should be adjusted to the characteristics of the selected substrate and the growth environment.

Spacing

The quality of ornamental plants is dependent upon plant spacing. Spacing is the balance between quality of the finished plant and profitability for the grower. Production of the largest number of plants per unit area is of little value if many of the plants are not marketable. Plants must have sufficient space to develop an acceptable shape without excessive and unnecessary elongation. Plants spaced too closely often have elongated shapes, sparse branching or breaks, abscission of lower leaves, excessive stem elongation and increased susceptibility to foliar diseases. Additional applications of plant growth regulators are required for closely spaced plants compared to plants produced with wider spacings. Closely spaced plants are also more susceptible to attacks by plant pests because of lack of adequate air movement, higher humidity among the plants, and difficulty in applying pesticides properly.

In the production of plants for foliage, small pots (less than 15 cm) have often been positioned at their final spacing at the beginning of the production phase to save on labor costs, whereas larger pots, in order to save on production space, are usually placed with the pots touching until the roots are well established and the plants are growing, before they are moved to their final spacing. Traditionally, a three-move spacing system has been used on some flowering potted crops to conserve production space and to match the environmental requirements during production with the stage of development. Labor costs for movement of the pots has restricted the use of this practice in recent years. However, the use of robotics, mechanical pot-moving systems, movable benches and computer control of the greenhouse environment, together with increased costs associated with greenhouse space, makes it possible to use this three-move system in modern greenhouses, for both flowering and foliage crops, at acceptable costs. Depending upon plant variety, spacing varies from one to six times the diameter of the pot. In many cases, spacing is determined by pot size. Many foliage plants grown in 10-cm pots are placed pot-to-pot, whereas flowering plants in containers of the same size are placed on 25-cm centers. Larger flowering plants, such as poinsettia and azalea grown in 15-cm pots, are placed on 30 to 38-cm centers. European flower growers producing plants in 11-cm or smaller pots often keep plants pot-to-pot throughout the production cycle, in view of the small size of the plants marketed throughout Europe. Minimum spacing should be used for all crops to maximize production efficiency, provided that crop quality is not jeopardized.

Fertilizer

Fertilizer type, the pH of the growing medium, production environment, and nutrient balance and form determine the effectiveness of the fertilizer program used for potted plants, whether grown for flowers or foliage. Many types and combinations of fertilizers can be used to produce ornamental plants of excellent quality with nutrient levels in the tissues within the ranges indicated in Table 7.5. Soluble fertilizers, organic or controlled release, applied frequently or over long durations are acceptable if applied and utilized properly. The ratio of N:P:K in fertilizers applied to plants also varies, with a "balanced" fertilizer (1:1:1, $N:P_2O_5:K_2O$) being one of the most popular for both foliage- and flowering-plant crops. Research has shown that a mixture in the ratio 3:1:2 (Poole and Conover, 1977, 1981d, 1985; Ben-Jaacov et al., 1982; Scharpf and Grantzau, 1985) supplies sufficient phosphorus and potassium for foliage plants. For some flowering-plant crops, a 1:0:1 formulation, provided phosphorus is incorporated in the substrate, supplies sufficient phosphorus and potassium (Nelson, 1978). Other ratios have provided optimum results on other flowering crops. For instance, *Begonia × hiemalis* (elatior begonias) develop more side shoots and grow faster with 2 parts nitrogen to 1 part potassium, whereas azaleas develop superior growth with a 3:1 (N:K) ratio (Nelson, 1978). A 1:1:1 fertilizer may supply excess phosphorus and potassium to plants that do not utilize them, which increases the cost of fertilizer and the electrical conductivity of the potting substrate without additional benefit to plant growth or quality, and can add unnecessary quantities of phosphorus and potassium to the groundwater.

The ratio of nitrate to ammonium in the fertilizer may affect growth and flowering in flowering potted plants, also. High levels of ammonium in the growth substrate can be toxic and may result in stunted growth, chlorosis, and poor and irregular flowering and, in severe cases, death of the plant. Ammonium toxicity is most commonly observed when fertilizer solutions contain more than 40% of the nitrogen in the urea or ammonium forms, and under cool, low-light conditions. Consequently, higher nitrate-nitrogen levels are used in the winter months when environmental conditions may lead to ammonium toxicity.

Some foliage plants have a fairly restricted fertilizer requirement (Poole and Conover, 1975) (Fig. 7.5), and are easily damaged by use of excess fertilizer. Other foliage plants produce optimum growth over a wide nutritional range (Ying and Joiner, 1960; Conover, 1969; Poole and Conover, 1986; Chase and Poole, 1987, 1989), but this should not encourage wasteful use of resources or damage to the environment. It is always best to utilize the least amount of fertilizer that will produce a quality crop. Suggested fertilizer levels for foliage plants can be found in Table 7.6. Levels of fertilizer application vary widely among the numerous flowering potted plants grown commercially, and references should be reviewed for specific species/cultivars to be grown (Nelson, 1978; Larson, 1992). High levels should not be used for foliage or flowering crops toward the end of the growing cycle because excess fertilizer in the substrate may cause problems with soluble salts

Table 7.5
Nutrient tissue levels (% dry weight) associated with high-quality foliage and flowering pot-plants[1]

Plant name	N	P	K	Ca	Mg
Foliage plants					
Aglaonema 'Fransher'	2.5–3.5	0.0–0.4	2.5–3.5	1.0–1.5	0.3–0.6
Brassaia actinophylla	2.5–3.5	0.2–0.4	2.5–3.5	1.0–1.5	0.3–0.6
Chamaedorea elegans	2.5–3.0	0.2–0.3	1.0–2.0	0.4–1.0	0.3–0.4
Dieffenbachia maculata	2.5–3.0	0.2–0.4	3.0–4.5	1.0–1.5	0.3–0.8
Dracaena 'Massangeana'	2.0–3.0	0.1–0.3	1.0–2.0	1.0–2.0	0.5–1.0
Epipremnum aureum	2.5–3.5	0.2–0.4	3.0–4.5	1.0–1.5	0.3–0.6
Ficus benjamina	1.8–2.5	0.1–0.2	1.0–1.5	2.0–3.0	0.4–0.8
Maranta leuconeura	2.0–3.0	0.2–0.3	3.0–4.5	0.5–1.5	0.5–1.0
Philodendron scandens ssp. *oxycardium*	2.0–3.0	0.1–0.3	3.0–4.5	0.5–1.5	0.3–0.6
Syngonium podophyllum	2.5–3.5	0.2–0.3	3.0–4.5	0.4–1.0	0.3–0.6
Flower plants					
Dendranthema (chrysanthemum)					
pot	1.8–2.2	0.4–1.0	0.8–2.4	1.1–1.5	0.4–1.1
standard	4.5–6.0	0.2–1.2	3.5–10.0	0.5–4.6	0.1–1.5
Dianthus (carnation)	3.4–3.9	0.3–0.4	2.8–3.3	1.3–1.5	0.2–0.6
Euphorbia (poinsettia)	4.51	0.97	2.78	0.74	0.57
Lilium (Easter lily)	2.8–4.0	0.1–0.3	1.5–4.1	0.2–2.2	0.2–0.5
Pelargonium (geranium)	3.3–4.8	0.4–0.7	2.5– 4.3	0.8–1.2	0.20–0.52
Rhododendron (azalea)	2.0–2.3	0.3–0.5	0.8–1.0	0.2–0.3	0.2–0.3
Rosa (rose)	3.0–5.0	0.2–0.3	1.8–3.0	1.0–1.5	0.25–0.35

[1] Adapted from Criley and Carlson (1970) and Poole et al. (1976).

Fig. 7.4. Effect of low, optimum and excess fertilizer application on *Chrysalidocarpus lutescens*.

when plants are moved into the low light of interior rooms where they frequently receive minimal irrigation (Fig. 7.5).

Foliage plants do not require large amounts of micronutrients, and can often obtain the required amount from irrigation water and the potting substrate. Palms require larger amounts than most other foliage plants, and supplemental micronutrients should be applied to the substrate. However, micronutrients are commonly applied to foliage plants as an insurance against deficiencies and will do no harm if not excessive, although boron can be phytotoxic in small amounts. Many fungicides contain copper, iron, manganese or zinc and may supply treated plants with both pathogen protection and these micronutrients. Micronutrients are provided for flowering potted plants as a substrate supplement before planting and as part of the standard daily/weekly fertilizer program (Nelson, 1978). Otherwise, micronutrient deficiencies can limit the marketability of these crops, as observed with molybdenum deficiency of poinsettia (Hammer, 1988).

Environmental conditions may restrict nutrient up-

Table 7.6
Suggested light and fertilizer levels for production of some potted plants in the greenhouse [1]

Foliage plants	Light intensity (klx)	Average fertilizer requirement ($g\,m^{-2}\,yr^{-1}$)		
		N	P	K
Aphelandra squarrosa cvs.	10–16	170	50	110
Brassaia spp.	53–64	200	70	130
Chamaedorea spp. (palms)	26–37	140	40	90
Dieffenbachia spp.	26–42	140	40	90
Dracaena spp.	32–37	140	40	90
Epipremnum spp.	32–42	140	40	90
Ficus spp.	53–64	200	70	130
Hedera spp.	32–42	140	40	90
Maranta spp.	16–21	100	30	60
Peperomia spp.	26–37	70	20	40
Philodendron spp.	26–37	170	50	110
Spathiphyllum spp.	16–26	170	50	110

[1] Adapted from Conover and Poole (1990).

take and result in a deficiency of certain nutrients even though adequate levels are available in the growing substrate. For instance, bract-edge burn in poinsettias has been associated with calcium deficiency, high humidity, high night temperatures and low light. Bract-edge burn is prevalent in the final 3 to 4 weeks of the production phase if light levels are low and temperatures are cool, thus restricting the translocation of calcium to the bracts. Applications of foliar sprays of calcium minimize, but do not totally eliminate, the problem. Recent information has suggested that cultivars sensitive to bract-edge burn have low calcium and high potassium in the bract margins as a result of low calcium uptake by the bracts (Stromme et al., 1994).

Fig. 7.5. Leaf deformity of *Dieffenbachia* caused by excess levels of soluble salts in the substrate.

Environmental factors

Genetic constitution and environmental factors each play a part in determining growth and quality of plants. Each plant has genetic characteristics that set the limits to its leaf and flower color, leaf shape, size and growth rate, lateral branching and time to flowering; but the display of these characteristics is controlled by the environment under which they develop.

Temperature

Optimum production temperature for most foliage and flowering plants ranges from 16°C to 32°C, depending on species, crop age, stage of development, other environmental conditions (light and CO_2) and the desired growth process – rooting, germination, floral initiation and development, and leaf unfolding, expansion and development. In general, foliage potted plants can be grown successfully with warmer temperatures than flowering potted plants, since floral initiation requires, in most cases, temperatures below 28°C under commercial procedures. Additionally, potted foliage plants will grow equally well at constant temperatures, probably because these correspond with their native habitats, whereas most flowering potted plants grow best with fluctuating day/night temperatures (Went, 1944). Went used the term "thermoperiodicity" to describe responses to fluctuating day and night temperatures.

Species respond in different ways to low air temperatures – both duration of exposure and temperature affect plant response. An extremely low temperature for a short duration or a less extreme temperature for a longer duration may cause chilling damage or death to certain plants (Marousky, 1980; Fooshee and McConnell, 1987); damage from chilling may be immediately apparent or may take several days to appear. Many species of ornamental plants are damaged by temperatures of 7°C, whereas others can tolerate temperatures below freezing (Hummel and Johnson, 1985). Growth rate of some plants, such as *Aglaonema*, *Dieffenbachia*, *Epipremnum*, *Euphorbia* and *Hibiscus*, is reduced by temperatures near 10°C, whereas *Ardisia*, *Cineraria*, *Cyclamen*, *Hedera*, *Pittosporum* and *Podocarpus* will grow satisfactorily at these lower temperatures. Also, floral initiation may be restricted

(a) (b)

Fig. 7.6. (a) Reduced growth and chlorosis of *Dracaena deremensis* cultivar "Janet Craig" due to high temperatures. (b) Leaf notching of *Dracaena deremensis* cultivar "Warneckii" due to high temperatures.

in some crops such as *Euphorbia* or *Hibiscus* whereas carnation and *Cyclamen*, for example, grow best at cool temperatures.

High air temperatures may limit growth and prevent or delay floral initiation. Maximum air temperatures of 40°C have been recorded in plastic and glass greenhouses in many areas of the world. Research (Conover and Poole, 1977b) has shown that some foliage plants tolerated the high temperatures with no change in plant quality, but the quality of other plants was greatly reduced. Reduced growth and chlorotic leaves (Fig. 7.6a) and notching of leaf margins (Fig. 7.6b) occurred due to high air temperatures. Production temperatures greater than 26°C can prevent floral initiation, delay floral development or result in abnormal bud development (Fig. 7.7) and, thus, limit production of marketable chrysanthemums, kalanchoes and other crops (Whealy et al., 1987b; Gehrke, 1988). Chrysanthemums and kalanchoes are most sensitive to high temperatures during the initial stages of the photoinduction period (de Jong and Smeets, 1982; Lepage et al., 1984; Whealy et al., 1987b; Gehrke, 1988). Responses vary among cultivars, some flowering without delay or development of abnormal flowers, while others produced at the same temperatures are unmarketable (Whealy et al., 1987a). Clearly, production temperature must be controlled and modified according to the crop being grown.

The high energy costs required to maintain optimum production temperatures have resulted in numerous techniques to reduce these costs. Most growing structures are heated by warming the air, but since substrate temperature is often more important in plant production than air temperature (Fig. 7.8) and heating the growing substrate can be achieved more economically than heating the air, growers in cold climates are choosing root-zone heating techniques to achieve good growth. Recent technological developments have allowed for the substrate in the pots to be heated independently of the air temperature, thus lowering the greenhouse heating costs and conserving energy. The medium is heated by placing pots on heated subirrigation floors or benches heated with steam pipes or hot air tubes or equipped with smaller tubes containing circulating hot water. Heating water prior to irrigation in cold climates is also effective in helping to maintain substrate temperatures. Maintaining warm soil temperatures promotes growth of foliage plants even in cool air temperatures. By keeping the substrate temperature elevated, air temperatures may drop as low as 16°C at night without significant crop damage to foliage plants (Bodnaruk et al., 1981; Poole and Conover, 1981a; Conover and Poole, 1987). Substrate temperatures maintained at a minimum of 21°C are satisfactory, but optimum substrate temperatures vary

Fig. 7.7. High temperatures cause abnormal bud development in *Dendranthema* cultivar "Orange Bowl". Outer florets are irregular and bracts form in the center of the inflorescence rather than florets. (Whealy et al., 1985).

Fig. 7.8. Substrate temperatures are more important than air temperatures for root growth. Air temperature for the two *Aglaonema* on the left was 16°C, and on the right 24°C; substrate temperatures were (left to right): 16°C, 27°C, 16°C and 27°C.

for different plants (Poole and Conover, 1981a; Johnson and Ingram, 1984; Vogelezang, 1988). Specific information on optimum root-zone temperatures is not available for most plants.

With flowering-plant crops, soil heating is dependent on the crop and stage of development. High root-zone temperatures reduced the time to flowering by 10% to 15% in *Saintpaulia* (African violet) and promoted flower development without affecting quality (Vogelezang, 1988). However, lowering root-zone temperature during the floral initiation phase did not improve flowering of *Begonia* × *hiemalis*. The number of inflorescences was reduced with a root-zone temperature of 23°C and air temperature of 20°C (Verberkt, 1988), and plants did not flower with high root-zone temperatures (29°C) regardless of air temperature (Vogelezang, 1990).

Temperature affects growth rate by influencing rates of photosynthesis and respiration (Went, 1953; Gates, 1968; Hadfield, 1968; Mortensen, 1992). In order to optimize plant growth, growers must provide proper temperature regimes to maximize carbohydrate production during the day and to control carbohydrate consumption through respiration at night (Beel and Schelstraete, 1986). In flowering-plant crops, temperature influences a number of processes related to growth and flowering (Moe and Heins, 1990). The rate of flower development is influenced primarily by average daily temperature within a certain temperature range (Moe, 1990). Conversely, the average daily temperature has only a small influence on internode length compared to the relative relationship of day and night temperature during a 24-h growth cycle, regardless of photoperiod response (Moe and Heins, 1990).

A differential between warm day and cool night temperatures (thermoperiodicity) is often used with flowering plants to enhance growth and development, and to avoid increased costs of heating at night when plants do not have light for photosynthesis. Stem elongation and internode length are reduced when plants are grown with warm nights and cool days (Went, 1944). Heins and his colleagues in the United States and Moe's research group in Norway have successfully refined and developed this physiological response into a commercial practice (Karlson et al., 1989; Moe and Heins, 1990). The use of high night and cool day temperatures offers the grower an excellent alternative to plant growth regulators for the control of plant height in flowering potted plants. The differential between the day/night temperatures affects the plant response – the greater the differential the larger the response. Normally a 5–10°C differential is used. Recently, it has been shown that high temperatures do not have to be maintained throughout the night. Raising the temperature 6–10 h before sunrise reduced the height of chrysanthemums by 12% to 17% as a direct result of shorter internodes (Jensen and Kresten, 1993). Also, plant height is reduced with a rapid reduction in temperatures for two hours before dawn (Heins and Erwin, 1990). In addition, Heins has combined this information with computer technology which allows for the modeling of plant growth (height) throughout the crop cycle. One limitation of the use of these reverse temperatures may relate to subsequent longevity. Recent work by Miller et al. (1993) related leaf yellowing in *Lilium longiflorum* to reduced carbohydrates when plants are grown with high night temperatures. Moe et al. (1992) showed a greater incidence of bract edge burn in poinsettias grown with warm night temperatures and cool days. Leaf drop was not affected by use of warm nights.

For a wide range of foliage plants, minimum night and day air temperatures of 18°C and 24°C are suitable, although particular types may require a more specific range for optimal growth. Most foliage plants will respond positively to greater than minimum temperatures for both day and night (Poole and Conover, 1981a,c; Chase and Poole, 1987; Mortensen, 1988). Additionally, growth rates of different species can be optimized by temperature control of air and substrate, but factors such as light intensity and irrigation frequencies must also be considered (Poole and Conover, 1981c; Chase and Poole, 1984b, 1987; Chase and Conover, 1987; Zieslin et al., 1987; Mortensen, 1988).

Light

Light intensity: Selecting the proper light intensity is an important aspect for growth of ornamental plants. The light energy received by the leaf activates chloroplasts and the radiant energy is converted to chemical energy, which is used to join carbon dioxide with water to form carbohydrates. The amount of light received within a broad spectral range determines the amount of carbohydrates synthesized. As light is decreased, carbohydrate production decreases until it is equal to the carbohydrates utilized by respiration (light compensation point). The optimal amount of light energy for best growth can vary from plant to plant (Horn and Huber, 1983).

In most flowering crops, the timing, uniformity of

flowering and time to flowering are directly influenced by light level, and plant quality is enhanced by increased light levels. In chrysanthemum, shoot length, number of leaves and growth of lateral shoots increase with supplemental light levels of 44–395 μmol m^{-2} s^{-1} (Mortensen and Moe, 1983). In northern production areas of the United States and Europe, light is a limiting factor in plant production during the winter months. In these cases, high-intensity discharge lights are used to increase the level of photosynthetically active radiation, thus increasing plant quality and reducing production time (Carpenter, 1974; Cathey et al., 1978; Hughes and Tsujita, 1981; Mortensen and Moe, 1983; Anderson, 1986; Gislerød et al., 1989; Mortensen and Grimstad, 1990), and offsetting the energy costs incurred. Generally, supplemental lighting is provided at levels of 6–20 W m^{-2} for up to 20 h per day, depending on the ambient light levels, from October to April. Supplemental lighting of chrysanthemums (17.2 W m^{-2}) reduced crop time by 6.4 days (Noble and Ellis, 1990). *Begonia* × *hiemalis* grown with 6 W m^{-2} from sodium vapor lamps had more lateral branches and a larger number of inflorescences and crop time was reduced by 10% to 15% compared to plants grown without supplemental lighting (Vogelezang and Verberkt, 1990). Similar results were obtained with several foliage plants. Not unexpectedly, results were dependent on the time of year. As the solar radiation integral increases during the first four weeks of short days, the time to flowering is reduced, the number of flowers is increased and the uniformity of flowering is greater (Cockshull and Hughes, 1971; Noble and Ellis, 1990). Supplemental lighting is most beneficial during the first half of the crop cycle when leaf expansion and stem elongation are most rapid (Carpenter, 1974; Vogelezang and Verberkt, 1990).

In a glasshouse, where excessive light energy can be received by plants unless shaded, the photosynthetic rate is frequently reduced because of limited carbon dioxide in the surrounding atmosphere and excessive heat build-up. Under conditions of excessive light, photo-oxidation can result because of reversal of the photosynthetic process. Light intensity affects growth rate, plant form and plant color; it influences internode length, width and height of plants, flower development and the color, variegation, size, shape and orientation of leaves (Cockshull and Hughes, 1971; Hughes and Cockshull, 1971; Conover and Poole, 1981; Chase and Poole, 1987). Basal branching of *Dracaena marginata* decreases as light intensity is reduced from full sun to 50 percent light exclusion (Donselman and Broschat, 1982). Leaves of *Ficus benjamina* grown under moderate light are usually smooth and dark green, but leaves produced under high light will be wavy and light green. Leaves of *Aphelandra squarrosa* produced under levels of photosynthetically active radiation of 100 μmol s^{-1} m^{-2}, are longer, narrower and flatter than leaves produced under levels of 200 μmol s^{-1} m^{-2}, while levels above 300 μmol s^{-1} m^{-2} cause excess puckering and reduce saleability (Poole et al., 1983). When light intensity is excessive, leaves of *Aglaonema* and *Dieffenbachia* will grow nearly vertical to the light source and, as a result, damage to the leaf surface is avoided. These vertically oriented leaves are usually long and somewhat chlorotic, even though high levels of fertilizer might be applied.

The ability of plants to adapt to varying light intensities has been known for years. This process of acclimatization can result in foliage plants of the same species having different anatomical and physiological features because of the light intensity received during the production phase (Mortensen and Olsen, 1987). Aphelandra plants grown under higher light levels have increased stomatal density and more palisade layers, but chloroplasts per palisade cell are lower (McConnell et al., 1984). Sun leaves of *Ficus benjamina* are smaller, thicker and lighter green, because of reduced chlorophyll levels (Table 7.7), than shade leaves and are often folded at the midrib (Conover and Poole, 1984). Chloroplasts of sun leaves will have vertical orientation and rudimentary grana often have this same orientation (Johnson et al., 1982).

Foliage color becomes deeper as shade level increases at constant fertilizer levels. Color intensity of plants, such as *Cordyline terminalis* which has red stripes on the leaves and *Codiaeum variegatum*, is

Table 7.7
Effect of light intensity on chlorophyll content of foliage plants

Light intensity (klx)	mg chlorophyll cm^{-2}		
	Brassaia actinophylla [1]	*Dracaena marginata* [2]	*Ficus benjamina* [3]
100	0.053	0.055	0.031
60	0.054	0.081	0.041
20	0.059	0.100	0.044

[1] Braswell et al. (1982).
[2] Conover and Poole (1975).
[3] Joiner et al. (1977).

reduced as light intensity is decreased (Conover and Poole, 1972). Variegated plants such as *Epipremnum aureum* have less yellow variegation when light levels are less than 40 klx, while *Dracaena sanderana* and *Dracaena deremensis* produced wider white stripes at less than 20 klx (Conover and Poole, 1972) and *Peperomia obtusifolia* variegation increased as light levels were reduced (Shen and Seeley, 1983).

Shape and size of foliage varies considerably as light intensity changes. *Philodendron scandens* subsp. *oxycardium* produces larger leaves and greater stem diameter under 40% shade than under 80% shade, but an intermediate light intensity produces a more desirable size and quality. Lance-shaped leaves, such as those on *Dracaena marginata*, are longer and narrower (Fig. 7.9) under high shade than under low shade (Conover and Poole, 1975), while petioles of plants such as *Brassaia actinophylla* will be lengthened, producing unsaleable plants, when grown under high shade.

Minimum light intensities are necessary for adequate flower production of some plants. *Aphelandra*, *Pachystachys*, *Pelargonium* and *Saintpaulia* must accumulate a required level of carbohydrates before flower initiation can occur. Failure to maintain optimum light levels can result in reduced floret development in chrysanthemum (Cockshull and Hughes, 1971; Hughes and Cockshull, 1971) and increased delay in flowering during high-temperature production periods (Whealy et al., 1987b).

Light levels for production of flowering pot-plants (31–54 klx) are higher than those for foliage plants. The light intensity suggested for foliage stock-plant production is frequently too high for plants in the production phase, since foliage might be of an unacceptable light-

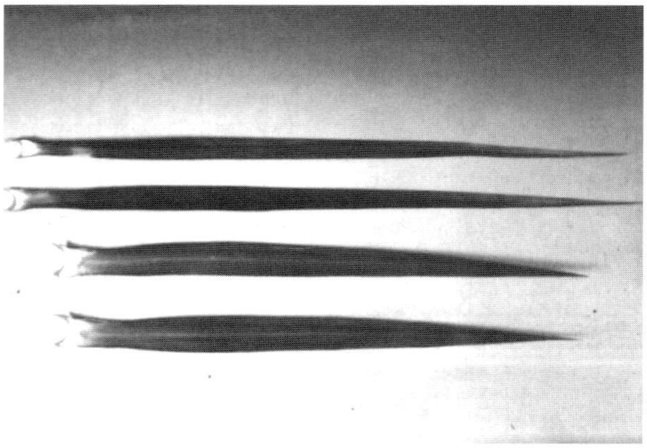

Fig. 7.9. Leaves produced under full sun (bottom) and 80% shade (top).

or yellow-green color because of reduced chlorophyll levels caused by the high light levels. In some cases increasing fertilizer levels will produce an acceptable shade of green, but build-up of soluble salts in the substrate and contamination of the aquifer might result; therefore, the best solution is proper selection of the light level.

Other problems related to light intensity also exist. For example, fluoride toxicity increases as light intensity is increased, because of the higher transpiration rate. Additionally, the method for controlling light intensity can influence plant quality. Plants grown under polypropylene shadecloth instead of slats are frequently more susceptible to diseases because of the higher humidity in the enclosed structures. An example is *Dracaena marginata,* which has more *Fusarium* problems when humidity is high (Knauss, 1973). Mite populations are usually higher under fully protected structures, due to lack of rainfall which would prevent the mites from proliferating rapidly. Light affects the distribution of roots in a pot. When plants are grown in dark pots, with the usual black or green color, roots will grow around the outside of the root ball adjacent to the pot, but if lighter colored, partially translucent pots are used, the roots are more uniformly distributed within the potting substrate (Conover and Poole, 1979).

Determination of proper light intensities for production depends on production techniques as well as the intended market for the crop (Conover and Poole, 1980, 1983). As stated previously, light level for production of foliage plants for sale is usually lower than for foliage stock plants. Light intensities and fertilizer requirements for the production of selected foliage crops grown in greenhouses are given in Table 7.6.

Light quality: Light quality is used commercially to promote flowering in photoperiodically sensitive plants. In addition, use of greenhouse covering of different spectral qualities may provide an opportunity to control plant morphogenesis. Potted roses grown with light filtered through cupric sulfate during production were shorter with reduced internode length and increased lateral shoots (Rajapakse and Kelly, 1994). This response to blue light appears to be related to the removal of far red light, thus increasing the red:far-red light ratio, suggesting the involvement of phytochrome in controlling plant height (Rajapakse et al., 1993). The commercial implications of using different greenhouse roof coverings are not understood at the present time, but it may be possible to modify plant morphogenesis without the use of plant growth regulators.

Photoperiod: Plant response to photoperiod, the relative length of the day and night, is related to the physiological activation/deactivation of phytochrome and is used as a method of controlling flowering in some commercial floriculture operations (Nelson, 1978). Plants are classified as long-day (LD), short-day (SD), or day-neutral (DN). Flowering is promoted by long-day conditions in LD plants, while long-day conditions inhibit flowering in SD plants. Similarly, short day lengths promote flowering of SD plants and inhibit floral initiation in LD plants. Day-neutral plants will flower regardless of the day length (Carpenter et al., 1972; Carpenter, 1974). Azalea does not respond to day length at temperatures below 18°C, but floral initiation is hastened under short-day conditions at temperatures of 18°C or above (Pettersen and Kristoffersen, 1969). Commercially, for production of a short-day plant (chrysanthemum, kalanchoe or poinsettia), a combination of long and short day lengths are used (Larson, 1992). To prevent floral initiation and to promote vegetative growth, long day lengths (provided by artificial lighting) are used for 1 to 6 weeks after planting to prevent floral initiation and to promote vegetative growth. The long-day period is followed by short days to promote floral initiation and development. The critical day length is dependent upon the plant being grown and the production temperature. Day lengths must be shorter for SD plants as the temperature increases.

Commercially, artificial lighting is used to extend the day length as a means of promoting flowering in LD plants or inhibiting flowering in SD plants. Lighting is provided by incandescent lamps (though other light sources may be used successfully) at levels exceeding 2 $\mu mol\, m^{-2}\, s^{-1}$. Plants do not need to be supplied with light throughout the entire dark period. Instead, artificial lighting breaks the dark period into two short periods which are shorter than the critical photoperiod. The greatest delay in flower initiation was observed when the night break was preceded by 4–8 h of darkness (Horridge and Cockshull, 1989). Thus, lighting is most often provided from 2200 h to 0200 h nightly. During naturally long-day periods, benches are covered with an opaque black cloth for 14–18 h per 24-h cycle. Plants (benches) may be covered manually, or automatic systems may be used to darken the entire house. Light levels in excess of 1 $\mu mol\, m^{-2}\, s^{-1}$ or less may prevent flowering in SD plants.

Moisture

Irrigation: Large amounts of water are utilized by plant producers. Greenhouse producers of foliage plants use as much as 20 000 kl $ha^{-1}\, yr^{-1}$ for crop production (Henley and Poole, 1981). Most research relating to irrigation frequency or volume shows that plants grow best when supplied with large volumes of water (Conover and Poole, 1977a; Spooner et al., 1978; Hipp et al., 1979; Blessington et al., 1980; Henley and Poole, 1981; Johnson et al., 1981; Poole and Conover, 1981c, 1986). Using large volumes of water results in leaching of fertilizer from the substrate and contamination of the ground water supply unless the leachate is reutilized. Decreasing the quantity of water applied at each irrigation can minimize nutrient leaching and quality plants can still be produced, provided that water stress does not occur. A few species, such as *Peperomia*, do not grow well with frequent watering (Chase and Poole, 1984a). During the winter, plant water requirements are reduced because of low light levels, even though heat is used to maintain the necessary temperatures. Watering frequency can vary from once a week during winter to daily during the peak of summer.

Irrigation frequency will depend not only on the season and the size of the pot and plant, but also on the plant's physiological and anatomical characteristics. Irrigation frequency is a direct function of plant leaf area, and environmental conditions promoting stomatal opening and an increase in the vapor pressure deficit. Large plants will lose more water than small plants in accordance with the leaf area. Under high-temperature production conditions and low relative humidity transpiration rates are higher, thus requiring more frequent irrigations. Irrigation requirements can be reduced by increasing the water-holding capacity of the growing substrate or by modifying the production environment. Antitranspirants (oils, waxes, plastic) sprayed on the foliage, and the addition of synthetic acrylic polymers or copolymers and starch-based polymers to the growing medium, have been used with very limited success on foliage and flowering pot-plants, and are not commonly included in commercial production systems.

Use of irrigation systems that water the potting substrate without wetting the foliage is desirable because foliar diseases and mineral residue problems from hard water are reduced (Fig. 7.10). A leader-tube system

Fig. 7.10. Efficient watering systems include (top to bottom): leader tubes, ebb-and-flow system, and troughs.

(a system with a small individual watering tube for each plant) is a good method of supplying water to the plants without wetting the foliage or using excess water. Capillary mats (absorbent materials placed directly on the bench) provide a simple irrigation system; water is applied to the mat and is then absorbed through the base of the pot by capillary action. Algal build-up may be a nuisance with capillary mats. Overhead irrigation, the most inefficient irrigation system, can lead to high moisture levels on the foliage and flowers, and result in high incidence of pathogen contamination.

Irrigation systems providing for recycling of excess irrigation water are used extensively throughout Europe and in some modern greenhouse operations in the United States, and the Dutch and Danish governments have required run-off in commercial flower and foliage nurseries to be reduced to zero by the year 2000. These systems help to reduce both water usage and nutrient run-off. The most promising closed systems appear to be those utilizing subirrigation either on bench, trough or concrete floors. In subirrigation systems, irrigation water with incorporated fertilizer – including unused solution from previous irrigations – is held in a reservoir tank and pumped to the bench, trough or floor. The surface is flooded with water 1.5–5 cm deep for several minutes. Plant pots or trays absorb the nutrient solution through capillary action, then the excess is drained into the reservoir tank until needed for subsequent irrigations. The fertilizer incorporated is carefully monitored to assure proper nutrient balance and levels of soluble salt at each irrigation. Although subirrigation systems are expensive to install, the initial costs are offset by lower labor, fertilizer and water requirements and, since foliage is kept dry, less incidence of disease.

Water temperature may be a cause for some concern, since cold water in contact with leaves can result in damage to plants such as *Dieffenbachia*, *Philodendron* and *Syngonium* and leaf spotting occurred on *Saintpaulia* leaves when water was 14°C cooler than the leaf temperature (Elliott, 1946). Tests with application of cold water (i.e. 5, 10, 15, or 20°C) to the substrate at each irrigation had no effect on growth or quality of *Aglaonema*, *Aphelandra*, *Dieffenbachia*, *Maranta* or *Nephrolepis* (fern) (Poole and Conover, 1981a). Irrigation with water at 2°C caused stomates in chrysanthemums and roses to close until the substrate temperature increased, whereas stomates remained open on plants irrigated with water at 18°C (Carpenter and Rasmussen, 1970).

Humidity: Humidity also influences plant growth. As relative humidity was increased from 55% to 95%, shoot length and number of leaves increased for some species (Mortensen, 1986b; Mortensen et al., 1988). High humidity can be a two-edged sword, however, since it may cause disease problems if proper air movement is not provided.

Carbon dioxide

Carbon dioxide must be present at adequate levels or photosynthesis will be reduced, regardless of light level or temperature. Low levels of carbon dioxide can become a limiting growth factor in northern greenhouses on cold days when ventilation with fresh air is minimal. Ambient carbon dioxide levels of 300 ppm are rapidly depleted unless supplemental carbon dioxide is added to the greenhouse atmosphere. Carbon dioxide generators are used to inject carbon dioxide into the atmosphere during the daylight period. Addition of carbon dioxide is not a common practice in warm production areas since ventilation is used throughout most of the day and supplemental carbon dioxide would be removed from the greenhouse by the ventilation before it could be used by the plants.

The level of supplemental carbon dioxide which is beneficial varies with light level, crop and temperature. Addition of carbon dioxide at levels up to 1000 $\mu\ell\,\ell^{-1}$ increased shoot dry weight by 27% to 60% with increased shoot length and number of leaves in chrysanthemum and *Begonia* × *hiemalis*; supplemental carbon dioxide was most beneficial with increased light levels. Levels of carbon dioxide above 1000 $\mu\ell\,\ell^{-1}$ did not increase plant growth (Mortensen and Moe, 1983; Mortensen and Ulsaker, 1985). Increasing the day temperature did not enhance the effect of supplemental carbon dioxide on rose or kalanchoe (Mortensen and Moe, 1992). Continuous supply of carbon dioxide throughout the day provided the greatest increases in growth and quality compared to intermittent periods throughout the day (Mortensen, 1986a). Commercially, carbon dioxide is added to greenhouses at levels between 1000 and 1200 ppm (Hanan et al., 1978; Anderson, 1991).

Growth regulators

Application of growth regulators (Table 7.8) has become a universal procedure in commercial production of flowering pot-plants, and, to a very limited extent, for foliage plants. Growth regulators are used to enhance the aesthetic value and marketability of ornamental crops by inducing flowering, controlling plant height and internode elongation, increasing branching, and overcoming natural dormancy. Several reviews of the use of growth regulator in floriculture have been compiled (Wittwer, 1968; Sachs and Hackett, 1977), and readers are directed to the most recent review by Larson (1985) for specific information on chemicals used, application procedures, and crops for which the materials are used commercially. It is difficult, if not impossible, to establish absolute guidelines for the use of growth regulators on account of the variability in production environments and cultural practices; the large variety of crops grown; the differences in cultivar responses to the growth regulators; and the effects desired (height control, induction of flowering, branching or delay of senescence). Each of these factors must be considered in selection of the chemical to be used, and the method, volume and time of application to maximize efficiency of the application and effectiveness of the chemical (Table 7.9).

The most common use of growth regulators is for control of height and internode elongation of flowering pot-plants. The control of plant size, especially height, is of concern in the production of quality flowering potted plants and bedding plants (Conover, 1986).

Because increased plant size is generally considered desirable in foliage plants, growth regulators have not been used traditionally in this segment of the ornamentals industry although their use is becoming more common to fill special niches in the marketplace, to increase branching of certain crops such as *Dieffenbachia*, and to promote flowering in crops such as bromeliads and spathiphyllum (Henny, 1980).

Control of plant height with growth regulators is dependent upon time of application, stage of plant development and, of paramount importance, selection of the particular chemical to be used. For example, chlormequat is effective as a foliar spray or a soil drench on poinsettias, whereas plants treated with daminozide are similar in height to non-treated plants. Paclobutrazol and uniconazole are both effective on poinsettias as a foliar spray or a soil drench. On chrysanthemums, daminozide is an effective plant growth regulator as a foliar spray, but has no effect as a soil drench (Barrett, 1992). Ancymidol and ethephon are the only chemicals effective against "stem topple" in potted *Hyacinthus orientalis* (hyacinth).

Cultural practices and production environment exert a primary influence on the effectiveness of growth regulators (Barrett and Nell, 1989); for instance, one may choose to apply ancymidol as a drench solution to the growing substrate or as a foliar spray. The effectiveness of drench applications was reduced 27% when applied to chrysanthemum plants growing in substrates containing bark (Barrett, 1982), and this loss of effectiveness was related to adsorption of ancymidol to the bark.

Table 7.8
Growth regulators used in cut-flower production[1]

Common name	Trade name	Chemical name
Abscisic acid (ABA)		5-(1-hydroxy-2,6,6-trimethyl 4-oxo 2-cyclohexenyl)-3-methyl-2,4-pentadienoic acid
Ancymidol	A-Rest Reducymol	α-cyclopropyl-α-(4-methoxyphenyl)-5-pyrimid methanol
AVG		aminoethoxyvinyl glycine
BA		6-benzylamino purine (or benzyladenine)
BOH		beta-hydroxyethyl hydrazine
Chlormequat	Cycocel	(2-chloroethyl) trimethylammonium chloride
Chlorphonium chloride	Phosfon D	2,4-dichlorobenzyl tributyl phosphonium chloride
Daminozide	Alar, B-Nine SP	butandedioic-acid mono-(2,2-dimethyl hydrazide)
Dikegulac	Atrinal	sodium salt of 2,3:4,6-bis-*O*-1-methyl ethylidene)-*x*-L-xylo-2-hexulofur-anosonic acid
Ethephon	Ethrel, Florel	(2-chloroethyl) phosphonic acid
Giberellins		
GA_3 GA_{4+7}	Gibrel Gibsol Gibtabs in Promalin	2,4*a*,7-trihydroxyl-1-methyl-1-methyl-8-methylenegibb-3-ene-1, 10-carboxylic acid 1-4 lactone
Methyl ester of fatty acids	Off-Shoot-O	methyl decanoate (*n*-decanol)
NAA		1-naphthaleneacetic acid
Paclobutrazol	Bonzi	(*2R,3R+2S,3S*)-1-(4-chlorophenyl-4,4-dimethyl-2-(1,2,4-triazol-1-yl)
PBA	Accel	6-benzylamino-9-(2-tetrahydro-pyranyl)-9*H*-purine
STS		silver thiosulfate

[1] Larson (1985).

Production temperature and light are the primary environmental factors affecting the marketable quality of flowers and foliage plants; production conditions are generally selected to produce and market plants in the shortest period of time. Under production conditions, plants such as poinsettia and chrysanthemum grow rapidly, developing strong stems, extended internodes and large leaves. Plants produced in warm, high-light areas, such as Florida or Texas in the United States and Spain, Portugal or southern France in Europe, have higher growth rates than plants produced in the cool, low light areas of the northern United States or Europe. Consequently, plants in the warm production areas must receive higher growth-regulator concentrations, or more applications, if height is to be controlled. Similarly, plants grown under suboptimal light or supraoptimal temperature conditions have increased stem elongation and internode lengths which will necessitate additional growth-regulator application.

As discussed earlier, recent work has demonstrated that modifying the greenhouse production environment by using higher temperatures by night than by day temperatures will effectively control plant height (Heins and Erwin, 1990). The use of warm nights and cool days may lead to a reduction in growth-regulator applications, especially in areas where day and night temperatures can be controlled accurately and consistently. In these areas, the greenhouse environment may be adequately modified to hasten growth while controlling plant height.

Modified greenhouse atmospheres and post-greenhouse environments may limit branching, fail to satisfy dormancy requirements, hasten senescence and/or limit flowering. Growth-regulator selection is dictated by the desired result. Under normal production conditions, some dieffenbachia varieties fail to develop lateral branches or axillary shoots, and, in such cases, foliar sprays of cytokinins (6-benzyl adenine) can increase branching by as much as 500% (Wilson and Nell, 1983). Moreover, increased branching is not the only

Table 7.9
Uses of growth regulators on floriculture crops[1]

Crop	Purpose of application	Growth regulator[2]
Azalea	Flower development	GA_3, GA_{4+7}
	Flower initiation	Chlormequat, Chlorphonium chloride, Daminozide
	Lateral shoot development	Dikegulac, Methyl esters of fatty acids
Bedding plants (annuals)	Growth control	Ancymidol, Chlormequat, Daminozide
	Growth promotion	Aminoethoxyvinyl glycine (AVG)
	Pollutant Protection	Ethylene diurea
Begonia × hiemalis	Growth control	Ancymidol, Chlormequat, Daminozide
	Lateral shoot development	Dikegulac
Bougainvillea spectabilis	Delay of abscission	Silver thiosulfate (STS)
	Flower induction	Chlormequat
Bromeliads	Flower induction	Beta-hydroxyethyl hydrazine (BOH), Ethephon, Naphthalene-acetic acid (NAA)
Calceolaria crenatifolia	Delay of abscission	Silver thiosulfate (STS)
	Growth control	Chlormequat
Carnation	Flowering	Abscisic acid (ABA)
	Rooting	Daminozide
Chrysanthemum	Disbudding	Alkylnaphthalene, Emulsifiable oils, Naphthalene, Oxathiin
	Growth control	Ancymidol, chlormequat, Chlorphonium chloride, Daminozide, Paclobutrazol, Piproctanyl bromide
	Lateral shoot	Methyl esters of fatty development acids, Oxathiin
	Rooting	Chlormequat, Daminozide
	Pollutant Protection	Ancymidol
Clerodendron	Flower initiation	Ancymidol, Chlormequat
Cyclamen persicum	Flowering	GA_3
Dahlia sp.	Growth control	Ancymidol, Chlormequat, Daminozide
	Propagation	Chlormequat
Easter lily	Growth control	Ancymidol, Chlorphonium chloride
Euphorbia fulgens	Flowering	Chlormequat
Foliage (interior) plants	Flowering	GA_3
	Growth control	Ancymidol, Ethephon, Paclobutrazol
	Lateral shoot development	Benzylamino purine (BA), Ethephon
Fuchsia sp.	Flowering	GA_3
	Growth control	Ancymidol, Ethephon
Geranium	Growth and Flowering	Ancymidol, Chlormequat, Chlorphonium chloride, Ethephon
	Petal abscission prevention	Silver thiosulfate (STS)
	Propagation	Chlormequat, Daminozide
Gladiolus × hortulanus	Propagation,	Ethephon
	Protection from disease	Ethephon
Gloxinia (*Sinningia hybrida*)	Growth control	Ancymidol, Daminozide
Hibiscus	Growth control	Chlormequat, Ethephon
Hyacinth	Prevention of topple	Ethephon

continued on next page

Table 7.9, *continued*

Crop	Purpose of application	Growth regulator[2]
Hybrid lilies	Growth control	Ancymidol, Ethephon, GA_3, Paclobutrazol
Hydrangea macrophylla	Growth control	Ancymidol, Daminozide, Ethephon, Paclobutrazol
	Flower abscission induction	Ethephon
Impatiens walleriana	Growth control	Ancymidol, Chlormequat, Daminozide
Iris	Reduction of flower abortion	Ethephon
	Flower induction	Ethephon
Justicia carnea	Flower induction	Ethephon
	Growth control	Ancymidol, Chlormequat, Daminozide, Ethephon
Kalanchoe blossfeldiana	Growth control	Ancymidol, Daminozide
Narcissus spp.	Growth control	Ethephon
Orchids	Growth control	Ethephon
Poinsettia (*Euphorbia pulcherrima*)	Growth control	Ancymidol, Chlormequat, Daminozide, Ethephon, Paclobutrazol
	Lateral shoot development	Benzylamino purine (BA)
	Propagation	Ancymidol, Chlormequat, Daminozide
	Protection from pests	Chlormequat
	Protection from pollutants	Ancymidol, Chlormequat
	Reduction of epinasty	Aminoethoxyvinyl glycine (AVG), Cycloheximide, Silver thiosulfate (STS)
Primula hybrida	Seed germination	GA_3
Rose	Lateral shoot development	Benzylamino purine (BA), Ethephon
Schlumbergera truncata	Growth and Flowering	Benzylamino purine (BA)
	Floral abscission retardation	Silver thiosulfate (STS)
Tulipa spp.	Growth control	Ancymidol, Ethephon, Paclobutrazol

[1] Larson (1985). [2] See Table 7.8.

known value of cytokinins. Potted flowers and foliage plants are frequently transported long distances from the production facility to the retail market, and this extended period of transport may lead to premature senescence, as is evidenced by leaf yellowing. In potted roses, the leaf yellowing associated with ethylene sensitivity during transit can be minimized by cytokinin application as an alternative to shipping plants at cooler temperatures (Halevy and Kofranek, 1976). Cytokinin also prevented discoloration of the lower leaves in *Dieffenbachia* "Rudolph Roehrs" during long-term storage conditions (Ben-Jaacov et al., 1985).

Other growth regulators are used to promote or induce flowering under environmental conditions in which flowering would not naturally occur. Gibberellic acid (GA) has been used to induce flowering in dieffenbachia (Henny, 1980), spathiphyllum (Henny and Rasmussen, 1981) and aglaonema (Henny, 1983), and ethylene is used to induce bromeliad flowering (Cathey and Downs, 1965). Flowering of spathiphyllum is used commercially to enhance marketability, while GA applications on dieffenbachia are valuable to breeders seeking to have plants flower more than once a year. In addition, GA has been used to overcome, partially or completely, the cold requirement in azaleas. Commercially produced azaleas require approximately six weeks of temperatures below 10°C after the flower buds have been initiated and partially developed. Six foliar applications of GA at weekly intervals can totally eliminate the need for the cold treatment (Larson and Sydnor, 1971; Nell and Larson, 1974). Comparable results may be obtained by cooling the plants for 3–4 weeks and then using 2 to 3 applications of GA in the greenhouse (Larson and Sydnor, 1971).

Growth regulators are important tools in the production of floriculture crops. It is anticipated that, until breeders can successfully develop genetically short varieties with good branching and post-production longevity, growth-regulator usage will continue to be an important part of the production scheme in ornamental horticulture.

Acclimatization

Acclimatization of foliage and flowering plants for interior use is necessary to ensure that they perform well indoors. Acclimatization is the adaptation of a plant to a new environment – in this instance, a building with low light and humidity, and warm temperatures. The production procedures and physiological systems involved in acclimatization vary between foliage and flowering pot-plants. For instance, reduction of the light compensation point in foliage plants by producing plants under low light levels has been an indicator of good interior longevity – that is, reduced leaf loss once plants are moved indoors. In flowering pot-plants, production under low light reduces the light compensation point, but longevity is not increased. However, it is clear that interior performance can be improved and plant longevity increased by maintaining not only good production practices, but also good production and post-production environments before the plants are placed indoors.

The most important aspect of the acclimatization process for foliage plants is development of "shade" foliage characterized by large, thin leaves with high chlorophyll levels, which allows proper "energy harvesting" in low-light situations. A second important factor in the process is the nutritional level which should be as low as possible while still producing a quality plant. Extensive research by Collard et al. (1977), Fonteno and McWilliams (1978), Conover and Poole (1985), Nell and Barrett (1989), and Fjeld (1992) has shown the importance of acclimatization.

Foliage plants such as *Brassaia actinophylla, Chrysalidocarpus lutescens* and *Ficus benjamina* are often grown in full sun and then acclimatized by placing them under suggested shade conditions and lowering fertilizer levels for 3 to 9 months. Although chloroplasts and grana are capable of reorientation within sun-grown foliage, the leaf anatomy – small size and thick cross section – prevents them from being as efficient after acclimatization as shade leaves; therefore, such plants are less tolerant of low or medium light levels indoors than plants acclimatized during production.

SHIPPING AND STORAGE

The environmental conditions for shipping and storage are similar because of the common nature of the two processes and the plant needs during these phases (Nell, 1990). During transport, plants are held in the dark at various temperature extremes, shaken due to road vibration and exposed to harmful gases. Plant preparation for shipping and/or storage does not start at the onset of this phase, but during the production period.

Pre-shipment factors affecting plant quality

Light

Pre-shipment light levels have a strong effect on the post-shipment quality of foliage plants. Foliage plants grown under high light will not ship so well as plants grown under low light because they are not properly acclimatized. High-light production results in excessive leaf drop or other loss of quality, either during shipment or after placement in a low-light interior environment. Also, plants grown in higher light are not so likely to tolerate cool shipping temperatures without sustaining chilling injury.

Flowering pot-plants exhibit decreased light compensation points and dark respiration similar to foliage plants at low light levels during the production phase, but the longevity inside buildings is reduced (Nell et al., 1990; Fjeld, 1986). Consequently, reduction of the light level during production offers few, if any, benefits in terms of interior performance and, thus, is not recommended for flowering potted plants.

Nutrition

Pre-shipment fertilizer levels also affect acclimatization and, in turn, subsequent quality for building interiors. Plants grown on higher than recommended nutritional regimes will not be well acclimatized, and will ultimately lose more leaves and be of lower quality than those properly acclimatized. Use of proper nutrition levels, optimum balance between ammonium and nitrate, and reduction of fertilizer prior to marketing, appear to be the most effective means of acclimatizing some flowering pot-plants, such as chrysanthemums (Nell et al., 1989; Roude et al., 1991). However, the reduction of fertilizer cannot be used effectively on all flowering pot-plants to increase longevity in interior situations. Easter lilies (*Lilium longiflorum*) exhibited leaf yellowing once plants were moved to interior conditions if fertilizer had been terminated prior to marketing, either at the stage when buds become visible or 2 weeks later (Prince and Cunningham, 1989).

Season

Foliage plants grown during high-light and high-temperature periods are less tolerant of the shipping environment and are more likely to decline in quality during shipment. This may be due to a lower degree of acclimatization, inability to tolerate lowered shipping temperatures or low carbohydrate reserves under high temperatures during the production phase. Seasonal differences have not been evaluated with potted flowering plants, but it appears that plants last longer during cool production periods than warm periods. However, low light levels during the winter may limit the longevity of flowering pot-plants in interior situations.

Environment factors during shipping

Several environmental factors affect the success of transport and storage of foliage and flowering plants. The effect of each factor individually has not been determined.

Light

Use of red light during simulated storage/transport reduced flower-bud drop in hibiscus from 74% to 31% (van Lieburg and Van Doorn, 1990). Although plants would tolerate shipping for longer periods if light could be supplied, this is not possible with present equipment.

Temperature

Shipping at temperatures lower than those used for production can be beneficial in maintaining quality of most plants, provided there is no chilling damage. Shipping at temperatures specified for the crop can be a major factor in maintenance of plant quality; however, the best shipping temperature for specific foliage plants changes with the season, and is lower in winter and higher in summer because summer-grown plants are more sensitive to chill damage. Duration of exposure to a specific temperature can also strongly affect plant quality; a temperature that might be optimum for 2-week shipments might cause damage when plants are exposed for 3 to 4 weeks (Table 7.10).

Shipping temperatures for flowering pot-plants (Table 7.11) are generally lower than those for foliage

Table 7.10
Shipping/storage temperatures (°C) for potted foliage plants [1]

Plant name	Duration [2]	
	1–14 days	15–28 days
Aglaonema cultivar 'Silver Queen'	15–18	15–18
Brassaia actinophylla	10–13	10–13
Chamaedorea seifrizii	13–15	13–15
Codiaeum variegatum cultivar 'Norma'	15–18	15–18
Dieffenbachia cultivar 'Tropic Snow'	13–18	13–18 [3]
Dracaena fragrans cultivar 'Massangeana'	15–18	15–18
Dracaena marginata	13–18	15–18 [3]
Epipremnum aureum	13–15	13–15 [4]
Ficus benjamina	13–15	13–15
Ficus lyrata	13–15	13–15
Hedera helix cultivar 'Eva'	10–15	10–13
Howea forsterana	10–18	10–18
Philodendron selloum	13–15	13–15
Schefflera arboricola	10–13	10–13
Spathiphyllum cultivar 'Mauna Loa'	10–13	13–15

[1] Adapted from Conover and Poole (1986).
[2] Plants shipped or stored for 1 to 7 days should be held at the highest temperature listed for that plant.
[3] Plants observed to have a loss in quality of about 25% each week beyond 2 weeks.
[4] Plants observed to have severe loss in quality beyond 2 weeks.

Table 7.11
Recommended shipping temperatures for some flowering pot-plants [1,2]

Shipping temperature of 2–4°C		Shipping temperature of 10–16°C
Calceolaria	*Oxalis*	*Begonia*
Crocus	*Pelargonium*	*Bougainvillea*
Cyclamen	*Rhododendron*	*Clerodendron*
Dendranthema	*Rosa*	*Crossandra*
Freesia	*Saintpaulia*	*Euphorbia*
Hippeastrum	*Senecio*	*Exacum*
Hydrangea	*Tulipa*	*Hibiscus*
Kalanchoe		*Saintpaulia*
Lilium		*Schlumbergera*
Muscari		*Sinningia*
Narcissus		*Streptocarpus*

[1] Adapted from Sterling and Molenaar (1985) and Nell (1993).
[2] Maximum shipping periods vary with plant species, cultivar, production environment and cultural practices (Nell, 1989).

plants except for plants such as hibiscus and poinsettia which are susceptible to chilling damage. Generally, flowering pot-plants will tolerate shipping and storage durations of 3–7 days without detrimental effects, provided proper temperatures are used and plants are not exposed to ethylene.

Substrate moisture

Plants should have adequate substrate moisture at the time of shipping, not too wet and not too dry. Plants shipped wet or with saturated substrates often suffer increased leaf drop and increased problems with disease on the flowers, whereas low substrate moisture during shipping periods increases desiccation of plant tissues and raises the potential for leaf drop or injury.

Humidity

Although all plants need to have adequate substrate moisture prior to shipment, this will not prevent them from desiccating if humidity is too low during shipment. A relative humidity level around 85% to 90% is best for maintenance of foliage plants shipped or stored for long durations. These levels can only be obtained by putting plants in boxes or by keeping air exchange in the shipping container to the minimum for unboxed plants or plants wrapped in paper or plastic ("sleeves") to protect them during shipment.

Gases

Research on foliage plants has shown that ethylene is probably the only contaminant that can occur in containers during shipping unless there is pollution from an outside source. However, foliage plants are somewhat tolerant of ethylene, and suffer damage only with fairly high levels (1–2 ppm) in relatively high temperatures (18°C or higher) for long durations (Marousky, 1979; Marousky and Harbaugh, 1979; Woltering, 1987). Experience has shown that ethylene does not seem to be a major problem when plants are shipped at cooler (16°C or lower) temperatures. In research on low oxygen storage, no benefit was observed over foliage plants stored in a normal atmosphere.

The presence of ethylene poses a major problem in the shipping and handling of flowering pot-plants (Nell, 1990). Injury due to ethylene includes leaf, bract and flower (bud) drop, flower wilting, failure of flowers to open, wilting of the sepals, petal drying and leaf yellowing. Sensitivity to ethylene varies among flowering potted plants, some species being sensitive to even low concentrations while other plants fail to exhibit any adverse symptoms to high concentrations (Table 7.12).

Table 7.12
Flowering pot-plants sensitive to ethylene[1,2]

Achimenes	*Euphorbia*
Begonia × *hiemalis*	*Exacum*
Bougainvillea	*Hibiscus*
Browallia	*Kalanchoe*
Calceolaria	*Pachystachys*
Clerodendron	*Pelargonium*
Crossandra	*Rhododendron*
Cyclamen	*Saintpaulia*
Cymbidium	*Sinningia*
Dianthus	*Streptocarpus*

[1] Adapted from Woltering (1987).
[2] Response varies with ethylene concentration, temperature, and duration of exposure.

Injury can be reduced by lowering the temperature during shipping provided injury will not result from the low temperatures. In addition, foliar applications of silver thiosulfate have been successful in minimizing the effects of ethylene exposure (Cameron and Reid, 1983).

A FINAL WORD

The true measure of a grower of ornamental foliage or flowering potted plant is the ability to consistently produce a high-quality, profitable crop using environmentally sound methods. This requires thorough understanding of crop requirements and refined ecosystem management in a consistent and purposeful manner in order to meet the environmental and cultural requirements of each plant species.

ACKNOWLEDGEMENT

The authors wish to acknowledge L.N. Satterthwaite, Publications Associate, for her contributions to this manuscript.

REFERENCES

Andersen, A.S. and Kirk, H.G., 1986. Low pressure storage of herbaceous cuttings. *Acta Hortic.*, 181: 305–310.
Anderson, N.E., 1986. Strategi for anvendelse af kunstyls. *Gartner Tid.*, 102(41): 1369–1373.

Anderson, N.E., 1991. The influence of constant and diurnally changing CO_2 concentrations on plant growth and development. *J. Hortic. Sci.,* 66(5): 569–574.

Barrett, J.E., 1982. Chrysanthemum height control by ancymidol, PP333 and EL-500 dependent on medium composition. *HortScience,* 17(6): 896–897.

Barrett, J.E., 1992. Mechanisms of action. In: H.K. Tayama, R.A. Larson, T.A. Hammer and T.J. Roll (Editors), *Tips on the Use of Chemical Growth Regulators on Floraculture Crops*. Ohio Florists' Association, Columbus, OH, pp. 12–18.

Barrett, J.E. and Nell, T.A., 1989. Factors affecting efficacy of paclobutrazol and uniconazole on petunia and chrysanthemum. *Acta Hortic.,* 272: 229–234.

Beel, E. and Schelstraete, A., 1986. Invloed van energiebesparing op de kwaliteit en de rendabiliteit van Warme Kasplanten (III). *Verbondsnieuws Belg. Sierteelt,* 30: 575–581.

Ben-Jaacov, J., Poole, R.T. and Conover, C.A., 1982. Effect of nutrition, soil water content and duration of storage on quality of *Dieffenbachia maculata* (Lodd.) G. Don 'Rudolph Roehrs'. *HortScience,* 17(4): 611–612.

Ben-Jaacov, J., Poole, R.T. and Conover, C.A., 1984. Tipburn of *Chlorophytum comosum* 'Vittatum'. *HortScience,* 19(3): 445–447.

Ben-Jaacov, J., Poole, R.T. and Conover, C.A., 1985. Long-term dark storage of dieffenbachia sprayed with cytokinin. *Gartenbauwissenschaft* 50(1): 19–22.

Blessington, T.M., Srithavaj, A.A. and Singletary, C.C., 1980. Influence of water methods and ancymidol on two tropical foliage species held in a simulated interior environment. *J. Am. Soc. Hortic. Sci.,* 105(6): 785–787.

Bodnaruk, W.H., Mills, T.W. and Ingram, D.L., 1981. Response of four foliage plants to heated soil and reduced air temperatures. *Proc. Fl. State Hortic. Soc.,* 94: 104–107.

Braswell, J.H., Blessington, T.M. and Price, J.A., 1982. Influence of production and postharvest light levels on the interior performance of two species of scheffleras. *HortScience,* 17(1): 48–50.

Broschat, T.K. and Donselman, H., 1986. Factors affecting storage and germination of *Chrysalidocarpus lutescens* seeds. *J. Am. Soc. Hortic. Sci.,* 111(6): 872–877.

Cameron, A.C. and Reid, M.S., 1983. Use of silver thiosulfate to prevent flower abscission from potted plants. *Sci. Hortic.,* 19(3/4): 373–378.

Carpenter, W.J., 1974. *High intensity lighting in the greenhouse,* Research Report 255. Michigan State University, East Lansing, MI, 16 pp.

Carpenter, W.J., 1988a. Temperature affects seed germination of four Florida palm species. *HortScience,* 23(2): 336–337.

Carpenter, W.J., 1988b. Seed after-ripening and temperature influence *Butia capitata* germination. *HortScience,* 23(4): 702–703.

Carpenter, W.J., 1989. Medium temperature influences the rooting response of *Hibiscus rosa-sinensis* L. *J. Environ. Hortic.,* 7(4): 143–146.

Carpenter, W.J. and Rasmussen, H.P., 1970. Temperature effect of greenhouse water on rose and chrysanthemum turgidity, stomate behavior, and flowering. *J. Am. Soc. Hortic. Sci.,* 95(5): 578–582.

Carpenter, W.J., Rodriguez, R.C. and Carlson, W.H., 1972. Effect of daylength on growth and flowering of roses (*Rosa hybrida*). *J. Am. Soc. Hortic. Sci.,* 97(1): 135–138.

Carpenter, W.J., Ostmark, E.R. and Ruppert, K.C., 1993. Promoting the rapid germination of needle palm seeds. *Proc. Fl. State Hortic. Soc.,* 106: 336–338.

Cathey, H.M., 1969. Guidelines for the germination of annual, pot plant and ornamental herb seed, Part 1. *Florists' Rev.* 144(3742): 21–23, 56, 60; Part 2. *Florists' Rev.* 144(3743): 18–20, 52–53; Part 3. *Florists' Rev.* 144(3744): 26, 29, 75–77.

Cathey, H.M. and Downs, R.J., 1965. Guidelines for regulating flowering of bromeliads. *The Exchange* 143(10): 27–29.

Cathey, H.M., Campbell, L.E. and Thimijan, R.W., 1978. Comparative development of 11 plants grown under various fluorescent lamps and different durations of irradiation with and without additional incandescent lighting. *J. Am. Soc. Hortic. Sci.,* 103(6): 781–791.

Chase, A.R. and Conover, C.A., 1987. Temperature and potting medium effects on growth of Boston fern infected with *Rhizoctonia solani*. *HortScience,* 22(1): 65–67.

Chase, A.R. and Poole, R.T., 1984a. Investigations into the role of fertilizer level and irrigation frequency on growth, quality and severity of *Pythium* root rot of *Peperomia obtusifolia*. *J. Am. Soc. Hortic. Sci.,* 109(5): 619–622.

Chase, A.R. and Poole, R.T., 1984b. Severity of acephate phytotoxicity on *Spathiphyllum* Schott. cv. Clevelandii as influenced by host nutrition and temperature. *J. Am. Soc. Hortic. Sci.,* 109(2): 168–172.

Chase, A.R. and Poole, R.T., 1987. Effect of fertilizer, temperature, and light level on growth of *Syngonium podophyllum* 'White Butterfly'. *J. Am. Soc. Hortic. Sci.,* 112(2): 296–300.

Chase, A.R. and Poole, R.T., 1989. Response of *Codiaeum variegatum* 'Gold Star' as influenced by slow-release fertilizer. *J. Environ. Hortic.,* 7(1): 21–23.

Cockshull, K.E. and Hughes, A.P., 1971. The effects of light intensity at different stages in flower initiation and development of *Chrysanthemum morifolium*. *Ann. Bot.* 35: 915–926.

Collard, R.C., Joiner, J.A., Conover, C.A. and McConnell, D.B., 1977. Influence of shade and fertilizer on light compensation point of *Ficus benjamina* L. *J. Am. Soc. Hortic. Sci.,* 102(4): 447–449.

Conover, C.A., 1969. Responses of pot-grown *Chrysanthemum morifolium* 'Yellow Delaware' to media, watering and fertilizer levels. *Proc. Fl. State Hortic. Soc.,* 82: 425–429.

Conover, C.A., 1986. Quality. *Acta Hortic.,* 181: 201–205.

Conover, C.A. and Poole, R.T., 1970. Foliage plant propagation. *Grounds Maintenance* 5: 27–29.

Conover, C.A. and Poole, R.T., 1971. Influence of fluoride on foliar necrosis of *Cordyline terminalis* cv Baby Doll during propagation. *Proc. Fl. State Hortic. Soc.,* 84: 380–383.

Conover, C.A. and Poole, R.T., 1972. Influence of shade and nutritional levels on growth and yield of *Scindapsus aureus*, *Cordyline terminalis* 'Baby Doll' and *Dieffenbachia exotica*. *Proc. Trop. Reg. Am. Soc. Hortic. Sci.,* 16: 277–281.

Conover, C.A. and Poole, R.T., 1973. Influence of shade level and soil temperature on forcing of *Caladium bicolor*. *Proc. Fl. State Hortic. Soc.,* 86: 369–372.

Conover, C.A. and Poole, R.T., 1974. Influence of shade nutrition and season on growth of *Aglaonema*, *Maranta* and *Peperomia*. *Proc. Trop. Reg. Am. Soc. Hortic. Sci.,* 18: 283–287.

Conover, C.A. and Poole, R.T., 1975. Influence of shade and fertilizer levels on production and acclimatization of *Dracaena marginata*. *Proc. Fl. State Hortic. Soc.,* 88: 606–608.

Conover, C.A. and Poole, R.T., 1977a. Influence of fertilization and watering on acclimatization of *Aphelandra squarrosa* Nees cv. Dania. *HortScience,* 12(6): 569–570.

Conover, C.A. and Poole, R.T., 1977b. *Influence of greenhouse design and cooling systems on temperature and energy requirements,* ARC-Apopka Research Report RH-77–3, 8 pp.

Conover, C.A. and Poole, R.T., 1979. Foliage plant responses to translucence of the growing container. *HortScience,* 14(5): 616–617.

Conover, C.A. and Poole, R.T., 1980. Effects of light and fertilizer levels on cut foliage production. *Proc. Fl. State Hortic. Soc.,* 93: 210–212.

Conover, C.A. and Poole, R.T., 1981. Environmental factors. In: J.N. Joiner (Editor), *Foliage Plant Production.* Prentice-Hall, Englewood Cliffs, NJ, pp. 273–274.

Conover, C.A. and Poole, R.T., 1983. Influence of shade and fertilizer levels on yield of croton stock plants. *Proc. Fl. State Hortic. Soc.,* 96: 261–263.

Conover, C.A. and Poole, R.T., 1984. Acclimatization of indoor foliage plants. In: J. Janick (Editor), *Horticultural Reviews*, Vol. 6. AVI Publishing, Westport, CT, pp. 119–154.

Conover, C.A. and Poole, R.T., 1985. *Acclimatization of foliage plants.* Horticultural Research Institute, Washington, DC, 7 pp.

Conover, C.A. and Poole, R.T., 1986. Shipping suggestions. *Interior Landscape Ind.,* 3(10): 32–37.

Conover, C.A. and Poole, R.T., 1987. Growth of *Dieffenbachia maculata* 'Perfection' as affected by air and soil temperatures and fertilization. *HortScience,* 22(5): 893–895.

Conover, C.A. and Poole, R.T., 1990. Light and fertilizer recommendations for production of acclimatized potted foliage plants. *Nursery Dig.,* 24 (10): 34–36, 58–59.

Conover, C.A., Sheehan, T.J. and McConnell, D.B., 1971. *Using Florida grown foliage plants,* Florida Agricultural Experiment Station Bulletin No. 746.

Criley, R.A. and Carlson, W.H., 1970. Tissue analysis standards for various floricultural crops. *Florists' Rev.* 146(3771): 19–21, 71–72.

de Jong, J. and Smeets, L., 1982. Effect of day and night temperature during long day photoperiods on the vegetative growth and flowering of *Chrysanthemum morifolium* Ramat. *Sci. Hortic.,* 17: 271–275.

Donselman, H. and Broschat, T.K., 1982. Light-induced basal branching of *Dracaena marginata. HortScience,* 17(6): 909–910.

Eisenberg, B.A., George, G.L. and Fretz, T.A., 1978. Low pressure and refrigerated storage of rooted and unrooted ornamental cuttings. *J. Am. Soc. Hortic. Sci.,* 103(6): 732–737.

Elliott, F.H., 1946. Saintpaulia leafspot and temperature differential. *Proc. Am. Soc. Hortic. Sci.,* 47: 511–514.

Eng, R.Y.N., Tsujita, M.J. and Grodzinski, B., 1985. The effects of supplemental HPS lighting and carbon dioxide enrichment on the vegetative growth, nutritional status and flowering characteristics of *Chrysanthemum morifolium* Ramat. *J. Hortic. Sci.,* 60: 389–395.

Fjeld, T., 1986. The effect of relative humidity, light intensity and temperature on keeping quality of *Begonia* × *Cheimantha* Everett. *Acta Hortic.,* 181: 251–255.

Fjeld, T., 1992. The effect of light acclimatization on photosynthesis, light compensation point, carbohydrate content, and keeping quality of Christmas begonia (*Begonia* × *Cheimantha* Everett). *Gartenbauwissenschaft,* 57(3): 115–120.

Flower Council of Holland, 1992. *Facts and Figures about the Dutch Horticultural Industry.* Leiden, the Netherlands.

Fonteno, W.C. and McWilliams, E.L., 1978. Light compensation points and acclimatization of four tropical foliage plants. *J. Am. Soc. Hortic. Sci.,* 103(1): 52–56.

Fooshee, W.C. and McConnell, D.B., 1987. Response of *Aglaonema* 'Silver Queen' to nighttime chilling temperatures. *HortScience,* 22(2): 254–255.

Gates, D.M., 1968. Transpiration and leaf temperature. *Ann. Rev. Plant Physiol.,* 19: 211–238.

Gehrke, P.W., 1988. *The effects of supraoptimal temperatures on the growth and flowering of Kalanchoe blossfeldiana (Poelln.),* M.S. Thesis, University of Florida, Gainesville, FL.

Gislerød, H.R., Eidsten, I.M. and Mortensen, L.M., 1989. The interaction of daily lighting period and light intensity on growth of some greenhouse plants. *Sci. Hortic.,* 38(3/4): 295–304.

Hadfield, W., 1968. Leaf temperature, leaf pose and productivity of the tea bush. *Nature,* 219: 282–284.

Halevy, A.H. and Kofranek, A.M., 1976. The prevention of flower bud and leaf abscission in pot roses during simulated transport. *J. Am. Soc. Hortic. Sci.,* 101(6): 658–660.

Hammer, P.A., 1988. Nutrition. In: H.K. Tayama (Editor), *Tips on growing poinsettias,* Ohio Cooperative Extension Service Bulletin FP764. The Ohio State University, Columbus, OH, pp. 7–10.

Hanan, J.J., Holley, W.D. and Goldsberry, K.L., 1978. *Greenhouse Management*. Springer, New York, 530 pp.

Hansen, J. and Mose, P., 1987. Formering af *Schefflera arboricola* ver ledstiklingler. Stiklingers position på moderplanten [Propagation of *Schefflera arboricola* by single-node cuttings. Position of cuttings on the mother plant]. *Gartner Tid.,* 103(17): 491–493.

Hansen, L., 1988. Experiences with a Grodan-rockwool based substrate for pot plants. *Acta Hortic.,* 221: 189–195.

Hartmann, H.T. and Kester, D.E., 1983. *Plant Propagation Principles and Practices,* 4th edition. Prentice Hall, Englewood Cliffs, NJ.

Heins, R. and Erwin, J., 1990. Understanding and applying DIF. *Greenhouse Grower,* 8(2): 73–78.

Henley, R.W., 1986. Tissue-cultured liner production. *Nurserymen's Dig.,* 20(1): 68–70.

Henley, R.W. and Poole, R.T., 1981. Water and foliage plants. In: J.N. Joiner (Editor), *Foliage Plant Production*. Prentice-Hall, Englewood Cliffs, NJ, pp. 203–228.

Henny, R.J., 1980. Gibberellic acid (GA3) induces flowering in *Dieffenbachia maculata* 'Perfection'. *HortScience,* 15(5): 613.

Henny, R.J., 1983. Flowering of *Aglaonema commutatum* 'Treubii' following treatment with gibberellic acid. *HortScience,* 18(3): 374.

Henny, R.J. and Rasmussen, E.M., 1981. Promotion of flowering in *Spathiphyllum* 'Mauna Loa' with gibberellic acid. *HortScience,* 16(4): 554–555.

Hipp, B.W., Colbaugh, P.F. and DiLeo, M., 1979. Influence of fertility and moisture level on growth of *Chlorophytum*. *HortScience,* 14(1): 65–66.

Horn, W. and Huber, J., 1983. Lichtversuche mit Grünpflanzen in Hydrokultur. *Dtsch. Gartenbau,* 37(26): 1191–1192.

Horridge, J.S. and Cockshull, K.E., 1989. The effect of timing of a night-break on flower initiation in *Chrysanthemum morifolium* Ramat. *J. Hortic. Sci.,* 64: 183–188.

Hughes, A.P. and Cockshull, K.E., 1971. The effects of light intensity and carbon dioxide on the growth of *Chrysanthemum morifolium* cv. Bright Golden Anne. *Ann. Bot.* 35: 899–914.

Hughes, B.R. and Tsujita, M.J., 1981. The effect of supplemental HPS lighting during propagation on rooting and quality of 'White Marble' cut chrysanthemums. *J. Am. Soc. Hortic. Sci.,* 106(5): 613–615.

Hummel, R.L. and Johnson, C.R., 1985. Freezing tolerance in the genus *Ficus*. *HortScience,* 20(2): 287–289.

Jensen, H.F. and Kresten, L., 1993. Influence of duration and placement of a high night temperature on morphogenesis of *Dendranthema grandiflora* Tzvelev. *Scientia,* 54: 327–335.

Johnson, C.R. and Ingram, D.L., 1984. *Pittosporum tobira* response to container medium temperatures. *HortScience,* 19(4): 524–525.

Johnson, C.R., Ingram, D.L. and Barrett, J.E., 1981. Effects of irrigation frequency on growth, transpiration and acclimatization of *Ficus benjamina* L. *HortScience,* 16(1): 80–81.

Johnson, C.R., Nell, T.A., Rosenbaum, S.E. and Lauritis, J.L., 1982. Influence of light intensity and drought stress of *Ficus benjamina* L. *J. Am. Soc. Hortic. Sci.,* 107(2): 252–255.

Joiner, J.N., Conover, C.A. and Poole, R.T., 1977. Factors affecting acclimatization of foliage plants. *Proc. Trop. Reg. Am. Soc. Hortic. Sci.,* 21: 41–43.

Karlson, M.G., Heins, R.D., Erwin, J.E., Berghage, R.D., Carlson, W.H. and Biernbaum, J.A., 1989. Temperature and photosynthetic photon flux influence chrysanthemum shoot development and flower initiation under short-day conditions. *J. Am. Soc. Hortic. Sci.,* 114(1): 158–163.

Knauss, J.F., 1971. *Suggestions for the control of some common diseases of foliage plants,* Mimeo 712. University of Florida, Agricultural Research Center, Apopka, FL, 20 pp.

Knauss, J.F., 1973. Common diseases of tropical foliage plants. *Florists' Rev.* 152(3937): 26–27, 55–58.

Knauss, J.F., Alfieri Jr., S.A., Marlatt, R.B. and Zettler, F.W., 1981. In: J.E. Joiner (Editor), *Foliage Plant Production*. Prentice-Hall, Englewood Cliffs, NJ, pp. 351–357.

Kristiansen, K., 1991. Post-propagation growth of cuttings from *in vitro* and *in vivo* propagated stock plants of *Ficus benjamina*. *Sci. Hortic.,* 46(3/4): 315–322.

Larson, R.A., 1985. Growth regulators in floriculture. In: J. Janick (Editor), *Horticultural Reviews*, Vol. 7. AVI Publishing, Westport, CT, pp. 399–481.

Larson, R.A. (Editor), 1992. *Introduction to Floriculture.* Academic Press, New York, 607 pp.

Larson, R.A. and Sydnor, T.D., 1971. Azalea flower bud development and dormancy as influenced by temperature and gibberellic acid. *J. Am. Soc. Hortic. Sci.,* 96(6): 786–788.

Lembke, E., 1992. Potteplante top-ti. *Gartner Tid.,* 31: 653.

Lepage, I., de Jong, J. and Smeets, J., 1984. Effect of day and night temperatures during short photoperiods on growth and flowering of *Chrysanthemum morifolium* Ramat. *Sci. Hortic.,* 22: 373–381.

Lyhne, M. and Hegnelt, J., 1988. Improved germination of seeds of *Chrysalidocarpus lutescens*. *Acta Hortic.,* 226: 263–270.

Marousky, F.J., 1979. Effects of ethylene in combination with light, temperature and carbon dioxide on leaf abscission in *Fittonia verschaffeltii* (Lem.) Coem. var. Argyroneura (Coem.) Nichols. *Proc. Fl. State Hortic. Soc.,* 92: 320–321.

Marousky, F.J., 1980. Chilling injury in *Dracaena sanderana* and *Spathiphyllum* 'Clevelandii'. *HortScience,* 15(2): 197–198.

Marousky, F.J., 1981. Symptomology of fluoride and boron injury in *Lilium longiflorum* Thunb. *J. Am. Soc. Hortic. Sci.,* 106(3): 341–344.

Marousky, F.J. and Harbaugh, B.K., 1979. Interactions of ethylene, temperature, light and CO_2 on leaf and stipule abscission and chlorosis in *Philodendron scandens* subsp. *oxycardium*. *J. Am. Soc. Hortic. Sci.,* 104(6): 876–880.

McConnell, D.B., Rugaber, P., Sheehan, T.J. and Henny, R.J., 1984. Light levels alter leaf anatomy of *Aphelandra squarrosa* 'Dania'. *J. Am. Soc. Hortic. Sci.,* 109(3): 298–301.

McConnell, D.B., Henley, R.W. and Kelly, C.B., 1989. Commercial foliage plants: twenty years of change. *Proc. Fl. State Hortic. Soc.,* 102: 297–303.

Mikorski, D.J. and White, J.W., 1977. Foliage plants – seed propagation and transplant research. *Florists' Review* 160(4153): 55, 99–101.

Miller, W.B., Hammer, P.A. and Kirk, T.I., 1993. Reversed greenhouse temperatures alter carbohydrate status in *Lilium longiflorum* Thunb. 'Nellie White'. *J. Am. Soc. Hortic. Sci.,* 118: 736–740.

Moe, R., 1990. Effect of day and night temperature alternations, and plant growth regulators on stem elongation and flowering of the long day plant *Campanula isophylla* Moretti. *Sci. Hortic.,* 43: 291–306.

Moe, R. and Heins, R., 1990. Control of leaf morphogenesis and flowering by light quality and temperature. *Acta Hortic.,* 272: 81–89.

Moe, R., Fjeld, T. and Mortensen, L.M., 1992. Stem elongation and keeping quality in poinsettia (*Euphorbia pulcherrima* Willd. ex Klotzsch.) as affected by temperature and supplemental lighting. *Sci. Hortic.,* 50: 127–136.

Molinar, J.M. and Williams, C.J., 1977. Response of *Cyclamen persicum* cultivars to different growing and holding temperatures. *Can. J. Plant Sci.,* 57: 93–100.

Mortensen, L.M., 1986a. Effect of intermittent as compared to continuous CO_2 enrichment on growth and flowering of *Chrysanthemum morifolium* Ramat. and *Saintpaulia ionantha* H. Wendl. *Sci. Hortic.,* 29: 283.

Mortensen, L.M., 1986b. Effect of relative humidity on growth and flowering of some greenhouse plants. *Sci. Hortic.,* 29(4): 301–307.

Mortensen, L.M., 1988. Hoye tempreaturer git okt tilvekst hos gronnplanter [High temperatures increase the growth of foliage plants]. *Gartneryrket* 78(11): 359.

Mortensen, L.M., 1992. Growth responses of three foliage plant species to temperature and photon flux density. *Sci. Hortic.,* 49(1/2): 159–166.

Mortensen, L.M. and Grimstad, S.O., 1990. The effect of lighting period and photon flux density on growth of six foliage plants. *Sci. Hortic.,* 41(4): 337–342.

Mortensen, L.M. and Moe, R., 1983. Growth response of some greenhouse plants to environment. VI. Effect of CO_2 and artificial light on growth of *Chrysanthemum morifolium* Ramat. *Sci. Hortic.,* 19(4): 141–147.

Mortensen, L.M. and Moe, R., 1992. Effects of CO_2 enrichment and different day/night temperature combinations on growth and flowering of *Rosa* L. and *Kalanchoe blossfeldiana* v. Poelln. *Sci. Hortic.,* 51: 145–153.

Mortensen, L.M. and Olsen, R., 1987. Light acclimatization of some foliage plants. *Gartenbauwissenschaft,* 52(4): 157–161.

Mortensen, L.M. and Ulsaker, R., 1985. Effect of CO_2 concentration and light levels on growth, flowering and photosynthesis of *Begonia × hiemalis* Fotsch. *Sci. Hortic.*, 27: 133–141.

Mortensen, L.M., Gislerød, H.R. and Bratberg, I., 1988. Liten effekt av luftfuktihet på dekorasjonsplanter [The small effect of relative humidity on ornamental plants]. *Gartneryrket* 78(23): 688.

Murakami, P.K. and Rauch, F.D., 1984. Effect of age and handling on subsequent growth and development of Areca palm, *Chrysalidocarpus lutescens*, seedlings. *J. Environ. Hortic.*, 2(3): 91–93.

Nell, T.A., 1989. The right variety adds more to your bottom line. *Grower Talks* 52(9): 48–55.

Nell, T.A., 1990. Commercial transport of flowering potted plants: keeping quality on the bench. *Grower Talks* 53(9): 24–39.

Nell, T.A., 1993. *Flowering Potted Plants: Prolonging Shelf Performance.* Ball Publishing, Batavia, IL, 96 pp.

Nell, T.A. and Barrett, J.E., 1989. Postproduction handling of bedding and potted plants. *Acta Hortic.*, 272: 311–317.

Nell, T.A. and Larson, R.A., 1974. The influence of foliar applications of GA3, GA4+7 and PBA on breaking flower bud dormancy on azalea cvs. Redwing and Dogwood. *J. Hortic. Sci.*, 49: 323–328.

Nell, T.A., Barrett, J.E. and Leonard, R.T., 1989. Fertilizer termination influences postharvest performance of pot chrysanthemum. *HortScience,* 24(5): 996–998.

Nell, T.A., Leonard, R.T. and Barrett, J.E., 1990. Production and postproduction irradiance affects acclimatization and longevity of potted chrysanthemum and poinsettia. *J. Am. Soc. Hortic. Sci.*, 115: 262–265.

Nelson, P.V., 1978. *Greenhouse Operation and Management*. Reston Publishing, Reston, VA, 518 pp.

Noble, R. and Ellis, S.R., 1990. Variation in chrysanthemums grown five to a pot. The influence of light. *J. Hortic. Sci.*, 65: 177–183.

Oglevee-O'Donovan, W., 1993. Culture-indexing for vascular wilts and diseases. In: J. White (Editor), *Geraniums IV*. Ball Publishing, Batavia, IL, pp. 277–286.

Ottosen, C.O., 1988. Selection of fast-growing clones among visually unseparable genotypes of *Ficus benjamina* L. *Gartenbauwissenschaft,* 53: 180–182.

Perry, L.P. and Boodley, J.W., 1980. Germination of foliage plant seeds in response to sowing media, depths of sowing, pH levels and medium temperatures. *HortScience,* 15(2): 194–196.

Pettersen, H. and Kristoffersen, T., 1969. The effect of daylength and temperature on flowering in the azalea cultivars 'Redwing' and 'Reinhold Ambrosius'. *Acta Hortic.*, 14: 27–38.

Pivot, D., 1988. Sludge and bark compost mixed or not with peat or rockwool for the cultivation of potted gerberas. *Acta Hortic.*, 221: 75–84.

Poole, R.T. and Conover, C.A., 1974. Germination of 'Neanthe Bella' palm seeds. *Proc. Fl. State Hortic. Soc.*, 87: 429–430.

Poole, R.T. and Conover, C.A., 1975. Media, shade and fertilizer influence production of the areca palm, *Chrysalidocarpus lutescens* Wendl. *Proc. Fl. State Hortic. Soc.*, 88: 603–605.

Poole, R.T. and Conover, C.A., 1977. Nitrogen and potassium fertilization of the *Aglaonema commutatum* Schott cvs. Fransher and Pseudobracteatum. *HortScience,* 12(6): 570–571.

Poole, R.T. and Conover, C.A., 1981a. Growth response of foliage plants to night and water temperatures. *HortScience,* 16(1): 81–82.

Poole, R.T. and Conover, C.A., 1981b. Influence of fertilizer, dolomite and fluoride levels on foliar necrosis of *Chamaedorea elegans* Mart. *HortScience,* 16(2): 203–205.

Poole, R.T. and Conover, C.A., 1981c. Influence of maximum air temperatures and irrigation frequencies during high temperature periods on growth of four foliage plants. *HortScience,* 16(4): 556–557.

Poole, R.T. and Conover, C.A., 1981d. Influence of N–P–K factorial fertilization on growth characteristics and foliar content of 4 foliage plants. *HortScience,* 16(6): 771–772.

Poole, R.T. and Conover, C.A., 1983. Establishment and growth of *in vitro*-cultured *Dieffenbachia*. *HortScience,* 18(2): 185–187.

Poole, R.T. and Conover, C.A., 1985. Nitrogen, phosphorus and potassium fertilization of *Brassaia actinophylla*, *Calathea makoyana* and *Chrysalidocarpus lutescens*. *J. Environ. Hortic.*, 3(1): 1–3.

Poole, R.T. and Conover, C.A., 1986. Growth of *Cissus*, *Dracaena* and *Syngonium* at different fertilizer, irrigation and soil temperatures. *Proc. Fl. State Hortic. Soc.*, 99: 268–269.

Poole, R.T. and Conover, C.A., 1988a. Propagation of single eye cuttings of *Philodendron scandens oxycardium* after storage. *Acta Hortic.*, 226: 587–590.

Poole, R.T. and Conover, C.A., 1988b. Storage of philodendron and pothos cuttings. *Proc. Fl. State Hortic. Soc.*, 101: 313–315.

Poole, R.T. and Waters, W.E., 1971. Soil temperature and development of cuttings and seedlings of tropical foliage plants. *HortScience,* 6(5): 463–464.

Poole, R.T., Conover, C.A. and Joiner, J.H., 1976. Chemical composition of good quality tropical foliage plants. *Proc. Fl. State Hortic. Soc.*, 89: 307–308.

Poole, R.T., Conover, C.A., Chase, A.R., Osborne, L.S., Henny, R.J. and Henley, R.W., 1983. *Aphelandra production guide,* University of Florida Ornamental Fact Sheet, 6 pp.

Prince, T.A. and Cunningham, M.S., 1989. Production and storage factors influencing quality of potted Easter lilies. *HortScience,* 24(6): 992–994.

Rajapakse, M.J., McMahan, M.J. and Kelly, J.W., 1993. End of day far red light reverses height reduction of chrysanthemum induced by $CuSO_4$. *Sci. Hortic.*, 53: 249–259.

Rajapakse, N.C. and Kelly, J.W., 1994. Influence of spectral filters on growth and postharvest quality of potted miniature roses. *Sci. Hortic.*, 56: 245–255.

Rishani, N. and Rice, R.P., 1988. Use of carob as a potting medium component. *HortScience,* 23(2): 334–336.

Roude, N., Nell, T.A. and Barrett, J.E., 1991. Longevity of potted chrysanthemum at various nitrogen and potassium concentrations and NH_4:NO_3 ratios. *HortScience,* 26: 163–165.

Sachs, R.M. and Hackett, W.P., 1977. Chemical control of flowering. *Acta Hortic.*, 68: 29–49.

Scharpf, H.C. and Grantzau, E., 1985. Correct N:K ratios guarantee successful culture. Plant tolerances should not lead to a salt enrichment in the substrate. *Gärtnerbörse und Gartenwelt,* 85(10): 403–405.

Shen, G.W. and Seeley, J.G., 1983. The effect of shading and nutrient supply on variegation and nutrient content of variegated cultivars of *Peperomia obtusifolia*. *J. Am. Soc. Hortic. Sci.*, 108(3): 429–433.

Spooner, J., Daley, L., Ware, G. and Vines, M., 1978. Determination of irrigation and fertilizer practices for Jade plant [*Crassula argentea* (L.) Thunb.]. *J. Am. Soc. Hortic. Sci.*, 103(3): 306–308.

Stamps, R.H., 1992. Commercial leatherleaf fern culture in the

United States of America. In: J.M. Ide, A.C. Jermy and A.M. Paul (Editors), *Fern Horticulture: Past, Present and Future Perspectives*. Intercept, Andover, UK, pp. 243–249.

Stamps, R.H., 1993. Reducing water use during cut-foliage production. *HortScience,* 28(4): 287–289.

Stamps, R.H. and Haman, D.Z., 1991. *Cold protection of leatherleaf fern in Lake, Putnam, and Volusia Counties, Florida,* Special Publication SJ 91SP 15. St. Johns River Water Management District, Palatka, FL, 72 pp.

Sterling, E.P. and Molenaar, W.H., 1985. *The influence of time and temperature during simulated shipment on the quality of pot plants,* Report No. 2286. Springer Institute, Wageningen, the Netherlands, 72 pp.

Stromme, E., Selmer-Olsen, A.R., Gislerød, H.R. and Moe, R., 1994. Cultivar differences in nutrient absorption and susceptibility to bract necrosis in poinsettia (*Euphorbia pulcherrima* Willd. ex Klotzsch). *Gartenbauwissenschaft,* 59: 6–12.

Tosi, D. and Tesi, R., 1987. Expanded clays as growing media for croton in pot culture. *Adv. Hortic. Sci.,* 1(1): 39–42.

U.S.D.A., 1994. *Floriculture crops, 1993 summary,* Sp Cr 61 (94). National Agriculture Statistics Service, United States Department of Agriculture, Washington, DC, 87 pp.

van Lieburg, M.J. and Van Doorn, W.G., 1990. Prevention of phytochrome-related postharvest loss of quality in ornamentals through red light emitted by diodes. *Acta Hortic.,* 272: 347–351.

Verberkt, H., 1988. Assimilatiebelichting verhoogt kwaliteit en verkort teeltduur. *Vakbl. Bloemisterij,* 41: 52–53.

Vogelezang, J.V.M., 1988. Effect of root-zone heating on growth, flowering and keeping quality of *Saintpaulia*. *Sci. Hortic.,* 34(1/2): 101–113.

Vogelezang, J.V.M., 1990. Effect of root zone and air temperatures on flowering, growth and keeping quality of *Begonia* × *hiemalis* 'Toral'. *Sci. Hortic.,* 44: 135–147.

Vogelezang, J.V.M., 1991. Effect of root-zone and air temperature on growth, ornamental value and keepability of *Ficus benjamina* and *Schefflera arboricola* 'Compacta'. *Sci. Hortic.,* 46(3/4): 301–313.

Vogelezang, J.V.M. and Verberkt, H., 1990. Supplemental lighting for potplant cultures. *Acta Hortic.,* 272: 159–163.

Wang, Y., 1989. Medium and hydrogel affect production and wilting of tropical ornamental plants. *HortScience,* 24(6): 941–944.

Welleman, J.C.C. and Smulders, P.M.J.A., 1988. Latest developments in soilless growing on Grodan rockwool in Holland. *Acta Hortic.,* 221: 347–356.

Went, F.W., 1944. Plant growth under controlled conditions. II. Thermoperiodicity in growth and fruiting of the tomato plant. *Am. J. Bot.,* 31: 135–150.

Went, F.W., 1953. The effect of temperature on plant growth. *Ann. Rev. Plant Physiol.,* 4: 347–362.

Whealy, C.A., Nell, T.A., Barrett, J.E. and Larson, R.A., 1985. Don't let heat delay your mums. *Greenhouse Manager* 4(4): 102–104, 106, 108.

Whealy, C.A., Nell, T.A., Barrett, J.E. and Larson, R.A., 1987a. Cultivar selection can minimize chrysanthemum heat delay. *Grower Talks* 50(9): 104–109.

Whealy, C.A., Nell, T.A., Barrett, J.E. and Larson, R.A., 1987b. High temperature effects on growth and floral development of chrysanthemum. *J. Am. Soc. Hortic. Sci.,* 112(3): 464–468.

Wilson, M.R. and Nell, T.A., 1983. Foliar applications of BA increase branching of 'Welkeri' dieffenbachia. *HortScience,* 18(4): 447–448.

Wittwer, S.H., 1968. Chemical regulators in horticulture. *HortScience,* 3(3): 163–167.

Woltering, E.J., 1987. Effects of ethylene on ornamental pot plants: classification. *Sci. Hortic.,* 31(3–4): 283–294.

Worrall, R.J., 1981. Comparison of composted hardwood and peat-based media for the production of seedlings, foliage and flowering plants. *Sci. Hortic.,* 15(4): 311–319.

Ying, H. and Joiner, J.N., 1960. The effect of sources and levels of nitrogen on the growth and flowering of potted chrysanthemums. *Proc. Fl. State Hortic. Soc.,* 73: 401–404.

Zieslin, N., Khayat, E. and Yogev, S., 1987. Response of plants of tropical origin to changes in the night temperature regime. *Sci. Hortic.,* 33(3/4): 299–305.

Chapter 8

PLANT DISEASES IN GREENHOUSES

Yigal ELAD

INTRODUCTION

The unique weather conditions which prevail in greenhouses make their ecosystem different from open growth habitats. In greenhouses, plants and their pathogens can develop during seasons which restrict their development in the open. As a consequence, the behavior of plant diseases differs according to the type of greenhouse.

High humidity, free leaf moisture, reduced intensity and changed quality of light, and temperatures higher than outdoors, at least during part of the day, promote the development of many plant diseases. Soil wetness and moderate temperatures affect soilborne plant pathogens. These conditions are often amplified by the development of a luxuriant plant canopy, which reduces aeration and illumination and facilitates development of diseases. Patterns of the same disease on the same host may vary in different types of greenhouses.

Greenhouse factors that affect the varying development of diseases comprise specific systems of heating, the architecture of the greenhouse and the covering material, systems of ventilation and irrigation, the growth medium, the general crop management, and factors which condition the interaction between pathogens and their hosts. For instance, ventilation or, conversely, restricted air movement and the concomitant increase in humidity, in addition to direct effects on disease, may affect the plant's development, its reproduction and yield, all of which may affect the disease indirectly. In this chapter are discussed the direct and indirect effects on plant diseases of conditions inside greenhouses.

COMMON PATHOGENS

Infectious diseases in greenhouse crops are caused by viruses, bacteria and fungi.

The main fungal disease agents that affect plants in greenhouses are classified into foliar pathogens and soilborne pathogens (see the Appendix for a listing of diseases and affected crops; see also Fletcher, 1984; Cline et al., 1988). The most frequent greenhouse airborne fungal pathogens are species of *Alternaria, Botrytis, Bremia, Cladosporium, Didymella, Erysiphe, Fulvia, Leveillula, Phoma, Phytophthora, Pseudoperonospora, Puccinia, Sclerotinia, Septoria* and *Sphaerotheca.* The most common soilborne fungal pathogens are species of *Fusarium, Phytophthora, Pyrenochaeta, Pythium, Rhizoctonia, Sclerotinia, Sclerotium* and *Verticillium.*

Bacterial pathogens cause leaf and fruit spots, soft rots, wilts, galls, hairy roots and scab. The main bacterial groups in greenhouse crops are *Agrobacterium, Clavibacterium, Erwinia, Pseudomonas, Streptomyces* and *Xanthomonas* (Appendix; see also Schaad, 1988).

The viruses which are often found in greenhouse crops cause mainly yellow mosaic and other symptoms on the leaves or fruits, as well as stunting. They may affect many plants in the greenhouse. Severe symptoms caused by virulent strains may end in stunting, leaf defoliation and even wilt. Major virus diseases (see Appendix) which are common in greenhouses are: tobacco mosaic virus (tomato and pepper strains) which is transmitted on seeds, soil and water; cucumber green mottle mosaic virus (cucumber and melon) and potato virus X (tomato), both transmitted by handling; tomato bushy stunt virus (tomato and eggplant); tobacco necrosis virus (cucumber); and melon necrotic spot virus (melon, cucumber and watermelon), transmitted by *Olpidium brassicae* and by contaminated water or soil. (See Table 8.1 for the scientific names of crops mentioned in this chapter.)

Table 8.1
Scientific names of crop plants mentioned in this chapter

Common name	Scientific name
Carnation	*Dianthus caryophyllus*
Chrysanthemum	*Dendranthema grandiflora*
Cucumber	*Cucumis sativus*
Eggplant	*Solanum melongena*
Gerbera	*Gerbera jamesonii*
Lettuce	*Lactuca sativa*
Melon (including muskmelon)	*Cucumis melo*
Pepper	*Capsicum* spp.
Rose	*Rosa* × *hybrida*
Tomato	*Lycopersicon esculentum*
Watermelon	*Citrullus lanatus*

CONDITIONS IN THE GREENHOUSE

Direct effect of high relative humidity and free leaf moisture

Conditions of high humidity (low vapor pressure deficit (VPD)) in greenhouses prevail mainly in non-heated greenhouses, and also sometimes in intensively heated greenhouses or where fuel is being saved in order to cut costs. The diseases that thrive under these conditions are gray mold (*Botrytis cinerea*), tomato late blight (*Phytophthora infestans*), downy mildew of lettuce and cucurbits (*Bremia lactucae* and *Pseudoperonospora cubensis*, respectively), tomato leaf mold (*Fulvia fulva*), chrysanthemum white rust (*Puccinia horiana*), and bacterial diseases. Powdery mildew in cucurbits (*Sphaerotheca fusca*), in roses (*S. pannosae*) and in other crops is favored by high humidity, but is negatively affected by leaf surface wetness (Jarvis, 1989). Conventional methods for the control of diseases promoted by wetness include reduction of atmospheric humidity by environmental manipulation (Morgan, 1984; Winspear et al., 1970).

Spores of most of the fungi causing major diseases will germinate only under high humidity or in free water. High humidity often leads to the condensation of moisture on aerial plant parts, and therefore the effect of free water is often difficult to separate from that of high humidity. The optimum levels of relative humidity to restrict the development of plant diseases are very difficult to define, because they are influenced by the temperature.

Effect of high humidity and nutrition on the physiological status of host plants and their susceptibility to diseases

Increased humidity and poor ventilation in the greenhouse have detrimental effects on plant development. Under such conditions there is reduced translocation of some ions and hormones from the roots to the shoots and leaves (O'Leary and Knecht, 1972). Transport of calcium and other divalent cations may be reduced, mainly because the calcium is translocated during the daytime and almost exclusively in the xylem by the transpiration stream. Such conditions may cause physiological disorders. Calcium is an important factor in many enzymatic processes. It interacts with plant hormones, and is a building block in the cell wall and middle lamella, where pectin is present. Thus, atmospheric humidity and the salinity in the plant root environment, which dictate the root pressure in the plant (at night), influence the level of calcium and its distribution within the plant. If stomatal resistance changes, increased damage due to air pollution may occur (Tibbits and Kobriger, 1983). Accelerated senescence due to hormonal changes has been reported under such conditions (Bakker, 1984). Physiological disorders of greenhouse plants are discussed in Chapter 5. However, if the physiological status of the plant is disturbed, it also enhances its susceptibility to pathogenic agents. Since calcium and hormones both affect membranes and meristematic tissues, interactions among hormones, calcium and microclimate, with respect to susceptibility of host organs to diseases, can be expected (Shear, 1975).

Increased calcium content in plant tissues inhibits the development of some diseases. Fertilizing with calcium reduced *Fusarium* wilt of tomato (Corden, 1965) and muskmelon (Spiegel et al., 1987) and also the susceptibility of bean hypocotyls to *Rhizoctonia solani* (Bateman and Lumsden, 1965). Liming of low-calcium soils caused a reduction in the incidence of tomato gray mold (Stall, 1963).

The use of soilless media for crop plants in greenhouses allows the precise control of fertilization in order to obtain healthy plants. An increased concentration of calcium in the fertilizer resulted in a significant reduction of gray mold of crops grown in perlite, rock-wool or volcanic gravel. Gray mold of cucumber grown in perlite was reduced 40–70% in fruits and stems by calcium nitrate treatment. Calcium content was increased considerably in the upper leaves of treated plants but the effect could be attributable to the high ratio of nitrate to ammonium nitrogen. Severity of ghost spots disease of tomato fruits was also decreased by Ca^+ fertilization. Similarly gray mold of pepper and eggplant was reduced by 50–60% and 30%, respectively. Tomato and pepper plants were grown in perlite or in soil amended with gypsum or superphosphate, both containing 21% calcium. Disease was reduced by these sources of calcium as well. Calcium nitrate sprayed on eggplants along with auxins decreased disease significantly (Elad et al., 1991).

Rose flowers picked from plants in commercial greenhouses grown in rock-wool or volcanic gravel and heavily fertilized with calcium were significantly less susceptible to gray mold than those which were not treated with additional calcium. The flowers from calcium plots had a higher calcium content and lower production of ethylene as compared with flowers from plants not treated with calcium. Higher calcium concentration is associated with lower activity of polygalacturonases produced by the pathogen (Volpin and Elad, 1991).

EPIDEMIOLOGY

Transfer of inoculum into greenhouses may be restricted because of unfavorable conditions which occur outside. However, once inoculum appears, the spread of an epidemic may be rapid. Epidemics start with the initial introduction of the pathogen, when the available inoculum meets a susceptible host in a favorable environment.

Inoculum

The inoculum may originate from the greenhouse itself or may be introduced from a distance. Some airborne fungi (e.g. rusts, powdery mildews and gray mold) can enter with streams of air carrying conidia. Soilborne inoculum may be carried into the greenhouse on pots or on the soles of shoes. However, with most soilborne diseases, the source of the pathogen is local, and originates from within the greenhouse. Pathogens may also be transmitted in irrigation water in droplets or aerosol. Greenhouses with hydroponic systems or detached growth media may be contaminated by pathogens which, due to the nature of the systems, may spread very fast and affect many plants. Other means of transmission are greenhouse tools, such as grafting implements or knives, that contaminate plants while being used. Several fungi, bacteria and viruses, bacteria and viruses can spread in seeds and in vegetatively propagated young plants. For instance, *Pseudomonas syringae* pv. *tomato* in tomato, *P. syringae* pv. *lachrymans* in cucurbits, *Xanthomonas campestris* in crucifers, and *X. vesicatoria* in pepper are seedborne diseases which have to be controlled in order that the farmer will receive pathogen-free seeds (Saettler et al. 1989).

Insect vectors may spread some fungi or bacteria into greenhouses. Soft-rot bacteria are commonly spread by flies, as are pathogens which infect cucurbit flowers. In cucurbits, bacteria may be carried in the vessels without producing symptoms, but are secreted in the guttation liquid, so that insects sucking it may infect leaves or flowers. Viruses are dispersed by vectors such as insects and nematodes, or by grafts of living plant material. Most viruses of perennial crops spread primarily by vegetative propagation. They are subsequently totally dependent on an insect vector as their sole means of dispersal and infection.

Epidemic development

Once established on plants in the greenhouse, the primary focus of infection provides inoculum for secondary spread. Spread through the soil is usually much slower than in air. The spores of airborne pathogens are produced in large quantities, and generally under wet conditions (e.g. downy mildew and gray mold), but are released most readily when the humidity drops. In some fungi a special mechanism is activated for spore release (Jarvis, 1980; Meredith, 1973). For several bacterial and fungal diseases a threshold level is required for the onset of an epidemic, but they subsequently spread rapidly, according to the conditions. Growers of commercial greenhouse crops tend to cultivate the same crop every year. In such monocultures pathogens may be established and cause increasing damage every year.

Survival

In between growing seasons, in the absence of major hosts, the pathogens may face severe conditions. Pathogens survive between periods of active parasitic stages in a variety of ways, depending on the type of organism involved. Fungal pathogens may survive in the active stage by producing dormant propagules such as resting spores, chlamydospores, aleariospores and sclerotia. In the absence of hosts, some fungi may undergo an active stage of saprophytic growth in soil or in organic materials such as plant debris, or they may grow on alternative hosts, including weeds.

Bacteria often survive in debris, or in soil and water, as saprophytes attached to organic particles, and in the rhizosphere of host and non-host plants. Non-active bacterial microcolonies were also proved to exist protected in polysaccharide slime, whereas their active forms are associated with different hosts. In general, under normal environmental conditions viruses are unable to persist for a long time outside the living plant cell or insects, but there are exceptions.

DISEASE CONTROL

For a commercial grower, cultivation must be profitable, and the net cost/benefit of disease control must be considered. On the other hand, the amateur grower or workers in botanical gardens may have other objectives, striving to achieve optimal disease control, or to cultivate plants without chemical treatments.

Control of plant diseases in the greenhouse can be on a preventive basis, or may start when the disease appears. Methods of prevention include the use of pathogen-free seeds and planting material, resistant cultivars or rootstocks, soil disinfestation by heat, steam, methyl bromide or formaldehyde, elimination of pathogen sources (sanitation), chemicals, pathogen-free soilless systems of culture and isolated beds, control of the environment to create conditions unfavorable for the development of diseases, or application of fungicides as protectants. One of the most reliable methods is the use of resistant or tolerant cultivars (Jarvis, 1989). After disease appears, hygienic practices and environmental control can slow down the dispersal of the pathogen. Fungicides may then be used to protect the plants and thus suppress the rate of disease development.

The most important production diseases in horticultural crops are those induced by the vascular wilt pathogens (see Appendix). Once introduced, these pathogens are difficult to eradicate even when beds are steamed and/or fumigated. Pathogen-free propagation programs are routine (Cline et al., 1988), employing clean sources of propagation material, and seeds treated by heat and chemicals.

Chemical control

For a given disease problem in the greenhouse, there may be several means of chemical control. Treatment before the pathogen enters the host plants is carried out mainly with "protectant" fungicides. However, due to the intensive growth of greenhouse crops, the protectants have to be applied at short intervals in order to cover new growth as well as old plant surfaces, on which residues of earlier applications are not effective. Systemic fungicides supposedly can provide a solution to this problem, but most of them are xylem-translocated and move upward into plant parts that are actively transpiring. There is only little movement into petals, sepals and fruits, but direct application to the target parts of the plant may solve the problem. Pesticides are usually applied to control vectors of viruses.

Surfactants (spreaders, wetters or stickers) can improve the performance of a fungicide or pesticide in the greenhouse by lowering the surface tension between the droplets of the spray and the waxy cuticle covering the host. These polymers themselves may have a fungistatic effect. Such effects were obtained by Elad et al. (1989, 1990) against gray mold and powdery mildew of cucumber grown in greenhouses.

Eradication of soilborne inoculum

Steam (80–95°C, 1–2 h) is a common means of control in northern countries. It gives effective sterilization down to a depth of 40 cm. In places where steam is too expensive to produce, methyl bromide fumigation may be used; however, in contrast to steam, it is associated with chemical residues (bromides) in the soil. Formaldehyde is replacing methyl bromide in certain greenhouses, as it has no residual effect. It is usually applied at the rate of 1–2 $m^3\ ha^{-1}$ against bacteria and 0.5 $m^3\ ha^{-1}$ against fungi and is activated by water. As it is biodegraded to alcohol and carbon dioxide, it does not pose hazards to the environment. "Structure solarization", carried out in summer (increasing the air temperature in the greenhouse dramatically by closing all openings for 1–2 months), can eradicate some pathogens, not only from the growth medium but also from all solid materials left in the greenhouse. This can be practiced in warm climate regions.

Soil sterilization (or pasteurization) is not immunization, but a method to eradicate pathogens. Afterwards, however, the introduction of an inoculum of a pathogen is followed by its rapid spread and the destruction of plants.

Water may carry pathogens, such as bacteria, into the greenhouses. Ultraviolet irradiation or chlorination is an effective treatment in such cases.

Application methods for foliar diseases

The application of fungicides and pesticides in the greenhouse requires special equipment. Sprayers can range from hand-pumped units carried as back-packs with a capacity of 1–15 ℓ to electric or gasoline-powered motor sprayers moved on wheels, with a capacity of 40–800 ℓ. Some larger greenhouses make use of a permanent plumbing system which provides spray from the mixing tank to any given section of the greenhouse. Apart from formulations for spray application, there are some fungicides which are formulated for evaporation (by heaters), to give applications in the form of dust, mist or smoke which are most suitable for greenhouses. Dusting or mist application of fungicides can also be conducted in greenhouses, but they as well as sprays pose a severe health problem to the workers. Because of the closed environment in the greenhouse, the use of respirators or gas masks along with suitable clothing is essential. For safety purposes, automatic sprayers which move on a rail between plant rows inside the greenhouse, and are remote-controlled, were developed. These systems can be expensive, but should be recommended to growers. It should be noted that crops may be more sensitive to phytotoxicity caused by fungicide and pesticides when grown in greenhouses, as compared with the open field.

The use of fungicides and pesticides in greenhouses is sometimes risky because of their toxicity to the plants under the conditions inside the greenhouse. The situation is more common in ornamentals and plants grown for their flowers, because in such crops there are many clones and varieties which were selected according to potential demands of the public, but their tolerance to fungicides and pesticides may differ from that of other varieties of the same species. In many countries, authorities approve particular chemical products for use on particular crops; this approval is known as "registration". Unfortunately, registration is not issued for all varieties but only the major ones in each species because of the cost, and farmers must be cautious when risking their crop with a non-registered product. Chemical agents are, however, the most common means employed for disease control in many places.

Effect of chemical control on yield

The ultimate goal of the grower is to achieve disease control which results in higher yield or better product quality. However, there may be cases where effective disease control does not result in yield increases. Such was the case with parthenocarpic cultivars of cucumber which were attacked by gray mold (Yunis et al., 1991). The reason for the lack of correlation between diseased senescing female flowers ("fruitlets") of cucumber and decreased yields is the physiology of the parthenocarpic cultivars. They produce such a large number of flowers that, even if most of them are aborted or infected by *Botrytis cinerea* or *Sclerotinia sclerotiorum,* others develop in compensation for these losses and there are enough left to give a high yield. This implies that fungicides can be applied less frequently. However, in some cases under severe epidemic conditions, usually at the end of the season, or in high-wire cropping systems where long bare stems are left spreading on the growth medium, gray mold can damage the stems of the plants and the upper part of the plants may die; this was observed in some greenhouses. A latent infection may also occur, following which fruits may begin to evince visible infection only during packaging or transport. Therefore, fungicide sprays should be employed only when a severe epidemic is suspected. Decisions can be made with the help of forecasting systems (described below, p. 202). However, on crops like tomato, eggplant, pepper and flowers where compensation as in cucumbers does not exist or where quality of the produce (flowers) is very important, fungicides are still the most important means for control.

Failures in chemical control

In greenhouses, fungicide-resistant strains of fungi have frequently appeared. This phenomenon occurs because the greenhouse is a closed system where the population of selected strains is not diluted by the outdoor wild population. Usually, the existence of epidemic conditions in greenhouses is a prerequisite for the development of resistant populations of pathogens. Greenhouse crops are usually grown during periods when relatively low temperatures prevail in the open fields, at times when the pathogens are inhibited, are absent in the open fields, or their hosts are not growing (e.g. *Botrytis cinerea, Pseudoperonospora cubensis,* and *Sphaerotheca fusca* syn. *fuliginea*). Moreover, conditions optimal for disease development in greenhouses prevail for long periods. The number of disease cycles is increased due to the optimal conditions or the extended time for which they prevail, and control necessitates frequent applications of fungicides. The latter result in high selection pressure towards resistance to the fungicides.

The benzimidazole fungicides are particularly likely to enable pathogens to develop resistance against pathogens because they have a specific mode of action. This has led to the selection of highly resistant strains when benzimidazoles are overused. Resistance to these fungicides is usually not associated with a significant loss of fitness of the pathogen. Mixtures and alternations with broad-spectrum contact fungicides may delay such selection, before resistance has become apparent. Acute problems of resistance to dicarboximide fungicides have arisen when they are used intensively and exclusively over many seasons. The resistance developed is moderate, and the resistant isolates tend to be almost as fit as sensitive strains in the absence of fungicides. It is recommended that the number of dicarboximide treatments be restricted to no more than three per crop in greenhouses where resistance is found; even in the absence of detectable resistant strains, the number of applications should not exceed four per season. When infection pressure is high, it is usually recommended to alternate these fungicides, or to apply them in a mixture with protectants such as chlorothalonil, captan or thiram, which usually do not select for resistance.

The main pathogens which are known to develop resistance to fungicides in greenhouses are *Botrytis cinerea, Didymella bryoniae,* (gummy stem blight of cucurbits), *Fusarium oxysporum* f.sp. *gladioli, Pseudoperonospora cubensis* (downy mildew of cucurbits), *Puccinia horiana, Sphaerotheca fusca* (powdery mildew of cucurbits) (Malathrakis, 1986) and *Uromyces dianthi.*

The failure of disease control in greenhouses is exemplified by the history of gray mold epidemics. The presence in Israel of *B. cinerea* isolates resistant to dicarboximides was first reported in 1982 (Katan, 1982a,b). Since then, failures in controlling gray mold were experienced in many Israeli greenhouses. A similar situation is common in Europe (Panayotakou and Malathrakis, 1983; Leroux and Clerjeau, 1985; Pappas, 1986). In surveys conducted following the detection of dicarboximide-resistant isolates in Israel (Katan, 1982a) it was noted that, at the beginning of each season, the proportion of resistant strains in the population of *B. cinerea* was lower than that present in greenhouses at the end of the previous season (Katan, 1982b). The frequency of resistant strains increased during the season in response to sprays with the dicarboximide fungicides iprodione or vinclozolin. This phenomenon has been reported also with other covered crops in areas in southern Europe (Panayotakou and Malathrakis, 1983; Pappas, 1986). In 1989 one witnessed failures to control gray mold in cucumber greenhouses even at the beginning of the season. Indeed, inoculum of *B. cinerea* in plant debris is able to survive high temperatures in semiarid countries (Yunis and Elad, 1989), or low winter temperatures in the temperate zone (Palti, 1981). These extreme seasonal conditions do not interfere with the survival of fungicide-resistant isolates. There is a possibility that resistant isolates of the pathogen are going through continuous selection towards greater fitness to survive under extreme conditions. Resistant populations of *B. cinerea* can be monitored by means of a selective medium (Elad and Shtienberg, 1995).

During the 1970s, *Pseudoperonospora cubensis* was controlled mainly with protective applications of dithiocarbamates and chlorothalonil. In the early 1980s metalaxyl was released, and soon resistant strains were selected. Metalaxyl-resistant strains seem to be more competitive than wild-type strains (Cohen et al., 1983). During the 1970s the use of benzimidazoles to control *D. bryoniae* led to the appearance of benzimidazole resistance in *D. bryoniae* populations (Malathrakis, 1986). Powdery mildew in greenhouses was controlled for several years by benzimidazoles, hydroxypyrimidines, pyrazophos, and ergosterol biosynthesis inhibitors (EBI). Resistance is known in populations of *Sphaerotheca fusca,* but alternation of fungicides, which is practiced in many countries, is helping to cope with the problem. It is generally recommended to rotate the use of EBI fungicides with fungicides from other groups, or to apply them in mixtures with other fungicides. It is usually better to limit the number of applications, especially against fungi with many cycles during the growing season. Moreover, it is recommended that non-chemical methods be used as well.

The limited efficiency of fungicidal treatments in roses and similar greenhouse crops is due to reasons other than resistance of pathogens to fungicides. Flowers which are closed at the time of treatment develop and are harvested before the next spray; meanwhile, infection of petals which were not exposed to fungicides may occur (Elad, 1988a). The use of vaporized fungicides distributed by thermal heaters at night may solve such problems of application, but requires a special formulation which may not be available. Similar problems of resistance are known with fungicide and pesticides aimed against insect vectors of viruses.

Some diseases have a latent stage, with symptoms becoming visible only after harvest. In such a case, infection occurs in the greenhouse but rot appears later, when the flowers or fruits mature. Infected plants may produce ethylene, which renders the tissues more susceptible to the development of disease and subsequent rot (Elad, 1988b).

Cultural control

Solarization

Soil disinfection can be done by steam and by fumigation or application of chemicals. Soil solarization was tested in warm regions and gave good results (Katan, 1981). This practice, which involves covering the soil with transparent polyethylene, is commonly used to raise the temperature of agricultural soils. Katan (1981) showed that mulching a moistened soil during the hot season in warm countries resulted in the control of many soilborne pathogens and the elimination of weeds. Soil solarization is now practiced in many hot countries, but it is not as effective in cooler regions or periods, because solar radiation is relatively low there, and the soil temperatures in the mulched soil do not reach levels which are lethal or suppressive for the pathogens and weeds. However, under marginal climatic conditions, although soil solarization might not be effective in the open field, the method may still be effective in greenhouses (Mahrer et al., 1987). This interesting approach was developed independently in Italy, Japan and other countries. Mulching soil with polyethylene or vinyl was done inside closed greenhouses to intensify soil heating. In northern Italy, soil solarization inside greenhouses controlled corky root of tomatoes in two successive crops (Garibaldi and Tamietti, 1984). This approach was successful in Japan for the control of *Fusarium* in strawberries, and of other diseases (Komada and Fukai, 1982). As mentioned above, greenhouse structure solarization can be carried out in summers in the absence of a crop.

Environmental control

Environmental control of disease can minimize the need to use chemicals. The factors which affect disease development in the greenhouse are temperatures of both soil and air, vapor pressure deficit (VPD), dew, soil water content, and light (quality, day length and intensity). All this can be controlled to a certain extent depending on the facilities available (Winspear et al., 1970; Morgan, 1984). However, it should be noted that there are interactions among air temperature, VPD, dew deposition on the canopy, physiological status of the host, saprophytic microflora and aggressiveness of the population of the pathogen in their effect on disease (Rotem et al., 1978; Elad et al., 1988). Interplay among these factors affects sporulation, dispersal, germination of conidia, penetration of the germ tubes, and lesion development.

Influence of cladding material on indoor temperature and humidity

Transparent plastic films are being used more and more in horticulture. In order to improve the heat retention there is a tendency to use polyethylene as the cladding material, and to reduce the far-red transmittance the polyethylene resin is enriched with vinyl acetate and/or aluminum silicate (or magnesium silicate). Such films have a "thermal" effect. Other cladding materials used are polyvinyl chloride (PVC), polyvinyl fluoride (PVF) and polyester films (PEST); these raise the crop temperature and decrease the leaf moisture. An example of such environmental control is given by cucumber gray mold development in non-heated polyethylene greenhouses covered with various types of cladding repelling infrared radiation compared with cladding which has no IR repellency (Elad et al., 1988). Non-persistence of dew on the foliage was the limiting factor in disease development in a relatively dry winter. Disease severity under the different IR sheets was correlated with the duration of dew. In a rainy winter dew periods were long, and gray mold severity was correlated with accumulated degree hours near the optimum for disease development (15–25°C). Yield was positively correlated with the number of infected fruits, probably because both it and disease are favored by temperatures around 20°C, and high humidity. This means that attempts to change environmental parameters for control of disease may have an effect on yield also. Plants generally grow better under thermal films. The tests also showed differences between infection of fruits and stems under the various brands of polyethylene, probably because these plant organs are influenced differently by the microclimatic factors.

Conditions in the greenhouse influence the physiological status of the host organs and thereby affect the latter's susceptibility to infection. Low night temperatures and high relative humidity in the greenhouse predispose cucumbers to disease, and optimal climatic conditions for crop production during the winter favor disease. Interactions probably exist among air temperature, relative humidity and dew deposition on the canopy in allowing disease development. In general, the thermal IR polyethylene covers reduce the duration of dew on plants but extend the duration of temperatures favorable for epidemics.

On the basis of disease records and microclimatic parameters, a model was developed for predicting outbreaks of gray mold in greenhouses. Outbreaks of gray mold occurred when on a weekly average the wetting period exceeded $7\,h\,d^{-1}$ and the duration of temperatures between 9°C and 21°C during the night (18:00–08:00) exceeded $9.5\,h\,d^{-1}$. It was suggested that the potential for outbreaks of gray mold epidemics could be reduced by measures which restrict the wetting period (Yunis et al., 1994). Control of these factors is easier in heated greenhouses, where the temperature is raised and the humidity is reduced by pumping fresh cold air from outside and heating it several times a day. As a result of the reduction of temperatures during the night time, water condenses on the greenhouse cover or fog forms. This results in the formation of drops, and dripping on the plant canopy. Wetting the plants makes them more vulnerable to disease development. This problem is now overcome by chemicals added to the plastic, which act as anti-foggers and anti-drippers, thereby leading to drier plants in the greenhouse.

Effect of light quality on diseases

It is well established that light stimulates the sporulation of some fungi in vitro (Marsh et al., 1959). Sporulation of certain fungi is stimulated by the ultraviolet, violet and blue regions of the light spectrum. Sporulation of *Botrytis cinerea* is known to be inhibited by blue and red light and stimulated by the near-ultraviolet and far-red regions of the spectrum (Tan, 1975).

Not only sporulation is affected by the quality of light. Lesion development in cucumber downy mildew is suppressed through decreasing photosynthetic activity of the host plant if light exposure is limited (Inaba and Kajiwara, 1971). Indeed, a reduction in cucumber downy mildew has been observed when ultraviolet irradiation was filtered by the greenhouse cover. Commercially available greenhouse covers filter the penetrating light selectively. For instance, PVC, glass and UV-absorbing-polyethylene all reduce ultraviolet irradiation.

Honda and Yuroki (1977) and Honda et al. (1977) reported that UV-absorbing vinyl film reduces the incidence of stem rot of eggplant and cucumber caused by *Sclerotinia sclerotiorum,* and of gray mold of tomato and cucumber caused by *Botrytis cinerea,* by filtering out wavelengths shorter than 390 nm. Reuveni et al. (1989) reported the control of tomato gray mold by using a polyethylene cover which reduces the ultraviolet irradiation significantly.

New custom-made polyethylene films with blockers in the far-red region of the spectrum were tested for their effect on sporulation of *B. cinerea*. A pink-pigmented polyethylene sheet partially screened visible light mainly in the range of 470–640 nm, with lowest transmissivity at 500–520 nm, and a green-pigmented sheet partially screened it mainly in the range of 570–800 nm, with lowest transmissivity at 620–670 nm. Both films screened out the UV radiation up to 380 nm. Light filtration inhibited sporulation of *B. cinerea* cultures on an artificial medium and on tomato leaves and cucumber fruits. However, the extent of inhibition varied with the fungus isolate, the quality of the light which reached the culture and the medium which supported the fungus colonies. Some isolates sporulated in darkness. It was speculated that an indirect effect on the host plant may be important in reducing the sporulation. The use in commercial greenhouses, of the green-pigmented polyethylene reduced gray mold by 35–75%, correlated with reduction of conidial load (Elad, 1997). White and leaf molds, downy and powdery mildews were also retarded (Elad, 1997). Recently it was shown that ultraviolet screening by greenhouse covers results in decreasing viral infection, because insects transmitting the viruses are discouraged from entering the greenhouse (Antignus et al., 1996a,b).

Soilless cultures

Soilless cultures are employed widely in many countries. The use of various media eliminates the need for regular soil sterilization, thus avoiding a very costly and laborious task. The plant roots are no longer exposed automatically to the pathogens which almost inevitably build up in intensively cropped glasshouse soils. Therefore there is a major advantage in soilless cultures.

Soilless culture systems, particularly those without

a solid substrate, can readily be made pathogen-free before planting, but there still remains the possibility of re-infestation during cropping. If, unluckily a root disease starts in a bed after the plants are grown, or is transmitted in propagation material, then the possibility exists of rapid spread of a pathogen throughout a hydroponic system. However, as Van Os (1982) claimed, in practice major disasters are rare. At present, knowledge is incomplete in regard to methods of controlling potential diseases (Runia et al., 1988), in terms of fungicides (dose and way of application) and other methods of control (UV irradiation of solutions), or in terms of factors which predispose crops to infection by pathogens (unbalanced nutrition and death of roots) (Daughtrey and Schippers, 1980; Vanachter et al., 1983).

There is an advantage in recirculating systems, in that pollution of water supplies by fungicides is minimized (Crum et al., 1985). Major fungal diseases which may spread in hydroponic beds are *Olpidium, Phytophthora* and *Pythium*. Control of these diseases can be achieved by slow dripping of the fungicides into the catchment tank, with thorough mixing (Price and Dickinson, 1980). Bacterial diseases of plants can in some situations represent a more serious problem than those caused by fungi, since it is virtually impossible to control them by addition of chemicals to the nutrient solution. In many cases, official clearance for the use of fungicides and pesticides on food crops grown in soilless media has not yet been obtained. This happens after special experiments in the new environment have been conducted, taking care to achieve effectiveness without reaching phytotoxic level.

Sand and gravel culture for re-use is sterilized after roots and plant debris have been removed from the heavy substrate. Where rock-wool or organic substances such as peat are to be re-used, steam sterilization is effective.

Soilless cultures offer an opportunity to employ non-conventional means of disease control, as discussed elsewhere in this chapter. Examples of such an opportunity are employing biocontrol agents against *Fusarium* spp. and deleterious pseudomonades (Van Peer and Schiperes, 1989; Rattink, 1991), including compost in the medium in order to increase its suppressiveness (Hoitink and Fahy, 1986) and the control of nutrients to reduce susceptibility of the host plants to diseases (Elad et al., 1991).

Resistant cultivars

Several problems common to greenhouse crops are overcome by growing plant cultivars resistant to diseases (Saccardo and Sonnino, 1988). Severe diseases, mainly *Fusarium* and *Verticillium* wilts of tomato, *Fusarium* crown and root rot of tomato, tomato corky root rot, *Fusarium* wilt of cucumber and melon, and *Meloidogyne* spp. on pepper and tomato, *Phytophthora capsici* and tobacco mosaic virus (TMV) on pepper, and leaf mold and TMV on tomato are effectively prevented either by employing resistant cultivars, or by grafting desirable cultivars on resistant rootstocks, or by both techniques (Jarvis, 1989). However, new problems appear in greenhouses as well as in the open fields as new races of pathogens (e.g. *Fusarium* race 3) are selected which threaten the resistant crops.

Other options for non-chemical disease control

Soilborne pathogens, such as *Pythium* spp. and *Rhizoctonia solani*, are of major importance in nurseries where plants are grown in container media with peat as the sole organic component. Application of fungicides, fumigation or steaming of the container medium before planting are used for controlling the diseases. However, such treated media may be recolonized by the pathogens. The use of composted wastes (tree barks, municipal solid wastes and sewage sludges) as substitutes to container media to control soilborne pathogens was suggested by Hoitink and subsequently reviewed by Hoitink and Fahy (1986). Most of the experiments in greenhouses were carried out with ornamental crops, and only a few with vegetables.

Control of diseases through composting includes kill of plant pathogens present in wastes during and after the high-temperature composting process. Specific chemical, physical and biological properties have a major effect on the ability of composts to suppress soilborne pathogens (Gouin, 1982; Hoitink and Fahy, 1986).

Other non-chemical methods which could potentially be used to decrease disease damage in greenhouses include (a) drip irrigation to reduce soil and foliage wetting; (b) soil mulching by black polyethylene sheeting to minimize water evaporation to the air and to extend the intervals between waterings; (c) installation of a thermal screen which prevents infrared irradiation from escaping into the atmosphere on bright nights, and thus increases the temperature of the canopy and minimizes the differences in temperature between the plant organs and the surrounding air; (d) growing plants in soilless media or hydroponic systems; (e) filtering and sterilizing the irrigation water, mainly to avoid pathogenic bacteria; (f) installing nets which can exclude vector insects from the greenhouse, in order to prevent virus infection (nets are, however, dangerous because they reduce ventilation and may increase high-humidity diseases, and should therefore be removed when not needed); and (g) covering the soil with shining reflective covers (aluminum foil or polyethylene) to deter vector insects, but the mechanism is not yet known.

Biological control

Biological control of infectious diseases in greenhouses has not yet come into use in many situations, although various attempts have been made (Jutsum, 1988). However, biological control may replace fungicides or provide solutions where the latter are ineffective.

For diseases of roots, foliage, flowers or fruit, glasshouse crops offer more attractive possibilities for biological control than field crops, because the population level of antagonists is easier to maintain (Elad and Chet, 1995).

Biocontrol of soilborne pathogens

Preparations of *Trichoderma* spp. have been used experimentally at several sites to control soilborne pathogens. Use of the biocontrol agents controlled effectively *Rhizoctonia solani, Pythium* spp. and *Fusarium* spp. in several greenhouse crops (Chet, 1987). Broadcast applications over the row were needed where the pathogens attacked mature plants (Elad et al., 1980, 1982). Crop losses caused by the pathogens were significantly reduced by maintaining a low inoculum potential for several months. The amount of *Trichoderma* preparation could be reduced if it was applied directly to the root zone of tomatoes and carnations (Elad et al., 1981, 1982). Seed coating by the antagonistic *Trichoderma* is aimed at the most susceptible sites of the plant and affects pre- and post-emergence diseases. The mature plant may be immune, in which case protection is needed for only a short period, at the seedling stage.

Harman et al. (1991) have developed a system to enhance the biocontrol efficacy of *Trichoderma harzianum* delivered on seeds. This system involves applying the bioprotectant in a thin layer before planting so that it colonizes the seed surface and can proliferate in the absence of competitive microflora; inclusion of amendments to the seed treatment to control pH; adding nutrients (food base) and toxicants. Another potential biocontrol agent was reported by Lumsden et al. (1991). This is a strain of *Gliocladium virens* used against damping-off caused by *Pythium* and *Rhizoctonia* on vegetable and ornamental bedding plants in greenhouses.

Synergism apparently is involved in integrated chemical, biological and physical control. The pathogen may be weakened by the chemical and therefore is controlled better and for longer periods by *Trichoderma.* In such a case the antagonist favors, and even multiplies on, dead cells of the pathogen. Combining *Trichoderma* with methyl bromide or soil solarization in the greenhouse prevented re-infestation of the fumigated soil with pathogens (e.g. *R. solani*), and had a long-term effect (Elad et al., 1982; Chet, 1987). Biocontrol agents applied to the soil may also increase plant growth, which may promote their use in greenhouses (Elad et al., 1987).

Fusarium wilt of carnations was effectively controlled by strains of *Pseudomonas putida* (Baker et al., 1986), the bacterium competing with the pathogen for availability of iron and carbon. Alabouvette (1986) and Baker et al. (1986) have emphasized the possibility of employing suppressive soils against soilborne *Fusarium* spp.

Crown gall caused by *Agrobacterium tumefaciens* endangers roses and several other ornamentals in nurseries, and disease may cause economic losses in greenhouses after transplanting. The biocontrol approach to crown gall control is based on the non-pathogenic *Agrobacterium radiobacter* strain Kd84 inoculant. This agent produces agromycin, which is effective against the pathogen (Moor and Warren, 1979). Commercialization of *A. radiobacter* has resulted in wide use of the preparation in greenhouses.

Biocontrol of foliar pathogens

Inoculation of plants with attenuated or avirulent strains of virus can prevent expression when the same plants are subsequently challenged with a virulent strain of the same virus. This has led to commercial development of TMV race M11-16 and a naturally occurring wild TMV strain to combat TMV in tomatoes (Ahoonmanesh and Shalla, 1981). There are other research projects under way on cross protection or induced resistance which may lead to commercial application in the future.

Hyperparasites may infect biotrophic pathogens such as powdery mildews and rusts. When used for biocontrol, hyperparasites must be applied after infection has occurred, rather than before (Krantz, 1981). In greenhouses Jarvis and Slingsby (1977) reduced infections of cucumber by *Sphaerotheca fusca* by spraying the plants with conidia of *Ampelomyces quisqualis,* and watering between inoculations to ensure establishment of the antagonist. The same workers and Sundheim (1982) obtained cucumber yield increases, although *A. quisqualis* did not completely retard the development of powdery mildew. Combining the hyperparasite with a reduced concentration of the fungicide triforine was also effective in achieving yield increases.

Some pathogens are partly controlled by micro-organisms which are found naturally on aerial surfaces of plants. Many attempts have been made to improve control by applying selected antagonists to such surfaces (Wood, 1950; Rishbeth, 1988). Fokkema (1976) demonstrated that antagonists often compete with the pathogen for nutrients, and inhibitory products may be formed that reduce germination and subsequent growth of the pathogen. Hyphae of fungal pathogens may be parasitized on contact with the antagonist. Apart from killing the pathogen, an antagonist may reduce its reproductive capacity. Competition for nutrients and interference with pathogenicity (Zimand et al., 1996) are regarded as the mechanisms of biological control of gray mold. A preparation of *Trichoderma harzianum* was also developed for the control of gray mold, and biocontrol is achieved by isolate T39 (TRICHODEX) in cucumber and tomato greenhouses (Elad et al., 1994a, Elad and Shtienberg, 1995). Average disease reduction by the *Trichoderma* preparations was >50%. A population of the biocontrol agent was established over the plants at a high level of 10^5–10^7 propagules per leaf or fruit. Alternation of a biocontrol preparation of *T. harzianum* T39 with fungicides proved to be effective in disease control, and reduced the exposure of the greenhouse environment and pathogen to fungicides (Elad et al., 1994a,b). Similar research has been carried out in Italy (Gullino et al., 1991) and elsewhere. Several commercial companies and organizations around the world are investing in projects with a biocontrol orientation.

Induced resistance

Plants possess an array of defense mechanisms against microbial challenges, preformed or induced by microbial challenge. Susceptible plants often contain the genetic information for resistance, and the nature of the phenomenon is non-specific (Dean and Kuc, 1987). Hypersensitivity is localized, whereas induced systemic resistance (immunity) involves a systemic effect in which challenge at one site triggers resistance at a second distant site (Sequeira, 1983). Resistance can be induced by chemicals, pathogens, injury and other physical stresses, products of hosts or infectious agents, or by races of micro-organisms that are non-pathogenic to the host cultivar.

The best known system of induced resistance is in cucurbits, where the duration of protection by a single immunization was found to be 4–5 weeks. Protection is achieved against anthracnose (*Colletotrichum lagenarium*), gummy stem, leaf blight (*Didymella bryoniae*) and powdery mildew (*Sphaerotheca fusca*) (Descalzo et al., 1990). The responses include defense compounds like phytoalexins, chitinases, β-1,3-glucanases, proteases, peroxidases, phenoloxidases, hydroxyproline-rich glycoproteins, lignin and callose (Kuc, 1990). Although crop immunization has high potential for use in greenhouses, it has not yet been practiced commercially (Elad, 1990).

Integrated pest management

Integrated pest management (IPM) involves the use of different methods and techniques (chemical, biological, physical and mechanical). It relies on using natural parasites of pests. As there have been fears that control measures against plant diseases may harm the beneficial agents integrated in the control schemes, several researchers have shown that disease control agents can be integrated in IPM against certain diseases, such as powdery mildew (wetting sulphur, tridimefon, peconazole, tridiminol, propioconazole, fenarimol); gray mold (chlorothalonil, vinclozolin, procimidone, iprodione); downy mildew (chlorothalonil, propamocarb); stem rot (*Phytophthora capsici*) (propamocarb, metalaxyl, benalaxyl); late blight (*P. infestans*) (chlorothalonil, copper, oxichloride); and leaf mold of tomato (*Fulvia fulva*) (chlorothalonil, penconazole) (Celli et al., 1991; Van de Veire, 1991). However, broader research is needed in order to cover the interaction between steps taken against diseases and IPM measures, as they may differ drastically from results observed in the open field. Greenhouse conditions may be responsible for the negative effect of fungicides on pest control agents. Examples of a negative effect of fungicide on natural enemies of pests can be found in publications of the IOBC (International Organization of Biological Control) working group "Pesticides and Beneficial Organisms". Generally, dusting is dangerous to the population development of natural enemies and it is not recommended. Wetters and spreaders are harmful for predatory mites. There are indications, for instance, that dichlofluanid, fenitrothion, formothion, polylfluanid, pyrazophos, and thiram are harmful to the parasite *Encarsia formosa* (Hymenoptera, Aphelinidae), and that benomyl, bupirimate, carbendazim, dichlofluanid, dinobuton, imazalil, metalaxyl and thiophanate methyl may be harmful to the predatory mite *Phytoseiulus persemilis* (Acarina, Phytoseiidae). Although much more research in IPM strategies is needed, it is already becoming an acceptable system in heated and non-heated greenhouses as growers realize that it is feasible and possible.

A decision-support system named BOTMAN (*Botrytis* manager) was developed for integration of chemical and biological controls of *B. cinerea* in greenhouse vegetables (Shtienberg and Elad, 1997). This was developed later to manage other diseases in the greenhouse and is named GREENMAN (greenhouse manager). Decisions whether to apply biological or chemical measures are taken before each spray according to a four-day weather forecast. The decision follows these lines: when weather is extremely favorable to pathogen development (e.g. humid, rainy weather), then a chemical fungicide is suggested according to the information regarding the resistance of the pathogen population in the greenhouse. On the other hand, when the weather is extremely unfavorable to disease development (e.g. hot and dry), then spraying is not recommended at all. In all other cases the biocontrol preparation TRICHODEX is recommended. GREENMAN was found as effective as the standard chemical control system (application of fungicides on a weekly basis) in spite of the fact that 4–9 chemical sprays were saved and replaced by a biocontrol preparation of *T. harzianum* T39. The system takes into consideration factors of pathogen resistance to fungicides, the presence of various diseases in the greenhouse, the structure and cover, the crop density and age, and the potential of fungicides or preparations of *Trichoderma* to control them (Shtienberg and Elad, 1997).

ORGANIC FARMING

Organic (biological) farming is practiced by farmers around the world. However, it is clear that definitions of this farming system vary widely. Organic farming may be defined as a method of husbandry aiming to optimize the production of biological products through the use and understanding of ecologically balanced systems. According to this view, the ecological balance is essential and the balance of the soil, a living medium, can be upset by the use of chemical fertilizers, fungicides, pesticides and other non-biological products. Sufficient nutrients can be returned to the soil through composting of vegetable and animal wastes to eliminate the need for chemical fertilizers. Compost also increases the population of micro-organisms in the soil and improves the soil texture. Such a soil is expected to produce healthy crops which are able to withstand infection by diseases.

Pathogens are antagonized by the established beneficial microflora in the soil. Some products which are used in organic farming have proved to be efficient in controlling diseases: sulphur, gypsum, compost extracts and soil extracts are evidently effective against root and foliar diseases. Other techniques which do not involve non-biological practices, and have been mentioned in this chapter, assist in the control (Jackson, 1988).

CONCLUDING REMARKS

Although already in use for a very long time, the agrosystem of greenhouses is facing new, as well as old, challenges to crop health. Greenhouse crops are expensive to produce and demand high quality standards. The special conditions which prevail for long periods in greenhouses, different from those outdoors, cause a direct increase in the incidence and severity of diseases, as well as an indirect one by rendering the crop physiologically more susceptible. As demonstrated in this chapter, greenhouses for the crops also are "greenhouses" for plant pathogens. However, as sources of high income, greenhouses justify taking expensive measures against plant diseases.

The new, highly capitalized, sophisticated facilities now being utilized for greenhouse crops have opened up new opportunities for the control of leaf, stem, flower and fruit diseases. These facilities include mechanized, computer-regulated growing systems; low-cost heating systems that incorporate overhead radiant heating as well as on-bench and under-bench heating, glazing materials that provide for the differential transmission of radiant energy; computer software for the prediction of epidemic outbreak of diseases and their management and for the modification of the environment to reduce the potential for epidemics, and the potential for the mechanized and computerized application of biocides. High priority thus should be given to innovative studies of the epidemiology and control of diseases, especially *Botrytis* blight, powdery mildew and bacterial diseases, that will utilize these technological advances. Even though the number of plant species and cultivars commercially produced in the greenhouses and the number of diseases is also large, the development of resistant cultivars using genetic engineering techniques certainly merits high priority for research.

More information is needed on physiology and susceptibility of protected crops and their effects on the development of epidemics. Research should focus also on the possibility of controlling the greenhouse environment, nutrition and light for the purpose of restricting disease development. The means to reduce the application of hazardous control agents for the production of healthy and high-quality crops in greenhouses are greatly desired and are being sought actively.

Appendix. IMPORTANT DISEASES OF MAJOR GREENHOUSE CROPS

This appendix lists important diseases of major greenhouse crops: in Table 8.2 the data are arranged by disease, in Table 8.3 the arrangement is by crop.

Diseases of local importance or diseases of same crops grown in the field may not be included in the list.

Symptoms of wilt diseases are observed on leaves; however, they are generally associated with root infections and vascular invasion by the pathogens.

Table 8.2
Important diseases of major greenhouse crops ordered by disease

Disease	Causal agent(s)	Crop(s)
Fungal diseases		
Botrytis flower rot	*Botrytis cinerea*	carnation, chrysanthemum, cucumber, eggplant, gerbera, melon, pepper, rose
Damping-off	*Phytophthora* spp., *Pythium* spp. [including *P. aphanidermatum, P. irregularae, P. ultimum*], *Rhizoctonia solani*	cucumber, eggplant, gerbera, lettuce, melon, pepper, tomato
Fruit gray mold	*Botrytis cinerea*	cucumber, eggplant, melon, pepper, tomato
Leaf gray mold	*Botrytis cinerea*	cucumber, eggplant, gerbera, lettuce, tomato
Powdery mildew	*Leveillula taurica*	cucumber, eggplant, gerbera, pepper, tomato
Root rot	*Botrytis cinerea, Phytophthora* spp., *Pythium* spp., *Rhizoctonia solani*	cucumber, eggplant, gerbera, tomato
Sclerotinia	*Sclerotinia sclerotiorum*	chrysanthemum, cucumber, eggplant, gerbera, lettuce, melon, tomato
Stem gray mold	*Botrytis cinerea*	carnation, chrysanthemum, cucumber, eggplant, gerbera, lettuce, melon, pepper, rose, tomato
Verticillium wilt	*Verticillium dahliae, V. albo-atrum* (rarely)	chrysanthemum, cucumber, gerbera, eggplant, melon, pepper, rose, tomato
Bacterial diseases		
Galls	*Agrobacterium tumefaciens*	many crops
Hairy roots	*Agrobacterium rhizogenes*	many crops
Soft rots of stems, leaves and fruits	*Erwinia carotovora, Pseudomonas fluorescens*	cucumber, eggplant, lettuce, melon, pepper, tomato
Viral diseases	Beet Pseudo Yellows Virus (BPYV)	cucumber, lettuce
	Cucumber Mosaic Virus (CMV)	carnation, cucumber, gerbera, lettuce, pepper, tomato
	Potato Virus X (PVX)	tomato
	Potato Virus Y Wilt (PVY)	pepper, tomato
	Tobacco Etch Virus (TEV)	pepper, tomato
	Tobacco Mosaic Virus (TMV)	pepper, tomato
	Tomato Bushy Stunt Virus (TBSV)	pepper, tomato
	Tomato Spotted Wilt Virus (TSWV)	gerbera, pepper, tomato
	Tomato Yellow Leaf Curl Virus (TYLCV)	pepper, tomato

Table 8.3
Important diseases of major greenhouse crops ordered by crops

Affected part of plant	Causal agent(s)	Disease	Pathogen(s)
Carnation			
Cuttings	Fungi	Basal rots	*Phytophthora* spp., *Pythium* spp.
		Black mold	*Alternaria dianthi*
		Fusarium basal rot	*Fusarium roseum*
		Rhizoctonia stem rot	*Rhizoctonia solani*
	Bacteria	Bacterial stem crack	*Pseudomonas caryophylli*
Stems	Fungi	Alternaria branch rot	*Alternaria dianthi*
		Fusarium stub rot	*Fusarium roseum*
	Bacteria	Bacterial stem crack	*Pseudomonas caryophylli*
		Galls	*Erwinia herbicola*
	Viruses	Carnation Etched Ring Virus (CERV) Carnation Italian Ringspot Virus (CRISV) Carnation Mottle Virus (CarMV) Carnation Necrotic Fleck Virus (CNFV) Carnation Yellow Fleck Virus (CYFV) Carnation Yellow Strip Virus (CYSV) Cucumber Mosaic Virus (CMV)	
Leaves	Fungi	Alternaria leaf spot or blight	*Alternaria dianthi*
		Fairy ring spot or ring spot	*Cladosporium echinulatum* [syn. *Didymellina dianthii*], *Heterosporium echinulatum, Mycosphaerella dianthi*
		Greasy blotch	*Zygophiala jamaicensis*
		Leaf rot	*Heteropatella valtellinensis*
		Powdery mildew	*Oidium* sp.
		Rust	*Uromyces dianthi*
		Septoria leaf spot	*Septoria dianthi*
	Wilt	Fusarium wilt	*Fusarium oxysporum* f.sp. *dianthi*
		Phialophora or Verticillium wilt, Fan mold	*Phialophora cinerescens* [syn. *Verticillium cinerescens*]
	Bacteria	Bacterial wilt, stem crack	*Pseudomonas caryophylli*
		Leafy gall	*Corynebacterium fascians*
		Slow wilt, Bacterial stunt	*Erwinia chrysanthemi*
Flowers	Fungi	Anther smut	*Ustilago violacea*
		Fusarium bud rot	*Fusarium tricinctum* f.sp. *poae*
		Powdery mildew	*Oidium* sp.
		Ring spot	*Cladosporium echinulatum*
Chrysanthemum			
Roots	Fungi	Phoma root rot	*Phoma chrysanthemicola* var. *chrysanthemicola*
Stems	Fungi	Pythium stem rot	*Pythium* spp.
		Ray blight	*Didymella chrysanthemi* [syn. *Mycosphaerella aligulicola*]
		Rhizoctonia stem rot	*Rhizoctonia solani*

Continued on next page

Table 8.3, *continued*

Affected part of plant	Causal agent(s)	Disease	Pathogen(s)
Leaves	Fungi	Blotch or leaf spot	*Septoria chrysanthemella*
		Brown rust	*Puccinia chrysanthemi*
		Fusarium wilt	*Fusarium oxysporum* f.sp. *chrisanthemi*
		Powdery mildew	*Oidium chrysanthemi*
		Ray blight	*Didymella chrysanthemi*
		Verticillium wilt	*Verticillium albo-atrum, V. dahliae*
		White rust	*Puccinia horiana*
	Bacteria	Bacterial wilt	*Erwinia chrysanthemi*
		Leafy gall	*Corynebacterium fascians*
	Viroids and viruses		Chrysanthemum Chlorotic Mottle Viroid Chrysanthemum Virus B (CVB) Impatiens Necrotic Spot Virus (INSV)
		Stunt	Chrysanthemum Stunt Viroid
Cucumber and melon			
Roots	Fungi	Black root rot	*Phomopsis sclerotioides*
		Corky root rot	*Pyrenochaeta lycopersici*
		Root rot	*Olpidium* spp.
Stems	Fungi	Basal rots	*Botrytis cinerea, Phytophthora* spp., *Pythium* spp., *Rhizoctonia solani* and non-pathogenic rots
		Black stem rot	*Didymella bryoniae* [syn. *Mycosphaerella melonis*]
		Fusarium wilt	*Fusarium oxysporum* f.sp. *cucumerinum* (cucumber) and *F. oxysporum* f.sp. *melonis* (melon)
		White rot	*Sclerotinia sclerotiorum*
	Bacteria	Bacterial speck	*Pseudomonas syringae* pv. *lachrymans*
Leaves	Fungi	Alternaria leaf spot	*Alternaria alternata, A. cucumerina*
		Cercospora leaf spot	*Cercospora melonis*
		Colletotrichum leaf spot or Anthracnose	*Colletotrichum lagenarium*
		Downy mildew	*Pseudoperonospora cubensis*
		Fusarium wilt	*F. oxysporum* f.sp. *cucumerinum* and *F. oxysporum* f.sp. *melonis*
		Powdery mildew	*Leveillula taurica, Sphaerotheca fusca* syn. *fuliginea*
		Ulocladium leaf spot	*Ulocladium atrum*
		Verticillium wilt	*Verticillium dahliae* and *V. albo-atrum* (rarely)
	Bacteria	Angular leaf spot	*Pseudomonas syringae* pv. *lachrymans*
		Bacterial wilt	*Erwinia tracheiphila*
Fruits	Fungi	Black stem rot	*Didymella bryoniae*
		Gummosis or scab	*Cladosporium cucumerinum*
		Pythium rot	*Pythium* sp.
		White rot	*Sclerotinia sclerotiorum*

Continued on next page

Table 8.3, *continued*

Affected part of plant	Causal agent(s)	Disease	Pathogen(s)
General	Viruses		Beet Pseudo Yellows Virus (BPYV) Cucumber Chlorotic Spot Virus (CCSV) Cucumber Green Mottle Mosaic Virus (CGMMV) Cucumber Mosaic Virus (CMV) Cucumber Vein Yellowing Virus (CVYV) Cucurbit Yellows Stunting Disorders (CYSDV) Melon Necrotic Spot Virus (MNSV) Squash Leaf Curl Virus (SLCV) Squash Mosaic Virus (SqMV) Watermelon Mosaic Virus 1 (WMV1) Watermelon Mosaic Virus 2 (WMV2) Zucchini Yellow Mosaic Virus (ZYMV)
Eggplant			
Roots	Fungi	Brown root rot	*Colletotrichum coccodes*
		Corky root rot	*Pyrenochaeta lycopersici*
		Phomopsis blight	*Phomopsis vexans*
Stems	Fungi	Phomopsis canker	*Phomopsis vexans*
Leaves	Fungi	Leaf spot	*Cercospora* spp. [including *C. melongena*], *Phyllosticta hortorum, Puccinia angivyi, P. paspalicola, Septoria melongenae*
		Phomopsis blight	*Phomopsis vexans*
	Bacteria	Leaf spot	*Pseudomonas solanacearum*
Fruits	Fungi	Phomopsis rot	*P. vexans*
		Fruit rot	*Colletotrichum melongenae*
General	Viruses		Eggplant mosaic virus (EMV)
Gerbera			
Roots	Fungi	Phytophthora crown rot	*Phytophthora cryptogea*
		Thielaviopsis root rot	*Thielaviopsis basicola*
Leaves	Fungi	Alternaria leaf spot	*Alternaria* spp.
		Ascochyta leaf spot	*Ascochyta gerberae*
		Cercospora leaf spot	*Cercospora gerberae*
		Phyllosticta leaf spot	*Phyllosticta gerbericola*
		Powdery mildew	*Erysiphe cichoracearum*
		Septoria leaf spot	*Septoria gerberae*
		White rust	*Albugo tragopogonis*
	Bacteria	Bacterial leaf spot	*Pseudomonas cichorii*
Flowers	Fungi	Alternaria flower spot	*Alternaria* sp.
		Downy mildew	*Peronospora radii*
		Flower shatter	*Pseudomonas* sp.
		Petal blight, Flower scorch	*Itersonilia perplexans*
		Ray blight	*Didymella chrysanthemi*

Continued on next page

Table 8.3, *continued*

Affected part of plant	Causal agent(s)	Disease	Pathogen(s)
	Viruses and	Chrysanthemum chlorotic mottle	Chrysanthemum Chlorotic Mottle Viroid
	viroids	Chrysanthemum stunt	Chrysanthemum Stunt Viroid Tomato Spotted Wilt Virus (TSWV)
		Flower distortion or aspermy	Tomato Aspermy Virus
General	Viruses		Cucumber Mosaic Virus (CMV) Gerbera Mosaic Virus (GMV)
		Mottle virus	Tobacco Mottle Virus
Lettuce			
Stems	Fungi	Bottom rot	*Rhizoctonia solani*
	Bacteria	Butt rot	*Pseudomonas marginalis*
Leaves	Fungi	Downy mildew	*Bremia lactucae*
		Pleospora leaf spot	*Pleospora herbarum*
		Powdery mildew	*Erysiphe cichoracearum*
		Ring spot	*Marssonina panattoniana*
		Rust	*Puccinia opizii*
		Septoria leaf spot	*Septoria lactucae*
	Bacteria	Bacterial spot	*Pseudomonas cichorii*
General	Viruses		Beet Pseudo Yellows Virus (BPYV) Beet Western Yellows Virus (BWYV) Cucumber Mosaic Virus (CMV) Lettuce Big-vein Virus (LBVV) Lettuce Infectious Yellows Virus (LIYV) Lettuce Mosaic Virus (LMV)
Pepper			
Root	Fungi	Brown root rot	*Colletotrichum coccodes*
		Corky root rot	*Pyrenochaeta lycopersici*
Stems	Fungi	Basal stem rots	*Phytophthora capsici* and *Pythium* spp.
Fruit	Bacteria	Bacterial wilt	*Erwinia carotovora, Pseudomonas corrugata, P. solanacearum, P. carotovora*
		Fruit scab (and leaf defoliation)	*Xanthomonas vesicatoria*
General	Viruses		Cucumber Mosaic Virus (CMV) Potato Virus Y (PVY) Tobacco Etch Virus (TEV) Tobacco Mosaic Virus (TMV) Tomato Bushy Stunt Virus (TBSV) Tomato Spotted Wilt Virus (TSWV)
Rose			
Roots	Fungi	Armillaria root rot	*Armillaria mellea*
Stems	Fungi	Brand canker or canker	*Leptosphaeria coniothyrium* [syn. *Coniothyrium fuckelii*]

Continued on next page

Table 8.3, *continued*

Affected part of plant	Causal agent(s)	Disease	Pathogen(s)
Leaves	Fungi	Blackspot	*Diplocarpon rosae*
		Downy or black mildew	*Peronospora sparsa*
		Powdery mildew	*Sphaerotheca pannosa*
		Rust	*Phragmidium mucronatum*
General	Viruses		Apple Mosaic Virus (APMV) Prunus Necrotic Ringspot Virus (PNRSV) Strawberry Latent Ringspot (SLRV)
Tomato			
Roots	Fungi	Brown and corky root rot	*Pyrenochaeta lycopersici*
		Brown root rot	*Phytophthora* spp. [basal stem rots]
		Calyptella root rot	*Calyptella campanula*
		Other root rots	*Colletotrichum coccodes, Rhizoctonia solani, Spongospora subterranea, Thielaviopsis basicola*
Stems	Fungi	Alternaria canker	*Alternaria alternata* f.sp. *lycopersici*
		Crown and foot rot	*Fusarium oxysporum* f.sp. *radicis-lycopersici*
		Didymella stem rot	*Didymella lycopersici*
		Foot rot	*Botrytis cinerea, Didymella lycopersici, Phytophthora* spp. [including *P. cryptogea, P. erythroseptica, P. megasperma, P. nicotianae* var. *parasitica* (syn. *P. parasitica*) and *P. verrucosa*] and *Rhizoctonia solani*
	Bacteria	Bacterial canker	*Clavibacter michiganense*
		Brown pith necrosis	*P. solanacearum*
		Pith necrosis	*Pseudomonas corrugata*
Leaves	Fungi	Early blight, target spot or Alternaria blight	*Alternaria solani*
		Fusarium wilt	*Fusarium oxysporum* f.sp. *radicis-lycopersici*
		Late blight	*Didymella lycopersici, Phytophthora infestans*
		Leaf mold	*Fulvia fulva* [syn. *Cladosporium fulvum*]
		Septoria leaf spot	*Septoria lycopersici*
		Stemphylium leaf blight	*Stemphylium botryosum* and *S. vesicarium*
	Bacteria	Bacterial speck	*Pseudomonas syringae* pv. *tomato*
		Bacterial wilt	*Clavibacter michiganense, Pseudomonas corrugata, P. solanacearum*
		Leaf spot	*Xanthomonas vesicatoria*
Fruits	Fungi	Blight	*Phytophthora infestans*
		Buckeye rot	*Phytophthora nicotianae* var. *parasitica*
		Didymella fruit rot	*Didymella lycopersici*
		Fruit spot	*Alternaria alternata* f.sp. *lycopersici, A. solani, Stemphylium botryosum*
		Ghost spots	*Botrytis cinerea*
		Phoma rot	*Phoma destructiva*

Continued on next page

Table 8.3, *continued*

Affected part of plant	Causal agent(s)	Disease	Pathogen(s)
	Bacteria	Bacterial canker	*Clavibacter michiganense*
		Bacterial speck	*Pseudomonas syringae* pv. *tomato*
		Fruit scab	*Xanthomonas vesicatoria*
General	Viruses		Cucumber Mosaic Virus (CMV) Potato Virus X (PVX) Potato Virus Y (PVY) Tobacco Etch Virus (TEV) Tobacco Mosaic Virus (TMV) Tomato Bushy Stunt Virus (TBSV) Tomato Infectious Chlorosis Virus (TICV) Tomato Spotted Wilt Virus (TSWV) Tomato Yellow Leaf Curl Virus (TYLCV)

REFERENCES

Ahoonmanesh, A. and Shalla, T.A., 1981. Feasibility of cross-protection for control of tomato mosaic virus in fresh market field grown tomatoes. *Plant Dis.,* 65: 56–58.

Alabouvette, C., 1986. Fusarium wilt suppressive soils from the Chateaurenard region: review of 10year study. *Agronomie,* 6: 275–284.

Antignus, J., Cohen, S., Mor, N., Masika, J. and Lapidot, M., 1996a. The effect of UV-blocking greenhouse covers on insects and insect-borne virus diseases. *Plasticulture,* 112: 15–20.

Antignus, J., Mor, N., Ben Joseph, R., Lapidot, M. and Cohen, S., 1996b. Ultraviolet-absorbing plastic sheets protect crops from insect pests and from virus diseases vectored by insects. *Pest Manage. Sampling*, 25: 919–924.

Baker, R., Elad, Y. and Sneh, B., 1986. Physical, biological and host factors in iron competition in soils. in: T.R. Swinburne (Editor), *Iron, Siderophores and Plant Diseases.* Plenum, New York, pp. 77–84.

Bakker, J.C., 1984. Physiological disorders in cucumber under high humidity conditions and low ventilation rates in greenhouses. *Acta Hortic.,* 56: 252–264.

Bateman, D.F. and Lumsden, R.D., 1965. Relation of calcium content and nature of the pectic substances in bean hypocotyls of different ages to susceptibility to an isolate of *Rhizoctonia solani. Phytopathology,* 55: 734–738.

Celli, G., Benuzzi, M., Maini, S., Mauzaroli, G., Antoniacci, L. and Nicoli, G., 1991. Biological and integrated control in protected crops of northern Italy's Po Valley, Overview and outlook. *WPRS Bull.,* 14: 2–12.

Chet, I., 1987. Trichoderma application, mode of action and potential as a biocontrol agent of soilborne plant pathogenic fungi. in: I. Chet (Editor), *Innovative Approaches to Plant Disease Control.* Wiley, New York, pp. 137–160.

Cline, M.N., Chastanger, G.A., Aragaki, M., Baker, R., Daughtrey, M.L., Lawson, R.H., MacDonald, J.D., Tammen, J.F. and Wort, G.L., 1988. Current and future research directions of ornamental pathology. *Plant Dis.,* 72: 926–934.

Cohen, Y., Reuveni, M. and Samoucha, Y., 1983. Competition between metalaxyl-resistant and -sensitive strains of *Pseudoperonospora cubensis* on cucumber plants. *Phytopathology,* 73: 1516–1520.

Corden, M.E., 1965. Influence of calcium nutrition on Fusarium wilt of tomato and polygalacturonase activity. *Phytopathology,* 55: 221–224.

Crum, S.J.H., Runia, W.T., Leistra, M. and Smelt, J.H., 1985. Behaviour of some pesticides in a nutrient film and in a rock-wool system. *Sci. Hortic.,* 25: 1–9.

Daughtrey, M.L. and Schippers, P.A., 1980. Root death and associated problems. *Acta Hortic.,* 98: 283–291.

Dean, R.A. and Kuc, J., 1987. Rapid lignification in response to wounding and infection as a mechanism for induced systemic protection in cucumber. *Physiol. Plant Pathol.,* 31: 69–81.

Descalzo, R.C., Rahe, J.E. and Mauza, B., 1990. Comparative efficacy of induced resistance for selected diseases of greenhouse cucumber. *Can. J. Plant Pathol.,* 12: 16–24.

Elad, Y., 1988a. Latent infection of *Botrytis cinerea* in rose flowers and combined chemical and physiological control of the disease. *Crop Prot.,* 7: 361–366.

Elad, Y., 1988b. Involvement of ethylene in the pathogenicity of *Botrytis cinerea* Pers. on rose and carnation flowers and the possibility of control. *Ann. Appl. Biol.,* 113: 589–598.

Elad, Y., 1990. Reasons for the delay in development of biological control of foliar pathogens. *Phytoparasitica,* 18: 99–105.

Elad, Y., 1997. Effect of filtration of solar light on the production of conidia by field isolates of *Botrytis cinerea* and on several diseases of greenhouse-grown vegetables. *Crop Prot.,* 16: in press.

Elad, Y. and Chet, I., 1995. Practical approaches for biocontrol implementation. In: R. Reuveni (Editor), *Novel Approaches to Integrated Pest Management.* Lewis Publishers, CRC Press, Boca Raton, FL, pp. 323–338.

Elad, Y. and Shtienberg, D., 1995. *Botrytis cinerea* in greenhouse vegetables; chemical, cultural, physiological and biological controls and their integration. *Integ. Pest Manag. Rev.,* 1: 15–29.

Elad, Y., Chet, I. and Katan, J., 1980. *Trichoderma harzianum*:

a biocontrol agent effective against *Sclerotium rolfsii* and *Rhizoctonia solani*. *Phytopathology,* 70: 119–121.

Elad, Y., Hadar, Y., Hadar, E., Chet, I. and Henis, Y., 1981. Biological control of *Rhizoctonia solani* by *Trichoderma harzianum* in carnation. *Plant Dis.,* 65: 675–677.

Elad, Y., Hadar, Y., Chet, I. and Henis, Y., 1982. Prevention with *Trichoderma harzianum* Rifai aggr., of reinfestation by *Sclerotium rolfsii* Sacc. and *Rhizoctonia solani* Kuhn of soil fumigated with methyl bromide, and improvement of disease control in tomatoes and peanuts. *Crop Prot.,* 1: 199–211.

Elad, Y., Chet, I. and Baker, R., 1987. Increased growth response of plants induced by rhizobacteria antagonistic to soilborne pathogenic fungi. *Plant Soil,* 98: 325–330.

Elad, Y., Yunis, H. and Mahrer, Y., 1988. Effect of climatic conditions in polyethylene-covered structures on gray mold disease of winter cucumbers. *Appl. Agric. Res.,* 3: 243–247.

Elad, Y., Ziv, O., Ayish, N. and Katan, J., 1989. The effect of film-forming polymers on powdery mildew of cucumber. *Phytoparasitica,* 17: 279–288.

Elad, Y., Ayish, N., Ziv, O. and Katan, J., 1990. Control of grey mold (*Botrytis cinerea*) with film-forming polymers. *Plant Pathol.,* 39: 249–254.

Elad, Y., Yunis, H., Volpin, H. and Pressman, E., 1991. Fertilization of plants grown in soilless media by calcium for the reduction of foliar diseases. *Phytoparasitica,* 19: 168.

Elad, Y., Gullino, L.M., Shtienberg, D. and Aloi, C., 1994a. Coping with tomato grey mold under mediterranean conditions. *Crop Prot.,* 14: 105.

Elad, Y., Shtienberg, D. and Niv, A., 1994b. *Trichoderma harzianum* T39 integrated with fungicides: Improved biocontrol of grey mold. *Brighton Crop Protection Conf. – Pests and Diseases.* British Crop Protection Council, Farnham, UK, pp. 1109–1113.

Fletcher, J.T., 1984. *Diseases of Greenhouse Plants.* Longman, London, 340 pp.

Fokkema, N.J., 1976. Antagonism between fungal saprophytes and pathogens on aerial plant surfaces. In: C.H. Dickinson and T.F. Preece (Editors), *Microbiology of Aerial Plant Surfaces.* Academic Press, London, pp. 487–506.

Garibaldi, A. and Tamietti, G., 1984. Attempts to use soil solarization in closed glasshouses in northern Italy for controlling corky root of tomato. *Acta Hortic.,* 152: 237–243.

Gouin, F.R., 1982. Using composted waste for growing horticultural crops. *BioCycle,* 23: 45–47.

Gullino, M.L., Aloi, C., Benzi, E. and Garibaldi, A., 1991. Biological and integrated control of grey mold of vegetable crops. *Petria,* 10: 149–150.

Harman, G.E., Taylor, A.G. and Jin, X., 1991. Seed treatment methods to enhance biocontrol efficacy of *Trichoderma harzianum. Petria,* 1: 137.

Hoitink, H.A.J. and Fahy, P.C., 1986. Basis for the control of soilborne plant pathogens with composts. *Annu. Rev. Phytopathol.,* 24: 93–114.

Honda, Y. and Yuroki, T., 1977. Control of Sclerotinia disease of greenhouse eggplant and cucumber by inhibition of development of apothecia. *Plant Dis. Rep.,* 61: 1036–1040.

Honda, Y., Toki, T. and Yuroki, T., 1977. Control of gray mold of greenhouse cucumber and tomato by inhibiting sporulation. *Plant Dis. Rep.,* 61: 1041–1044.

Inaba, T. and Kajiwara, T., 1971. Effect of light on lesion development in cucumber downy mildew. *Ann. Phytopathol. Soc. Jpn.,* 37: 340–347.

Jackson, A.W., 1988. Organic farming – crop protection implications. *Brighton Crop Protection Conf.* British Crop Protection Council, Farnham, UK, 3A-3, pp. 103–109.

Jarvis, W.R., 1980. Epidemiology. in: J.R. Coley-Smith, K. Verhoeff and W.R. Jarvis (Editors), *The Biology of Botrytis.* Academic Press, London, pp. 219–249.

Jarvis, W.R., 1989. Managing diseases in greenhouse crops. *Plant Dis.,* 73: 190–194.

Jarvis, W.R. and Slingsby, K., 1977. The control of powdery mildew of greenhouse cucumber by water sprays and *Ampelomyces quisqualis. Plant Dis. Rep.,* 61: 728–730.

Jutsum, A.R., 1988. Commercial application of biological control: status and prospects. *Phil. Trans. R. Soc. London B,* 318: 357–373.

Katan, J., 1981. Solar heating (solarization) of soil for control of soilborne pests. *Annu. Rev. Phytopathol.,* 19: 211–236.

Katan, T., 1982a. Resistance to 3,5-dichlorophenyl-*N*-cyclicimide ('Dicarboximide') fungicides in the grey mould pathogen *Botrytis cinerea* on protected crops. *Plant Pathol.,* 31: 133–141.

Katan, T., 1982b. Persistence of dicarboximide-fungicide resistance in populations of *Botrytis cinerea* in a warm, dry temperate agroclimate. *Phytoparasitica,* 10: 209–211.

Komada, T. and Fukai, T., 1982. Solar heating in closed plastic house for control of soilborne diseases. V. Application for control of Fusarium wilt of strawberry. *Ann. Phytopathol. Soc. Jpn.,* 48: 570–577.

Krantz, J., 1981. Hyperparasitism of biotrophic fungi. in: J.P. Blakeman (Editor), *Microbial Ecology of the Phylloplane.* Academic Press, London, pp. 327–392.

Kuc, J., 1990. A case for self defense in plants against disease. *Phytoparasitica,* 18: 3–8.

Leroux, P. and Clerjeau, M., 1985. Resistance of *Botrytis cinerea* Pers. and *Plasmopara viticola* (Berk. & Curt.) Berl and de Toni to fungicides in French vineyards. *Crop Prot.,* 4: 137–160.

Lumsden, R.D., Locke, E. and Walter, J.F., 1991. Approval of *Glioclodium virens* by the US Environmental Protection Agency for biological control of *Pythium* and *Rhizoctonia* damping off. *Petria,* 1: 138.

Mahrer, Y., Avissar, R., Naot, O. and Katan, J., 1987. Intensified soil solarization with closed greenhouses: numerical and experimental studies. *Agric. For. Meteorol.,* 41: 325–334.

Malathrakis, N.E., 1986. Fungicide resistance in diseases of vegetables growing under plastic in Crete. *EPPO Bull.,* 16: 429–435.

Marsh, P.B., Taylor, E.E. and Bassler, L.M., 1959. A guide to the literature on certain effects of light on fungi: reproduction, morphology, pigmentation and prototrophic phenomena. *Plant Dis. Rep.,* Suppl., 261: 251–312.

Meredith, D.S., 1973. Significance of spore release and dispersal mechanisms in plant disease epidemiology. *Annu. Rev. Phytopathol.,* 11: 313–342.

Moor, L.W. and Warren, G.W., 1979. *Agrobacterium radiobacter* strain 84 and biological control of crown gall. *Annu. Rev. Phytopathol.,* 17: 163–179.

Morgan, W.M., 1984. Integration of environmental and fungicidal control of *Bremia lactucae* in glasshouse lettuce crops. *Crop Prot.,* 3: 349–361.

O'Leary, J.W. and Knecht, G.N., 1972. Salt uptake in plants grown at constant high relative humidity. *Ariz. Acad. Sci.,* 7: 125–128.

Palti, J., 1981. *Cultural Practices and Infectious Crop Diseases.* Springer, Berlin, 243 pp.

Panayotakou, M. and Malathrakis, N.E., 1983. Resistance of *Botrytis cinerea* to dicarboximide fungicides in protected crops. *Ann. Appl. Biol.,* 102: 293–299.

Pappas, A.C., 1986. Seasonal fluctuations in the benzimidazole- and dicarboximide-resistant population of *Botrytis cinerea* in greenhouse crops. *ISPP Chem. Control Newsl.,* 7: 10–11.

Price, D. and Dickinson, A., 1980. Fungicides and the nutrient film technique. *Acta Hortic.,* 98: 269–275.

Rattink, H., 1991. Possibilities of biocontrol of soilborne fungi by *Trichoderma harzianum* in Dutch glasshouse crops. *Petria,* 1: 144.

Reuveni, R., Raviv, M. and Bar, R., 1989. Sporulation of *Botrytis cinerea* as affected by photoselective polyethylene sheets and filters. *Ann. Appl. Biol.,* 115: 417–424.

Rishbeth, J., 1988. Biological control of air-borne pathogens. *Phil. Trans. R. Soc. London B,* 318: 265–281.

Rotem, J., Chen, Y. and Bashi, E., 1978. Host and environment influences on sporulation *in vivo. Annu. Rev. Phytopathol.,* 16: 83–101.

Runia, W.Th., Van Os, E.A. and Bollen, G.J., 1988. Disinfection of drain water from soilless cultures by heat treatment. *Neth. J. Agric. Sci.,* 36: 237–238.

Saccardo, F. and Sonnino, A., 1988. Energy saving through breeding for disease resistance. *Acta Hortic.,* 229: 21–29.

Saettler, A.W., Schaad, N.W. and Roth, D.A. (Editors), 1989. *Detection of Bacteria in Seed and Other Planting Material.* APS Press, St. Paul, MN, 122 pp.

Schaad, N.W. (Editor), 1988. *Laboratory Guide for Identification of Plant Pathogenic Bacteria.* APS Press, St. Paul, MN, 164 pp.

Sequeira, L., 1983. Mechanisms of induced resistance in plants. *Annu. Rev. Microbiol.,* 37: 51–79.

Shear, C.B., 1975. Calcium related disorders of fruits and vegetables. *HortScience,* 10: 361–365.

Shtienberg, D. and Elad, Y., 1997. Incorporation of weather forecasting in integrated, biological–chemical management of *Botrytis cinerea. Phytopathology,* 87: 332–340.

Spiegel, Y., Netzer, D. and Kafkafi, Y., 1987. The role of calcium nutrition on Fusarium wilt in muskmelon. *J. Phytopathol.,* 118: 220–226.

Stall, R.E., 1963. Effects of lime on incidence of Botrytis gray mold of tomato. *Phytopathology,* 53: 149–151.

Sundheim, L., 1982. Control of cucumber powdery mildew by the hyperparasite *Ampelomyces quisqualis* and fungicides. *Plant Pathol.,* 31: 209–214.

Tan, K.K., 1975. Interaction of near-ultraviolet, blue, red, and far-red in sporulation of *Botrytis cinerea. Trans. Br. Mycol. Soc.,* 64: 215–222.

Tibbits, J.W. and Kobriger, J.N., 1983. Mode of action of air pollutants in injuring horticultural plants. *HortScience,* 18: 675–680.

Van de Veire, M., 1991. Progess in IPM in glasshouse vegetables in Belgium. *WPRS Bull.,* 14: 22–32.

Van Os, E.A., 1982. Dutch developments in soilless culture. *Outlook on Agric.,* 11: 165–171.

Van Peer, R. and Schiperes, B., 1989. Bacteria in hydroponic systems: Effects on plant growth. In: *Proc. 7th Int. Congr. on Soilless Culture,* pp. 375–384.

Vanachter, A., Van Wambeke, E. and Van Assche, C., 1983. Potential danger for infection and spread of root diseases of tomatoes in hydroponics. *Acta Hortic.* 133: 119–128.

Volpin, H. and Elad, Y., 1991. Influence of calcium nutrition on susceptibility of rose flowers to *Botrytis* blight. *Phytopathology,* 81: 1390–1394.

Winspear, K.W., Postlethwaite, J.D. and Cotton, R.F., 1970. The restriction of *Cladosporium fulvum* and *Botrytis cinerea* attacking glasshouse tomatoes by automatic humidity control. *Ann. Appl. Biol.,* 65: 75–83.

Wood, R.K.S., 1950. The control of diseases of lettuce by the use of antagonistic organisms. I. The control of *Botrytis cinerea* Pers. *Ann. Appl. Biol.,* 38: 203–216.

Yunis, H. and Elad, Y., 1989. Survival of dicarboximide-resistant strains of *Botrytis cinerea* in plant debris during summer in Israel. *Phytoparasitica,* 17: 13–21.

Yunis, H., Elad, Y. and Mahrer, Y., 1991. Influence of fungicidal control of cucumber and tomato grey mould (*Botrytis cinerea*) on fruit yield. *Pestic. Sci.,* 31: 325–335.

Yunis, H., Steinberg, D., Elad, Y. and Mahrer, Y., 1994. Qualitative approach for modelling outbreaks of grey mold epidemics in non-heated cucumber greenhouses. *Crop Prot.,* 13: 99–104.

Zimand, G., Elad, Y. and Chet, I., 1996. Effect of *Trichoderma harzianum* on *Botrytis cinerea* pathogenicity. *Phytopathology,* 80: 1255–1260.

Chapter 9

ARTHROPOD FAUNA

Michael P. PARRELLA

INTRODUCTION

The greenhouse ecosystem represents a unique environment for the development of arthropod populations. Pest colonization in this system is initiated through importation of infested plant material, migration into greenhouses through open vents or doorways, movement from one crop to the next, or survival through a non-cropping period in the greenhouse. During crop production in the greenhouse, mortality from abiotic factors such as rain, wind, heat, etc., is dramatically reduced and there is usually an abundance of food. This artificial environment favors the rapid development of pest populations. Furthermore, in many greenhouses there are often multiple crops, crops of different ages, and weeds present in the same or adjacent areas which provide a conduit for pests to be a continual problem.

Both ornamental and vegetable crops are widely grown in greenhouses, and the demand for aesthetic quality is high in these products. This marketing reality and the consistent supply of inexpensive, effective pesticides has led to the wide-scale dependence on the use of these products; this is particularly true in the production of ornamentals. It is more common to find biological control an integral part of vegetable production in greenhouses. The success achieved with the use of biological control agents for managing vegetable pests is often cited as one of the major achievements in practical biological control. Despite this outstanding success, pesticides still dominate in the greenhouse. Of the approximate 280 000 ha of glass and plastic houses around the world only a relatively small percentage (15 000 ha) is under biological control.

Today, as never before, there is an increasing stance by the general public against the use of pesticides. This cuts across all uses, from traditional chemical applications made to agricultural crops to spraying urban areas for eradication of introduced, exotic pests. This strong bias has its roots in concerns over, among others, direct contamination of the general public, contact with residues in foods, contamination of groundwater, general impact on the environment, and the health and safety of agricultural workers. These concerns have manifested themselves in political arenas all over the world where, among other activities, there are now very strict requirements governing the registration of new pesticides, re-evaluation and removal of previously registered materials, and tougher rules and regulations governing the use of pesticides. Rules and regulations enacted in one country are likely to have an impact on other importing and exporting countries. The cost of registering new pesticides also varies considerably from country to country. For example, direct registration costs in England are approximately 100 times greater than in England's three major horticultural competitors – France, Italy and the Netherlands (Vale, 1995). This, coupled with the facts that there are few new insecticide registrations, insecticide resistance is on the rise, and many commercially available materials are losing registration status, magnifies the importance of alternatives to pesticides.

Many of the characteristics of the greenhouse environment that favor pests also favor the natural enemies of these pests. Successful biological control in the greenhouse has been under continual siege over the past ten years as "new" pests have appeared in greenhouse systems all over the world. Pest movement on plant material used in greenhouse production was likely the cause of the worldwide spread of *Liriomyza trifolii* – a serpentine leafminer, probably originating in North America; *L. huidobrensis* – the pea leafminer, originating in South America; *Spodoptera exigua* – the beet armyworm, originating in North America; *Frankliniella occidentalis* – the western flower thrips, originating in North America; and *Bemisia argentifolii* – the silverleaf whitefly, originating in the Middle East or Southeast Asia.

These new pests pose difficult challenges to research on biological control for several reasons: (1) for some of these pests, such as the western flower thrips, biological control agents that could be used in greenhouses are largely unknown; (2) until biological control agents are found and put into effect, the biological control programs currently in place for other pests are compromised by chemical treatments for the new pests; (3) these new pests may attack crops already beset by a serious pest complex, further complicating biological control; and (4) some of these pests, such as western flower thrips, feed directly on vegetables, fruits and flowers for which the threshold for unacceptable damage is lower, making effective biological control more difficult to achieve. Pests that transmit diseases to plants make heavy demands on any biological control programs in place on that crop. Such is the case with western flower thrips which transmit tomato spotted wilt virus to vegetables and impatiens necrotic spot virus to ornamentals. The threat of virus transmission makes it important to keep the population of thrips as low as possible.

Integrated pest management (IPM) is a logical vehicle through which successful pest control can be achieved in the greenhouse. Integrated pest management can be defined as the intelligent selection of measures which keep pests from exceeding the population at which economic injury results. Various pest-control options making up an IPM program are selected on the basis of economic, ecological, and political considerations in an attempt to balance the need for pest control with the sensitivities of a more environmentally aware general public. A new era in pest control is beginning as pesticides become gradually reduced. This change will have an impact on the greenhouse industry. Some may argue that this is already occurring through the widespread adoption of IPM programs for many agricultural crops, where presumably the role of insecticides has substantially changed (Metcalf, 1980). However, there are differences in what is occurring today with respect to the use of pesticides, and what will be required tomorrow. The majority of IPM programs in operation today have the use of one or more pesticides as vital components. In many cases, the entire IPM program has been built around the use of these materials. For example, an important aspect of biological control in many vegetable greenhouses is the use of the insecticide pirimicarb[1] for aphid control. Aphids can be brought under control with pirimicarb without destroying natural enemies used for control of other pests. If this material is lost due to registration problems, as occurred in the United States in the late 1970s, or because of the development of insecticide resistance, which has recently been discovered in Europe (Jacobson, 1990), biological control programs may be in jeopardy.

If current trends continue, IPM programs must be developed which rely on other methods besides the use of pesticides. The concept and practice of IPM necessitates a thorough understanding of the pest–natural enemy–plant–environment interaction, so that the full arsenal of tools necessary to combat pest problems can be brought into action. For natural enemies to be effective, they must be factored into this overall system; this, too, requires considerable fundamental data about the pest, natural enemy, plant and environment. This is important given the polyphagous nature of many of the most common pests. In addition, the recent focus on the use of abiotic factors within the greenhouse to control aspects of crop production (Heins, 1990) may have a profound impact on the development of populations of pests or use of natural enemies. Tools such as controlling emigration/immigration, host-plant resistance, managing insecticide resistance, and biological control, among others, must be brought into concerted action.

At the present time there is considerable evidence, based primarily on practical experience and trial-and-error experimentation in the greenhouse, to suggest that IPM is very effective (van Lenteren and Woets, 1988; Heinz and Parrella, 1994a). However, there are few good scientific studies documenting the impact of specific control programs on pest populations. While such studies are not crucial to the outcome of current IPM programs, their lack reduces the repeatability and predictability of IPM programs which hinders their wide-scale adoption. For example, too little is known about how to use effectively many of the commercially available natural enemies in IPM programs for most ornamental crops grown in the greenhouse. Until more scientific studies are done and additional data are in place, fully workable IPM programs will not be a reality for these crops in the greenhouse.

Much of the literature on pests in the greenhouse

[1] 2-dimethylamino-5,6-dimethylpyrimidin-4-yl dimethylcarbamate.

environment has taken a narrow approach when examining growing systems, crops, or pests. The prevalent attitude is that what applies in one location is unlikely to apply in another. It can be argued that the greenhouse ecosystem must be viewed in its totality, and examined with respect to similarities rather than differences. In this chapter an attempt is made to do this by providing a global perspective of the greenhouse ecosystem.

The overall focus of this review is to examine the greenhouse environment with respect to the populations therein of pests and natural enemies. The greenhouse ecosystem can be viewed as less complex than field agriculture because many biotic and abiotic factors can be controlled. Therefore, the greenhouse ecosystem may offer the potential to understand complex aspects of plant–pest–natural-enemy interactions which exist and function in general agriculture but understanding of which has heretofore been unattainable through experimentation in more complex field ecosystems.

Overview of the arthropod fauna

There have been some thorough reviews which have listed the diverse pest complexes found in greenhouses in various parts of the world (Pritchard, 1949; Wilson and Becker, 1960; Hussey et al., 1963; Wilson et al., 1977; Penna et al., 1983; Sunderland et al., 1993). While these have been significant contributions, they have focused only on regional pest complexes, or they have become out of date. Here an attempt is made to list the major and minor pests occurring in greenhouses throughout the world, and to provide information concerning which crops (see Table 9.1 for the scientific names of crops mentioned in this chapter) are primarily attacked and in which areas the problem exists (Table 9.2). While the list is quite extensive, it does not cover all the incidental pests which may occur in greenhouses throughout the world.

One aspect of the greenhouse environment that has not been adequately addressed by researchers is the fauna present in the growing media. Many different types of growing media are used throughout the world and run from the full range of soil types to soilless mixes and to completely artificial substrates such as rock-wool. Reviews of the pest complex in soil are available and major pests have been documented (Ascerno, 1990); however, the diverse fauna of predaceous and phytophagous invertebrates occurring in various growing media used in greenhouses has not really been examined. The fauna varies enormously with the crop grown, the length of the cropping cycle, the type of soil, the watering regime, cultural practices, and pesticide use, among other factors. However, information is accumulating on the fauna occurring in

Table 9.1
Scientific names of crop plants mentioned in this chapter

Common name	Scientific name
African violet	*Saintpaulia ionantha*
Asparagus fern	*Asparagus plumosus*
Azalea	*Rhododendron* spp.
Beans	*Phaseolus* spp.
Boston fern	*Nephrolepis exaltata*
Cabbage	*Brassica oleracea*
Cala lily	*Zantedeschia*
Carnation	*Dianthus caryophyllus*
Chrysanthemum	*Dendranthema grandiflora*
Celery	*Apium graveolens*
Cineraria	*Senecio cruentis*
Cucumber	*Cucumis sativus*
Dock	*Rumex* spp.
Easter lily	*Lilium* spp.
English ivy	*Hedera helix*
Geranium	*Pelargonium* spp.
Gerbera	*Gerbera jamesonii*
Grapevine	*Vitis vinifera*
Holly	*Ilex* spp.
Lettuce	*Lactuca sativa*
Lily	*Lilium* spp.
Maidenhair	*Adiantum* spp.
Melon	*Cucumis melo*
Nasturtium	*Tropaeolum* spp.
Pansy	*Viola tricolor*
Parsley	*Petroselinum crispum*
Pepper	*Capsicum* spp.
Poinsettia	*Euphorbia pulcherrima*
Rose	*Rosa* × *hybrida*
Rubber plant	*Ficus elastica*
Snapdragon	*Antirrhinum majus*
Strawberry	*Fragaria* spp.
Sweet pea	*Lathyrus odoratus*
Sweet pepper	*Capsicum annuum*
Tomato	*Lycopersicon esculentum*
Violet	*Viola* spp.

Table 9.2
Major and minor greenhouse pests

Scientific name	Common name	Crops attacked	Country/Area
Ants			
Arthropoda: Insecta			
Hymenoptera: Formicidae			
Formica cinerea Mayr	California brown field ant	All crops	USA
Linepithema humile (Mayr)	Argentine ant	All crops	USA, South America
Prenolepis imparis californica Wheeler	California honey ant	All crops	USA
Solenopsis eloni maniosa Wheeler	California fire ant	All crops	USA
Tetramorium pacificum Mayr	Western pavement ant	All crops, especially orchids	USA
Defoliators			
Arthropoda: Crustacea: Malacostraca			
Isopoda			
Armadillidium spp.		Cucumbers	Europe: UK
Oniscus spp.		Cucumbers	Europe: UK
Porcellio spp.		Cucumbers	Europe: UK
Trichoniscus spp.		Cucumbers	Europe: UK
Arthropoda: Insecta			
Dermaptera: Forficulidae			
Forficula auricularia Linnaeus	European earwig	Carnation, many crops	Cosmopolitan
Lepidoptera: Gelechiidae			
Phlyctaenia rubigalis (Guenee)	Celery leaf tier	Many ornamental plants	USA
Phthorimaea operculella Zell.	Potato tuber moth	Chrysanthemum	USA, Europe
Lepidoptera: Geometridae			
Sabulodes caberata Guenee	Omnivorous looper	Many ornamental plants	USA
Lepidoptera: Noctuidae			
Agrotis segetum		Tomato	Israel
Autographica gamma	Silvery moth	Chrysanthemums, lettuce, peppers	Europe: UK
Carpitarsia spp.		Chrysanthemums, etc.	South America: Colombia
Choristoneura rosaceana (Harris)	Oblique-banded leafroller	Many ornamental crops	USA
Chrysodeixis chalicites Esper		Tomato	Europe
Euplexia lucipara (Linnaeus)	Fern cutworm	Ferns, other ornamentals	USA, Europe
Heliothis armigera Hubner	African armyworm	Tomato	Europe: Greece
Heliothis virescens (Fabricius)	Tobacco budworm	Geranium, pelargonium, etc.	USA
Heliothis zea (Boddie)	Corn earworm	Many ornamental plants	USA
Lacanobia oleracea (L.)	Tomato moth	Tomato	Europe
Mamestra brassicae	Cabbage moth	Chrysanthemums, peppers	Europe: UK
Naenia typica	Gothic moth	Chrysanthemums	Europe: UK
Noctua pronuba	Large yellow underwing	Chrysanthemums, peppers	Europe: UK
Peridroma saucia Hubner	Variegated cutworm	Many ornamental crops	USA
Phlogophora meticulosa	Angleshades moth	Chrysanthemum	Europe: UK

continued on next page

Table 9.2, *continued*

Scientific name	Common name	Crops attacked	Country/Area
Lepidoptera: Noctuidae (*cont'd*)			
Plusia gamma		Rose	Europe: France
Spodoptera exigua (Hubner)	Beet armyworm	Chrysanthemum	USA, Europe: Netherlands
Spodoptera littoralis Boisduval		Tomato	Middle East: Israel, Europe: Greece
Spodoptera litura	Tobacco cutworm	Strawberries	Far East: Japan
Lepidoptera: Pterophoridae			
Platyptilia antirrhina Lange	Snapdragon plume moth	Snapdragons	USA
Platyptilia pica Walsingham	Geranium plume moth	Geranium	USA
Lepidoptera: Tortricidae			
Adoxophyes orana F.V.R.	Summer fruit tortrix moth		Europe: UK
Argyrotaenia citrana (Fernald)	Orange tortrix	Rose, many ornamental plants	USA
Argyrotaenia franciscana (Walsingham)	Apple skinworm	Many ornamental plants	USA
Cacoecimorpha pronuba	Carnation tortrix	Carnation, ferns, rose	Europe: UK, France
Choristoneura rosaceana (Harris)	Oblique-banded leafroller	Many flowering plants	USA
Clepsis buskana Keifer	Cyclamen leafroller	Cyclamen, other ornamental plants	USA
Clepsis spectrana Tr.	Garden leafroller	Azalea, other ornamental plants	USA, Europe
Epichoristodes acarbella		Carnation	Europe: France, UK
Platynota stultana Walsingham	Carnation leafroller	Carnation, other ornamental plants	USA
Hymenoptera: Tenthredinidae			
Ametastegia pallipes (Spinola)	Violet sawfly	Pansies, violets	USA
Hartigia cressoni (Kirby)	Western rose stem sawfly	Roses	USA
Leafminers/Gall Formers/Borers			
Arthropoda: Insecta			
Coleoptera: Scolytidae			
Xyleborus affinis Eichoff., *X. ferrugineus* (Fabricius)	Bark beetles	Dracaena	Florida
Diptera: Agromyzidae			
Amauromyza maculosa (Malloch)		Chrysanthemum, dock, etc.	USA, South America
Chromatomyia syngenesiae (Hardy)	Chrysanthemum leafminer	Chrysanthemum, cineraria, others	USA, Europe
Liriomyza bryoniae Kaltenbach	Tomato leafminer	Tomato	Europe: UK
Liriomyza huidobrensis Blanchard	Pea leafminer, etc.	Gyposophila, lettuce, chyrsanthemum	USA: California; Europe; South America
Liriomyza sativae Blanchard	Tomato leafminer	Tomato	Middle East: Israel; USA
Liriomyza strigata (Blanchard)	Serpentine leafminer	Tomato	Europe: Spain
Liriomyza trifolii (Burgess)	Serpentine leafminer	Many vegetables and ornamentals	Cosmopolitan
Diptera: Cecidomyiidae			
Dasineura rhodophaga (Coquillett)	Rose midge	Rose	USA
Rhopolmyia chyrsanthemi (Ahlberg)	Chrysanthemum midge	Chrysanthemum	USA

continued on next page

Table 9.2, *continued*

Scientific name	Common name	Crops attacked	Country/Area
Diptera: Tephritidae			
Philophylla heraclei	Celery leafminer	Celery and other vegatable crops	Europe: UK
Lepidoptera: Gracillaridae			
Gracilaria azaleella Brants	Azalea leafminer	Azalea	USA
Hymenoptera: Eurytomidae			
Eurytoma orchidaerium (Westwood)	Cattleya fly	Cattleya orchids	Cosmopolitan
Root feeders/Soil inhabitants			
Mollusca: Gastropoda			
Stylommatophora: Arionidae			
Arion circumscriptus Johnston	Banded slug	Many greenhouse crops	Cosmopolitan
Arion hortensis Ferussac		Many greenhouse crops	Cosmopolitan
Stylommatophora: Helicidae			
Helix aspersa Muller	Brown garden snail	Many ornamental crops	USA (California), Europe
Stylommatophora: Limacidae			
Agriolimax agrestis (Linnaeus)	Gray garden slug	Orchids, many other ornamental plants	USA
Limax flavus Linnaeus	Tawny garden slug	Many greenhouse crops	USA, Europe
Limax marginatus Muller	Striped garden slug	Many greenhouse crops	USA, Europe
Limax maximus Linnaeus	Spotted garden slug	Many greenhouse crops	USA, Europe
Milax gagetes (Muller)	Greenhouse slug	Many ornamental crops	USA
Stylommatophora: Zonitidae			
Oxychilus draparnaldi (Beck)	Draparnald snail	Many greenhouse crops	USA, Europe
Oxychilus alliaris		Many greenhouse crops	USA, Europe
Oxychilus cellarius (Muller)	Cellar snail	Many greenhouse crops	USA, Eruope
Zonitoides arboreus (Say)	Bush snail	Orchids, many other greenhouse crops	USA, Cosmopolitan
Arthropoda: Crustacea			
Malacostraca: Isopoda			
Armadillidium vulgare (Latreille)	Greenhouse pillbug	Bedding plants, asparagus fern	USA
Porcellio laevis Koch	Dooryard pillbug	Many greenhouse plants	USA
Arthropoda: Arachnida			
Acari: Tarsonemidae			
Rhizoglyphus echinopus (Fumouze & Robin)	Bulb mite	Bulb crops	Cosmopolitan
Arthropoda: Symphyla			
Scutigerellidae			
Scutigerella immaculata (Newport)	Garden centipede	Bedding plants, cut flowers	USA, South America: Colombia
Arthropoda: Diplopoda			
Diploiulus luscus (Meinert)	Common millipede	Begonia	USA
Nopoiulus minutus (Brandt)	Bulb millipede	Bulb crops	USA
Orthomorpha gracilis Koch	Greenhouse millipede	Many ornamental crops	USA
Oxidis gracilis	Glasshouse millipede	Cucumbers	Europe: UK

continued on next page

Table 9.2, *continued*

Scientific name	Common name	Crops attacked	Country/Area
Arthropoda: Collembola			
Onychiuridae			
Archorutes armatus (Nicolet)	Armed springtail	Ferns, seeds, many ornamentals	USA
Bourletiella hortensis (Fitch)	Garden springtail	Ferns, seeds, many ornamentals	UK, South America
Collembola spp.	Springtails	Cucumbers, melons	Europe: UK
Hypogastura armata Nic.		Orchids	Europe
Onychiurus spp.		Carnations, chrysanthemums	USA, Australia, South America??
Onychiurus stachianus Bagn.		Many crops, young seedlings	Europe
Orchesella spp.		Orchids	Europe
Tomocerus longicornis Mull.		Many crops, young seedlings	Europe
Sminthuridae			
Folsomia candida		Carnations, chrysanthemums	Australia, South America
Arthropoda: Insecta			
Isoptera: Termitidae			
Reticulotermes hesperus Banks	Western subterranean termite	Geranium, poinsettia, etc.	USA
Homoptera: Aphididae			
Aphis middletonii Thomas	Aster root aphid	Roots of many composites	USA
Homoptera: Pseudococcidae			
Geococcus coffeae Green	Root mealybug	Foliage plants	USA
Rhizoecus americanus	Root mealybug	Nursery plants	USA
Rhizoecus cacticans	Cactus mealybug	Cactus	Europe
Rhizoecus falcifer Kunckel d'Herculais	Ground mealybug	Many potted ornamentals; succulents	Cosompolitan
Rhizoecus floridanus Hambleton		Bromeliads	USA
Coleoptera: Curculionidae			
Anthonomus signatus	Strawberry weevil	Cyclamen, strawberries, potted plants	USA
Cholus cattleya	Cattleya weevil	Orchids	Cosmopolitan
Diorymerellus laevimargo Champion	Orchid weevil	Cattleya, dendrobium orchids	USA
Listroderes obliquus Klug.	Vegetable weevil	Cyclamen, numerous ornamental plants	USA
Otiorhynchus sulcatus (Fabricius)	Black vine weevil	Cyclamen, strawberries, potted plants	Cosmopolitan
Pantomorus godmani (Crotch)	Fuller rose weevil	Gardenias, roses, etc.	USA
Coleoptera: Elateridae			
Agriotes obscurus	Wireworms	Lettuce, beans	Europe: UK
Diptera: Anthomyiidae			
Delia platura (Meigen)	Bean seed fly	Freesia	Europe: UK
Psilla rosae (Fabricius)	Carrot fly	Celery and other vegetables	Europe: UK
Diptera: Cecidomyiidae			
Clinodiplosis cattleyae (Molliard)	Cattleya midge	Orchids	Europe
Diptera: Ephydridae			
Scatella spp.	Shore flies	Potted ornamental plants	USA

continued on next page

Table 9.2, *continued*

Scientific name	Common name	Crops attacked	Country/Area
Diptera: Sciaridae			
Braydesia coprophila (Lintner)	Fungus gnats	Potted ornamental plants	USA
Braydesia impatiens (Johannsen)	Fungus gnats	Potted ornamental plants	Cosmopolitan
Braydesia paupera Toum.	Fungus gnats	Potted ornamental plants	Cosmopolitan
Sciara spp.	Fungus gnats	Cucumbers, etc.	Cosmopolitan
Diptera: Syrphidae			
Citibaena narcissi (Smith)	Least bulb fly	Bulb crops	USA
Citibaena tuberculata (Rondani)	Lesser bulb fly	Bulb crops	USA
Merodon equestris (Fabricius)	Narcissus bulb fly	Bulb crops	USA
Diptera: Tipulidae			
Tipula spp.	Leatherjackets	Hydrangea	Europe: UK
Diptera: Asilidae			
Psilla nigricornis	Stool miner	Chrysanthemum	Europe: UK
Sucking pests			
Arthropoda: Arachnida			
Acari: Eriophyidae			
Aculops (Vasetes) *lycopersici* (Massee)	Tomato russet mite	Tomato	Middle East: Israel
Acari: Pyemotidae			
Pyemotes gramineum (Reuter)	Grass mite	Carnation	USA
Acari: Tarsonemidae			
Polyphagotarsonemus latus Banks	Broad mite	Many ornamental and vegetable crops	Cosmopolitan
Steneotarsonemus laticeps (Halbert)	Bulb scale mite	Narcissus, hippeastrum	Europe: UK
Phytonemus pallidus (Banks)	Cyclamen mite	Many ornamental and vegetable crops	Cosmopolitan
Tarsonemus confusus		African violet	Europe: UK
Tyrophagus longior	French fly	Cucumbers	England
Acari: Tenuipalpidae			
Tenuipalpus pacificus Baker	Phalaenopsis mite	Phalaenopsis orchids	USA
Brevipalpus inornatus (Banks)	Privet mite	Azaleas, fuchsias, etc.	USA
Acari: Tetranychidae			
Bryobia spp.	Clover mites	Cucumbers, ornamentals	Cosmopolitan
Schizotetranychus asparagi (Oudemanns)	Asparagus spider mite	Asparagus ferns	USA
Septanychus tumidus (Banks)	Palm spider mite	Palms, maranta	USA
Tetranychus cinnabarinus (Boisduval)	Carmine spider mite	Tomato	Europe, Middle East: Israel
Tetranychus ludeni Zacher	Luden spider mite	Lantana, etc.	USA
Tetranychus urticae (Koch)	Two-spotted spider mite	Many ornamentals and vegetables	Cosmopolitan

continued on next page

Table 9.2, *continued*

Scientific name	Common name	Crops attacked	Country/Area
Arthropoda: Insecta			
Hemiptera: Miridae			
Calocoris norvegicus		Many ornamental crops	Europe: UK
Lygcoris pabulinus		Many ornamental crops	Europe: UK
Lygus hesperus		Many ornamental crops	USA
Lygus lineolaris Palisot de Beauvois	Tarnished plant bug	Many ornamental crops	USA
Poecilocapsus lineatus (Fabricius)	Four-lined plant bug	Many ornamental crops	USA
Homoptera: Aleyrodidae			
Aleurotulus nephrolepidis Quaintance	Fern whitefly	Ferns	USA
Aleyrodes azaleae Baker & Moles	Azalea whitefly	Azalea	USA
Aleyrodes spiraeoides Quaintance	Iris whitefly	Iris, Gloxinia, etc.	USA
Bemisia argentifolii Bellows and Perring	Silverleaf whitefly	Poinsettia, hibiscus, many vegetables, etc.	Cosmopolitan
Bemisia tabaci (Gennadius)	Sweetpotato whitefly	Many vegetables	
Trialeurodes vaporariorum (Westwood)	Greenhouse whitefly	Many vegetables and ornamentals	Cosmopolitan
Homoptera: Aphididae			
Amorphophora tulipaella (Theobald)	Tulip leaf aphid	Tulips	Europe, USA
Aphis fabae (L.)	Black bean aphid	Many ornamentals and vegetables	Cosmopolitan
Aphis gossypii Glover	Cotton or melon aphid	Melon, chrysanthemums, etc.	Cosmopolitan
Aphis hederae Kaltenbach	Ivy aphid	English ivy	USA
Aphis rumicis Linnaeus	Bean or dock aphid	Dock, Nasturtium, many ornamental crops	USA
Aulacorthum circumflexum	Mottle arum aphid	Bulb crops (Lilies), chrysanthemum, etc.	Europe: UK
Brachycaudus helichrysi Kaltenbach	Leaf curling plum aphid	Chrysanthemum	Cosmopolitan
Brevicoryne brassicae (L.)	Cabbage aphid	Cabbage	Europe
Ceratophis lataniae (Boisduval)	Latana aphid	Orchids, palms and ferns	USA, Europe: UK
Dysaphis tulipae (Boyer de Fonscolombe)	Tulip aphid	Bulb crops	Europe: UK
Idiopterus nephrolepidis Davis	Fern aphid	Maidenhair, Boston ferns	USA
Macrosiphoniella sanborni (Gillete)	Chrysanthemum stem aphid	Chrysanthemum	Cosmopolitan
Macrosiphum euphorbiae (Thomas)	Potato aphid	Tomatoes	Europe: Italy
Macrosiphum pelargonii (Kaltenbach)	Geranium aphid	Geranium	USA
Macrosiphum pisi (Kaltenbach)	Pea aphid	Sweet peas, other legumes	USA
Macrosiphum rosae (Linnaeus)	Rose aphid	Rose	Cosmopolitan
Micromyzus violae (Pergande)	Violet aphid	Violets	USA

continued on next page

Table 9.2, *continued*

Scientific name	Common name	Crops attacked	Country/Area
Homoptera: Aphididae (*cont'd*)			
Myzaphis rosarum Walker	Small green rose aphid	Rose	USA
Myzus ascalonicus Doncaster		Bulb crops (lilies)	Europe: UK
Myzus circumflexus (Buckton)	Lily aphid	Easter lilies, tulips, etc.	USA
Myzus ornatus Laing	Ornate aphid	Many ornamental crops	USA
Myzus persicae (Sulzer)	Green peach aphid	Many ornamentals and vegetables	Cosmopolitan
Myzus solani (Kaltenbach)	Foxglove aphid	Many ornamental crops	USA
Nasanovia ribisnigri		Lettuce	Europe: UK
Rhopalosiphonius latysiphon (Davidson)	Periwinkle aphid	Tulips	USA
Rhopalosiphum prunifoliae (Fitch)	Apple grain aphid	Tulips, strelitzia	USA
Toxoptera aurantii (Fonscolombe)	Black citrus aphid	Gardenia, camellias	USA
Homoptera: Cicadellidae			
Hauptidia maroccana (Melichar)	Glasshouse leafhopper	Tomato	Europe: UK
Zygina pallidifrons Edw.		Calceolaria, cucumbers	Europe: UK
Homoptera: Coccidae			
Coccus elongatus (Signoret)	Elongate soft scale	Palms, other ornamentals	USA
Coccus hesperidum L.	Brown soft scale	Many ornamentals	Cosmopolitan
Coccus pseudohesperidum (Cockerell)	Orchid soft scale	Dendrobium, other orchids	USA
Eucalymnatus tessellatus (Signoret)	Tesselated scale	Many ornamentals	USA
Parthenolecanium cornii (Bouche)	Brown scale	Grapevine, foliage plants	Cosmopolitan
Pulvinaria floccifera Westwood	Cottony camellia scale	Camellias, ferns, podocarpus, etc.	USA
Pulvinaria hydrangea Steinweden	Cottony hydrangea scale	Hydrangea	USA
Pulvinaria phaiae Lull.	Cottony orchid scale	Orchids	USA
Pulvinaria psidii Maskell	Cottony guava or green shield scale	Many woody ornamentals	USA
Saisetia oleae (Bernard)	Black scale	Many ornamentals	USA
Saisettia coffea (Targioni)	Hemispherical scale	Foliage plants	Cosmopolitan
Saisettia nigra (Nietner)	Nigra scale	Ivy, holly, aralia, etc.	Cosmopolitan
Homoptera: Diaspididae			
Aonidiella aurantii (Maskell)	California red scale	Many ornamental plants	USA
Aonidiella citrina (Coquillett)	Yellow scale	Many ornamental plants	USA
Aspidiotus hederae (Vallot)	Ivy scale, oleander scale	Many ornamentals	Cosmopolitan
Aulacaspis rosae (Bouche)	Rose scurfy scale	Rose, other ornamentals	Cosmopolitan
Chromphalus ficus Ashmead	Florida red scale	Palms, rubber plants, etc.	USA
Chrysomphalus bifasciculatus Ferris	False Florida red scale	Aspidistra, many other ornamental plants	USA
Chrysomphalus dictyospermum (Morgan)	Dictyospermum scale, Spanish red scale	Palms, many ornamental plants	USA
Diaspis boisduvalii Signoret	Boisduval scale	Orchids, bromeliads, etc.	USA

continued on next page

Table 9.2, *continued*

Scientific name	Common name	Crops attacked	Country/Area
Homoptera: Diaspididae (*cont'd*)			
Diaspis echinocacti (Bouche)	Cactus scale	Cactus	USA
Fiorinia fioriniae (Targioni)	Palm fiornia scale	Palms, other ornamental plants	
Furcadiaspis zamiae (Morgan)	Cycad scale	Cycads	USA
Hemiberlesia lataniae (Signoret)	Latania scale	Palms and many other plants	USA
Hemiberlesia rapax (Comstock)	Greedy scale	Many woody ornamentals	USA
Hemiberlesia cyanophylli (Signoret)	Cyanophyllum	Many ornamental plants	USA
Howardia biclavis (Comstock)	Mining scale	Many ornamental plants	USA
Lepidosaphes beckii (Newman)	Purple scale	Many ornamental plants	USA
Lepidosaphes machili (Maskell)	Cymbidium scale	Cymbidium orchids	USA
Lepidosaphes mackiana McKenzie	Mackie scale	Dendrobium orchids	USA
Lepidosaphes noxia McKenzie	Noxious scale	Dendrobium scale	USA
Parlatoria pergandei Comstock	Chaff scale	Many ornamental plants	USA
Pinnaspis aspidistrae (Signoret)	Fern scale, aspidistra scale	Ferns, aspidistra	USA
Temnaspidiotus excisus		Foliage plants	USA
Homoptera: Eriococcidae			
Eriococcus azaleae Comstock	Wooly azalea scale	Azaleas and other rhododendrons	USA
Eriococcus coccineus Cockerell	Wooly cactus scale	Cacti and other succulents	USA
Homoptera: Margorodidae			
Icerya purchasi Maskell	Cottony cushion scale	Croton, geranium, ivy, etc.	USA
Orthezia insignis Douglas	Greenhouse orthezia	Many ornamentals	USA
Homoptera: Pseudococcidae			
Phenococcus gossypii (Townshend & Cockerell)	Mexican mealybug	Pelargonium, fuchias, etc.	USA
Phenococcus solani Ferris		Foliage plants	Florida
Planococcus (Pseudococcus) *adonidum* (L.)	Long tailed mealybug	Dracaena, begonia, many ornamentals	Cosmopolitan
Planococcus spp.		Many ornamentals	Cosmopolitan
Pseudococcus citri		Many ornamentals	
Pseudococcus calceolariae (Maskell)	Citrophilus mealybug	English ivy, rubber plant, etc.	USA
Pseudococcus maritimus (Ehrhorn)	Grape or Baker mealybug	Boston fern, etc.	USA
Pseudococcus nipae (Maskell)	Palm or coconut mealybug	Palms, etc.	USA
Trionymus peregrinus (Green)	Amaryllis mealybug	Clivia, etc.	USA
Thysanoptera: Thripidae			
Anaphothrips orchidaceus (Bagnall)	Orchid thrips	Many orchids	Cosmopolitan
Anaphothrips orchidii (Moulton)	California orchid thrips	Many orchids	USA
Echinothrips americanus Morgan		Impatiens	USA
Frankliniella bispinosa (Morgan)		Chrysanthemums	Southeastern USA

continued on next page

Table 9.2, *continued*

Scientific name	Common name	Crops attacked	Country/Area
Thysanoptera: Thripidae (*cont'd*)			
Frankliniella intonsa Trybom		Rose	Europe: France; Southeast Asia
Frankliniella occidentalis Pergande	Western flower thrips	Many ornamentals and vegetables	USA
Frankliniella spp.		Chrysanthemums	South America
Frankliniella tritici (Fitch)	Eastern flower thrips	Many ornamental crops	Eastern USA
Heliothrips haemorrhodalis (Bouche)	Greenhouse thrips	Foliage plants	Cosmopolitan
Hercinothrips femoralis (Reuter)	Banded greenhouse thrips	Cala lily, foliage plants	Europe: UK
Liothrips vaneeckei Priesner	Lily thrips	Lily bulbs	USA
Microcephalothrips abdominalis (Crawford)	Composite thrips	Many flowering plants	USA
Parthenothrips dracaena Heeger	Dracaena thrips	Foliage plants	USA
Scirtothrips longipennis (Bagnall)	Long-winged thrips	Begonia, cyclamen	USA
Taeniothrips atratus (Haliday)		Carnation	England
Taeniothrips xanthius (Williams)	Yellow orchid thrips	Cattleya orchids	USA
Thrips dianthi Moulton		Carnation	France
Thrips fuscipennis Haliday		Cucumbers and melons	Europe: UK
Thrips nigropilosus (Uzel)	Chrysanthemum thrips	Chrysanthemum, gloxinia, etc.	Cosmopolitan
Thrips palmi Karny	Palm thrips	Many ornamentals and vegetables	Far East; Hawaii, Florida
Thrips simplex (Morison)	Gladiolus thrips	Gladiolus	USA
Thrips tabaci Lindeman	Onion thrips	Many flowers and ornamentals	Cosmopolitan

the greenhouse as researchers focus on the interaction of root pathogens and invertebrates, and on biological control by means of soil-dwelling arthropods and nematodes (Londoño and Vañegas, 1990; Nedstam and Burman, 1990). In addition, the fact that biological control agents exist whose life cycles include a soil-dwelling stage necessitates that more attention be focused on the growing-media environment to enhance overall effectiveness in controlling pest populations (Gilkeson, 1990a). The increasing popularity of artificial substrates for crop production may reduce the diversity of arthropods involved within the root zone of plants, but this does not make the relationship of pests to the artificial growing media, plants and natural enemies any less complex.

The pest and natural enemy complex found in the greenhouse occasionally mimics what is present in a field situation, but this usually occurs only where the greenhouses are not totally enclosed and similar crops grow outdoors in adjacent areas (Avilla et al., 1990). In most situations, there is a dearth of natural enemies in the greenhouse. If migration takes place from outside, this usually occurs too late to provide successful pest control. Exceptions to this include the migration of *Diglyphus isaea* during the summer which helps to control *Liriomyza* spp. in greenhouses in western Europe (Minkenberg and van Lenteren, 1986).

Most often, successful biological control of greenhouse pests occurs through manipulation of the natural enemies by inoculative or inundative releases (van Lenteren and Woets, 1988). In comparison to the number of pests, natural enemies used for biological control are relatively few (Table 9.3). However, this will likely increase as problems with insecticides intensify and more natural enemies are sought to combat new pests (Osborne et al., 1990; Loomans and van Lenteren, 1990; van der Linden, 1990).

Table 9.3
Natural enemies of greenhouse pests

Pest grouping	Natural enemy	Common name//Trade name
Aphids	**Fungi**	
	Verticillium lecanii (Vertelac)	
	Beauveria bassiana	BotaniGard; Naturalis O
	Paecilomyces fumosoroseus	PFR-97
	Hemiptera: Anthocoridae	
	Orius spp.	
	Neuroptera: Chrysopidae	
	Chrysoperla spp.	Green lacewings
	Coleoptera: Coccinellidae	
	Hippodamia convergens Guerin-Meneville	Convergent lady beetle
	Diptera: Cecidomyiidae	
	Aphidoletes aphidimyza (Rondani)	Aphid midge
	Hymenoptera: Aphelinidae	
	Aphelinus spp.	
	Hymenoptera: Braconidae	
	Aphidius matricariae Hal.	
	Aphidius colemani Viereck	
	Diaeretiella rapae (Curtis)	
	Ephedrus cerasicola Stary	
	Lysiphlebus tetraceipes (Cresson)	
	Moncrotonus spp.	
	Praon spp.	
	Trioxys spp.	
Fungus gnats	**Bacteria**	
(Mycetophilidae, Sciaridae)	*Bacillus thuringiensis* Berliner var. *israelensis*	Gnatrol
	Nematoda: Steinernematidae	
	Steinernema carpocapsae (Weiser)	
	Steinernema feltiae (Filipjev)	X-GNAT
	Acari: Laelapidae	
	Hypoaspis (Stratiolaelaps) sp.	
Lepidopterous larvae	**Bacteria**	
	Bacillus thuringiensis var. *kurstaki*	Dipel etc.
	Viruses	
	SeMNPV (*S. exigua* multiply-embedded nuclear polyhedrosis virus)	
	SIMNPV (*S. litura* multiply-embedded nuclear polyhedrosis virus)	
Leafhoppers (Cicadellidae)	**Hymenoptera: Mymaridae**	
	Anagrus atomus Haliday	
Leafminers (Agromyzidae)	**Hymenoptera: Braconidae**	
	Dacnusa sibirica Telenga	
	Opius dissitus Muesebeck	
	Opius pallipes Wesmael	

continued on next page

Table 9.3, *continued*

Pest grouping	Natural enemy	Common name//Trade name
	Hymenoptera: Eulophidae	
	Chrysocharis parksi Crawford	
	Diglyphus begini Ashmead	
	Diglyphus intermedius Girault	
	Diglyphus isaea (Walker)	
	Hymenoptera: Pteromalidae	
	Cyrtogaster vulgaris Walker	
	Pgnalio proximus	
Mealybugs (Pseudococcidae)	**Coleoptera: Coccinellidae**	
	Cryptolaemus montrouzieri Mulsant	Mealybug destroyer
	Nephus reunioni Chazeau	
	Hymenoptera: Encyrtidae	
	Lepomatidea abnormis (Girault)	
	Leptomastix dactylopii Howard	
	Pseudaphycus maculipennis (Mercet)	
	Tetracnemoidea peregrina (Compere)	
	Hymenoptera: Mymaridae	
	Anagrus pseudococci (Girault)	
Mites (Acari)	**Acari: Phytoseiidae**	
	Amblyseius aurescens Athias-Henriot	
	Amblyseius barkeri (Hughes)	
	Amblyseius cucumeris (Ouds.)	
	Amblyseius fallacis	
	Amblyseius potentillae (Garman)	
	Amblyseius rademacheri Dosse	
	Amblyseius aberrans (Ouds.)	
	Euseius finlandicus (Ouds.)	
	Iphiseius degenerans	
	Metaseiulus longipilus (Nesbitt)	
	Metaseiulus occidentalis Nesbitt	
	Phytoseiulus longipes	
	Phytoseiulus persimilis Athias-Henriot	
	Phytoseius macropilis (Banks)	
	Seiulus tiliarum (Ouds.)	
	Typhlodromus pyri Scheuten	
	Coleoptera: Coccinellidae	
	Stethorus sp.	
	Diptera: Cecidomyiidae	
	Therodiplosis persicae (Kieffer)	
Root weevils (Curculionidae)	**Nematoda: Heterorhabditidae**	
	Heterorhabditis heliothidis (Kahn, Brooks and Hirschman)	
	Steinernema carpocapsae (Weiser)	
	Steinernema feltiae (Filipjev)	

continued on next page

Table 9.3, *continued*

Pest grouping	Natural enemy	Common name//Trade name
Scale insects (Coccoidea)	**Coleoptera: Coccinellidae**	
	Chilocorus nigritus	
	Cryptolaemus montrouzieri Mulsant	Mealybug destroyer
	Lindorus lophanthae	
	Hymenoptera: Encyrtidae	
	Aphytis sp.	
	Diversinervus elegans Silvestri	
	Encyrtus infelix (Embleton)	
	Encyrtus lecaniorum (Mayr)	
	Metaphycus flavus (Howard)	
	Metaphycus helvolus (Compere)	
	Metaphycus lichtensiae (Howard)	
	Metaphycus bartletti (Howard)	
	Scutellista cyanea Motschulsky	
Thysanoptera	**Fungus**	
	Verticillium lecanii	Thriptal
	Beauveria bassiana	BotaniGard; Naturalis O
	Acari: Laelapidae	
	Hypoaspis (Stratiolaelaps) sp.	
	Hemiptera: Anthocoridae	
	Orius spp.	
	Hymenoptera: Eulophidae	
	Ceranisus menes (Walker)	
	Thripobius semiluteus Boucek	
Whiteflies (Aleyrodidae)	**Fungi**	
	Aschersonia aleyrodis	
	Paecilomyces fumosoroseus	PFR-97
	Verticillium lecanii	Mycotal
	Beauveria bassiana	BotaniGard; Naturalis O
	Hemiptera: Miridae	
	Dichphus errans L.	
	Dichphus tamaninii Wagner	
	Macrolphus caliginosus Wagner	
	Coleoptera: Coccinellidae	
	Delphastus pusillus Casey	
	Hymenoptera: Aphelinidae	
	Encarsia formosa Gahan	
	Encarsia pergandiella Howard	
	Encarsia luteola Howard	
	Encarsia transvena (Timberlake)	
	Encarsia tricolor Foerster	
	Eretomocerus californicus	

THE GREENHOUSE ENVIRONMENT

Types of greenhouses

Movement within the greenhouse, emigration, and immigration

The basic reason for the existence of greenhouses is to produce plants under controlled environmental conditions throughout the year. The degree of control can vary considerably from a completely controlled and enclosed glasshouse in northern and western Europe and Japan, to the simpler structures in the Caribbean, the Mediterranean, Central and South America, southern and central California, and Florida. In the simplest of situations greenhouses are merely constructed of wooden frames designed to hold polyethylene plastic film over a crop. In some situations the sides are permanently open and cannot be sealed. Details of greenhouse types are covered more fully elsewhere (see Chapter 2), but certainly the nature of the greenhouse often influences the arthropod fauna associated with it. For example, a greenhouse with open sides or side vents may allow greater immigration of pests from outside. In addition, the nature of such a greenhouse suggests a mild climate which may provide a better year-round habitat for pests in nearby weeds or crops.

Completely enclosed greenhouses constructed of glass are usually found in areas of the world with a relatively harsh climate and less sunshine. In these areas a tight structure made of glass is required to allow light penetration and to insulate against inclement abiotic conditions during certain times of the year. Around such a greenhouse there are likely to be diapausing populations of pests and their natural enemies, and decreased activity of natural enemies as a result of low light levels. This can often be countered by supplemental lighting, maintaining night-time temperature above a certain minimum, and/or using a non-diapausing strain of natural enemy (Elliot, 1993). Clearly, the environment must be taken into account when deciding on using biological control agents. Furthermore, completely enclosed greenhouses often experience smaller overwintering pest populations and reduced movement from outside. However, in some areas of the world where intense production occurs in greenhouses side by side, migration of pests into the greenhouse can present a problem regardless of how tightly enclosed a greenhouse is. This is especially true if greenhouses for vegetables are located adjacent to greenhouses for ornamentals, and pests are polyphagous. Since tolerable pest populations in vegetable production are often much greater than in ornamental production, migration of even a few into ornamentals can be a serious problem when the vegetable crop terminates.

In an enclosed greenhouse where migration of pests and natural enemies from outside is reduced, a small pest population on the plants at the start of production may gradually build up and reach damaging levels. In this situation the population dynamics are less complex, there may be a more discrete age-class structure during pest development (at least for the first few generations) and overall development is more predictable. This makes control with pesticides or natural enemies more predictable. Population monitoring is important in these situations, but becomes paramount if migration into the production area occurs.

Greenhouses found in areas with a moderate to subtropical climate are usually more open, and are therefore confronted with the possible threat of migration from outside. Pests can migrate from nearby weeds, bordering agricultural production, or adjacent greenhouses. Where migration is a continual problem, pest control becomes extremely complicated and far more difficult. In such situations the use of screening to exclude pests is gaining popularity, even in areas of the world (such as southern California and Bogota, Colombia) where it has been considered impossible due to concomitant reduction in air flow and increasing temperatures (Berlinger et al., 1991; Robb, 1995). In addition, with regular movement of pests into open greenhouses, pest numbers have the potential to increase more quickly. Growers must have some way of monitoring this migration so that they can respond, either with chemical or biological control. Greenhouse structure, and hence migration, must for each system be factored into the framework for decision-making for overall pest management (Parrella, 1990b). Models of the leafminer-parasite interaction on chrysanthemums have been developed for the overall purpose of making biological control more predictable; but for simplicity migration was assumed to be zero (Parrella et al., 1992). If more sophisticated and realistic models are to take migration into account, an accurate method of quantifying this factor must be developed.

Movement and spread of pests and natural enemies within individual glasshouses has been studied in very few instances. This is unfortunate because these data can be very useful in designing effective

IPM-strategies. Smits and Vlak (1988) found that larvae of the beet armyworm (*Spodoptera exigua*) initially moved vertically on chrysanthemum, but that, once the pest was in the third instar, horizontal spread occurred. This was limited to an area of approximately 1 m^2 around the egg mass. Such information is vital when developing useful strategies for an entomopathogen which must be ingested by the pest insect to be effective (Moed et al., 1990).

Probably more is known about the vertical and horizontal movement of the greenhouse whitefly, *Trialeurodes vaporariorum,* and its parasite, *Encarsia formosa,* than about any other pest–natural-enemy system. It has been estimated that the greenhouse whitefly disperses at the rate of 3.5 cm h^{-1} during the first few days of its life. After three days this dispersal ceases, and the average distance from hatching to the point where adult emergence occurs is 1.25 m (van Lenteren and Noldus, 1990). Data are not as precise for *E. formosa,* but it is believed that they are capable of finding patches of whiteflies in large greenhouses (Eggenkamp-Rotteveel Mansveld et al., 1982; Noldus and van Lenteren, 1990).

Research done on the movement of individual *Liriomyza trifolii* using mark-recapture techniques (Jones and Parrella, 1986) demonstrated that adult leafminers could fly the full distance of a large greenhouse within one week; 90% of the females were captured within 46 m from the release point. This has implications for the development of insecticide resistance in the greenhouse, because it suggests a thorough mixing of individuals and further reduces the feasibility of utilizing untreated refugia in a resistance management program (Leeper et al., 1986). Based on the distances flown, suggestions were made for the optimal number of yellow traps[2] to use per unit area (Jones and Parrella, 1986). Studies done with whiteflies (*Bemisia argentifolii*) in poinsettia (*Euphorbia pulcherrima*) have demonstrated that they too would fly the full distance of a large greenhouse in a one-week period (Robb and Parrella, unpublished). It is dangerous to extrapolate this data to other species of whiteflies; this information may or may not be applicable to *T. vaporariorum.* Additional experimentation is needed to decide if the information can be used in a generalized way.

If movement is restricted after emergence, as demonstrated by Noldus and van Lenteren (1990) for greenhouse whitefly on tomato (*Lycopersicon esculentum*), then yellow traps may be useful for detecting some of the initial foci of infestations. This could lead to effective spot treatments by insecticides or suggest areas for parasite release, but would be limited by the cost and number of traps required. However, such data on movement and dispersal must be collected for each crop and pest. Crop height, canopy structure, and preferences of the pest insect, among other factors, will influence movement in the vertical and horizontal planes (Sutterlin et al., 1990). It may be difficult, for example, to extrapolate movement patterns of whiteflies from work done in tomatoes to other greenhouse crops such as gerbera (*Gerbera jamesonii*) and poinsettias. In addition to differences in architecture and duration of these crops, they are grown under different environmental conditions. The distance over which western flower thrips *Frankliniella occidentalis* dispersed was much greater in a poinsettia greenhouse than in a rose greenhouse (Robb, 1989). Host and cultivar preferences, and size and timing of the crop, could be responsible for the differences observed. Data on movement and dispersal in both vertical and horizontal planes is needed for most pests and their natural enemies in greenhouses. Such information is important, too, when trying to understand the mathematical distributions of pests and how these distributions change over time for the purpose of developing sampling plans.

Life history and adaptations

Isolating effect

The physical structure and crop-production characteristics of greenhouses favor the establishment of reproductively isolated arthropod populations. In addition, the ideal climate, shortage of naturally occurring parasites and predators, and a plentiful food supply all year long allow for multiple pest generations in a year and high individual survival rates. Consequently, one might expect the development of pest biotypes – members of a population or populations with a specific set of biological characteristics – within the greenhouse environment.

Work on the development of insecticide resistance provides the strongest evidence of the formation of greenhouse biotypes. Because application of pesticides in the greenhouse is a common occurrence in both

[2] Yellow traps (and other colors) coated with a sticky material are used in greenhouses to monitor a wide range of insect pests.

ornamental and vegetable production, such intense selection pressure has led to the development of high levels of resistance in key pests. In addition, development of resistance varies from greenhouse to greenhouse (Keil and Parrella, 1990), this is a result of disparate patterns of pesticide use in different greenhouses. The development of resistance in greenhouse pests began soon after pesticides were first used in the 1950s, and has continued to be a serious problem. Pesticide resistance has been documented in *Aphis gossypii* (Blackman and Eastop, 1985), *Bemisia argentifolii* (Sanderson and Roush, 1995), *Frankliniella occidentalis* (Robb, 1990; Immaraju et al., 1992), *Liriomyza trifolii* (Keil et al., 1985), *Myzus persicae* (Blackman and Eastop, 1985), *Spodoptera exigua* (Yoshida and Parrella, 1987), *Tetranychus urticae* (Overmeer and Helle, 1980; Ramdev et al., 1988; Gough, 1990) and *Trialeurodes vaporariorum* (Wardlow et al., 1972). Although insecticide failure has led to the suspicion of resistance development in many more species, confirmational studies in the laboratory have not been done. In many of the pests listed above, higher levels of resistance were found in greenhouse populations compared to resistance levels observed in the field. Overmeer and Helle (1980) believed that genes for resistance to tetradifon[3] and dicofol[4] in greenhouse populations of *Tetranychus urticae* could be used as genetic markers to separate greenhouse from field populations.

The potential isolating effect of the greenhouse environment has been investigated using the leafroller, *Clepsis spectrana* (van den Bos and van de Vrie, 1981). The greenhouse biotype, unlike the field biotype, did not have a facultative diapause in the larval stage, allowing continuous breeding throughout the year. A similar situation was found with *Tetranychus urticae* on roses (*Rosa* spp.) where populations in greenhouses did not enter diapause in response to short days as did their field counterparts (Overmeer et al., 1980). In rose greenhouses heated during the winter, diapause was rare.

There has been little research into the interaction of greenhouse pests and their host plants, so that the formation of biotypes in response to host-plant preference cannot be fully addressed. It was found that populations of the melon aphid, *Aphis gossypii,* developing on cucumber (*Cucumis sativus*) for several generations, developed specificity to cucumber and were found to be unable to develop on chrysanthemums. The reverse situation of *A. gossypii* developing host specificity for chrysanthemums has also been documented (Blackman and Eastop, 1985). Very rapid development of host-plant specialization has also been shown for *Trialeurodes vaporariorum* developing on sweet pepper (*Capsicum frutescens*) and gerbera (van Lenteren and Noldus, 1990). Although sweet pepper and gerbera are normally considered to be poor host plants for the greenhouse whitefly, each population exhibited faster development, longer adult life, and reduced mortality on its second host compared to its original host after several generations on the second host. This local adaptation of greenhouse whitefly to relatively poor host plants is of concern in terms of expansion onto other crops. It is of major concern with the silverleaf whitefly, *Bemisia argentifolii,* a relatively new pest in greenhouses. This pest's ability to cause serious damage to a number of greenhouse crops is enhanced by its ability to act as a vector for many plant viruses (Byrne et al., 1990).

Development of resistance to insecticides and acaricides

Resistance to insecticides and acaricides is extremely serious in greenhouse situations and is an artifact of the greenhouse environment, the desire to produce a perfect product, and the biology of most of the major pests encountered. Many greenhouse pests have relatively short development times and high reproductive rates, which have been shown, in some situations, to foster the development of resistance (Sanderson and Roush, 1995). Of course, in the greenhouse many pests breed all year (as opposed to their field counterparts which may have fewer generations) and are exposed to insecticides throughout each generation. In addition, as stated previously, the greenhouse fosters the isolation of arthropod populations. Insecticides are often applied at short intervals resulting in very thorough coverage, and thus few individuals are left untreated. Consequently, survivors of applications are likely to be resistant, and will breed with other resistant survivors to produce a homozygous strain of resistant pests. Gradually, an insecticide-resistant insect population is selected. Insecticide applications are made quite frequently in many greenhouses (this is especially true for ornamentals) to produce a commodity free of pests and

[3] 4-chlorophenyl 2,4,5-trichlorophenyl sulfone.
[4] 2,2,2-trichloro-1,1-bis(4-chlorophenyl) ethanol.

their damage. In the Netherlands, approximately 223 and 14.4 kg ha^{-1} yr^{-1} of pesticides (active ingredient) are used in chrysanthemum and cucumber greenhouses, respectively (van Lenteren, 1990). While these figures are high when contrasted with field agriculture, similar amounts are probably used in ornamental and vegetable greenhouses throughout the world. Such intensive pesticide application exacerbates the resistance problem.

The general failure of insecticides to control the two-spotted spider mite (*Tetranychus urticae*) and the greenhouse whitefly (*Trialeurodes vaporariorum*) was an important incentive for growers in western Europe to reconsider biological control in greenhouses (Noldus and van Lenteren, 1990). The success, safety, and economic competitiveness of biological control compared with conventional control by insecticides has made this a preferred method for many greenhouse vegetable growers all over the world, irrespective of concerns about pesticide resistance.

In greenhouse production of ornamentals, where the use of chemicals is still the dominant method of pest control throughout the world, the phenomenon of resistance development is playing a similar role in getting growers to consider alternatives. Together with legislation enforcing reductions in pesticide use, this will force growers to take another approach to pest control (Parrella, 1990a; van Lenteren, 1990). Such programs involving biological control are being developed (Sanderson and Ferrentino, 1990), but practical applications such as the program for spring bedding plants in England (Jacobson, 1993) are rare. Ideally, programs for biological control in ornamentals should be developed that are both successful and economic, so that growers have other incentives to change.

Another serious problem associated with the spread of new resistant species and new biotypes is the movement of infested plant material from country to country. No other agricultural industry moves so much plant material so frequently as the greenhouse industry. A considerable proportion of vegetable and ornamental production in the greenhouse relies on propagative material produced elsewhere to initiate crop production. This material is grown by large propagators, who strive to produce a pest-free product. However, in the production of millions of cuttings or transplants of a particular commodity, it becomes impossible to eradicate all of the species. Furthermore, the fact that many of these large propagators use pesticides quite frequently suggests that any pests sent out with plant material may be resistant to one or more materials.

Research on the development of pesticide resistance in the greenhouse has focused on documenting resistance (Keil et al., 1985; Ramdev et al., 1988) and on the development of techniques to assess pesticide resistance rapidly (Haynes et al., 1986; Sanderson et al., 1989). While these techniques should enable producers to assess resistance levels within their greenhouses, traditional strategies for managing insecticide resistance in field situations may have applicability in the greenhouse (Sanderson and Roush, 1989, 1995). A thorough understanding of resistance mechanisms and the genetics of resistance are necessary before good management plans can be put into action. These data are available for only a few pests (Overmeer and Helle, 1980; Keil and Parrella, 1990). Attempts to manage resistance in the greenhouse through traditional approaches such as rotation of classes of insecticides, switching to a new class of material, and low-dose strategy among others, will, at best, slow the development of resistance; total prevention is unlikely (Parrella et al., 1987).

CURRENT PEST MANAGEMENT STRATEGIES

Insecticide use in the greenhouse

Despite the development of resistance rather early in the history of insecticide use in greenhouses, most growers embrace insecticides as their major method of pest control. When insecticides work, there are few alternatives that are as easy or convenient. The use of insecticides in the greenhouse promoted an environment free of pests which was a desired goal of many cut-flower and vegetable growers. In addition, because new insecticides became available frequently during the 1960s and 1970s, a continual supply of materials was available. Furthermore, requirements for registering pesticides were then far less demanding than they are today; many insecticides could be made available very quickly for greenhouse use on ornamentals because the requirement to specify tolerance levels for nonfood crops has often been waived. Also, in the United States, a special interregional project ("IR-4 project"), funded at the national level, supports the registration of pesticides for minor crops (such as those grown in greenhouses) which has further enlarged the number of materials available (Elson, 1988). Finally, the enclosed greenhouse is conducive to the application of insecticides as aerosols, fogs, and smokes, enhancing the ease and convenience of their use.

Today, the situation is changing dramatically with respect to pesticide availability, pesticide effectiveness, and the rules and regulations governing usage. Applications of materials two and three times per week are no longer sufficient in some cases to produce a product that is pest-free at harvest. However, many growers must still produce a crop virtually free of pests and their damage if the commodity is destined for export to another country. Many countries have intensified quarantine and stepped up inspections of imported plant material. Thus growers are expected to produce material with no visible pests or damage whatever, and yet are expected to reduce their reliance on pesticides to accomplish this goal (Parrella, 1990a; van Lenteren, 1990).

Although this may seem to be putting the grower in an awkward situation, there is indisputable evidence that insecticides are not providing as good control as they once were. Because insecticide resistance has rendered many materials ineffective, it is certainly not economical nor convenient to make multiple applications of such a product. This is especially true when increasing labor costs in insecticide application are taken into consideration.

There are no easy answers to this problem, and speculation as to the future is difficult. It is likely that insecticides will not be rendered totally ineffective or lost altogether, despite problems of resistance and legal restrictions, among others. Rather, the number of insecticides may be drastically reduced and the type of materials available may change. Materials with inherently lower toxicity to fish and mammals, and fewer potential problems with ground-water contamination or long-lasting residues (e.g. insect growth regulators, insecticidal soaps), and materials of natural origin (e.g. botanicals) may dominate the marketplace. In addition, the conventional use of wet sprays requiring volumes of water and applications heavy enough to cause runoff may gradually disappear (Lindquist and Powell, 1990). Low- and ultra-low-volume techniques which make more efficient use of pesticide and water may become increasingly popular (Kabashima et al., 1995). Because of continual problems with infested propagative material either purchased or produced on site and the challenge associated with obtaining acceptable control during production phases of the crop, dipping propagative cuttings or transplants into various pesticide solutions may become more popular (Osborne, 1986).

The use of specific botanical insecticides and insect growth regulators may create a greenhouse ecosystem which allows more effective use of natural enemies. This generates excitement and optimism over the utility of biological control in conjunction with the use of pesticides. Thus the full integration of insecticides and natural enemies may be used in the future to produce export-quality greenhouse vegetables and ornamentals. [Compatibility of insecticides with natural enemies is discussed later in this chapter, p. 243].

Frequent pesticide application to flowers or vegetables in greenhouses can have profound effects on the plants themselves, in the form of obvious physical injury, phytotoxicity, more subtle effects on overall plant physiology (Jones et al., 1986b) and effects on yield (Heinz and Parrella, 1991). In addition, insecticide application may cause the plants to become more sensitive to infection by pathogens (Chase and Osborne, 1983; Osborne and Chase, 1984). During the winter when plants are generally more sensitive, pesticides can cause abortion of flowers and small fruits (van Lenteren et al., 1980). It is also suspected that pesticide application to cut flowers may reduce overall yield and vase life after cutting (Jones and Parrella, 1984).

Some cut flowers such as roses may be harvested several times a day, and every day during the week. Some vegetable commodities, such as tomato, require harvesting 2–3 times per week. If frequent contact of workers with the crop is necessary, there are difficulties in using pesticides. An interval ("re-entry interval") must elapse after pesticide application before it is safe for workers to re-enter the greenhouse. This interval is often embodied in regulations. For many pesticides commonly used in the United States, the required interval is increasing to as much as three days after application (Parrella, 1988), which makes their legal use by many growers impossible. Toxicological data concerning surface and airborne residues of pesticides in the greenhouse are just becoming available (Stephanou and Zourari, 1989) and stricter re-entry intervals are likely in the future. Finally, in a crop such as tomato where bumblebee colonies are brought into the glasshouse to enhance pollination (Anonymous, 1990), alternatives to pesticide application must be sought.

Host-plant resistance

The utilization of host-plant resistance to manage

arthropod pests is one of the cornerstones of integrated pest management (Beck, 1965). However, many more varieties resistant to pathogens have been bred than varieties resistant to insects and other arthropod pests. As speculated by de Ponti et al. (1990), this may be due, in part, to a lack of mutual interest and cooperation between entomologists and plant breeders. Research on host-plant resistance in greenhouse vegetable crops has focused primarily on whiteflies and spider mites (de Ponti, 1980a; de Ponti and Hogenboom, 1980; de Ponti et al., 1990). In floriculture, research has spanned a greater number of crops and pests, but the effort has focused on evaluating host-plant preferences of different arthropods, and ranking various cultivars or host plants as to their resistance to a particular pest.

In floriculture the range of different crops and the variety of cultivars per crop is much wider than in food crops and is changing more rapidly. In fact, diversity in the product being offered to the public is an essential requirement, stemming from the function of flowers as ornaments in the home, as gifts for special occasions, or expressions of sympathy. Fashion plays an important role, especially where color is considered, and novelties are introduced regularly to stimulate demand. Consequently, ornamental value is the principal criterion of most breeding programs, and resistance to arthropods and diseases is almost always a secondary selection criterion (Sparnaaij, 1991). Breeding for resistance is a slow process which cannot always keep pace with rapid changes. By the time an adequate level of resistance has been reached, the types of cultivar used in the development of resistant plant material may no longer be in demand. The introduction of resistance in a crop can only have an impact on pest management through less dependence on pesticides if the resistance is broadly based and can be transferred rapidly to the majority of cultivars. Resistance in chrysanthemum cultivars to viruses transmitted by western flower thrips may be developed more quickly by genetic manipulation (Urban et al., 1992). However, it is not practical to transform all chrysanthemum cultivars – a more traditional approach must be found to transfer this resistance to other cultivars.

Research on host-plant resistance in greenhouse crops has focused on evaluating the susceptibility of plants to aphids (Markkula et al., 1969), leafminers (Ibrahim and Madge, 1977; Alverson and Gorsuch, 1982; Oetting, 1982; Parrella et al., 1983; Broadbent and Blom, 1984), Lepidoptera (Parr and Thurston, 1968; Yoshida and Parrella, 1991), spider mites (Markkula et al., 1969; Osborne and Chase, 1985; Craig et al., 1986; Al-Abassi et al., 1987), thrips (Hargreaves and Cooper, 1980; de Jaeger and Butot, 1993) and whiteflies (Fisher and Shanks, 1974; Bilderback and Mattson, 1977). In only a few instances have the mechanisms of this resistance been explored (Craig et al., 1986; Al-Abassi et al., 1987; Yoshida and Parrella, 1991; Tomczyk and Kropczynska, 1989). Such data are required for a breeding program which tries to incorporate resistant traits into plant material.

The unqualified success of the biological control of whiteflies by *Encarsia formosa* in glasshouse vegetable crops has prompted studies aimed at understanding the relationship between the pest, its primary parasite and several host plants (van Lenteren et al., 1980). Such studies led to predictions concerning the ability of the parasite to control whiteflies successfully on new glasshouse crops or cultivars. A general conclusion from this work was that biological control can be improved by using varieties of plants with few or no hairs. For example, *E. formosa* walks 3.5 times faster and kills a greater number of prey on a hairless cucumber variety than on a hairy variety. Similar results were found on poinsettia cultivars with a variety of natural enemies (Heinz and Parrella, 1994b). Such data suggest the possibility of breeding crop plants for improved parasite efficiency rather than for pest resistance (de Ponti, 1980b; de Ponti et al., 1986). In addition, where pest resistance is the goal of a breeding program, such studies emphasize the need for assessing the potential impact of the new cultivar on the control capability of natural enemies.

The greenhouse environment and biological control

Arthropod natural enemies

Biological control within a protected environment has a long history; details can be found in reviews by Hussey and Scopes (1985) and van Lenteren and Woets (1988). This control method within greenhouses has been enthusiastically adopted by vegetable producers on more than 12 000 hectares worldwide (van Lenteren, 1990). Biological control in ornamentals has lagged far behind that in vegetables because of greater constraints in production, which include aesthetic quality and strict quarantine requirements (Parrella and Jones, 1987; Parrella, 1990a). Considerable attention is now being focused on the development and implementation of a biological control strategy for ornamentals (van Lenteren and Wardlow, 1993; Gilkeson, 1990b).

The greenhouse environment clearly influences the success of any biological control effort. Successes have been achieved in the closed systems found in northerly latitudes (45–53°N) in Europe. In the warmer climates of more southerly regions (30–45°N) greenhouses are more open and success has been less pronounced, perhaps because migration seems to be a continual problem. Also, higher temperatures usually translate into faster population development. Smith and Webb (1977) speculated that the continuous land mass and the north–south orientation of mountain chains in North America provided little in the way of natural barriers to pest migration. In order for biological control to be successful, Smith and Webb believed that biological control would be limited to certain regions, and that there must be changes in the crops grown and adherence to very strict guidelines. Adherence to regimented guidelines and concomitant changes in crop production practices were also emphasized by Hussey and Scopes (1977), who indicated that successful biological control requires more than simply releasing the correct number of natural enemies and choosing pesticides compatible with those natural enemies. Unfortunately, until there are no other alternatives, crop production scenarios will be dictated by market forces, rather than what is best for biological control.

Some of the concepts favoring successful biological control were reviewed by Parrella and Jones (1987) and Parrella (1990b), and are actually part of a complete IPM program. These include: 1) starting with clean propagative material and mother block[5]; 2) sterilizing soil between crops; 3) controlling weeds in and around the greenhouse; 4) use of insect-screening where and when appropriate; 5) immediate removal of plant material such as discarded leaves and small buds from the greenhouse; 6) intelligent scheduling of crops to avoid continuous cropping; 7) elimination of highly susceptible cultivars; 8) grouping of susceptible cultivars together, so that monitoring and pest-control activities can be disproportionately concentrated in these areas; 9) monitoring pests and natural enemies regularly, with particular attention to early detection of migration into the greenhouse; 10) avoidance of excessive fertilizer levels in the crop; and 11) contingency plans to deal quickly with new pests migrating into the crop without disturbing natural enemies. These concepts and others were utilized in delivering an effective IPM/biological control program to producers of spring bedding plants in England (Jacobson, 1993).

At the start of a crop in a fully enclosed greenhouse, the environment usually has few or no insects and mites present. When a crop is planted, pests may become a problem for any number of reasons already discussed. Within this context, augmentative releases of natural enemies must be used for biological control to be effective. Van Lenteren and Woets (1988) use the term "seasonal inoculative release" to describe a release of natural enemies into vegetable greenhouses (discussed below). The methodology employed in the United States, where parasites have been released into ornamental greenhouses for biological control of leafminers (Jones et al., 1986a; Parrella and Jones, 1987; Heinz and Parrella, 1990c) falls between the classical definitions of inundative and inoculative releases. In these studies large numbers of parasites were released on a regular basis, and yet control exerted by the offspring of these releases was crucial to obtaining success.

Outstanding success with biological control in greenhouses has been achieved for whiteflies on vegetable crops using seasonal inoculative releases (van Lenteren and Woets, 1988). This method is intended to have a long-term effect over a number of generations of the pest, and is actually similar to an inoculative release in a classical program of biological control. However, the culture method used for most greenhouse crops prevents the effects of a single release from lasting many years. At crop termination, the pest and natural enemy are removed from the greenhouse with the crop. Hence, with each season comes a new crop requiring new releases of natural enemies. In the Netherlands, the seasonal inoculative release method for control of whiteflies consists of placing parasitized whitefly pupae on cards in the tomato crop at the first appearance of susceptible whitefly stages. This is generally repeated three times at two-week intervals; a minimum of 1 pupa per plant is introduced at each release (Woets, 1978; Vet et al., 1980). The density of whiteflies at the time of initial parasite release is critical: there can be no more than 1 adult whitefly per 10 m^2 (Ekbom, 1977). Biological control of whiteflies in European glasshouses is often in the hands of the private sector

[5] Mother block generally refers to a high density planting used primarily for the production of cuttings. These cuttings are then planted and result in the marketed crop.

which, through experience, may modify release rates and frequency.

Inundative releases also have a place in the greenhouse, though there is a divergence of opinion regarding the utility of this method. As early as 1951 it was believed that inundative release would be limited to the control of pests infesting crops of high value where pest infestations were very light (Flanders, 1951). Debach (1964) considered this technique to be most effective with univoltine pests or, if multivoltine, where injurious numbers were reached in only a single generation each year. Van Lenteren and Woets (1988) believed that it had potential only against minor pests in greenhouses. The exact reasons for these different viewpoints are not clear, although economics may play a role in the development of these ideas. A general consensus (Debach, 1964; Stinner, 1977; King et al., 1985) is that, if multiple releases of natural enemies are required, then the inundative release technique would probably not be economically feasible. Unfortunately, there are very few hard data to support this contention in a field or greenhouse situation (King et al., 1985). As data are gathered, especially for high-value ornamental crops (Heinz and Parrella, 1990c; Parrella et al., 1991; Flint et al., 1995), inundative releases seem very promising.

The overall goal of any biological control program is the suppression of host numbers below the threshold for economic damage. While long-term stability is a desirable characteristic in a classical biological control program, this is usually not a requirement in the greenhouse given the short-term nature of many crops. However, predaceous mites have been shown to persist in rose greenhouses for more than two seasons (Field and Hoy, 1984, 1986; Gough, 1991). There is a similarity between inoculative releases made into the greenhouse system against a multivoltine pest, and inoculative releases of natural enemies made against univoltine pests in a perennial field agricultural system. An analysis of the underlying mechanisms of biological control may be more feasible in an enclosed greenhouse where multiple pest generations, and control of abiotic conditions and extrinsic mortality factors, allow research which might otherwise require many years to be condensed into one season. However, only one study (van Lenteren, 1986) has attempted to utilize the greenhouse ecosystem as a model system to determine the utility of several theoretical explanations of biological control.

Theoretically, biological control may result from one of two situations:

(1) a stable pest–enemy equilibrium is achieved at low host densities, with aggregation of natural enemies in patches of high host density. Here, positive density-dependent parasitism may occur and lead to overall pest suppression (Beddington et al., 1978). Extinction of pests is prevented by spatial heterogeneity, and the selective discovery of pests by natural enemies which leaves refugia (Huffaker, 1958);

(2) large-scale suppression and stability of the pest population may occur where parasites searching in a random manner cause local extinction of pests and therefore locally unstable population dynamics; but when random search is combined with pest movement on a larger scale, stability of the pest population occurs. Implicit in (2) is that a stable equilibrium between the pest and natural enemy is not a requisite for biological control.

More information on these theories as they relate to biological control in the greenhouse can be found in papers by van Lenteren and Woets (1988), Noldus and van Lenteren (1990) and Waage (1990).

Van Lenteren (1986), using some of the fundamental ideas presented above, tried to explain why biological control of greenhouse whitefly by *Encarsia formosa* was regularly achieved in large greenhouses, whereas failures were often recorded in smaller greenhouses. It was speculated that spatial heterogeneity in large greenhouses would lead to stability below the economic threshold, but that in smaller greenhouses extinction of the pest and parasite would occur. Migration of whiteflies into the smaller greenhouses would then lead to uncontrolled and damaging pest populations. However, data were insufficient to support these ideas. The asynchronous dynamics of individual whitefly-infested foci may serve as a stabilizing mechanism at the greenhouse level, and may explain the success and failure in large and small greenhouses, respectively (Noldus and van Lenteren, 1990).

The use of pathogens in a greenhouse environment has tremendous potential and as we move into the next century it is likely that they will play a greater and greater role in pest control. Depending on the pathogen, they may be used via inundative releases (such as *Bacillus thuringiensis* for control of Lepidoptera) or inoculative/inundative releases (such as *Paecilomyces fumosoroseus* for whitefly control) which may kill after contact and continue to cycle in the greenhouse. A thorough review of fungi with biological control potential in glasshouses is provided by Fransen (1990).

When total greenhouse hectares on a worldwide basis are examined with respect to biological control, more hectares are treated with *Bacillus thuringiensis* var. *kurstaki* than with any other biological control agent (van Lenteren, 1990). This reflects the effectiveness of the toxin produced by this bacterium, and underscores the importance of Lepidoptera as pests in the greenhouse. The application of this pathogen is likely to increase with the discovery of new strains or isolates (e.g. *B. thuringiensis* var. *israelensis* against fungus gnats) which will expand the host range of this pathogen, and their registration as officially approved. In addition, *B. thuringiensis* is currently one of the major focal points of molecular biologists; the material itself is the target for alterations, and a great effort is being made to incorporate into plants portions of the genetic structure of *B. thuringiensis* coding for the toxin.

The use of fungi in the greenhouse has received considerable attention with much of the focus on *Verticillium lecanii* (Hall, 1985a). This fungus was very appealing commercially because it could be easily mass-reared and because several biotypes existed which attacked important greenhouse pests, including whiteflies, aphids, and western flower thrips. The greenhouse whitefly becomes infected in all developmental stages except for eggs. *Verticillium lecanii* was slow to be adopted by European growers, and its popularity gradually declined towards the end of the 1980s, at which time the material was no longer available commercially. The restrictive set of environmental conditions under which epizootics develop probably explains the inconsistent performance of *V. lecanii.* This fungus has a rather narrow range of environmental conditions under which it functions effectively (15–25°C, 85–90% RH, with high humidity for at least 10–12 h per day). Furthermore, the very effective aphidicide, pirimicarb (2-dimethylamino-5,6-dimethylpyrimidin-4-yl dimethylcarbamate), was readily available and did not require specific environmental conditions to be effective. To counter problems with environmental constraints, Helyer and Wardlow (1987) developed a multiple low-dose strategy which provides a greater opportunity for the fungus to encounter the right environmental conditions. *Verticillium lecanii* is making a strong comeback in the 1990s, in part due to pirimicarb resistance in aphid populations throughout Europe (Jacobson, 1990), and commercial formulations are now available. However, other fungi may supplant it as biological control agents (see discussion below).

Verticillium lecanii has the greatest chance of success where environmental conditions necessary for infection are satisfied; such locations include rooting areas with regular misting, and areas of production chrysanthemums after black shade cloth is pulled over them to initiate flowering (Vehrs and Parrella, 1991). In more open greenhouses these conditions will rarely be met, and the use of the product is probably not worthwhile. For example, in Bogota, Colombia, the endemic fungus *Zoophthora erinacea* appears to have much more potential than *V. lecanii* for aphid control on chrysanthemums because it is adapted locally to growing conditions in the savannah (Hall, 1985b). However, in this area shading of chrysanthemums with black cloth is not necessary to induce flowering, which further reduces the possibility of correct environmental conditions for *V. lecanii.* Because of the strict environmental requirements necessary to maintain pathogenicity, Fransen (1990) suggested that, when one is searching for and evaluating biotypes of fungi for biological control, studies should be done on intact plants under fluctuating environmental conditions typical of actual production areas.

Fungi in the genus *Aschersonia* are specific for whiteflies and do not require the strict environmental conditions for infection outlined above, though they have the disadvantage of being more difficult to rear *en masse* (Fransen, 1990). Successful whitefly control has been achieved at 50% RH and 20°C. The conidia of *A. aleyrodis* may survive up to 28 days in the greenhouse (as opposed to 5 days for *V. lecanii*), but a major drawback is that these conidia must be dispersed by rain in order to infect new whiteflies and cause the desired epizootics (at low prey densities or early in the crop). Hence, good coverage of new foliage is essential. Frequent low-dose and low-volume applications of *A. aleyrodis,* as suggested for *V. lecanii* (Helyer and Wardlow, 1987), may have great potential.

The fungus *Paecilomyces fumosoroseus*[6] is being developed commercially both in the United States and Europe. The fungus was originally targeted for whitefly control, but the broad spectrum of its activity will probably lead to additional recommendations for control of aphids and western flower thrips (Osborne et al., 1990).

Several other fungi have been examined with respect

[6] PFR-97™ 20% WDG Biological/Microbial Insecticide, W.R. Grace & Co., Conn. 7379 Route 32, Columbia, MD 21044, USA.

to their control potential in greenhouses. Two of the more common ones are *Beauveria bassiana* and *Metarhizium anisopliae,* the primary activity of which is against soil-dwelling larvae. These materials have potential, and there are methods which can be used to isolate them from artificial potting media (Chase et al., 1986). *Beauveria bassiana* has received registration for greenhouse use in the USA, and may be useful in controlling whiteflies, aphids and western flower thrips (Murphy et al., 1998).

Nematodes are currently exempt from the safety requirements mandated by the United States Environmental Protection Agency for other pathogens, so there is considerable interest in exploiting these as biological control agents in greenhouses. Commonly used species are *Steinernema feltiae* (=*bibionis*), *S. carpocapsae* and *Heterorhabditis bacteriophora* (=*heliothidis*), with major emphasis on control of fungus gnats and root weevils (Bedding and Miller, 1987; Nedstam and Burman, 1990; Lindquist, 1995). Application rates and frequency of application vary, but generally 120 000 nematodes per square meter one to nine times during a crop can provide good control. Control of fungus gnats require multiple applications. Nematodes have also been used successfully against the larval stage of leafminers in chrysanthemum leaves (Harris, 1988; Williams, 1993; Sher and Parrella, 1996), although this is not recommended commercially. Similar results were reported by the Vineland Agricultural Station in Ontario, Canada (Bruce Broadbent, personal communication), but the strain of nematode used was critical to obtaining successful control of leafminers.

Selection of arthropod natural enemies

The two most widely used arthropods for biological control in greenhouses are *Encarsia formosa* for biological control of whiteflies and *Phytoseiulus persimilis* for biological control of the two-spotted spider mite. Neither of these was discovered as a result of rigorous laboratory and greenhouse studies; rather, they were found by serendipity. In 1926, a tomato grower in England brought some parasitized whitefly scales to the attention of a researcher in the United Kingdom, and the era of biological control in greenhouses was born. In 1960, a German scientist found *P. persimilis* in Germany in a shipment of oranges from Chile (Hussey and Scopes, 1985). There is no doubt that detailed studies on the biology and ecology of these natural enemies has contributed enormously to their successful use in greenhouses. However, little has been learned from the successes of *E. formosa* and *P. persimilis* about what attributes characterize a successful natural enemy. As new pests invade greenhouses and integration with chemical control becomes more critical, there is a greater emphasis on searching for other effective natural enemies.

Considerable controversy exists over the appropriate selection criteria to use when searching for and screening natural enemies. This is understandable considering that the mechanisms leading to successful biological control are not understood.

Van Lenteren (1980) suggested that the trial-and-error approach for selecting and releasing natural enemies would probably lead to successful selection of a good natural enemy rapidly and with minimum cost. In addition, this approach would probably be as accurate as conducting more detailed biological and ecological studies designed to compare natural enemies and select the best one. Van Lenteren (1980) proposed a three-step approach to evaluating natural enemies which eliminates the need for detailed laboratory evaluations prior to making trial releases directly into the greenhouse: 1) can a parasitoid species discover pest-infested patches throughout the greenhouse before economic damage is reached?; 2) is it able to reduce pest numbers sufficiently after locating a patch?; and 3) is it able to keep the pest density below the economic threshold throughout the growing season? However, such studies are clearly very time-consuming and, as discussed earlier, they must be done in large commercial-size greenhouses, because of the differences reported for biological control in small and large greenhouses (van Lenteren, 1986). A more promising approach may be a combination of laboratory studies together with actual greenhouse releases; these may provide the necessary data to evaluate different natural enemies efficiently (Heinz and Parrella, 1994a,b).

A thorough review of criteria for selecting parasitoids, based on the type of natural-enemy release, can be found in van Lenteren (1986), and references therein. Many of these criteria fall under the reductionist approach to selection (Waage, 1990), where agents are chosen on the basis of particular biological attributes or life-history traits. However, many of these characteristics are not independent, so attempting to measure them in isolation is of questionable value. It is more important to evaluate and measure these characteristics as a function of what they contribute to reducing pest populations in the greenhouse. One example is the calculation of intrinsic rate of increase for the natural enemy, which may include fecundity, longevity, handling time, rate of egg production, oviposition rate, host-stage preferences, larval survival, etc. All of these are integrated to provide an estimate of the potential of a natural enemy to control a pest.

One of the most widely measured features of all biological control agents is the functional response. Typical functional-response experiments are conducted by confining the natural enemy with a known number of pests and by recording selected natural-enemy characteristics (e.g., attack rate) under specific criteria. Such studies are usually not very difficult to conduct, and may yield valuable data. For example, when a natural enemy is unable to kill most of the prey in the test arena when the density is similar to what might be encountered in the greenhouse, then this natural enemy may not be worth examining under more challenging circumstances. Thus, as noted by Minkenberg (1990), one of the goals of a preselection study is to identify those natural enemies likely to fail as biological control agents. Predicting which biological control agent will be the best to use is far more difficult. More detailed studies of the functional response, where measurements are taken to determine how well a natural enemy is able to find patches of prey, and whether it avoids leaving these areas after an initial encounter could prove valuable as evaluation criteria for effective natural enemies (Kareiva, 1990). These aspects are inherent in the three evaluation criteria discussed earlier (van Lenteren, 1986). However, Kareiva (1990) stressed that the issue is not whether a natural enemy aggregates to pest outbreaks (i.e. generates spatially density-dependent mortality differing from place to place), but how rapidly it does so. Such a criterion is of vital concern in an ornamental crop where the acceptable level of pests and their damage are very low. Inundative releases may render this a moot point, but where the actual release spans the definitions of inoculative and inundative, response time may still be important. Noldus and van Lenteren (1990) tried to explain observed levels of parasitism in the *Encarsia formosa*–whitefly system in tomato greenhouses taking into account the spatial distribution of the prey and behavior of the parasite. They found that *E. formosa* was not attracted to infested plants over long distances, but did aggregate in areas of high host density because honeydew associated with these whiteflies increased parasitoid search and residence times. However, density-dependent mortality of whiteflies did not occur, regardless of the scale used (greenhouse, plant, or leaf).

A mechanistic approach to understanding natural-enemy foraging in a greenhouse, and how it responds to the pest population in time and space needs to be done for most of the natural enemies used in biological control programs. An understanding of whether or not a natural enemy responds to chemical cues in the greenhouse environment is an important aspect of its foraging behavior. This is currently an active area of research – integration of chemical ecology with population biology provides a powerful tool to evaluate the characteristics of successful natural enemies. Roland (1990) proposed two approaches for experimentation: 1) observe a pattern of search and host mortality over several generations when a chemical cue is present or absent; and 2) identify and utilize chemical cues themselves in order to manipulate search behavior and pattern of attack. The greenhouse environment has been used as a test arena for some of this research, especially the latter experimentation method (Roland, 1990), but little work has been done with natural enemies of economically important greenhouse pests.

The use of mathematical simulation models to evaluate natural enemies as potential biological control agents is gaining wider acceptance. As suggested by Ehler (1990), modeling has the potential to explain what will be observed if a particular release strategy is adopted or a particular natural enemy is released. The utility of these simulation models can be expressed in several ways:

(1) they may tell whether random search is sufficient for pest suppression over the growing season. If this is true, then detailed studies on searching behavior would not be a priority in the evaluation process (van Lenteren, 1986);
(2) the relative potential of two biological control agents can be contrasted by running the model with the known biological attributes of each natural enemy and then examining the impact on the pest population over several generations. Depending on the system and definition of successful biological control, the superior natural enemy can be determined (Waage, 1990). Simulation models can also be used to enhance the predictability and economic feasibility of biological control (Parrella et al., 1992; Heinz et al., 1993).

Mass-rearing, availability and quality of natural enemies

These factors continue to be major impediments

to the wide-scale adoption of biological control in greenhouse ecosystems. The broad area of mass-rearing is beyond the scope of this chapter, and more information can be found in Hussey and Scopes (1985); van Lenteren (1986, 1996) and van Lenteren and Woets (1988). Excellent methods for mass-production of many of the species listed (Table 9.3) have been developed. Mass-rearing of natural enemies has become commercially profitable, making any innovation in older rearing techniques or the development of rearing methodology for new natural enemies proprietary information. Although this is inevitable and a sign that biological control has attained significant success, it clearly stifles the exchange of information and slows down expansion into new areas. This is especially true as new pests continue to appear and others develop resistance to selective insecticides [e.g., *Aphis gossypii* to pirimicarb: Jacobson (1990)]. To be judged a success, any natural enemy must be available in large quantities and on almost a year-round basis. Because of this, the ease with which a natural enemy can be mass-reared must be considered a priority in evaluating potential natural enemies (Waage, 1990).

Although availability and supply of natural enemies have long been limiting factors in the widespread adoption of biological control in greenhouse and other agroecosystems, there has been tremendous growth in the number and size of commercial insectaries in many parts of the world, especially in North America. This growth should increase the supply of many species of natural enemies, and potentially decrease their prices. The number of North American producers of natural enemies increased from 53 in 1985 to 142 in 1997 and, more importantly, the number of species available increased from 45 to 130 during this same time period (Hunter, 1997). There are at least ten producers of *Encarsia formosa* in Europe alone (Onillion, 1990).

Considerable attention has been focused on developing realistic quality-control criteria for commercial insectaries (van Lenteren, 1993). A concern over quality control was one of the reasons for the formation of the "Association of Natural Biocontrol Producers" in California in 1990. The purpose of this association is to provide a unified approach for the protection and improvement of the natural-enemy production industry on a regional, national, and international basis.

With the universal problems associated with packaging and shipping, the arrival of natural enemies in good condition is something that should not be taken for granted. When natural enemies are shipped as active adults, a cursory glance can reveal whether the predators or parasites are alive upon receipt. However, this may be insufficient to reveal quality-control problems such as were experienced with predaceous mites in Finland (Markkula et al., 1987). The assessment of quality control of natural enemies becomes more complicated if other stages, such as pupae, are shipped and received by the grower. For example, in a trial using *Encarsia formosa* for whitefly control, the parasitized pupae might appear normal upon receipt, but examination of the pupae after placement in the greenhouse could reveal problems with adult emergence. In a biological control trial using *E. formosa* against sweet-potato whitefly (*Bemisia tabaci*), less than 40% emergence of adult parasitoids from cards was observed on some dates (Parrella et al., 1991).

The scientific measurement of overall quality of mass-reared insects has received considerable attention from a number of researchers [Boller and Chambers (1977), King et al. (1985), and Leppla and Ashley (1989), among others]. The division of quality into four categories – adaptability, sexual activity, host selection, and mobility (King et al., 1985) – combines many of the other suggested evaluation criteria. These categories should serve as useful guidelines for evaluating natural-enemy quality in large mass-rearing programs.

Introduction methods and release rates

Considerable variation exists among methods of introduction and rates of release for most of the commercially available natural enemies. With respect to the whitefly parasite *Encarsia formosa,* the only stage released in the greenhouse is the pupa (within a parasitized whitefly), but there are more than ten different recommendations describing release options (Onillion, 1990). These include:

(1) the "pest in first" method (Hussey and Scopes, 1985) where a uniform release of whiteflies (10 adults per 100 tomato plants) is followed by a release of parasitoids (400 in three waves). It is difficult to win grower acceptance of this method, and since comparisons of this method with releases of natural enemies soon after the pest is detected, have yielded similar results (Markkula and Tiittanen, 1976), the "pest in first" method is rarely used. However this method, which artificially creates a uniform distribution of the prey in space, offers intriguing possibilities to examine characteristics of natural-enemy foraging behavior;

(2) a technique in which "storage" plants which contain the pest (or an alternate host for the natural enemy) and natural enemy in a stable equilibrium are introduced into the greenhouse. Stacy (1977) suggested using 50 to 90 of these plants per hectare, with each plant producing 8000 *E. formosa* over a period of 8 weeks;
(3) a preventative or multiple-introduction method, which is a true inundative release as described by Debach (1964). Because this method is based on repetitive releases before any whiteflies are detected, the objective is not to establish the parasite in the greenhouse. No less than six releases of 35 000 *E. formosa* per hectare are necessary;
(4) a method in which releases of *E. formosa* are made at precise dates over several weeks either ignoring whitefly density (Nilsson, 1976) or taking whitefly density into consideration (Onillion et al., 1976). The latter method requires 6–40 or 1–16 *E. formosa* per m^2 or per plant, respectively;
(5) releases of *E. formosa* against predetermined whitefly populations are made on a per plant basis (five parasitoids per plant if the pest density is between 0.3 and 1 adult per plant [Stenseth, 1983]) or in relation to overall density of adult whiteflies in the greenhouse. In all situations, if more than five adult whiteflies are present per plant, then chemical control is recommended (Onillion, 1990).

With so many different recommendations, it is difficult to determine which predominates in practice. It seems that each individual country and natural-enemy producer has its own formula for success. In all situations where biological control is in the hands of the private sector, prescribed recommendations are subject to change based on the conditions in specific grower operations and on the past experience of the practitioner.

Almost all other release methods for natural enemies used in greenhouses fall under one of the categories discussed above for *E. formosa.* For example, the banker plant or rearing-unit system is employed to release aphid parasites and predators (Hofsvang and Hagvar, 1979; Hansen, 1983) and has recently been used to release predaceous mites (*Amblyseius* spp.) for thrips control (Jacobson, 1990). A very common method is to release natural enemies in numbers per plant or per unit area when the pest population is at low levels; often some attempt is made to establish a pre-determined natural-enemy/pest ratio. This methodology has been used to release predaceous mites for control of spider-mites (Hamlen and Lindquist, 1981; Smitley, 1990) and green lacewing larvae (*Chrysoperla* spp.) for control of aphids (Tulisalo et al., 1977). Inundative releases are being explored for the control of leafminers by parasites (Heinz and Parrella, 1990c) and for control of thrips by predaceous mites (Gilkeson et al., 1990).

It is difficult to determine accurately the dosage, frequency, and distribution of natural-enemy releases in the greenhouse to effect successful biological control. As indicated by van Lenteren and Woets (1988), most programs rely on empirical data rather than on calculations based on dispersal behavior and reproductive capacity of the natural enemy. Results from studies designed to investigate the mechanisms of population regulation of whiteflies by *Encarsia formosa* prompted Noldus and van Lenteren (1990) to suggest a re-evaluation (downward) of the numbers of parasites needed. When a density-dependent numerical response for *E. formosa* attacking greenhouse whitefly was not observed and successful biological control still occurred, it suggested that release rates could be reduced. More detailed studies such as this will likely change recommendations for releasing natural enemies, and place biological control on a more scientific basis.

An attempt has been made to upgrade the predictability of inundative releases through the development of a mechanistic mathematical model which provides answers to some basic questions regarding biological control of the leafminer *Liriomyza trifolii* by the parasite *Diglyphus begini* (Parrella et al., 1992). The model itself incorporates data collected over ten years on the biology of the leafminer and parasite. Of particular importance is the realistic depiction of parasite impact on the leafminer population through inclusion of age structure in the model. This permits the utilization of specific behavioral parameters for the leafminer and parasitoid, which are extremely important to population dynamics. Age-specific survivorship, fecundity and development rates for *L. trifolii* have been included in the model (Parrella, 1987). These parameters, as well as age-specific utilization parameters and sex ratio for *Diglyphus begini,* were added to the model (Heinz and Parrella, 1989, 1990a,b).

Details of the model can be found elsewhere (Parrella et al., 1992; Heinz et al., 1993), but its overall purpose is to predict the number of parasites which must be released weekly to bring the leafminer population under control within 40 days of crop initiation. This time interval coincides with the period when the chrysanthemum plant does not yet have marketable foliage (Parrella and Jones, 1987). Successful biological control is defined as the maintenance of a leafminer population of no more than one susceptible leafminer larva per 1000 leaves from day 40 after crop initiation. The intent is to keep marketable foliage on the plant free of leafminer injury. It is unlikely that the predictive model will ever be able to estimate the exact number of parasites to release in order to obtain effective control; however, it offers an additional tool in the decision-making process, and has distinct advantages over the subjective criteria currently used.

Evaluation methods after introductions of natural enemies

The efficacy and economics of inoculative and inundative releases of natural enemies for biological control on most greenhouse crops have not been demonstrated. Although recommendations can be found for almost all the natural enemies found in Table 9.3, quite often there have been no scientific studies on which to base these recommendations. This is not to imply that the recommendations do not work; on the contrary, there is ample evidence from practice that success can be achieved regularly. Nonetheless, the lack of scientific studies is still very surprising. Whether such successes can continue without a good scientific approach is questionable, especially in light of the spread of new greenhouse pests, many of which (e.g. the western flower thrips, *Frankliniella occidentalis*) feed directly on marketable vegetables and flowers.

The logistics of conducting large-scale greenhouse studies in which the impact of natural-enemy releases are evaluated are formidable, even with substantial fundamental knowledge about the natural enemy. The following elements must be in place for studies to be scientifically valid: 1) studies should be replicated using similar greenhouses or isolated greenhouse sections, which should be chosen carefully to avoid the problem of statistically significant plot×result interactions; 2) statistically accurate sampling plans must be in place for the pest and natural enemy; 3) some method must be available to exclude the natural enemy in subplots (perhaps by the use of cages) so that population development of the pest can be followed in the absence of the natural enemy [see Luck et al. (1988) for other methods]; and 4) methods must be available to separate mortality due to other causes from that caused by the natural enemy released (Carey and Krainacker, 1988). Satisfying these requirements can be extremely difficult. For example, if the natural enemy released is already present in the greenhouse from previous crops, or similar genera are present, separating the impact of what is released from what is already there becomes difficult. In all cases the experimental design is critical, and the work will require considerable time, manpower, and expense. These factors undoubtedly contribute to the lack of research in this area.

One general concern is the lack of statistically accurate sampling plans for greenhouse pests and their natural enemies. This area seems to have been ignored by most researchers, even though it is often considered the fulcrum on which a program for integrated pest management pivots. On ornamental crops grown for their aesthetic value, where any pest density is considered unacceptable, it is understandable that only limited sampling plans have been developed (Parrella et al., 1989). However, it is surprising that, even on vegetable crops, statistically accurate sampling plans for most pests and natural enemies do not exist.

As summarized by Ekbom and Xu (1990), more sampling data have been developed for whiteflies than for any other greenhouse vegetable pest. Understanding the spatial pattern of the whiteflies is an important first step in developing a sampling plan. It has become standard practice to utilize Taylor's power law (Taylor, 1961) and Iwao's patchiness regression (Iwao, 1968) to describe the relationship between the variance and the mean of a population. These data are lacking for most greenhouse pests. However, even with the data on whiteflies, few authors have used these regressions to describe their data; often distributional data have not been collected over a wide enough range of densities, host plants or cultivars, and involving enough sampling sites or localities, to be of general utility. The size of the sampling area is also an important factor (Noldus et al., 1986), and this has made it difficult to compare different studies which have developed distributional data on whiteflies in greenhouse tomatoes.

Although some data have suggested that random sampling for greenhouse whiteflies is not feasible, and the use of yellow sticky traps too coarse a sampling method (Noldus et al., 1986), there are other interpretations which are more optimistic (Hassan, 1983; Yano, 1987; Ekbom and Xu, 1990). Nonetheless, considerable research remains to be done for both silverleaf and greenhouse whiteflies. Studies with other pests, such as western flower thrips (Shipp and

Zariffa, 1990), have only recently been initiated. The possible development of stratified random sampling plans, binomial sampling plans, and studies into the utility of yellow traps for monitoring adults, are a few of the areas worth exploring for most greenhouse pests.

Recent work with spider mites has demonstrated that it is feasible to utilize generic coefficients from Taylor's power law derived from many different crops to develop plans for binomial sampling with constant precision on crops where mites are a problem (Jones, 1990b). This was done with roses and carnations, and the results were validated (Jones, 1990b; Zhang and Sanderson, 1995). The use of generic coefficients may be possible for other pests such as aphids and leafminers; this could rapidly increase the development of sampling plans on greenhouse crops.

Options in the development of integrated pest management programs

Much of what has already been covered in this review can be combined to form components of an IPM program. For example, breeding for host-plant resistance, managing insecticide resistance, biological control, development of sampling plans, etc., all form important parts of a pest-management program. Although full IPM programs for greenhouse-grown crops are rare (Lindquist et al., 1979; Costello et al., 1992; Cross et al., 1983; Parrella and Jones, 1987; Fransen, 1993), much of what is published under the guise of biological control often includes the components of a good IPM program (Hussey and Scopes, 1985). For biological control to be successful it must be done within the context of an integrated pest management program.

Cultural controls

Cultivation of most crops can be manipulated with the overall intent of reducing pest populations, as is attempted with host-plant resistance; however, these are far too extensive to cover in detail here. Methods which have been shown to have a dramatic effect on pest populations include specific irrigation regimes, soil type, misting systems, and fertilizer applications (Poe et al., 1976; Lindquist et al., 1985, 1987; Minkenberg and Ottenheim, 1990). A different spatial arrangement of plants on a bench may provide better penetration and coverage of pesticide sprays, or allow more efficient movement by a natural enemy. Because this may also mean fewer plants per bench and therefore less production, the cost-effectiveness of such a strategy has yet to be determined. Keeping the greenhouse clean and free of wet areas, and prompt removal of discarded plant material and weeds, will provide an environment less conducive to pest problems (Parrella and Jones, 1987; Ascerno, 1990). Proper crop scheduling, avoiding planting a crop adjacent to where an older crop is being harvested, and infrastructural organization can dramatically reduce pest populations on successive crops. As greenhouse production systems become more sophisticated – for example, rearing plants without soil in substrates such as rock-wool, utilizing nutrient-film techniques, or employing ebb-and-flow watering systems (among others) – they will have an impact on pest population development and on IPM and biological control strategies. Crop production systems must be taken into account when examining pest population development and in developing control strategies.

Physical and mechanical controls

Sterilizing the soil between crops can substantially reduce the movement of pests from one crop to the next. Much higher populations of leafmining pests have been observed in chrysanthemums where growers did not sterilize between crops (Parrella and Jones, 1987). The options for practical soil sterilization are diminishing with the gradual abandonment, because of environmental concerns, of methyl bromide and other effective soil sterilants. In some countries, the complete elimination of preventative chemical soil disinfection is being mandated by law (van Lenteren, 1990). Other options for growers include safer chemicals, steam sterilization, or producing in a soilless substrate. Initiation of a crop in a greenhouse free of pests is very important to overall management of pest populations. In the Caribbean, Central and South America many growers are unable to sterilize between crops because of the lack of necessary equipment and chemicals; as a result there is more of a problem with pests moving from one crop to the next.

The physical structure of a greenhouse does afford some protection from movement of pests, but complete exclusion may be possible with the use of insect screens. The idea of screening greenhouses has been a concept for many years, and early studies suggested treating the screens with a residual insecticide (such as dieldrin[7]) to kill insects trying to pass through.

[7] (1*R*,4s,4a*S*,5*R*,6*R*,7*S*,8*S*,8a*R*)-1,2,3,4,10,10-hexachloro-1,4,4a,5,6,7,8,8a-octahydro-6,7-epoxy-1,4:5,8-dimethanonaphthalene.

However, practical application of screens had not been addressed seriously until the problem with western flower thrips and tomato spotted wilt virus emerged in greenhouses (Baker and Jones, 1989; Jones, 1990a; Robb, 1990). The size of the widest portion of the thorax of the western flower thrips is approximately 0.25 mm; the flexibility of the insect's body enables it to pass through even smaller openings. However, 100-mesh screen (pore size of 0.19 mm) and larger screening (pore size 0.28 mm) where the holes are extruded and male side is positioned to face the migration source are capable of excluding western flower thrips. Such small screens greatly reduce air flow (Jones, 1990a; Sase and Christianson, 1990). In addition, an efficient method of cleaning these screens must be available because they very easily become dirty and blocked.

The use of screening may be difficult for many growers when the entire greenhouse is to be screened. However, it becomes much more feasible when only one section of the greenhouse is considered for screening (e.g., mother-block production, rooting areas, etc.). Thus, screening can be used to prevent movement from one portion of the greenhouse to another without greenhouse-wide reduction in air flow.

Very ingenious methods of blocking the entrance of insects into the greenhouse have been developed (Kuack, 1989). One such method involves modifying an evaporative cooling pad to screen out pests. Prior to the air passing through the pad it is drawn through a cloth prefilter and then through a foam-rubber filter which is kept wet with water delivered from high-pressure mist nozzles. The resistance of the filter, cooling pad and screen was accounted for when the total area through which the air was to pass was determined. Clearly such a modification requires major expense and considerable engineering skill. The production of thrips- and virus-free plant material is the potential reward, but this is only possible when used in the context of the greenhouse as a specialized quarantine facility with strictly controlled admittance, and mandatory change of clothing for workers before entry, among other factors.

Chemical and biological control

Chemical and biological control has each been discussed separately, but it is the integration of the two that is critical to the success of an IPM program in greenhouses. In ornamental production where pesticide use is so intensive, the integration of pesticides with biological control agents is especially important for successful IPM. The development of data on the compatibility of pesticides with biological control agents is fraught with problems, largely because of the differences in the methodology of testing methods from researcher to researcher. This was recognized early in Europe, and in 1974 a working group, "Pesticides and Beneficial Arthropods"', of the International Organization of Biological Control, Western Palearctic Regional Section, was formed. This group has been developing methodology for rearing and testing, evaluating important pesticides for their compatibility with natural enemies (Hassan and Oomen, 1985; Samsoe-Petersen et al., 1989), and publishing information jointly (Hassan et al., 1983).

Other information is published routinely by scientists not in the working group (Croft, 1990); many of the private commercial insectaries also have publications detailing their own data on compatibility. With considerable information already available, with fewer new pesticides being registered, and with the abandonment of existing materials, one might expect that data on compatibility are nearly complete. However, this is not the case, and many materials remain to be evaluated. In addition, many of these studies must be repeated as new natural enemies with control potential appear. Furthermore, as new application methodology develops and is adopted by the industry (such as low-volume application), direct extrapolation of older data on compatibility may not be possible. One positive point is that many of the pesticides obtaining registration in the future will have less impact on the environment (lower mammalian and fish toxicity, less persistence in soil and water, shorter residual times) than many of the traditional pesticides registered today such as organophosphates and pyrethroids. While this does not mean that compatibility with natural enemies is automatic, such compatibility may occur and is very encouraging for the future integration of pesticides and natural enemies. However, compounds which have little or no impact on the natural enemy in laboratory assays may have an impact when used in the greenhouse. An actively searching natural enemy may respond in a way that cannot be predicted from laboratory studies. Effects on searching behavior, possible sublethal effects due to pesticides, and repellency toward natural enemies are extremely difficult to determine in laboratory bioassays. Continued research in this area is critical to the successful implementation of IPM in greenhouses.

Utilization of natural enemies resistant to one

or more pesticides would increase their chances of survival and hence improve compatibility. Success has been achieved in this area with *Chrysoperla* sp. strains which have exhibited a resistance factor seven times the field application rates of certain carbamates and pyrethroids (Grafton-Cardwell and Hoy, 1986). Extensive research has been focused on the development and field release of predaceous mites showing resistance to one or more pesticide materials (Schulten, 1980; Field and Hoy, 1984, 1986). These predators provided control of two-spotted spider mite in roses, and resistant *Metaseiulus occidentalis* were found in a rose greenhouse two years after initial releases (Field and Hoy, 1984, 1986).

Weapons in the IPM arsenal

Many of the pest-management options discussed for greenhouses have counterparts in field agriculture. There are only a few options, like screening, that take advantage of the unique greenhouse environment. In contrast, current pest-management technology in the field often is only tested and practiced in greenhouses after many years.

Although members of the order Lepidoptera are major pests in greenhouses, the possible use of pheromones as a control strategy through male confusion has hardly been evaluated. This technique has been successful against a number of major Lepidopteran pests in fruit and nut orchards. Van den Bos and van de Vrie (1981) suggested that the possibilities for success of this technique are much greater in the greenhouse than in the field, provided that moth density is not so high that random encounters and matings occur. Although promising, this idea has not been tested in the greenhouse.

The possible use of mass-trapping of insect pests together with the release of natural enemies offers promise in certain situations (Webb and Smith, 1979; van de Veire and Vacante, 1984). Although severely limited in its application (Parrella et al., 1989), this methodology deserves further study to define conditions under which such control strategies are feasible.

Genetic control of greenhouse pest populations has received scant attention. Van den Bos and van de Vrie (1981) suggested releasing males of the field strain of *Clepsis spectrana* into greenhouses with the objective of introducing diapause characteristics into the greenhouse population. Genetic control of *Tetranychus urticae* has been contemplated in the Netherlands for many years. Feldman (1980) speculated that incompatibility resulting from a cross between a translocated hemizygous male and a wild-type female could dramatically reduce fertility of the population. This effect would even be more dramatic if the translocated male was resistant to an acaricide to which the greenhouse population was still susceptible. Obviously some serious safeguards would be needed, but clearly this is an exciting approach; the control potential of this strategy still remains to be tested.

Modifying the greenhouse environment to benefit biological control of *Trialeurodes vaporariorum* and *Planococcus citri* has been suggested (Helgesen and Tauber, 1976; Copeland and Varley, 1987). A great deal of information must be known about the thermal requirements of the pest and natural enemy, but once in place, manipulation of abiotic conditions to benefit the natural enemy may be possible. Although rarely used in practice today, this is a method which truly takes advantage of the unique greenhouse ecosystem. Considerably more research in this area and practical application of this concept are likely in the future.

CONCLUSIONS

Considerable progress has been made toward understanding the arthropod complex occurring in greenhouses throughout the world. However, the importation of new pests and reduced reliance on chemicals to control pest populations will be a major challenge to researchers as growers demand answers well before any data are available. Thus, short-term crisis solutions must be sought while more long-term and permanent control strategies are being developed.

The unique greenhouse environment affords the opportunity to examine fundamental areas such as population genetics in relation to insecticide resistance, local adaptation to host plants or the environment, and population dynamics (e.g., invasion, migration and dispersal, population structure), behavior (e.g., foraging, mate selection, etc.), novel examples for biological control and plant-insect interactions (e.g., host-plant resistance, induced response, compensation, etc.), and sampling theory (e.g., dispersion indices, horizontal and vertical stratification, etc.). These are exciting areas of research which have both applied and basic components and clear parallels in field agriculture.

REFERENCES

Al-Abassi, S.H., Wiegle, J.L. and Hart, E.R., 1987. Biological interactions between New Guinea impatiens and the two-spotted spider mite (Acari: Tetranychidae). *J. Econ. Entomol.*, 80: 47–50.

Alverson, D.R. and Gorsuch, S.R., 1982. Evaluation of chrysanthemum cultivars and insecticides for control of damage by a leafminer, *Liriomyza trifolii* (Diptera: Agromyzidae). *J. Econ. Entomol.*, 75: 888–891.

Anonymous, 1990. Bumble bee factory. *Grower,* 114: 15–17.

Ascerno, M.E., 1990. Arthropod problems in the soil. In: M.P. Parrella and J. Hall (Editors), Proc. Sixth Conf. on Insect and Disease Management on Ornamentals. Society of American Florists, Alexandria, VA, pp. 119–124.

Avilla, J., Artigues, M., Sarasua, M.J. and Albages, R., 1990. A review of the biological characteristics of *Encarsia tricolor* and their implications for biological control. *Bull. Int. Org. Biol. Integr. Control of Noxious Plants and Animals/West Palearctic Regional Section,* XIII: 14–18.

Baker, J.R. and Jones, R.K., 1989. Screening as part of insect and disease management in the greenhouse. *N.C. Flower Growers Bull.*, 34: 1–9.

Beck, S.D., 1965. Resistance of plants to insects. *Annu. Rev. Entomol.*, 10: 207–232.

Bedding, R.A. and Miller, L.A., 1987. Use of a nematode, *Heterorhabditis heliothidis,* to control black vine weevil, *Otiorhynchus sulcatus,* in potted plants. *Ann. Appl. Biol.*, 99: 211–216.

Beddington, J.R., Free, C.A. and Lawton, J.H., 1978. Characteristics of successful natural enemies in models of biological control of insect pests. *Nature,* 273: 513–519.

Berlinger, M.J., Mordechi, S. and Leeper, A., 1991. Application of screens to prevent whitefly penetration into greenhouses in the Mediterranean basin. *Bull. IOBC/WPRS,* 1991/XIV/5: 105–110.

Bilderback, T. and Mattson, R., 1977. Whitefly host preference associated with selected biochemical and phenotypic characteristics of poinsettias. *J. Am. Soc. Hortic. Sci.,* 102: 327–331.

Blackman, R.L. and Eastop, V.F., 1985. *Aphids on the World's Crops.* Wiley, New York, 466 pp.

Boller, E.F. and Chambers, D.L., 1977. Quality aspects of mass-reared insects. In: R.L. Ridgway and S.B. Vinson (Editors), *Biological Control by Augmentation of Natural Enemies. Insect and Mite Control by Parasites and Predators.* Plenum Press, New York, pp. 219–235.

Broadbent, A.B. and Blom, T.J., 1984. Comparative susceptibility of chrysanthemum cultivars to *Liriomyza trifolii* (Diptera: Agromyzidae). *Proc. Entomol. Soc. Ont.*, 114: 91–93.

Byrne, D.N., Bellows Jr., T.S. and Parrella, M.P., 1990. Whiteflies in agricultural systems. In: D. Gerling (Editor), *Whiteflies: Their Bionomics, Pest Status and Mangement.* Intercept, Ltd, Andover, Hants, UK, pp. 227–261.

Carey, J.R. and Krainacker, D., 1988. Demographic analysis of mite populations: Extensions of stable theory. *Exp. Appl. Acarol.*, 4: 191–210.

Chase, A.R. and Osborne, L.S., 1983. Influence of insecticidal soap on several diseases of foliage plants. *Plant Disease,* 67: 1021–1023.

Chase, A.R., Osborne, L.S. and Ferguson, V.M., 1986. Selective isolation of the entomopathogenic fungi *Beauveria bassiana* and *Metarhizium anispoliae* from an artificial potting medium. *Fl. Entomol.*, 69: 285–292.

Copeland, M.J.W. and Varley, M.J., 1987. Progress in developing a controlled glasshouse environment to promote biological pest control. *Bull. Int. Org. Biol. Integr. Control of Noxious Plants and Animals/West Palearctic Regional Section,* X: 41–45.

Costello, R.A., Elliot, D.P., Gilkeson, L.A. and Gillespie, D.R., 1992. *Integrated control of greenhouse pests.* Province of British Columbian Ministry of Agriculture, Fisheries and Food, 19 pp.

Craig, R., Mumma, O., Gerlhold, D.L., Winner, B.L. and Snetsinger, R., 1986. Genetic control of a biochemical mechanism for mite resistance in geraniums. In: M.B. Green and P.A. Hedin (Editors), *Natural Resistance of Plants to Pests,* American Chemical Society Symposium Series Number 296, 243 pp.

Croft, B.A., 1990. *Arthropod Biological Control Agents and Pesticides.* Wiley, New York, 723 pp.

Cross, J.V., Wardlow, L.R., Hall, R., Saynor, M. and Bassett, P., 1983. Integrated control of chrysanthemum. *Bull. Int. Org. Biol. Integr. Control of Noxious Plants and Animals/West Palearctic Regional Section,* VI: 181–185.

de Jaeger, C.M. and Butot, R.P.T., 1993. Chrysanthemum resistance to two types of thrips (*Frankliniella occidentalis* Pergande) feeding damage. *Proc. Exp. Appl. Entomol. Neth. Entomol. Soc.,* 4: 27–32.

de Ponti, O.M.B., 1980a. Breeding cucumber for resistance to two-spotted spider mite (*Tetranychus urticae*). In: A.K. Minks and P. Gruys (Editors), *Integrated Control of Pests in the Netherlands.* Centre for Agricultural Publishing and Documentation, Wageningen, pp. 191–195.

de Ponti, O.M.B., 1980b. Breeding glabrous cucumber (*Cucumis sativus*) varieties to improve the biological control of the greenhouse whitefly (*Trialeurodes vaporariorum*). In: A.K. Minks and P. Gruys (Editors), *Integrated Control of Pests in the Netherlands.* Centre for Agricultural Publishing and Documentation, Wageningen, pp. 197–199.

de Ponti, O.M.B. and Hogenboom, N.G., 1980. Breeding tomato (*Lycopersicon esculentum*) for resistance to greenhouse whitefly (*Trialeurodes vaporariorum*). In: A.K. Minks and P. Gruys (Editors), *Integrated Control of Pests in the Netherlands.* Centre for Agricultural Publishing and Documentation, Wageningen, pp. 187–190.

de Ponti, O.M.B., van Lenteren, J.C. and Sabelis, M.W., 1986. Breeding cucumbers and tomatoes for improved biological control of whiteflies and spider mites. *Hortscience,* 21: 603.

de Ponti, O.M.B., Romanow, L.R. and Berlinger, M.J., 1990. Whitefly–plant relationships: plant resistance. In: D. Gerling (Editor), *Whiteflies: Their Bionomics, Pest Status and Management.* Intercept, Ltd, Andover, Hants, UK, pp. 91–106.

Debach, P., 1964. *Biological Control of Insect Pests and Weeds.* Chapman and Hall, London, 844 pages.

Eggenkamp-Rotteveel Mansveld, M., Ellenbroek, F.J.M. and van Lenteren, J.C., 1982. The parasite–host relationship between *Encarsia formosa* (Hymenoptera: Aphelinidae) and *Trialeurodes vaporariorum* (Homoptera: Aleyrodidae). XII. Population dynamics of parasite and host in a large commercial glasshouse and test of the parasite introduction method. *Z. angew. Entomol.*, 93: 113–130.

Ehler, L.E., 1990. Some contemporary issues in biological control of insects and their relevance to the use of entomopathogenic nematodes. In: R.G. Gaugler and H. Kaya (Editors), *Entomopathogenic Nematodes in Biological Control.* CRC Press, Boca Raton, FL, pp. 1–19.

Ekbom, B.S., 1977. Development of a biological control program for greenhouse whiteflies (*Trialeurodes vaporariorum* Westwood) using its parasite *Encarsia formosa* (Gahan) in Sweden. *Z. angew. Entomol.,* 84: 145–154.

Ekbom, B.S. and Xu, Rumei, 1990. Sampling and spatial patterns of whiteflies. In: D. Gerling (Editor), *Whiteflies: Their Bionomics, Pest Status and Management.* Intercept, Ltd, Andover, Hants, UK, pp. 107–121.

Elliot, D., 1993. *Biological Technical Manual.* Applied Bionomics Ltd, Westgrow Sales, Inc., 49 pp.

Elson, J.E., 1988. IR-4: Clearing the way for new pesticides. *Soc. Am. Florists Business News for the Floral Industry,* 5: 28, 46.

Feldman, A.M., 1980. Genetic control of greenhouse spider mite *Tetranychus urticae.* In: A.K. Minks and P. Gruys (Editors), *Integrated Control of Pests in the Netherlands.* Centre for Agricultural Publishing and Documentation, Wageningen, pp. 137–140.

Field, R.P. and Hoy, M.A., 1984. Biological control of spider mites on greenhouse grown roses. *Calif. Agric.,* 38: 29–32.

Field, R.P. and Hoy, M.A., 1986. Evaluation of genetically improved strains of *Metaseiulus occidentalis* (Nesbitt) (Acarina: Phytoseiidae) for integrated control of spider mites on roses in greenhouses. *Hilgardia,* 54: 1–31.

Fisher, S.J. and Shanks, J.B., 1974. Host preference of greenhouse whitefly: Effect of poinsettia cultivars. *J. Am. Soc. Hortic. Sci.,* 99: 260–262.

Flanders, S.E., 1951. Mass culture of California red scale and golden chalcid parasites. *Hilgardia,* 21: 1–42.

Flint, M.L., Driestadt, S.H., Rentner, J. and Parrella, M.P., 1995. Lady beetle release controls aphids on potted plants. *Calif. Agric.,* 49: 5–8.

Fransen, J.J., 1990. Natural enemies: Fungi. In: D. Gerling (Editor), Whiteflies: *Their Bionomics, Pest Status and Management.* Intercept, Ltd, Andover, Hants, UK, pp. 187–210.

Fransen, J.J., 1993. Development of integrated protection in glasshouse ornamentals. *J. Pesticide Sci.,* 36: 329–333.

Gilkeson, L.A., 1990a. Biological control of aphids in greenhouse sweet peppers and tomatoes. *Bull. Int. Org. Biol. Integr. Control of Noxious Plants and Animals/West Palearctic Regional Section,* XIII: 64–70.

Gilkeson, L.A., 1990b. Biological control in ornamentals: A producer's perspective. In: M.P. Parrella and J. Hall (Editors), *Proc. Sixth Conf. on Insect and Disease Management on Ornamentals.* Society of American Florists, Alexandria, VA, pp. 65–72.

Gilkeson, L.A., Morewood, W.D. and Elliot, D.E., 1990. Current status of biological control of thrips in Canadian greenhouses with *Amblyseius cucumeris* and *Orius tristicolor. Bull. Int. Org. Biol. Integr. Control of Noxious Plants and Animals/West Palearctic Regional Section,* XIII: 71–75.

Gough, N., 1990. Evaluation of miticides for the control of two-spotted mite *Tetranychus urticae* Koch on field roses in southern Queensland. *Crop Prot.,* 9: 119–127.

Gough, N., 1991. Long term stability in the interaction between *Tetranychus urticae* and *Phytoseiulus persimilis* producing successful integrated control on roses in southern Queensland. *Exp. Appl. Acarol.,* 12: 83–101.

Grafton-Cardwell, E.E. and Hoy, M.A., 1986. Selection of the common green lacewing for resistance to carbaryl. *Calif. Agric.,* 40: 22–24.

Hall, R.A., 1985a. Whitefly control by fungi. In: N.W. Hussey and N. Scopes (Editors), *Biological Pest Control. The Glass House Experience.* Cornell University Press, Ithaca, NY, pp. 116–118.

Hall, R.A., 1985b. Aphid control by fungi. In: N.W. Hussey and N. Scopes (Editors), *Biological Pest Control. The Glasshouse Experience.* Cornell University Press, Ithaca, NY, pp. 138–141.

Hamlen, R.A. and Lindquist, R.K., 1981. Comparison of two *Phytoseiulus* species as predators of two spotted spider mites on greenhouse ornamentals. *Environ. Entomol.,* 10: 524–527.

Hansen, L.S., 1983. Introduction of *Aphidoletes aphidimyza* (Rond.) (Diptera: Cecidomyiidae) from an open rearing unit for control of aphids in glasshouses. *Bull. Int. Org. Biol. Integr. Control of Noxious Plants and Animals/West Palearctic Regional Section,* VI: 146–150.

Hargreaves, J.R. and Cooper, L.P., 1980. Differential susceptibility of some gladiolus cultivars to gladiolus thrips, *Taeniothrips simplex* (Morison). *Queensl. J. Agric. Anim. Sci.,* 37: 161–163.

Harris, M., 1988. Biological control. *Soc. Am. Florists Business News for the Floral Industry,* 5: 24–25.

Hassan, S.A., 1983. A practical method to monitor pests and natural enemies in integrated control experiments under glass. *Bull. Int. Org. Biol. Integr. Control of Noxious Plants and Animals/West Palearctic Regional Section,* VI: 186–193.

Hassan, S.A. and Oomen, P.A., 1985. Integration of biological and chemical control of diseases and minor pests. In: N.W. Hussey and N. Scopes (Editors), *Biological Pest Control. The Glasshouse Experience.* Cornell University Press, Ithaca, NY, pp. 145–152.

Hassan, S.A., Bigler, F., Bogenschultz, H., Brown, J.V., Firth, S.I., Huang, P., Ledieu, M.S., Naton, E., Oomen, P.A., Overmeer, W.P.J., Rieckman, W., Samsoe-Peterson, L., Viggiani, G. and van Zon, A.Q., 1983. Results of the second joint pesticide testing programme by the IOBC/WPRS working group 'Pesticides and Beneficial Arthropods.' *Z. angew. Entomol.,* 95: 151–158.

Haynes, K.F., Parrella, M.P., Trumble, J.T. and Miller, T.A., 1986. Monitoring insecticide resistance with yellow sticky cards. *Calif. Agric.,* 40: 11–12.

Heins, R.D., 1990. Where are we in taking the guesswork out of growing? *Am. Floral Endowment Res. Rep.,* 1: 1–6.

Heinz, K.M. and Parrella, M.P., 1989. Attack behavior and host size selection by *Diglyphus begini* on *Liriomyza trifolii* in chrysanthemum. *Entomol. Exp. Appl.,* 53: 147–156.

Heinz, K.M. and Parrella, M.P., 1990a. The influence of host size on sex ratios in the parasitoid *Diglyphus begini* (Hymenoptera: Eulophidae). *Ecol. Entomol.,* 15: 391–399.

Heinz, K.M. and Parrella, M.P., 1990b. Holarctic distribution of the leafminer parasitoid *Diglyphus begini* (Ashmead) (Hymenoptera: Eulophidae) and notes on its life history attacking *Liriomyza trifolii* (Burgess) (Diptera: Agromyzidae) in chrysanthemum. *Ann. Entomol. Soc. Am.,* 83: 916–924.

Heinz, K.M. and Parrella, M.P., 1990c. Biological control of insect pests on greenhouse marigolds. *Environ. Entomol.,* 19: 825–835.

Heinz, K.M. and Parrella, M.P., 1991. The effect of leaf-mining by *Liriomyza trifolii* on seed set in greenhouse marigolds. *Ecol. Appl.,* 2: 139–146.

Heinz, K.M. and Parrella, M.P., 1994a. Biological control of *Bemisia argentifolii* (Homoptera: Aleyrodidae) infesting

Euphorbia pulcherrima: Evaluations of releases of *Encarsia luteola* (Hymenoptera: Aphelinidae) and *Delphastus pusillus* (Coleoptera: Coccinellidae). *Environ. Entomol.,* 23(5): 1346–1353.

Heinz, K.M. and Parrella, M.P., 1994b. Poinsettia (*Euphorbia pulcherrima* Willd. ex Koltz) cultivar-mediated differences in performance of five natural enemies of *Bemisia argentifolii* Bellows and Perring, n.sp. (Homoptera: Aleyrodidae). *Biol. Control,* 4: 305–318.

Heinz, K.M., Nunney, L. and Parrella, M.P., 1993. Toward predictable biological control of *Liriomyza trifolii* (Diptera: Agromyzidae) infesting greenhouse cut chrysanthemums. *Environ. Entomol.,* 22: 1217–1237.

Helgesen, R.G. and Tauber, M.J., 1976. Biological control of greenhouse whitefly *Trialeurodes vaporariorum* (Aleyrodidae: Homoptera), on short-term crops by manipulating biotic and abiotic factors. *Can. Entomol,* 106: 1175–1188.

Helyer, N.L. and Wardlow, L.R., 1987. Aphid control of chrysanthemums using frequent, low dose applications of *Verticillium lecanii. Bull. Int. Org. Biol. Integr. Control of Noxious Plants and Animals/West Palearctic Regional Section,* X: 62–65.

Hofsvang, T. and Hagvar, E.B., 1979. Different introduction methods of *Ephedrus cerasicola* Stary to control *Myzus persicae* (Sulzer) in small paprika greenhouses. *Z. angew. Entomol.,* 88: 16–23.

Huffaker, C.B., 1958. Experimental studies on predation: Dispersion factors and predatorprey oscillations. *Hilgardia,* 27: 343–383.

Hunter, C.D., 1997. *Suppliers of Beneficial Organisms in North America.* California Environmental Protection Agency, Department of Pesticide Regulation, Environmental Monitoring and Pest Management Branch, Sacramento, CA, 32 pp.

Hussey, N.W. and Scopes, N. (Editors), 1985. *Biological Pest Control. The Glasshouse Experience.* Cornell University Press, Ithaca, NY, 240 pp.

Hussey, N.W. and Scopes, N.E.A., 1977. The introduction of natural enemies for pest control in glasshouses: Ecological considerations. In: R.L. Ridgway and S.B. Vinson (Editors), *Biological Control by Augmentation of Natural Enemies. Insect and Mite Control by Parasites and Predators.* Plenum Press, New York, pp. 349–377.

Hussey, N.W., Read, W.H. and Hesling, J.J., 1963. The Pests of Protected Cultivation – The Biology and Control of Glasshouse and Mushroom Pests. American Elsevier, New York, 404 pp.

Ibrahim, A.G. and Madge, D.S., 1977. The susceptibility of eighteen chrysanthemum cultivars to the chrysanthemum leaf miner, *Phytomyza syngenesiae. Plant Pathol.,* 26: 103–108.

Immaraju, J.A., Paine, T.D., Bethke, J.A., Robb, K.L. and Newman, J.P., 1992. Western flower thrips (Thysanoptera: Thripidae) resistance to insecticides in coastal California greenhouses. *J. Econ. Entomol.,* 85: 9–14.

Iwao, S., 1968. A new regression method for analyzing the aggregation pattern of animal populations. *Res. Popul. Ecol.,* 10: 1–20.

Jacobson, R., 1990. Battle tactics. *Grower,* 114: 15, 16, 18.

Jacobson, R., 1993. Integrated pest management in spring bedding plants; a successful package for commercial crops. *Bull. SROP/WPRS,* 16(8): 105–109.

Jones, R.K., 1990a. Control of tomato spotted wilt virus. In: M.P. Parrella and J. Hall (Editors), *Proc. Sixth Conf. on Insect and Disease Management on Ornamentals.* Society of American Florists, Alexandria, VA, pp. 29–35.

Jones, V.P., 1990b. Developing sampling plans for spider mites: Those who don't remember the past are bound to repeat it. *J. Econ. Entomol.,* 83: 1656–1664.

Jones, V.P. and Parrella, M.P., 1984. The effects of selected pesticides on photosynthetic rates of chrysanthemum. In: S.L. Poe (Editor), *Proc. Fourth Annual Industry Conf. on the Leafminer.* Society of American Florists, Alexandria, VA, pp. 4–9.

Jones, V.P. and Parrella, M.P., 1986. The movement and dispersal of *Liriomyza trifolii* (Diptera: Agromyzidae) in a chrysanthemum greenhouse. *Ann. Appl. Biol.,* 109: 33–39.

Jones, V.P., Parrella, M.P. and Hodel, D.R., 1986a. Biological control of leafminers in greenhouse chrysanthemums. *Calif. Agric.,* 40: 10–12.

Jones, V.P., Toscano, N.C., Johnson, M.W., Welter, S.C. and Youngman, R.R., 1986b. Pesticide effects on plant physiology: Integration into a pest management program. *Bull. Entomol. Soc. Am.,* 32: 103–109.

Kabashima, J., Giles, K. and Parrella, M.P., 1995. Analysis of pesticide application in the greenhouse. *Calif. Agric.,* 49(4): 31–35.

Kareiva, P., 1990. The spatial dimensions in pest–enemy interactions. In: L.E. Ehler, M. Mackauer and J. Roland (Editors), *Critical Issues in Biological Control.* Intercept, Ltd, Andover, Hants, UK, pp. 213–227.

Keil, C.B. and Parrella, M.P., 1990. Characterization of insecticide resistance in two colonies of *Liriomyza trifolii* (Diptera: Agromyzidae). *J. Econ. Entomol.,* 83: 18–26.

Keil, C.B., Parrella, M.P. and Morse, J.G., 1985. Method for monitoring and establishing baseline data for resistance to permethrin by *Liriomyza trifolii. J. Econ. Entomol.,* 78: 419–422.

King, E.G., Hopper, K.R. and Powell, J.E., 1985. Analysis of systems for biological control of crop arthropod pests in the U.S. by augmentation of predators and parasites. In: M.A. Hoy and D.C. Herzog (Editors), *Biological Control in Agricultural IPM Systems.* Academic Press, Orlando, FL, pp. 201–227.

Kuack, D.L., 1989. Pest-free greenhouses are possible. *Greenhouse Grower,* 7: 78, 80.

Leeper, J.R., Roush, R.T. and Reynolds, H.T., 1986. Preventing or managing resistance in arthropods. In: E.H. Glass (Chairman), *Pesticide Resistance. Strategies and Tactics for Management.* National Academy Press, Washington, DC, pp. 335–346.

Leppla, N.C. and Ashley, T.R., 1989. Quality control in insect mass production: A review and model. *Bull. Entomol. Soc. Am.,* 35: 33–44.

Lindquist, R.K., 1995. Control of fungus gnats and shore flies. In: A. Bishop, M. Hausbeck, and R. Lindquist (Editors), *Proc. Eleventh Conf. on Insect and Disease Management. Fort Meyers, FL.* Society of American Florists, Alexandria, VA, pp. 69–4.

Lindquist, R.K. and Powell, C.C., 1990. Pesticide application techniques for disease and insect management. In: M.P. Parrella and J. Hall (Editors), *Proc. Sixth Conf. on Insect and Disease Management on Ornamentals.* Society of American Florists, Alexandria, VA, pp. 98–117.

Lindquist, R.K., Frost, C. and Wolgamott, M.L., 1979. *Integrated control of insects and mites on greenhouse crops,* Research Circular 245. Ohio Agricultural Research and Development Center, 20 pp.

Lindquist, R.K., Faber, W.R. and Casey, M.L., 1985. Effect of various soilless root media and insecticides on fungus gnats. *Hortscience,* 20: 358–360.

Lindquist, R.K., Casey, M.L., Bauerle, W.L. and Short, T.L., 1987. Effects of an overhead misting system on thrips populations and spider mite-predator interactions on greenhouse cucumber. *Bull. Int. Org. Biol. Integr. Control of Noxious Plants and Animals/West Palearctic Regional Section,* X: 97–100.

Londoño, M.E.A. and Vañegas, J.A.B., 1990. *Busqueda y reconocimiento de los enemigos naturales de las principales plagas en floras bajo invernaderos en la sabana de Bogota,* Master's Thesis. Universidad Nacional de Colombia, Facultad de Ciencias, Departamento de Biologia, 197 pp.

Loomans, A.J.M. and van Lenteren, J.C., 1990. Hymenopterous parasites as biological control agents of *Frankliniella occidentalis* (Perg.)? *Bull. Int. Org. Biol. Integr. Control of Noxious Plants and Animals/West Palearctic Regional Section,* XIII: 109–114.

Luck, R.F., Shepard, B.M. and Kenmore, P.E., 1988. Experimental methods for evaluating arthropod natural enemies. *Annu. Rev. Entomol.,* 33: 367–391.

Markkula, M. and Tiittanen, K., 1976. "Pest in first" and "natural infestation" methods in the control of *Tetranychus urticae* Koch with *Phytoseiulus persimilis* Anthias-Henriot on glasshouse cucumbers. *Ann. Agric. Fenn.,* 15: 81–85.

Markkula, M., Roukka, K. and Tiittanen, K., 1969. Reproduction of *Myzus persicae* (Sulz.) and *Tetranychus telarius* (L.) on different chrysanthemum cultivars. Annales Agriculturae Fenniae, 8: 175–183.

Markkula, M., Tiittanen, K. and Hokkanen, H.M.T., 1987. Failures in biological control of spider mites – due to predatory mites or their users. *Bull. Int. Org. Biol. Integr. Control of Noxious Plants and Animals/West Palearctic Regional Section,* X: 108–110.

Metcalf, R.L., 1980. Changing role of insecticides in crop protection. *Annu. Rev. Entomol.,* 25: 219–256.

Minkenberg, O.P.J.M., 1990. *On seasonal inoculative biological control,* Ph.D. Thesis. Agricultural University, Wageningen, Netherlands, 230 pp.

Minkenberg, O.P.J.M. and Ottenheim, J.J.G.W., 1990. Effect of leaf nitrogen content of tomato plants on preference and performance of a leafmining fly. *Oecologia,* 83: 291–298.

Minkenberg, O.P.J.M. and van Lenteren, J.C., 1986. *The leafminers Liriomyza bryoniae and Liriomyza trifolii (Diptera: Agromyzidae), their parasites and host plants: A review,* Agriculture University, Wageningen Papers, 86–2, 50 pp.

Moed, G.H.D., van der Werf, W. and Smits, P.H., 1990. Modelling the epizootiology of *Spodoptera exigua* nuclear polyhedrosis virus in a spatially distributed population of *Spodoptera exigua* in greenhouse chrysanthemums. *Bull. Int. Org. Biol. Integr. Control of Noxious Plants and Animals/West Palearctic Regional Section,* XIII: 135–141.

Murphy, B.C., Morisawa, T.A., Newman, J.P., Tjosvold, S.A. and Parrella, M.P., 1998. Fungal pathogen provides control of western flower thrips in greenhouse flowers. *Calif. Agric.* (in press).

Nedstam, B. and Burman, M., 1990. The use of nematodes against sciarids in Swedish greenhouses. *Bull. Int. Org. Biol. Integr. Control of Noxious Plants and Animals/West Palearctic Regional Section,* XIII: 147–148.

Nilsson, B.G., 1976. Progress in biological control in Sweden. *Bull. Int. Org. Biol. Integr. Control of Noxious Plants and Animals/West Palearctic Regional Section,* IV: 25–29.

Noldus, L.P.J.J. and van Lenteren, J.C., 1990. Host aggregation and parasitoid behavior: Biological control in a closed system. In: L.E. Ehler, M. Mackauer and J. Roland (Editors), *Critical Issues in Biological Control.* Intercept, Ltd, Andover, Hants, UK, pp. 229–262.

Noldus, L.P.J.J., Xu, R., Eggenkamp-Rotteveel Mansveld, M.H. and van Lenteren, J.C., 1986. The parasite–host relationship between *Encarsia formosa* Gahan (Hymenoptera, Aphelinidae) and *Trialeurodes vaporariorum* (Westwood) (Homoptera, Aleyrodidae). XX. Analysis of the spatial distribution of greenhouse whiteflies in a large greenhouse. *J. Appl. Entomol.,* 102: 484–498.

Oetting, R.D., 1982. Susceptibility of selected chrysanthemum cultivars to *Liriomyza trifolii. J. Entomol. Soc. Georgia,* 17: 552–558.

Onillion, J.C., 1990. The use of natural enemies for biological control of whiteflies. In: D. Gerling (Editor), *Whiteflies: Their Bionomics, Pest Status and Management.* Intercept, Ltd, Andover, Hants, UK, pp. 287–313.

Onillion, J.C., Onillion, J. and Pietro, J.P., 1976. Resultats preliminaires due controle biologique de l'Aleurode des serres *T. vaporariorum* (Westwood) (Homot., Aleurodidae) par *E. formosa* (Gahan) (Hymenoptera, Aphelinidae) en serres d'aubergine. *Bull. Int. Org. Biol. Integr. Control of Noxious Plants and Animals/West Palearctic Regional Section,* 4: 138–150.

Osborne, L.S., 1986. Dip treatment of tropical ornamental foliage cuttings in fluvalinate to prevent spread of insect and mite infestations. *J. Econ. Entomol.,* 79: 465–470.

Osborne, L.S. and Chase, A.R., 1984. Influence of acephate and oxamyl on *Alternaria* leaf spot of schefflera and the pathogen, *Alternaria panax. Plant Dis.,* 68: 870–872.

Osborne, L.S. and Chase, A.R., 1985. Relative susceptibility of twelve cultivars of *Hedera helix* to two-spotted spider mite and *Xanthomonas* leaf spot. *Hortscience,* 20: 269–271.

Osborne, L.S., Hoelmer, K. and Gerling, D., 1990. Prospects for biological control of *Bemisia tabaci. Bull. Int. Org. Biol. Integr. Control of Noxious Plants and Animals/West Palearctic Regional Section,* XIII: 153–160.

Overmeer, W.P.J. and Helle, W., 1980. Genetics of resistance in spider mites from rose houses. In: A.K. Minks and P. Gruys (Editors), *Integrated Control of Pests in the Netherlands.* Centre for Agricultural Publishing and Documentation, Wageningen, pp. 125–127.

Overmeer, W.P.J., van Zon, A.Q. and Helle, W., 1980. Isolation mechanisms in populations of spider mite in rose houses. In: A.K. Minks and P. Gruys (Editors), *Integrated Control of Pests in the Netherlands.* Centre for Agricultural Publishing and Documentation, Wageningen, pp. 133–136.

Parr, J.C. and Thurston, R., 1968. Toxicity of *Nicotiana* and *Petunia* species to larvae of the tobacco hornworm. *J. Econ. Entomol.,* 61: 1525–1531.

Parrella, M.P., 1987. Biology of *Liriomyza. Annu. Rev. Entomol.,* 32: 201–224.

Parrella, M.P., 1988. Rethinking the war on pests. The writing is on the wall: Growers can no longer rely on pesticides alone. *Soc. Am. Florists Business News for the Floral Industry,* 5: 20–22, 27.

Parrella, M.P., 1990a. Biological control in ornamentals: Status and perspectives. *Bull. Int. Org. Biol. Integr. Control of Noxious Plants and Animals/West Palearctic Regional Section,* XIII: 161–168.

Parrella, M.P., 1990b. The development of IPM strategies in

chrysanthemums and gerberas with an emphasis on biological control. In: V.O. Sing, M. Maldonado and M. Pizano (Editors), *Proc. Latin American Symp. Flowers and Ornamental Plants, Bogota, Colombia,* pp. 43–59.

Parrella, M.P. and Jones, V.P., 1987. Development of integrated pest management strategies for floriculture crops. *Bull. Entomol. Soc. Am.,* 33: 28–34.

Parrella, M.P., Robb, K.L. and Bethke, J.A., 1983. Influence of selected host plants on the biology of *Liriomyza trifolii* (Diptera: Agromyzidae). *Ann. Entomol. Soc. Am.,* 76: 112–115.

Parrella, M.P., Baker, J.R., Sanderson, J.P., Lindquist, R.K. and Ali, A.D., 1987. Resistance to insecticides: What a grower can do. *GrowerTalks,* 50: 136, 138–142.

Parrella, M.P., Jones, V.P., Malais, M.S. and Heinz, K.M., 1989. Advances in sampling in ornamentals. *Fl. Entomol.,* 72: 394–403.

Parrella, M.P., Paine, T.D., Bethke, J.A., Robb, K.L. and Hall, J., 1991. Evaluation of *Encarsia formosa* (Hymenoptera: Aphelinidae) for biological control of sweetpotato whitefly (Nomoptera: Aleyrodidae) on poinsettia. *Environ. Entomol.,* 20: 713–719.

Parrella, M.P., Heinz, K.M. and Nunney, L., 1992. Biological control through augmentative releases of natural enemies: A strategy whose time has come. *Am. Entomol.,* Fall: 172–179.

Penna, R.J., Morgan, W.M. and Ledieu, M.S., 1983. Chapter 12. Pests and diseases of protected crops. In: N. Scopes and M. Ledieu (Editors), *Pest and Disease Control Handbook.* The Lavenham Press Limited, Lavenham, Suffolk, United Kingdom, pp. 375–472.

Poe, S.L., Green, J.L. and Shih, C.I., 1976. Cultural practices affect damage to chrysanthemum by *Liriomyza sativae* Blanchard. *Proc. Fl. State Hortic. Soc.,* 89: 299–301.

Pritchard, A.E., 1949. *California greenhouse pests and their control,* Bull. 713. California Agricultural Experiment Station, The College of Agriculture, University of California, Berkeley, CA, 71 pp.

Ramdev, Y.P., Lindquist, R.K. and Hall, F.R., 1988. Evaluation of resistance to Pentac and Bifenthrin in nine spider mite populations from Ohio greenhouses. *Ohio Florists' Association Bull.,* 704: 6–7.

Robb, K.L., 1989. *Analysis of Frankliniella occidentalis (Pergande) as a pest in floricultural crops in California greenhouses,* Ph.D. Thesis. University of California, Riverside, CA, 135 pp.

Robb, K.L., 1990. Western flower thrips; their biology and control. In: M.P. Parrella and J. Hall (Editors), *Proc. Sixth Conf. on Insect and Disease Management in Ornamentals.* Society of American Florists, Alexandria, VA, pp. 19–28.

Robb, K.L., 1995. Controlling thrips and the use of exclusion devices. In: A. Bishop, M. Hausbach and R. Lindquist (Editors), *Proc. Eleventh Conf. on Insect and Disease Management on Ornamentals, Fort Meyers, FL.* Society of American Florists, Alexanderia, VA, pp. 113–120.

Roland, J., 1990. Parasitoid aggregation: Chemical ecology and population dynamics. In: L.E. Ehler, M. Mackauer and J. Roland (Editors), *Critical Issues in Biological Control.* Intercept, Ltd, Andover, Hants, UK, pp. 185–212.

Samsoe-Petersen, L., Bigler, F., Bogenschutz, H., Brun, J., Hassan, S.A., Helyer, N.A., Kuhner, C., Mansour, F., Naton, E., Oomen, P.A., Overmeer, W.P.J., Polgar, L., Reickmann, W. and Staubli, A., 1989. Laboratory rearing techniques for 16 beneficial arthropod species and their prey/hosts. *J. Plant Dis. Prot.,* 96: 289–316.

Sanderson, J.P. and Ferrentino, G.W., 1990. Beat greenhouse whiteflies with a proven biological – *Encarsia formosa. GrowerTalks,* 54: 108–114.

Sanderson, J.P. and Roush, R.T., 1989. Managing insecticide resistance in the greenhouse. In: M. Daughtrey and J. Begley (Editors), *Proc. Fifth Conf. on Insect and Disease Management in Ornamentals.* The Society of American Florists, Alexandria, VA, pp. 93–102.

Sanderson, J.P. and Roush, R.T., 1995. Managing insecticide resistance in the greenhouse. In: A. Bishop, M. Hausbach and R. Lindquist (Editors), *Proc. Eleventh Conf. on Insect and Disease Management on Ornamentals, Fort Meyers, FL.* Society of American Florists, Alexandria, VA, pp. 23–36.

Sanderson, J.P., Parrella, M.P. and Trumble, J.T., 1989. Monitoring resistance using yellow sticky traps. *J. Econ. Entomol.,* 82: 1011–1018.

Sase, S. and Christianson, L.L., 1990. Screening greenhouses – Some engineering considerations. *Northeast Agriculture/Biological Engineering Conference,* Paper No. NABEC 90201. American Society of Agricultural Engineers.

Schulten, G.G.M., 1980. A strain of *Phytoseiulus persimilis* (Acari: Phytoseiidae) resistant to organophosphorus compounds. In: A.K. Minks and P. Gruys (Editors), *Integrated Control of Pests in the Netherlands.* Centre for Agriculural Publishing and Documentation, Wageningen, pp. 119–120.

Sher, R.B. and Parrella, M.P., 1996. Integrated biological control of leafminers, *Liriomyza trifolii,* on greenhouse chrysanthemums. *Bull. Int. Org. Biol. Integr. Control of Noxious Plants and Animals/ West Palearctic Regional Section,* 19: 147–150.

Shipp, J.L. and Zariffa, N., 1990. Developing a sampling program for western flower thrips on greenhouse peppers. *Bull. Int. Org. Biol. Integr. Control of Noxious Plants and Animals/West Palearctic Regional Section,* XII: 194–197.

Smith, F.F. and Webb, R.E., 1977. Biographic and agronomic problems relating to the utilization of biocontrol organisms in commercial greenhouses in the continental United States. In: F.F. Smith and R.E. Webb (Editors), *Pest Management in Protected Culture Crops in the United States. XV Int. Congr. of Entomology, Washington, DC.* Agricultural Research Service, US Department of Agriculture, ARS-NE-85, pp. 89–93.

Smitley, D.R., 1990. Biological control of spider mites in greenhouse rose production. *Roses, Inc., Bull.,* August: 47–52.

Smits, P.H. and Vlak, J.M., 1988. Biological activity of *Spodoptera exigua* nuclear polyhedrosis virus against *Spodoptera exigua* (Lepidoptera: Noctuidae) larvae. *J. Invertebr. Zool.,* 51: 107–114.

Sparnaaij, L.D., 1991. Breeding for disease and insect resistance in flower crops. In: J. Harding, F. Singh and J.M.M. Mol (Editors), *Genetics and Breeding of Ornamental Species.* Kluwer Academic Publishers, Dordrecht, Netherlands, pp. 179–211.

Stacy, D.L., 1977. 'Banker' plant production of *Encarsia formosa* (Gahan) and its use in control of glasshouse whitefly on tomatoes. *Plant Pathol.,* 23: 63–66.

Stenseth, C., 1983. The use of the parasite *E. formosa* as a part of pest management on cucumbers. *Entomophaga,* 28: 17–26.

Stephanou, E. and Zourari, M., 1989. Exposure to residues in greenhouses: Determination of airborne residues and surface deposition. *Toxicol. Environ. Chem.,* 25: 17–27.

Stinner, R.E., 1977. Efficacy of inundative releases. *Annu. Rev. Entomol.,* 22: 515–531.

Sunderland, K.D., Chambers, R.J., Helyer, N.L. and Sopp, P.I., 1993. Integrated pest management of greenhouse crops in northern Europe. In: J. Janick (Editor), Horticulture Reviews, Vol. 14. Horticultural Research International, Worthing Road, Littlehampton, UK, 66 pp.

Sutterlin, S., van Lenteren, J.C. and Fransen, J.J., 1990. Dispersal of the greenhouse whitefly, *Trialeurodes vaporariorum* (Westwood), on the rosette plant *Gerbera. Bull. Int. Org. Biol. Integr. Control of Noxious Plants and Animals/West Palearctic Regional Section,* XIII: 205–208.

Taylor, L.R., 1961. Aggregation, variance and the mean. *Nature,* 189: 732–735.

Tomczyk, A. and Kropczynska, D., 1989. Some aspects of influence of two-spotted spider mite feeding on chrysanthemum plants. *Bull. Int. Org. Biol. Integr. Control of Noxious Plants and Animals/West Palearctic Regional Section,* XII: 15–17.

Tulisalo, U., Tuovinen, T. and Kurppa, S., 1977. Biological control of aphids with *Chrysopa carnea* on parsley and green pepper in the greenhouse. Ann. Agric. Fenn., 43: 97–100.

Urban, L.A., Speck, J., Moyer, J.W. and Daub, M.E., 1992. Transformation of chrysanthemum with the nucleocapside gene of tomato spotted wilt virus. *Phytopathology,* 82: 1147.

Vale, R., 1995. Agrochemicals industry issues crisis warning. *Grower,* 123(11): 5.

van de Veire, M. and Vacante, V., 1984. Greenhouse whitefly and leafminer control by the combined use of the color attraction system with the parasite wasp, *Encarsia formosa* (Gahan). *Meded. Fac. Landbouwwet. Rijksuniv. Gent,* 49: 107–114.

van den Bos, J. and van de Vrie, M., 1981. The isolating effect of greenhouse on the leafroller moth, *Clepsis spectrana*; prospects of non-chemical control of the greenhouse populations. *Meded. Fac. Landbouwwet. Rijksuniv. Gent,* 46: 443–456.

van der Linden, A., 1990. Prospects for the biological control of *Liriomyza huidobrensis* (Blanchard), a new leafminer for Europe. *Bull. Int. Org. Biol. Integr. Control of Noxious Plants and Animals/ West Palearctic Regional Section,* XIII: 100–103.

van Lenteren, J.C., 1980. Evaluation of control capabilities of natural enemies: Does art have to become science? *Neth. J. Zool.,* 30: 369–381.

van Lenteren, J.C., 1986. Parasitoids in the greenhouse: Successes with the seasonal inoculative release systems. In: J. Waage and D. Greathead (Editors), *Insect Parasitoids.* Academic Press, London, pp. 341–374.

van Lenteren, J.C., 1990. Integrated pest and disease management in protected crops: The inescapable future. *Bull. Int. Org. Biol. Integr. Control of Noxious Plants and Animals/West Palearctic Regional Section,* XIII: 91–99.

van Lenteren, J.C., 1993. Quality control for natural enemies used in greenhouses. *Bull. Int. Org. Biol. Integr. Control of Noxious Plants and Animals/West Palearctic Regional Section,* 16: 89–92.

van Lenteren, J.C., 1996. *Quality control tests for natural enemies used in greenhouse biological control. Bull. Int. Org. Biol. Integr. Control of Noxious Plants and Animals/West Palearctic Regional Section,* 19: 83–86.

van Lenteren, J.C. and Noldus, L.P.J.J., 1990. Whitefly plant relations: Behavioural and ecological aspects. In: D. Gerling (Editor), *Whiteflies: Their Bionomics, Pest Status and Management.* Intercept, Ltd, Andover, Hants, UK, pp. 47–89.

van Lenteren, J.C. and Wardlow, L.R. (Editors), 1993. IPM in greenhouse ornamentals. *Bull. Int. Org. Biol. Integr. Control of Noxious Plants and Animals/West Palearctic Regional Section,* 16: 166 pp.

van Lenteren, J.C. and Woets, J., 1988. Biological and integrated pest control in greenhouses. *Annu. Rev. Entomol.,* 33: 239–269.

van Lenteren, J.C., Ramakers, P.M.J. and Woets, J., 1980. Integrated control of vegetable pests in greenhouses. In: A.K. Minks and P. Gruys (Editors), *Integrated Control of Pests in the Netherlands.* Centre for Agricultural Publishing and Documentation, Wageningen, pp. 109–118.

Vehrs, S. and Parrella, M.P., 1991. Aphid problems increase on ornamentals. *Calif. Agric.,* 45: 28–29.

Vet, L.E.M., van Lenteren, J.C. and Woets, J., 1980. The parasite host relationship between *Encarsia formosa* (Hymenoptera; Aphelinidae) and *Trialeurodes vaporariorum* (Homoptera; Aleyrodidae). IX. A review of the biological control of the greenhouse whitefly with suggestions for future research. *Z. angew. Entomol.,* 90: 26–51.

Waage, J., 1990. Ecological theory and the selection of biological control agents. In: L.E. Ehler, M. Mackauer and J. Roland (Editors), *Critical Issues in Biological Control.* Intercept, Ltd, Andover, Hants, UK, pp. 135–157.

Wardlow, L.R., Ludlam, F.A.B. and French, N., 1972. Insecticide resistance in glasshouse whitefly. *Nature,* 139: 164–165.

Webb, R.E. and Smith, F.F., 1979. Greenhouse whitefly control of an integrated regimen based on adult trapping and nymphal parasitism. *Bull. Int. Org. Biol. Integr. Control of Noxious Plants and Animals/ West Palearctic Regional Section,* III: 235–46.

Williams, E.C., 1993. Entomopathogenic nematodes for leafminer control. *Bull. WPRS/SROP,* 16: 158–162.

Wilson, G.F. and Becker, P., 1960. *Horticultural Pests. Detection and Control.* Crosby Lockwood & Son, Ltd, London, 240 pp.

Wilson, M.C., Schuder, D.L. and Provonsha, A.V., 1977. *Insects of Ornamental Plants.* Waveland Press, Prospect Heights, IL, 134 pp.

Woets, J., 1978. Development of an introduction scheme for *Encarsia formosa* Gahan (Hymenoptera; Aphelinidae) in greenhouse tomatoes to control the greenhouse whitefly, *Trialeurodes vaporariorum* (Westwood) (Homoptera; Aleyrodidae). *Meded. Fac. Landbouwwet. Rijksuniv. Gent,* 43: 379–385.

Yano, E., 1987. Quantitative monitoring techniques for the greenhouse whitefly. *Bull. Int. Org. Biol. Integr. Control of Noxious Plants and Animals/West Palearctic Regional Section,* X: 198–202.

Yoshida, H.A. and Parrella, M.P., 1987. The beet armyworm, a serious pest of floricultural crops. *Calif. Agric.,* 41: 13–15.

Yoshida, H.A. and Parrella, M.P., 1991. Chrysanthemum cultivar preferences exhibited by *Spodoptera exigua* (Lepidoptera: Noctuidae). *Environ. Entomol.,* 20: 160–165.

Zhang, Zhi-Giang and Sanderson, JP., 1995. Two-spotted spider mite (Acari: Tetranychidae) and *Phytoseiulus persimilis* (Acari: Phytoseiidae) on greenhouse roses: spatial distribution and predator efficacy. *J. Econ. Entomol.,* 88: 352–357.

Chapter 10

HUMANS – HEALTH, PSYCHOLOGICAL AND SOCIOLOGICAL FACTORS

Peter LUNDQVIST

INTRODUCTION

In this chapter the major factors affecting the health, safety and well-being of people working in greenhouses will be discussed. The information presented is mainly extracted from selected studies carried out in Sweden.

The greenhouse is not only a system for protecting cultivated crops, it is also the working environment for humans such as growers and greenhouse workers, female and male, all over the world. The shape, size and interior of the greenhouse can range from large-scale, highly mechanized modern greenhouses to very simple small and old greenhouses totally manually operated. The humans are influenced by the greenhouse environment in various ways. They can enjoy the environment as a result of such factors as interesting work tasks, being able to see the result of their own work, to see the crop grow and mature, to follow the changes in the seasons, to make a profit and the like. They can also be affected by the environment in a negative way; they may suffer occupational injuries and diseases, may be exposed to occupational accidents and to physiological, physical, biological, chemical, psychological and sociological factors.

Official statistics from Sweden show that greenhouse workers have many work-related accidents due to falling, overexertion of body parts, and handling of materials. The most common occupational disease was caused by monotonous or strenuous movements and working postures (about 60%), followed by chemical substances, noise and biological factors – allergies, etc. (Lundqvist, 1984).

What do the greenhouse workers feel about their work themselves? Inquiries in Sweden and Germany show that the body postures and the climate are the worst factors. Other problems include stress, pesticides and lengthy work operations (Table 10.1).

Table 10.1
Feelings of greenhouse workers about their work. The question was "Which is the worst factor in your working environment?"

Ranking according to:		
Nilsson (1975)	Stoffert (1986)	Lundqvist (1988a)
1. Climate	1. Body postures	1. Body postures
2. Stress	2. Pesticides	2. Climate
3. Body postures	3. Lengthy work operations	3. Pesticides

OCCUPATIONAL ACCIDENTS

According to the Swedish definition, occupational injuries can be of three kinds: occupational accidents, occupational diseases, and accidents on the way to or from the work site (commuting accidents). In 1979, a new system for collecting information about occupational injuries was instituted by the Swedish National Board of Occupational Safety and Health to provide basic data for accident prevention in industry and other areas. To obtain more detailed information about the horticulture sector, a special analysis of the information in the system was made (Lundqvist, 1984): The total Swedish horticultural labour force consisted of about 15 000 persons working half-time or more. In 1979, 1980 and 1981, there were 742, 751, and 761 injuries, respectively, excluding those occurring when travelling to or from work. Of the 761 injuries during 1981, 695 were occupational accidents and

66 occupational diseases. One-third of the accidents resulted in more than 14 days of compensation (sick leave). The average number of compensation days per injury was slightly more than 20 for the entire horticulture industry. Accidents due to overexertion, falls, and materials-handling dominated in both 1980 and 1981. Injuries due to falls, stepping on an uneven surface, slipping due to mis-steps, fires and explosions, or some type of overexertion, often required more than three weeks recovery (compensation days). Materials, structural parts of greenhouses, vehicles, machines and hand tools and equipment were the most common external agencies involved. The back, fingers, hip joints, legs, knees and feet were the body parts most affected. The most common injuries were sprains, strains, wounds, and tissue injuries due to crushing. Long periods of compensation were common for skeletal injuries and amputations. During 1980 and 1981, no fatal injuries were reported. The numbers of injuries per million man-hours by industry group are presented in Table 10.2, which shows that the number of injuries in the horticultural industry is quite high.

Near-accidents occurring in greenhouses were studied (Lundqvist, 1982) to gather more detailed information about the risks involved in greenhouse work than was available from official statistics. It has been shown (Carter and Menckel, 1985; Gustafsson et al., 1991) that there are more near-accidents than actual accidents, and that they have the same causes. Furthermore, a person who had a near-accident is more likely to discuss it than one who has had an accident requiring medical care or sick leave. Therefore, a study of near-accidents would be expected to generate more information for purposes of accident prevention than would a study of accidents.

The data were obtained by interviewing a large number of selected commercial growers and greenhouse workers about the occurrence of near-accidents in their work. Every person was interviewed by a research leader every second day during a 14-day period. Before the start of the interview period, the commercial growers and greenhouse workers received oral instruction about the study with special reference to what should be considered as a near-accident. After the study was completed and the data analyzed, every person received an individual summary of the results and a list of suggested preventive measures.

Twenty greenhouse operations and 86 persons working in these greenhouses took part in these investigations. There were a total of 55 near-accidents observed during the study period. Two of these incidents were evaluated as being accidents. The occurrence of near-accidents was spread relatively evenly over the week, and over the working day, with a slightly greater number occurring during the morning hours.

Most of the near-accidents occurred during harvest or transport work (about 40%) (Table 10.3). Quite a few near-accidents were observed when a hose was used to water the plants. About 70% of the near-accidents occurred in the greenhouse passages or alleys. The main causes of these near-accidents were slippery and uneven surfaces, obstacles and sharp objects; these were involved in up to 70% of the near-accidents (Table 10.4). In older greenhouses there tended to be

Table 10.2
Occupational injuries per million man-hours in different types of industry, 1983 [1]

Type of injury	Industry division [2]			
	Hortic.	Agric.	Manuf.	All
Occupational accidents	51,3	46,5	30,2	19,0
Occupational diseases	5,1	5,1	6,5	3,8
All occupational injuries	56,4	51,6	36,7	22,8

[1] Source: National Swedish Board of Occupational Safety and Health (1987a).
[2] Abbreviations: Hortic., horticulture; Agric., agriculture; Manuf., manufacture; All, all industries.

Table 10.3
Circumstances causing near-accidents [1]

Main cause	Near-accident	
	Number	%
Slippery surface	14	25.5
Uneven surface, obstacles	13	23.6
Sharp object	11	20.0
Projecting object	5	9.1
Falling object	4	7.3
Lack of space	4	7.3
Climate	2	3.6
Chemicals	1	1.8
Electricity	1	1.8
Total	55	100.0

[1] Source: Lundqvist (1982).

Table 10.4
Distribution of near-accidents according to type of work [1]

Type of work	Near-accidents	
	Number	%
Harvesting	14	25.5
Transporting	9	16.4
Watering	5	9.1
Planting/sowing	5	9.1
Repairing/maintenance	5	9.1
Sorting	4	7.3
Clipping, pruning, etc.	3	5.4
Cleaning	3	5.4
Cultivation	2	3.6
Other work	5	9.1
Total	55	100.0

[1] Source: Lundqvist (1982).

a relatively higher incidence of near-accidents than in more modern greenhouses.

OCCUPATIONAL DISEASES

According to the National Swedish Board of Occupational Safety and Health (1987a), occupational diseases are injuries that have arisen through the adverse action of an agency other than accident. Examples of other harmful factors are hazardous substances, unusually strenuous motions or work postures (ergonomic factors), vibrations from machinery and tools, noise, and infection. Under certain conditions, moreover, adverse effects may come from low or high temperatures, sudden fluctuations in temperature, moisture or strong draughts. Insurance cover may extend to disorders arising from mentally trying conditions which can be directly traced to the job.

The statistics from the Swedish horticultural industry (Lundqvist, 1984) showed that the most common occupational diseases were due to monotonous or strenuous movements and working positions (about 60%), followed by chemicals, noise, and biological factors (for instance allergies). The average period of compensation days per disease was about one month. Diseases of the locomotor system were the most frequent, followed by those of the respiratory system.

ENVIRONMENTAL FACTORS

Physiological factors

Complaints affecting the back, neck, shoulders, knee-joints and other joints are among the commonest causes of illness among the employed population. That was the major reason why Sweden was one of the first countries in the world to issue a standard concerning working postures and working movements (National Swedish Board of Occupational Safety and Health, 1983). Persons with physically demanding jobs involving the lifting of heavy loads, repetitive and monotonous movements and unsuitable work postures have higher rates of sick leave than others. Between them, "overexertion of body part" and "handling accidents" accounted for some 30% of all reported occupational accidents in Sweden in 1980. The official occupational injury statistics for 1980 included just over 18 000 reports of presumptive occupational diseases, and in 50% of these cases physical stress factors were the suspected cause. These conditions cause widespread and difficult problems for community and individual alike. A high rate of sick leave, transfers to alternative duties and rehabilitation involve the community in heavy expenditure, and many people incur pain and suffering.

It should however be made clear that a certain amount of physical stress is beneficial to the individual. Some measure of physical activity is needed to keep the organs of the body in trim and to maintain physical fitness. Work involving the right amount of mobility can be good for the health.

Capacity for physical exertion varies from one person to another, depending on size, sex, age, experience, fitness, motivation and possible disabilities. The balance between job requirements and human characteristics is achieved primarily by adapting work to suit the individual.

There is no distinct line between acceptable and harmful physical stress. The risk of injury is influenced both by the type of stress and by the characteristics of the individual. Even apparently "light" work may involve operations demanding the constant tension of certain muscles. Operations of this kind, if repeated frequently or sustained for long periods, may cause not only fatigue but also, eventually, injury as well.

The individual person's subjective perception of exertion is a relatively good yardstick of overexertion.

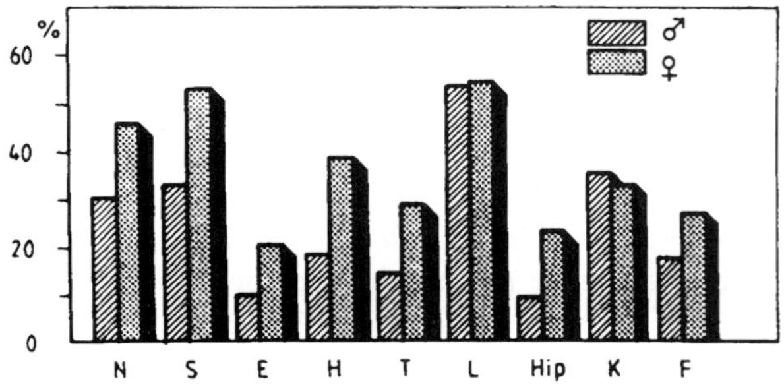

Fig. 10.1. Percentage of males ($n=134$) and females ($n=150$), reporting musculo-skeletal troubles in different body regions during the past 12 months: N, neck; S, shoulders; E, elbows; H, wrists/hands; T, upper back (thoracic spine); L, low back (lumbar spine), Hip, hips; K, knees; F, ankles/feet. Source: Lundqvist and Gustafsson (1988).

Subjective perception can furnish guidance for efforts to improve working conditions.

A study of musculo-skeletal problems was carried out among horticultural workers in Scania in southern Sweden (Lundqvist and Gustafsson, 1988), using a standardized questionnaire. A 70% response (284 replies) was achieved. The lower back (both males and females) was the area most reported. Females noted more problems than males, although they had slightly less problems with the knees. More than 70% of those reporting had musculo-skeletal problems in at least one body region, indicating that the working conditions in horticulture need to be improved greatly (Fig. 10.1).

Stoffert and Wildt (1987) have made work-physiological investigations of pot-plant handling on the ground. In greenhouses, workers have to do much manual work on or near the ground in many different working positions. The physical workload of three methods of handling pot-plants on the ground was assessed by recording the heart rate, heart arrhythmia and oxygen uptake in resting and working situations. With these ergonomic data it was possible to rank the physical workloads of the handling methods. German anthropometrical studies on women working in a bent position (Stoffert and Timme, 1987) show, among other things, that women are not better suited to work in a bended position than men. Studies on working postures in greenhouse work were also carried out in the Netherlands. Van der Schilden (1989) has described a method for recording and evaluation of working postures during pot-plant handling. This method has also been used by Lundqvist (1988b), Zegers and van Dieën (1987), Wildt (1988), and the Finnish inventors Karhu, Kansi and Kuorinka (Karhu et al., 1977). In a study by Maeda et al. (1980) the working postures of workers engaged in strawberry and eggplant culture in greenhouses were observed and recorded (see Table 10.5 for the scientific names of crop plants mentioned in this chapter). They found that more than 50% of the workers complained of fatigue in the lower back and shoulders. The prevalence of low back pain was significantly higher among strawberry workers than among eggplant workers, probably due to the deep stooping postures during picking strawberries. Other studies of ergonomical problems in greenhouses have been presented by van Dieën (1989) and Hildebrandt (1989).

Table 10.5
Scientific names of crop plants mentioned in this chapter

Common name	Scientific name
Carnation	*Dianthus caryophyllus*
Chrysanthemum	*Dendranthema grandiflora*
Cucumber	*Cucumis sativus*
Daffodil	*Narcissus pseudo-narcissus*
Egg plant	*Solanum melongena*
Strawberry	*Fragaria×ananassa*

Physical factors

In this section the physical factors are divided into: climate, light, noise and vibration. The emphasis is put on the climate as a factor in the working environment which also includes the effect of radiation (mainly from the sun).

Climate

The climate is one of the major problems of the working environment for man in greenhouses, and it can include troubles with high temperatures, as well as with radiation from the sun, humidity, low temperatures and draught during the cold season. In order to describe the climate in greenhouses during the summer in Sweden, daily measurements of the air temperature, relative humidity and radiant temperature were made. Radiant temperature were measured with a globe thermometer, which provides an expression of how a normally dressed person experiences the combined effect of air temperature, air movement and heat radiation of the environment. The measurements were made every hour between 7.00 and 16.00 hours. It was found that in greenhouses for carnations, tomatoes and cucumbers, 7, 24 and 52% of the air temperature observations, respectively, exceeded 27°C. On some occasions radiant temperatures of 50°C and over were measured during May, June and August (Brunsson et al., 1978). The occasional high radiant temperatures were considered to provide ample reason for a continual recording of radiant temperatures in a conventional-type greenhouse over a long period of time. For this study (Ekström and Gustafsson, 1978), a greenhouse of conventional type equipped with automatic climate control and a hot-water heating system was used. Its ventilation was controlled by a thermostat regulating air circulation through windows at the roof ridge and at the bottom of the side-walls. The thermostat also controlled a fan system. No artificial shade system (sunblinds or shading paint) was used. The aim of the study was to test the ability of the ventilation and heating systems to maintain a greenhouse air temperature in the spring, summer and autumn that did not exceed 20°C, and in the winter did not fall below 12°C. It was observed that the radiant temperature from October to February was only insignificantly higher than the air temperature. However, from May to August a significant difference was found. The monthly mean for the radiant temperature during the four summer months was nearly 35°C, whereas the average air temperature during the same period was just below 25°C. This investigation showed that the possibility of controlling greenhouse climate during the summer, using only a ventilation system, was limited (Gustafsson and Ekström, 1980). Additional observations made in greenhouses covered with hard or soft plastic gave similar results (Ekström et al., 1979). With the present level of technology these working conditions were likely to remain unaltered for several years. What remained to be determined was how the greenhouse climate and workload affected the workers. In horticulture, the major part of the work force are women, often employed in subordinate positions and with male supervisors. Since women have a lower heat tolerance than men (Burse, 1979), it was decided to study the effect of heat loads equivalent to the greatest one occurring in greenhouses, on female and male subjects carrying out light lifting work in a climate chamber, while using working positions and movements which as closely as possible resembled those in greenhouse work (Lundqvist et al., 1987; Lundqvist, 1988b). A secondary problem is that of protective clothing. This is often worn when pesticides are used in greenhouses, and the air intakes are closed in order to achieve high air concentrations of the pesticide quickly. Wearing protective clothing may be supposed to increase the workload and heat stress on the worker. Thus, two series of climate chamber trials were carried out in this study, the first with subjects wearing light clothing appropriate to warm temperatures, and the second with a number of the same subjects wearing protective clothing (Gustafsson and Lundqvist, 1981). Parallel with these studies, the metabolism (as oxygen usage) during routine work operations in greenhouse cultivation was measured on location in order to determine the true workload.

The general results of the Swedish studies (Gustafsson et al., 1989) showed that:

(1) High environmental temperatures mean extra work for the heart. It increases parallel with the increase in environmental temperature.
(2) Women are more sensitive to an increase in environmental temperatures than men. If exposed to the same temperatures, they showed a greater increase in pulse rate than men.
(3) During work periods after rest, when the participants had chosen their own work pace, the number of work operations per minute was almost the same as in periods with a metronome-controlled work pace. In the subsequent work periods the work pace increased greatly. No proportional increase in pulse rates, however, was observed at climate-chamber temperatures up to and including 32°C.
(4) The deep body temperature and the skin temperature of the subjects increased mainly during the first half-hour spent in the climate chamber and remained practically constant for the last two hours of the three-hour period.
(5) The higher the environmental temperature, the less was the difference observed between deep body temperature and skin temperature.
(6) Sweat production increased with an increase in environmental temperature. Women perspired less than men.
(7) As shown by the metabolic investigations, most of the routine work in greenhouse cultivation required little energy. The exceptions included hoeing walkways and flower beds, cutting edges, and shovelling soil. Skin temperature at rest and at work, and sweat production, were higher with the use of suits than without.

As a result of the Swedish studies, there have been some recommendations about work in greenhouses during warm summer periods (Gustafsson, 1982):

(1) avoid physically strenuous body movements and working positions;

(2) have frequent pauses during the working day;
(3) avoid working in the greenhouse at midday (when the temperature is high);
(4) drink frequently, and only small amounts (but not cold);
(5) replace body salt; and
(6) wear light, roomy clothing.

It was also pointed out that:
(1) working in high temperature increases the risk of accidents;
(2) working in high temperature results in a loss of effective working time; and
(3) heat tolerance decreases with age.

Outside Sweden, there have been few studies of the working climate in greenhouses. Löbel (1977) in former East Germany prepared a thesis on factors of the working environment in greenhouses, in which he also found that the heat load in the warm summer months could be high in different types of greenhouses. When he compared the heat load with that in other industries, he found that the fact that the heat load in greenhouses is discontinuous (with respect to the time of the day, season and type of weather) made it much more difficult for greenhouse workers to get acclimatized. In another thesis by Schulz (1975), limits for human work in greenhouses were set at 25°C under 90–100% atmospheric humidity, 26°C under 80% atmospheric humidity or 28°C under 60% atmospheric humidity. According to Stoffert (pers. commun., 1992) there is an agreement in Italy with the trade union to the effect that all work in greenhouses must be stopped when the air temperature exceeds 28°C. In the Netherlands the legal limit is a maximum air temperature of 29°C. Warmer working places need special permission.

Laitinen (1980) reported on measurements of the working climate in Finnish greenhouses. In Czechoslovakia, Visnovsky and Stehlo (1982) studied the heat load in plastic-foil greenhouses. They found that the pulse rate rose with increasing temperatures, and that special heat recovery periods ought to be allowed during warm summer days. In a small Danish study by Lander et al. (1985) another aspect of the warm climate in greenhouses was discussed, namely the stimulation of the growth of fungi that might cause respiratory problems among the workers.

Light

Light or illumination in greenhouses is often thought of as daylight, with the sun as the actual source of lighting. Most of the activities in the greenhouses are completed during daylight, but this can also involve problems for the workers. When human activities are performed in the greenhouse at late hours or in special types of greenhouses without any natural light (without windows) one has to depend on artificial lighting. Working in daylight in a greenhouse does not normally cause any problems due to lack of light. Instead, the very high intensity of sunlight can generate occasional inconvenience from glare. Glare is produced by brightness within the field of vision that is sufficiently greater than the luminance to which the eyes are adapted, to cause annoyance, discomfort, or loss in visual performance and visibility (McCormick, 1976). Direct glare is caused by light sources (mostly the sun in greenhouse work) in the field of view. Reflected glare is caused by reflections of high brightness from polished or glossy surfaces (glass, interior fittings, etc.) that are reflected toward an individual. To prevent these problems at work during sunny periods or in sunny districts, it might be advisable to use some kind of shading, such as screens or shading paints on the roof.

Today it is becoming more and more common to use artificial lighting in greenhouses for the regulation of growth. A study by Pettersson (1987) has indicated that there could be problems of glare, if people are working in this type of light. Workers have complained about eye irritation and difficulties in selecting products according to colour. The recommendation was proper location of the light sources to avoid glare, in addition to the use of eye protectors, and, if possible, to cut off the lighting during work in any particular section of the greenhouse.

Noise and vibration

Noise is not a serious problem of the working environment in most greenhouses. When it occurs, it is due to the use of compressors, cultivators, potting machines or other type of machinery (Table 10.6). The occupational injuries in horticultural work caused by noise and vibrations are limited (about 10–12%) according to Swedish statistics (Lundqvist, 1984). The increasing use of different types of machinery in greenhouse work may raise the risk. On the other hand, modern machines are often more user-friendly than older ones. Still, it is important to choose and demand machinery and equipment with a low noise level and with a construction that eliminates the risks of injuries from vibrations. The accepted levels of noise differ from country to country. In Sweden the limit is set at 85 dB(A) as a mean during an 8-hour

Table 10.6
Summary of the range of sound levels from different sources in greenhouses [1]

Noise source	Sound level, dB(A)
Rotary cultivator	80–100
Engine-driven portable sprayer	70–90
Pot machine	50–60
Heating boiler	60–70
Electricity generating set	110
Compressor	70–85

[1] Source: Lundqvist (1982).

working day, according to the National Swedish Board of Occupational Safety and Health (1986). There are also standards on vibrations from hand-held machines (1987b). It is important to know that each person reacts differently to noise, and it is therefore imperative to use ear protection when noisy work operations are performed.

Biological factors

The most common negative effects of biological factors in the greenhouse environment are allergic reactions to certain crops, such as *Primula obconica,* daffodils, chrysanthemum and the like. In official statistics (Lundqvist, 1984) the frequency of occupational disorders such as skin diseases is about 10–15% of the total number of qualified disorders among greenhouse workers in Sweden. In a study by Lundqvist (1982) it was shown that more than 20% of the greenhouse workers interviewed had problems with allergic skin diseases. In the greenhouse environment, allergies may also be caused by pesticides such as pyrethrum and formalin. The best way of preventing allergies and skin diseases is to avoid direct contact with the causative agents. It might be possible to grow another crop or to change the type of pesticide. It could also help if protective gloves are used. Finally, it is very important to maintain a high level of personal hygiene.

Chemical factors

Pesticides

In horticulture, there is a potential exposure to a number of pesticides. In Sweden, some of the most harmful pesticides for man and/or the environment have been banned.

Working in greenhouses involves exposure in enclosed spaces to high temperature, high application rates of pesticides, continuous work-time exposure during almost the whole year, and a considerable amount of manual work, which all enhance the risk for significant uptake of chemical substances.

Pesticide concentrations in the air are correlated not only with the amount of substance applied, but also with temperature and air circulation (Williams et al., 1980, 1982). Certain pesticides in greenhouses have been shown to remain detectable in the air for long periods after application (Williams et al., 1987).

Moreover, it has been shown that the main route of exposure to pesticides is through the skin (Knaak, 1980; Kolmodin-Hedman et al., 1983). This means that manual work other than pesticide application can be at least as important as mixing and spraying operations (Skördare, 1989). A special concern, then, arises with those pesticides which are slowly metabolized in the plant, so that residues remain in the foliage for a long time (Chiba and Veres, 1981; Wallin and Olevall, 1987; Adolfsson et al., 1989). There is some evidence that some pesticides may be mutagenic in humans. It has been reported that workers, after spraying pesticides in greenhouses, had mutagenic urine (Shane et al., 1988).

Epidemiological literature in the field of horticulture is scarce. According to Littorin et al. (1990) the proportion of deaths due to respiratory cancer among orchardists was 30% higher than among other men (Milham, 1983), though no special exposure could be blamed for this increased risk (Wicklund et al., 1988). One of many risk factors for lip cancer in Sicily was working in greenhouses (Dardanoni et al., 1984). In a group of farmers, farm managers and market gardeners in the United Kingdom there was a moderately increased risk of soft-tissue sarcoma (STS), but for gardeners and groundsmen there was no such risk (Balarajan and Acheson, 1984). A census-based Swedish study of sub-cohorts of agriculturists, among them market gardeners, did not show an excessive risk of STS (Wiklund and Holm, 1986), or malignant lymphomas (Wiklund et al., 1988). In a United States study of non-Hodgkin's lymphomas, there was, however, an increased risk among orchard workers (Pearce et al., 1987). Proportional mortality ratios in British Columbia for melanomas were significantly elevated for, among others, gardeners and nursery workers (Gallagher et al., 1986).

In summary, many pesticides are mutagenic, and

some are animal carcinogens. The epidemiology of pesticide-exposed workers is conflicting. One explanation is geographical variations in exposure, due to different substances, work practices and climate.

In order to investigate possible effects of exposure to pesticides (mainly fungicides, and insecticides), a research group (Littorin et al., 1990) studied a cohort of 2363 subjects, who had been members of a horticulturists' trade association (market gardeners/nurserymen and orchardists), for at least one year in the period 1903–82. In the cohort, there was a significantly reduced mortality in the period 1964–86, compared to a local reference population. Moreover, there were significant under-representations of circulatory and respiratory deaths. There was no statistically significant deviation of deaths from intoxication and violence, from mental disorders, nor from nervous system diseases. During the period 1964–86, there were no statistically significant deviations in morbidity from tumours of the stomach, male genital organs or lymphomas or leukemias. The risks for all brain tumours, and for gliomas, were raised, but the difference was not statistically significant. For menindomas, however, a significantly elevated risk was recorded. Further, there was a significantly reduced risk for tumours of the large bowel. The mortality and tumour morbidity patterns were similar in two sub-cohorts of market gardeners/nursery-men and orchardists. In a group of 1936 association members who had horticulture as a full-time occupation, there was an increase in total brain tumour morbidity. The risk of gliomas and meningiomas was increased, but significantly only for the menindomas. The increase in risk of total brain tumours was particularly evident in the age groups below 60, irrespective of other cohort restrictions, especially in middle-aged full-time orchardists. The risk was again increased for both gliomas and meningiomas, but reached statistical significance only for the menindomas in full-time horticulturists, as well as in market gardeners/nursery-men and orchardists.

"Re-entry" intervals

There is a potential for acute effects when highly toxic fumigants are used in closed systems such as greenhouses. It is common practice to fumigate the greenhouse, and ventilate it for some hours, before permitting workers to enter. Exposure is aggravated when labor involves prolonged and continuous contact with foliar surfaces bearing pesticide residues. The dislodgeable foliar residues are transferred to workers' skin and clothing. This health problem is currently referred to as the "re-entry problem", in reference to worker re-entry into a field or a greenhouse after pesticide application (Iwata et al., 1979). Re-entry intervals have been established only in some countries. The California Department of Food and Agriculture has established re-entry intervals for 21 organophosphate insecticides (Knaak, 1980). A re-entry interval of 48 hours has been set by the US Environmental Protection Agency for 11 organophosphates. Intervals are normally calculated for pesticide application in fruit orchards, and they vary from 1 to 30 days depending on the pesticide and fruit. Hungary has a re-entry interval of 3 days, for instance, for a number of pesticides in greenhouses (Adamis et al., 1985). The aim of a Finnish study (Liesivuori et al., 1988) was to measure the diminution of fumigant concentrations in greenhouse air and the dissipation of toxic foliar residues on the plant leaves, so that practical re-entry intervals could be established. According to the results, some practical instructions could be given to prevent excessive exposure of greenhouse workers to pesticides or their degradation products. Fumigations should take place in the evening, and the greenhouse must be ventilated for 2 hours before work in the morning. Pickers, harvesters, and packers must wear protective gloves when handling treated flowers.

Fertilizers, fuels and other chemical substances

In greenhouses, liquid fertilizers of many different kinds and brands are often being used. Some of these are explosive, and have to be handled with great care. Others are very irritating and dangerous for the eyes. Fuels of petroleum type used in greenhouses must also be handled with care, because of risks of explosion, and harmful effects if they get into the eyes. People handling all types of dangerous chemical substances should have personal protective equipment according to the warning text on the packages of the products. If the package has no warning, it is always recommended to use personal protective equipment if there is any suspicion that the substance could be dangerous.

Psychological and sociological factors

Working in greenhouses can be very interesting and rewarding for some people. Others might consider it the worst job they could think of. In an interview study, 86 Swedish greenhouse workers were asked how satisfied they were with their present work

(Lundqvist, 1982). Nearly one-third (28%) said that they were very satisfied, 70% were quite satisfied and only two persons were dissatisfied. No one was very dissatisfied. When problems were discussed, matters like stressful time limits for deliveries, problems in relations with the management, and worries about losing their jobs, were mentioned. The owners and the management felt sometimes depressed due to financial problems. Some people in this group also complained about pain in the stomach, sleeping problems and problems with family relations due to their work situation. In 1984–85 a survey was made among 70 greenhouse operations in the Netherlands (Belt, 1988). The aim of the survey was to gain an insight into the (excessive) mental load, how the mental load is perceived, and what the causes are of any excessive loads in commercial greenhouse operations. By means of two lines of approach it was attempted to get an idea of the mental load: (1) short-term load (data-processing, making decisions) related to a large number of operations; and (2) long-term load, in the literature often referred to as "stress."

In the survey it was found that there are great differences in mental loads originating from the performance of operations. Operations with a high level of mental load are characterized by:

(1) constant concentration during performance;
(2) complexity (multitude of simultaneous processes);
(3) little manual labour;
(4) sometimes: work under pressure of time;
(5) sometimes: no direct feedback with results of work performed; and
(6) large consequences for business, or personal risk, in case mistakes are made.

Activities indicated as having a high level of mental load are: soil disinfection, plant protection, management, and keeping business and financial records.

The long-term level of mental load was determined according to the consequences, which can be physical and psychological complaints. A comparison was made with other occupational groups. In this comparison the grower appeared to occupy a low position regarding mental load. The survey revealed a relationship between high mental load and level of education. Growers with a high level of education (but who since school had attended few or no continuation classes) appear to be confronted with problems in the field of management and business administration which they could not solve themselves. A different cause of a high mental load is the fact that one is restricted to one's business (few days off, long working hours). This specially refers to small-sized greenhouse operations, or those growers who have difficulty in detaching themselves from business. Finally, it was found that the group of ambitious growers and those whose state of health forbade them to continue cooperating in the greenhouse operation, had relatively more physical and/or psychological complaints.

Greenhouse enterprises are very often family businesses, that engage the husband, the wife, children and a varying number of employees. This type of enterprise can socially have both positive and negative values. Barling (1990) is one of the authors who have seriously dealt with employment, stress and family functioning. He pointed out that neither family nor work roles exist in a vacuum. On the contrary, there is a growing realization that they affect each other, and the need to understand their interaction is critical.

PREVENTIVE MEASURES

Safety and preventive health organizations

Occupational health hazards have been long recognized in Sweden, and a well-developed system has evolved for industrial safety and health surveillance of workers. This is a result of far-reaching agreements between labour unions and employers' organizations, with supporting legislation in specific areas (Höglund, 1989). In agriculture and horticulture, however, health and safety efforts have been less aggressive. In 1975 a trial program was established to evaluate different models of organization of an occupational health service mainly for farmers. It was soon found that the model with farmers' health centers organized by Lantbrukarnas riksförbund (the Swedish farmers federation) was the most effective. The staff of the centers consists of a safety engineer, a physiotherapist, an occupational nurse and a physician who concentrates on the specific work-related health problems affecting persons working on farms, in greenhouses and related work-places.

The farmers, who have to pay a considerable amount of the costs for occupational health care themselves, are very decided about what kind of service they wish to receive in return. Among farmers, as in many other groups, there is a strong belief in health checkups. There is less consideration about workplace visits and preventive work. Of course, this view to a great extent depends on lack of information. Thus, one important task for the association is to inform and teach the farmers about the objectives of occupational health care and the benefits of the prevention of work injuries.

The current service program offers some kind of service to all members every year. Every second year there is a health checkup, which may be designed in different ways depending on what kind of exposure the farmer might have. In the years between the health checks the farmers are offered a menu of different activities. It includes farm visits and short courses on different topics – for instance, how to reduce hazardous organic dust, how to improve the working conditions in forestry, and the like. There may also be courses on such topics as first aid, and care of the back and neck. Beside this, the health centres are always ready to take care of any problem that the members may have concerning health and work environment. The physiotherapists perform ergonomic investigations on the farms, and also investigate and treat back problems. The physicians are mainly occupied with work-related diseases; investigation, diagnosis and proper treatment are carried out.

Besides Sweden, several other countries have occupational health and safety service for farmers and greenhouse managers, workers and similar categories. In Finland, there has been an occupational health service since 1979 (Vohlonen et al., 1985). The Farm Safety Associaton in Ontario, Canada, is another example of activities where the main efforts are devoted to advisory and educational services related to safety and health in the workplace. The main purpose is to reduce the number of injuries and fatalities occurring on Ontario farms. The Association's objective is to provide a complete education service to make all persons engaged in agricultural work in Ontario aware of occupational safety and health hazards. The Association's aim is to make people involved in agriculture aware of the proper procedures and precautions necessary in order to preserve people, equipment, materials and the environment, all of which are an integral part of today's workplace. Safety management is a necessary component in good business management. The Association encourages people in agriculture, horticulture and landscaping industries to take action before accidents occur (Farm Safety Association Inc., 1990).

Other measures

To cope with the problems that managers, owners and workers might have with their work environment or work situation, it seems necessary to give them support in different ways. The people involved in the greenhouse industry need active support from their own organizations, and from private and or governmental agencies like the advisory and preventive health services. The manager's decision-making process must be facilitated. He must be helped to collect and analyse new technical and economic information needed for planning the greenhouse operation. He also needs somebody to turn to when he is unable, mentally or physically, to handle the "stress" caused by economic uncertainty or similar factors that he cannot influence. In such situations family members, neighbours and other friends as well as professional counselors can be of great help.

NEW TECHNOLOGY FOR THE FUTURE OR BACK TO NATURE?

The use of technology in greenhouses could mean many different types and levels of solutions, from manually opening and closing of air inlets to fully automated and computerized climate control. Another example is from manual transport, placing, spacing and harvesting of pot-plants to the use of robots and automation of the whole process. The greenhouse industry is approaching a new era where one can expect two types of development:

(1) The intensive high-technological industry with increasing use of computerized, automatized and robotized processes in climate control, fertilization, pesticide treatment, transport, harvesting and packing.
(2) The small company with a low level of technology, small investments and a lot of manual work, often with an "ecological" touch, involving a drastic reduction of chemical substances such as pesticides and fertilizers used in the growing process.

In many places in the world, a lot of automatization, computerization and robotization is used. One example of high technology, which would have been unbelievable just 10 years ago, is a harvesting robot for cucumbers (Amaha et al., 1989). This new technology could have very positive effects in eliminating working procedures involving poor and harmful working positions and movements, and heavy lifting. As regards the accident risks, there might be increased problems – perhaps not in the number of accidents, but in the seriousness of injuries. In other industries with increased automated production, it has been shown that serious accidents have occurred during work operations such as maintenance work, adjustments and the like. The use of high technology will also lead to a decreased demand for workers, which could lead to unemployment. The new types of jobs that will be established could on the other hand be more interesting and need higher qualifications. The few aspects mentioned show that new technology could cause a lot of possibilities as well as problems, and indicate the importance of evaluating all aspects of new technology before it is put into practice on a large scale in many workplaces. This new development in the greenhouse sector goes along with the increasing problems of recruiting new students for agricultural and horticultural education. During the 1990s it is becoming more and more difficult to get the new generation interested in jobs in this sector. It is therefore very important that the greenhouse industry concentrates on a good working environment, and on the production of high-quality products in balance with the environment. It also has to offer meaningful jobs and a decent level of income. The next generation will critically compare work in greenhouses with jobs in other fields.

REFERENCES

Adamis, Z., Antal, A., Füzesi, I., Molnar, J., Nagy, L. and Susan, M., 1985. Occupational exposure to organophosporus insecticides and synthetic pyrethroid. *Int. Arch. Occup. Environ. Health,* 56: 299–305.

Adolfsson, M., Skördare, R. and Wettermark, K.G., 1989. *Organiskt fosfor i växthus. En pilotstudie gällande personalens exponering för bekämpningsmedel.* Lantbrukshälsan, Södertälje, 10 pp. In Swedish.

Amaha, K., Shono, H. and Takakura, T., 1989. A harvesting robot of cucumber fruit. *The 1989 International summer meeting. Quebec, Canada, June 25–28, 1989.* Canadian Society of Agricultural Engineering. ASAE Paper no. 897053.

Balarajan, R. and Acheson, E.D., 1984. Soft tissue sarcomas in agriculture and forestry workers. *J. Epidemiol. Community Health,* 38: 113–116.

Barling, J., 1990. *Employment, Stress and Family Functioning,* Wiley series on studies in occupational stress. Wiley, Toronto, 286 pp.

Belt, A.H.M., 1988. *Een verkennend onderzoek naar de mentale belasting in de glastuinbouw,* Publikatie 222. Wageningen. Instituut voor mechanisatie, arbeid en gebouwen (IMAG), Wageningen, 34 pp. In Dutch, with English summary.

Brunsson, A., Lindgren, G., Ekström, B., Gustafsson, B., Lundqvist, P. and Löfstedt, B., 1978. *Working environment in greenhouses. Part 1. Technical and physiological observations,* Report 4. Swedish University of Agricultural Sciences, Department of Farm Buildings, Lund, 90 pp.

Burse, R., 1979. Sex differences in human thermoregulatory response to heat and cold stress. *Human Factors,* 21: 687–699.

Carter, N. and Menckel, E., 1985. Near-accident reporting: A review of Swedish research. *J. Occup. Accidents,* 7: 41–64.

Chiba, M. and Veres, D.F., 1981. Fate of benomyl and its degradation compound methyl 2-benzimidazolecarbamate on apple foliage. *J. Agric. Food Chem.,* 29: 588–590.

Dardanoni, L., Gafa, L., Patemo, R. and Pavone, G., 1984. A case-control study on lip cancer risk factors in Ragusa (Sicily). *Int. J. Cancer,* 34: 335–337.

Ekström, B. and Gustafsson, B., 1978. *Working environment in greenhouses. Part 2. Globe temperature measurements in greenhouses,* Report 5. Swedish University of Agricultural Sciences, Department of Farm Buildings, Lund, 30 pp.

Ekström, B., Gustafsson, B. and Lundqvist, P., 1979. *Working environment in greenhouses. Part 3. Long term observations of air temperature and relative humidity in greenhouses,* Report 6. Swedish University of Agricultural Sciences, Department of Farm Buildings, Lund, 37 pp.

Farm Safety Association Inc., 1990. *Annual report 1989.* Farm Safety Association Inc., Guelph, Canada, 14 pp.

Gallagher, R.P., Elwood, J.M., Threlfall, W.J., Band, P.R. and Spinelli, J.J., 1986. Occupation and risk of cutaneous melanoma. *Am. J. Industr. Med.,* 9: 289–294.

Gustafsson, B., 1982. Future developments in horticulture. In: *Future Hazards in Agriculture,* Report from an international symposium at Svaneholms slott 22–23 May 1980, Report 9: 1627. Lantbrukshälsan, Stockholm, 62 pp.

Gustafsson, B. and Ekström, B., 1980. *Working environment in greenhouses. Part 4. Computerized recording of observations, processing and analysis of climate data obtained from greenhouses,* Report 12. Swedish University of Agricultural Sciences, Department of Farm Buildings, Lund, 22 pp.

Gustafsson, B. and Lundqvist, P., 1981. *Working environment in greenhouses. Part 5. Heat and work load in subjects wearing two types of protective suits,* Report 15. Swedish University of Agricultural Sciences, Department of Farm Buildings, Lund, 26 pp.

Gustafsson, B., Lundqvist, P. and Lindgren, G., 1989. Climatic effects on the greenhouse worker. Physiological measurements in a climate chamber and in greenhouses. *Swed. J. Agric. Res.,* 10: 217–225.

Gustafsson, B., Lindgren, G. and Lundqvist, P., 1991. Near-accidents in agriculture. A survey of Swedish studies. *Swed. J. Agric. Res.,* 21: 85–93.

Hildebrandt, V.H., 1989. *Preventie aandoeningen bewegingsapparaat in de land- en tuinbouw. Gezondheidsproblematiek van het bewegingsapparaat bij mannelijke ondernemers en werknemers werkzaam in veertien agrarische sektoren,* NIPG89104. Nederlands Instituut voor Praeventieve Gezondheidszorg, TNO, Leiden, 125 pp. In Dutch.

Höglund, S., 1989. Occupational health service for farmers in Sweden. *J. Occup. Med.,* 31(9): 767–770.

Iwata, Y., Dusch, M.E., Carman, G.E. and Gunther, F.A., 1979. Worker environment research: Residues from carbaryl, chlorobenzilate, dimethoate, and trichlorfon applied to citrus trees. *J. Agric. Food Chem.,* 27: 1141–1145.

Karhu, O., Kansi, P. and Kuorinka, I., 1977. Correcting working postures in industry: a practical method for analysis. *Appl. Ergon.*, 8(4): 199–201.

Knaak, J.B., 1980. Minimizing occupational exposure to pesticides: Techniques for establishing safe levels of foliar residues. *Res. Rev.*, 75: 81–96.

Kolmodin-Hedman, B., Höglund, S., Swensson, A. and Åkerblom, M., 1983. Studies on phenoxyacid herbicides. II. Oral and dermal uptake and elimination in urine of MCPA in humans. *Arch. Toxic.*, 154: 267–273.

Laitinen, A., 1980. Ergonomiska problem vid växthusarbete. Paper presented at NJF-seminar *Arbetsmiljön i lantbruket,* 2–4 December 1980. Lillehammer, 6 pp. In Swedish.

Lander, F., Skytthe, A. and Jeune, B., 1985. *Gartnernes arbejdsmiljö og helbred. Arbejdsmiljöfondet,* Rapport. Köbenhavn, 44 pp. In Danish, with English summary.

Liesivuori, J., Liukkonen, S. and Pirhonen, P., 1988. Reentry intervals after pesticide application in greenhouses. *Scand. J. Work Environ. Health,* 14(suppl. 1): 35–36.

Littorin, M., Attewell, R., Olson, M., Skerfving, S., Andersson, C. and Möller, T., 1990. *Mortality and tumor morbidity among Swedish market gardeners and orchardists,* Report. Department of Occupational and Environmental Medicine, University Hospital, Lund, 25 pp.

Löbel, H., 1977. *Arbeitshygienische Untersuchungen im Zierpflanzen- und Gemüseanbau in Gewächshäusern,* Diplomarbeit. Institut für Arbeitshygiene der Medizinischen Akademie Erfurt, 95 pp.

Lundqvist, P., 1982. *Near-accidents in greenhouses,* Report 22. Swedish University of Agricultural Sciences, Department of Farm Buildings, Lund, 46 pp.

Lundqvist, P., 1984. *Occupational accidents and diseases in horticulture 1980/81,* Special Report 136. Swedish University of Agricultural Sciences, Department of Farm Buildings, Lund, 53 pp.

Lundqvist, P., 1988a. *Människan och Tekniken i Växthusproduktionen.* Sveriges lantbruksuniversitet. Trädgårdskonferensen 1988. Konsulentavdelningen Trädgård. Alnarp, 149 pp. In Swedish.

Lundqvist, P., 1988b. *Working environment in farm buildings. Results of studies in livestock buildings and greenhouses,* Dissertation, Report 58. Swedish University of Agricultural Sciences, Department of Farm Buildings, Lund, 235 pp.

Lundqvist, P. and Gustafsson, B., 1988. Working environment in greenhouses – a review of Swedish research. *HortScience,* 23(3): 446–448.

Lundqvist, P., Löfstedt, B., Gustafsson, B., Lindgren, G., Brunsson, A. and Persson, M., 1987. *Working environment in greenhouses. Part 6. Physiological data from trained workers in a climate chamber and in greenhouses,* Report 56. Swedish University of Agricultural Sciences. Department of Farm Buildings, Lund, 103 pp.

Maeda, K., Okazaki, T., Suenaga, T., Sakurai, T. and Takamatsu, M., 1980. Low back pain related to bowing posture of greenhouse farmers. *J. Hum. Ergology,* 9: 117–123.

McCormick, E.J., 1976. *Human Factors in Engineering and Design.* McGraw-Hill, London, 491 pp.

Milham, S., 1983. *Occupational mortality in Washington State 1950–1979.* DHHS (NIOSH) publication no. 83116. Cincinnati, 18 pp.

National Swedish Board of Occupational Safety and Health, 1983. *Ordinance (AFS 1983:6) concerning Working postures and working movements.* Solna, 26 pp.

National Swedish Board of Occupational Safety and Health, 1986. *Ordinance (AFS 1986:15) concerning Noise.* Solna, 22 pp. In Swedish.

National Swedish Board of Occupational Safety and Health, 1987a. *Occupational injuries in Sweden. 1983,* Report. Solna, 193 pp.

National Swedish Board of Occupational Safety and Health, 1987b. *Ordinance (AFS 1987:7) concerning Vibrations from hand-held machines.* Solna, 12 pp.

Nilsson, K., 1975. *Arbetsmiljö i växthus. En intervjuundersökning i Malmö. Lunds universitet,* Meddelande nr 15. Institutionen för socialmedicin, Jordbrukets socialmedicin, Lund, 16 pp. In Swedish.

Pearce, N.E., Sheppard, R.A., Smith, A.H. and Teague, C.A., 1987. Non-Hodgkin's lymphoma and farming: An expanded case-control study. *Int. J. Cancer,* 39: 155–161.

Pettersson, S.-E., 1987. *Lux till max. Assimilationsbelysning i växthus sett ur arbetsmiljösynpunkt,* Examensarbete. Sveriges lantbruksuniversitet, Skyddsingenjörsutbildningen, Skinnskatteberg, 9 pp. In Swedish.

Schulz, D., 1975. *Arbeitshygienische Probleme der Frauenarbeit bei Industriemässiger Produktion unter Glas – dargestellt am beispiel des Arbeitsklimas,* Dissertation. Humboldt Universität, Berlin, 152 pp.

Shane, B.S., Scarlett-Kranz, J.M., Shaw Reid, W. and Lisk, D.J., 1988. Mutagenicity of urine from greenhouse workers. *J. Toxicol. Environ. Health,* 24: 429–437.

Skördare, R., 1989. *Resultat från mätning av bekämpningsmedel Orthene 75 i luft applicerat med kalldimningsaggregat i växthus.* Lantbrukshälsan AB. Södertälje, 6 pp. In Swedish.

Stoffert, G., 1986. Welche Arbeiten als besonders unangenehm empfunden werden. *Gartnerbörse Gartnerwelt,* 31: 1158–1162.

Stoffert, G., 1992. Personal communication.

Stoffert, G. and Timme, U., 1987. Women and work in a bended position: An anthropometrical study. Paper presented at IXth Joint Int. Ergonomics Symp. *Working Postures in Agriculture and Forestry,* CIGR/IAAMRH/IUFRO, June 9–12, 1987. Kuopio, Finland.

Stoffert, G. and Wildt, U., 1987. Work physiological investigations of potplant handling on the ground. *Acta Hortic.,* 203: 95–100.

van der Schilden, M., 1989. The OWAS system for analysing working postures. *Acta Hortic.,* 237: 129–136.

van Dieën, J.H., 1989. *Preventie aandoeningen bewegingsapparaat in de land- en tuinbouw. Ergonomische analyse agrarische sektoren,* IMAG-nota 447. Instituut voor mechanisatie, arbeid en gebouwen (IMAG), Wageningen, 237 pp. In Dutch.

Visnovsky, J. and Stehlo, P., 1982. Labour efforts in plastic foil greenhouses. In: H. Dupois (Editor), *Ergonomics of Indoor Working Places, Proc. 6th Int. Joint Ergonomic Symp.,* 9–11 September 1981, Mainz. pp. 69–72.

Vohlonen, I., Husman, K., Kalimo, E., Nuutinen, J., Tupi, K. and Virolainen, R., 1985. *Occupational health services for farmers: a study based on an experiment in 1979–83.* Publications of the Social Insurance Institution, Finland, A:21, Helsinki, 26 pp. In Finnish, with English summary.

Wallin, S. and Olevall, O., 1987. *Studie över "resthalter" i olika exponeringskällor vid användning av Alar i växthus,* Rapport 870429. Lantbrukshälsan AB, Södertälje, 8 pp. In Swedish.

Wicklund, K.G., Daling, J.R., Allard, J. and Weiss, N.S., 1988. Respiratory cancer among orchardists in Washington State, 1968 to 1980. *J. Occup. Med.,* 30: 561–564.

Wiklund, K. and Holm, L-E., 1986. Soft tissue sarcoma risk in Swedish agricultural and forestry workers. *J. Natl. Cancer Inst.,* 76: 229–234.

Wiklund, K., Lindefors, B-M. and Holm, E., 1988. Risk of malignant lymphoma in Swedish agricultural and forestry workers. *Br. J. Industr. Med.,* 45: 19–24.

Wildt, U., 1988. *Belastung und Beanspruchung des Menschen bei Gartenbaulichen Arbeiten und die hieraus resultierenden Erholungszeiten,* Dissertation, Forschungsberichte zur Ökonomie im Gartenbau, Heft 65. Institut für Gartenbau-ökonomie der Universität Hannover, Hannover und Weihenstephan, 309 pp.

Williams, D.T., Denley, H.V. and Lane, D.A., 1980. On site determination of sulfotep air levels in a fumigating greenhouse. *Am. Industr. Hyg. Assoc. J.,* 41: 647–658.

Williams, D.T., Denley, H.V., Lane, D.A. and Quan, E.S.K., 1982. Real time monitoring of methomyl air levels during and after spraying in a greenhouse. *Am. Industr. Hyg. Assoc. J.,* 43: 190–195.

Williams, D.T., Shewchuck, C., Lebel, G.L. and Muir, N., 1987. Diazinon levels in indoor air after periodic application for insect control. *Am. Industr. Hyg. Assoc. J.,* 48: 780–785.

Zegers, D. and van Dieën, J.H., 1987. The application of the OWAS and NIOSH working posture analysing system in the Dutch agriculture. Paper presented at IXth Joint Int. Ergonomics Symp. *Working Postures in Agriculture and Forestry,* CIGR/IAAMRH/IUFRO, June 9–12, 1987. Kuopio, Finland.

Chapter 11

ENERGY CYCLE

A. BAILLE

INTRODUCTION

General context of greenhouse production

Agriculture is the human activity devoted to converting natural solar energy into useful chemical energy by means of plants and animals. In recent years it has become apparent that considerable amounts of other energy sources are also used in the process. In some cases, and particularly in the case of protected crops, this energy consumption may be so important that modern agriculture could, perhaps with some exaggeration, be defined as the human activity that converts fossil energy into useful chemical energy by means of plants, animals and sun (Dekkers et al., 1974; de Wit, 1986).

Precisely in the case of the greenhouse-crop industry, which has faced (and will face in the foreseeable future) problems connected with the large fluctuations in the price of fossil energy, an overview of the energy cycle, including the main processes of the system, is useful in order to assess the relative weight of these different processes (Enoch, 1978a).

The present chapter covers the different aspects of the greenhouse energy cycle. Basically, a greenhouse can be considered as a "physical" solar collector which contains many other smaller "biological" solar collectors (plant leaves). The environmental conditions created by the physical collector have to be the most appropriate for the growth and development of the small biological collectors.

Unfortunately, the "spontaneous" environment created by the greenhouse is seldom in concordance with the requirements for plant-physiological comfort. So, human intervention by means of auxiliary equipment and artificial sources of energy has to be provided in order to create adequate environmental conditions. This intervention is mainly needed for supplementary energy which is used to compensate for an insufficient supply of natural energy (in winter) or to remove the excessive energy supply which enters the greenhouse (in summer). The second case is the worse in terms of energy efficiency: artificial energy is used in order to waste some natural energy.

Another characteristic of the greenhouse system is the wide range of technical solutions which can be applied in order to control the environmental conditions. From a physiological point of view, the lower and upper limits of the system, as quoted by Enoch (1986), may be defined as:

- a "minimal" environment modification which permits the survival of the crop during a given critical period, when outside conditions are particularly adverse;
- a "maximal" environment modification, enabling the plant to grow in the optimal conditions for maximum yield and quality.

Between these two limits, many other possibilities are available to the grower, who generally will adopt a compromise based on technical, biological and economic criteria, in which energy parameters will play an important part, in the investment cost as well as the running cost.

Thus, energy inputs may vary to a large extent, following the level of environmental control which is required for the crop by the grower. Generally, two environmental sub-systems may be distinguished: (1) the aerial part of the crop, which is affected by the climate control facilities: heating, ventilation, carbon dioxide enrichment, (de)humidification, supplementary lighting, etc.; and (2) the root part of the crop, which is affected mainly by the facilities for supplying water and minerals: irrigation, fertilization, soil heating, soil sterilization, etc.

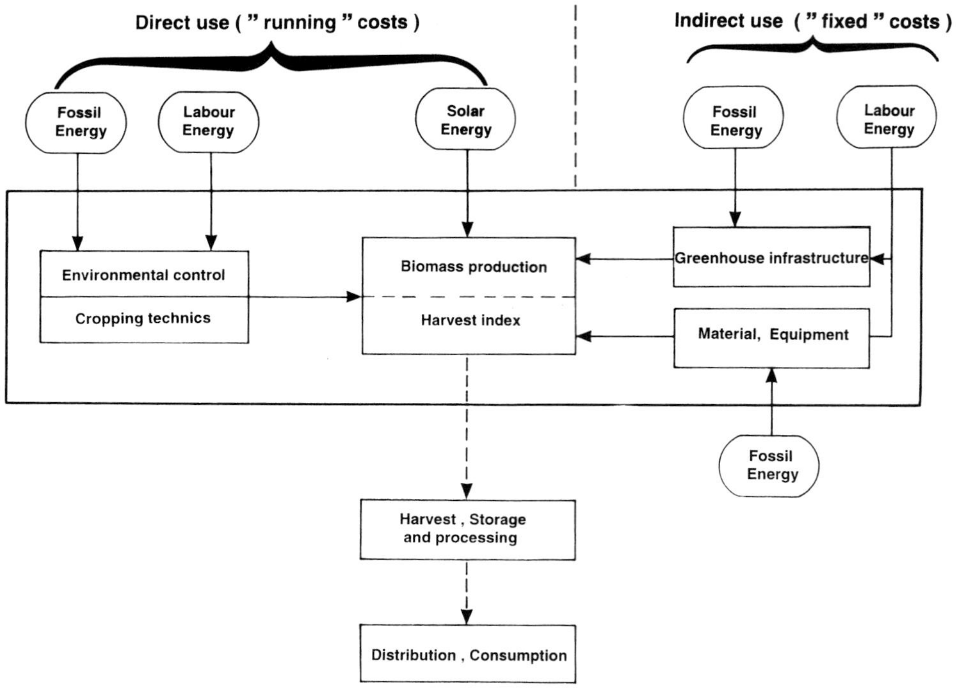

Fig. 11.1. Global scheme of the energy cycle in greenhouse ecosystems.

Apart from these two main environmental-control inputs (the first one being specific to greenhouse crops), the other "direct" or "indirect" inputs which are currently considered in the energy analysis of agricultural systems are: (1) chemical products for plant protection; (2) energy input comprised in the greenhouse infrastructure, covering material and equipment (insulation devices, heating systems, mechanization, etc.); and (3) human labor for cultural practices such as pruning, plant training and harvest.

The global scheme of the energy cycle in the greenhouse system

In the literature, three forms of energy are generally recognized: (a) Solar energy and groundwater: this first form is considered as a free subsidy in the economic analysis of agricultural systems (Slesser, 1973). (b) Human energy and energy for life support: the value of this term was preponderant in pre-modern agriculture, but is nowadays only a small part of the whole energy input to modern intensive agricultural systems. (c) Direct and indirect fossil-energy subsidy: this sort of energy input combines many direct inputs: fossil fuels (oil, gas, coal), nutrients, fertilizers, pesticides and herbicides, which may be considered as *direct* inputs, as they are used directly in the course of the production process. The *indirect* energy input is, for example, the energy needed for manufacturing and building the farm-housing infrastructure, irrigation, machinery, equipment, etc.

In terms of economic analysis, direct energy input will be taken into account in the "running" costs, and indirect energy input will be included in the "investment" or "fixed" costs (Manby, 1984). The two last forms (b) and (c) may also be qualified as "support" energy or "cultural" energy.

Another classification is to distinguish between inputs which are *yield increasing* and *yield protecting* (de Wit, 1986). In work which contributes to yield, substitution is possible between human labour and mechanical equipment with their associated energy requirements. On the other hand, other necessary "yield-increasing" inputs, such as water, minerals and organic materials, cannot be replaced by labour. "Yield-protecting" inputs are mainly for disease and pest control.

Figure 11.1 gives the global scheme for the energy cycle in the greenhouse system where the different energy inputs are separated: (1) into direct and indirect inputs; and (2) into solar, fossil and labour inputs.

In the following, I shall consider only the production

system – that is, the processes up to the harvest stage. The storage and processing stages, as well as the distribution process, up to the moment when the product is served on the consumer's plate (Enoch, 1978a), will not be discussed here.

I shall successively deal with three aspects:

(1) the solar energy input and biomass production output;
(2) the direct fossil-energy input used for environmental control processes;
(3) other direct and indirect energy inputs into the greenhouse ecosystem.

At the end of each section, the energy efficiencies (or ratios) which are related with these inputs/outputs will be discussed.

SOLAR ENERGY INPUT INTO THE GREENHOUSE

Solar energy availability

Solar energy is a natural energy source which is theoretically available and harvestable in considerable amounts if one considers the earth surface (seas and lands) and the yearly average global radiation incident on it. Clearly, some parts of the earth are more favored in this aspect than others.

Table 11.1 gives some examples of monthly average global radiation at four places in the Northern Hemisphere where greenhouse crops represent an important component of the local agricultural production and economy. It may be seen that, in January, Israel or southern Spain receive 4 to 5 times more radiation than the Netherlands. Differences are much smaller

Table 11.1
Monthly average global radiation in January, March and June for 4 locations in the Northern Hemisphere

Location	Lat.	Global radiation ($MJ\,m^{-2}\,day^{-1}$) monthly average		
		January	March	June
The Netherlands (de Bilt)	52°N	2.5	8.1	18.9
Southern France (Avignon)	44°N	6.3	13.0	24.0
Southern Spain (Almeria)	37°N	9.8	15.4	26.0
Central Coast of Israel (Bet Dagan)	32°N	11.4	19.5	28.2

in summer, the surplus of energy in more southerly latitudes (30°N) being only 1.5 times greater than towards the north (50°N).

Another comparison can be made on an annual basis: 4000 $MJ\,m^{-2}$ are received in the Netherlands and southern England, 6000 $MJ\,m^{-2}$ in southern France and about 7000 $MJ\,m^{-2}$ in Israel.

These climatic differences will, of course, influence significantly the supplementary input of energy required for heating, but also the possibilities of carbon dioxide enrichment, the choice of internal equipment for the greenhouse, climate control, crop management, etc.

Coping with the large variations between minimum winter and maximum summer radiation levels (tenfold for 50°N latitude, threefold for 30°N latitude) represents the major challenge to continuous crop production in greenhouses. Insufficient light for winter production is the main challenge for growers in northern latitudes (also in more southerly latitudes, but to a lesser extent), whereas sufficient cooling of the greenhouse environment in summer, which is absolutely necessary to achieve good-quality production, is hard to realize in southerly latitudes.

In fact, the greenhouse environment from winter to summer is a compromise (Short and Breuer, 1985). Most greenhouse growers from most latitudes would prefer to have all external conditions similar to their own springtime, when heat or water stresses may be avoided without costly and sophisticated equipment, and without a large requirement for fossil energy: heating and cooling loads depend strongly on global solar radiation and on outside air temperature (cf. Chapter 3).

A proper understanding of the energy behavior of a greenhouse demands that the various heat fluxes be quantified for given climatic conditions. The input of solar energy is the basic forcing function which will determine the rate of energy dissipation or gain. The following analysis, using a global energy balance of the greenhouse, is intended to stress the main characteristics which prevail in this behavior, and the role played by solar input in the energy balance of the greenhouse.

The global energy balance of greenhouses

The energy balance of greenhouses is formulated by writing that heat input is equal to heat output plus a term including the inertial thermal behavior of the greenhouse, which is zero if steady-state conditions are assumed. A simple and straightforward approach is to consider the greenhouse as a solar collector, and to use the same form of equations (Duffie and Beckman, 1974) to describe the greenhouse energy balance:

$$Q = H - U(T_i - T_o), \qquad (11.1)$$

where: H, solar input to the greenhouse per unit floor area (W m^{-2}); U, overall heat loss per unit floor area and per unit temperature difference (W m^{-2} K^{-1}); T_i, T_o, inside and outside air temperature (K); Q, residual term (W m^{-2}), or enthalpy gain (or loss) of the greenhouse.

If Q is positive, then heat is in excess and increases T_i, unless vents are operated. Estimates of Q will indicate the cooling requirements of the greenhouse. In hot climates, venting only is not sufficient to extract all the excess energy, and high values of T_i can result, unless other means for cooling are available (evaporative cooling, shading screens, etc.).

If Q has negative values, then heat supply is necessary to keep the desired temperature T_i. In this case, estimates of Q will indicate the heat requirements of the greenhouse.

In this section, I shall focus on the analysis of H, the heat input into a greenhouse resulting from outside incident solar radiation, G_o. This heat input may be expressed as (Garzoli and Shell, 1984):

$$H = G_o \tau \alpha \frac{[1 - (\varphi + F_s)]}{[1 - (-\alpha)\rho]}, \qquad (11.2)$$

where: G_o, outside global solar radiation per unit floor area (W m^{-2}); τ, transmissivity of the greenhouse; α, solar absorptivity of the internal surface (soil, canopy); ρ, reflectivity of the greenhouse to global radiation; φ, fraction of internal solar radiation which is used in the photosynthetic process; F_s, fraction of internal solar radiation which is transferred by conduction in the soil. The term τG_o represents the global solar radiation G_i incident on the canopy.

For estimating H, it is necessary to know, apart from G_o, the values of the different parameters included in eq. (11.2), and which characterize the greenhouse (ρ,τ), crop (α,φ) and soil (F_s).

Radiative properties of greenhouses

Transmissivity

The fraction of solar radiation which is transmitted into a greenhouse is a complex function of the sun's position in the sky, the greenhouse orientation and the radiative characteristics of the greenhouse structure and contents. Simple methods for estimating transmissivity have been given by Smith and Kingham (1971), Takakura et al. (1971) and Baille et al. (1989), while Morris (1972) has stressed the difficulties in measuring this quantity, caused by considerable sampling problems. More sophisticated numerical models have also been proposed (Kozai et al., 1973; Critten, 1983a,b), mostly for purposes of design optimization.

Values of greenhouse global transmissivity higher than 0.7 are the exception. Standard values for single-wall greenhouses range between 0.55 to 0.70, with the highest values in summer and the lowest in winter. For double-wall greenhouses, values from about 0.5 to 0.6 are currently measured. Variability in transmissivity (and, likewise, in reflectivity) is mainly due to the direct component of solar radiation. Transmissivity for diffuse radiation may be considered as constant for a given greenhouse (van de Kieboom and Stoffers, 1985).

Reflectivity and absorptivity

The reflectivity of the greenhouse, ρ, may be taken as 0.20–0.25 on average, and absorptivity of the cover and outer structures (ridges, glazing bars) ranges from 0.15 for single glazing to 0.25 for double glazing. A significant fraction of the radiant energy which is absorbed by the greenhouse structure is dissipated in the form of sensible heat in the air within the greenhouse, and contributes in fact to the solar input, H.

In the following, values of τ are chosen as 0.70 for a single-glazed greenhouse and 0.60 for a double-glazed greenhouse. Reflectivity is 0.15 in the two cases and absorptivity 0.15 and 0.25, respectively (the sum of the three parameters is 1.0).

Conductive flux into the soil

The conductive flux into the soil is the main storage term (if one neglects the inertia thermal mass constituted by air, canopy and internal equipment such as heating pipes). This term is positive during daytime (storage) and negative during night-time (restitution). Recently, some work has been focused on the estimation of this flux. Albright et al. (1985) have analyzed

Table 11.2
Scientific names of crop plants mentioned in this chapter

Common name	Scientific name
Alfalfa [lucerne]	*Medicago sativa*
Chrysanthemum	*Dendranthema grandiflora*
Cucumber	*Cucumis sativus*
Lettuce	*Lactuca sativa*
Lucerne [alfalfa]	*Medicago sativa*
Maize	*Zea mays*
Rose	*Rosa × hybrida*
Sorghum	*Sorghum bicolor*
Sugarcane	*Saccharum officinarum*
Tomato	*Lycopersicon esculentum*
Wheat	*Triticum aestivum*

the energy balance of the greenhouse by adding an inertial term to eq. (11.1):

$$Q = H - U(T_i - T_o) - C\left(\frac{dT_m}{dt}\right), \quad (11.3)$$

where: C (kJ m^{-2} °C^{-1}) is the heat capacitance of the thermal mass (soil, or soil plus canopy), and T_m its temperature; H, the solar input, is expressed as γG_o. Estimation of γ, U and C by multiple regression led to values of C ranging from 14 kJ m^{-2} °C^{-1} for a dry soil, to 137 kJ m^{-2} °C^{-1} for a wet cultivated soil.

Boulard and Baille (1987) have slightly modified eq. (11.3), approximating the daily stored energy by the term $C'(T_d - T_n)$, where T_d and T_n are the average daytime and night-time air temperatures, respectively, and C' is an equivalent heat transfer coefficient for heat storage. From measurements in a tomato greenhouse, they found values of C' ranging from 0.8 to 3.6 W m^{-2} °C^{-1}, with an average value of 2.2 W m^{-2} °C^{-1} (corresponding to a C value of about 140 kJ m^{-2} °C^{-1}).

Thus, the order of magnitude of storage flux may be estimated as 20 W m^{-2} for $T_d - T_n = 10$°C. For a daytime period of 12 hours, the energy stored into the soil is about 0.70 MJ m^{-2}.

Jolliet (1988), by another method, has estimated the energy stored in the soil for two glasshouse crops (Fig. 11.2): (1) a chrysanthemum canopy, completely covering the soil, for which the maximum value of stored energy is 0.50 MJ m^{-2} day^{-1}; and (2) a tomato crop, with paths between plant rows, for which the maximum value is 1.0 MJ m^{-2} day^{-1}.

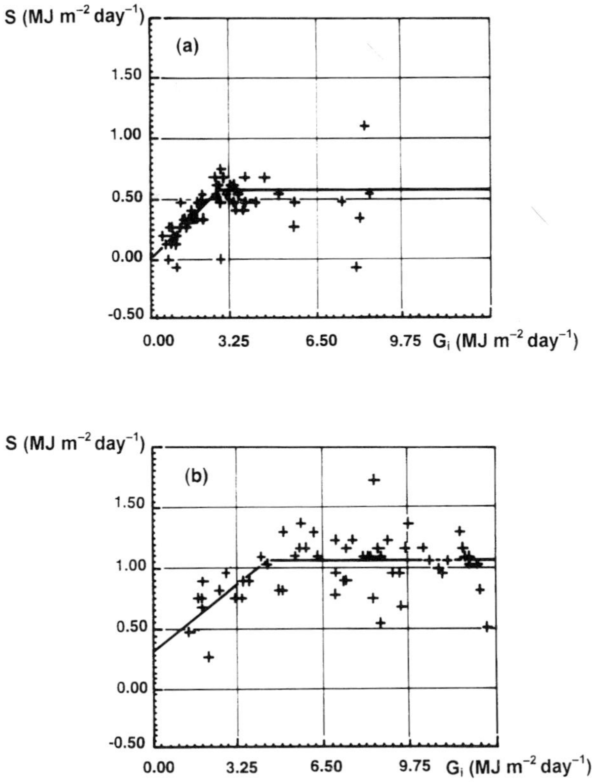

Fig. 11.2. Energy stored in the ground (S) against incident global radiation (G_i) for two greenhouse crops: (a) chrysanthemum; (b) tomato. From Jolliet (1988).

Considering also the value of 0.53 MJ m^{-2} day^{-1} given by Bailey (1977) for a tomato glasshouse crop, one may adopt for the soil storage term a value of 0.5 MJ m^{-2} day^{-1} for closed canopy and 0.7 MJ m^{-2} day^{-1} for row canopy. Thus, taking an average value of incident radiation flux of 10 MJ m^{-2} day^{-1}, one obtains values of F_s ranging from 0.05 to 0.07.

This storage term can be increased by adding some additional thermal mass inside the greenhouse. Passive systems, such as plastic tubes filled with water, known as "water sleeves" (Graffiadellis, 1986), allow storage of 1 to 1.5 MJ m^{-2} day^{-1} during sunny winter days in northern Greece. Active systems, using fans which force the hot air of the greenhouse into a storage medium, such as a rock pile (Garzoli and Shell, 1984) or the greenhouse soil itself (Boulard and Baille, 1986a,b), have also been tested and proved to be efficient. In these cases, the greenhouse acts like a true solar collector.

Radiation absorption by the canopy

To allow estimation of α and of the fraction φ of solar input which is used for the photosynthetic process, the amounts of global radiation and photosynthetically active radiation (PAR) absorbed by the crop have to be calculated.

The crop absorption coefficient for global radiation, α_g, may be defined (Gallagher and Biscoe, 1978) as:

$$\alpha_g = \frac{G_{abs}}{G_i} = \frac{(G_i - G_r - G_t)}{G_i}, \tag{11.4}$$

where: G_i is the incident global radiation at the level of canopy top; G_r the reflected global radiation; G_t the transmitted global radiation at the ground level; and G_{abs} the global radiation absorbed by the crop.

Other authors (Varlet-Grancher and Bonhomme, 1974; Hodge and Kanemasu, 1977) have used the relationship:

$$G_{abs} = G_i - G_r - G_t + G_{t,r}, \tag{11.5}$$

where $G_{t,r}$ is the reflected fraction of G_t (Fig. 11.3). Equations (4) and (5) are very close to each other, as $G_{t,r}$ is generally negligible, except in the case of sparse canopies.

Some quantitative studies on the absorption of global and PAR radiation by greenhouse crops are available in the literature. The first comprehensive study was performed by Stanhill et al. (1973), who estimated the absorption of a rose row crop in Israeli conditions. Using several solarimeters for the measurement of the different components of G_{abs}, they found values of α_g ranging from 0.45 when the canopy has reached half its final height, to 0.51 when the canopy development was complete.

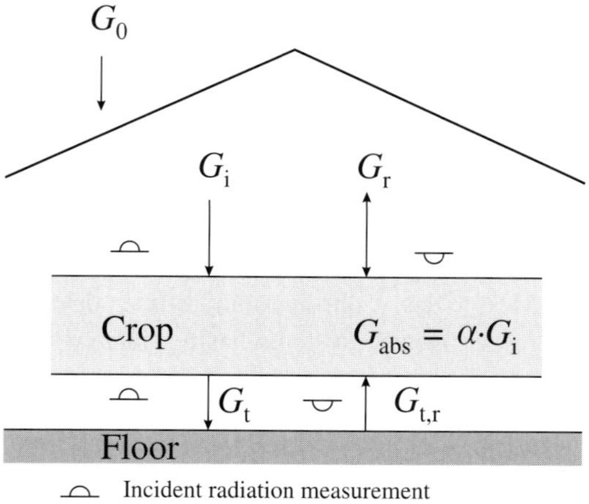

Fig. 11.3. Determination of the crop absorption coefficient, α_g.

Using hemispherical photographs, estimation of the absorption coefficient for the direct component gave values between 0.61 (at the equinox) and 0.69 (at the summer solstice). This example shows that canopy absorption of diffuse radiation is significantly lower than absorption of direct radiation, and confirms a general characteristic of crop radiative behavior: direct radiation is more efficiently and rapidly intercepted by crop canopies than diffuse radiation. It may also be concluded that seasonal changes in direct-component absorption and effects of greenhouse structure, orientation and cover material on the ratio diffuse/direct are the most important non-plant factors determining solar absorption by the crop canopy (Stanhill et al., 1973).

For the same rose crop, reflection from the surface of the crop canopy ranged from 0.14 to 0.20 for PAR radiation (<700 nm) and from 0.19 to 0.41 for near-infrared (IR) radiation (>700 nm). These differences are due to the higher reflectivity of rose leaves in the near-IR waveband.

These results lead, for this rose crop, to the following conclusions: (1) about 20% of the solar radiation reaching a fully developed rose canopy is reflected in the PAR waveband, and up to 40% in the near-infrared. (2) 30% of the solar radiation reaching the top of the fully developed rose canopy is transmitted through the canopy, either through gaps in the canopy or through the pathways between the rows. (3) 50% of the incident solar radiation is absorbed by the rose canopy, and the absorption coefficient α_g increases with the size of the direct component and the horizontal homogeneity of foliage.

Similar studies for other row canopies such as the tomato crop (Acock et al., 1978; Kurata et al., 1988; Papadopoulos and Ormrod, 1988a,b) or cucumber crops (Nederhoff, 1984) lead to analogous results. The effects of spacing and planting density of greenhouse tomato crops have been extensively studied by Papadopoulos and Ormrod. They found a deeper penetration of PAR in widely spaced canopies than in more closely spaced canopies. This is of significant interest, as several investigators (Charles-Edwards, 1982) have demonstrated a positive correlation between the uniformity of PAR distribution in the canopy and the efficiency of PAR utilization in photosynthesis. The effect of plant spacing may be considered in terms of light exposure, as with wide spacing there is less

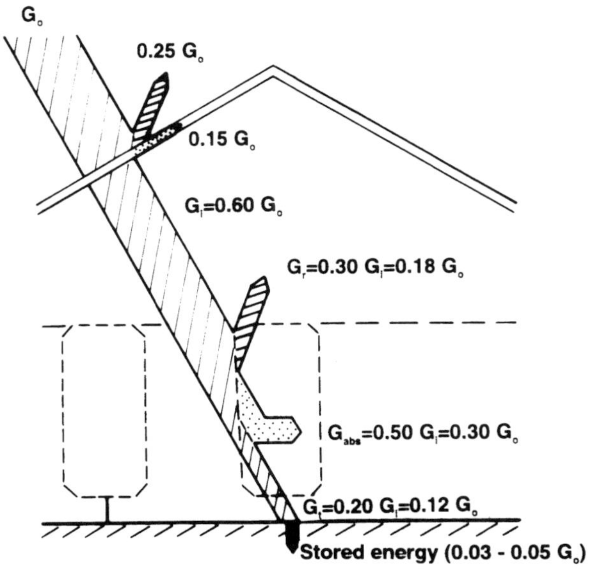

Fig. 11.4. Example of solar radiation partition in a typical greenhouse crop.

overlapping and mutual shading of leaves, and less competition for light.

For greenhouse row crops, values of α_g may range from very low values at the beginning of the crop to maximum values of 0.8 to 0.9 when the crop is fully developed.

Inside the crop, complete absorption of PAR at the bottom of the crop is commonly observed when the leaf area index (LAI) is about 4 or above (Papadopoulos and Ormrod, 1988a).

The light absorption by the crop can be improved by adding reflective surfaces on the ground or at the north wall in the Northern Hemisphere (in the case of single or twin span greenhouses). Covering the floor with a highly reflective surface enhances the radiation absorption significantly, especially when a substantial part of the radiation is transmitted by the crop to the ground – for instance, in the case of widely spaced crops and crops with low LAI, increases of 3% were measured at the top of a rose canopy, which means an increase of 10% (or more) in absorbed radiation (Stanhill et al., 1975). Lining the north wall with a reflective material has been studied by several authors, either theoretically (Kozai and Sugi, 1972; Kurata, 1983) or experimentally (Stanhill et al., 1975; Thomas, 1978). This design increases the incident radiation at the top of the canopy, in the northern part of the greenhouse in the Northern Hemisphere, by about 10% or more.

Other devices have also been proposed to create a more homogeneous distribution of radiations in greenhouses: venetian blinds (Critten, 1985, 1988; Cave and Cockshull, 1989), and partially reflective screens between row crops (Aikman, 1989). In the last case, effects on crop production can be significant (17%), as demonstrated theoretically by Aikman.

An example of the partition of solar radiation in a typical greenhouse crop, assuming $\tau = 0.7$ and $\alpha_g = 0.5$, is presented in Fig. 11.4.

Energy used in the photosynthetic process and biomass production

The proportion φ of absorbed solar radiation which is used in the photosynthetic process can be estimated by the "radiation use" efficiency, RUE, also named "conversion" efficiency or "biological" efficiency, and defined as:

$$\mathrm{RUE} = \frac{\mathrm{ES}}{\text{Radiation Integral}}, \tag{11.6}$$

where ES ($\mathrm{MJ\,m^{-2}}$) is the equivalent energy stored in dry matter (d.m) production due to a given radiation integral ($\mathrm{MJ\,m^{-2}}$) received during a given reference period (crop duration, year, etc.). Several RUE values can be calculated depending on the radiative parameter that is used: G or PAR, incident or absorbed. The type of reference radiation has to be specified when using such efficiencies.

The RUE is directly related, via the calorific equivalent C of dry matter (taken here as $16\,\mathrm{MJ\,kg^{-1}}$ d.m.), to the "yield" ratio YR ($\mathrm{g\,MJ^{-1}}$):

$$\mathrm{YR} = \frac{\mathrm{RUE}}{C} = \frac{\text{d.m.}}{\text{Radiation Integral}}. \tag{11.7}$$

This simple physiological framework, based on amount of radiation and its efficiency of conversion to dry matter, has been proposed by Monteith (1977) and used by several authors (Charles-Edwards, 1982; Varlet-Grancher et al., 1982; Gosse et al., 1986) to either predict or analyze crop growth. The results since published on this subject mainly relate to open-field C_3 or C_4 crops, for which RUE depends mainly on crop structure, genotype and cultural practices. In greenhouse crops, RUE will depend also on glasshouse structure (transmissivity) and on environmental control strategy.

Table 11.3 presents the successive stages of radiation loss and use in a typical greenhouse ecosystem, as hypothesized by Warren-Wilson (1972).

Table 11.3
Radiation use in a typical greenhouse ecosystem [1]

Hypothesis	Available energy in units of G_o	
Outside radiation	1	
PAR proportion in G_o: 0.5	0.50	(= PAR_o)
PAR transmissivity of the greenhouse: 0.7	0.35	(= PAR_i)
Absorbed PAR: 0.5	0.175	(= PAR_a)
Dry matter fixed in photosynthesis: 0.143	0.025	(gross photosynthesis)
Less photorespiration and maintenance cost: 0.4	0.015	(net photosynthesis)
Less growth respiration cost: 0.333	0.010	(total d.m. energy content)
Harvest index: 0.5	0.005	(harvest d.m. energy content)

[1] Warren-Wilson (1972).

Table 11.4
Comparisons of RUE and YR (based on PAR_a) given in the literature

Crop	RUE (%)	YR ($g\,MJ^{-1}$)	Source
Open-field crops			
Wheat (C_3)			
old cultivars	3.7	2.35	Siddique et al. (1989)
new cultivars	4.8	2.9	Siddique et al. (1989)
Alfalfa (C_3)	2.9	1.8	Gosse et al. (1986) (aerial d.m. only)
Maize (C_4)	1.0	2.5	Gosse et al. (1986)
Sugarcane (C_4)	4.3	2.65	Gosse et al. (1986)
Sorghum (C_4)	3.6–4.9	2.2–3.0	Hammer and Vanderlip (1989)
Glasshouse crops			
Tomato	7.1	4.4	Aikman (1989)
Rose	3.9	2.4	Stanhill et al. (1973)

With these hypotheses, one obtains RUE values of 0.5% for harvested d.m., when based on G_o; 1% for total d.m., when based on G_o; 2% for total d.m., when based on PAR_o; 3% for total d.m., when based on PAR_i; 6% for total d.m., when based on PAR_a.

If absorbed PAR is near 100%, then a RUE value (based on PAR_a) of about 12% may be expected. Some values of RUE for greenhouse crops are available in the literature: Stanhill et al. (1973), for a glasshouse rose crop (cv. Baccara), estimated the annual dry matter production as $2.14\,kg\,m^{-2}\,yr^{-1}$, which converted to its calorific equivalent represents an energy flux of $35\,MJ\,m^{-2}\,yr^{-1}$. Compared to the PAR incident on the canopy (PAR_i: $1965\,MJ\,m^{-2}\,yr^{-1}$), RUE is 1.8%. When expressed on the basis of absorbed flux ($PAR_a = 0.45\,PAR_i$), RUE is 3.9%, the corresponding Yield Ratio (YR) values are $1.1\,g\,MJ^{-1}$ and $2.4\,g\,MJ^{-1}$, respectively.

Aikman (1989), for southern England, gave figures of $4000\,MJ\,m^{-2}\,yr^{-1}$ for G_o. Assuming a greenhouse transmissivity of 0.60, PAR/G ratio of 0.5 and PAR canopy absorption of 0.75, leads to the PAR absorbed by the canopies being $900\,MJ\,m^{-2}\,yr^{-1}$. The maximum yield obtained by growers in this region with a year-round tomato crop is about $45\,kg\,m^{-2}\,yr^{-1}$ (fresh matter) with carbon dioxide enrichment and accurate climate control. Taking into account the harvest index (0.66) and the d.m. content (6%) leads to a total d.m. production of $4\,kg\,m^{-2}\,yr^{-1}$. In this case, the PAR_a use efficiency would be:

$$RUE\,(PAR_a) = \frac{4^*16}{900} \approx 7\%$$

and:

$$YR\,(PAR_a) = \frac{4000}{900} \approx 4.4 gMJ^{-1}.$$

Referred to the outside global radiation, RUE(G_o) is 1.6% and YR(G_o) = 16 g (fresh matter) MJ^{-1}, or 1 g (d.m) MJ^{-1}.

Bailey (1988) has given, for a glasshouse tomato crop in southern England, YR(G_o) = 16 g (fresh matter) MJ^{-1}, or 1 g (d.m.) MJ^{-1} for the harvested part, which means about 1.4 g (d.m) MJ^{-1} for the total dry matter. This last value, 40% higher than the previous value, would correspond, if one adopts the same assumptions as Aikman, to RUE(PAR_a) values of about 10%.

From these examples, one can assume that the proportion of absorbed PAR which is used for dry matter production varies from 4% up to 10%, which is probably the maximum now attained by the most effective growers. Thus, plausible values of φ in eq. (11.2) may be taken as ranging from 0.04 to 0.10 if related to PAR_a, and 0.02 to 0.05 when related to G_i.

In Table 11.4, comparisons of RUE and YR values based on PAR_a are presented for greenhouse crops and

Table 11.5
Comparisons of G_o, PAR_i, PAR_a and respective values of RUE for one greenhouse crop and two C_4 open-field crops

Inputs	GTC [1]	MF [1]	SFW [1]
G_o ($MJ\,m^{-2}\,yr^{-1}$)	4000	4500	7500
Crop duration (days)	330	200	340
Integral of PAR_i($MJ\,m^{-2}$) during the crop duration	1200	1300	3500
Integral of PAR_a ($MJ\,m^{-2}$)	900	550	2200
Total d.m. production ($kg\,m^{-2}$)	3.0 [2] 4.0 [3]	1.4	6.0
RUE (G_o)	1.3 [2] 1.6 [3]	0.5	1.3
RUE (PAR_i)	4.0 [2] 5.3 [3]	1.7	2.7
RUE (PAR_a)	5.3 [2] 7.1 [3]	4.0	4.3

[1] Abbreviations: GTC, greenhouse tomato crop (Southern England); MF, maize (France, near Paris); SFW, sugarcane (French West Indies).
[2] Average value.
[3] Maximum value.

several open-field crops (all values refer to the PAR absorbed radiation integral). One may note that the efficiencies of greenhouse crops are similar to or higher than those estimated for the best-performing open-field C_4 crops (sorghum, sugarcane, maize). These high efficiencies are consequences of optimal environmental control, which allows high levels of photosynthesis although radiation levels are 30% lower than outside. Table 11.5 presents more detailed comparisons between two C_4 crops: maize (in France) and sugarcane (in the French West Indies) and a glasshouse tomato crop (in southern England).

Solar heat input and its partition between sensible and latent heat

Returning to eq. (11.2), where values of parameters α, ρ, φ and F_s are chosen as: $\alpha = 0.80$; $\rho = 0.15$; $\varphi = 0.05$; $F_s = 0.03$; one obtains:

$$H = 0.77\tau G_o = 0.77 G_i.$$

For a single-glazed greenhouse ($\tau = 0.7$), the available solar input would be about 0.55 G_o, while for a double-glazed greenhouse ($\tau = 0.6$), $H = 0.46\,G_o$.

Taking into account the fact that a significant part of the radiation absorbed by the structure is dissipated into the greenhouse, one may deduce that about 60 to 65% of the outside radiation is available in the single-glazed greenhouse for either sensible or latent heat exchanges, and about 55% in a double-glazed greenhouse. These values may be compared favourably to experimental evaluations of H given in the literature in the form

$$H = \gamma G_o,$$

with values of γ: for solar collectors, 0.6 to 0.8 (Beckman et al., 1977); for single-glazed greenhouses, 0.65 (Duncan et al., 1981); 0.64 (Jolliet, 1988); 0.66–0.79 (Boulard and Baille, 1987).

The fraction of H which is used for evapotranspiration, ET, is a complex function of a number of microclimatic and plant parameters which are difficult to measure, especially in the non-homogeneous situations which commonly occur in greenhouses.

The partition of H between sensible and latent heat may be estimated from measurements of variations in the dry and wet enthalpy of the greenhouse air, when extracted at a known rate by extractors (Boulard, 1985). Figure 11.5 shows changes over the year in the ratio of latent heat enthalpy to H, for a glasshouse tomato crop in the south of France. This ratio is about 0.5 at the beginning of the crop (low LAI) and increases up to 0.7 at the end (July). Garzoli and Blackwell (1973) found experimentally that transpiration from a fully developed crop varies between 48 and 75% of G_i, with an average of 57%.

Other authors have shown that, for a given growth stage and set of environmental conditions, the relationship between evapotranspiration (ET) and either G_i or G_o is quite constant (Stanhill and Scholte Albers, 1974).

These results have led to the use of simple regression formulas relating ET to G_o (or G_i) for estimating the daily water requirements of greenhouse crops (de Villèle, 1974; Michels, 1984):

$$\mathrm{ET} = K_c\left(m + n\frac{G_o}{L}\right), \tag{11.8}$$

with ET in mm day^{-1} and G_o in $MJ\,m^{-2}\,day^{-1}$, m and n are constants for a specific greenhouse and climate, L is the latent heat of vaporization ($MJ\,kg^{-1}$), and K_c is a

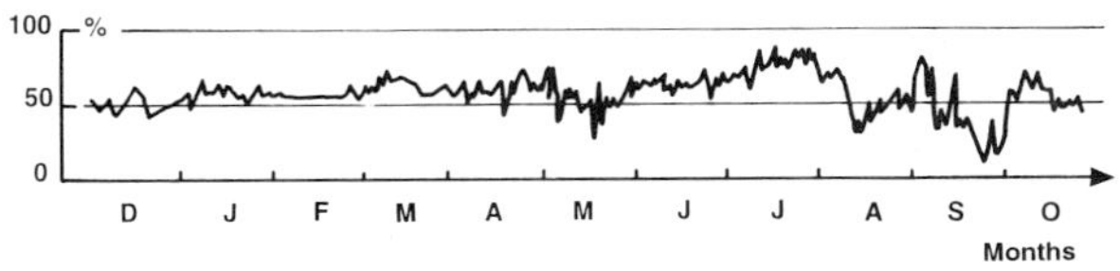

Fig. 11.5. Seasonal changes in the ratio (latent heat enthalpy)/H, for a tomato greenhouse crop. From Boulard (1985).

"crop" coefficient, depending on the crop development stage. The m constant is generally negligible, unless nocturnal heating is intensive (de Graaf and van den Ende, 1981).

Other rough approximations which are sometimes used are:

$$ET \approx G_{abs} \tag{11.9}$$

and

$$ET \approx Rn, \tag{11.10}$$

where Rn is the greenhouse net radiation (solar plus long-wave net balance) measured above the crop. The long-wave radiative exchanges between the canopy and the greenhouse roof (or roof and sky in the case of transparent cover material like polyethylene) do not represent a large amount of energy, as usually during most of the day the difference between roof temperature and canopy temperature is not important. So, neglecting the long-wave element, the daytime net radiation may be approximated by the net solar radiation balance, thus:

$$Rn \approx (1 - A)G_i, \tag{11.11}$$

where A is the albedo of the canopy. As the value of $(1 - A)$ is near the absorption coefficient of the crop, the approximation $Rn \approx G_{abs}$ may be acceptable in most cases.

However, it has to be noted that G_{abs}, for a freely transpiring canopy, provides only a part of the latent energy equivalent to the crop transpiration (Stanhill et al., 1973). If the crop is freely transpiring, the canopy temperature is lower than the ambient temperature, and consequently there is a positive sensible heat flux from the air to the canopy surface. This sensible heat is converted into latent heat by the transpiration of the crop, which then contributes to the cooling of the greenhouse air. This observation stresses the importance in summer of supplying enough water to greenhouse crops and encouraging water absorption by the roots, in order to maintain a high level of transpiration, and thus a significant cooling effect by means of crop transpiration. On the other hand, if the plants do not transpire freely and are not able to dissipate the energy absorbed, leaf temperature will be higher than air temperature and the air saturation deficit will increase. These conditions will enhance thermal and water stresses, leading to important losses in production and quality.

Solar energy input, recovery and storage (solar greenhouses)

Greenhouse collector efficiency

Whatever the type of greenhouse, solar energy supplies an important part of the heating demand of the crop. "Solar" greenhouses, a phrase which sounds somewhat pleonastic, are greenhouses designed to allow a better efficiency in the use of this natural and free source of energy. The greenhouse efficiency, GE, in the sense of solar efficiency, is defined as the amount of useful energy derived by the greenhouse, H_u, divided by the incident solar radiation intensity (Garzoli, 1989):

$$GE = \frac{H_u}{G_o}. \tag{11.12}$$

The efficiency of a greenhouse then depends (1) on the potential solar input H to the greenhouse (Eq. 11.2), as H_u is the useful part of H; and (2) on the performances of the systems which are used for collection, storage and restitution of the useful fraction.

As seen above, latent heat constitutes an important part of H, and one of the main problems in maximizing GE (i.e. H_u) is to collect and efficiently convert the latent energy fraction (mainly by condensation and/or absorption).

Main types of solar greenhouses

Solar greenhouses may be classified following two main criteria: (1) the way in which radiative solar energy is converted into heat energy; and (2) the type of heat storage.

Conversion system

Three main conversion systems are generally distinguished (Von Zabeltitz, 1986): (1) Separate solar collectors (air or water solar collectors, solar ponds) are built near the greenhouse and provide most of the heating (Roberts and Mears, 1978; Montero and Short, 1984; Rheinlaender and Photiades, 1984; Bartok and Aldrich, 1984). (2) Solar collectors are integrated into the greenhouse structure, such as the fluid-roof solar greenhouse with a selective cover material (glass, plastic) absorbing the infrared fraction of the solar spectrum and transmitting PAR. Experimentation (Chiapale, 1981; Nilsen et al., 1984) and modelling (Chiapale et al., 1983; Sadler and Van Bavel, 1984; Van Bavel et al., 1985) have provided evidence of the very good performance of such a system, in winter (heating) as well as in summer (cooling) conditions. However, the high investment cost and short working life of the equipment are severe limitations. (3) The greenhouse itself is used as the solar collector (Eq. 11.1). It is clear that conventional greenhouses are solar collectors of poor efficiency: solar radiation is absorbed by the crop, floor and internal structures and is converted to heat which raises the temperature above that outside. But once the chosen daytime temperature is reached, the energy in excess is vented outside. Instead, in solar greenhouses, this energy is extracted at a given rate and stored for utilization in the near future (generally, for heating the following night). This type of greenhouse therefore acts as its own solar collector, generating all or part of its own energy requirements (Garzoli, 1989). This type of solar greenhouse has been studied extensively (Takami and Uchijima, 1977; Blackwell and Garzoli, 1981; Yamamoto, 1985; Mavroyanopoulos and Kyritsis, 1986; Boulard and Baille, 1986a,b).

Heat storage

Three storage materials may be distinguished (Baille, 1987): (1) Storage in a solid medium, such as rock beds, gravel beds, or greenhouse soil. Generally, buried pipes are used as heat exchangers between the fluid (air, water) and the storage medium. However, compared to water, the heat capacity of these materials is lower, and the volume involved is limited to a small distance from the pipes (Boulard and Baille, 1986b). (2) A liquid medium, which is generally water, or salt water in the case of solar ponds where an inverse temperature gradient is created by a salt concentration gradient. (3) Phase-change materials, such as sodium sulphate (Nishina and Takakura, 1985), or hydrated calcium chloride ($CaCl_2 + 6H_2O$), studied by Jaffrin and Cadier (1982) and Boulard et al. (1990). The interest of this storage material is its very high thermal capacity when the phase-change phenomenon is occurring. This allows a small storage volume. A limitation is the lack of flexibility of the system, due to the fixed temperature of fusion (21°C to 26°C in the case of hydrated calcium chloride).

Some examples of greenhouse efficiency

Garzoli (1985), using average climatic data for Griffith (Australia, 34°S) during the heating season (May to November), established the values for H in single and double glazed greenhouses and compared these maximum (or potential) heat gains to daily heat losses (Table 11.6).

Table 11.6
Potential heat gain, heat loss and greenhouse efficiency[1]

Parameters	Single glazing	Double glazing
Transmissivity τ	0.7	0.6
Incident radiation G_i ($G_o = 500\ W\ m^{-2}$)	350	300
Daytime heat loss ($W\ m^{-2}$)	163	73
H ($W\ m^{-2}$)[2]	110	167
GE (%) for $G_o = 500\ W\ m^{-2}$	22	33
6 hours heat gain ($MJ\ m^{-2}$)	2.3	3.6
Night-time heat loss ($W\ m^{-2}$)	169	92
10 hours heat loss ($MJ\ m^{-2}$)	6.1	3.3
% Supplied by daytime heat gain	39	109

[1] From Garzoli (1985).
[2] Average incident radiation during the heating season.

Table 11.7
Seasonal solar contribution (%) to heating requirements in Griffith (Australia) and Hamburg (northern Germany)[1]

Location	Summer	Autumn	Winter	Spring	Year
Griffith (34°S)	None[2]	100	74	100	84
Hamburg (50°N)	100	67	0	21	23

[1] From Garzoli (1985). [2] No heating requirements.

Thus, if all the solar energy input H, for times when the air temperature exceeds the desired one, could be stored and used, GE would be about 0.2 for single-glazed and 0.3 for double-glazed greenhouses, allowing supplementary storage of 2.3 and 3.6 $MJ\ m^{-2}\ day^{-1}$ (compared with soil flux values of 0.5 to 1.0 $MJ\ m^{-2}\ day^{-1}$ in a conventional greenhouse). From this example, it can be seen that a double-glazed greenhouse is about 50% more efficient than a single-glazed greenhouse, and that heating requirements are covered to about 100% by the solar contribution in double-glazed houses, but only to 40% with single glazing. Table 11.7 shows the solar contributions calculated for Griffith and Hamburg, in a double-glazed greenhouse with rock-pile storage and 22/14° (day/night) temperatures. Clearly, outside climatic conditions have a significant influence on the solar contribution to heating requirements.

In terms of absolute energy savings, 640 $MJ\ m^{-2}\ yr^{-1}$ in total is saved in Griffith, and 380 $MJ\ m^{-2}\ yr^{-1}$ in

Table 11.8
Performance of a soil heat-storage system for two heating seasons, in the south of France[1]

Parameters	Lettuce			Tomato			
	Nov	Dec	Jan	Feb	Mar	Apr	May
Night temperature (°C)	6	6	4	12	14	14	16
Day temperature (°C)	15	15	15	22	26	26	26
Outside solar radiation ($MJ\,m^{-2}\,day^{-1}$)	6.7	4.7	6.1	7.9	12.9	16.0	21.7
Mean energy stored ($MJ^{-2}\,day^{-1}$)	0.12	0.26	0.58	1.32	1.40	1.43	1.86
Mean energy extracted ($MJ^{-2}\,day^{-1}$)	0.46	0.27	0.53	1.26	1.54	1.30	1.42
Mean energy auxiliary heating ($MJ^{-2}\,day^{-1}$)	0	0	0.18	0.92	0.41	0	0
Mean power in storage ($MJ\,m^{-2}$)	33	65	67	106	95	96	95
Mean power in extraction ($MJ\,m^{-2}$)	100	94	79	49	48	57	64
Mean ratio [latent energy extracted/total energy extracted]	0.28	0.30	0.31	0.30	0.31	0.34	0.49
Mean ratio [useful heat/fan consumption]	5.5	3.9	2.7	2.0	3.2	3.7	4.4

[1] From Boulard et al. (1989a).

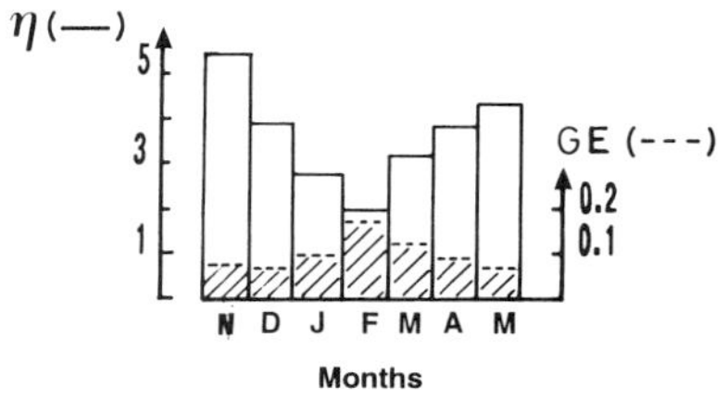

Fig. 11.6. Average monthly values of GE from November to May for a greenhouse with an underground heat-storage system, in southern France. From Boulard et al. (1989a). Dashed line, GE; solid line, ratio η = stored energy/fan consumption.

Hamburg (with 127 and 73 $MJ\,m^{-2}\,yr^{-2}$, respectively, for fan consumption).

Experimental studies in southern France (Boulard et al., 1989a) in a greenhouse with double polycarbonate cladding equipped with soil storage (buried pipes) have given results tabulated in Table 11.8. They show that: (1) Values of stored energy depend on the daytime temperature: for a lettuce crop with 15°C day temperature, the stored energy was 1.5 $MJ\,m^{-2}\,day^{-1}$ in January; for a tomato crop, with a 22°C day temperature, the stored energy was 1.32 $MJ\,m^{-2}$ in February and did not increase very much until May. (2) Optimal periods for heat exchange occurred during March and April, when about 1.5 $MJ\,m^{-2}\,day^{-1}$ was stored during daytime and the same amount could be extracted for heating during the night. (3) Only 30% auxiliary heating is needed for these two months, and the system is self-sufficient by the end of March. (4) The latent heat exchange represents about 30% of the total exchange. (5) The highest greenhouse efficiency GE is obtained in February with an average value of about 0.2, whereas GE is about 0.1 in March and April (Fig. 11.6), compared with GE = 0.30 in the case where all the solar input H is collected and used (Table 11.6). (6) The performance ratio of the storage system [(useful extracted heat)/(fan consumption)] ranges from the lowest value of 2 in February to about 4 in April/May. These values could surely be increased if management of the soil storage system is optimized.

DIRECT ENERGY INPUTS FOR ENVIRONMENTAL CONTROL

The major environmental factors that are controlled in greenhouses at present are: (1) the air temperature, T_i (°C); (2) the air humidity, q_i ($kg\,kg^{-1}$); (3) the concentration of carbon dioxide, C_i (ppm); and (4) the flux density of photosynthetically active radiation, PAR_i ($W\,m^{-2}$).

These aspects are developed in more detail in other chapters of this book. So, I shall consider here only the general knowledge which is available on the energy requirements for environmental control in order to give some orders of magnitude for these requirements.

Air temperature control

Heating

Heating normally constitutes the major energy requirement for greenhouses. The amount of supplementary heat Q_h ($W\,m^{-2}$) which is necessary to maintain the greenhouse air temperature at fixed daytime and night-time temperatures can be estimated from simple formulas such as Eq. (11.1), where the solar input H is reduced to its useful sensible part, H_s:

$$Q_h = U(T_i - T_o) - H_s = U(T_i - T_o) - fG_o. \qquad (11.13)$$

This formula shows that, for a given greenhouse characterized by parameters U and f, the heating demand is mainly determined at night by the difference $(T_i - T_o)$ and, during daytime, by this difference and the outside solar radiation, G_o. The factor f is generally taken as being between 0.20 and 0.30 (van de Braak et al., 1984). These values come from the order of magnitude of H (~0.60–G_o) and of its sensible/latent heat partition, which varies between 0.3 and 0.5, as stated in the previous section. For example, Short and Breuer (1985) used Eq. (11.13) with $H_s = \tau G_o/3$ (i.e., $f = \tau/3$) when calculating heat requirements for single-glazed and double-glazed greenhouses in a maritime climate (de Bilt, Netherlands) and a continental climate (Wooster, Ohio, USA). Furthermore, their model takes into account two common grower practices: (1) a lower limit for pipe temperature limit in a hot-water heating system, regardless of heat requirements; and (2) a dehumidification process, by heating with the roof vents slightly open.

Cumulative frequency curves obtained with this model (Fig. 11.7) present the predicted number of hours for each level of heating requirement, which range from 0 to a maximum of $300\,W\,m^{-2}$. The predicted heating requirement for Wooster was found to be about 20–30% higher than those for de Bilt in winter, and 15–40% lower in spring and autumn. The predicted annual gas consumption for a rose or tomato crop ($T_i \approx 15°C$) with single glass and a thermal screen was $62.6\,m^3\,m^{-2}$ for de Bilt and $65.1\,m^3\,m^{-2}$ for Wooster. Using the calorific equivalent of one cubic meter of natural gas ($36\,MJ\,m^{-3}$), this gives an annual heat-energy expenditure of about $2000\,MJ\,m^{-2}\,yr^{-1}$.

If double glazing was used, the predicted annual gas consumption was $40.6\,m^3\,m^{-2}$ for de Bilt and $42\,m^3\,m^{-2}$ for Wooster. The heat consumption, about $1300\,MJ\,m^{-2}$, is 35% less than with single glazing. The predicted heat requirements below $70\,W\,m^{-2}$ are quite similar for single glazing and for double glazing. Above $70\,W\,m^{-2}$, significant energy savings were predicted with double glazing.

These values for heat consumption have to be

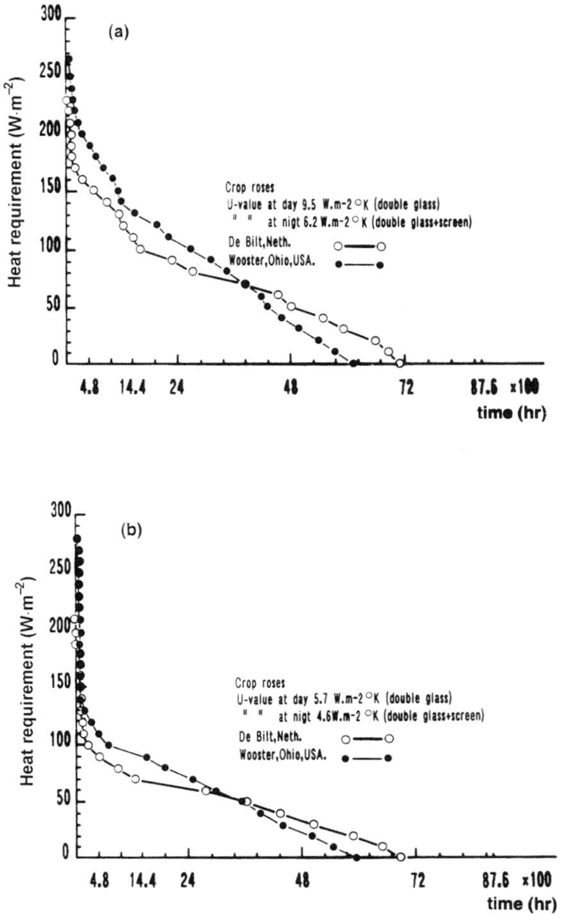

Fig. 11.7. Average number of hours in a year that each level of heating is reached or exceeded at de Bilt (Netherlands) and Wooster (Ohio, USA). (a) Single-glazed greenhouse; (b) double-glazed greenhouse (Short and Breuer, 1985).

multiplied by a factor 1.2 to 1.3 (efficiency of the gas boiler and heat distribution losses), leading to values of $2500\,MJ\,m^{-2}$ for single glazing with a thermal screen and $1600\,W\,m^{-2}$ for double glazing.

If these values are compared, for example, to the annual outside global radiation G_o in the Netherlands (about $4000\,MJ\,m^{-2}$), the proportions of non-solar input for heating are about 60% and 40%, respectively – that is, approximately the values of the annual solar heat input into the greenhouse, H, for single-glazed and double-glazed greenhouses. Thus, in this particular example, we have $Q_h \approx H$ on an annual basis.

Enoch (1978a) mentioned that, under United Kingdom glasshouse conditions, heating consumes about

58 tons of oil $ha^{-1} yr^{-1} °C^{-1}$. If an average inside–outside temperature difference of 10°C is assumed during the heating season, then the annual heating requirement amounts to about 3000 MJ $m^{-2} yr^{-1}$. Thus, without a thermal screen, values of 3000–3500 MJ $m^{-2} yr^{-1}$ may be taken as sensible approximations for single-glazed greenhouses with heat-demanding crops such as rose, tomato or cucumber, in climates like those of northern Europe (lat. >45°N) or the central United States (lat. >40°N).

For coastal countries nearer the equator, heating requirements for the same greenhouse crops are significantly lower – about 1200 MJ $m^{-2} yr^{-1}$ for a rose glasshouse in central Israel, and 1600 MJ $m^{-2} yr^{-1}$ for the same crop in southern France (Nice).

Table 11.9
Heat requirements for different areas of Europe (in litres of light oil per $m^{-2} yr^{-1}$)[1]

Area	Indoor temperature		
	6°C	12°C	18°C
Nice	0.1	13.6	42.7
Paris	8.1	32.6	70.3
London	4.8	32.9	81.1
The Hague	12.0	44.0	94.1
Hamburg	17.0	52.0	103.7
Copenhagen	20.2	55.2	106.7
München	20.2	50.1	116.8

[1] From Storck (1978).

Table 11.9 presents estimates for set-point temperatures of 6°C, 12°C and 18°C in several areas of western Europe (Storck, 1978). For an inside temperature of 18°C, consumption of light oil ranges from 42.7 $\ell\, m^{-2} yr^{-1}$ for Nice up to 117 $\ell\, m^{-2} yr^{-1}$ for München, 1560 MJ $m^{-2} yr^{-1}$ and 4300 MJ $m^{-2} yr^{-1}$, respectively.

For the global energy input, indirect energy inputs required in processing and transport of oil or natural gas have to be added. The energy equivalent to one litre of oil would be 45 MJ l^{-1} instead of 36 MJ l^{-1}, thus increasing direct energy input by about 25%.

Cooling by ventilation

Cooling by ventilation is not really an important component of direct energy requirement in greenhouse climate control. Natural ventilation by opening motorized vents needs a very small amount of electric power to operate the motors. Only forced ventilation (fan, cooling-pad systems) requires some significant amount of electricity: estimates for greenhouses in southern France give values of about 10 kWh $m^{-2} yr^{-1}$, or 36 MJ $m^{-2} yr^{-1}$. Using an efficiency of 29% for conversion from the energy of fossil fuel to electric energy yields a value of about 120 MJ $m^{-2} yr^{-1}$ for the usage of fossil fuel.

Humidity control

The major energy requirement for control of air humidity in greenhouses occurs in the dehumidification process, mainly during winter periods. Well-insulated greenhouses frequently have excessive levels of humidity, mainly during night-time and at sunrise, for the two following reasons: (1) if the greenhouse is airtight, only very small exchanges of water vapour occur between the humid greenhouse air and the dry outside air; and (2) in double-glazed greenhouses, the high temperature of the inner wall reduces its effectiveness as a natural condensation sink.

Possible techniques for reducing humidity levels in greenhouses include the traditional one of ventilation and heating to maintain the desired temperature, ventilation with heat recovery, and refrigerative dehumidification (Bailey, 1984).

In the common practice of simultaneous heating and ventilation, the heat required by this process may be calculated as (Short and Breuer, 1985):

$$Q_{dh} = VNC_p (T_r - T_o), \tag{11.14}$$

where Q_{dh} ($W m^{-2}$) is the heat loss; T_r is the average temperature of the desired temperature T_i on the one hand and the temperature at which vents are opened on the other; V is the average air volume per unit area of greenhouse floor ($m^3 m^{-2}$); N is the number of air changes per second; and C_p is the specific heat of air ($J m^{-3} °C^{-1}$).

From results of van de Braak et al. (1984), this dehumidification process may be estimated to constitute about 5–10% of the total heating requirements, that is about 100 to 200 MJ $m^{-2} yr^{-1}$. Other estimates (Boulard et al., 1989c) give, in southern France, an average value (from January to May) of 94 MJ $m^{-2} yr^{-1}$ for a double-skin polyethylene greenhouse with a spring crop of tomatoes – that is, about 15% of the heating requirements during the same period.

Other dehumidification systems which are nowadays in use, such as heat pumps or latent-heat converters using a brine solution, require a high investment cost. They allow the greenhouse to remain closed, thus allowing longer periods of enrichment with carbon dioxide and reduced running costs compared to the heating/ventilation process. In the case of the heat pump, measurements of electric consumption by Boulard et al. (1989b) in a double-wall polyethylene greenhouse give a value of 37 MJ m^{-2} yr^{-1}. This quantity of heat is recovered at the heat pump condenser and serves to heat the greenhouse. In latent-heat dehumidification with a brine solution, energy (gas) is used in a small boiler for brine regeneration. Energy input, estimated at 1.2 kWh kg^{-1} (water), is also released in the greenhouse together with the energy from water condensation, thus leading to a net energy saving during heating periods (Peiper et al., 1987).

Carbon dioxide enrichment

Enrichment with carbon dioxide is now a common practice in advanced horticulture. After heating and irrigation, carbon dioxide enrichment is generally the next investment that growers will choose, in order to increase crop production (vegetables) or to improve quality and yield (pot plants and cut flowers).

Carbon dioxide can be supplied from several sources: (1) decay of organic material (straw, etc.); (2) burning organic compounds (parafin, propane, etc.); or (3) using compressed carbon dioxide from bottles or containers. One of the commonest ways, when the heating system burns natural gas, is to use the carbon dioxide contained in the exhaust gases, which are emitted either directly (air heaters) or indirectly (stack-gas condensers) into the greenhouse.

Some of the added carbon dioxide will be absorbed by the plants and used in the photosynthetic process, but a significant part (up to 75%) will be lost to the exterior through leakages or ventilation. Thus, the carbon dioxide consumption will vary widely with greenhouse tightness, climate (wind-speed and radiation), crop, and enrichment strategy (level, enrichment periods, etc.).

Some figures for carbon dioxide consumption are available in the literature. For glasshouses with tomatoes under winter conditions in the United Kingdom, Slack and Calvert (1972) gave an average of 416 kg CO_2 ha^{-1} yr^{-1} to raise the carbon dioxide concentration by 1 p.p.m.: if the greenhouse carbon dioxide concentration is double that outside (350 p.p.m.) – that is, 700 p.p.m. – the consumption of carbon dioxide would be about 130 t ha^{-1} yr^{-1}. For the same crop in France, Wacquant (personal communication, 1985) estimated the supply from January to May, when carbon dioxide enrichment was stopped, as about 8 kg CO_2 m^{-2} or 80 t ha^{-1}. Other estimations give values around 100 t ha^{-1}. One may adopt this last value for an average increase in carbon dioxide level of 350 p.p.m. As 1 m^3 natural gas yields about 2 kg CO_2 and its heat of combustion is 35 MJ m^{-3}, the energy input corresponding to 100 t ha^{-1} would require 50 000 m^3 natural gas or 175 MJ m^{-2} yr^{-1}. Computation with propane (1 kg propane gives 3 kg CO_2, for a combustion heat of 50 MJ kg^{-1}) leads to approximately the same value. Consequently, direct energy cost for carbon dioxide enrichment may be estimated as about 200 MJ m^{-2} yr^{-1}. If the indirect energy cost for producing and transporting natural gas or propane is added – Enoch (1978a) has estimated an indirect requirement of 57.6 MJ kg^{-1} for propane – the total energy requirement for carbon dioxide enrichment at about 700 p.p.m. would be about 400–500 MJ m^{-2} yr^{-1}.

Artificial lighting

Photosynthetically active radiation (PAR) (400–700 nm) is provided by solar radiation. As for global radiation, the PAR transmissivity of greenhouse covers ranges from 0.5 to 0.7. During winter months, greenhouse crops in high latitudes receive very low PAR levels: a PAR integral inside the greenhouse of 135 MJ m^{-2} from November to February is given by Bailey (1985) for southern England. In December, glasshouse crops receive an average PAR flux of 25 W m^{-2} during 8 hours. Thus, using artificial light to supplement natural illumination is a common practice with seedlings, flowers and bulbous crops, and sometimes is used for vegetables, but to a much lesser extent.

In the electrical sources used for supplementary lighting, the relative proportions of PAR (400–700 nm), far red (700–800 nm) and near-infrared (800–2800 nm) differ from those in natural light. Several peaks at given wave-lengths are typical of lamp spectral distributions (Sager, 1984). They also contain a significant fraction of thermal infra-red (2800 nm–50 000 nm), which can act as a radiant-heat source for direct crop heating. Some special types of equipment ("total energy" units) which generate electricity and heat (Giniger and Mears, 1984; Huijs, 1989) are presently available.

Two main types of lamps can be distinguished

(Enoch, 1978a): (1) incandescent lamps, mainly used for photoperiod control, with only 6–12% of the electrical energy emitted in the PAR waveband; and (2) fluorescent sodium, mercury and metal halide lamps, whose efficiencies range from 20 to 25%, and which are mainly used for photosynthesis. High Pressure Sodium (HPS) lamps, which have a PAR efficiency of 25%, are the most commonly used (McAvoy and Janes, 1984). Generally, one unit of 400–450 W power load covers about 8 to 10 m^2 greenhouse floor, leading to an electric power requirement of about 50 $W m^{-2}$ and a PAR level of about 10 $W m^{-2}$, if it is assumed that 70% of the flux is directed towards the canopy by the ballast reflector. With a value of 29% efficiency for electricity production, this leads, for HPS lamps, to an overall lamp efficiency of 5%. For incandescent lamps, the overall lamp efficiency is only 1–2%.

The following example concerns a crop of pot plants in the Val de Loire (west central France, 45°N), with one 450 W HPS unit for 9 m^2 floor, supplying 10 $W m^{-2}$ PAR flux. The following lighting strategy was used (Baille et al., 1987): by night: 7 hours (0h–7h U.T.) in December–January; 6 hours (1h–7h U.T.) in February; and 5 hours (2h–7h U.T.) in March; and by day whenever PAR_i level is lower than 10 $W m^{-2}$.

This strategy led to an annual use of 650 hours lighting at night and 550 hours lighting during the daytime – that is a total of 1200 hours. The electric consumption was 60 $kWh m^{-2}$, or 740 $MJ m^{-2}$ if the efficiency is 29%, for a PAR integral supply of 43.2 $MJ m^{-2}$ annually if the lamp efficiency is 5.8%.

Another example concerns a rose crop continuously lighted for 24 hours per day during 6 months (October to March) in Ontario (Canada), with HPS lamps giving a PAR flux of 5.4 $W m^{-2}$ and needing 45 $W m^{-2}$ power load (Blom and Ingratta, 1984). For 4400 hours of lighting, electric consumption amounts to 200 $Kwh m^{-2}$ or 2480 $MJ m^{-2}$ with 29% efficiency. The PAR integral is 85 $MJ m^{-2}$ and lamp efficiency is 3.4%.

OTHER ENERGY INPUTS IN THE GREENHOUSE ECOSYSTEM

Direct inputs other than those used for environmental control

The other direct inputs which have to be considered in the greenhouse crop production system include some related to the soil system: fertilizers (mainly nitrogen, phosphorus and potassium), irrigation water supply, soil conditioners, artificial substrates (in soilless culture), a nutrient-film treatment system, etc., and soil sterilization.

Other direct inputs include pesticides, human labour (for planting, pruning, harvest, etc.), and such miscellaneous items as plants, overheads, plant supports and plastic film for ground covering.

Details and references about the estimation of the equivalent energy input corresponding to these components were assembled by Stanhill (1980), who compared six different tomato cropping systems, of which three were protected.

System (A): semi-intensive production protected by low plastic tunnels, winter–spring, on the central coast of Israel;

System (B): intensive production in an unheated glasshouse, winter–spring, on the southern coast of Israel;

System (C): intensive production in a heated Venlo-type glasshouse, spring–summer, on the southern coast of England.

Table 11.10 presents the results of all the direct inputs in terms of annual energy equivalent in

Table 11.10
Energy equivalent ($MJ m^{-2} yr^{-1}$) of direct inputs for three protected tomato-cropping systems [1]

Inputs	System		
	A	B	C
Chemical fertilizers	1.6	11.9	14.7
Organic soil dressings	29.0	0.0	8.0
Soil sterilants	0.0	5.3	261.9
Cultivation and transport	0.5	0.5	0.7
Water supply	4.5	6.7	10.0
Pesticides	1.0	1.8	1.8
Plants	0.1	0.6	2.5
Labour	0.3	0.7	1.4
Others	0.6	0.9	4.7
Sub total	37.6	28.5	305.7
Input for environmental control			
Heating	0.0	0.0	2357.0
Ventilation	0.0	0.0	106.8
CO_2 enrichment	0.0	0.0	54.8
Sub total	0.0	0.0	2518.6
Total direct inputs	37.6	28.5	2824.3

[1] From Stanhill (1980).

MJ m^{-2} yr^{-1}. For the first two cropping systems, which do not need any environmental-control input, the total energy content of direct inputs is relatively low (38 and 28.5 MJ m^{-2} yr^{-1}) and comparable to open-field cropping systems. The third system requires ten times more energy input (306 MJ m^{-2} yr^{-1}) without considering the environmental-control inputs, and 100 times more (2820 MJ m^{-2} yr^{-1}) when these inputs are included.

Energy equivalent of indirect inputs

In the same paper, Stanhill (1980) estimated the indirect energy inputs which include the energy sequestered in the amortization of the protective structures, machinery, irrigation equipment, environmental control design, etc.

Table 11.11 gives estimates of these indirect inputs for the three systems. The total direct and indirect inputs are of the same order of magnitude, but for System (C), the indirect component represents only 4% of the direct energy requirements.

Adding direct and indirect inputs give the following estimates for the gross energy requirements (GER): System (A) = 55 MJ m^{-2} yr^{-1}; System (B) = 69 MJ m^{-2} yr^{-1}; System (C) = 2930 MJ m^{-2} yr^{-1}.

If one compares non-solar input to solar input (here G_i, global radiation incident on crop, is taken as the reference), the ratios of non-solar to solar input are: System (A) = 55/3030 ≈ 2%; System (B) = 69/3030 ≈ 2%; System (C) = 2930/2256 ≈ 130%.

It may be seen that, for the last system, the non-solar input is of the same order of magnitude as the global solar radiation incident on the crop.

Table 11.11
Energy equivalent (MJ m^{-2}) of indirect inputs for three protected cropping systems [1]

Inputs	System		
	A	B	C
Machinery	0.1	0.1	0.1
Irrigation equipment	1.0	1.7	2.6
Protective structures	15.9	38.7	38.7
Environmental control system	0.0	0.0	62.5
Plant support	0.0	0.2	0.2
Miscellaneous	0.1	0.1	0.3
Total	17.1	40.8	104.4

[1] From Stanhill (1980).

Another example for glasshouse crops is given by Leach (1975), who analyzed only the direct inputs for a winter lettuce crop (using artificial light for seedlings) in southern England. Environmental control inputs were estimated as 400 to 536 MJ m^{-2} for heating, 31 MJ m^{-2} for carbon dioxide enrichment and 5 to 20 MJ m^{-2} for lighting seedlings, the total input thus ranging from 436 to 587 MJ m^{-2}. Other direct inputs amounted to 20 MJ m^{-2}. Indirect inputs may be approximated by the same value as found by Stanhill for System (C), that is 104 MJ m^{-2} yr^{-1}, a value which has to be divided by 3 as the winter crop lasts 4 months. This leads to an indirect input of 35 MJ m^{-2} and a ratio of indirect to direct inputs of about 6%.

From these results, it can be seen that environmental control, and mainly heating, is by far the largest component of GER, as it can represent up to 90% of the GER in the case of heated glasshouse crops in climates of northern Europe or the central United States. This environmental input will be halved in coastal Mediterranean countries such as the south of France, or reduced to zero when no environment control is used ("cold greenhouses"). In this case, GER is of an order of magnitude similar to values for open-field crops.

ENERGY RATIO OF THE GREENHOUSE ECOSYSTEM

The two energy-related parameters that are commonly used to describe an agricultural ecosystem are (Pimentel and Pimentel, 1979; Lorber et al., 1984):

(1) the energy ratio, ER, defined as:

$$\mathrm{ER} = \frac{\text{caloric content of crop (MJ)}}{\text{energy sequestered in its production (MJ)}} = \frac{\mathrm{EC}}{\mathrm{GER}}.$$

(2) the energy productivity, EP, defined as:

$$\mathrm{EP} = \frac{\text{mass of dry matter production (kg)}}{\text{energy sequestered in its production (MJ)}} = \frac{\text{d.m.}}{\mathrm{GER}}.$$

The sequestered energy in these definitions is the total non-solar input, GER, estimates of which for some greenhouse crops have been presented in Table 11.10. The energy ratio and energy productivity correspond to the efficiency measures RUE and YR related to solar input, respectively, and detailed on pp. 272–273.

Table 11.12
Values of energy ratio (ER) and energy productivity (EP) for some open-field and greenhouse crop systems

Product	ER	EP	Source
Maize grain	4.1	0.26	Lorber et al. (1984)
Alfalfa	16.1	0.91	Lorber et al. (1984)
Woodchips [1]	4.3–5.5	0.22–0.28	Lorber et al. (1984)
Sugarcane	4.8–26.7	0.23–1.32	Lorber et al. (1984)
System A [2]	0.08	5.10^{-3}	Stanhill (1980)
System B [3]	0.27	0.017	Stanhill (1980)
System C [4]	7.10^{-3}	$4.4.10^{-4}$	Stanhill (1980)
Lettuce [5]	2.10^{-3}	$1.2.10^{-4}$	Leach (1975)

[1] From forests ranging from natural to intensively cultured.
[2] Low tunnels, tomato cropping.
[3] Glasshouse tomato crop, non-heated.
[4] Greenhouse tomato crop, heated.
[5] Glasshouse winter lettuce crop.

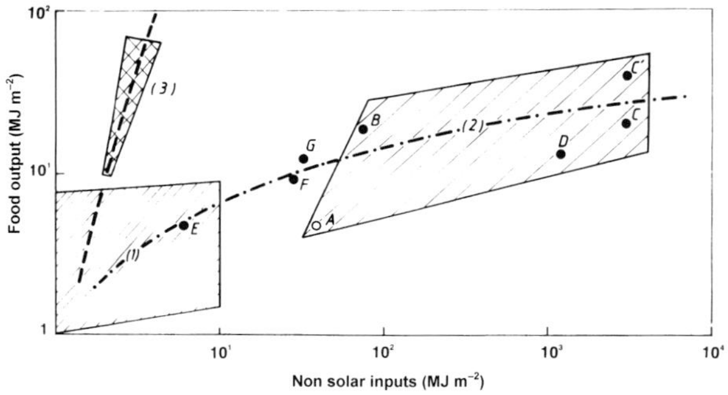

Fig. 11.8. Relationships between food yield output and non-solar inputs for several cropping systems. (a) A, B, C, greenhouse systems discussed in this chapter; C′, same as C, with a twofold yield (42 kg m^{-2} yr^{-1}); E, F, G, intensive mechanized open-field tomato crop (California). From Stanhill (1980). Hatched area (1) encloses points representative of 17 fully industrialized crops. From Leach (1975). (b) Hatched area (2), greenhouse cropping area. (c) Hatched area (3), points representative of several sugarcane cropping systems (Lorber et al., 1984).

Data on ER and EP for several open-field agricultural systems are presented in Table 11.12, together with estimates for greenhouse crops. This Table shows that the least efficient greenhouse crop (ER = 2.10^{-3} for winter lettuce) and the most efficient open-field crop (ER ≈ 20 for sugar cane) are separated by a factor 10^4!

In terms of energy "intensity", 230 MJ are necessary to produce 0.046 kg d.m., or 1 kg (fresh matter) of lettuce, whereas only 23 kJ are needed for 1 kg (f.m.) of sugar cane. Values of energy intensity for glasshouse tomatoes range from 3.4 MJ kg^{-1} (f.m.) for System (A) to 137 MJ kg^{-1} (f.m.) for System (C), as estimated by Stanhill (1980). For greenhouse crops, the energy ratios for the most efficient crop (unheated glasshouse in Israel) and the least efficient system (glasshouse winter lettuce in England) differ by a factor 100 (ER = 0.2 and 2.10^{-3}, respectively).

It is clear from these estimates that increased production per unit area and time will be associated with increases in the inputs. Moreover, the relationships between yield output and GER (Fig. 11.8) show that the production system becomes less energy-efficient as the output rises and the harvest is earlier (Stanhill, 1980).

CONCLUSION

In this chapter, some quantitative estimates of greenhouse efficiency (collector efficiency, solar radiation use efficiency, non-solar energy ratio) have been presented. They allow the following conclusions about energy use in greenhouse crops.

A. Greenhouse efficiency, in the sense of solar collection and storage, is very low for conventional greenhouses. An increase in this efficiency by a better use of the solar input H would lead to saving of significant amounts of heating inputs, even in high-latitude climates (von Zabeltitz and Damrath, 1984). Utilization of passive or active solar systems is now possible. Takakura (1989) reports that there are about 64 ha of soil-storage greenhouses in Japan; and this system should be promoted whenever possible. The success of "solar" greenhouses will depend mainly on oil prices and on the supplementary investment cost needed for the solar system. Moreover, greater awareness by growers of these aspects related to a better efficiency of their solar collectors is necessary, as at present motivation for alternative energy use is not really high among them.

B. Radiation use efficiencies by greenhouse crops are among the highest observed in agricultural ecosystems: they are of the same order as or higher than, the most efficient open-field C_4 crops, such as sugar cane or maize. The obvious main causes for these high efficiencies are, first, the possibility of continuous cropping, with harvests all the year round, and second, the optimization of aerial and soil environment by various supplemental equipment requiring input of non-solar energy, which overcomes the main limiting factors, that is, low carbon dioxide concentration, and low light intensity during the winter months. With new tools like computers and software (models, expert systems, etc.) for optimizing climate control, water and nutrient supply, and the use of pesticides, RUE (on the basis of absorbed PAR) will continue to increase in coming years and probably will be very near to the maximum theoretical value.

C. When GER and energy ratio are considered, heated greenhouses clearly show the worst efficiency of all the agricultural ecosystems. In spite of significant improvements in methods for saving energy (double glazing, thermal screens, etc.) and cultural practices (selection of crop rotations with less energy requirements, use of cultivars with lower temperature requirements), the high GER values estimated for heated greenhouses cast doubt on the economics of producing out-of-season crops in high latitudes (Storck, 1978). A comparative study of energy costs incurred by two alternatives to winter greenhouse production – transport of products from lower latitudes, or long-term storage of the products (summer to winter) – has been made by Enoch (1978b). His results prove that these two alternative solutions demand much less energy than winter greenhouse production in high latitudes.

We may conclude this chapter with the following statement from Stanhill (1980): "it would appear unwise to expand energy intensive forms of protected cropping which require large annual energy inputs, whether direct, e.g. heating fuel or indirect, e.g., polyethylene film. Future development should rather seek to increase the exploitation of the natural advantages of the region, i.e., high winter solar radiation and plentiful human labour." This statement clearly indicates that low-latitude greenhouse areas offer the greatest promise in the future, if drastic changes occur in world oil production and prices, thus reducing the gap between the results of energy and economic analyses.

GLOSSARY OF SYMBOLS, ABBREVIATIONS, ETC.

A	albedo of the crop
C	CO_2 concentration (p.p.m.)
C_d	calorific equivalent of dry matter ($MJ\,kg^{-1}$)
C_m	heat capacitance of the thermal mass ($kJ\,m^{-2}\,°C^{-1}$)
C_p	specific heat of air ($J\,m^{-3}\,°C^{-1}$)
d.m.	dry matter (kg)
EC	Energy Content (MJ)
EP	Energy Productivity ($kg\,MJ^{-1}$)
ER	Energy Ratio
ES	Energy Stored in dry matter ($MJ\,m^{-2}$)
ET	evapotranspiration rate ($mm\,day^{-1}$)
f.m.	fresh matter (kg)
F_s	fraction of internal solar radiation which is transferred by conduction in the soil
G_o	outside global solar radiation per unit floor area ($W\,m^{-2}$)
GE	Greenhouse Efficiency
GER	Gross Energy Requirements ($MJ\,m^{-2}$)
H	solar input to the greenhouse per unit floor area ($W\,m^{-2}$)
L	latent heat of vaporization ($MJ\,kg^{-1}$)
LAI	Leaf Area Index ($m^2\,m^{-2}$)
N	number of air changes per unit time
q	air humidity ($kg\,kg^{-1}$)
Q	enthalpy gain (or loss) of the greenhouse ($W\,m^{-2}$)
Q_h	energy requirements for heating ($W\,m^{-2}$)
Q_{dh}	energy requirements for dehumidification ($W\,m^{-2}$)
PAR	Photosynthetic Active Radiation ($W\,m^{-2}$)
Rn	net radiation ($W\,m^{-2}$)
RUE	Radiation Use Efficiency
T	air temperature (°C)
T_m	temperature of the thermal mass (°C)
U	overall heat loss coefficient per unit floor area ($W\,m^{-2}\,°C^{-1}$)
U.T.	Universal time
V	air volume per unit area of greenhouse floor ($m^3\,m^{-2}$)
YR	Yield Ratio
α	solar absorptivity of the internal surface (soil, canopy)
α_g	crop absorption coefficient for global radiation
ρ	reflectivity of the greenhouse to global radiation
τ	transmissivity of the greenhouse

φ fraction of internal solar radiation which is used in the photosynthetic process

Subscripts

o outside
i inside
a absorbed
r reflected
t transmitted
d daytime
n nighttime
s sensible heat

REFERENCES

Acock, B., Charles-Edwards, A., Fitter, D.J., Hand, D.W., Ludwig, J.L., Warren-Wilson, J. and Withers, A.C., 1978. The contribution of leaves from different levels within a tomato crop to canopy net photosynthesis: an experimental examination of two canopy models. *J. Exp. Bot.,* 29: 815–827.

Aikman, D.P., 1989. Potential increase in photosynthetic efficiency from the redistribution of solar radiation in a crop. *J. Exp. Bot.,* 40: 855–864.

Albright, L.D., Seginer, I., Marsh, L.S. and Oko, A., 1985. In situ thermal calibration of unventilated greenhouses. *J. Agric. Eng. Res.,* 31: 265–281.

Bailey, B.J., 1977. *The calculation of glasshouse fuel requirements using degree-day data corrected for solar heat gain,* Dept. Note DN/6/798/040 13. National Institute of Agricultural Engineering, Silsoe.

Bailey, B.J., 1984. Limiting the relative humidity in insulated greenhouses at night. *Acta Hortic.,* 148: 411–420.

Bailey, B.J., 1985. Microclimate, physical processes and greenhouse technology. *Acta Hortic.,* 174: 35–42.

Bailey, B.J., 1988. Improved control strategies for greenhouse thermal screen. *Acta Hortic.,* 230: 485–492.

Baille, A., 1987. Transferts thermiques et systèmes énergétiques en agriculture: connaissances actuelles et prospectives. In: *Energies Renouvelables et Utilisation Rationnelle de l'Energie,* Vol. 2. INSA, Lyon, France, pp. 361–382. In French.

Baille, M., Laury, J.C., Granger, J., Lajoux, C. and Baille, A., 1987. Eclairage artificiel et apports de CO_2 en cultures de plant d'ornement. *Les cahiers du CNIH,* 10: 15–22. In French.

Baille, M., Baille, A. and Tchamitchian, M., 1989. A simple model for the estimation of greenhouse transmission: influence of structures and internal equipment. *Acta Hortic.,* 281: 35–46.

Bartok, J.W. and Aldrich, R.A., 1984. Low cost solar collectors for greenhouse water heating. *Acta Hortic.,* 148: 771–774.

Beckman, W.A., Klein, S.A. and Duffie, J., 1977. *Solar Heating Design.* Wiley Interscience, New York.

Blackwell, J. and Garzoli, K.V., 1981. Australian development of a low energy greenhouse based on thermal storage rockpile. *Acta Hortic.,* 115: 113–123.

Blom, T.J. and Ingratta, F.J., 1984. The effect of high pressure sodium lighting on the production of tomatoes, cucumbers and roses. *Acta Hortic.,* 148: 905–914.

Boulard, T., 1985. *Etude du gisement de chaleur récupérable sur l'air d'une serre. Comparaison avec les besoins en chauffage,* Internal Report M/85/1. INRA, Station de Bioclimatologie d'Avignon. In French.

Boulard, T. and Baille, A., 1986a. Simulation and analysis of soil heat storage systems for a solar greenhouse. I. Analysis. *Energ. Agric.,* 5: 175–184.

Boulard, T. and Baille, A., 1986b. Simulation and analysis of soil heat storage systems for a solar greenhouse. II. Simulation. *Energ. Agric.,* 5: 285–293.

Boulard, T. and Baille, A., 1987. Analysis of thermal performance of a greenhouse as a solar collector. *Energ. Agric.,* 6: 17–27.

Boulard, T., Razafinjohany, E. and Baille, A., 1989a. Heat and water vapour transfer in a greenhouse with an underground heat storage system. Part I. Experiment results. *Agric. Forest Meteorol.,* 45: 175–184.

Boulard, T., Baille, A., Mermier, M., Lagier, J. and Vanderschitt, E., 1989b. Water vapour transfer in a plastic house equipped with a dehumidification heat pump. *J. Agric. Eng. Res.,* 44: 191–204.

Boulard, T., Baille, A., Lagier, J. and Mermier, M., 1989c. Water vapour transfers and dehumidification in an inflatable plastic greenhouse. *Acta Hortic.,* 245: 462–469.

Boulard, T., Razafinjohany, E., Baille, A., Jaffrin, A. and Favre, B., 1990. Performances of a greenhouse heating system with a phase change material. *Agric. Forest Meteorol.,* 52: 303–318.

Cave, C.R.J. and Cockshull, K.E., 1989. The effect of using a venetian reflector assembly to improve winter light levels in a single span glasshouse on the growth of a pot chrysanthemum crop. *Sol. Energ.,* 42: 319–325.

Charles-Edwards, D.A., 1982. *Physiological Determinants of Crop Growth.* Academic Press, London.

Chiapale, J.P., 1981. La serre solaire: resultats physiques. *Acta Hortic.,* 115: 387–400.

Chiapale, J.P., van Bavel, C.H.M. and Sadler, E.J., 1983. Comparison of calculated and measured performance of a fluid roof and a standard greenhouse. *Energ. Agric.,* 2: 75–89.

Critten, D.L., 1983a. A computer model to calculate the daily light integral and transmission of a greenhouse. *J. Agric. Eng. Res.,* 28: 61–76.

Critten, D.L., 1983b. The evaluation of a computer model to calculate the daily light integral and transmissivity of a greenhouse. *J. Agric. Eng. Res.,* 28: 545–563.

Critten, D.L., 1985. The use of reflector in venetian blinds to enhance irradiance in greenhouses. *Sol. Energ.,* 34: 83–92.

Critten, D.L., 1988. Light enhancement using E.W. aligned long prismatic arrays at high latitude. *Sol. Energ.,* 41: 583–592.

de Graaf, R. and van den Ende, J., 1981. Transpiration and evapotranspiration of the glasshouse crops. *Acta Hortic.,* 119: 147–158.

de Villèle, O., 1974. Besoins en eau des cultures sous serre. Essai de conduite des arrosages en fonction de l'ensoleillement. *Acta Hortic.,* 35: 123–129. In French.

de Wit, C.T., 1986. Introduction. In: H. Van Keulen and J. Wolf (Editors), *Modelling of Agricultural production: Weather, Soils and Crops.* PUDOC, Wageningen, the Netherlands, pp. 3–10.

Dekkers, W.A., Lange, J.M. and de Wit, C.T., 1974. Energy production and use in Dutch agriculture. *Neth. J. Agric. Sci.*, 22: 107–118.

Duffie, J.A. and Beckman, W.A., 1974. *Solar Energy Thermal Processes.* Wiley Interscience, New York, 324 pp.

Duncan, G.A., Loewer, D.J. and Colliver, D.G., 1981. Simulation of energy flows in a greenhouse: magnitude and conservation potential. *Trans. Am. Soc. Agric. Eng.*, 24: 197–201.

Enoch, H.Z., 1978a. A theory for optimalization of primary production in protected cultivation. I. Influence of aerial environment upon primary plant protection. *Acta Hortic.*, 76: 31–43.

Enoch, H.Z., 1978b. A theory for optimalization of primary production in protected cultivation. II. Primary plant production under different outdoor light regimes. *Acta Hortic.*, 76: 45–57.

Enoch, H.Z., 1986. Climate and protected cultivation. *Acta Hortic.*, 176: 11–20.

Gallagher, J.N. and Biscoe, P.V., 1978. Radiation absorption, growth and yield of cereals. *J. Agric. Sci. Cambridge,* 91: 47–60.

Garzoli, K., 1985. A simple greenhouse climate model. *Acta Hortic.*, 174: 393–400.

Garzoli, K., 1989. Efficient energy greenhouses. *Acta Hortic.*, 245: 53–62.

Garzoli, K. and Blackwell, J., 1973. The response of a glasshouse to high solar radiation and ambient temperature. *J. Agric. Eng. Res.*, 18: 205–216.

Garzoli, K. and Shell, G.S.G., 1984. Performance and cost analysis of an Australian solar greenhouse. *Acta Hortic.*, 148: 723–730.

Giniger, M.S. and Mears, D.R., 1984. Cogeneration of heat and electricity for greenhouse use. *Acta Hortic.*, 148: 151–160.

Gosse, G., Varlet-Grancher, C., Chartier, M., Bonhomme, R., Allirand, J.M. and Lemaire, G., 1986. Production maximale de matière sèche et rayonnement solaire intercepté par un couvert végétal. *Agronomie,* 6: 47–56. In French, with English summary.

Graffiadellis, M., 1986. Development of a passive solar system for heating greenhouses. *Acta Hortic.*, 191: 245–252.

Hammer, G.L. and Vanderlip, R.L., 1989. Genotype-by-environment interaction in grain sorghum. I. Effects of temperature on radiation use efficiency. *Crop Sci.*, 29: 370–376.

Hodge, T. and Kanemasu, E.T., 1977. Modeling daily matter production of winter wheat. *Agron. J.*, 69: 974–978.

Huijs, J.P.G., 1989. Cogeneration of heat and power in combination with supplementary light. *Acta Hortic.*, 245: 185–192.

Jaffrin, A. and Cadier, P., 1982. Latent heat storage applied to horticulture: La Baronne solar greenhouse. *Sol. Energ.*, 28: 313–323.

Jolliet, O., 1988. *Modélisation du comportement thermique d'une serre horticole,* Ph.D. Dissertation. Ecole Polytechnique de Lausanne, Switzerland, 244 pp. In French, with English summary.

Kozai, T. and Sugi, J., 1972. Studies on the solar irradiation in glasshouses. 3. Use of reflective mirror. *J. Agric. Meteorol. Jpn.*, 28: 79–88.

Kozai, T., Goudriaan, J. and Kimura, M., 1973. *Light Transmission and Photosynthesis in Greenhouses,* Simulation Monograph. PUDOC, Wageningen.

Kurata, K., 1983. Studies on improvement of the light environment in greenhouses. II. Utilization of reflective blinds. *J. Agric. Meteorol. Jpn.*, 39: 219–223.

Kurata, K., Okada, M. and Sase, S., 1988. Row orientation and direct solar radiation absorptance of tomato canopy. Fisheye photography analysis. *J. Agric. Meteorol. Jpn.*, 44: 15–22.

Leach, G., 1975. *Energy and Food Production.* International Institute for Environment and Development, London, 151 pp.

Lorber, M.N., Fluck, R.C. and Mishoe, J.W., 1984. A method for analysis of sugarcane (*Saccharum* sp.) biomass production system. *Trans. Am. Soc. Agric. Eng.*, 27: 146–152.

Manby, T.C.D., 1984. Selected aspects of efficient crop production. *J. Agric. Eng. Res.*, 29: 275–293.

Mavroyanopoulos, G.N. and Kyritsis, S., 1986. The performances of a greenhouse heated by an earth–air exchanger. *Agric. Forest Meteorol.*, 36: 263–268.

McAvoy, R.J. and Janes, H.W., 1984. The use of high pressure sodium lights in greenhouse tomato crop production. *Acta Hortic.*, 148: 877–884.

Michels, P., 1984. *Comparison of soil, plant and climate measuring procedures for scheduling irrigation of greenhouse tomatoes,* Dissertationes de Agricultura. Louvain Catholic University, 185 pp.

Monteith, J.L., 1977. Climate and the efficiency of crop production in Britain. *Philos. Trans. R. Soc. London B,* 281: 277–294.

Montero, J.I. and Short, T.H., 1984. A comparison of solar heated greenhouses in Wooster (Ohio) and Malaga (Southern Spain). *Acta Hortic.*, 145: 805–812.

Morris, J.I., 1972. Solar radiation in greenhouses: a brief review. *Isr. J. Agric. Meteorol. Res.*, 22: 85–97.

Nederhoff, E.M., 1984. Light interception of a cucumber crop at different stages of growth. *Acta Hortic.*, 148: 525–537.

Nilsen, S., Dons, C. and Pettersen, H., 1984. A method for reducing the infrared radiation or changing the spectral distribution of sunlight in greenhouses. *Biotronics,* 13: 1–10.

Nishina, H. and Takakura, T., 1985. Greenhouse heating by means of latent heat storage systems (3): heating experiment in a greenhouse. *J. Agric. Meteorol. Jpn.*, 40: 313–322.

Papadopoulos, A.P. and Ormrod, D.P., 1988a. Plant spacing effects on light interception by greenhouse tomatoes. *Can. J. Plant Sci.*, 68: 1197–1208.

Papadopoulos, A.P. and Ormrod, D.P., 1988b. Plant spacing effects on photosynthesis and transpiration of the greenhouse tomato. *Can. J. Plant Sci.*, 68: 1209–1218.

Peiper, U.M., Levav, N., Geoola, F. and Seginer, I., 1987. Experimental evaluation and mathematical model of an hygroscopic greenhouse dehumidifier, Final Research Report. BARD Project I39082, Bet-Dagan, Israel, April 1987.

Pimentel, D. and Pimentel, M., 1979. *Food, Energy and Society.* Edward Arnold, London, 165 pp.

Rheinlaender, J. and Photiades, I., 1984. A comparison of fossil and solar heating of greenhouses in Cyprus. *Acta Hortic.*, 148: 775–784.

Roberts, N.J. and Mears, D.R., 1978. A solar heating system for a 0.54 ha. greenhouse. *Acta Hortic.*, 87: 311–319.

Sadler, E.J. and van Bavel, C.H.M., 1984. Simulation and measurement of energy partition in a fluid roof greenhouse. *Agric. Forest Meteorol.*, 83: 1–14.

Sager, J.C., 1984. Spectral effects on the growth of lettuce under controlled envrionment conditions. *Acta Hortic.*, 148: 889–896.

Short, T.H. and Breuer, J.J.G., 1985. Greenhouse energy demand comparisons for the Netherlands and Ohio, USA. *Acta Hortic.*, 174: 145–153.

Siddique, K.H.M., Belford, R.K., Perry, M.W. and Tennant, D., 1989. Growth, development and light interception of old and modern wheat cultivars in a mediterranean type environment. *Aust. J. Agric. Res.,* 40: 473–87.

Slack, G. and Calvert, A., 1972. Control of carbon dioxide concentration in glasshouses by use of conductimetric controllers. *J. Agric. Eng. Res.,* 17: 107–115.

Slesser, M., 1973. Energy subsidy as a criterion in food policy planning. *J. Sci. Food Agric.,* 24: 1193–1207.

Smith, C.V. and Kingham, H.G., 1971. A contribution to glasshouse design. *Agric. Meteorol.,* 8: 447–468.

Stanhill, G., 1980. The energy cost of protected cropping: a comparison of six systems of tomato production. *J. Agric. Eng. Res.,* 25: 145–154.

Stanhill, G. and Scholte Albers, J., 1974. Solar radiation and water loss from glasshouse roses. *J. Am. Soc. Hortic. Sci.,* 99: 107–110.

Stanhill, G., Fuchs, M., Bakker, J. and Moreshet, S., 1973. The radiation balance of a glasshouse rose crop. *Agric. Meteorol.,* 11, 385–404.

Stanhill, G., Moreshet, S., Jurgrau, M. and Fuchs, M., 1975. The effect of reflecting surfaces on the solar radiation regime and carbon dioxide fixation of a glasshouse rose crop. *J. Am. Soc. Hortic. Sci.,* 100: 112–115.

Storck, H., 1978. Towards an economics of energy in horticulture. *Acta Hortic.,* 76: 15–30.

Takakura, T., 1989. Protected cultivation in Japan. *Acta Hortic.,* 230: 29–37.

Takakura, T., Jordan, K.A. and Boyd, L.L., 1971. Dynamic simulation of plant growth and environment in the greenhouse. *Trans. Am. Soc. Agric. Eng.,* 15: 964–971.

Takami, S. and Uchijima, Z., 1977. A model of the greenhouse with a storage type exchanger and its verification. *J. Agric. Meteorol. Jpn.,* 33: 155–166.

Thomas, R.B., 1978. The use of specularly reflecting back walls in greenhouse. *J. Agric. Eng. Res.,* 23: 85–97.

van Bavel, C.H.M., Damagnez, J. and Sadler, E.J., 1985. The fluid roof solar greenhouse: energy budget analysis by simulation. *Agric. Meteorol.,* 23: 61–76.

van de Braak, N.J., Breuer, J.J.G. and Heyna, B.J., 1984. Cumulative frequency distribution curves of greenhouse heat requirements. *Acta Hortic.,* 148: 337–344.

van de Kieboom, A.M.G. and Stoffers, J.A., 1985. Light transmittance under diffuse radiation circumstances. *Acta Hortic.,* 174: 67–74.

Varlet-Grancher, C. and Bonhomme, R., 1974. Utilisation de l'énergie solaire par une culture de *Vigna sinensis* III. Evolution du rendement énergétique pendant la phase végétative. *Ann. Agron.,* 25: 811–819. In French, with English summary.

Varlet-Grancher, C., Bonhomme, R., Chartier, M. and Artis, P., 1982. Efficience de la conversion de l'énergie solaire par un couvert végétal. *Acta Oecol./Oecol. Plant.,* 3: 3–26. In French, with English summary.

von Zabeltitz, C., 1986. Greenhouse heating with solar energy. *Energ. Agric.,* 5: 111–120.

von Zabeltitz, C. and Damrath, J., 1984. Greenhouse heating with sun energy. The greenhouse as collector. *Acta Hortic.,* 148: 763–769.

Warren-Wilson, J., 1972. Control of crop processes. In: A.R. Rees, K.E. Cockshull, D.W. Hand and R.G. Hurd (Editors), *Crop Processes in Controlled Environment.* Academic Press, London, pp. 7–30.

Yamamoto, Y., 1985. Greenhouses with underground heat exchange system. *J. Agric. Meteorol. Jpn.,* 41: 273–280.

Chapter 12

THE GREENHOUSE CARBON CYCLE

C. GARY and A. BAILLE

INTRODUCTION

The greenhouse ecosystem contains, on a smaller scale, part of the dynamic and constant features of the carbon cycle in natural or agricultural ecosystems. Its specificity arises from the nature and purpose of a greenhouse. A greenhouse is designed to increase crop productivity by increasing temperature, thanks to a transparent enclosure which reduces convective and radiative heat losses. But this same enclosure also limits the availability of photosynthetic radiation and the carbon dioxide (CO_2) replenishment of the atmosphere around the plants. While higher temperatures increase potential productivity by increasing the potential growth rate of sinks and extending the growing season, lower light and CO_2 availability may reduce photosynthetic rate and prevent the realization of a higher yield. The balance between the activities of the sources and sinks of carbon in the crop is altered. The skill in the design and management of a greenhouse is to come to a balance which leads to a better production (quantity, quality, timing) than in outside conditions.

The greenhouse is a semi-closed system, as there are exchanges with the outside atmosphere, with artificial CO_2 sources and with biomass consumers (through harvests or prunings if removed). The main carbon pools are (1) the greenhouse atmosphere; (2) the crop; and (3) either the soil or artificial root medium. Various sources of energy are necessary for the carbon to move between these different compartments. For a given greenhouse structure, the turn-over of its internal atmosphere mainly depends on wind speed and on the setting of the vents. In the greenhouse ecosystem, a main forcing variable is photosynthetically active radiation (PAR), which drives the assimilation of CO_2 in photosynthetic organs. Part of the carbohydrates produced are transformed into plant structures, while another part is oxidized back into CO_2 to provide energy for biosynthesis and other energy-requiring processes. Heterotrophic micro-organisms living in the rhizosphere acquire energy from the oxidation of organic compounds in roots and soil. The crop subsystem is a carbon trap, which would quickly exhaust the greenhouse atmosphere if there was no renewal from the big outside atmospheric reservoir, or from CO_2 enrichment. Another important forcing variable is temperature, as it drives most of the carbon flows: gross photosynthesis, translocation, respiration, growth, and rhizosphere activity. The cycle of other elements may interact with the carbon cycle. For instance, the water status in the root medium and in the atmosphere controls stomatal closure and consequently resistance to the diffusion of gases, CO_2 as well as water vapor. If some mineral is in short supply, production of fundamental structures of the carbon metabolism can decrease (for example synthesis of chlorophyll under magnesium deficiency).

In order to improve carbon exportation from the greenhouse ecosystem, different methods are available, such as providing more mineral carbon (CO_2), supplying more energy for its assimilation (light), or increasing flow rates in the system (temperature). Knowing all sources and sinks of carbon and estimating their activity is of interest to provide not only a case study in ecology but also a rationale for greenhouse management.

After an overview of the carbon cycle in the greenhouse ecosystem, each of its flux variables will be described here, with some emphasis on its importance in the whole greenhouse system, on the possible interactions with other flux and state variables, and on the controls by external parameters. Finally, an attempt at quantifying the different terms of the carbon cycle will be given for typical commercial greenhouses. All flow and state variables will be expressed per square meter of greenhouse, CO_2 concentrations in $\mu\ell\,\ell^{-1}$ (ppm), and CO_2 and biomass fluxes in $g\,C\,m^{-2}\,h^{-1}$.

OVERVIEW OF THE GREENHOUSE CARBON CYCLE: COMPARTMENTS AND FLUXES

The components of the carbon cycle are compartments of different sizes (Fig. 12.1). A global description of the carbon cycle is possible through the balance of the various inputs and outputs of each carbon compartment. These balances will give clues for understanding and managing the carbon cycle in the greenhouse.

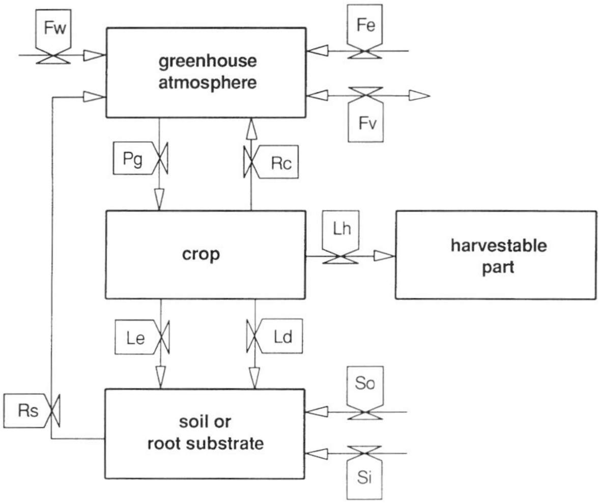

Fig. 12.1. Compartments and fluxes in the greenhouse carbon cycle. For the meanings of symbols, see the Glossary, p. 299.

The outside atmosphere is regarded as an infinite environment with which the greenhouse system exchanges gases. Its mean CO_2 concentration (C_o) is around 350 $\mu\ell\,\ell^{-1}$ in the early 1990s, and is increasing slowly due to human activities. C_o presents diurnal and seasonal fluctuations of some tens of $\mu\ell\,\ell^{-1}$, due to photosynthesis and respiration in the biosphere. Moreover, greenhouses located near urbanized or industrial areas are exposed to higher and more variable CO_2 concentrations.

The greenhouse atmosphere

The atmosphere inside the greenhouse is a very small carbon compartment. Its mass of carbon (M_a, in $g\,m^{-2}$) is:

$$M_a = C_a \cdot 10^{-6} \cdot V_g \cdot d \cdot \tfrac{12}{44}, \qquad (12.1)$$

where C_a is the inside CO_2 concentration ($\mu\ell\,\ell^{-1}$), V_g is the greenhouse volume per area unit ($m^3\,m^{-2}$), and d is the density of carbon dioxide (1.83 $g\,\ell^{-1}$ at 20°C) at ambient temperature and pressure (12 is the atomic mass of carbon and 44 is the molar mass of CO_2). Being small, the size of the greenhouse atmospheric carbon pool varies constantly according to the balance of different fluxes (all expressed in $g\,C\,m^{-2}\,h^{-1}$):

(1) air exchange with the outside atmosphere through leakage or ventilation (F_v);
(2) CO_2 exchanges with the crop, i.e., gross photosynthesis (P_g) and respiration (R_c);
(3) CO_2 release by the soil (R_s);
(4) artificial CO_2 supply by injection of pure CO_2 or use of the products of combustion of fossil fuels (F_e); and
(5) respiration of human beings (F_w).

This balance may be written

$$\frac{dM_a}{dt} = 10^{-6} \cdot V_g \cdot d \cdot \tfrac{12}{44} \cdot \frac{dC_a}{dt}$$

or

$$\frac{dM_a}{dt} = 10^{-6} \cdot V_g \cdot d \cdot (F_v - P_g + R_c + R_s + F_e + F_w). \qquad (12.2)$$

This last equation may be used as a basis in algorithms for the control of CO_2 concentration in greenhouses (Jones et al., 1984).

The crop

The developed crop represents a much larger carbon compartment than the greenhouse atmosphere. Its growth is the result of various processes (all expressed in $g\,C\,m^{-2}\,h^{-1}$):

(1) CO_2 exchanges with the inside atmosphere (P_g and R_c);
(2) harvest of organs of economic interest, and pruning (L_h);
(3) death of shoot or root organs after normal senescence or as a consequence of pests and diseases (L_d); and
(4) root exudation (L_e).

Three important components of the crop compartment are the shoot, root and harvestable parts (which may include part of the shoot and/or part of the root). They interact with the atmospheric compartment, the soil compartment and the outside atmosphere. Therefore, partitioning of carbon among the crop compartments is an important process in the greenhouse carbon cycle. The size of the whole crop compartment, M_c, varies as follows:

$$\frac{dM_c}{dt} = (P_g - R_c) - L_h - L_d - L_e. \tag{12.3}$$

The soil

The last compartment is the soil, the components of which differ according to cultivation method. In production on natural soil, a pool of organic matter is mineralized by micro-organisms. Its origin may be the original organic matter present before the crop was planted (S_i), dead tissues from the crop (L_d) or organic materials supplied for fertilization or mulching (S_o). In soilless culture, the rhizosphere will develop only around the root system, as a result of root exudation (L_e). The balance of the soil compartment (in $g\,C\,m^{-2}\,h^{-1}$) is:

$$\frac{dM_s}{dt} = S_i + L_d + L_e + S_o - R_s. \tag{12.4}$$

CARBON DIOXIDE ASSIMILATION BY THE CROP

Photosynthesis is basically a process of energy conversion for carbon acquisition. The interception of photons in the visible part (400–700 nm) of the solar spectrum by the foliar pigments allows a movement of electrons from water (transformed into O_2) to nicotinamide adenine dinucleotide phosphate ($NADP^+$, changing into its reduced form, NADPH). In the process, a gradient of protons is created across the thylakoid membrane which allows the formation of energetic compounds (adenosine triphosphate, ATP). The reducing power and energy are available for the reduction of atmospheric CO_2 into carbohydrates, but may also be used for the reduction of O_2 with loss of CO_2 (photorespiration), which reduces the efficiency of the use of radiant energy for carbon assimilation. Potential photosynthesis will then depend on the availability of light and CO_2, and on the ability of the photosynthetic apparatus to intercept light and assimilate CO_2. Actual photosynthesis will be modulated by the demand for photoassimilates by sinks.

Light interception and canopy morphogenesis

Light harvesting depends on leaf area index (leaf area per unit soil area, LAI), and on the spatial distribution and duration of green leaf area. The physics of light interception is detailed elsewhere (Chapters 3 and 11), so only the determination of canopy morphogenesis will be considered here. Patterns of canopy morphology differ greatly among species. Furthermore, they can be affected by temperature, day-length, light quantity and quality, and water stress. The rates of initiation of new leaves and of leaf-area expansion depend mainly on temperature; they are relatively independent of light and of CO_2 concentration except when these factors are very limiting (Verkerk, 1955; Picken et al., 1986). As light and CO_2 concentration increase, the ratio of leaf area to dry weight (specific leaf area) decreases (Thornley and Hurd, 1974). The rate of leaf aging also accelerates with higher temperatures (Picken et al., 1986). Water stress limits light interception in several ways: leaf wilting or rolling, reduction of leaf area expansion and acceleration of ageing and senescence (Amthor and McCree, 1990).

Carbon dioxide assimilation at leaf level

The assimilation of CO_2 in leaves is determined by: (1) the availability of energy and reducing power; (2) the CO_2 concentration at the carboxylation site; (3) the activation of ribulose biphosphate (RuBP) carboxylase, and (4) the availability of RuBP (regenerated in the Calvin cycle). Conditions (1) and (3) depend on light level. Condition (2) depends on the CO_2 concentration in the atmosphere and on the various resistances to diffusion, and condition (4) on the ability of the cell to transform assimilated carbon into storage compounds (starch in the chloroplast) or translocated compounds (soluble carbohydrates). At low light levels photosynthesis is limited by quantum yield, whereas at high light levels it is limited by internal CO_2 concentration – that is, by external CO_2 concentration and resistances to diffusion. In non-limiting light conditions, at low CO_2 concentrations photosynthesis is limited by CO_2 availability, and at high CO_2 concentration by RuBP regeneration. The increase of photosynthesis under high CO_2 concentrations is due not only to a mass effect but also to a change in relative affinity of the RuBP carboxylase/oxygenase enzyme for CO_2 and O_2.

Whereas major greenhouse crops are C_3 species[1], some have crassulacean acid metabolism (CAM) (*Kalanchoë*, cacti and many orchids).

Carbon dioxide assimilation at canopy level

Photosynthesis at canopy level results from the integration of photosynthetic activity of individual leaves. Light, temperature, CO_2 concentration and humidity are heterogeneous inside the canopy and vary with the geometry of the canopy and with its activity. Each leaf or even leaf element is under the influence of different microclimatic conditions, which determine the instantaneous photosynthetic response, and may also lead to acclimation. The mathematical description of PAR penetration into the canopy and therefore of photosynthesis is possible, from the knowledge of incident PAR intensity and angle distribution (ratio of diffuse to direct radiation, and direction of direct beams), and a description of the foliage (geometry and photosynthetic characteristics). Such an approach has been applied to row crops such as cucumber (*Cucumis sativus*) (Gijzen and ten Cate, 1988) and tomato (*Lycopersicon esculentum*) (Tchamitchian and Longuenesse, 1991). In this last reference, a map of the photosynthetic activity across a crop row showed the contribution of the leaves to the crop photosynthesis according to their position in the canopy; the most influential region moves as the sun angle changes, and includes the lower leaves near noon. This behavior is typical of row canopies.

More empirical approaches have led to the determination of whole-plant or canopy response curves to light and/or CO_2, for flower crops: roses (*Rosa* spp.) (Hand and Cockshull, 1975), carnations (*Dianthus caryophyllus*) (Enoch and Hurd, 1977), chrysanthemum (*Dendranthema grandiflora*) (Acock et al., 1978a), and other flower species (e.g. Berninger, 1989). Vegetable crops such as tomato (Acock et al., 1978b) and green pepper (*Capsicum annuum*) (Acock et al., 1976) have also been studied. In these models, the diffusion of light in the canopy is generally considered as isotropic.

Validation of a gross photosynthesis model is not easy, except at leaf level with short pulses of $^{14}CO_2$. Most results reported in the literature come from measurements of net CO_2 exchange, at leaf, canopy or greenhouse level. Gross photosynthesis rate is computed by adding an estimated respiration rate. *In situ* techniques are convenient for studying the greenhouse carbon balance. For example, a frameless, slightly pressurized, assimilation chamber has been designed to enclose several whole plants (Daudet, 1987). Its size may increase with the canopy development, and it may be easily moved from one sample to another. The same estimation of net assimilation has been carried out at the greenhouse scale (Dayan et al., 1985; Nederhoff et al., 1988). To the extent that all other CO_2 flows are known (ventilation, CO_2 injection, soil respiration, and changes in internal concentration), the canopy net assimilation can be deduced. The difficulties in estimating ventilation rate can be avoided by maintaining internal CO_2 concentration equal to the outside concentration (Matthews et al., 1987).

CARBON TRANSFERS INSIDE AND OUT OF THE CROP

Carbon pools and flows

The various carbon pools inside the plant consist of recently assimilated carbohydrates (soluble sugars, starch) in the leaf, structural biomass and long-term storage, while the carbon flows are translocation, respiration, structural growth and breakdown of degradable structures.

The trioses emerging from the Calvin cycle may be used either to synthesize starch in the chloroplast itself or sucrose in the cytoplasm. During a 24 hour period, the starch content typically increases by day and decreases by night (Challa, 1976). This carbon pool buffers possible differences between assimilation and translocation rates. The translocation rate varies according to the activities of sources and sinks (Ho, 1979), which are determined by their intrinsic potential and by external parameters such as light, CO_2 (which modulate source activity) and temperature (acting on both sources and sinks) (Verkleij and Challa, 1988).

Once unloaded in the sink tissues, part of the

[1] In C_3 plants, RuBP carboxylase is the initial CO_2-fixing enzyme which catalyzes the reduction of CO_2 into three-carbon molecules. CAM plants open their stomata at night to allow the fixation of CO_2 by PEP (phosphoenol pyruvate) carboxylase, and the production of malate, a four-carbon molecule; the process of carbon assimilation goes on by day, when stomata are closed, to limit water losses. In C_4 plants, fixation of CO_2 by PEP carboxylase is a first step which concentrates CO_2 close to RuBP carboxylase, located in inner tissues; consequently, photorespiration – that is, the fixation of O_2 by RuBP carboxylase, is suppressed (Osmond et al., 1982).

assimilates are oxidized into CO_2 for energy production through respiration. The respiratory metabolism is under feed-forward and feedback controls. Carbohydrate availability is limiting in young tomato plants when it is lower than about 10% of the dry weight (Gary, 1988). This may explain differences in responses of respiration rate to temperature – for instance, the Q_{10} factor measured on the same plant material may decrease from 1.5 at dusk to 1.1 after 40 hours darkness (Gary, 1988). The respiration rate also depends on the activity of energy sinks. These include the production of structural biomass (structural growth) – that is, more or less reduced compounds synthesized from carbon skeletons. The ratio between carbon losses and structural growth, the CO_2 production factor (CPF) (Penning de Vries et al., 1974), depends on the energy cost of the synthesized compounds. About 1.6 g of carbon are lost for the production of 1 g lipids, 0.4 g for 1 g proteins and 0.1 g for 1 g carbohydrates.

A second energy sink is the maintenance of existing structures. This sink arises from the constant turnover of some structural compounds such as enzymes, and leakage across membranes with ionic gradients. The maintenance coefficient (m) would be between 0.4 and 1.1 $\mathrm{mg\,C\,g^{-1}\,C\,h^{-1}}$ (Penning de Vries, 1975). Therefore, total respiration can be expressed as:

$$R_c = \mathrm{CPF}\frac{dM_c}{dt} + mM_c, \qquad (12.5)$$

without considering harvest or death of organs nor exudation in dM_c/dt.

The last energy sink, the assimilation of minerals, is not considered yet in whole-plant respiration models, but is implicitly included in growth respiration. The uptake of anions needs the activity of ion pumps, with consumption of ATP. The most costly process is the reduction of oxidized minerals, such as nitrates and sulfates. When it occurs in leaves, reductive power is available in excess in the chloroplasts under non-limiting light conditions. In darkness and in roots, however, respiratory metabolism has to provide this energy.

All assimilates are not metabolized immediately. Some may be diverted to long-term carbohydrate reserves in specific organs such as stems and roots of perennials or bulbs. These reserves may be mobilized later.

Carbon partitioning

Carbon partitioning depends on the number and relative strength of the various sinks for assimilates. Warren-Wilson (1972) considered sink strength of a growing organ to be expressed by its absolute growth rate. Sink strength is the product of sink size (dry weight) and sink activity (relative growth rate). Modelers often give a different meaning to this expression, sink strength being considered as potential growth rate (Jones et al., 1989; Marcelis et al., 1989). This potential growth can be quantified in non-limiting assimilate conditions – for instance, by pruning sinks which would compete with the organ studied. Potential growth generally follows a bell-shaped function of the organ age. Thus, knowing the number, quality and age of sinks, dry-matter distribution has been well simulated for tomato (Jones et al., 1989) and cucumber plants (Marcelis et al., 1989). A more empirical approach has been developed for the growth of flowering rose shoots, with the use of a set of equations giving the time-course of partitioning coefficients (Lieth and Pasian, 1991).

Sink activity responds positively to temperature; as a consequence, competition among sinks increases at high temperatures. In this situation, the number of sinks may be reduced by abortion of young organs. This is often the case in winter with greenhouse conditions of low light and high temperature; the activity of assimilate sources is low, while sink activity is high. In tomato, this imbalance leads to flower abortion, except when the number of flowers per inflorescence is artificially limited by pruning (Bertin and Gary, 1992). Reproductive organs are generally major sinks; for instance in tomato the growth of leaves and fruits is markedly depressed when fruit growth begins (Hurd et al., 1979), and root/shoot ratio is higher when the first inflorescences have been pruned (Bertin, personal communication, 1991). The pattern of carbon allocation can also be disturbed by a limitation of a growth factor such as light, water or minerals. The organ that can acquire whichever growth factor is most limiting is often given priority, thereby maintaining a "functional equilibrium" (Brouwer, 1983). For example, limitation in water or mineral supply lowers the shoot/root ratio, whereas low light conditions increase it.

Carbon export

Biomass leaves the greenhouse as a result of harvest, pruning or death of organs during growth, or removal at the end of the growth cycle. The ratio of dry weight of the harvested product to the total plant dry weight, the harvest index (HI, here in $\mathrm{g\,C\,g^{-1}\,C}$) is generally high and varies from about 60% for tomato (Ho and Hewitt, 1986) to almost 100% for ornamentals and leaf crops like lettuce (*Lactuca sativa*):

$$L_h = HI(P_g - R_c). \quad (12.6)$$

The time-course of this carbon flux may vary greatly, with a rather constant export rate in species with indeterminate fruiting like cucumber or tomato, and more variable export rates for flower or ornamental crops.

CARBON OUTPUT FROM GREENHOUSE SOIL

Of all CO_2 fluxes, respiration of the root medium is certainly that which is least quantified. In all studies of the CO_2 balance in the greenhouse, this component was measured or estimated in a cruder way than any other. Measured CO_2 production rates range from 50 to 220 mg C m^{-2} h^{-1} (e.g. Matthews et al., 1987). The carbon sources may be the root system, the rhizosphere, and the soil microflora. Many kinds of root media may be used in greenhouses and this will affect the origin and the flux of CO_2.

As stated before, the respiration rate of the root is correlated with its dry weight, its growth rate and the rate of mineral uptake. It depends on temperature, and may be limited by the availability of the respiratory medium which reflects the rate of carbon assimilation by shoots. Another loss of carbon from roots is the exudation of carbohydrates (Lambers, 1987). The flux of carbon from the root system to its microflora has been estimated at from 0.017 to 0.150 mg C (g d.wt. roots)$^{-1}$ h^{-1}. Heterotrophic micro-organisms live in the rhizosphere, partly using these products, and, consequently, their growth and respiration rates will be closely linked with root activity. In soilless culture, the only source of carbon is the plant (except for some photosynthetic algae the contribution of which is not significant), and root and microflora activities may be considered as a whole. Experimental estimations of the activity of the tomato root system range from 1 to 3 mg C (g d.wt. roots)$^{-1}$ h^{-1} (Gary, 1988).

In soils, biological degradation of organic matter also makes a contribution. The carbon sources are either plants through senescence of roots or shoot organs, or organic materials (peat, manure, etc.) imported from outside the greenhouse or present in the original soil. Carbon dioxide evolution therefore depends on the organic matter content, increasing from about 70 mg C m^{-2} h^{-1} with 2% carbon content up to 200 mg C m^{-2} h^{-1} with 4% carbon content (Berninger, personal communication, 1980). Carbon mineralization in greenhouse soils is mainly driven by temperature. It increases exponentially from about 5°C up to about 35°C (Mathes and Schriefer, 1985). Temperature affects not only microbial activity but also the composition of soil microflora, as each species has its own optimal temperature range. Bacterial development needs a sufficient level of soil moisture, but not excessive amounts that would limit the diffusion of oxygen. At lower pH values, fungi will dominate over bacteria. As soil micro-organisms need all kinds of minerals as well as carbon, the mineral composition of the organic material and particularly its nitrogen content, together with the nature and amount of fertilizers, will affect the evolution of CO_2 from the soil.

Different methods are available to measure CO_2 evolution from soil: open or closed vessels with gas analysis, or absorption by soda lime or alkali. Large discrepancies were observed among different techniques used on the same samples (Minderman and Vulto, 1973). Part of this variability may be due to the effect of the vessel pressure on CO_2 diffusion and mass flow from the soil (Kanemasu et al., 1974). All of these methods have more value for comparative studies than for absolute measurements of CO_2 evolution. In the case of soilless culture, enclosing the root medium in a cuvette for measurement of CO_2 exchange is easier and less uncertain.

The contribution of CO_2 evolution from the soil to the greenhouse carbon balance depends greatly on the nature of the root medium. In the case of soil, the carbon input through organic materials used as fertilizer, root medium or mulch may not be negligible. Provided temperature, moisture and mineral conditions are not limiting, about one-third of the assimilatory activity of the canopy may be covered (Berninger, 1989). But the time-course of this CO_2 production remains out of control.

CARBON DIOXIDE ENRICHMENT AND ITS EFFECTS ON CROP GROWTH

Increasing the CO_2 concentration in the atmosphere enhances growth and yield of crops, as reported by many authors. Reviews by Kimball (1983, 1986), Cure and Acock (1986) and Enoch (1990), among others, have summarized the results of the most significant work on the effects of atmospheres enriched with CO_2. Current interest in these effects has been renewed, as several scenarios on anthropogenic CO_2 emissions predict a doubling of the atmospheric concentration of CO_2 in the near future (2025–2050). In these predictions, the carbon cycle of the biosphere is the main frame supporting computations and simulation models. Carbon dioxide enrichment of the earth's atmosphere could increase crop photosynthesis, yield and water-use efficiency, as it does in greenhouses.

Effects of CO_2 enrichment on the crop carbon balance

The carbon balance of CO_2-enriched crops involves the same compartments (pools, sources and sinks) and exchange fluxes as for non-enriched crops, as described previously. The main effect of CO_2 enrichment is to modify the intensity of carbon fixation, in both the short term and the long term. It is beyond our scope here to detail all the aspects related to these changes in the physiological behavior of the crops. Only the main effects will be summarized below.

Photosynthesis

The short-term response of net photosynthesis rate to CO_2 enrichment is generally a significant increase (about 50% when the ambient CO_2 concentration is doubled), but this increase in photosynthetic rate is only approximately 30% after acclimation to a new level of CO_2 concentration (Cure and Acock, 1986). Greenhouse species, which are mainly C_3-type, respond relatively well to CO_2 enrichment, as shown by photosynthesis measurements on cucumber and tomato (Gaastra, 1959; Ito, 1971; Frydrych, 1976; Ho, 1977; Slack, 1986), on rose (Hand and Cockshull, 1975) and carnation (Enoch and Hurd, 1977).

Respiration

Some recent work indicates that dark respiration is reduced at high CO_2 levels (Reuveni and Gale, 1985; Spencer and Bowes, 1986; Gary and Veyrès, 1992). Therefore, a part of the increase in net assimilation rate and yield of plants grown at elevated CO_2 might be attributed to this phenomenon. But this behavior does not seem to be general, as other authors (Hrubec et al., 1985) observed a higher respiration rate in CO_2-enriched soybean (*Glycine max*) plants, associated with a higher carbohydrate content.

Net assimilation and acclimation

Net assimilation rate (NAR, $g\,m^{-2}\,d^{-1}$) averages CO_2 exchanges over days or weeks. As night respiration is included in NAR, short- and long-term NAR responses are smaller than photosynthesis responses (Cure and Acock, 1986): the initial increase of NAR is on average +23%, whereas the long-term increase of NAR is about +18%. The decrease in carbon exchange rate and growth rate observed after continuous long-term enrichment is well known (Aoki and Yabuki, 1977; Sage et al., 1989), but the explanation of this behavior has not been clearly established. It is possible that the maximum rate of carbon assimilation may be fixed genetically, or may be determined by the number and strength of existing sinks (Mott, 1990). When NAR exceeds this limit, feedback mechanisms reduce photosynthesis to preserve a steady state. Accumulation of starch (Madsen, 1968) and sucrose in the leaves of CO_2-enriched plants, and the reversibility of this process when plants return to ambient CO_2 concentration, support this hypothesis (Sasek et al., 1985).

Biomass production and allocation

Many studies report a significant increase in dry-matter production by CO_2-enriched greenhouse crops: chrysanthemum (Hughes and Cockshull, 1972), rose (Holley, 1970; Hand and Cockshull, 1975), tomato (Knecht and O'Leary, 1974; Hicklenton and Jolliffe, 1978; Nielsen et al., 1983), and cucumber (Frydrych, 1976; Ito, 1978). Yield increases for CO_2-enriched plants are roughly similar to increases in total biomass (Kimball, 1983), suggesting that the allocation pattern in CO_2-enriched plants is more or less the same as for unenriched plants. The response of biomass accumulation and yield is roughly similar to that for the rate of carbon fixation. Cure and Acock (1986) calculated that the average increase in biomass accumulation for a doubling of CO_2 concentration is about 28% for C_3 grasses. In the same review, results were more erratic for C_3 broadleaf species: yields increased on average by 40%. Kimball (1983) found in a literature review that the mean yield increase for C_3 crops with CO_2 enriched to $600\,\mu\ell\,\ell^{-1}$ was 34%. Based upon this, and noting a decline in incremental response as CO_2 concentration increased from 660 to $1000\,\mu\ell\,\ell^{-1}$, Bellamy and Kimball (1986) assumed a 50% increase in yield for a CO_2 concentration of $1000\,\mu\ell\,\ell^{-1}$.

To summarize, a variety of physiological and morphological changes have been reported for plants grown at elevated CO_2. The general trend is a significant increase (for C_3 types) in carbon fixation rate, biomass accumulation and yield. For greenhouse crops one may take an average of 20% increase in carbon fixation and yield for CO_2-enriched conditions.

SOURCES OF CO_2 ENRICHMENT

Carbon dioxide is available from a number of natural and industrial processes. Before 1960, most greenhouse growers were using manure to produce CO_2. In the 1960s, many began to use either liquid compressed CO_2, or CO_2 produced by burning of liquid fuel or gas (propane, butane, methane). Detailed descriptions of these various sources of CO_2 are available in papers by van Berkel (1986), Hand (1986) and Bauerle et al. (1986). A brief summary is presented here.

Decomposition of organic matter

Microbiological decomposition of organic matter produces considerable amounts of CO_2, particularly at the beginning of crop growth on large volumes of organic soil and fertilized with organic matter (e.g., manure, or straw-balls). Bierhuizen (1973) gives values of about 0.5 g CO_2 $m^{-2}\,h^{-1}$ for straw. However, later in the season the demand increases and the CO_2 production declines.

Pure CO_2

A purified form of CO_2 can be obtained from various by-product sources (distilleries, chemical industry, etc.). Problems of phytotoxic impurities can be avoided when this form is used. The gas is generally stored in cylinders or bottles (small holdings) at 7 MPa or in tanks containing several cubic meters, fully insulated and cooled to −18°C, at about 2 MPa.

Natural gas, propane and kerosene

Another commonly used source of CO_2 is the combustion of natural gas, propane, butane or kerosene (low-sulfur grade) with release of flue gases into the greenhouse atmosphere. In some countries (the Netherlands, for example) natural gas is widely used to enrich crops, while in others (Denmark) nearly all CO_2 enrichment is with compressed CO_2 from large cooled tanks. Available systems (van Berkel, 1986) include the following: (1) using CO_2 enrichment from gas-fired central systems; the flue gases can be directed (totally or partially, depending on the CO_2 concentration desired) via ducts outside or inside the greenhouse; or (2) using air heaters located throughout the greenhouse which burn natural gas or kerosene.

The chemical equations for calculating the equivalent production of CO_2 for a given combustion gas are the following:

propane: $C_3H_8 + 5O_2 \rightarrow 3CO_2 + 4H_2O$,

butane: $2C_4H_{10} + 13O_2 \rightarrow 8CO_2 + 10H_2O$,

methane: $CH_4 + 2O_2 \rightarrow CO_2 + 2H_2O$.
(main component of natural gas)

Gas consumption is often expressed in volume units: for example, 1 m^3 (NTP) natural gas produces about 1.8 kg CO_2.

The use of such combustion gases has a drawback: the production of toxic compounds such as CO (toxic for humans), C_2H_4, sulfur and nitrogen oxides. In the presence of N_2 and O_2 from the air, various nitrogen oxides such as NO, N_2O, N_2O_2 and NO_2 will always be produced, the amount depending on flame temperature, geometry, etc. Even at relatively low concentrations, NO_x compounds can be phytotoxic for some sensitive species (Hand, 1990).

Respiration of human beings (F_w)

When people are working in greenhouses, some supplementary CO_2 is added to the internal atmosphere, due to human breathing (about 1.6 g C h^{-1} per person). This amount is generally negligible and does not change significantly the CO_2 balance, except in some periods of heavy activity (cropping, picking tasks): if ten people are working simultaneously in a 1 ha greenhouse, the CO_2 release will be about 1.6 mg C $m^{-2}\,h^{-1}$.

CARBON DIOXIDE EXCHANGE WITH OUTSIDE ATMOSPHERE

Importance of CO_2 exchanges with the outside atmosphere

The outside atmosphere can be considered either as a CO_2 source or a CO_2 sink, depending on the concentration inside the greenhouse, C_a. As the outside CO_2 concentration, C_o, is relatively constant with time (about 350 $\mu\ell\,\ell^{-1}$), the CO_2 transfer will be mainly governed by the time-course of C_a and by the intensity of air exchanges (leakage, ventilation) between the inside and outside atmospheres.

The outside atmosphere is a CO_2 sink: 1) during the night, when crop and soil respiration create a CO_2-enriched atmosphere: C_a can reach values around 500 to 600 $\mu\ell\,\ell^{-1}$ (Wittwer and Robb, 1964); and 2) during the day, when artificial CO_2 is supplied from a tank of pure CO_2 or from gas burners and C_a is significantly increased above C_o. The outside atmosphere is a CO_2 source during the day, when the photosynthetic activity of the crop depletes the inside concentration C_a below C_o (down to 150 $\mu\ell\,\ell^{-1}$ in tight greenhouses with a well-developed crop, high light intensity and no CO_2 enrichment).

From a practical point of view, CO_2 transfers between inside and outside atmospheres are a key mechanism in greenhouse CO_2 control, whether or not the atmosphere is enriched. A close analogy with sensible and latent heat transfers is noticeable: to estimate the transfers of CO_2, water and sensible heat, knowledge of the air exchange rate is absolutely necessary. It is generally expressed in volumes per greenhouse volume per hour ($m^3\,m^{-3}\,h^{-1}$) or in flow rate per unit ground area ($m^3\,m^{-2}\,h^{-1}$).

Greenhouse air exchange rate

In the case of mechanical ventilation (fans), the air flow rate can be known with high accuracy. However, air-exchange rates in natural ventilation systems (windows, openings, etc.) are difficult to estimate. Natural ventilation systems are widely used in agricultural buildings (greenhouses, animal housing, etc.) to control energy and mass transfers. They require less energy and maintenance than mechanical ventilation systems, but are more difficult to control and manage. Furthermore, they are strongly dependent on external factors which are out of the grower's control: wind speed and wind direction (pressure-driven forces), differences between inside and outside temperature and humidity (buoyancy forces), and global radiation (heating load).

Thermal buoyancy has little effect on natural ventilation in windy conditions. At low wind speed, thermal effects prevail, but there is no clear demarcation between conditions where wind speed or thermal effects dominate (Brockett and Albright, 1987). Theoretical analysis of wind and thermal actions, considered separately or jointly, are available in the literature (Kozai and Sase, 1978; Kozai et al., 1980; Bot, 1983). Wind-tunnel measurements (Sase et al., 1980) and numerical models of flow around and inside greenhouse (Okushima et al., 1989) have supplied more information on streamlines, turbulence levels and mass transfers, and show the high complexity of the exchange processes, influenced by numerous parameters (greenhouse shape and size, window location, obstacles such as crop rows, etc.).

For practical purposes, simple empirical models, derived from experimental results, are generally used. In the case of Venlo-type glasshouses (a widespread Dutch glasshouse), with vents in the roof, several experimental studies (Nederhoff et al., 1984; Goedhart et al., 1984) using the tracer gas technique (Barber and Ogilvie, 1984) showed that the ventilation rate could be satisfactorily approximated by a simple linear relationship:

$$N = K(r + l)\,\upsilon, \tag{12.7}$$

where N is the ventilation rate ($m^3\,m^{-3}\,h^{-1}$), K is a constant factor depending on the height of the greenhouse, r is the relative window aperture (%), l is the leakage component, expressed in equivalent window aperture (%), and υ is the wind speed ($m\,s^{-1}$). The constant K is about 0.07, and l depends on the tightness of the greenhouse (l ranges from 1 to 2% for Venlo-type greenhouses). This simple equation will be used in the following to estimate the ventilation rate of greenhouses. For $\upsilon = 4\,m\,s^{-1}$ and $r = 10\%$, N is about $3\,m^3\,m^{-3}\,h^{-1}$. For the same wind speed, and $r = 100\%$ (maximal opening), N is about $30\,m^3\,m^{-3}\,h^{-1}$. These values can be considered as typical for natural ventilation in moderate wind conditions.

Estimates of CO_2 exchanges by ventilation

The equation giving the amount of CO_2 exchanged by leakage and natural ventilation, F_v (in $g\,C\,m^{-2}\,h^{-1}$), is

$$F_v = N \cdot \tfrac{12}{44} \cdot d \cdot V_g \cdot 10^{-3} \cdot (C_o - C_a), \tag{12.8}$$

where d is the CO_2 density ($1.83\,g\,\ell^{-1}$ at 20°C), V_g is the ratio volume/ground area of the greenhouse ($m^3\,m^{-2}$), and $(C_o - C_a)$ is the difference in CO_2 concentration between outside and inside ($\mu\ell\,\ell^{-1}$).

Some numerical examples are presented below, giving ranges of values for F_v in a greenhouse with $V_g = 3\,m^3\,m^{-2}$.

(a) Unventilated greenhouse, during a sunny winter day, with a depletion $C_o - C_a = 100\,\mu\ell\,\ell^{-1}$ due to photosynthesis. Assuming $N = 1$ (well-sealed greenhouse), F_v is ~$0.5\ g\,m^{-2}\,h^{-1}$, or $5\ kg\,ha^{-1}\,h^{-1}$ (CO_2 supply to the greenhouse).

(b) Vented greenhouse during the day, with $C_o - C_a = 50\ \mu\ell\,\ell^{-1}$, and $N = 10\ m^3\,m^{-3}\,h^{-1}$. The resulting F_v is $2.7\ g\,m^{-2}\,h^{-1}$.

(c) Unventilated greenhouse, with a CO_2-enriched atmosphere at $650\ \mu\ell\,\ell^{-1}$. With the assumption $N = 1$, F_v amounts to $-1.6\ g\,m^{-2}\,h^{-1}$ CO_2 loss. If the greenhouse windows are slightly opened ($r = 10\%$) in order to control temperature or humidity, then $N = 3\ m^3\,m^{-3}\,h^{-1}$ (Eq. 12.8), and the CO_2 efflux out of the greenhouse is about $-5\ g\,m^{-2}\,h^{-1}$. If the average value for F_v during enrichment periods at $650\ \mu\ell\,\ell^{-1}$ is $-3\ g\,m^{-2}\,h^{-1}$, and the total enrichment duration is 1500 h during the whole year, the CO_2 efflux is 45 tons CO_2 ha^{-1} $year^{-1}$. This represents the quantity of artificial CO_2 lost by leakage and ventilation during the 1500 hours of CO_2 enrichment.

(d) Unventilated greenhouse during the night, with an average C_a of $450\ \mu\ell\,\ell^{-1}$, assuming $N = 1$, $F_v = -0.5\ g\,m^{-2}\,h^{-1}$. If this value is considered an annual average for a continuously grown crop (rose, for example), then the yearly nocturnal loss of CO_2 is approximately 25 tons CO_2 ha^{-1} $year^{-1}$.

Prediction of internal CO_2 concentration

The internal CO_2 concentration may be predicted by the following equation, which represents the steady-state CO_2 balance of the greenhouse atmosphere ($dM_a/dt = 0$ in Eq. 12.2):

$$C_a = C_o + \frac{F_e - P_g + R_c + R_s}{N \cdot V_g \cdot d} \cdot \frac{44}{12} \cdot 10^3, \qquad (12.9)$$

where F_e is the supply from an artificial source, P_g is crop gross assimilation rate, and R_c and R_s are crop and soil respiration, respectively (all fluxes in $g\,C\,m^{-2}\,h^{-1}$).

If these fluxes and N are known (measured or estimated by models), then C_a can be determined. Figure 12.2 presents the isolines ($C_a - C_o$) versus ventilation rate, N, and the net CO_2 exchange ($P_g - R_c - R_s$), in the case of non-enriched greenhouses ($F_e = 0$). During the night ($P_g = 0$, $R_c + R_s = -0.15\ g\,C\,m^{-2}\,h^{-1}$), $C_a - C_o$ can be positive, with values from +50 to $+100\ \mu\ell\,\ell^{-1}$ for $N = 1$ to $2\ m^3\,m^{-3}\,h^{-1}$. During the day, ventilation rates under $5\ m^3\,m^{-3}\,h^{-1}$ induce depletions greater than $-50\ \mu\ell\,\ell^{-1}$ for $P_g - R_c - R_s = 0.3\ g\,m^{-2}\,h^{-1}$ and $-100\ \mu\ell\,\ell^{-1}$ for $P_g - R_c - R_s = 0.8\ g\,m^{-2}\,h^{-1}$.

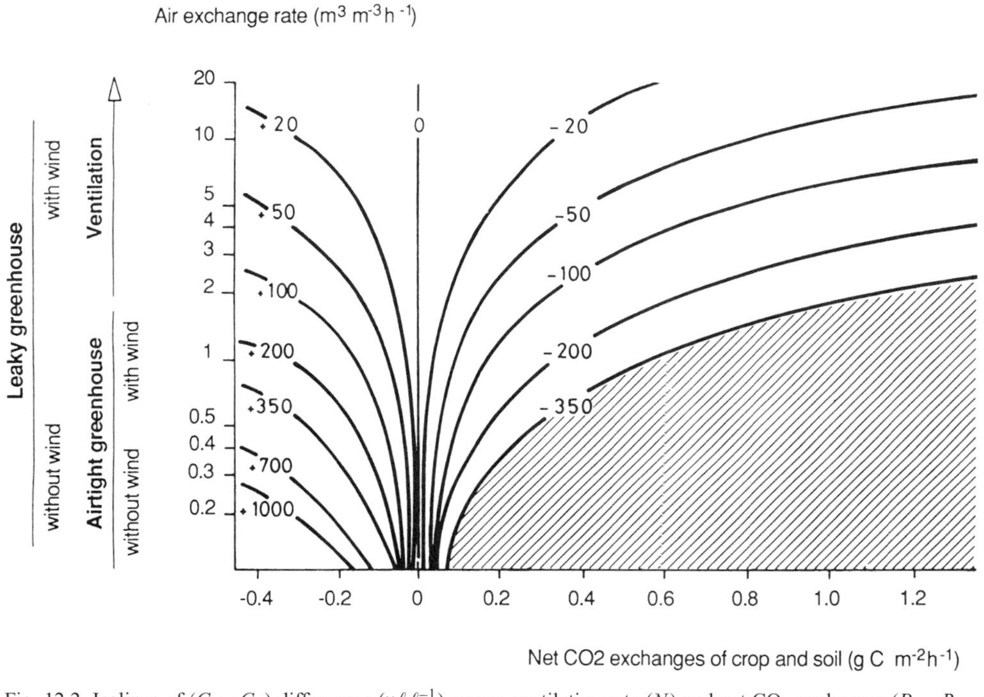

Fig. 12.2. Isolines of ($C_a - C_o$) differences ($\mu\ell\,\ell^{-1}$) versus ventilation rate (N) and net CO_2 exchanges ($P_g - R_c - R_s$). The so-called "airtight greenhouse" is a relatively well-sealed greenhouse, designed for energy saving.

Other aspects of the problem are (1) determining the ventilation rate necessary to maintain a given concentration C_a inside a non-enriched greenhouse (Challa and Schapendonk, 1986):

$$N = \frac{P_g - R_c - R_s}{d\,(C_o - C_a)\,V_g} \cdot \frac{44}{12} \cdot 10^3, \quad (12.10)$$

and (2) calculating the CO_2 injection rate (F_e) maintaining a given CO_2 concentration for a given ventilation rate:

$$F_e = P_g - R_c - R_s + (C_a - C_o) \cdot d \cdot N \cdot V_g \cdot \tfrac{12}{44} \cdot 10^{-3}. \quad (12.11)$$

These equations can be used in control algorithms for optimization of CO_2 concentration in greenhouses – for example, to calculate optimal CO_2 concentrations, considering energy and CO_2 consumption and additional dry-matter production due to increased photosynthesis (Schapendonk and Tilburg, 1984; Bakker, 1985; Nederhoff, 1988; Houter, 1990).

A simple approach to optimization of CO_2 enrichment in greenhouses has been proposed by Enoch (1984), with the concept of carbon dioxide uptake-rate efficiency (CDUE), defined as the ratio of CO_2 assimilated by the crop to CO_2 supplied to the crop. This dimensionless parameter, which can take values between 0 and 1, can be determined from (1) estimates of CO_2 uptake ($P_g + R_c$) and CO_2 loss rate (F_v) and (2) measurements of the ratio of the isotopes ^{14}C and ^{13}C in the crop (Enoch et al., 1984). However, other information must be ascertained to optimize the economic return from CO_2 enrichment. In sum, it is necessary to calculate the economic benefits and costs of enrichment, and identify the level with the highest difference between the two. This approach requires accurate estimations of CO_2 usage (assimilation + ventilation) and a good simulation of expected harvest. Moreover, in order to compare costs and yields, the price of CO_2 and the expected price for the harvest must be taken into account (Challa and Schapendonk, 1986; Seginer et al., 1986; Nederhoff, 1988).

CARBON CYCLE IN COMMERCIAL GREENHOUSES: CASE STUDIES

The purpose of the following is to quantify carbon pools and fluxes in some typical commercial greenhouse systems. Formulae (Eqs. 2, 3, 5, 6 and 8) and ranges of values have been given for the different variables, together with a discussion of the greenhouse carbon cycle; but it may be helpful to remember the relative orders of magnitude of these values in a real system. The carbon cycle may be described in terms of daily or yearly balances.

Case studies and hypothesis

Three greenhouse cropping systems will be considered; calculations have been made for a tomato crop in production.

System I: culture on natural soil, in an unheated and unenriched greenhouse (common system in the Mediterranean area); average yield is about 12 kg fruits m^{-2} $year^{-1}$, for a crop duration of 7 to 8 months.

System II: soilless culture in heated but unenriched glasshouse, with an average yield of about 35 kg m^{-2} $year^{-1}$ for a whole-year crop.

System III: soilless culture in heated and CO_2-enriched glasshouse (common system in northern Europe), with a surplus yield of about 30% compared to system II – that is, an average yield of 45 kg m^{-2} $year^{-1}$.

All flow values are expressed in g C m^{-2} day^{-1}. The volume of the greenhouse is 3 m^3 m^{-2}. Without added CO_2, the crop has a total dry weight of 730 g m^{-2}, with 10% roots, 30% vegetative shoots and 60% fruits. Its mean carbon content is 45% dry weight. Carbon dioxide enrichment doubles the concentration of CO_2 in the greenhouse and increases all state and flux variables in the crop sub-system by 20%. The size of the different carbon compartments is: – for the greenhouse atmosphere, about 0.5 g C m^{-2} at 350 $\mu\ell\,\ell^{-1}$ and 1.0 g C m^{-2} if the CO_2 concentration is doubled; – for the whole crop (including shoots and roots), 330 g C m^{-2} at 350 $\mu\ell\,\ell^{-1}$ and 396 g C m^{-2} under CO_2 enrichment; – for the soil, when present, 12 kg C m^{-2} in the upper 50 cm containing 2.3% organic carbon. The estimations of flux variables have been made from statements detailed previously in this chapter.

Carbon cycle in systems I and II

The lower productivity of system I is mainly due to the shortness of the production period. In periods without heating, system II is equal to system I, except for the exchanges with the soil. The calculated fluxes and pools are given in Fig. 12.3a. The special feature of system I is the existence of a big carbon compartment in the soil. But fluxes into and out of the soil are

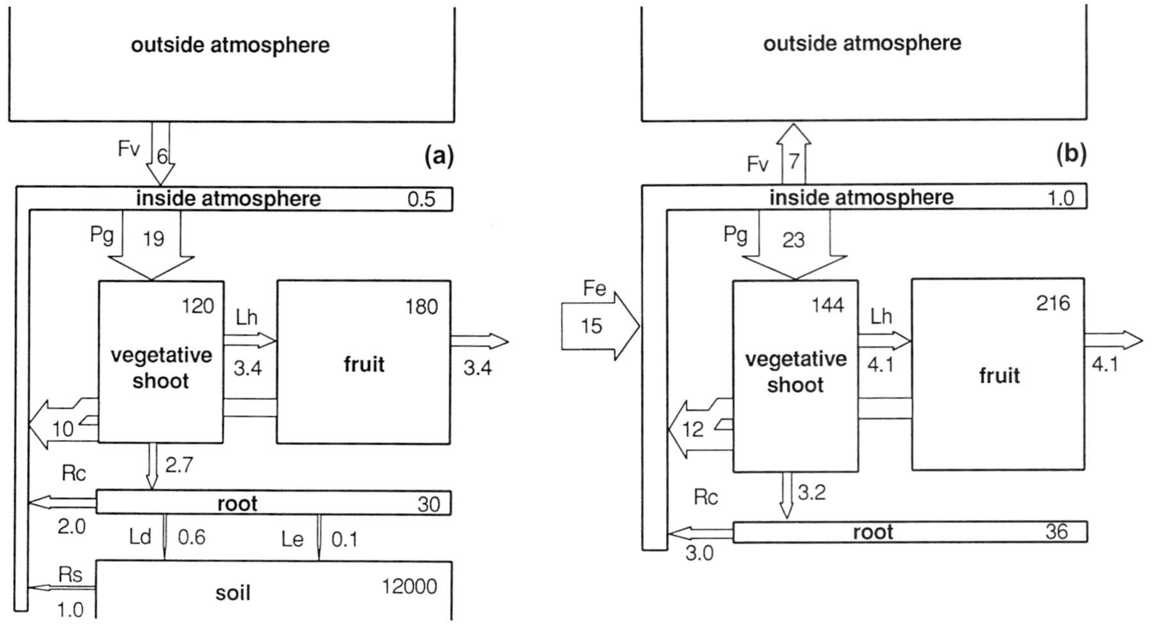

Fig. 12.3. The carbon cycle in a greenhouse. Inside the carbon compartments, numerical values are masses of carbon in $g\,m^{-2}$; the fluxes are expressed in $g\,C\,m^{-2}\,day^{-1}$ (flux symbols as in Fig. 12.1). (a) Cultivation on soil, without CO_2 enrichment; (b) soilless cultivation, with CO_2 enrichment.

small. Mean values of R_s (soil respiration minus root respiration) have been used. Where there is an important supply of manure or organic mulch, R_s can increase to $5\,g\,C\,m^{-2}\,day^{-1}$ and almost compensate for carbon assimilation by the crop. Carbon flux from the outside atmosphere would then be reduced to $1\,g\,C\,m^{-2}\,day^{-1}$. In both systems, the main supply of carbon comes from air exchanges. Because of the small size of the inner atmospheric carbon pool, any limitation in air movement from the outside will reduce carbon assimilation by the crop and productivity. The absence of any buffering capacity stresses the importance of a good and rational climate control, and indicates the necessity of CO_2 enrichment.

Carbon cycle in system III

The outside atmosphere is not a carbon source in periods of CO_2 enrichment. This external source feeds the crop, but half of the injected CO_2 is lost through air exchanges with the outside atmosphere (Fig. 12.3b). The limitation of carbon assimilation due to the low CO_2 concentrations in system II does not exist in this case, and productivity may increase.

One can observe in these examples that, due to the limited size of the internal atmosphere, the turnover of carbon in this compartment is high: each day, the internal atmosphere exchanges its carbon content several times with the crop. In contrast, the soil compartment, when present, has a much lower carbon turnover; the flow of released CO_2 is significant only when the soil carbon compartment is large (e.g., culture on soil with organic fertilization).

CONCLUSION

This chapter presents a set of expressions or estimations of the various fluxes in the greenhouse carbon cycle. These afford at least a theoretical description of all kinds of carbon-cycle scenarios for various greenhouse designs and methods of crop management. For reasons of clarity, each component of the greenhouse carbon cycle has been described separately, but it should not be forgotten how strong the interactions and controls are in such a small ecosystem.

Balancing the activities of carbon sources and sinks is certainly a key to the optimization of the growth (Gary, 1989) and yield (Challa and Schapendonk, 1986) of greenhouse crops. The greenhouse modifies the balance between temperature, light and CO_2 around the crop. Higher temperatures increase the demand for carbon (both the number and activity of sinks): case studies demonstrate how carbon supply is critical to obtain the high levels of production a grower can expect from the greenhouse farming system.

Models developed during the two last decades have provided valuable descriptions of the different functions of the greenhouse carbon cycle, and make the optimization of greenhouse management possible. Further attention to less well-known processes would benefit greenhouse management. Respiration in the root medium may be important in the case of soil cultivation and use of organic medium or fertilizer, but its estimation remains difficult. Carbon assimilation and its partitioning to harvested organs present a large range of patterns among greenhouse crops; a lot remain to be done for predictive models to be available for the major vegetable, fruit and ornamental greenhouse crops. Finally, interactions between the carbon cycle, water status and mineral availability are still poorly taken into account in growth models. More integrated approaches including carbon, water and mineral cycles would help address important issues such as the extension of the cultivation period to summer months in the Mediterranean area, or the increasing concern about pollution by mineral releases to the environment.

GLOSSARY OF SYMBOLS

C_o mean CO_2 concentration outside the greenhouse ($\mu\ell\,\ell^{-1}$)

C_a CO_2 concentration inside the greenhouse ($\mu\ell\,\ell^{-1}$)

CPF CO_2 production factor ($g\,C\,g^{-1}\,C$)

d density of carbon dioxide ($g\,\ell^{-1}$)

F_e artificial CO_2 supply ($g\,C\,m^{-2}\,h^{-1}$)

F_v air exchange with the ouside atmosphere through leakage or ventilation ($g\,C\,m^{-2}\,h^{-1}$)

F_w respiration of human beings ($g\,C\,m^{-2}\,h^{-1}$)

HI harvest index ($g\,C\,g^{-1}\,C$)

l leakage component, expressed in equivalent window aperture (%)

L_d death of shoot or root organs ($g\,C\,m^{-2}\,h^{-1}$)

L_e root exudation ($g\,C\,m^{-2}\,h^{-1}$)

L_h harvest of organs of economic interest and pruning ($g\,C\,m^{-2}\,h^{-1}$)

m coefficient of maintenance respiration ($g\,C\,g^{-1}\,C\,h^{-1}$)

M_a mass of carbon of the atmosphere inside the greenhouse ($g\,C\,m^{-2}$)

M_c mass of carbon in the crop ($g\,C\,m^{-2}$)

M_s mass of carbon in the soil ($g\,C\,m^{-2}$)

N ventilation rate ($m^3\,m^{-3}\,h^{-1}$)

P_g crop gross photosynthesis ($g\,C\,m^{-2}\,h^{-1}$)

r relative window aperture (%)

R_c crop respiration ($g\,C\,m^{-2}\,h^{-1}$)

R_s CO_2 release by the soil ($g\,C\,m^{-2}\,h^{-1}$)

S_i organic materials present before the crop was planted ($g\,C\,m^{-2}\,h^{-1}$)

S_o organic materials supplied for fertilization or mulching ($g\,C\,m^{-2}\,h^{-1}$)

V_g greenhouse volume per area unit ($m^3\,m^{-2}$)

υ wind speed ($m\,s^{-1}$)

REFERENCES

Acock, B., Hand, D.W., Thornley, J.H.M. and Warren-Wilson, J., 1976. Photosynthesis in stands of green pepper. An application of empirical and mechanistic models to controlled-environment data. *Ann. Bot.,* 40: 1293–1307.

Acock, B., Charles-Edwards, D.A., Fitter, D.J., Hand, D.W. and Warren-Wilson, J., 1978a. Modelling canopy net photosynthesis by isolated blocks and rows of Chrysanthemum plants. *Ann. Appl. Biol.,* 90: 255–263.

Acock, B., Charles-Edwards, D.A., Fitter, D.J., Hand, D.W., Ludwig, L.J., Warren-Wilson, J. and Withers, A.C., 1978b. The contribution of leaves from different levels within a tomato crop to canopy net photosynthesis: an experimental examination of two canopy models. *J. Exp. Bot.,* 29: 815–827.

Amthor, J.S. and McCree, K.J., 1990. Carbon balance of stressed plants: a conceptual model for integrated research results. In: R.G. Alscher and J.R. Cumming (Editors) *Stress Responses in Plants: Adaptation and Acclimation Mechanisms.* Wiley-Liss, New York, pp. 1–15.

Aoki, M. and Yabuki, K., 1977. Studies on the CO_2 enrichment for plant growth. 7. Changes in dry matter production and photosynthetic rate of cucumber during CO_2 enrichment. *Agric. Meteorol.,* 18: 475–485.

Bakker, J.C., 1985. A CO_2 control algorithm based on simulated photosynthesis and ventilation rate. *Acta Hortic.,* 174: 387–392.

Barber, E.M. and Ogilvie, J.R., 1984. Interpretation of tracer gas experiments in ventilation research. *J. Agric. Eng. Res.,* 30: 57–63.

Bauerle, W.L., Kretchman, D.W. and Tucker-Kelly, L.L., 1986. CO_2 enrichment in the U.S. In: H.Z. Enoch and B.A. Kimball (Editors), *CO_2 Enrichment of Greenhouse Crops sources,* Vol. I, *Status and CO_2 sources.* CRC Press, Boca Raton, FL, pp. 49–53.

Bellamy, L.A. and Kimball, B.A., 1986. CO_2 enrichment duration and heating credit as determinated by climate. In: H.Z. Enoch and B.A. Kimball (Editors), *CO_2 enrichment of Greenhouse Crops,* Vol. II,

Physiology, Yield and Economics, CRC Press, Boca Raton, FL, pp. 167–198.

Berninger, E., 1989. *Cultures florales de serre en zone méditerranéenne française. Eléments climatiques et physiologiques.* INRA-PHM Revue horticole, Paris, 206 pp.

Bertin, N. and Gary, C., 1992. Tomato fruit set and competition for assimilates during the early production period. *Acta Hortic.,* 303: 121–126.

Bierhuizen, J.F., 1973. CO_2 supply and net photosynthesis. *Acta Hortic.,* 32: 119–126.

Bot, G.P.A., 1983. *Greenhouse climate: from physical processes to a dynamic model,* PhD Thesis. Agricultural University, Wageningen, 240 pp.

Brockett, B.L. and Albright, L.D., 1987. Natural ventilation in single airspace buildings. *J. Agric. Eng. Res.,* 37: 141–154.

Brouwer, R., 1983. Functional equilibrium: sense or non-sense? *Neth. J. Agric. Sci.,* 31: 335–348.

Challa, H., 1976. *An analysis of the diurnal course of growth, CO_2 exchange and carbohydrate reserve content of cucumber.* Agricultural Research Reports, Wageningen.

Challa, H. and Schapendonk, A.H.C.M., 1986. Dynamic optimalization of CO_2 concentration in relation to climatic control in greenhouses. In: H.Z. Enoch and B.A. Kimball (Editors), *CO_2 enrichment of Greenhouse Crops,* Vol. I, *Status and CO_2 sources.* CRC Press, Boca Raton, FL, pp. 147–160.

Cure, J.D. and Acock, B., 1986. Crop responses to carbon dioxide doubling: a literature survey. *Agric. Forest Meteorol.,* 38: 127–145.

Daudet, F.A., 1987. Un système simple pour la mesure in situ des échanges gazeux de couverts végétaux de quelques mètres carrés de surface foliaire. *Agronomie,* 7: 133–139.

Dayan, E., Zipori, I., Enoch, H.Z., Shmuel, D. and Gefen, P., 1985. A system for measuring photosynthesis and respiration rate crops in commercial size greenhouses. *Acta Hortic.,* 174: 505–512.

Enoch, H.Z., 1984. Carbon dioxide uptake efficiency in relation to crop intercepted solar radiation. *Acta Hortic.,* 162: 387–392.

Enoch, H.Z., 1990. Crop responses to aerial carbon dioxide. *Acta Hortic.,* 268: 17–29.

Enoch, H.Z. and Hurd, R.G., 1977. Effect of light intensity, CO_2 concentration and leaf temperature on gas exchanges of spray carnation plants. *J. Exp. Bot.,* 28: 84–95.

Enoch, H.Z., Carmi, I., Rounick, J.S. and Magaritz, M., 1984. Use of carbon isotopes to estimate incorporation of added CO_2 by greenhouse tomato plants. *Plant Physiol.,* 76:1083–1085.

Frydrych, J., 1976. Photosynthesis characteristics of cucumber seedlings grown under two levels of carbon dioxide. *Photosynthetica,* 10: 335–338.

Gaastra, P., 1959. Photosynthesis of crop plants as influenced by light, carbon dioxyde, temperature and stomatal diffusion resistance. *Meded. Landbouwhogesch. Wageningen,* 59(13): 1–68.

Gary, C., 1988. Relation entre température, teneur en glucides et respiration de la plante entière chez la tomate en phase végétative. *Agronomie,* 8: 419–424.

Gary, C., 1989. Interest of a carbon balance model for on-line growth control: the example of a daylight-dependent night temperature control. *Acta Hortic.,* 248: 265–268.

Gary, C. and Veyrès, P., 1992. CO_2 concentration and the photosynthesis–respiration balance of vegetative tomato plants. In: J.T. Lambers and L.H.W. van der Plas (Editors), *Molecular, Biochemical and Physiological Aspects of Plant Respiration.* SPB Academic Publishing, The Hague, pp. 553–557.

Gijzen, H. and ten Cate, J.A., 1988. Prediction of the response of greenhouse crop photosynthesis to environmental factors by integration of physical and biochemical models. *Acta Hortic.,* 229: 251–258.

Goedhart, M., Nederhoff, E.M., Udink Ten Cate, A.J. and Bot, G.P.A., 1984. Methods and instruments for ventilation rate measurements. *Acta Hortic.,* 148: 393–400.

Hand, D.W., 1986. CO_2 sources and problems in burning hydrocarbon fuels for CO_2 enrichment. In: H.Z. Enoch and B.A. Kimball (Editors), *CO_2 enrichment of Greenhouse Crops,* Vol. I, *Status and CO_2 sources.* CRC Press, Boca Raton, FL, pp. 99–121.

Hand, D.W., 1990. CO_2 enrichment in greenhouses: problems of CO_2 acclimation and gaseous air pollutants. *Acta Hortic.,* 268: 81–102.

Hand, D.W. and Cockshull, K.E., 1975. The effects of CO_2 concentration on the canopy photosynthesis and winter bloom production of the glasshouse rose "Sonia". *Acta Hortic.,* 51: 243–252.

Hicklenton, P.R. and Jolliffe, P.A., 1978. Effects of greenhouse CO_2 on the yield and photosynthetic physiology of tomato plants. *Can. J. Plant Sci.,* 58: 801–817.

Ho, L.C., 1977. Effects of CO_2 enrichment on the rates of photosynthesis and translocation of tomato leaves. *Ann. Appl. Biol.,* 87: 191.

Ho, L.C., 1979. Regulation of assimilate translocation between leaves and fruits in the tomato. *Ann. Bot.,* 437–448.

Ho, L.C. and Hewitt, J.D., 1986. Fruit development. In: J.G. Atherton and J. Rudich (Editors), *The Tomato Crop. A Scientific Basis for Improvement.* Chapman and Hall, London, pp. 201–239.

Holley, W.D., 1970. CO_2 enrichment for flower production. *Trans. ASAE,* 3: 257–258.

Houter, G., 1990. Simulation of CO_2 consumption, heat demand and crop production of greenhouse tomato at different CO_2 strategies. *Acta Hortic.,* 268: 157–164.

Hrubec, T.C., Robinson, J.M. and Donaldson, R.P., 1985. Effect of CO_2 enrichment and carbohydrate content on the dark respiration of soybeans. *Plant Physiol.,* 79: 684–689.

Hughes, A.P. and Cockshull, K.E., 1972. Further effects of light intensity, CO_2 concentration and day temperature on the growth of *Chrysanthemum morifolium,* c.v. Bright Golden Anne, in controlled environments. *Ann. Bot.,* 36: 533–550.

Hurd, R.G., Gay, A.P. and Mountifield, A.C., 1979. The effect of partial flower removal on the relation between root, shoot and fruit growth in the indeterminate tomato. *Ann. Appl. Biol.,* 93: 77–89.

Ito, T., 1971. Photosynthetic activity of vegetable plants and its horticultural significance. XI. The time course of photosynthesis in the tomato plants as influenced by some external and internal factors, especially by water and starch contents in the leaf. *J. Jpn. Soc. Hortic. Sci.,* 40: 41.

Ito, T., 1978. Physiological aspects of CO_2 enrichment to cucumber plants grown in greenhouses. *Acta Hortic.,* 87: 139–146.

Jones, J.W., Dayan, E., Van Keulen, H. and Challa, H., 1989. Modeling tomato growth for optimizing greenhouse temperatures and carbon dioxide concentrations. *Acta Hortic.,* 285–294.

Jones, P., Jones, J.W., Allen, L.H. and Mishoe, J.W., 1984. Dynamic computer control of closed environmental plant growth chambers. Design and verification. *Trans. ASAE,* 879–888.

Kanemasu, E.T., Powers, W.L. and Sij, J.W., 1974. Field chamber measurements of CO_2 flux from soil surface. *Soil Sci.*, 233: 118.

Kimball, B.A., 1983. CO_2 and agricultural yield: an assemblage and analysis of 430 prior observations. *Agron. J.*, 75: 779–788.

Kimball, B.A., 1986. Influence of elevated CO_2 on crop yield. In: H.Z. Enoch and B.A. Kimball (Editors), *CO_2 enrichment of Greenhouse Crops,* Vol. II, *Physiology, Yield and Economics.* CRC Press, Boca Raton, FL, pp. 105–116.

Knecht, G.W. and O'Leary, J.W., 1974. Increased tomato fruit development by CO_2 enrichment. *J. Am. Soc. Hortic. Sci.*, 99: 214–216.

Kozai, T. and Sase, S., 1978. A simulation of natural ventilation for a multi-span greenhouse. *Acta Hortic.*, 87:39–50.

Kozai, T., Sase, S. and Nara, M., 1980. A modeling approach to greenhouse ventilation control. *Acta Hortic.*, 106:125–136.

Lambers, H., 1987. Growth, respiration, exudation and symbiotic associations: the fate of carbon translocated to the roots. In: P.J. Gregory, J.V. Lake and D.A. Rose (Editors), *Root Development and Function, SEB Seminar Series,* Vol. 30. Cambridge University Press, Cambridge, pp. 125–145.

Lieth, J.H. and Pasian, C.C., 1991. A simulation model for the growth and development of flowering rose shoots. *Sci. Hortic.*, 46: 109–128.

Madsen, E., 1968. Effect of CO_2 concentration on the accumulation of starch and sugar in tomato leaves. *Phys. Plant.*, 21: 168–175.

Marcelis, L.F.M., Heuvelink, E. and De Koning, A.N.M., 1989. Dynamic simulation of dry matter distribution in greenhouse crops. *Acta Hortic.*, 248: 269–276.

Mathes, K. and Schriefer, T., 1985. Soil respiration during secondary succession: influence of temperature and moisture. *Soil Biol. Biochem.*, 17: 205–211.

Matthews, R.B., Marshall, B., Saffell, R.A. and Harris, D., 1987. Computer control of carbon dioxide concentration in experimental glasshouses and its use to estimate net canopy photosynthesis. *Agric. Forest Meteorol.*, 40: 279–292.

Minderman, G. and Vulto, J.C., 1973. Comparison of techniques for the measurement of carbon dioxide evolution from soil. *Pedobiologia,* 13: 73–80.

Mott, K.A., 1990. Sensing of atmosphere CO_2 by plants. *Plant, Cell Environ.*, 13: 731–737.

Nederhoff, E.M., 1988. Dynamic optimization of the CO_2 concentration in greenhouses: an experiment with cucumber (*Cucumis sativus* L.). *Acta Hortic.*, 229: 341–348.

Nederhoff, E.M., Van de Vooren, J. and Udink Ten Cate, A.J., 1984. A method to determine ventilation in greenhouses. *Acta Hortic.*, 148: 345–353.

Nederhoff, E.M., Gijzen, J.G. and Vegter, J., 1988. Measurement and simulation of crop photosynthesis of cucumber (*Cucumis sativus* L.) in greenhouses. *Neth. J. Agric. Sci.*, 36: 253–264.

Nielsen, S., Holland, K., Dons, C. and Sletten, P.S., 1983. Effect of CO_2 enrichment on photosynthesis growth and yield of tomato. *Sci. Hortic.*, 20: 1–14.

Okushima, L., Sase, S. and Nara, M., 1989. Modeling natural ventilation design of greenhouses based on computation aerodynamics. *Acta Hortic.*, 248: 129–136.

Osmond, C.B., Winter, K. and Ziegler, H., 1982. Functional significance of different pathways of CO_2 fixation in photosynthesis. In: O.L. Lange, P.S. Nobel, C.B. Osmond and H. Ziegler (Editors), *Physiological Plant Ecology. 2. Water relations and carbon assimilation.* Springer, Berlin, pp. 479–547.

Penning de Vries, F.W.T., 1975. The cost of maintenance processes in plant cells. *Ann. Bot.*, 39: 77–92.

Penning de Vries, F.W.T., Brunsting, A.H.M. and Van Laar, H.H., 1974. Products, requirements and efficiency of biosynthesis: a quantitative approach. *J. Theor. Biol.*, 45: 339–377.

Picken, A.J.F., Stewart, K. and Klapwijk, D., 1986. Germination and vegetative development. In: J.G. Atherton and J. Rudich (Editors), *The Tomato Crop.* Chapman and Hall, London, pp. 111–166.

Reuveni, J. and Gale, J., 1985. The effect of high levels of carbon dioxide on dark respiration and growth of plants. *Plant, Cell Environ.*, 8: 623–628.

Sage, R.F., Smackey, T.D. and Seeman, J.R., 1989. Acclimation of photosynthesis to elevated CO_2 in five C3 species. *Plant Physiol.*, 89: 590–596.

Sase, S., Kozai, T., Nara, M. and Negishi, H., 1980. Ventilation of greenhouses. Wind tunnel measurements of pressure and discharge coefficients for a single span greenhouse. *J. Agric. Meteorol. Japan,* 36: 3–12.

Sasek, T.W., Delucia, E.M. and Strain, B.R., 1985. Reversibility of photosynthetic inhibition in cotton after long-term exposure in elevated CO_2 concentrations. *Plant Physiol.*, 78: 619–622.

Schapendonk, A.H.C.M. and Tilburg, W., 1984. The CO_2 factor in modelling photosynthesis and growth of glasshouse crops. *Acta Hortic.*, 162: 83–92.

Seginer, I., Angel, A., Gal, S. and Kantz, D., 1986. Optimal CO_2 enrichment strategy for greenhouses. A simulation study. *J. Agric. Eng. Res.*, 34: 265–304.

Slack, G., 1986. CO_2 enrichment of tomato crops. In: H.Z. Enoch and B.A. Kimball (Editors), *CO_2 Enrichment of Greenhouse Crops,* Vol. II, *Physiology, Yield and Economy.* CRC Press, Boca Raton, FL, pp. 151–163.

Spencer, W. and Bowes, G., 1986. Photosynthesis and growth of water hyacinth under CO_2 enrichment. *Plant Physiol.*, 92: 528–533.

Tchamitchian, M. and Longuenesse, J.J., 1991. Photosynthèse d'une culture en rangs de tomates sous serre. Modélisation analytique et cartographie de l'activité du feuillage. *Agronomie,* 11: 17–26.

Thornley, J.H.M. and Hurd, R.G., 1974. An analysis of the growth of young tomato plants in water culture at different light integrals and CO_2 concentrations. II. A mathematical model. *Ann. Bot.*, 38: 389–400.

van Berkel, N., 1986. CO_2 enrichment in the Netherlands. In: H.Z. Enoch and B.A. Kimball (Editors), *CO_2 Enrichment of Greenhouse Crops,* Vol. I, *Status and CO_2 Sources.* CRC Press, Boca Raton, FL, pp. 17–34.

Verkerk, K., 1955. Temperature, light and the Tomato. *Meded. Landbouwhogesch. Wageningen,* 55: 175–224.

Verkleij, F.N. and Challa, H., 1988. Diurnal export and carbon economy in an expanding source leaf of cucumber at contrasting source and sink temperature. *Physiol. Plant.*, 74: 284–293.

Warren-Wilson, J., 1972. Control of crop processes. In: A.R. Rees, K.E. Cockshull, D.W. Hand and R.G. Hurd (Editors), *Crop Processes in Controlled Environment.* Academic Press, London, pp. 7–30.

Wittwer, S.H. and Robb, W., 1964. CO_2 enrichment of greenhouse atmospheres for food crop production. *Econ. Bot.*, 18: 34–56.

Chapter 13

THE WATER CYCLE

O. JOLLIET

INTRODUCTION

Water is essential to life and plant growth, and is therefore one of the most important substances in any ecosystem. In the greenhouse ecosystem, the water cycle also plays a very important role, and this chapter provides an overview of its functions, variations and control:

(1) The function of water and particular features of the greenhouse ecosystem are described, together with the strong interaction between the water cycle and heat transfer.
(2) The physical principles and the order of magnitude of the different water flows are given.
(3) Models for calculating transpiration and inside humidities are reviewed in detail.
(4) A sensitivity study shows how location, crop stage, greenhouse type and control system affect the water cycle.
(5) Humidity control is discussed. In a warm climate, transpiration and evaporative systems are studied in respect to their cooling efficiency. Optimal values (set-points) are presented for humidity control for tomatoes in low-transpiration conditions.
(6) Practical solutions for irrigation systems are discussed.

Function and effect of water in plants

In plants, water ensures a wide range of functions (Fig. 13.1). (1) It is the largest component, typically 70% to 95% of the fresh plant weight. It also gives shape and rigidity to the plant by maintaining a positive turgor pressure. (2) As solvent and source of ions, water is the medium for the transport of nutrient elements from soil to plants and for many biochemical reactions. (3) Finally, because of its high latent heat, water cools leaves by transpiration and prevents overheating.

The water content of plants must remain within a narrow range to provide good developmental conditions: a decrease of 10% will already greatly affect activity and growth – indeed, may be deadly. Qualitative and quantitative data on how plant growth, development and yield are related to transpiration and humidity have recently been provided for greenhouse crops.

The water cycle affects dry matter production first of all by its effect on crop photosynthesis. On a short-term basis, when vapor pressure deficit increases (corresponding to a decreasing relative humidity: Fig. 13.2), leaf stomata are progressively closed in order to reduce water stress (Choudury and Monteith, 1986). As stomata are the points of interaction between the carbon and water cycles, photosynthesis response to light also decreases. Bakker (1991) showed that this effect is limited, and that crop photosynthesis

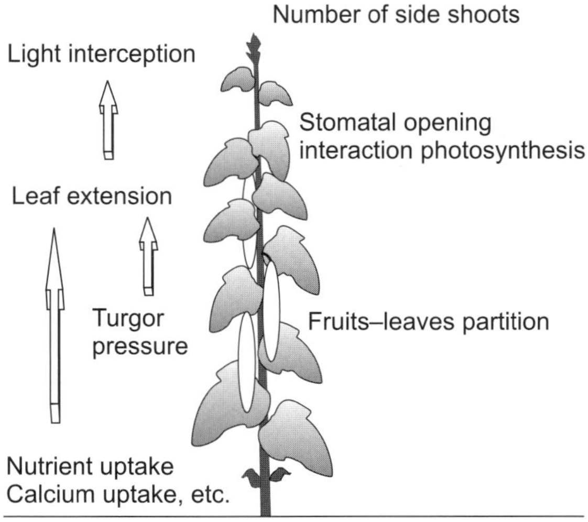

Fig. 13.1. Effect of water and humidity on plant growth and yield.

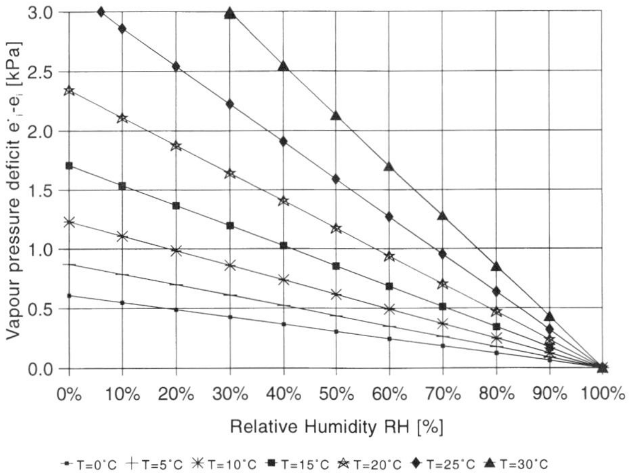

Fig. 13.2. Vapor pressure deficit (in kPa) as a function of relative humidity (RH in %) for different air temperatures (T in °C).

is reduced by less than 3% as long as the vapor pressure deficit remains lower than 1 kPa. This will be a critical limiting factor at low humidities, or in the case of water stress (mainly in dry hot climates like the Mediterranean). If higher levels of water stress are reached, roots cannot supply enough water, plants lose their rigidity, leaf expansion is reduced and irreversible damage to tissues may even be caused.

Low humidity also affects the distribution of water and nutrients within plants. At high transpiration rates, most of the water is drawn to the leaves and the water flow to fruits is reduced, limiting fruit growth. This effect is partially compensated by osmotic adjustments, where the concentration of solute sugars in fruits is increased to maintain the water flow into them. In some cases, water stress is desirable in order to control the vegetative growth of young plants, flower initiation or fruit size (Graves and Hurd, 1983).

In addition, humidity and transpiration have an indirect effect on foliage temperature, the evaporation contributing to reduce plant overheating and therefore damage to its growth.

In the long term, humidity affects leaf growth and development, and therefore light interception and dry matter production. High humidities tend to increase the number of leaves, especially on the side shoots. This may be the mechanism for the positive response of cucumber (*Cucumis sativus*) to high levels of humidity, because lateral branches are usually allowed to grow after removal of the main growing point (Bakker, 1991). On the other hand, high humidities combined with low radiation levels induce low transpiration rates and low rates of nutrient uptake. This may cause calcium deficiency, which may lead to leaf area reductions of up to 50% (Holder and Cockshull, 1990) and to a significant yield reduction (Bakker, 1990). This effect is critical for tomatoes (*Lycopersicon esculentum*), at high humidities and low light levels, for example in northern Europe during winter. Jolliet et al. (1993) have shown that, in case of dehumidification, tomato yield could be related to the increase in transpiration at the top of the crop, together with the duration of dehumidification.

Root health also depends on the water quality and its oxygen content. In addition, the spread of fungal diseases, and the development of fungi used for biological pest control, can be strongly affected by humidity. This is a major effect of bad humidity control on greenhouse crop. Chapter 8 discusses this subject in detail.

The greenhouse ecosystem

The greenhouse ecosystem has been introduced by man in order to provide appropriate growth conditions for plants. Therefore, greenhouse design and control should also match the water needs of plants. In practice this should be achieved while taking into account the other needs of plant growth together with energetic, ecological and economic considerations. Compared to open fields, the greenhouse affects the water cycle in several ways.

It obviously does not rain in the greenhouse, and water has to be provided by an irrigation system. In general, the irrigation system is controlled to provide enough easily available water to plants, and thus transpiration is close to its potential value.

The microclimate within the greenhouse affects and is affected by plant transpiration. As for outside crops, water exchanges are strongly coupled to heat transfer, evaporation and condensation being directly linked to latent heat. In addition, physical processes involved in the water cycle depend on temperature and air movements, and therefore on radiative and convective exchanges. Moreover, physiological processes, such as the control of stomatal opening, also depend on incident radiation, for instance. In greenhouses, heating, shading, ventilation, and dehumidification or humidification modify transpiration and offer possibilities of controlling the water use of plants.

Concepts like "Water Use Efficiency" (Dry-weight growth/water uptake) or the percentage of solar radiation transformed into latent heat, must be considered very cautiously in greenhouses, where there can be auxiliary heating and where vapor pressure deficit can be controlled independently of solar radiation. In intensive open-field cultivation of tomatoes, the water use per ton of fruit is about 50 m^3, under a plastic-net roof 40 m^3, whereas the production in a glasshouse would only require 30 m^3. For regions with water shortages, water use per unit yield might be reduced by as much as 50%, given the larger output from protected cultivation (Stanghellini, 1993).

If there are more control possibilities, it does not automatically mean that water use will be better managed. A higher degree of complexity implies the necessity of more knowledge, if optimal control is to be ensured. In particular, the complex interactions between transpiration and humidity in the greenhouse need to be understood and modelled in a simple way.

This chapter provides the basis for an overall understanding of the water cycle in a greenhouse: after a short overview of the different water fluxes, the main factors affecting each flux are discussed. Considering humidity and transpiration in interaction, the effects of location, outside climate, leaf area and greenhouse type on the water cycle are then presented. For the first time optimal set-points to control humidity and transpiration are calculated for tomato crops in high-humidity and low-light conditions. Finally, irrigation and recirculation systems are treated.

WATER FLUXES AND WATER BALANCE

Figure 13.3 presents the different water fluxes. The water balance could be established at any time step. For the air within the greenhouse, three water sources must be considered: the crop transpiration (E_t), the evaporation from the soil to the air (E_e), and the water added by a fogging system (E_{ad}), if any. The water can be removed from the air either by condensation on the cladding and drainage to the outside (E_c), by leakage and ventilation (E_v) or by a dehumidification heat pump (negative values of E_{ad}). For water condensed on the cladding, the fraction which has not been removed could be re-evaporated (negative value of E_c). For the air within the greenhouse, the law of water vapor conservation states that the sources of water are equal to the water sinks plus the variation in the water vapor stored in the air (E_{sa}):

$$E_t + E_e + E_{ad} = E_c + E_v + E_{sa} \quad [\text{kg m}^{-2}\,\text{d}^{-1}]. \qquad (13.1)$$

Several of these fluxes depend directly on the humidity of the inside air. As a matter of fact, the inside humidity evolves according to the conservation of water vapor and could therefore be determined by

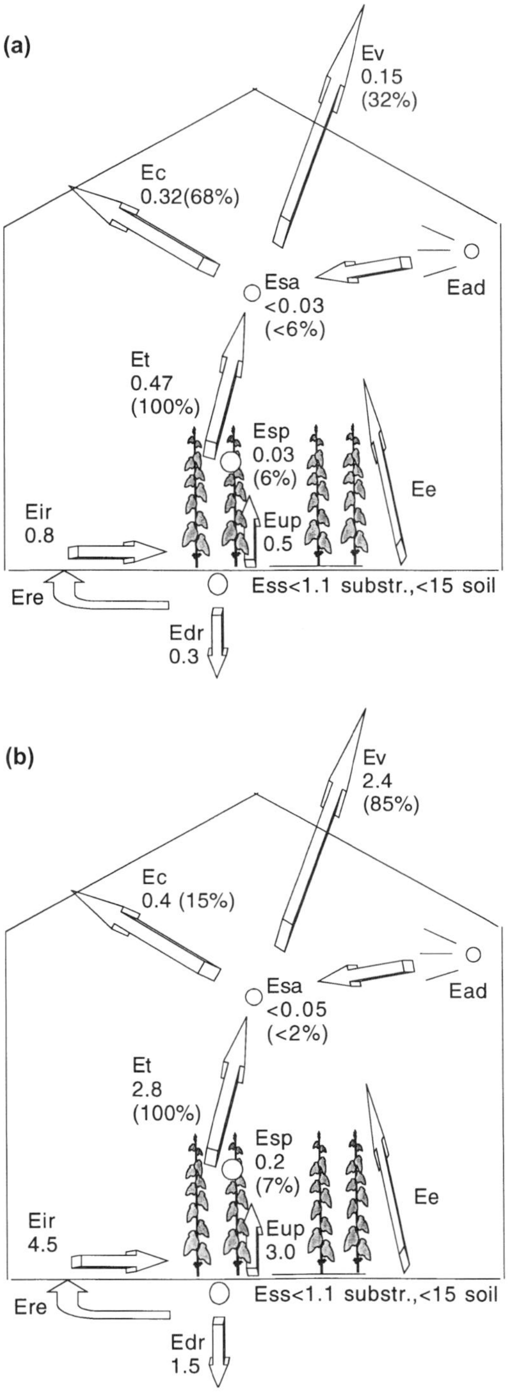

Fig. 13.3. Water fluxes in the greenhouse: average of the water flows in January (a) and June (b) in a 500 m^2 single glazed greenhouse in London.

solving this equation. To bring a clear understanding of how transpiration and humidity vary with outside conditions, each flux should be explicitly expressed as a function of the inside humidity. The points of interaction with the heat balance, namely the air and cladding temperatures, should also be clearly characterized.

This water balance could be established at different time steps (hourly, daily). Control problems and predictive models usually operate at an hourly time step. In this chapter, all flows, whether established on an hourly, daily or monthly basis, are expressed in $kg\,m^{-2}\,d^{-1}$ in order to be able to compare the different graphs ($1\,kg\,m^{-2}\,d^{-1} = 11.57\,mg\,m^{-2}\,s^{-1} = 41.7\,g\,m^{-2}\,h^{-1}$).

The water uptake of the plant (E_{up}) compensates for the transpiration and the variation of water stored within the plant (E_{sp}):

$$E_{up} = E_t + E_{sp} \quad [kg\,m^{-2}\,d^{-1}]. \tag{13.2}$$

The amount of water drained out of the greenhouse (E_{dr}) can be related to the water supplied by the irrigation system (E_{ir}), the change in water stored in soil or substrate (E_{ss}), together with the plant uptake, and the soil evaporation (E_e):

$$E_{dr} = E_{ir} - (E_{up} + E_e + E_{ss}) \quad [kg\,m^{-2}\,d^{-1}]. \tag{13.3}$$

In a closed system this drained water can be partly or fully recirculated.

Figure 13.3 gives typical values of the daily water fluxes for a tomato crop. The water storage in the air is limited and can be neglected for long-term calculation. It could be fully described (Jolliet, 1994a), but because of its limited influence it has been ignored in the rest of this chapter. The water stored in plants is limited, but not negligible as it amounts to nearly 10% of the total uptake. The storage capacity in soil depends strongly on the growing medium; compared to the daily uptake, it is high in real soil but considerably less for artificial substrates or systems using nutrient-film techniques (NFT). Ventilation losses are dominant during summer (Fig. 13.3b), whereas condensation is the largest water sink for winter conditions (Fig. 13.3a). Figure 13.7a (p. 313) also shows that the transpiration varies greatly during the day: as expected, transpiration is limited at night, when condensation is the main water sink.

Water storage in plants and in soil could also play an important role when measuring transpiration or water uptake: continuous weighing methods can measure transpiration directly, without including water stored in the plant, whereas measurements of the amount of water supplied include the variations of water stored in plants and in the growing medium.

The detailed study follows the order of the three equations above. It starts with transpiration, the driving force, and the study of its relation with air humidity. Water uptake and irrigation systems are then discussed.

MODELLING TRANSPIRATION, WATER FLUXES AND INSIDE HUMIDITY

Transpiration

Experimental correlations and factors of influence

As indicated by Morris et al. (1957), de Villèle (1972) and van der Post et al. (1974) for tomatoes or by Yang et al. (1990) for cucumbers, there is a high correlation between plant transpiration and the inside solar radiation above the crop. The adjusted parameters of this correlation depend on the crop stage, as shown by de Graaf and van den Ende (1981) for lettuce (*Lactuca sativa*), cucumber and tomato crops: for a young crop they are about one-fifth of the value for a mature one.

The air humidity inside the greenhouse has also a smaller but significant influence on transpiration. This influence is characterized by the vapor pressure deficit of the inside air, which is the difference between the vapor pressure at saturation (e_i^*) and the effective vapor pressure (e_i). It is therefore a characteristic of the air dryness and, at a given temperature, it is inversely proportional to the relative humidity. Figure 13.2 shows how it varies with air temperature and relative humidity. Lake et al. (1966) found an influence of the outside-air vapor pressure deficit on transpiration; Okuya and Okuya (1988) related tomato transpiration to inside-air vapor pressure deficit and solar radiation. Jolliet and Bailey (1992) confirmed in the case of tomatoes that transpiration can be expressed as a linear function of solar radiation (R_i, in $MJ\,m^{-2}\,d^{-1}$) and inside vapor pressure deficit ($e_i^* - e_i$, in kPa):

$$E_t = \alpha R_i + \beta(e_i^* - e_i) \quad [kg\,m^{-2}\,d^{-1}]. \tag{13.4}$$

The vapor pressure deficit itself is generally correlated with the inside solar radiation. For a single-glazed

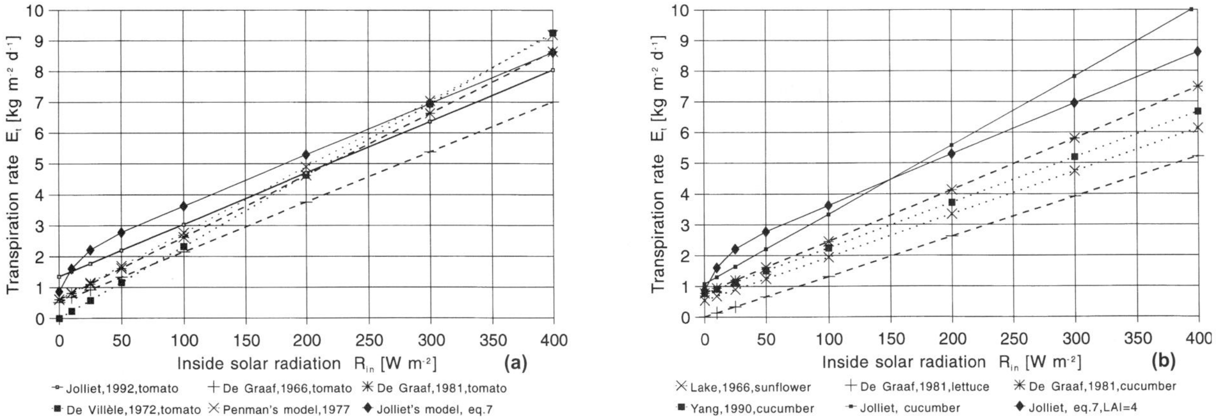

Fig. 13.4. Transpiration rate of a mature crop as a function of solar radiation, for a vapor pressure deficit (in kPa), given by Eq. (13.5). Comparison between the following experimental correlations: (a) $E_t = 0.12\,R_i + 0.21\ (e_i^* - e_i)$, tomato, Jolliet and Bailey (1992); $E_t = 0.19\,R_i + 0.52$, tomato, de Graaf (1966–1968), July–November in van der Post et al. (1974); $E_t = 0.23\,R_i + 0.60$, tomato, de Graaf and van den Ende (1981, Table 1, mature crop); $E_t = 0.27\,R_i$, tomato, de Villèle (1972); $E_t = 0.22\,R_i + 1.1\ (e_i^* - e_i)$, tomato, Penman's model (FAO, 1977, mid-season); Eq. (13.7) Jolliet (1994a, simplified model, LAI = 4). (b) $E_t = 0.13\,R_i + 1.3\ (e_o^* - e_o) - 0.17$, sunflower (*Helianthus annuus*), Lake et al. (1966); $E_t = 0.15\,R_i + 0.01$, lettuce, de Graaf and van den Ende (1981); $E_t = 0.19\,R_i + 0.78$, cucumber, de Graaf and van den Ende (1981); $E_t = 0.17\,R_i + 0.77$, cucumber, Yang et al. (1990); $E_t = 0.26\,R_i + 1.08$, cucumber, Jolliet and Bailey (1992); Eq. (13.7), Jolliet (1994a, simplified model, LAI = 4), where: E_t is the transpiration rate (kg m^{-2} d^{-1}); R_i is the incoming global solar radiation on a horizontal surface inside the greenhouse (MJ m^{-2} d^{-1}); $e_i^* - e_i$ is the air vapor pressure deficit (kPa).

greenhouse in May ($r^2 = 0.61$), Jolliet and Bailey measured

$$e_i^* - e_i = 0.56 + 0.0218 R_i \quad \text{[kPa]}. \tag{13.5}$$

Therefore, in most cases, solar radiation alone accounts for most of the variation of the transpiration and hides the influence of the vapor pressure deficit. It is however essential to take the inside vapor pressure deficit into account if a heating, fogging or dehumidification system is used.

Concentration of carbon dioxide, pipe temperature or the nutrient-solution conductivity were found to have little influence on transpiration. Air movements inside the greenhouse could potentially increase transpiration (Jolliet and Bailey, 1992). However, trials on a full-size greenhouse at the Silsoe Research Institute showed that air circulation forced by fans inside a tomato crop was not sufficient to affect its transpiration significantly. It also appears that the increase in transpiration with a heating system observed by de Graaf and van den Ende (1981) was due to an increase of the vapor pressure deficit of the air inside the greenhouse (see p. 312).

Figures 13.4a and b compare different experimental correlations by plotting transpiration against inside solar radiation, for an air vapor pressure deficit varying itself with solar radiation, according to Eq. (13.5). Most of the correlations for a mature tomato crop agree reasonably well. Therefore, such experimental correlations can be used for irrigation control as long as only rough estimations are required. Their validity and precision is however limited and they cannot be used for different climate conditions, crop stages or crop configurations without determining coefficients for the particular situation.

Predictive intermediary models

For predictive purposes, more general and more accurate models are needed, which will predict the value of the parameters α and β in Eq. (13.4) as a function of the crop stage or the Leaf Area Index. Once developed, such models should not require any parameter adjustment.

The experimental correlations have shown that crop stage, solar radiation and vapor pressure deficit are the three main factors affecting transpiration. Based on these factors the Penman model (after FAO, 1977) is a simplified model which has been developed for outside crops and includes evaporation. For typical greenhouse conditions (air speed of 0.15 m s^{-1} during summer or during the heating period, reflectivity of the crop surface of 25%, net radiation equal to solar radiation) this model can be expressed as follows:

$$E_t = k_{cr}[0.80\,W\,R_i + 3.2(1 - W)(e_i^* - e_i)] \quad [\text{kg m}^{-2}\,\text{d}^{-1}], \tag{13.6}$$

where R_i is the inside solar radiation in MJ m^{-2} d^{-1},

$e_i^* - e_i$ is the vapor pressure deficit in kPa, W is the temperature-related weighting factor for the effect of radiation and vapor pressure deficit on transpiration [$W = 0.52$ at 8ºC, 0.66 at 18ºC and 0.77 at 28ºC], and k_{cr} characterizes the crop stage and crop type. For mature crops, k_{cr} amounts to 1.05 for tomatoes and sunflowers (*Helianthus annuus*), 0.9 for cucumbers, and 0.95 for lettuces, aubergines (*Solanum melongena*), sweet peppers (*Capsicum frutescens*) and melons (*Cucumis melo*).

For young tomato crops just after planting (crop height of 0.4 m) k_{cr} amounts typically to 20% of the value of a mature crop and increases almost linearly with height until maturity; k_{cr} is equal to 45% at 0.8 m., 70% at 1.3 m and 100% above 1.7 m (extrapolated from de Graaf and van den Ende, 1981).

Figure 13.4a shows that this model of Penman agrees with experimental correlations for a mature tomato crop. Jolliet and Bailey (1992) showed that this simplified Penman model gives a good prediction on average, but does not account accurately for variation in the air vapor pressure deficit. There is therefore a need for a more general relationship predicting transpiration.

For tomatoes, Jolliet (1994a) developed a linearized relation based on the Stanghellini detailed model (see below). This linear relation proved as accurate as the Stanghellini model even on a short-term basis (typically hourly), despite being simpler. The total transpiration rate is expressed as follows:

$$E_t = \frac{a}{\lambda} R_i + \frac{h_t}{\lambda \gamma \phi}(e_i^* - e_i) \quad [\mathrm{kg\,m^{-2}\,d^{-1}}]. \tag{13.7}$$

with

$$a = 0.154 \ln(1 + 1.1\mathrm{LAI}^{1.13}), \tag{13.8}$$

$$h_t = 1.65\mathrm{LAI}(1 - 0.56\mathrm{e}^{-R_i/1.12}) \quad [\mathrm{W\,m^{-2}\,K^{-1}}], \tag{13.9}$$

λ the latent heat of vaporization of water (2.45 MJ kg^{-1} at 20ºC), γ the psychrometric constant (0.066 kPa K^{-1}), and ϕ a dimensional factor which allows for the conversion of seconds to days: 11.57 d Ms^{-1}.

Parameter a expresses how the increase in transpiration with solar radiation depends on the Leaf Area Index, with saturation at high LAI. The influence of the vapor pressure deficit on transpiration, expressed by h_t, increases linearly with LAI. This parameter h_t also depends on the solar radiation, because stomata close at night or at very low light levels and reduce transpiration. As the crop stage is characterized by the Leaf Area Index (LAI), the method for estimating LAI for practical applications is described in the discussion of optimal control (p. 320).

Transpiration of young leaves at the top of the crop (Jolliet et al. 1993) can also be expressed according to Eq. (13.7), where:

$$a = 0.296, \qquad h_t = 1.95(1 - 0.63\mathrm{e}^{-R_i/0.864}) \quad [\mathrm{W\,m^{-2}\,K^{-1}}]. \tag{13.10}$$

Predictive detailed models and stomatal conductance

For most practical applications, the intermediate models are accurate enough. However, it can be interesting to model and understand in detail the processes involved, particularly if the interaction between the water and carbon cycles is to be determined. Most of the detailed transpiration models follow the pattern of the Penman–Monteith equation (Monteith, 1973: see Eq. (3.31) in chapter 3, p. 94). This equation confirms that solar radiation, vapor pressure and crop stage are the three main factors affecting transpiration, and it is also similar to Eq. (13.4). However, a and b are generally calculated as a function of the Leaf Area Index (LAI), the aerodynamic conductance ($1/r_a$) and the stomatal resistance ($1/r_p$). Models differ mainly in the way the coefficients of absorption for radiation, r_a and r_p, are calculated.

The Penman–Monteith equation is very sensitive to the stomatal conductance, and it is essential to predict it accurately. Moreover, stomatal conductance is the point of interaction with photosynthesis and with the carbon cycle. It is mainly affected by solar radiation and vapor pressure deficit (Choudury and Monteith, 1986; Stanghellini, 1987; Jolliet and Bailey, 1993), the relative response to vapor pressure deficit being equal for different glasshouse crops (Bakker, 1991). In an experiment on tomato crops where climatic variables such as carbon dioxide concentration, air speed, air temperature and radiation were controlled independently, Jolliet and Bailey (1993) observed that leaf temperature and carbon dioxide concentration had little effect on the stomatal conductance. Soil water potential can also influence the stomatal conductance. Turner (1991) suggested that this influence is through direct changes in the level of phytohormones produced by the roots in response to soil dehydration, but this explanation is still disputed. In greenhouses, water supply is usually not limiting and this effect can often be neglected.

As observed by van Bavel et al. (1985), different hypotheses can lead to large differences in predicted humidity and transpiration, and the validity of these models need to be checked. In a comparison of six different models with measurements, Jolliet and Bailey (1992) have shown that the single-layer model of Stanghellini (1987) provides an accurate prediction of tomato transpiration, because it accounts for the effect of both solar radiation and vapor pressure deficit on stomatal conductance. On the other hand, models assuming constant day and night values of the stomatal conductance (van Bavel et al., 1981; Aikman and Houter, 1990) have poor accuracy.

It is therefore better to use a simple experimental correlation or intermediate simplified models with approximate coefficients such as Eqs. (13.6) or (13.7) above, than the more complex Penman–Monteith equation with incorrect values for the stomatal conductance. *In situ* identification of transpiration model parameters could offer another possibility (Boulard and Jeema, 1992), but so far it has mainly been applied in research projects.

Other sources and sinks of water vapor

Air infiltration and ventilation

When air is exchanged by infiltration and ventilation between inside and outside, wet inside air is replaced by drier outside air. The exchange of water vapor is therefore proportional to the difference between the vapor pressure of the inside air (e_i, in kPa) and that of the outside air (e_o):

$$E_v = \frac{h_v}{\lambda\gamma\phi}(e_i - e_o) \quad [\mathrm{kg\,m^{-2}\,d^{-1}}], \tag{13.11}$$

where h_v is the coefficient of sensible heat transfer due to air renewal, calculated as follows:

$$h_v = \rho_a C_{pa}\frac{q}{A} \quad [\mathrm{W\,m^{-2}\,K^{-1}}], \tag{13.12}$$

with ρ_a the air density in $\mathrm{kg\,m^{-3}}$, C_{pa} the heat capacity of the air in $\mathrm{J\,kg^{-1}\,K^{-1}}$, q the air flow between inside and outside in $\mathrm{m^3\,s^{-1}}$, and A the greenhouse soil area in $\mathrm{m^2}$.

The air flow exchanged between the inside and the outside of the greenhouse depends on the wind speed (ω, in $\mathrm{m\,s^{-1}}$), the temperature difference between the inside (T_i) and outside air (T_o), the area of the cladding (A_c), the area of the ventilators (A_v) and their angle of opening (ζ, in °). It is calculated using a physical model of a building adapted to greenhouses by Jolliet et al. (1991):

$$q = A_f\left[f_\omega^2\omega^2 + f_T^2(T_i - T_o)\right] \quad [\mathrm{m^3\,s^{-1}}] \tag{13.13}$$

with

$$A_f = A_c f_f + 0.00588 A_v \zeta e^{-\zeta/50} \quad [\mathrm{m^2}]. \tag{13.14}$$

f_ω and f_T define the relative influence of wind and stack effect; they amount to 0.22 and 0.16 $\mathrm{m{-}s^{-1}\,K^{-1/2}}$, respectively. The value A_f is the effective area for air infiltration ($\mathrm{m^2}$), and f_f characterizes the tightness of the cladding to air infiltration. The factor f_f amounts to 2.5×10^{-4} for a very tight cladding with rubber tape, to 5×10^{-4} for tight cladding without apparent holes, and to between 10×10^{-4} and 20×10^{-4} for leaky cladding with cracks and holes. With a screen, this factor is reduced by 25%. This model was successfully checked for air infiltration by Jolliet (1988) and was also found valid for natural ventilation by Fernandez and Bailey (1992).

Condensation on the cladding

A considerable quantity of water is condensed on the cladding and removed from the greenhouse by capillary action, through glass overlaps or via a condensation channel. Physically, warm wet inside air is brought by natural convection towards the cladding and is cooled down to the cladding temperature. When the air temperature is reduced, its relative humidity increases and water is condensed as soon as the saturation level is achieved.

The amount of water condensed is proportional to the difference between the vapor pressure of the inside air and the vapor pressure at saturation at the inside surface of the cladding (e_c^*, in kPa):

$$E_c = \frac{h_c}{\lambda\gamma\phi}(e_i - e_c^*)[\mathrm{kg\,m^{-2}\,d^{-1}}]. \tag{13.15}$$

h_c is the coefficient of heat transfer by convection at the inside surface of the cladding. It is affected by air movement within the whole greenhouse (forced convection at the surface of the cladding) and amounts to approximately 3.5 $\mathrm{W\,K^{-1}}$ per square metre of cladding, multiplied by the ratio of ground to cladding area.

The *saturation vapor pressure* e_c^* is directly related to the temperature of the inside surface of the cladding (see Glossary of Symbols (p. 324) under e^* for the approximate equation). Without condensation, this temperature is independent of the humidity and can be predicted on the basis of the inside air, outside air and apparent sky temperature, together with the cladding heat-exchange coefficients (Jolliet, 1994a, Appendix B). During condensation periods, there is however a small interactive effect of the inside humidity, which affects and is affected by the cladding temperature.

Jolliet (1994a) showed that this interaction is limited to less than 1.3°C, and that this problem can be solved in a simple way by calculating first the temperature that the cladding would have if the inside air were saturated (for 100% relative humidity of the inside air). A small corrective factor, depending linearly on the inside vapor pressure deficit, is then introduced to account explicitly for the feedback effect of inside humidity on cladding temperature.

If the vapor pressure at saturation of the cladding is greater than the vapor pressure of the inside air ($e_c^* > e_i$), there is no water condensed on the cladding. In glasshouses most of the condensed water is usually drained away at the bottom of the glass panes. The fraction which has not been drained away could be re-evaporated (negative value of E_c), until the cladding is dry.

Fogging system and heat pump

The quantity of water vapor added to the air by a fogging system (E_{ad}, in $kg\,m^{-2}\,d^{-1}$) depends on its size and period of operation. It is therefore determined by the control procedure (see below p. 315).

The water extracted by the condenser of a dehumidification heat pump depends slightly on the inside humidity, but the main factors determining the amount of water condensed are still its size and operating period. Therefore, the heat pump can be dealt with like the fogging system, but with E_{ad} taking negative values.

Soil evaporation

In the past, soil evaporation played an important role in greenhouses. However, nowadays with new growing methods in which the substrate is contained in bags (rockwool, peat, etc.), with plastic sheets covering the soil, with closed channels (NFT), or with trickle irrigation, the moist surface is small and evaporative loss is minimal (Cooper, 1979, p. 94). It is only in the case of irrigation by spraying that soil evaporation becomes an essential part of the water cycle and has to be included in calculations. It could then be estimated using the method proposed by Monteith (1981).

Air humidity and transpiration predicted from the outside climate

The main limitation of a transpiration model like that of Stanghellini is that it uses the humidity inside the greenhouse as data input. In practice, there is a feedback effect of transpiration on inside humidity, which affects transpiration, etc. For climate-control purposes, a transpiration submodel needs therefore to be incorporated into a greenhouse climate model, to predict humidity and transpiration directly as a function of the outside conditions and of the greenhouse characteristics.

Until the 1980s, only complex dynamic models (e.g., Bot, 1983; Kimball, 1986) were able to achieve that aim. Based on the conservation of heat and water vapor for each element of the greenhouse (soil, crop, air, cladding, etc.), these models are difficult to use for optimization purposes, because of their complexity and lack of transparency. Boulard et al. (1989) determined water sources and sinks within a greenhouse, but his model does not predict the inside humidity directly. Fuchs (1990) proposed a useful simplified approach, but valid only for a highly ventilated greenhouse, without condensation on the cladding. For humidity control in mediterranean countries, Fuchs (1993), Boulard and Baille (1993) and Seginer (1994) have presented models allowing computation of inside-air temperature and humidity together with crop transpiration as a function of control parameters of ventilation and fogging.

In the present chapter, all water-vapor sources and sinks have been expressed as linear functions of the inside-air vapor pressure. By introducing Eqs. (13.7, 11, 15) into the water-balance Eq. (13.1), the actual vapor pressure in the greenhouse can be determined explicitly:

$$e_i = \frac{\lambda\gamma\phi E_{ad} + \gamma\phi a R_i + h_t e_i^* + h_c e_c^* + h_v e_o}{h_t + h_c + h_v} \quad [\text{kPa}]. \tag{13.16}$$

This equation brings a clear understanding of the influence of each process and variable on humidity; the inside humidity varies with solar radiation, inside air temperature (e_i^* is a direct function of T_i), outside temperature and humidity, or ventilation rate (taken into account by h_v). Replacing the inside vapor pressure calculated by Eq. (13.16) in Eq. (13.7) enables also the actual crop transpiration to be determined as a function of these variables.

Jolliet (1994a) presented a slightly more accurate and elaborate but similar linear relationship, where the feedback effect of inside humidity on cladding condensation and the storage of water vapor in the air are also taken into account.

A validation check against measurements has shown that such a simple model is accurate in predicting transpiration, relative humidity or vapor pressure deficit, with an average difference lower than 8% and an r^2 value between 0.79 and 0.93. Such a model is also able to determine the quantity of water vapor and energy to be extracted from or added to the greenhouse, in order to reach a given humidity or transpiration set-point.

For growers, horticultural advisors and researchers, a dynamic programme "HORTITRANS" is available on IBM or IBM-compatible personal computers (Jolliet, 1992). It allows quick calculations of the greenhouse humidity and transpiration and can determine optimal humidity set-points for a tomato crop.

Water uptake and variation of the plant water content

As described by Eq. (13.2), the water uptake of the plant is strongly related to transpiration, the difference between these two flows corresponding to the variation of water stored in plant tissues (E_{sp}). This amount of water stored in the plant is limited, but not negligible, as it amounts to nearly 10% of the total water uptake (Fig. 13.3). It could be evaluated by estimating or measuring the increases in fresh and dry matter production. Basically, this relationship between water uptake and transpiration is specific to greenhouse crops only through the special environmental conditions and the additional control possibilities available.

Hunt et al. (1991) proposed to estimate directly the amount of water transferred from the soil to the plant, as a function of the soil and leaf water potential and as a function of the soil, root and stem resistivities to water flow. In practice, this approach is very difficult to apply for predictive purposes, because of the uncertainty in estimating root and stem resistances as well as the difficulty in measuring and predicting the leaf water potential continuously (Batta, 1989).

More research is needed to understand the relation between nutrient and water uptake, together with water distribution within the plant and competition between organs. Aikman and Houter (1990) provided a first theoretical model for calculating nutrient transport mainly in the xylem, such as calcium uptake in tomato plants. Several recent models aim to model water and nutrient distribution within the plant together with organ growth, using the concepts of sinks and sources (Marcelis, 1993). However, a comprehensive understanding of these phenomena (active or passive filtering of the root system, etc.) is far from being achieved.

SENSITIVITY STUDY: VARIATIONS OF HUMIDITY AND TRANSPIRATION

Using the HORTITRANS software (Jolliet, 1992: see left-hand column), a sensitivity study was performed in order to quantify the effects on humidity and transpiration of climatic conditions, crop stage, period of the day and the year, location and greenhouse type.

Outside and inside climatic conditions

Figures 13.5a–c show how transpiration and humidity vary with outside air temperature and solar radiation. Calculations are performed for a tight single-glazed greenhouse of 500 m^2 with a mature tomato crop (LAI = 4) and an inside air temperature set-point of 20°C, except for the unheated scenario where inside temperature varies freely. The outside relative humidity is assumed to be 84%. Ventilators were fully opened if temperatures were to reach 25°C. Consequently, the air renewal rate at $T_o = 18$°C varied between 0.4 h^{-1} if inside solar radiation was below 200 W m^{-2}, and up to 20 h^{-1} for 400 W m^{-2}.

At high radiation levels there is overheating, and transpiration and vapor pressure deficit increase linearly with solar radiation. At night, it is only in the unheated greenhouse that transpiration and vapor pressure deficit decrease down to zero. The use of the heating system results in an increase of vapor pressure deficit and thus of transpiration. The higher the difference between inside and outside air temperatures, the higher are the vapor pressure deficit and transpiration, and thus the lower the relative humidity inside the greenhouse. This explains the increase in transpiration during heating periods observed by de Graaf and van den Ende (1981).

At night, stomata are closed, transpiration is reduced accordingly and the vapor pressure deficit is therefore relatively high. At low radiation levels, vapor pressure deficit reaches its minimum (relative humidity maximum) when stomata are open and no overheating occurs. This also implies that with artificial lighting at night humidity levels are very high.

For warm climates, the water cycle contributes to reduce high canopy temperature, a crucial issue

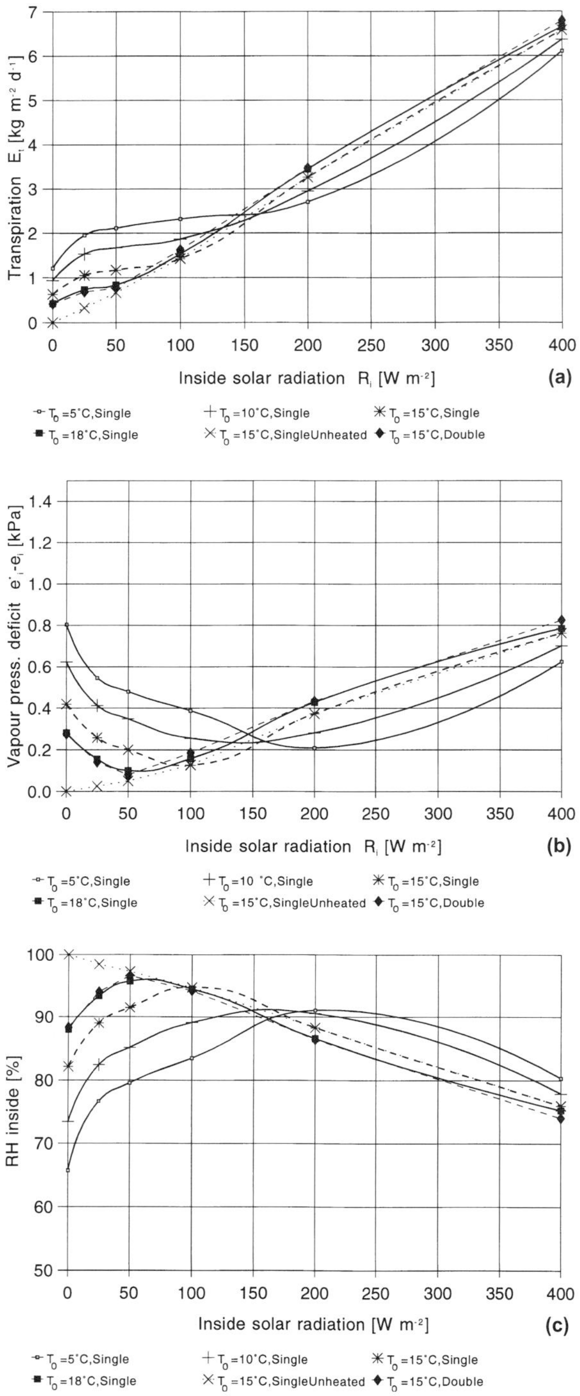

Fig. 13.5. Transpiration (a), vapor pressure deficit of the inside air (b) and relative humidity (c) calculated by the HORTITRANS model as a function of the solar radiation above the crop, for different outside air temperatures. 500 m^2 single-glazed greenhouse, for LAI = 4 and T_i = 20°C. In addition, double-glazed and unheated greenhouse, for T_o = 15°C.

for summer production. The combined effects of varying air exchange rate and fog intensities on cooling effectiveness are discussed below (Fig. 13.10, p. 315).

Crop stage

After solar radiation, crop stage, characterized by the Leaf Area Index of the canopy, is the main factor affecting the water cycle (Fig. 13.6): as expected, the higher the leaf area, the higher is the transpiration and the lower is the vapor pressure deficit (the higher the relative humidity). For a tomato crop, LAI can be approximately evaluated on the basis of leaf number or crop height (see Eq. 13.19, p. 320).

Daily changes, seasonal changes and location

Transpiration and humidity vary strongly within a day (Fig. 13.7). The lowest transpiration levels are measured at night, but the lowest vapor pressure deficits (highest humidity levels) are recorded just after sunrise when solar radiation is limited (Fig. 13.5b, p. 312). Figure 13.7a shows that the condensation flow also reaches its maximum just after sunrise and decreases quickly to zero when overheating occurs and the ventilators open.

These quick variations of water flows and humidities imply that an optimal humidity control must be operated on a very short-term basis.

Figure 13.8 presents the yearly changes in water flow in different locations covering a wide range of climatic conditions: Northern Europe (London; transpiration in Holland is practically identical); very cold winter (Montréal); temperate Mediterranean (Marseilles); and arid conditions (Cairo). Simulations were performed for a crop planted in December (LAI = 0.5) and growing until March (LAI = 2 in January, 3.5 in February and 4 from March to November). On an annual basis, the transpiration varied in proportion to the radiation level of the month and was always higher in Cairo. Humidities in a greenhouse in Cairo during the winter can be higher than in a colder climate, where the heat demand is higher.

Moreover, at night the lowest transpiration and highest humidities were generally observed for Cairo, where the greenhouse is unheated and the difference between inside and outside temperatures is small.

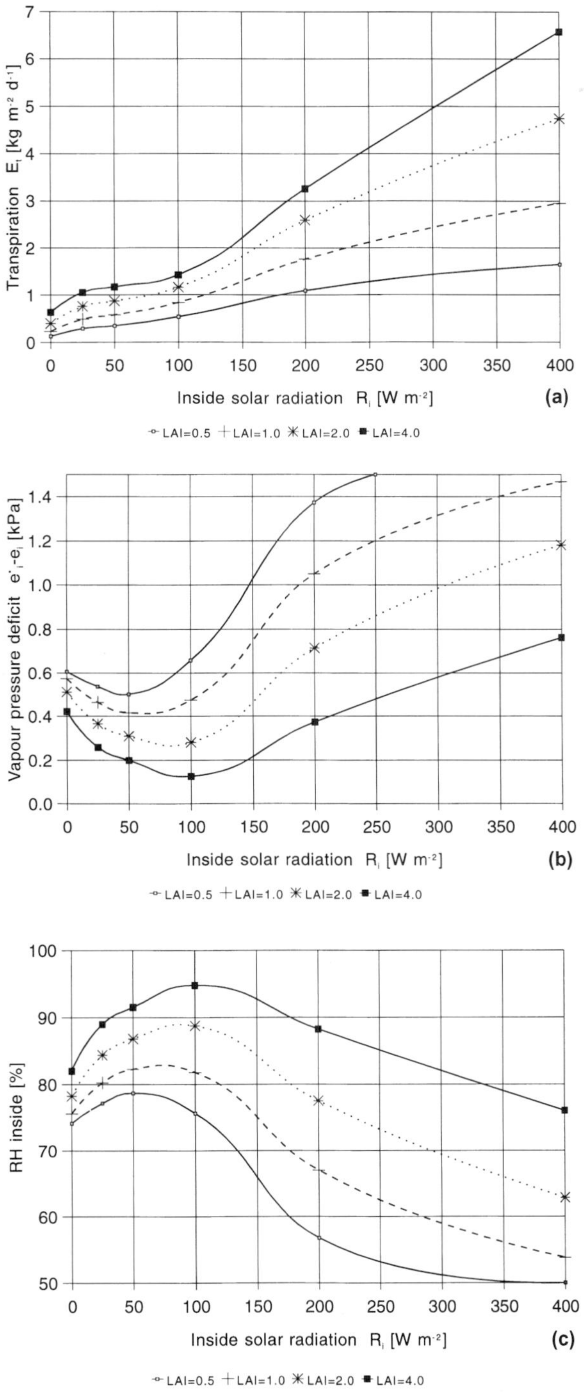

Fig. 13.6. Transpiration (a), vapor pressure deficit of the inside air (b) and relative humidity (c) calculated by the HORTITRANS model as a function of the solar radiation above the crop, for different Leaf Area Index. 500 m^2 single glazed greenhouse, for $T_o = 15$°C, $T_i = 20$°C and LAI = 4 ($1\,kg\,m^{-2}\,d^{-1} = 11.57\,mg\,m^{-2}\,s^{-1} = 41.7\,g\,m^{-2}\,h^{-1}$).

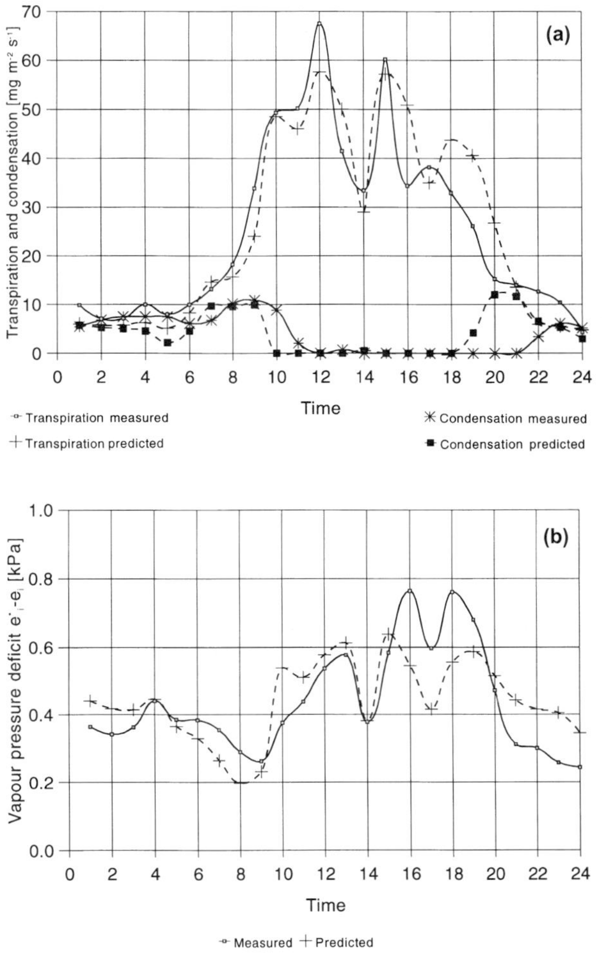

Fig. 13.7. Hourly variations of transpiration and condensation on the cladding (a) and vapor pressure deficit (b). Measurements and predicted values by the HORTITRANS model on the 23rd May 1991, at Silsoe, England, in a 427 m^2 single-glazed glasshouse ($1\,kg\,m^{-2}\,d^{-1} = 11.57\,mg\,m^{-2}\,s^{-1} = 41.7\,g\,m^{-2}\,h^{-1}$).

Greenhouse type, size and tightness

Figure 13.5 also allows single- and double-glazed greenhouses to be compared for an outside temperature of 15°C. The general behavior is very similar; but, with a shift in humidity and transpiration, double glazing with a temperature of 15°C corresponds to single glazing with an outside temperature of 18°C. Differences in transpiration due to glazing type are much smaller than variation with location or time of year (Fig. 13.8a). However, significant differences are observed between single- and double-glazed greenhouses for vapor pressure deficit in winter (Fig. 13.8b). During winter in London, a double-glazed cladding

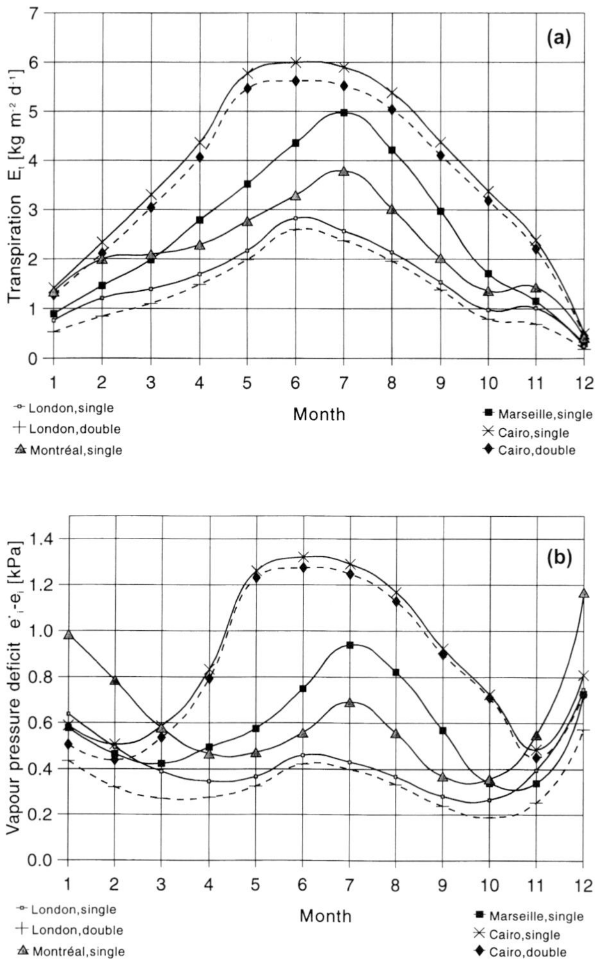

Fig. 13.8. Annual variation of transpiration (a) and vapor pressure deficit of the inside air (b) calculated by the HORTITRANS model for London, Montreal, Marseille and Cairo, for 500 m^2 single- and double-glazed greenhouses.

results in a 25% reduction in transpiration and vapor pressure deficit, compared to single glazing.

Some additional simulations were performed for London climatic conditions to determine the influence of greenhouse size and air-tightness. The largest effects are also observed in winter and can be summarized as follows: compared to a greenhouse of 500 m^2, transpiration and vapor pressure are increased by about 15% in a greenhouse of 100 m^2 and decreased by 10% in one of 5000 m^2. If the glazing bars are not fitted with rubber seals, ventilation increases and causes an increase in transpiration and vapor pressure deficit of about 9%.

OPTIMAL CONTROL OF HUMIDITY AND TRANSPIRATION

Generalities

Similarly to general climate control (see Chapter 16) optimal control of humidity and transpiration is a compromise between ensuring the best growing conditions for the crop and the financial costs to achieve these conditions (Fig. 13.9). Energy, ecological or other criteria could also be included in the choice of the control policy. To determine an optimal control strategy, it is therefore necessary on the one hand to know how crop growth, disease development and yield are affected by changes in transpiration and humidity. On the other hand, it is necessary to know how the climate can be controlled, and which investments and which quantities of energy and water are required to maintain a given humidity set-point.

Technically, control systems for the water cycle aim to control the main variables affecting transpiration: solar radiation and vapor pressure deficit (or relative humidity, see Fig. 13.2, p. 304, and Fig. 13.5, p. 312). In addition, the water cycle could be controlled through irrigation. There is, however, little room in greenhouses for control by irrigation, as the optimum for growth always requires enough easily available water to be provided to the plant.

For the *command* parameter, it has been suggested that humidity control in greenhouses should be based not on humidity but on transpiration set-points (Stanghellini and van Meurs, 1992). As discussed by Jolliet et al. (1993), it is the transpiration of the

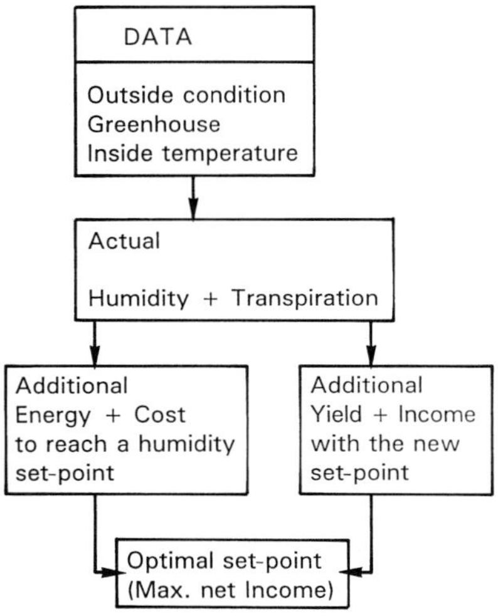

Fig. 13.9. Strategy to determine the optimal humidity set-points.

top growing leaves which is the most important, and the effect of transpiration on crop growth could also be related to vapor pressure deficit. Therefore, vapor pressure deficit and solar radiation are also suitable command variables for humidity control.

Control in warm climates

In warm climates, production is limited by high canopy temperatures. In this situation, the water cycle itself can help to reduce high canopy temperature. In addition the cooling system, generally based on evaporative principles, can enhance plant well-being.

Effect on crops

As discussed in the introduction (p. 303), at low humidities the leaf stomata close progressively to avoid water stress, and therefore photosynthesis is reduced. This is a critical limiting factor in dry hot climates. If higher levels of water stress are reached, the roots cannot supply enough water, plants lose their rigidity, leaf expansion is reduced and even irreversible damage could be caused to the leaves. At low humidity, most of the water goes to the leaves, and fruit growth is also limited. In addition, damage could occur directly with foliage burns. Typically, leaf temperature should not exceed 28°C to 30°C to avoid important adverse effects (Fuchs, 1993).

Cooling systems

To reduce high canopy temperatures and water stress, all practical cooling methods involve some sort of evaporative cooling (Seginer, 1994). The first and most efficient method is transpiration, in which evaporation takes place inside the stomata. Transpiration and cooling effects may be enhanced by proper ventilation. Boulard and Baille (1993) showed how air and leaf temperatures vary with the air exchange rate (Fig. 13.10c: $E_{ad} = 0$): leaf temperature is strongly reduced by an increasing rate of air renewal and falls below the air temperature if the air-renewal rate exceeds $25\,h^{-1}$. Inside temperature is also reduced, but it remains above the outside temperature. Proper ventilation rate depends mainly on wind speed, but it can be enhanced for example by increasing the stack effect through combination of side and roof vents.

Further temperature and transpiration reductions involve water supply and greenhouse air humidification, in addition to ventilation. Traditionally, wetted pads

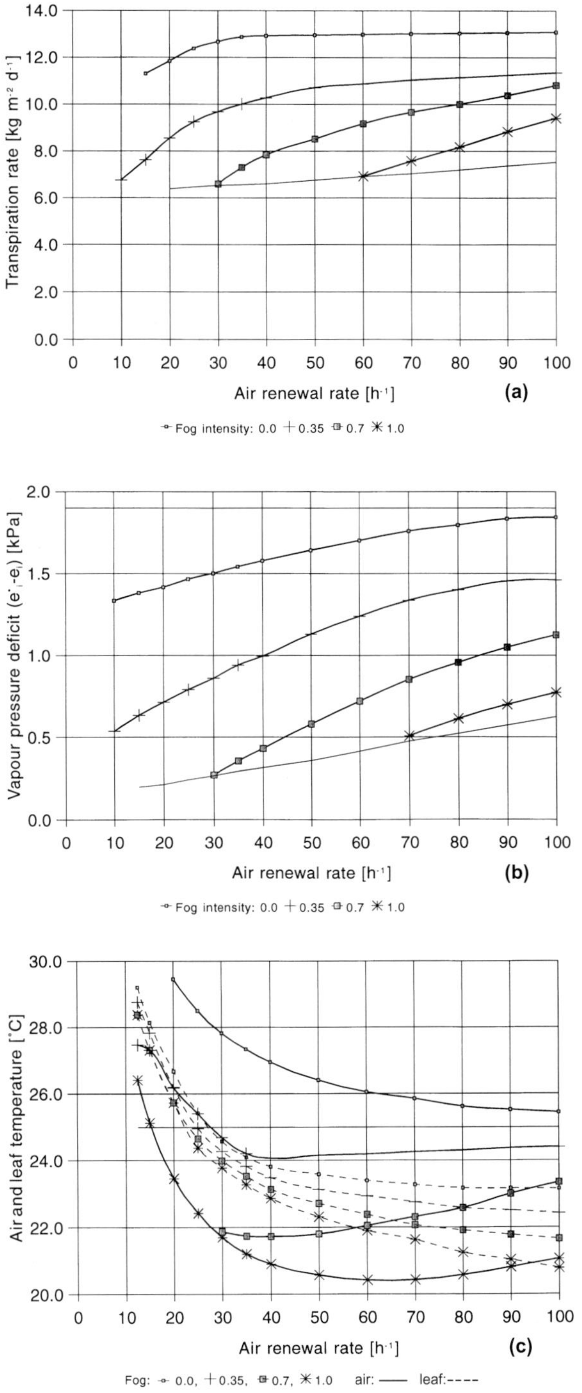

Fig. 13.10. Simulated variation of greenhouse-crop climate as a function of air exchange rate and fog intensities, according to Boulard and Baille (1993). Crop transpiration (a), air vapor pressure deficit (b), air and tomato leaf temperature (c) as a function of air exchange rate (1/h), for different fogging intensities: $E_{ad} = 0.0$, 0.35, 0.7, $1.0\,kg\,h^{-1}\,m^{-2}$. Boundary conditions: $R_o = 820\,W\,m^{-2}$; $T_o = 25°C$; $e_o^* - e_o = 1.8\,kPa$; LAI = 3.0.

have been used with forced ventilation by fans. The main disadvantage of this system is the gradient in temperature and humidity created in the greenhouse, together with the limited greenhouse width, the distance between pad and fans amounting usually to less than 30 meters (Montero and Segal, 1993). Nowadays, fogging systems allow more uniform conditions throughout the greenhouse and eliminate the need for forced ventilation, while having a higher cooling efficiency. Disadvantages are the high investment, and the need for very good water quality and filters to avoid problems with the nozzles. Air temperature and relative humidity are simultaneously reduced, leading to an important reduction of vapor pressure deficit and transpiration (Fig. 13.10). However, reduction in leaf temperature is limited, unless evaporation takes place at the surface of the leaves with misting or with irrigation sprays. On the other hand, the risks of rapid development of disease would be enhanced in these cases, because of the wet leaves.

Cooling the inside air without modifying the quantity of water in the air would reduce the vapor pressure at saturation and therefore also limit vapor pressure deficit and transpiration.

Shading can also reduce these adverse effects. Ideally, the inside solar radiation should be reduced to the saturation level of the photosynthetic response to light (see Chapter 12). Boulard and Baille (1993) indicated that a shading screen significantly reduces crop transpiration through the decrease of light transmission, while temperature is only slightly reduced.

Control strategies

Optimal values for aeration and fog parameters differ, according to whether the objective is to cool the air, to cool the foliage or to maximize/minimize the transpiration rate. Figure 13.10 gives a first basis for choosing aeration rate and fogging intensity in order to obtain the desired air or leaf temperature. Boulard and Baille (1993) (table 1), have given a detailed, interesting, more general but more complex climate-control algorithm.

For most applications, there is however agreement between the range of efficient air-renewal rate in Fig. 13.10 and the recommendations of different authors (Fuchs, 1993; Montero and Segal, 1993): an air-renewal rate of $20\,h^{-1}$ as a minimum should be ensured; an exchange rate of approximately $30\,h^{-1}$ achieves most of the potential cooling, allowing in most conditions for the temperature not to exceed the critical value of 28°C. Above this rate, little additional cooling of the foliage is obtained. Therefore, the fogging system must be designed to provide sufficient water to maintain high humidities in well-ventilated greenhouses (air-renewal rate of $30\,h^{-1}$).

Optimal control at high humidities

With the introduction of tighter and well insulated greenhouses, higher humidity levels occur more often nowadays.

Effect on crops

The occurrence and severity of fungal diseases, a major economic factor, depends on the humidity level; the spores of pathogens are produced in large quantities under wet conditions, but are released most readily when the humidity drops. As discussed in chapter 8, air temperature, dew deposition on the canopy, physiological status of the host and the aggressiveness of the pathogen population have interactive effects on disease development. More research is still needed to understand, quantify and predict these effects.

High humidities also modify leaf development, either by increasing the number of side shoots or by causing low transpiration and therefore calcium deficiencies, which may lead to reduction in leaf area of up to 50% (Holder and Cockshull, 1990; see also chapter 8).

For cucumber, the resulting response to high humidities is positive, whereas no significant influence was found for aubergines (Bakker, 1991). Therefore, there is no reason to apply dehumidification with these crops, unless fungal diseases cause a major problem.

For tomatoes, high levels of humidity lead to yield losses (Bakker, 1991). The response of yield to humidity was modelled by Jolliet et al. (1993) as a function of the increase in vapor pressure deficit, weighted by solar radiation. This work showed how yield of greenhouse tomato crops relates both to inside-air humidity and to transpiration. Yield data were obtained during three multifactorial experiments (Holder and Cockshull, 1990) at the Horticulture Research Institute in Efford, England in 1987, 1988 and 1991. Tomato plants were grown under different day and night humidity set-points (air vapor pressure deficit between 0.1 and 0.8 kPa).

Humidity treatments had a significant effect on total tomato yield. As day and night air vapor pressure deficits had a similar influence on yield but a different effect on transpiration, differences in yield were not directly related to the total absolute transpiration during the treatment period. In fact, yield response to transpiration was higher at low levels of transpiration and the increase in yield proved proportional to the *relative* increase in transpiration, 89% of the yield variability being accounted for by such a model.

For prediction purposes, a simplified relationship was developed to predict the effect of a new vapor pressure set-point (e_{pi}) achieved by dehumidification. The increase in yield is a function of the integrated increase in vapor pressure deficit [$(e_{pi} - e_i)\Delta t$, in kPa d], weighted by the inside solar radiation (R_i, in MJ m^{-2} d^{-1}):

$$\Delta Y = 0.064 \left[\frac{(e_{pi} - e_i)\Delta t}{\left((e_i^* - e_{pi}) + cR_i\right)_{day}} + \frac{(e_{pi} - e_i)\Delta t}{\left(e_i^* - e_{pi}\right)_{night}} \right] \quad [\mathrm{kg\,m^{-2}}], \qquad (13.17)$$

where $c = 0.116/(1 - 0.63 \exp(-R_i/0.864))$.

This relationship was validated by comparison with long-term yields ($r^2 = 0.90$. Standard error of the estimate of ΔY: 0.34 kg m^{-2}) and a relationship was suggested for short time periods. Combined with the HORTITRANS model (Jolliet, 1994a, see also Eq. (13.16) which predicts inside humidity, transpiration and the energy necessary for dehumidification, all data necessary to operate an optimal humidity control are provided.

Ways of dehumidification and energy use

Before determining optimal humidity set-points, the technique used must be considered. Classically, dehumidification is performed by opening the ventilators while the heating system is operating. This leads to a very high increase in energy consumption (20% to 60%). The lower the external temperature, the more energy is required to dehumidify the greenhouse.

Table 13.1 compares the energy use for alternative ways of increasing transpiration: an increase in vapor pressure deficit of 0.1 kPa results in an increase in tomato yield of 0.010 kg m^{-2} d^{-1}. The additional heat consumption amounts to 28 W m^{-2} for dehumidification by heating and ventilation, and to 13 W m^{-2} by increasing the inside temperature by 2°C. The use of a dehumidifying heat pump leads to an electricity consumption of 5 W m^{-2} but to a reduction of 19 W m^{-2} in the heat consumption. However its investment cost cannot be recovered at the present energy price (Boulard et al., 1989; Bailey, 1991). Artificial lighting, besides its effect on photosynthesis rate, also increases transpiration and therefore yield.

Table 13.1
Energy consumption with alternative ways of increasing transpiration[1]

Method of dehumidification	Addheat[2] (W m^{-2})	Electricity (W m^{-2})	Addyield[2] (g m^{-2} d^{-1})
Ventilation + heating	+28	0	+10
Heat pump	−19	+5	+10
2°C increase in air temperature	+13	0	+10

[1] Calculation using HORTITRANS software for a tomato crop in a double-glazed 5000 m greenhouse. $T_o = 0$°C, at night. Dehumidification of +0.1 kPa (−4% RH).
[2] Addheat, Additional heating applied; Addyield, increase in yield.

Air movement within the greenhouse could also potentially increase transpiration. In practice however, the use of fans cannot ensure a sufficient air speed through the whole vegetation.

A control strategy

For disease development, the present state of knowledge does not allow one to predict optimal humidity set-points. Due to its potentially strong economic influence on yield, the control of a declared disease is certainly a priority for short-term humidity control. Additional factors of influence can be identified: (1) direct foliage wetting should be avoided. Therefore, additives which act as antifoggers and antidrippers are used for plastic cladding, to prevent the dripping of water condensed on the cladding onto the canopy. Trickle irrigation rather than sprays reduces soil and foliage wetting. (2) Condensation of water on leaves can also be reduced by the use of thermal plastics or even thermal screens, which reduce the radiative losses from the top of the crop. It therefore increases its temperature and reduces the risk of direct condensation on the leaves. Radiative heating or heating at crop level also contributes to reduce this effect.

Regarding the effect of humidity and transpiration on plant growth and yield, the optimal humidity set-points were determined for a tomato crop by Jolliet (1994a). The study was performed using the HORTITRANS model and software, and assuming that dehumidification was performed by ventilation and heating (Fig. 13.9):

(a) Actual inside humidity and transpiration are predicted from outside conditions.
(b) The energy necessary to increase the vapor pressure deficit by 0.02 kPa using simultaneous ventilation and heating is then determined.

(c) The increase in yield is estimated according to Jolliet et al. (1993); by introducing the prices of energy and tomatoes, the additional income resulting from dehumidification is determined as the difference between the additional yield income and the additional energy cost.
(d) Operations (b) and (c) are repeated until the maximum income is achieved; the corresponding vapor pressure deficit gives the optimal set-point.

As the growth model was developed on the basis of humidity treatment applied to a young tomato crop in winter conditions (LAI between 0.5 and 3, between December and March at Efford, England), the calculations are essentially valid for the same conditions.

Optimal set-points

The optimal set-point of vapor pressure deficit decreases practically linearly with solar radiation (Fig. 13.11a). This set-point is higher for low leaf area index (LAI = 1) and high inside temperature ($T_i = 20$°C). It is only at very high levels of humidity that dehumidification by heating and ventilation can be justified – for relative humidity higher than 80% at night ($T_i = 20$°C) and higher than 90% as soon as the inside solar radiation exceeds 50 W m^{-2} (Fig. 13.11b). Greenhouse type and size do not significantly affect the optimum set-point.

Dehumidification is worthwhile *only when the present vapor pressure deficit is lower than the optimal set-point* given in Eq. (13.18). Figure 13.12 compares the present and optimal vapor pressure deficits and shows that this condition is most likely to occur at very low light levels, between 20 and 40 W m^{-2}, whatever the crop stage.

There is, however, little need for dehumidification in a single-glazed greenhouse, as the actual vapor pressure deficit is nearly always higher than or equal to the optimal set-point. On the other hand, dehumidification could be justified for a double-glazed greenhouse when inside radiation falls below 75 W m^{-2}. The increase in vapor pressure deficit required to reach the optimum does not exceed 0.15 to 0.2 kPa (6 to 8% RH at 20°C).

Similarly, In terms of relative humidity (Fig. 13.13), it is when the present relative humidity is *higher* than the optimum that dehumidification is worthwhile.

Jolliet (1994a) showed that the optimal set-point is a function of inside solar radiation (in MJ m^{-2} d^{-1}), the vapor pressure of the inside air (in Pa), the vapor pressure of the outside air (in Pa), the temperature difference between inside and outside air (in K) and the Leaf Area Index [$r^2 = 0.94$, standard error on individual predictions = 50 Pa (for 262 points)]:

$$(e_i^* - e_i)_{opt} = c_0 + c_1 R_i + c_2\sqrt{e_i - e_o} + c_3(T_i - T_o) \quad [\text{Pa}], \tag{13.18}$$

where c_0, c_1, c_2 and c_3 are given in Table 13.2 as a function of Leaf Area Index and of the ratio of the crop sale price to the price of energy (δ, in MJ/kg). The validity of this equation is directly related to the validity of the yield model (see above, p. 316): it has been established for the first three months after planting and for winter high humidities together with low light levels. Extrapolation to summer conditions or to a mature crop should therefore be undertaken only with care.

The present relation (13.18) could be directly used by a climate controller to check if the present humidity is higher than the optimal value and to decide whether or not dehumidification is needed for a tomato crop.

Table 13.2
Coefficients used to calculate the optimal humidity set-point in Eq. (13.18)[1]

Coefficient	Units	General relationship	Particular values (δ = 320 MJ/kg)	
			LAI = 1	LAI = 4
c_0	Pa	−42.2 · (1 − 0.59 LAI) (±48%) (±28%)	−17.3	57.4
c_1	s m^{-1}	−10.5 · (1 + 0.53 LAI) · (1 − 0.53[ln(δ) − 5.77]) (±18%) (±28%) (±10%)	−16.1	−32.4
c_2	Pa$^{1/2}$	−20.6 · (1 − 0.108 LAI) · (1 + 0.46[ln(δ) − 5.77]) (±5%) (±7%) (±6%)	18.4	11.7
c_3	Pa K^{-1}	−19.7 · (1 − 0.093 LAI) · (1 + 0.21[ln(δ) − 5.77]) (±9%) (±19%) (±47%)	−17.9	−12.4

[1] General relationship determined by nonlinear regression analysis as a function of Leaf Area Index (LAI) and the ratio of yield to energy price (δ). Particular values for δ = 320 MJ/kg (ln(320) = 5.77) and for LAI = 1 and 4.

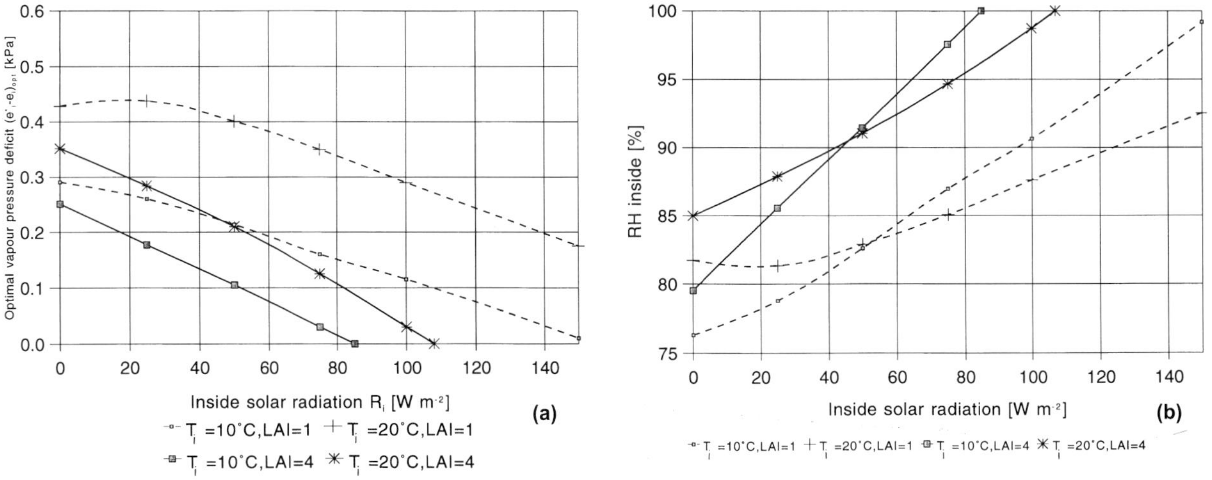

Fig. 13.11. Optimal vapor pressure deficit (a) and optimal relative humidity (b) of the inside air as a function of the inside solar radiation. For $T_i = 10$°C and $T_i = 20$°C, LAI = 1 and LAI = 4.

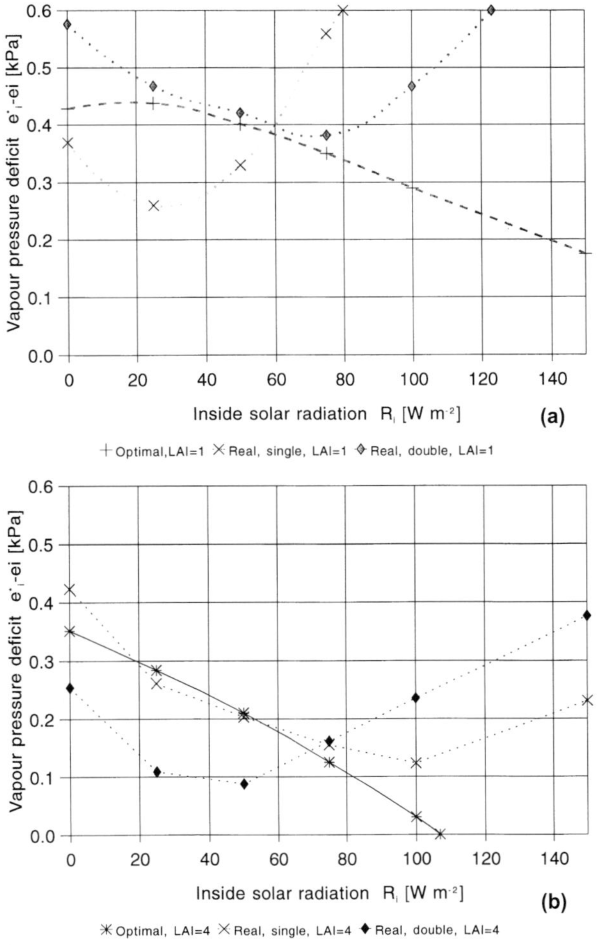

Fig. 13.12. Comparison between optimal vapor pressure deficit and the effective vapor pressure deficit (for $T_o = 15$°C, $T_i = 20$°C) in a 500 m^2 single-glazed and a 5000 m^2 double-glazed greenhouse. (a) LAI = 1; (b) LAI = 4.

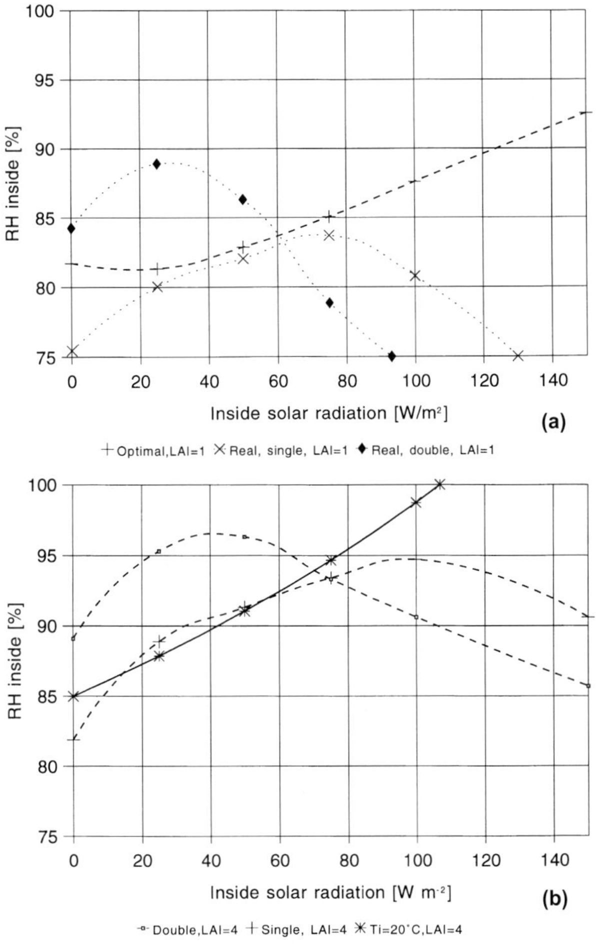

Fig. 13.13. Comparison between optimal and effective relative humidities (for $T_o = 15$°C, $T_i = 20$°C) in a 500 m^2 single-glazed and a 5000 m^2 double-glazed greenhouse. (a) LAI = 1; (b) LAI = 4.

Evaluation and measurements of Leaf Area Index (LAI)

For the optimal humidity-control scenarios and more generally for transpiration studies, crop stage is characterized by the Leaf Area Index (LAI), which is the ratio of leaf to soil area. How can this parameter be estimated or measured? For tomatoes, the Leaf Area Index could be approximated as a function of the plant density (ρ_p, in m^{-2}), the mean plant height (H, in m) or the mean leaf number per plant (N):

$$\text{LAI} = \rho_p(0.054N - 0.38) \qquad (13.19)$$

and

$$\text{LAI} = \rho_p \sum_{k=1}^{N} 0.35 l_k w_k. \qquad (13.20)$$

These two correlations were measured on the cultivar Counter at Silsoe in England between February and June 1990, for a plant density of 2 to 2.65 plants per square metre.

A more accurate determination of LAI can be made by measuring non-destructively the length (l: from the first small leaflet to the extremity of the leaf k) and width (w: distance between the extremities of the two largest leaflets) of each leaf of 5 to 12 randomly chosen plants. In these conditions the Leaf Area Index can be calculated to within 10 to 20% as:

$$\text{LAI} = \rho_p \sum_{k=1}^{N} 0.35 l_k w_k. \qquad (13.21)$$

Compared to planimeter measurements, this method proved accurate [±3%; $r^2 = 0.94$ (for 80 measurements)], the major error being due to the variability between plants.

IRRIGATION SYSTEMS AND RECIRCULATION

Generalities

The irrigation system aims to meet the plant needs for water and nutrients. It must also provide a favorable environment to ensure root health and development – that is, a sufficient supply of oxygen and a limited build-up of salinity in the root medium. As mentioned by Cooper (1979), maintaining the best compromise between the supply of water and the supply of air is the essence of skill in irrigation management. These objectives are met by adjusting the quantity and frequency of irrigation, together with the nutrient and oxygen concentrations in water. While meeting them, pollution of the environment through drainage needs to be avoided.

In addition, the irrigation system can be used in special cases for indirect control of flower initiation, vegetative growth or fruit size by imposing a water stress (Graves and Hurd, 1983).

This section discusses the different possibilities for irrigation control and compares the available irrigation systems from a practical point of view. Recirculation of water is finally treated as a way of limiting fertilizer use and environmental pollution.

Irrigation control

Quantity control

The total quantity of water supplied must correspond to the water uptake – that is, the amount transpired and the increase in the water stored in plants [Eqs. (13.2, 13.3)]. Figure 13.8a shows that transpiration of a mature tomato crop amounts to $0.5\,\text{kg}\,\text{m}^{-2}\,\text{d}^{-1}$ on a very dull day, and up to $6\,\text{kg}\,\text{m}^{-2}\,\text{d}^{-1}$ on a very sunny day. In addition to water uptake, adequate drainage must be ensured to avoid a build-up of salinity in the root medium, and to account for irregularities in water distribution (minimum drainage 25% of the water uptake for a rock-wool medium).

In practice, growers tend to reduce risks by providing a generous water supply. This results in a very high run-off (between 50% and 100% of the water uptake, or even more, and high pollution), particularly for young crops. Closed systems on the one hand greatly simplify irrigation control, as water can be applied in excess without a problem. On the other hand, it makes fertilization control more complex, requiring on-line control of the solution.

The quantity of water is now commonly controlled by computer on the basis of on-line measurements of solar radiation. As solar radiation is the main factor affecting transpiration (see p. 306), this system provides a good control as long as the amount applied is adapted to the crop stage or the Leaf Area Index. In summer, any of the correlations presented in Fig. 13.4a could be used as a basis for calculations, adding a 25% run-off percentage when necessary. For young crops or crops other than tomatoes, amounts could be corrected proportionally to the k_{cr} factors of Eq. (13.6).

If air is artificially humidified or dehumidified, the vapor pressure deficit should be taken into account. In

Table 13.3
Standard concentrations, total input and amount of drainage for nutrient elements for greenhouse (and field) crops[1]

	Elements				
	N	P	K	Mg	Ca
Standard concentration in irrigation water[2] (mg ℓ^{-1})	200	46	360	42	184
Total input, kg ha^{-1} yr^{-1}					
Greenhouse crops	2100	480	3780	440	1930
Field crops	280	140	700	98	–
Nutrients in drainage (kg ha^{-1} yr^{-1})					
Greenhouse crops	700	160	1260	150	640
Field crops	50	0.1–0.5	–	–	–
Maximum permitted concentration (mg ℓ^{-1})	5.66	0.08	–	–	–
Volume of surface water polluted up to maximum permitted concentration by nutrients in drainage					
Per unit area (m^3 ha^{-1} yr^{-1})	124000	2000000	–	–	–
Per unit crop (tomato) (m^3 kg^{-1})	3700	60000			
Equivalent area polluted during 1 year (m^2 year ha^{-1})	125500	551000			

1 Maximal admissible concentration of element in water: 25 mg NO_3 ℓ^{-1} (5.66 mg N ℓ^{-1}) and 0.08 mg P ℓ^{-1} for open waters. Volume of water polluted up to that maximum, ratio of the drainage divided by the maximum concentration. Equivalent area polluted according to the critical surface-time method (total volume of dilution: 4.5 m^3 m^{-2}; residence time of 4.6 years for N, 1.2 years for P; Jolliet, 1994b).
2 Sonneveld (1988).

winter, transpiration and inside vapor pressure deficit are increased by the heating system (see p. 311); therefore, correlations which explicitly account for the vapor pressure deficit should also be used (e.g. Eq. 13.7). If not, a minimum amount of water at least should be applied: between 0.5 and 1.5 kg m^{-2} d^{-1} for a tomato crop, depending on the difference between inside and outside temperature and the Leaf Area Index (Figs. 13.5a, 13.6a).

An irrigation controller based only on time leads generally to large losses to drainage, in order to avoid water stress in sunny days. The set-point must also be regularly adjusted.

Some other systems are commercially available: a) Tensiometer control for crops grown in soil, where the soil water potential is maintained close to field capacity (−10 kPa to −40 kPa). Irrigation will be commanded as soon as the suction exceeds its upper limits. This method is interesting, but problems can be encountered, as the water column in the tensiometer breaks if suction exceeds −70 kPa. b) Drainage measurements using tipping buckets or lysimeter. c) Evaporimeter measurements. d) Weighing lysimeter (mostly for research purposes).

These methods are rather difficult to apply for on-line control and the choice of a sampling position representative of the whole crop is also difficult. However, they provide efficient ways of checking occasionally whether the irrigation supply is suitable, and they could be used in parallel with a solar-based controller.

Frequency control

Irrigation frequency depends mainly on the type of rooting medium (see also Chapter 4) and on its buffer capacity: about 15 kg m^{-2} of readily available water in soil, down to 1.1 kg m^{-2} in rock-wool, or less for aeroponics. Frequency must be chosen to avoid both water depletion and permanent saturation of the root system, which would hinder its oxygenation.

Generally, the lower the buffer capacity, the more accurate the system can be, but the more difficult it is to manage (higher frequency and short time to react in case of breakdown). Frequency is often regulated by solar radiation, an irrigation being delivered every 1 to 4 MJ m^{-2} (trickle irrigation for rock-wool) or every 14 MJ m^{-2} ("tide" system, see below, p. 322) of incoming outside solar radiation (for a single-glazed greenhouse with a transmission of 0.65).

Quality control

Water quality strongly affects yield. Table 13.3 gives standard values for fertilizer concentration in solutions (see also Baets et al., 1985). For most crops, pH should be maintained close to 6 (Cooper, 1979)

Table 13.4
Properties and comparison of irrigation systems

Irrigation system	Root medium	Reaction time (hours)	Recycling possible	Control system [1]	Other comments
Overhead spray	soil	24–100	No	T, S, P	High effect on air humidities. With nutrients, risks of sprayer obstruction. Deposits on flowers and fruits.
Drip/trickle	soil	24–100	No	S, T, P	Positioning of capillaries needs to be checked regularly.
	rockwool, etc.	4–12	Yes	S, T	
	pot plants	12–24	Yes	S, T	
Tide	pot plants	12–48	Yes	T, S	
NFT	pot plants	12–48	Yes	continuous	Use double pumps + electricity
	channel	2–24	Always	level of reservoir	
Nutrient mist	channel	0.25–2	Yes	nearly continuous	Risks of nozzle obstruction

[1] Abbreviations: S, solar; T, time; P, tensiometer.

and the electrical conductivity (salinity measurement) between 2 and 6 mS m^{-1}. At the beginning of crop growth, higher conductivities, up to 10 mS m^{-1}, can be maintained to initiate flowering (Baets et al., 1985, p. 36).

While supplying the fertilizer needs (see Chapter 14), build-up of toxicity in the irrigation water should be avoided, especially for chlorine (maximum 90 mg l^{-1}: ADAS, 1981, p. 14) and sodium (maximum 60 mg l^{-1}). In closed systems, the build-up of ion concentrations should be regularly checked by weekly analyses of the nutrient solution. If these limits are reached, the solution should be changed.

Choice of irrigation system

Table 13.4 gives an overview of the different irrigation systems which are commercially available. Properties, advantages and disadvantages are described in the following.

Overhead spray

This system is widely applied for crops with little foliage [e.g. radishes (*Raphanus sativus*), lettuces]. As it greatly increases the inside-air humidity, it is also used with crops like cucumbers, which require high humidity levels. Because water is sprayed on foliage, the risk of disease is great with this system. Finally, care must be taken to ensure a reasonable uniformity in the water distribution.

Trickle/drip

Water is brought to the foot of individual plants, through capillary tubes linked to a distribution pipe. It is usually used for crops in artificial substrates (rockwool) or in soil, with a limited density (tomatoes, etc.). It can also be used for pot plants or other crops, but it requires a lot of work and is more expensive because of the large number of capillary tubes.

Depending on the soil medium, an irrigation will be ordered for every 1 to 4 MJ m^{-2} of outside solar radiation on a horizontal surface. In order to ensure an equal flow in all capillary tubes, distribution pipes must be horizontal, with a maximum slope of 0.2% and their length must not exceed 35 m [see Baets et al. (1985) for more practical details]; irrigation duration should be long enough, at least 5 times the time required to fill the pipes.

Water has usually to be filtered (sand filter, etc.) to avoid obstruction of the capillary tubes. It is also necessary to check the outlet positions regularly and to clear any blocked outlet.

"Tide"

The "tide" system is mainly suitable for pot-plant irrigation. Pots are set on water-tight tables with raised edges, where a water height of 1 to 3 cm (Steffen, 1988) can be maintained for short periods. It is controlled through duration and frequency of irrigation. According to Volkening and Strauch (1988), the duration is typically 15 minutes. It can be increased to 30 minutes, as long as the roots are able to sustain a high water content in the substrate. In winter, at low radiation levels, a daily irrigation is sufficient most of the time. For higher levels of solar radiation, an additional irrigation should be given for every 14 MJ m^{-2} (sometimes every 29 MJ m^{-2} in midsummer) of outside solar radiation.

A mat can be set under the pots in order to provide higher surface evaporation and local air humidity, together with a slightly lower temperature. A manual irrigation can be performed once a week to avoid a build-up of salinity.

Flow in channels and Nutrient Film Technique (NFT)

A thin layer of water runs freely in a channel in which the roots are partly immersed. The solution drains into a reservoir and is recirculated. This allows a very accurate control of the nutrient supply to the roots. Pot plants can also be grown in narrow closed channels. Cooper (1979), Graves (1983) and Baets et al. (1985) have made very good reviews of NFT. Channel length must not exceed 30 m, with a slope of 1 to 1.5% and a flow of 2 to 3 $\ell\,min^{-1}$ at the top of the channel. About 90% of water circulates in the system against only 10% in the reservoir. As there is little time available to react in case of breakdown, water pumps should be doubled.

To limit root diseases and death, *oxygenation* of the solution is a very important issue and a minimum of 20 ml oxygen per hour and per plant (Graves, 1983, p. 31) should be provided for a tomato crop. In addition, the roots are only partly submerged in the gully to enable direct oxygen exchange with the air.

Nutrient mist supply and aeroponics

Water and nutrients are provided as a fine mist projected directly onto the roots. A very precise control of the root environment can be achieved, all roots being in contact with the solution and aeration being ensured. It is however a rather expensive system, with potential technical problems due to nozzle obstruction. It needs high-pressure pumps and also a system to hold the roots. As a result of these characteristics, it is mainly suitable for the multiplication phase in plant micropropagation (Kurata et al., 1991).

Water pollution and recycled solutions

Water pollution

For crops grown in rock-wool or peat bags, a minimal drainage of 25% of the water supplied must be maintained throughout the cropping period. In practice, it amounts often to more than 50%. Assuming a yearly water usage of 1050 $\ell\,m^{-2}$ with 700 $\ell\,m^{-2}$ taken by plants and 350 $\ell\,m^{-2}$ drained, Table 13.3 presents the amount of nutrient contained in the run-off and the corresponding pollution, according to the life-cycle assessment of greenhouse tomato production carried out by Jolliet (1993).

The amount of fertilizer applied is 5 to 7 times higher than for field crops. The potential pollution level is very high, especially for phosphate: compared to field crops where most of the phosphate remains in the soil, the potential pollution per hectare is 300 to 1500 times higher in the greenhouse. Furthermore, nitrogen run-off is probably even higher in reality than is indicated in Table 13.3 (about 1350 $kg\,ha^{-1}$).

The level of potential pollution is thus very high. Therefore, water drainage should be almost totally recycled. In the coming years, this issue will become a condition for the survival of the horticultural industry, as environmental criteria become more and more important in public opinion. In the Netherlands, total recycling of nutrient solutions will be compulsory from the year 2000 onwards.

New methodologies developed for Life Cycle Impact Assessment, such as the Critical Surface-Time method (Jolliet, 1994b; Jolliet and Crettaz, 1997), will help to determine the most critical pollution loads, for which measures are needed in priority.

Closed systems

In new production systems, where crops are grown on artificial substrates instead of in soils, drainage can easily be collected and then recycled. However, apart from NFT systems, the nutrient solutions are rarely recycled for fear of the spread of root diseases.

The experience gained in twenty years of NFT closed systems should now be applied to other growing methods. As mentioned by Cooper (1979, p. 119), the early fear that a disease organism would rapidly spread in recirculating solution and wipe out the crop has not materialized. Soilless systems could even contribute to the control of fungal diseases (see Chapter 8). Disease organisms have not established themselves unless the root system has been affected by lack of aeration or other factors (phytotoxic elements, etc.), or unless disease has been artificially introduced. Good oxygenation of the solution and roots is certainly a central issue in closed systems. The build-up of toxic elements (sodium, chloride: see p. 322) must be avoided.

For crops grown in rockwool or perlite, recent experiments have demonstrated that recycled solution can be used with little adverse consequence on the growth of crops such as tomatoes or chrysanthemums (Hardgrave, 1991; A. Reist, personal communication).

Partial recirculation of the solution (e.g. no recirculation until the crop is mature) also gives good results but is not very satisfactory, leading to a high level of environmental pollution.

For crops very sensitive to root diseases, many ways are now available for solution disinfection, as long as the amount of drainage is not too high: pasteurization, fungicides, ozone, UV-irradiation or slow filtration through quartz sand. Good studies were carried out by Reist et al. (1991) and by Wohanka (1991).

An interesting alternative to closed systems is the storage of the drainage water in tanks and its distribution in contiguous fields. The field area must be 3 to 10 times greater than the greenhouse area, and the required storage volume is about 35 $\ell\,m^{-2}$ for one month's storage. The cost of such storage can be estimated as US\$ 4 m^{-2}. If the cost of nutrients in the recycled solution amounts to about US\$ 0.8 $m^{-2}\,year^{-1}$, the initial investment is paid back in 5 years (Reist et al., 1991).

CONCLUSION

Solar radiation is the main factor affecting and controlling transpiration and the water cycle. To a lesser extent, transpiration can be controlled by modifying the inside humidity, characterized by the vapor pressure deficit.

As the existing models of transpiration provide good accuracy, there is little need for more detailed models (multi-level, etc.). On the other hand, much more understanding is needed on the uptake of individual nutrients and their distribution within the plant, and on disease development.

Optimal set-points were calculated in the case of a tomato crop for the first time. This enables the operation of optimal humidity control in the high-humidity range. The optimization study shows that there is little need for dehumidification in a single-glazed greenhouse. For double glazing, dehumidification could be financially justifiable for a tomato crop at very low levels of solar radiation, in particular just before sunset.

For other crops such as cucumbers or aubergines, there is no positive effect of dehumidification on crop growth and only serious risks of fungal diseases could justify it.

When properly managed, new irrigation systems enable very accurate control of plant nutrition, but tend to involve more knowledge and more risks in case of breakdowns. Environmental pollution is a very important issue and complete recycling of nutrient solutions is becoming a survival condition for the greenhouse industry.

GLOSSARY OF SYMBOLS

A	greenhouse soil area (m^2)
A_f	effective area for air infiltration (m^2)
A_v	ventilator area (m^2)
a	dimensionless parameter characterizing the influence of solar radiation on transpiration
C_{pa}	heat capacity of the air ($J\,kg^{-1}\,K^{-1}$)
E	flow of water added, extracted, condensed, transpired, exchanged by ventilation or stored ($kg\,m^{-2}\,d^{-1}$) ($1\,kg\,m^{-2}\,d^{-1} = 11.57\,mg\,m^{-2}\,s^{-1} = 41.7\,g\,m^{-2}\,h^{-1}$)
e	vapor pressure (kPa)
e^*	vapor pressure at saturation, function of the temperature (T, in K) of the body considered (kPa): approximated as $e^* = 1.41\,10^7 \exp[3928.5/(T-41.5)]$
f_f	dimensionless factor characterizing the tightness of the cladding to air infiltration
f_T	factor characterizing the relative influence of temperature on ventilation and air infiltration ($m\,s^{-1}\,K^{-1/2}$)
f_ω	dimensionless factor characterizing the relative influence of wind on ventilation and air infiltration
g	aerodynamic or stomatal conductance ($m\,s^{-1}$)
h	coefficient of transfer by sensible heat ($W\,m^{-2}\,K^{-1}$)
h_t	influence of vapor pressure deficit on transpiration ($W\,m^{-2}\,K^{-1}$)
H	mean plant height (m)
k_{cr}	factor characterizing the crop stage and the crop type (according to FAO, 1977), dimensionless
l	leaf length (m)
LAI	Leaf Area Index, dimensionless
N	leaf number per plant
q	air flow between inside and outside ($m^3\,s^{-1}$)
r_a	aerodynamic resistance ($s\,m^{-1}$)

r_p stomatal resistance ($s\,m^{-1}$)

R_i incoming solar radiation inside the greenhouse ($W\,m^{-2}$ or $MJ\,m^{-2}\,d^{-1}$)

R_n net radiation exchange between the canopy and the environment ($W\,m^{-2}$)

T temperature (K)

t time (s)

w leaf width (m)

W temperature-related weighting factor (according to FAO, 1977), dimensionless

Y yield ($kg\,m^{-2}$)

α parameter characterizing the influence of solar radiation on transpiration ($kg\,MJ^{-1}$)

β parameter characterizing the influence of vapor pressure deficit on transpiration ($S^2\,d^{-1}\,km^{-1}$)

δ ratio of the crop sale-price to the price of energy inputs ($MJ\,kg^{-1}$)

λ latent heat of vaporization of water ($2.45\,MJ\,kg^{-1}$ at 20°C)

γ psychrometric constant ($0.066\,kPa\,K^{-1}$)

ϕ dimensional factor for the conversion of seconds to days ($11.57\,d\,Ms^{-1}$)

ζ angle of opening of the ventilators (°)

ρ_a air density ($kg\,m^{-3}$)

ρ_p plant density (m^{-2})

ω wind speed ($m\,s^{-1}$)

Subscripts

ad added to or extracted from the greenhouse air

c cladding

dr drained

e soil evaporation

i inside air

ir supplied by irrigation

l leaf

o outside air

n ordinal position of a leaf

p plant

sa stored in the air

sp stored in the plant

sso stored in the soil

t transpiration

up uptake

v ventilation

REFERENCES

ADAS, 1981. *Guide notes on the Nutrient Film Technique for tomatoes,* Booklet G1. Ministry of agriculture, fisheries and food, Great Britain, 23 pp.

Aikman, D.P. and Houter, G., 1990. Influence of radiation and humidity on transpiration: Implications for calcium levels in tomato leaves. *J. Hortic. Sci.,* 65 (3): 245–253.

Baets, W., Bleyaert, P., Craemers, J., Jansen, G. and Pasture, E., 1985. *La culture sur substrat et l'hydroculture des tomates.* Ministère de l'agriculture Belge, 1210 Bruxelles, pp. 46.

Bailey, B.J., 1991. *Economic assessment of dehumidification for greenhouse tomatoes,* Contract Report CR/445/91/8037, Silsoe Research Institute, Silsoe, Beds MK45 4HS, UK.

Bakker, J.C., 1990. Effects of day and night humidities on yield and fruit quality of glasshouse tomatoes. *J. Hort. Sci.,* 65: 323–331.

Bakker, J.C., 1991. *Analysis of humidity effects on growth and production of glasshouse fruit vegetables,* Doctoral Thesis. Proefstation voor Tuinbouw onder Glas (PGT), NL-2670 AA Naaldwijk, Netherlands.

Batta, L.G.G., 1989. Modelling of water uptake rate of greenhouse tomato plants. *Acta Hortic.,* 248: 355–360.

Bot, G.P.A., 1983. *Greenhouse climate: from physical processes to a dynamic model,* Doctoral Thesis. Landbouwhogeschool, Wageningen, Netherlands, 240 pp.

Boulard, T. and Baille, A., 1993. Greenhouse ventilation and evaporative cooling in warm climate. In: M. Fuchs (Editor), *Proc. Workshop on Cooling System for Greenhouse,* Tel Aviv, 2–6 May 1993. Department of Agricultural Meteorology, Agricultural Research Organization,Bet-Dagan 50-250, Israel, pp. 89–98.

Boulard, T. and Jeema, R., 1992. Greenhouse tomato crops transpiration, application to irrigation control. *Acta Hortic.,* 335: 381–387.

Boulard, T., Baille, A., Lagier, J., Mermier, M. and Vanderschmitt, E., 1989. Water vapour transfer in a plastic house equipped with a dehumidification heat pump. *J. Agric. Eng. Res.,* 44: 191–204.

Choudury, B.J. and Monteith, J.L., 1986. Implications of stomatal responses to saturation deficit for the heat balance of vegetation. *Agric. For. Meteorol.,* 36: 215–225.

Cooper, A., 1979. *The ABC of Nutrient Film Technique.* Grower Books, London, 225 pp.

de Graaf, R. and van den Ende, J., 1981. Transpiration and evapotranspiration of the glasshouse crops. *Acta Hortic.,* 119: 147–158.

de Villèle, O., 1972. Besoins en eau des cultures sous serres, essai de conduite des arrosages en fonction de l'ensoleillement. *Acta Hortic.,* 35.

FAO, 1977. Crop water requirements. *FAO Irrig. Drainage Pap.,* 24: 15–44.

Fernandez, E. and Bailey, B.J., 1992. Measurement and prediction of greenhouse ventilation rates, *Agric. For. Meteorol.,* 58: 229–245.

Fuchs, M., 1990. Effect of transpiration on greenhouse cooling. In: I. Segal (Editor) *Proc. Int. Seminar and British–Israel Workshop on Greenhouse Technology.* Agricultural Research Organization, Bet Dagan, Israel, pp. 155–181.

Fuchs, M., 1993. Transpiration and foliage temperature in a greenhouse. In: M. Fuchs (Editor), *Proc. Workshop on Cooling System for Greenhouse,* Tel Aviv, 2–6 May 1993. Department of Agricultural Meteorology, Agricultural Research Organization, Bet-Dagan 50-250,Israel, pp. 89–98.

Graves, C., 1983. *The Nutrient Film Technique, Horticultural Reviews,* Vol. 5, 48 pp.

Graves, C. and Hurd, R.G., 1983. Intermittent solution circulation in the nutrient film technique. *Acta Hortic.,* 133: 47–52.

Hardgrave, M., 1991. *Substrates and Recirculation,* Proc. Tomato Conference, 25 September. Horticulture Research International, Littlehampton, West Sussex, England, 14 pp.

Holder, R. and Cockshull, K.E., 1990. Effects of humidity on the growth and yield of glasshouse tomatoes. *J. Hortic. Sci.,* 65: 31–39.

Hunt Jr., E.R., Running, S.W. and Federer, C.A., 1991. Extrapolating plant water flow resistances and capacitances to regional scales. *Agric. For. Meteorol.,* 54: 169–195.

Jolliet, O., 1988. *Modélisation du comportement thermique d'une serre horticole,* Thèse de doctorat n°713. Département de physique de l'Ecole Polytechnique Fédérale de Lausanne, Lausanne, Switzerland, 247 pp.

Jolliet, O., 1992. *HORTITRANS, a greenhouse dynamic program for optimizing humidity and transpiration. User Manual.* Horticulture Development Council (HDC), Petersfield, Hants, England, 60 pp.

Jolliet, O., 1993. Bilan écologique de la production de tomates en serre. *Rev. Suisse Vitic. Arboric. Hortic.,* 25(4): 261–267.

Jolliet, O., 1994a. HORTITRANS, a model for predicting and optimizing humidity and transpiration in greenhouses. *J. Agric. Eng. Res.,* 57: 23–37.

Jolliet, O., 1994b. Critical Surface-Time: an evaluation method for life cycle assessment. In: *Integrating Impact Assessment into LCA.* SETAC, av. E. Mounier 83, 1200 Brussels, pp. 133–142.

Jolliet, O. and Bailey, B.J., 1992. The effect of climate on tomato transpiration in greenhouses: measurements and models comparison. *Agric. For. Meteorol.,* 58: 43–62.

Jolliet, O. and Bailey, B.J., 1993. *Climate effect on stomatal conductance of greenhouse tomato plants,* Report 1/93. EPFL-IATE/HYDRAM, CH-1015 Lausanne, 22 pp.

Jolliet, O. and Crettaz, P., 1997. *Critical Surface-Time 95, an Impact Life Cycle Assessment methodology, including exposure and fate.* Swiss Federal Institute of Technology, Institute of Soil and Water Management, EPFL, IATE-HYDRAM, CH-1015, Lausanne.Initially prepared for the workshop on impact assessment of the concerted action on LCA in agriculture.

Jolliet, O., Danloy, L., Gay, J.-B., Munday, G.L. and Reist, A., 1991. HORTICERN, an improved static model for predicting the energy consumption of a greenhouse. *Agric. For. Meteorol.,* 55, 265–294.

Jolliet, O., Bailey, B.J., Hand, D.J. and Cockshull, K.E., 1993. Tomato yield related to humidity and transpiration. In: *Workshop on Greenhouse Crop Models, 26–29 August 1991, INRA Bioclimatologie, Avignon. Acta Hortic.,* 328: 115–124.

Kimball, B.A., 1986. *Modular energy balance program including subroutines for greenhouse and the other latent devices,* A.R.S. 33. US Agricultural Research Service, Phoenix, AZ, 360 pp.

Kurata, K., Ibaraki, Y. and Goto, E., 1991. Micropropagation by nutrient mist supply. *Trans. ASAE,* 34(2): 621–624.

Lake, J.V., Postlethwaite, J.D., Slack, J. and Edwards, R.I., 1966. Seasonal variation in the transpiration of glasshouse plants. *Agric. Meteorol.,* 3: 187–196.

Marcelis, L., 1993. Simulation of biomass allocation in greenhouse crops – A review. *Acta Hortic.,* 328: 49–67.

Monteith, J.L., 1973. *Principles of Environmental Physics.* Contemporary biology. Edward Arnold, London, 241 pp.

Monteith, J.L., 1981. Evaporation and surface temperature. *Q. J. R. Meteorol. Soc.,* 107(451): 1–27.

Montero, J.I. and Segal, I., 1993. Evaporative cooling of greenhouses by fogging combined with natural ventilation and shading. In: M. Fuchs (Editor), *Proc. Workshop on Cooling System for Greenhouse,* Tel Aviv, 2–6 May 1993. Department of Agricultural Meteorology,Agricultural Research Organization, Bet-Dagan 50-250, Israel, pp. 89–98.

Morris, L.G., Neale, F.E. and Postlethwaite, J.D., 1957. The transpiration of glasshouse crops and its relationship to the incoming solar radiation. *J. Agric. Eng. Res.,* 2(2): 111–122.

Okuya, A. and Okuya, T., 1988. The transpiration of greenhouse tomato plants in rockwool culture and its relationship to climatic factors. *Acta Hortic.,* 230: 307–311.

Reist, A., Gysi, C. and Corbaz, R., 1991. Cultures hors-sol en circuit fermé: recommandations pour l'horticulture. *Le Maraîcher (CH),* 18: 7–18.

Seginer, I., 1994. Transpiration cooling of a greenhouse crop with partial ground cover. *Agric. For. Meteorol.,* 71: 265–281.

Sonneveld, C., 1988. Rockwool as a substrate in protected cultivation. *Horticulture in high technology era,* May 10–11.

Stanghellini, C., 1987. *Transpiration of greenhouse crops, an aid to climate management,* Doctoral Thesis. Agricultural University, Wageningen, Netherlands, 150 pp.

Stanghellini, C., 1993. Evapotranspiration in greenhouses with special references to mediterranean conditions. *Acta Hortic.,* 335: 295–304.

Stanghellini, C. and van Meurs, W.Th.M., 1992. Environmental control of greenhouse crop transpiration. *J. Agric. Eng. Res.,* 51: 297–311.

Steffen, K., 1988. Erfahrungen mit der Ebbe- und Flut-Bewässerung. *Dtsch. Gartenb.,* 21: 1308–1310.

Turner, N.C., 1991. Measurements and influence of environmental and plant factors on stomatal conductance in the fields. *Agric. For. Meteorol.,* 54: 103–106.

van Bavel, C.H.M., Damagnez, J. and Sadler, E.J., 1981. The fluid-roof solar greenhouse: energy budget analysis by simulation. *Agric. Meteorol.,* 23: 61–76.

van Bavel, C.H.M., Takakura, T. and Bot, G.P.A., 1985. Global comparison of three greenhouse climate models. *Acta Hortic.,* 174: 21–33.

van der Post, C.J., van Schie, J.J. and de Graaf, R., 1974. Energy balance and water supply in glasshouses in the West-Netherlands. *Acta Hortic.,* 35: 13–22.

Volkening, M. and Strauch, K.H., 1988. Steuerung der Bewässerung bei Topfpflanzen. *Dtsch. Gartenb.,* 21: 1316–1319.

Wohanka, W., 1991. Wasserentkeimung bei der Hydrokultur von Zierpflanzen. *Gartenbau,* 38(1): 43–46.

Yang, X., Short, T.H., Robert, D.F. and Bauerle, W.L., 1990. Transpiration, leaf temperature and stomatal resistance of a cucumber crop. *Agric. For. Meteorol.,* 51: 197–209.

Chapter 14

THE NUTRIENT CYCLE

S. DASBERG

INTRODUCTION

A greenhouse can be defined as a man-made ecosystem, although this definition is somewhat out of context. Ecosystems are usually recognized as natural systems, as parts of nature, where organisms interact as influenced by the natural environment. In a greenhouse, the organisms (the horticultural plants) are grown artificially and all competing organisms (weeds, fungi, etc.) are eradicated. The environment is artificially controlled by enclosure, heating, cooling, irrigation, fertilization, drainage, addition of carbon dioxide, etc. Therefore, to describe a greenhouse as an ecosystem seems far-fetched. On the other hand, there are some advantages in defining a greenhouse as an ecosystem. In a natural ecosystem, the boundaries are diffuse in many cases, while in the greenhouse the boundaries are well defined by the enclosure. Moreover, since plant growth is usually limited to one variety, the production can be monitored fairly easily; and the energy flow and its components are also much better defined and easier to quantify than in a natural ecosystem. Nutrient flow, the subject of this chapter, is also quantified much better than in natural systems, except for the case when the greenhouse is erected on natural soil and no artificial drainage system is installed. In that case, with natural drainage of excess water and nutrients into the vadose zone, the lower boundary of the system is diffuse and ill-defined. The inputs of chemicals in fertilizers and irrigation water and the output in harvested material are well-defined.

The greenhouse differs from other man-made agricultural systems by the high value of the product. Gross income from flower culture or high-quality vegetables may reach values of 20–40 dollars per square metre. The cost of fertilization is in most cases only 1–2% of this value (de Jong and Maurits, 1990). In the case of field crops, the cost of fertilization may reach 10–20% of the gross value of the product. This means that in greenhouse culture there is little incentive for the grower to economize on fertilization. Wasteful application, as long as the effects are not harmful, makes more sense economically, since the added production caused by added fertilizer will always cover the cost. Moreover, because of the high production per unit area, fertilizer application rates are in any case very high. Because of these reasons, the excess of fertilizer application over plant uptake tends to be high in protected cultivation.

GREENHOUSES ON NATURAL SOILS

Nitrogen cycle

The nitrogen cycle in soils has been described and reviewed many times during the last decades (for a recent review, see Keeney, 1989). Numerous sources of nitrogen exist, and there are several storage pools (in soil organic matter, in soil solution, exchangeable and fixed ammonium ions, in plants and plant residues and in microorganisms) and several processes occur simultaneously which transform nitrogen from one form to another (mineralization, immobilization, nitrification, denitrification, N_2 fixation, and nitrate leaching). The relative importance of these processes varies according to the agriculture system and according to the environmental conditions. Nitrogen can be easily leached from the soil in the form of nitrate by excess water. This nitrate has been added as fertilizer, and transformed from soil organic matter and NH_4^+ by the processes of mineralization and nitrification (brought about by soil micro-organisms).

In horticulture, and in particular in greenhouse culture, there is more nitrate leaching than in agricultural systems (Juergens-Geschwind, 1989). There are several reasons for this:

(1) Higher rates of nitrogen application because of the high yields obtained, and because of the relatively low cost of fertilizer in the total production costs, as stated above.

(2) Large quantities of organic material in the form of manure, compost or crop residues are added to the soil. This organic matter eventually decomposes, the end product being nitrate in the soil solution prior to leaching.

(3) The tendency to apply more irrigation water than needed for evapotranspiration, for the purpose of leaching of salt and in order to prevent osmotic stress.

The excess nitrate may reach surface water directly, or leach into the subsoil and eventually reach the aquifer, depending on the hydrological conditions. Many cases are documented in the literature showing high nitrate contents of ground water caused by intensive irrigated crops (Pratt, 1984; Keeney, 1989). As an example: a detailed study of an irrigated area of 5000 ha in Spain showed that, with an average nitrogen fertilizer application of 291 kg ha^{-1} yr^{-1}, 66 kg ha^{-1} was leached as nitrate into the drainage water (Bellot and Golley, 1989). Some of these losses are inevitable. Plant roots are not able to use all the nitrate in the soil solution, and even with nitrogen fertilization below requirement, some leaching occurs, as was shown for citrus orchards (Dasberg et al., 1984).

The environmental concern for this phenomenon has increased in the developed countries. Management practices are available to increase the efficiency of water and fertilizer use. However, the incentive for the grower to use these practices is not great. The profit which can be obtained from over-irrigation and over-fertilization of a high-value crop can be large. Until recently, no enforceable public policy existed for the prevention of nitrate pollution. The possible measures to be adopted are regulation (problems of monitoring); effluent charges or taxes; or subsidies to farmers who implement practices that reduce leaching (Follett and Walker, 1989).

The hazard of nitrate pollution from greenhouses is large. If only a small part of the total area is used for greenhouses, the damage to the environment is relatively small. With large concentrated areas of greenhouses, the effect on ground or surface water may be large. A possible way to decrease nitrate leaching is by using high-quality irrigation water, with a very low salt content. This water can be re-used, and no leaching for removal of excess salt is required. This can only be practiced in soilless cultures, which are discussed in detail below (p. 329).

Phosphorus cycle

Only a small part of the soil phosphorus is in the soil solution (0.05–0.3 mg ℓ^{-1}), partly in mineral and partly in organic form. In greenhouses, higher concentrations are often measured, up to 5 mg ℓ^{-1}. Most of the phosphorus reserves in the soil are relatively inert. Moreover, a large part of the phosphorus supplied by fertilizers is fixed by the soil (Ozanne, 1980). Successful crop-growing depends on continued replenishment of phosphorus in soluble form for plant uptake. The fixing power of most soils is very large. Excess leaching of phosphorus from the soil may occur only in very special cases: on shallow soils, or on sandy soils, with relatively low capacity for fixing phosphorus, or when large amounts of animal manure have been applied continuously for many years. No direct health hazards are reported as a result of excess phosphorus in drinking water. However, ecological hazards, because of the prevalence of certain algae in surface waters or lakes related to excess phosphorus, have been reported (Rigler, 1973).

Potassium cycle

The potassium phases in the soil and their interactions are less complicated than those of the nitrogen cycle. Potassium is active only in the ionic form, K^+, and the governing phase for potassium transformations is K^+ in the soil solution. Plant uptake is from the soil solution; excess potassium is leached into the soil solution and there is a dynamic chemical equilibrium between the potassium in the soil solution and the exchangeable potassium. Potassium applied in fertilizer is highly soluble and enters the soil solution. There is also a reverse process of potassium moving from the soil solution to nonexchangeable form, which is called fixation. It is difficult to generalize and quantify these processes of potassium transformation, since much depends on soil properties (clay content, nature of soil minerals) and climatic conditions (temperature, rainfall).

Leaching of potassium depends, first of all, on the

potassium-fixing power of the soil. In soils with a high clay content most of the excess potassium is fixed by the soil and little leaching occurs. In sandy soils, higher losses of potassium below the root zone are prevalent. Much depends on the rate of fertilizer application and on the crop grown. Leaching of potassium is of less environmental concern than leaching of nitrate, since no health hazard is known. The annual loss by leaching may vary from negligible amounts in loamy soils to 10–20 kg K ha^{-1} in sandy soils with arable crops (Bertsch and Thomas, 1985). High-value crops use large amounts of potassium. Average-yielding vegetable crops require 100–500 kg K ha^{-1} (Geraldson, 1985). An average crop of greenhouse tomatoes in the Netherlands may yield 500 t ha^{-1} (Sonneveld, 1989), requiring more than 2000 kg K ha^{-1}. Under these conditions of highly fertile soils saturated with potassium, the leaching of excess potassium may reach an order of magnitude of 100 kg K ha^{-1} yr^{-1}.

Table 14.1
Salinity tolerance of greenhouse crops[1]

Crop	Threshold EC ($dS\ m^{-1}$)	Salinity yield decrease (% dS^{-1} m)
Beans (*Phaseolus* spp.)	0.5	20
Carnation (*Dianthus caryophyllus*)	1.0	6
Chrysanthemum (*Dendranthema grandiflora*)	0.6	16
Cucumber (*Cucumis sativus*)	1.5	15
Gerbera jamesonii	0.6	25
Lettuce (*Lactuca sativa*)	0.6	5
Pepper (*Capsicum* spp.)	0.5	17
Tomato (*Lycopersicon esculentum*)	1.8	9

[1] As based on data of Sonneveld (1988).

NUTRIENT CYCLES IN SOILLESS CULTURES

Originally, most soilless cultures (for details see Chapter 4) were implemented with open drainage systems. This means that the excess irrigation water and part of the dissolved nutrients flow directly into the soil. The nutrient cycle in this case is very similar to the nutrient cycle on natural soils, as detailed in the previous section. More advanced growing systems have artificial drainage features – the excess water is collected and may be reused for irrigation. This re-use of drainage water, although attractive for reasons of economics (less waste of water and fertilizer) and environment (less pollution of surface and ground water), has some drawbacks related to problems of salinity, phytopathology and the need for intensive monitoring of the concentrations of nutrient elements.

Salinity problems in greenhouses

Greenhouse crops, like other agricultural crops, show different tolerance to salinity. Salinity of the irrigation water is expressed in units of electrical conductivity (EC, $dS\ m^{-1}$). Salt tolerance is usually denoted by two values – the threshold EC below which no damage occurs, and the percentage yield decrease per unit increase of EC above the threshold value. Table 14.1 gives some data on greenhouse crops, as summarized by Sonneveld (1988). Based on these data, and on the quality of the irrigation water, a leaching fraction (drainage water as a percentage of applied water) can be estimated. This estimate has to be taken into account together with the non-uniformity of water application and plant water use (not all plants are alike in their development and transpiration).

In Westland, the glasshouse district located in the vicinity of Naaldwijk, the Netherlands, water of three different qualities is potentially available for irrigation: river water, currently in use ($EC = 2\ dS\ m^{-1}$); high-quality drinking water ($EC = 1\ dS\ m^{-1}$) or rainwater collected from the roofs of the glasshouses and stored. Detailed calculations of the necessary leaching fractions were carried out by P. Hamaker (personal communication) based on the above assumptions for the local climatic conditions, on the effect of water quality on the volume of drainage needed from one hectare of greenhouse. The results are given in Table 14.2. These data show that, with a moderately sensitive crop like tomato, virtually no drainage is necessary if high-quality irrigation water is used. With free drainage, large amounts of water are wasted, even when rainwater is applied. If the drainage water is re-used, even a relatively small rainwater reservoir (1000 m^3 ha^{-1}) can greatly decrease the drainage needed for the conditions prevailing in Westland.

These data are adduced in order to illustrate the importance of primary water quality for minimizing the drainage necessary. The higher the water quality, the higher the cost to the farmer. The benefit can be measured as less waste of fertilizer, probably much less than the actual cost. There is, however, a large benefit to

Table 14.2
Drainage requirement[1] ($m^3\ ha^{-1}\ yr^{-1}$) for growing tomatoes or pepper on rock-wool, assuming an annual water requirement of $6000\ m^3\ ha^{-1}$

Drainage system	Leaching fraction	Drainage requirement		
		No rainwater collection	$1000\ m^3$ rainwater collection	$2000\ m^3$ rainwater collection
Tomato				
Free drainage	0.3[2]	2625	1850	1750
Re-use	0.14[2]	1000	700	75
	0[3]	0	0	0
Pepper				
Free drainage	0.4[2]	3300	2450	2200
Re-use	0.27[2]	2275	450	175
	0.09[3]	600	125	50

[1] Based on data of Hamaker (personal communication, 1989).
[2] Sodium in irrigation water: $2\ mmol\ \ell^{-1}$.
[3] Sodium in irrigation water: $1\ mmol\ \ell^{-1}$.

the environment. The data in Table 14.2 are based on an annual water application of $6000\ m^3\ ha^{-1}$. In the worst case, more than 50% of this amount has to be added for necessary drainage in order to sustain production. With proper re-use of high-quality water, the drainage requirement becomes negligible. The drainage water is not only more concentrated in salts not taken up by the plant, but also contains nutrients and pesticides, posing a hazard to the environment.

Pathogens in drainage water from soilless cultures

Re-use of drainage water from soilless cultures poses the hazard of rapid spread of plant diseases. Proper re-use of drainage water implies its freedom from pathogens of viral, bacterial or fungal nature. Several methods are available and have been developed for the treatment of drinking water, or waste water. Not all methods are suitable for the treatment of drainage water because of the large volumes involved and because of the cost. Recently, some disinfection methods were tested on laboratory and semi-commercial scales (Runia, 1988). Several root pathogens were tested – cucumber mosaic virus, tomato mosaic virus, *Fusarium* and *Verticillium.* Only partial disinfection was achieved by chlorination, iodination, copper and silver ions, which pose environmental problems. Ozone treatment, ultrafiltration and heat treatments were effective. Ultrafiltration has the disadvantage that only 70–80% of the drainage water is recovered, while 20–30% is rejected, whereas with proper heat treatment all the drainage water is recovered. More detailed studies showed that treatment for 10 s at 90ºC was sufficient to eliminate tobacco mosaic virus and *Verticillium dahliae.* For *Fusarium* species, longer exposure times were needed (Runia et al., 1988). This system is suitable for disinfection of drainage water from rock-wool culture (25% drainage, $30\ m^3\ ha^{-1}\ day^{-1}$). In the nutrient-film technique, the rate of water flow reaches $1000\ m^3\ ha^{-1}\ day^{-1}$, which is beyond the capacity of the system. Lately, it has been shown that ultraviolet radiation is also an effective method of water disinfection. The water has to be pretreated by filtration in order to achieve good penetration of the radiation (Runia, 1992).

Control of water and nutrient supply

In order to minimize the waste of water and nutrients, both irrigation and fertilization should be applied with high precision according to the actual demands of the plant based on its growth stage. The question is whether there is enough information available for accurate systems of irrigation and fertilization. The technology for precise application is available: accurate water meters activated by solenoid valves and controlled by microprocessors; pressure-compensated accurate drippers and sprayers which apply water at a constant rate in spite of variations in water pressure; on-line control of total salt content (EC) of nutrient solutions or concentration of individual ions by specific electrodes. However, despite many experiments with specific electrodes, no system using it is currently in practical use.

The theory of evaporation and transpiration, and the effects of environmental and crop factors, have been thoroughly researched. The question whether this research is sufficient to provide accurate water-scheduling for specific crops during the growth cycle for specific environmental conditions is dealt with in Chapter 13 (pp. 314–322).

Research on nutrient uptake by horticultural crops is less extensive. A review by van Goor et al. (1988) concluded that growth, water absorption and mineral uptake are closely connected. The plant is a complicated organism, consisting of roots, stems, leaves and fruits. Not only uptake is important, but

Table 14.3
Daily nitrogen, phosphorus and potassium consumption rates of tomatoes (kg ha^{-1} day^{-1}) grown under drip irrigation[1]

Time after emergence (days)	Processing tomatoes[2]			Greenhouse tomatoes[3]			Fresh tomatoes[4]		
	N	P	K	N	P	K	N	P	K
1–10	0.10	0.00	0.10	0.15	0.00	0.10	0.30	0.01	0.40
11–20	0.50	0.05	0.30	0.35	0.07	0.15	0.30	0.02	0.50
21–30	1.00	0.16	2.00	0.75	0.13	0.15	0.30	0.03	0.50
31–40	2.80	0.19	2.30	1.25	0.15	0.30	0.40	0.03	0.50
41–50	4.50	0.75	8.00	2.10	0.35	6.00	0.40	0.03	0.55
51–60	6.50	0.80	8.50	2.50	0.47	6.00	0.45	0.04	0.55
61–70	7.50	1.80	9.00	2.60	0.50	1.90	0.50	0.04	0.60
71–80	3.50	0.50	4.50	2.85	0.53	2.50	1.70	0.18	2.20
81–90	5.00	0.50	9.20	3.65	0.60	6.00	1.80	0.22	4.80
91–100	8.00	0.89	9.00	6.15	0.96	12.50	1.30	0.10	2.90
101–110	–	–	–	7.70	1.06	13.00	2.70	0.30	5.70
111–120	–	–	–	6.35	1.28	8.20	4.60	0.60	7.80
121–130	–	–	–	0.10	0.50	0.50	3.90	0.45	7.00
131–150	–	–	–	0.05	0.12	0.50	2.70	0.17	2.00
151–200	–	–	–	1.70	0.52	7.00	–	–	–
Total (kg ha^{-1})	393	59	520	450	95	943	250	24	37

[1] Bar-Yosef (1991).
[2] Variety VF H82-1-2; density (ha^{-1}), 50 000; soil, clay; yield (t ha^{-1}), 160.
[3] Variety Grenadier; density (ha^{-1}), 28 000; soil, sandy; yield (t ha^{-1}), 153.
[4] Variety B_{75}; density (ha^{-1}), 12 000; soil, sandy; yield (t ha^{-1}), 127.

also translocation between organs at different growth stages. There are large differences in the uptake ratios of different elements. This means that the composition of the nutrient solutions must be adapted to the growth stage of the crop. This topic of accurate plant nutrition based on the actual demand of the plant is still in its infancy, and much more research is needed.

Data on this topic were published by Bar-Yosef (1991) based on detailed field experiments (Table 14.3). These may serve as a first approximation for the minimal amounts of nitrogen, phosphorus and potassium that must be added to the irrigation water to satisfy plant demand. Extrapolation of these data to different environmental conditions – ambient temperatures, radiation intensities, root temperatures – should be done with caution.

REFERENCES

Bar-Yosef, B., 1991. Fertilization under drip irrigation. In: D.E. Palgrave (Editor), *Fluid Fertilizer Science and Technology.* Marcel Dekker, New York, pp. 285–329.

Bellot, J. and Golley, F.B., 1989. Nutrient input and output of an irrigated agro-ecosystem in an arid Mediterranean landscape. *Agric. Ecosys. Environ.,* 25: 175–185.

Bertsch, P.M. and Thomas, G.W., 1985. Potassium status of temperate region soils. In: R.D. Manson (Editor), *Potassium in Agriculture.* ASA, Madison, WI, pp. 131–163.

Dasberg, S., Erner, Y. and Bielorai, H., 1984. Nitrogen balance in a citrus orchard. *J. Environ. Qual.,* 13: 353–361.

de Jong, J. and Maurits, J., 1990. *Quantitatieve informatie voor de glastuinbouw 1989/1990.* Consulentschappen Aalsmeer en Naaldwijk.

Follett, R.F. and Walker, D.J., 1989. Ground water quality concerns about nitrogen. in: R.F. Follett (Editor), *Nitrogen Management and Ground Water Protection.* Elsevier, Amsterdam, pp. 1–20.

Geraldson, C.M., 1985. Potassium nutrition of vegetable crops. In: R.D. Manson (Editor), *Potassium in Agriculture.* ASA, Madison, WI, pp. 915–929.

Juergens-Geschwind, S., 1989. Ground water nitrate in Europe; relationships to land use patterns. In: R.F. Follett (Editor), *Nitrogen Management and Ground Water Protection.* Elsevier, Amsterdam, pp. 75–139.

Keeney, D.R., 1989. Sources of nitrate to ground water. in: R.F. Follett (Editor), *Nitrogen Management and Ground Water Protection.* Elsevier, Amsterdam, pp. 23–35.

Ozanne, P.B., 1980. Phosphate nutrition of plants, a general treatise. In: F.E. Khasawneh, E.C. Sample and E.J. Kamprath (Editors), *The*

Role of Phosphorus in Agriculture. ASA, Madison, WI, pp. 559–590.

Pratt, P.F., 1984. Nitrogen use and nitrate leaching in irrigated agriculture. In: R.D. Hauck (Editor), *Nitrogen in Crop Production.* ASA, Madison, WI, pp. 204–250.

Rigler, F.H., 1973. A dynamic view of the phosphorus cycle in lakes. In: E.J. Griffith (Editor), *Environmental Phosphorus Handbook.* Wiley, New York, pp. 539–572.

Runia, W.Th., 1988. Elimination of plant pathogens in drainage water from soilless cultures. In: *Proc. 7th Int. Congress on Soilless Culture.* ISOSC, Wageningen, Netherlands, pp. 428–443.

Runia, W.Th., 1992. Drain water disinfection with ultra-violet light gives good results. *Vakbl. Bloemisterij,* 10: 44–47. In Dutch.

Runia, W.Th., Van Os, E.A. and Bollen, G.J., 1988. Disinfection of drainwater from soilless cultures by heat treatment. *Neth. J. Agric. Sci.,* 36: 231–239.

Sonneveld, C., 1988. The salt tolerance of greenhouse crops. *Neth. J. Agric. Sci.,* 36: 63–73.

Sonneveld, C., 1989. Rockwool as a substrate in protected cultivation. *Chron. Hortic.,* 29: 33–37.

van Goor, R.J., de Jager, A. and Voogt, W., 1988. Nutrient uptake by some horticultural crops during the growing period. In: *Proc. 7th Int. Congr. on Soilless Culture.* ISOSC, Wageningen, Netherlands, pp. 163–175.

Chapter 15

POTENTIAL PRODUCTION WITHIN THE GREENHOUSE ENVIRONMENT

Hugo CHALLA and Menno BAKKER

INTRODUCTION

Potential crop production

The annual production per square meter of different crops in greenhouses at different locations and in different cultivation systems varies greatly. Also there has been, for many years, a general trend towards higher productivity through improvement of crop varieties, cultivation techniques and equipment (van der Velden et al., 1990). Against the background of these general observations the question arises how to compare productivity of different crops and crop varieties at different locations, and what maximum production levels may be attained in greenhouse cultivation.

In order to answer these questions reference can be made to the work in Wageningen of de Wit and his co-workers (de Wit et al., 1979), who introduced the concepts of *potential production* and *production levels*. According to Penning de Vries (1982) production systems may be classified into four levels, depending on the factors that limit growth and production:

Level 1: radiation, temperature;
Level 2: as Level 1, + availability of water;
Level 3: as Level 2, + nitrogen;
Level 4: as Level 3, + other minerals, particularly phosphate.

The philosophy behind this classification is that in modern agriculture only radiation and temperature, at least for crops growing outdoors, cannot be controlled and that in more extensive agriculture water, nitrogen and finally other nutrients, respectively, may limit the level of production.

According to de Wit et al. (1979) the potential production rate of a crop may be defined as "the growth rate of a closed, green crop surface, optimally supplied with water and nutrients, in a disease-free environment under the prevailing weather conditions". In this definition, potential production is related to the prevailing weather conditions, and thus linked to the particular conditions during a given period. Potential production, however, may also be considered as an abstraction, independent of the instantaneous weather, but rather as a property or characteristic of a specific site (de Wit and van Heemst, 1976; de Wit et al., 1979; Versteeg and van Keulen, 1986), related to the average annual weather conditions.

In the latter context the concept of potential crop production may be considered as a reference value, which makes it possible to compare production of different crops and production at different locations of the world in relation to climate. The concept may, in addition, be used to qualify an agricultural system in terms of the discrepancy between (average) actual and potential production, and as such provides a target for agricultural research.

Potential crop production in greenhouses

In order to quantify the theoretical maximum production level in greenhouse cultivation for a given location the concepts of de Wit and his co-workers can be used, but should be worked out to a certain extent in relation to the particular conditions that play a role in greenhouse cultivation. Obviously, these production systems are at Production Level 1, where nutrients and water are non-limiting. For outdoor crops allowance is usually made for the length of the growing season as determined by temperature. In modern greenhouses temperature is a controlled factor, and therefore low temperature does not play a role.

High temperature, however, may limit the length of the growing season in warm climates. In northern latitudes, low irradiance limits the growing season in winter for specific crops. In this chapter, however, potential production is quantified in complete annual cycles, without considering the practical implications of temperature or radiation extremes: the crop is supposed to continue to produce at its potential rate at constant temperature. Predictions and comparisons with actual yield can subsequently be made by selecting the periods when production of a particular crop is feasible (see below, p. 342).

In the definition of potential production by de Wit et al. (1979) previously quoted, the crop referred to has a green, closed canopy. Obviously, greenhouse crops often are not fully closed: many crops are grown in rows, and particularly in the young stage plants are often widely spaced. Potential production therefore should be considered as an abstraction which, in the majority of cases, will become reality only over a certain period of time. It can best be imagined by thinking of a non-growing crop, starting to produce each day of the year at the same fixed initial state.

Potential production as discussed above is primarily determined by diurnal radiation, because of its impact on crop photosynthesis. Because the properties of the greenhouse cover greatly affect radiation inside the greenhouse, it is necessary to define a reference greenhouse. The transmission of light through the greenhouse cover and the reference greenhouse that is used in the present study will be discussed below (p. 338).

Another detail of potential production in greenhouses that needs further elaboration is the concentration or partial pressure of carbon dioxide in the greenhouse air, because of its significance for crop photosynthesis. In this chapter the term carbon dioxide (partial) pressure will generally be used rather than the term carbon dioxide concentration, because this is what crops respond to and what is measured with infrared gas analysers. Depending on the rate of air exchange, the carbon dioxide pressure inside a greenhouse often will deviate from the pressure outside, due to crop photosynthesis and carbon dioxide enrichment, and therefore it is hard to provide a representative value for it. In order to account for such variations two pressures of carbon dioxide were considered, a "normal" atmospheric pressure ($34\,\mathrm{Pa} \approx 340\,\mu\ell\,\ell^{-1}$) and a higher value ($100\,\mathrm{Pa} \approx 1000\,\mu\ell\,\ell^{-1}$). The higher carbon dioxide pressure approaches saturation for photosynthesis (Challa and Schapendonk, 1986), and is quite often considered as a safe maximum level for crop growth; the lower level may be considered a minimum that could and should be maintained under all conditions in greenhouses provided with adequate facilities for carbon dioxide enrichment (Nederhoff, 1990).

Another difficulty with the choice of a suitable reference for potential production is the question whether characteristics of C_3 or C_4 crops are to be selected. Because the vast majority of greenhouse crops are C_3 species, C_4 crops were not considered. Other complications result from the wide range of crops cultivated in greenhouses with, probably, widely diverging photosynthetic properties, and from activities such as application of whitewash on the cover, the use of screens for reduction of light or day length control, and the application of artificial light. Such cases could be dealt with in more detailed studies for specific situations, but are not considered here. In comparisons of actual with potential production these factors, however, should be considered because they may interfere strongly.

Objectives and general set-up of the study

The objective of this study is to extend the concept of potential production to greenhouse cultivation, and to make it useful – for example to horticulturists and plant breeders as a tool to quantify performance in greenhouse culture, to research leaders and policy makers to provide insight into the progress that could be obtained in increasing yield per unit area, and in the evaluation of new locations for greenhouse production. The theoretical background is treated superficially, but is well referenced; the emphasis is on the practical implementation of the approach at different locations in the world. These locations are selected rather arbitrarily from regions in the world, at different latitudes and with different climates, where greenhouse production plays a role (Table 15.1). An example of how the simulation output could be used is presented for tomato (*Lycopersicon esculentum*) in the Netherlands (p. 342), where also a comparison with actual production is presented. Furthermore a greatly simplified model derived from our calculations, a so-called meta-model (Kleijnen, 1987), is presented, which enables reasonably accurate calculations of potential production in greenhouses based on daily global radiation outside the greenhouse and its distribution between direct and diffuse radiation.

Table 15.1
Selected locations, their latitude, annual radiation and estimated potential dry matter production with carbon dioxide pressure of 34 Pa ($\approx 340\,\mu\ell\,\ell^{-1}$) and 100 Pa ($\approx 1000\,\mu\ell\,\ell^{-1}$)

Location	Country	Latitude (°N)	Annual radiation ($MJ\,m^{-2}$)	Annual production ($g\,m^{-2}$)	
				34 Pa CO_2	100 Pa CO_2
Yangambi	Zaire	1	6254	5594	7326
Miami	USA	26	7091	6086	7998
Gilat	Israel	31	8081	6479	8578
Los Angeles	USA	34	6860	5857	7696
Tokyo	Japan	35	4033	4004	5182
Avignon	France	49	5367	4754	6216
de Bilt	The Netherlands	52	3450	3284	4274
Stockholm	Sweden	59	3085	2907	3773

Scheme 15.1
Procedure for the calculation of daily crop photosynthesis and related dry matter production

Input	Operation	Output example
Latitude	Generate annual course of extra-terrestrial daily radiation integral ↓	Fig. 15.1a
Day number	Generate diurnal course of extra-terrestrial radiation, solar elevation and azimuth ↓	Fig. 15.1b
Average daily radiation (Fig. 15.2)	Generate diurnal course of diffuse and direct PAR outside the greenhouse ↓	Fig. 15.5
Greenhouse properties (Fig. 15.4)	Generate diurnal course of diffuse and direct radiation above the canopy ↓	Fig. 15.5
Leaf Area Index, extinction coefficient	Extinction of diffuse and direct radiation inside the canopy; absorbed radiation of sunlit and shaded leaves ↓	Fig. 15.1c
Photosynthesis–light response curve of individual leaves	Rate of CO2 assimilation of sunlit and shaded leaves ↓	
	Integration of instantaneous CO_2 assimilation of individual leaves over canopy height ↓	Fig. 15.7
	Integration over the day to obtain total daily CO_2 assimilation of the canopy ↓	
Conversion yield, crop structural dry weight, temperature	Daily dry matter production after correction for maintenance respiration	Fig. 15.8

[1] Modified after Kropff and Spitters (1987).

The theoretical background for the calculation of potential crop production is essentially the same as that followed by de Wit and co-workers (de Wit, 1965; Penning de Vries and van Laar, 1982; Spitters et al. 1989), but was extended by taking into account the light transmission through the greenhouse cover (Bot, 1983), and the effect of carbon dioxide concentration on leaf photosynthesis (Challa and Schapendonk, 1986). The method followed can be summarized in four steps (see Scheme 15.1 for details):

(1) An average diurnal course of photosynthetically active radiation (PAR, direct and diffuse) is generated, for each location and for each day, from average daily global radiation figures throughout the year (Fig. 15.1).
(2) From this and the diurnal transmission properties of the greenhouse (separately for direct and diffuse radiation), diurnal PAR inside the greenhouse is generated for a standard multi-span Venlo glasshouse (see Table 15.2 for characteristics).

Fig. 15.1. Calculated curves for (a) annual course of extra-terrestrial daily radiation integral, (b) diurnal course of extra-terrestrial radiation, and (c) extinction of diffuse radiation inside the canopy.

(3) Light distribution inside a standard crop canopy (a crop with spherical distribution of leaf angle and leaf area index (LAI, defined as one-sided leaf area divided by greenhouse area) of 3 is derived for the radiation conditions obtained in the previous step and used for the calculation of photosynthesis at different levels within the crop; daily crop photosynthesis is subsequently derived by integration over the total leaf area and over the whole day.

(4) Daily crop photosynthesis is converted to dry matter, after subtraction of a fixed rate of maintenance respiration (based on the selected leaf area index, and leaf area ratio).

RADIATION

Introduction

In order to predict potential production in greenhouses in relation to daily radiation, account has to be taken of spatial and temporal variations in leaf irradiation, because of the non-linear response of leaf photosynthesis to incident radiation. For many locations in the world values of average global radiation per day, per five or per ten days is available, but information on its diurnal course is scarcer. Likewise information about the diurnal distribution of radiation between direct and diffuse radiation is also scarce. Because diffuse and direct radiation are transmitted differently through the greenhouse cover and within the crop, this information is essential. In the following sections a procedure for the calculation of daily photosynthesis of a crop within a greenhouse, which takes these aspects into account, will be discussed.

Daily radiation as related to latitude and season

The annual course of average daily global radiation for the locations considered (the inputs used for simulation) is given in Fig. 15.2. The higher the latitude, the larger, in general, is the annual variation. Near the equator the minimum radiation is about 80% of the maximum daily radiation; in the subtropics this ratio is about 30%, in the Netherlands about 10% and in Sweden only 5%. Not only the ratio, but also the absolute radiation levels, are low in the winter season. Of course these differences are not only caused by the average radiation level, but also by differences in day length. In some cases (Israel, Japan) the annual radiation pattern depicted is less smooth than in other countries. This is, at least partly, caused by the availability of more detailed data, covering shorter intervals, and as a result more random variation in the average daily radiation over that period.

Diurnal radiation patterns can be generated from daily totals of global radiation, as was done by Spitters et al. (1989). In this procedure radiation at each moment of the day is derived from the theoretical maximum radiation that would be obtained if there were no absorption by the atmosphere (extra-terrestrial radiation). The ratio of average global daily radiation to the extra-terrestrial radiation integral of that day is defined as the average daily atmospheric transmission. Taking into account the change in average instantaneous atmospheric transmission as a function of sun elevation, an average diurnal course of global radiation outside the greenhouse can be generated.

As has been pointed out before, the proportion of diffuse radiation (the ratio of diffuse to total

Table 15.2
Characteristics of the reference greenhouse, a multi-span Venlo glasshouse with a N–S orientation

Property	Value	Property	Value
Roof slope	26°	Width of gutters	
Refraction index	1.5	At lower side	.130 m
Glass thickness	4 mm	At upper side	.220 m
Power absorbtion coefficient	$5\,m^{-1}$	Height of bars	.030 m
		Width of bars	.020 m
Height of ridges		Separation of bars	.730 m
Lower half	.097 m	Height of girders	.440 m
Upper half	.100 m	Width of girders	.040 m
Width of ridges		Separation of girders	3.000 m
At lower side	.057 m		
At upper side	.025 m	Diffuse transmissivity	.639
Height of gutters			
Lower half	.035 m		
Upper half	.050 m		

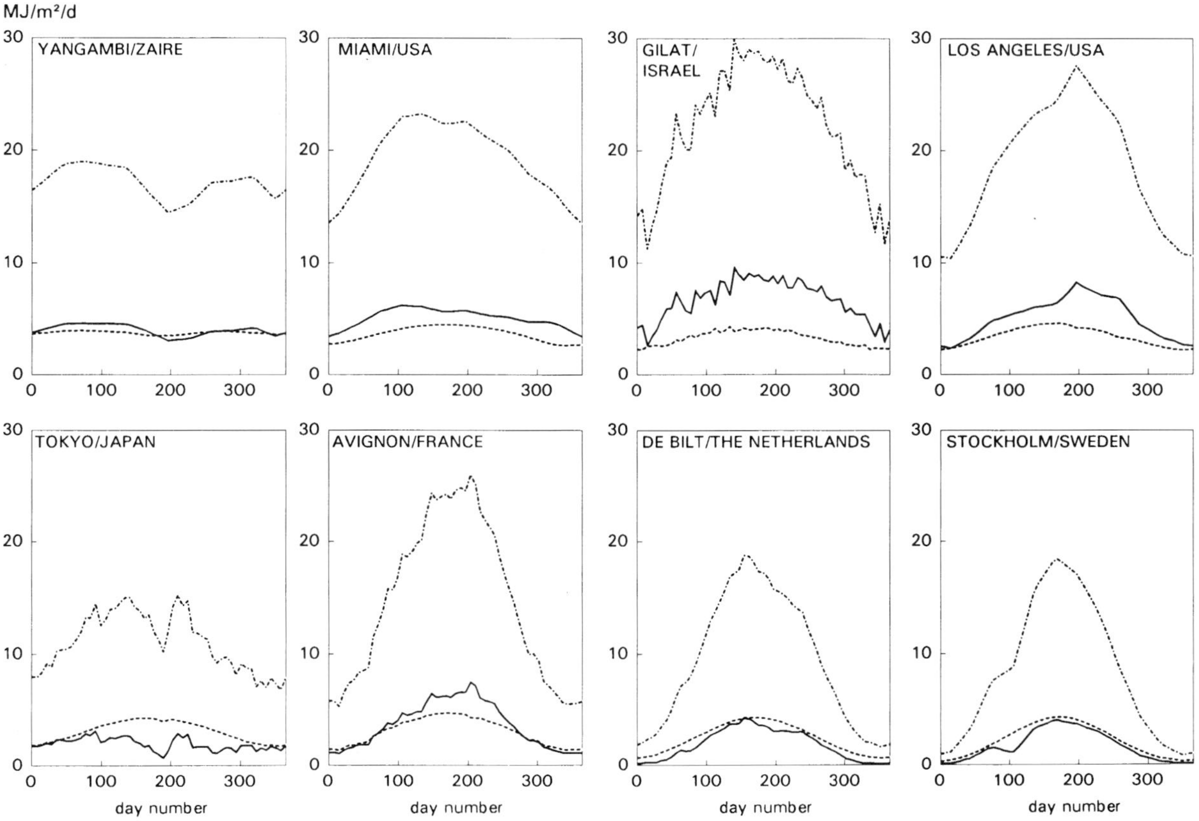

Fig. 15.2. Annual course of daily average global radiation (dash–dotted curve) and calculated diffuse (dashed) and direct (solid) photosynthetically active radiation outside the greenhouse, for the locations considered.

radiation) at different times of the day is an important characteristic because direct and diffuse radiation are transmitted differently by the greenhouse cover and inside the crop canopy. Spitters et al. (1986) described an empirical relation between atmospheric transmission and the proportion of diffuse radiation for hourly radiation values (Fig. 15.3). Their relationship can be used for hourly data that have actually been measured, but not for climatic means (compare the relation of Spitters et al. in Fig. 15.3 with observed monthly means according to Iqbal (1983)).

This discrepancy was remedied by the following procedure to add simulated variability to our data for climatic means. We worked with two atmospheric transmissivities, one increased by 0.23 and one decreased by 0.23 in comparison with the observed climatic mean. After using the procedure of Spitters et al., incoming diffuse and direct radiation were averaged, and the mean proportion of diffuse radiation was calculated. Atmospheric transmission cannot vary beyond certain limits, and therefore atmospheric transmission was allowed to vary only between a minimum of 0.18 and a maximum of 0.75. To ensure agreement with the observed means, we altered the relative weights of the two values.

The relation obtained in this way between atmospheric transmission and the proportion of diffuse radiation for long-term average radiation values differs greatly from that for hourly radiation values, except for high and low atmospheric transmission, where both curves come together (Fig. 15.3). The results obtained with this method agree reasonably with those observed by Iqbal (1983).

Among the locations included in this study, the highest calculated proportions of diffuse radiation are for Tokyo, Zaire and the northern countries, the Netherlands and Sweden. The highest proportions of direct radiation are found in Israel and Florida. Later in this chapter it will become clear that these differences affect the relation between global radiation and potential production (p. 340).

For potential production only photosynthetically

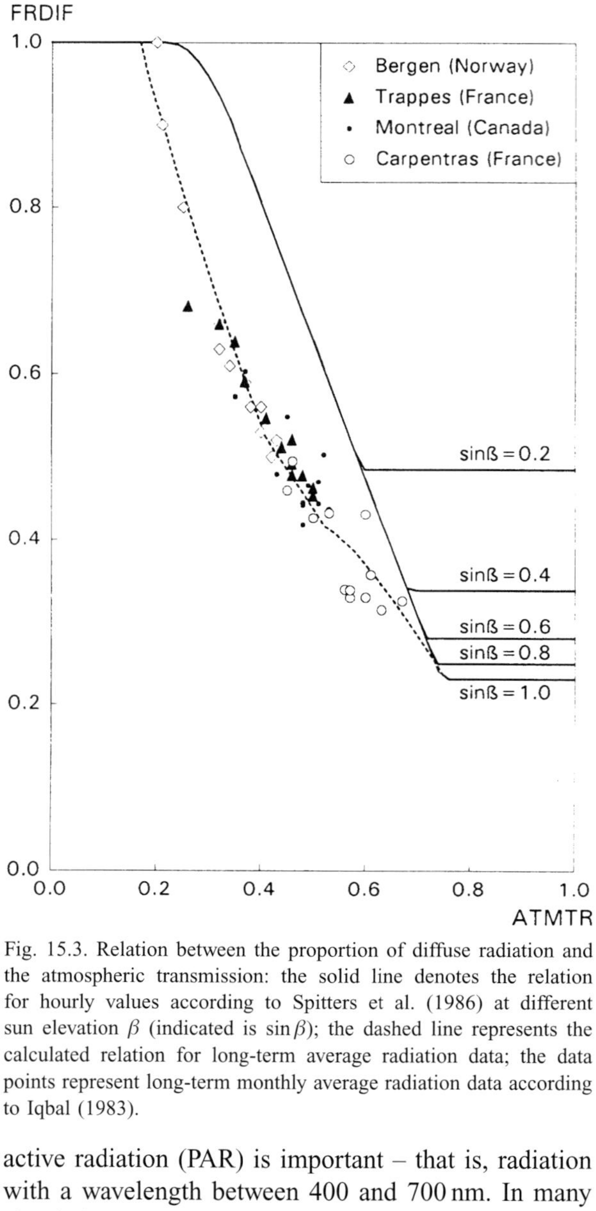

Fig. 15.3. Relation between the proportion of diffuse radiation and the atmospheric transmission: the solid line denotes the relation for hourly values according to Spitters et al. (1986) at different sun elevation β (indicated is $\sin\beta$); the dashed line represents the calculated relation for long-term average radiation data; the data points represent long-term monthly average radiation data according to Iqbal (1983).

active radiation (PAR) is important – that is, radiation with a wavelength between 400 and 700 nm. In many simulation studies PAR is estimated as 50% of global radiation. This percentage, based on Szeicz (1974), however, is for radiation between 300 and 700 nm. A better estimate for the percentage PAR (by definition 400–700 nm) is 45% (Moon, 1940; Meek et al., 1984), a value adopted in our calculations.

Transmission of radiation through the greenhouse cover

The radiation level inside the greenhouse is lower than outside, due to absorption and reflection by the greenhouse cover. For diffuse radiation daily and annual variations in the transmission coefficient of a given greenhouse may be ignored, but for direct radiation the situation is more complex. Several researchers have analyzed this problem in general (Critten, 1989; Kurata, 1990), or more specifically, in the case of a multi-span Venlo greenhouse covered with glass (Stoffers, 1971; Kozai and Kimura, 1977; Kozai et al., 1978; Bot, 1983). On account of the diurnal and annual variations in sun position with respect to the greenhouse cover, there are variations in light interception by the construction parts. In addition the different angles at which direct radiation reaches the glass panes cause variations in reflection (particularly at low angles of incidence) and absorption (due to an increase in path length). At low sun elevation (lower than the slope of the roof) and azimuth more or less perpendicular to the orientation of the greenhouse, direct radiation has to pass the cover of more than one bay until it reaches the crop. When the angle between greenhouse orientation and azimuth of the sun approaches zero, shadows of the ridges tend to overlap at low sun elevation.

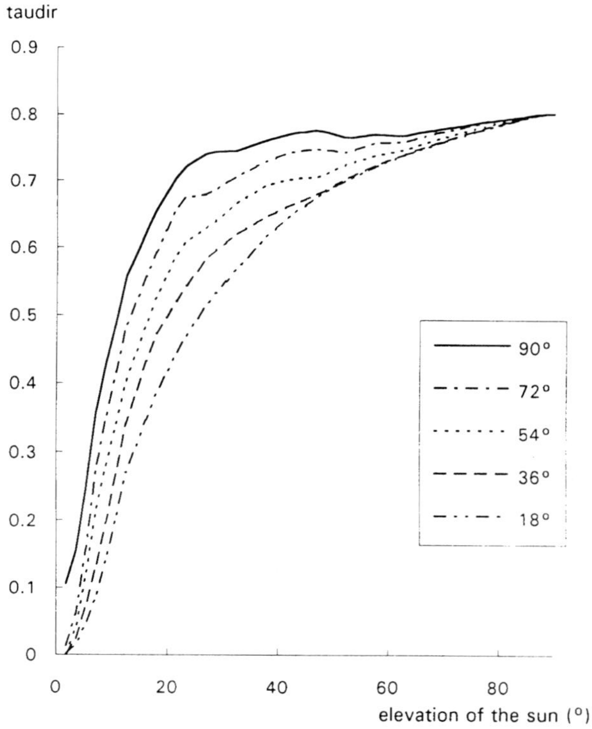

Fig. 15.4. Transmission coefficient for direct radiation (τ_{dir}) of the reference glasshouse of Table 15.2 as related to elevation of the sun and different angles between glasshouse orientation and azimuth of the sun.

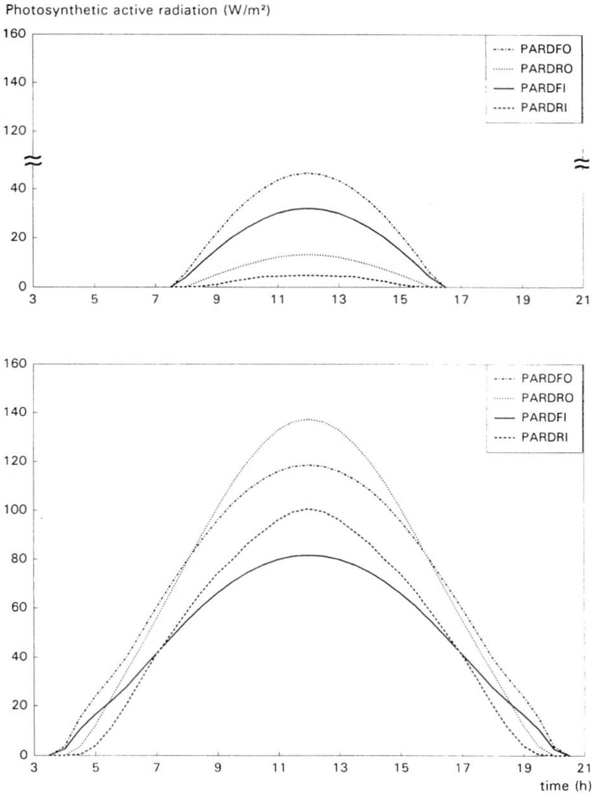

Fig. 15.5. Calculated diurnal pattern of diffuse and direct photosynthetically active radiation (PAR) inside and outside the reference greenhouse in January (Day 26, top) and in June (Day 156, bottom) in the Netherlands. PARDFO, diffuse PAR outside; PARDFI, diffuse PAR inside; PARDRO, direct PAR outside; PARDRI, direct PAR inside.

The theory for transmission of direct and diffuse radiation of a multi-span Venlo glasshouse has been treated extensively by Bot (1983). The calculations used in this chapter are based on his approach, which takes all the effects mentioned into account. Some relevant properties of the reference glasshouse are presented in Table 15.2. The transmission for diffuse radiation has been included in this table but the transmission for direct radiation varies, depending on the elevation of the sun and the angle between its azimuth and the orientation of the greenhouse. The transmission for direct radiation of the reference greenhouse has been summarized in Fig. 15.4. Of the two variables considered, the elevation of the sun has by far the greatest effect on transmission of direct radiation, but the effect of the angle between the sun's azimuth and greenhouse orientation cannot be neglected. Diffusion of direct radiation by the greenhouse cover is ignored in our calculation.

Daily radiation inside a greenhouse

During the day, elevation and azimuth of the sun change. As a result the transmission for direct light changes within the day, and this effect will be stronger in winter than in summer, particularly at higher latitudes. As a result the diurnal pattern of radiation inside a greenhouse will differ from that outside (Fig. 15.5). Likewise, the annual course of daily light integrals inside deviates from that outside. An example is the elimination of nearly all direct radiation inside the greenhouse in winter at higher latitudes, in the Netherlands and in Sweden (Fig. 15.6).

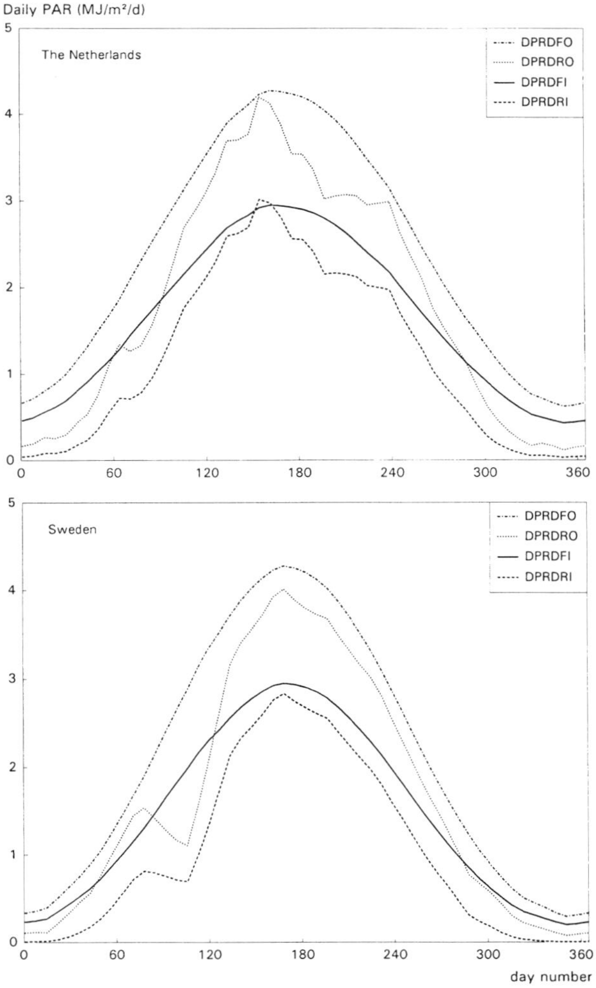

Fig. 15.6. Calculated annual pattern of daily totals of diffuse and direct PAR inside and outside the greenhouse in the Netherlands and in Sweden. DPRDFO, daily diffuse PAR outside; DPRDFI, daily diffuse PAR inside; DPRDRO, daily direct PAR outside; DPRDRI, daily direct PAR inside.

POTENTIAL PHOTOSYNTHESIS, GROWTH AND YIELD

Introduction

Since a general approach to the question of potential greenhouse production is the primary objective, it is obvious that attention should be given to those processes that are common to most, if not all, greenhouse crops, and that differences have to be ignored. We, therefore, adopted standard crop parameters for the simulation of crop growth (Table 15.3).

A point that should be clarified is the choice of the leaf area index (LAI), the ratio of leaf area to greenhouse area. According to the definition of potential production discussed above (p. 333), this production level is obtained with a closed green crop surface. However, the choice of a really saturating LAI (maximum absorption of radiation) would imply a very high dry weight per unit greenhouse area and consequently a high rate of maintenance respiration, and low or negative yields at low radiation. In greenhouse cultivation, where extremely low daily radiation integrals may occur in the winter season, high values of LAI are not common. As a compromise a value of 3 $m^2\,m^{-2}$ was chosen, where the influence of variations in LAI on light interception is limited, but maintenance respiration is not excessive.

Table 15.3
Parameters of the standard crop used to calculate potential production at crop temperature of 20°C

Quantity	Value	Units
Light use efficiency at 34 Pa CO_2	0.0131	$mg\,CO_2\,J^{-1}$
Light use efficiency at 100 Pa CO_2	0.0155	$mg\,CO_2\,J^{-1}$
Maximum rate of CO_2 uptake at 34 Pa CO_2	0.949	$mg\,m^{-2}\,s^{-1}$
Maximum rate of CO_2 uptake at 100 Pa CO_2	1.400	$mg\,m^{-2}\,s^{-1}$
Leaf area index	3.0	$m^2\,m^{-2}$
Leaf area ratio	0.0182	$m^2\,g^{-1}$
Scattering coefficient, leaves	0.2	
Extinction coefficient of the canopy for diffuse radiation	0.8	
Maintenance respiration (glucose units)	2.3	$g\,m^{-2}\,d^{-1}$
Total crop dry weight	165.0	$g\,m^{-2}$
Conversion efficiency for crop growth (structural dry weight/glucose units)	0.724	$g\,g^{-1}$

In the previous section the radiation conditions inside the greenhouse have been described. In this section the distribution of radiation over leaves, and the integration of photosynthesis over the whole canopy and of instantaneous rates of crop photosynthesis to daily totals are considered. Subsequently, daily dry matter production is calculated from daily crop photosynthesis. Production figures in this chapter are based on total dry matter production rather than fresh weight of the harvested product – which, however, is the information more commonly available in horticultural practice. Therefore the procedure to relate yield to dry matter production will be discussed taking into account the dry matter distribution in the production phase and the dry weight, or dry matter content, of the product.

Daily crop photosynthesis and potential dry matter production

Since by far the majority of greenhouse crops are C_3 crops, leaf photosynthesis parameters were estimated accordingly for both carbon dioxide pressures considered (Table 15.3).

The rate of crop photosynthesis is not only determined by the radiation level inside the greenhouse. The ratio between direct and diffuse radiation, as well as the sun elevation, are important; both factors affect the extinction and hence the distribution of radiation over different leaf layers, and therefore affect the rate of crop photosynthesis (Goudriaan, 1982).

Though many greenhouse crops are grown in rows, we have assumed a uniform crop. The implications of this assumption, at a LAI of 3, however, are not very great (Gijzen and Goudriaan, 1989). An example of the effect that the proportion of diffuse radiation may have on the response of crop photosynthesis to light is represented in Fig. 15.7 for two different sun elevations. Thanks to a better distribution of radiation over the leaves, a high proportion of diffuse radiation gives rise to a high rate of crop photosynthesis. The effect is most pronounced at low sun elevation because penetration of direct radiation in the canopy is poor and the top leaf layers are exposed to higher, and hence more saturating, irradiance.

The calculation of potential dry matter production follows the procedure given by Spitters et al. (1989) for their simulation in SUCROS87:

$$GTW = \frac{(GPHOT \cdot 30/44 - MAINT)}{ASRQ}, \qquad (15.1)$$

where GTW = daily dry matter production ($g\,m^{-2}\,d^{-1}$),

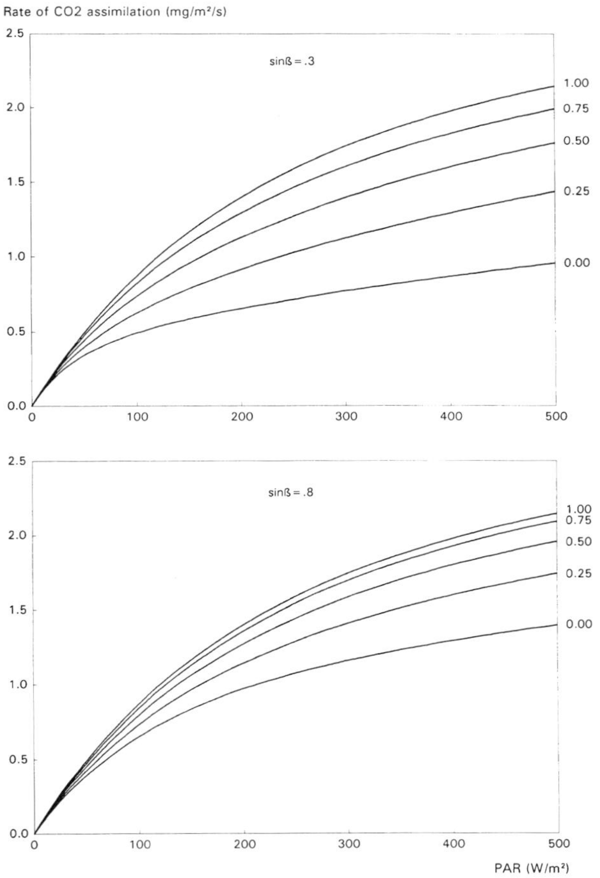

Fig. 15.7. Demonstration of the influence of the proportion of diffuse radiation (indicated at the right end of the curves) on the photosynthesis–light response curve of the standard crop at sun elevations β of 17.5° (top) and 53.1° (bottom). The light compensation point is taken as the origin.

GPHOT = daily gross CO_2 assimilation ($g\,m^{-2}\,d^{-1}$), 30/44 makes the conversion from CO_2 to equivalent glucose weight, MAINT = daily loss of substrate (glucose) due to maintenance respiration ($g\,m^{-2}\,d^{-1}$), ASRQ = assimilate requirement, that is the weight of glucose required to form 1 g of structural dry matter (dimensionless). Values adopted for the rate of maintenance respiration and the conversion efficiency of assimilates to structural dry matter ($\alpha = 1/\text{ASRQ}$) are given in Table 15.3.

The calculations of potential dry matter production have been performed at both levels of carbon dioxide for all the locations considered. The results are presented as daily rates (Fig. 15.8) and cumulatively (Fig. 15.9). The daily rates of potential dry matter production clearly reflect the annual course of radiation for the locations considered (Fig. 15.2). In Sweden the standard crop is not able to grow for part of the winter. Total annual potential production is in the same order of magnitude for Israel, the United States and Zaire, whereas it is much smaller in France, Japan, the Netherlands and Sweden. Obviously, production patterns observed in practice may differ greatly because of the suitability of the greenhouse environment and/or the local market situation.

The cumulative curves enable an easy estimation of total dry matter production over a given interval of time (see Fig. 15.10), whereas the dry matter production per day shows more clearly the annual variation and the actual level.

Potential yield of greenhouse crops

The total dry weight produced is only indirectly related to marketable yield: in most cases only part of the crop is harvested; and besides it should be noted that horticultural products often have a low dry matter content. The efficiency of a crop to utilize its photosynthetically fixed dry matter for the formation of harvestable products is therefore an important crop characteristic, and forms in the present approach the only basis for crop-to-crop variations in yield. In order to obtain yield from dry matter production the production efficiency, EFF, has to be known, defined as:

$$\text{EFF} = \frac{\text{HI}}{\text{DMC}}, \qquad (15.2)$$

where EFF is the ratio between marketable yield and structural dry weight ($g\,g^{-1}$, or number g^{-1}), HI, the harvest index, the ratio between dry weight of the harvestable part and that of the whole plant, and DMC, dry matter content, which, depending on the type of crop, is either the ratio of dry weight over fresh weight of the product, or the amount of dry weight contained in 1 unit of product (cut flowers, cucumbers (*Cucumis sativus*), etc.).

One of the simplifications that have been made in this chapter is the assumption that differences in photosynthetic properties and assimilate requirement (ASRQ) among crops and/or cultivars may be ignored in the calculation of potential production. As a result of this assumption and of the fact that standard values have been adopted for all crop and greenhouse parameters, the curves of Figs. 15.8 and 15.9 are common to all crops. In individual cases great deviations of these standard curves may occur, due to different type of greenhouses, different photosynthetic parameters, etc.

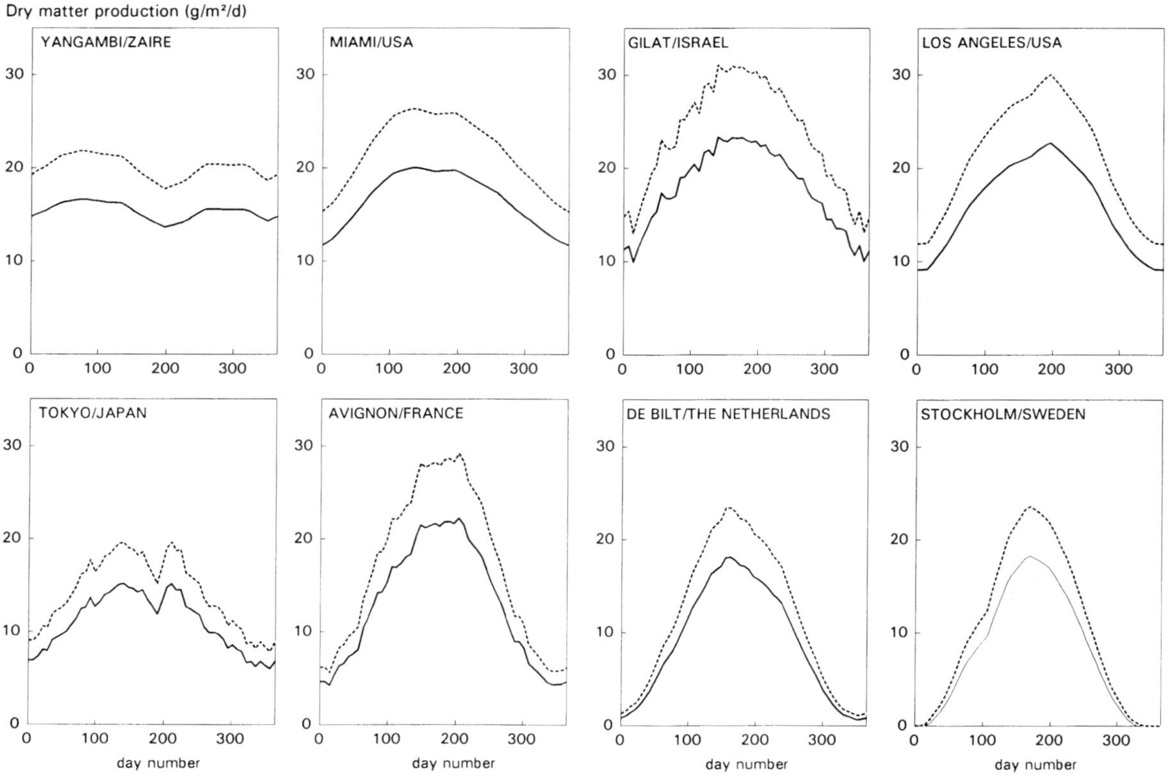

Fig. 15.8. Annual course of the daily rate of potential dry matter production at the locations considered, with carbon dioxide pressure of 34 Pa (solid) and 100 Pa (dashed).

For a limited number of vegetable crops values of HI and DMC were presented by Heuvelink and Challa (1989), which make possible the estimation of EFF (Table 15.4). For flower crops it is more difficult to obtain such data. The value of HI for crops that are not harvested as a whole is in general of the order of 0.7 to 0.8; DMC of ornamentals is often, however, much higher than for vegetables. By using the EFF values thus obtained, the curves in Figs. 15.8 and 15.9 can be used to obtain potential yield figures for a given location and a given crop. Obviously there may be great differences in EFF – differences that have substantial consequences for potential yield of different crops.

Example: tomato

In order to demonstrate the procedure for the calculation of potential yield from the potential dry matter figures presented in this chapter, one may consider the cultivation of tomato in the Netherlands. The example is derived from an official publication in the Netherlands (Anonymous, 1989), where round tomatoes planted in Week 50 are cultivated on rock-wool in a high-wire system. The production figures are average yields feasible in reasonably modern and well-managed nurseries.

The first problem to be solved is in which period production was taking place. Harvesting in this example starts in March (Day 75) and ends in November (Day 320). Photosynthates contributing to fruits harvested in that period had already been formed before the first harvest. The period from flowering until harvest of a cluster is approximately 8 to 9 weeks (Hurd and Graves, 1985). Because a stable harvest index is not immediately reached, however, it is not realistic to include this period fully in the calculations; therefore, rather arbitrarily, only half of this period is included. The starting point for the annual production consequently is approximately $75-60/2=$ Day 45. The total duration of the production phase is then $320-45=275$ days. In Fig. 15.10 for the Netherlands at Day 45 a cumulative dry matter production of

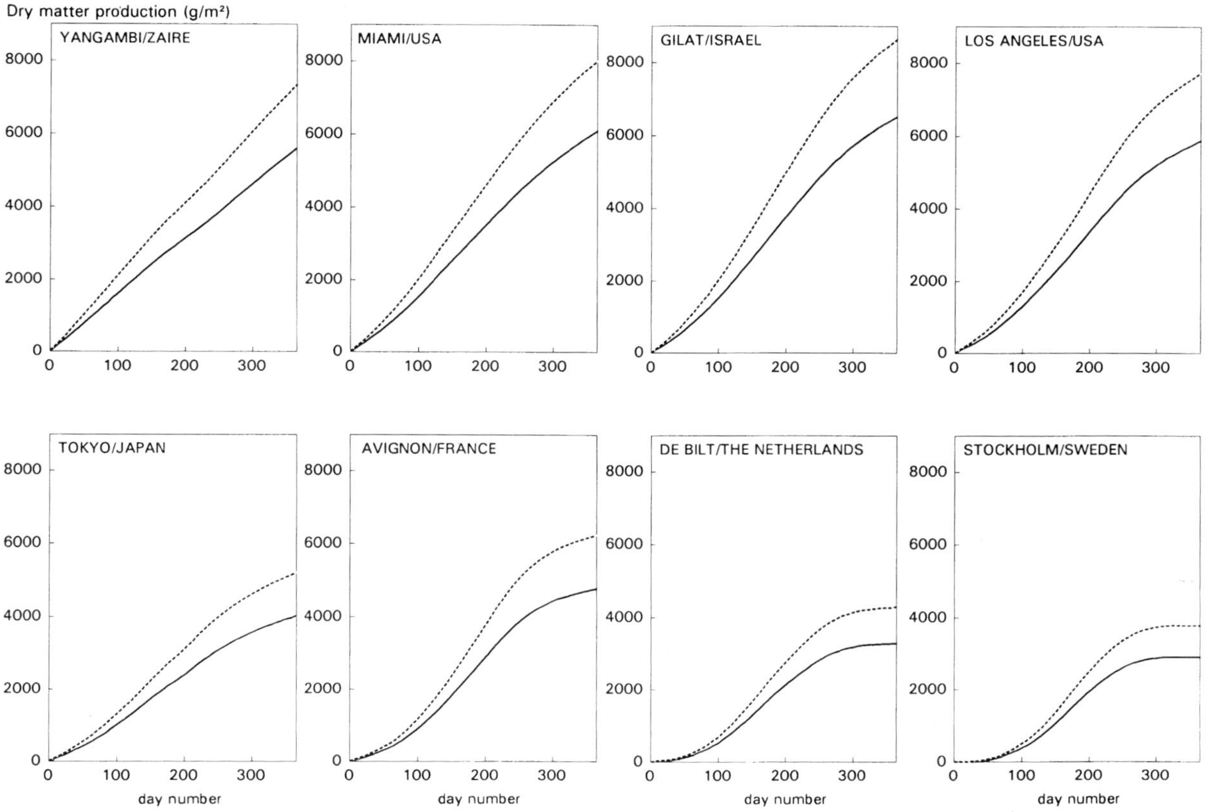

Fig. 15.9. Cumulative annual potential dry matter production at the locations considered, with carbon dioxide pressure of 34 Pa (solid) and 100 Pa (dashed).

Table 15.4
Estimated values of HI (harvest index), DMC (dry matter content of harvestable product) and EFF (ratio of the amount of fresh product formed to total plant dry matter produced, HI/DMC) for a number of crops

Crop	HI	DMC	EFF
Tomato (*Lycopersicon esculentum*)	0.7	0.06	12.0
Cucumber (*Cucumis sativus*)	0.7	0.035	20.0
Sweet pepper (*Capsicum annuum*) red	0.8	0.085	9.4
Eggplant (*Solanum melongena*)	0.8	0.07	11.4
Radish (*Raphanus sativus*)	0.9	0.05	18.0
Butterhead lettuce (*Lactuca sativa*)	0.85	0.05	17.0

90 and 126 is shown for carbon dioxide pressure of 34 and 100 Pa, respectively, and at Day 320 these values are respectively 3243 and 4210. Thus at low carbon dioxide pressure potential dry matter production in the period considered is $3243-90=3153\,g\,m^{-2}$ and at high CO_2 $4210-126=4084\,g\,m^{-2}$. During the production phase, HI is about 70% and DMC of the fruit is about 6% (Heuvelink and Challa, 1989), so that the corresponding yields are respectively $3.153 \cdot 0.7/0.06=37\,kg\,m^{-2}$ at ambient carbon dioxide pressure and $4.084 \cdot 0.7/0.06=48\,kg\,m^{-2}$ at high carbon dioxide pressure.

In the Netherlands yields of about $44\,kg\,m^{-2}$ are normal in well-managed greenhouses (Anonymous, 1989), where carbon dioxide enrichment is common practice, and where transmission figures are usually better than for our standard greenhouse. An exact comparison is difficult to make, because it would require a detailed analysis of the production figures used and the prevailing conditions, but with some caution the conclusion may be drawn that yields obtained in practice are not far from potential values over the period considered.

Preliminary conclusions may be drawn from a comparison of the annual patterns of potential and actual average production rate (Fig. 15.11). The actual

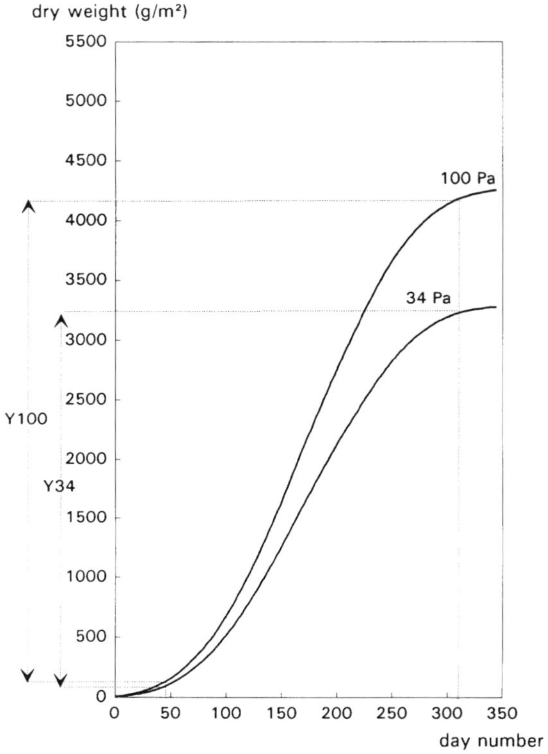

Fig. 15.10. Cumulative potential dry matter production in the Netherlands with carbon dioxide pressure of 34 and 100 Pa (the final yields are indicated on the vertical axis by Y34 and Y100, respectively) and its use for the derivation of potential dry matter production of a tomato crop over a given period of time.

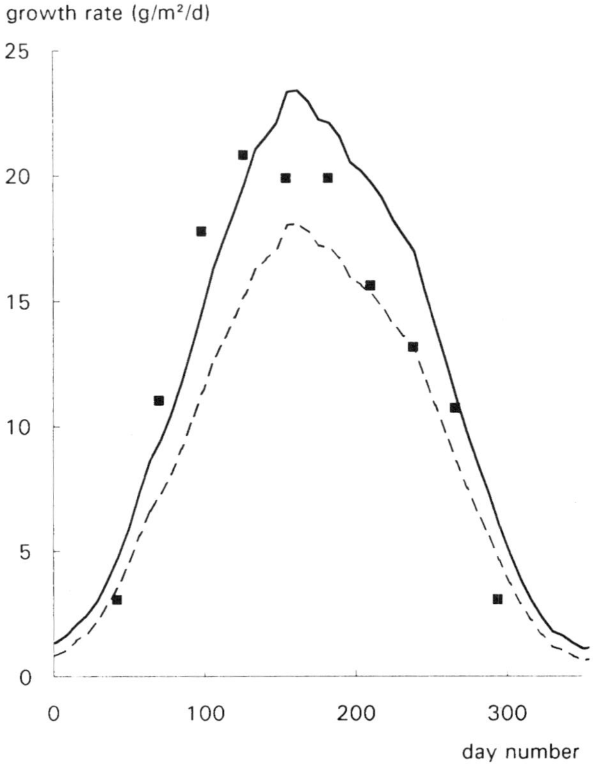

Fig. 15.11. Annual course of daily potential production rate with carbon dioxide pressure of 34 Pa (solid) and 100 Pa (dashed) in the Netherlands, compared with average production figures for tomato in the Netherlands (solid squares), converted to total dry weight per m^2.

production rates are available only over periods of about 28 days. Conversions to total dry weight were made by dividing fresh weight by EFF, the fresh weight of tomatoes produced per gram of total structural dry weight formed. Average rates were obtained by dividing the total production by the duration of the period. As a consequence the first and last actual production rates are underestimations, since the crop was not fully productive during the whole period. Furthermore all observed values were advanced 4 weeks, in order to account for the average delay between formation of assimilates and harvest.

The seasonal variations in actual average production compare well with those of potential production, supporting the assumption that photosynthesis and dry matter production are strongly related to actual yield. In the second part of the growing season, there is a clear tendency towards a lower actual average production compared to potential production, suggesting that yield could be increased in that phase, although other interpretations are also possible.

A simple procedure for estimating potential yield

The calculations presented in this chapter cover only a limited number of locations. In order to facilitate the estimation of daily potential dry matter production (GTW) and yield for other locations, a so-called meta-model (Kleijnen, 1987) was derived on the basis of the results of the simulation model. To this end, the linear relation observed by Versteeg and van Keulen (1986) between daily crop photosynthesis, and thus GTW, and AVRAD, the daily light integral, with separate parameters for low and for high irradiances, was investigated. When simulated GTW was plotted as a function of AVRAD, however, large variations occurred.

Deviations from a simple relationship between GTW and AVRAD may be anticipated, since variations in AVRAD are brought about by variations in day length, sun elevation and atmospheric transmission, which may affect GTW in different ways. Yet it seemed worthwhile

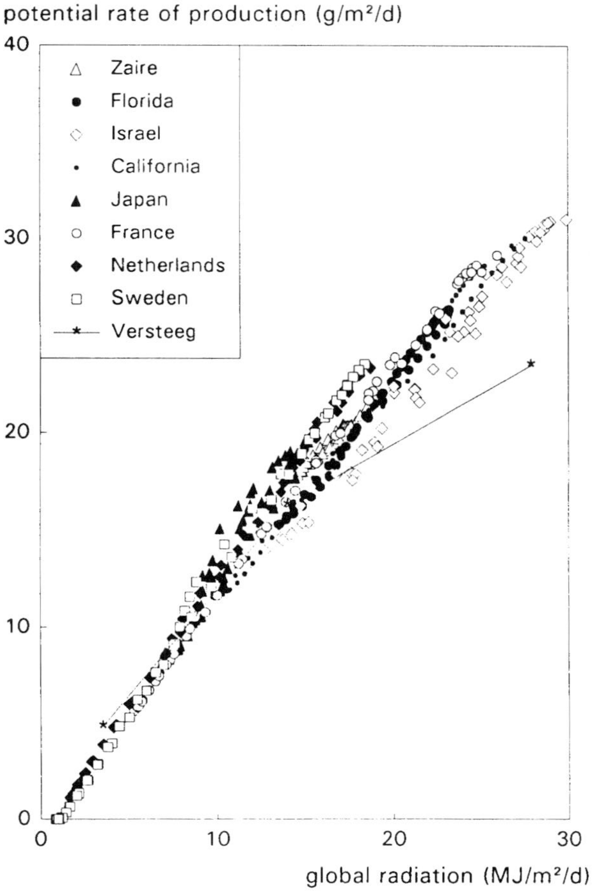

Fig. 15.12. Relation, for all the locations considered, between daily rate of potential dry matter production and daily global radiation, at different times of the year, with carbon dioxide pressure of 34 Pa. The line denoted as "Versteeg" shows the relation proposed by Versteeg and van Keulen (1986), converted to the units of this graph to facilitate comparison.

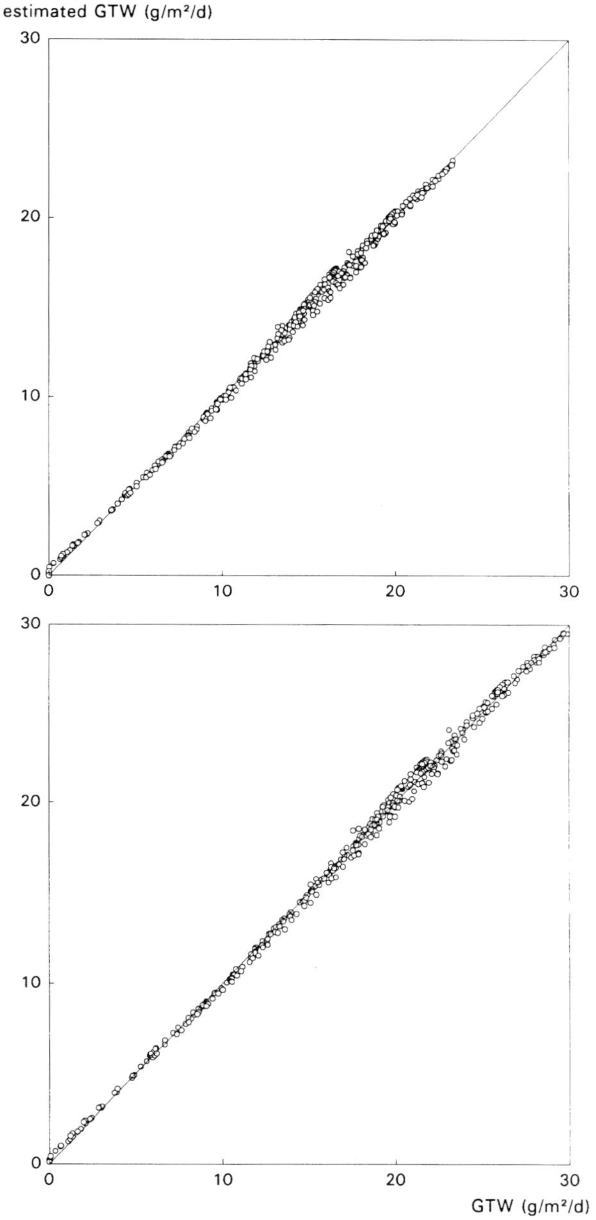

Fig. 15.13. Comparison of the daily rate of potential dry matter production calculated with the simulation model (GTW) and with the meta-model (estimated GTW) with carbon dioxide pressure of 34 Pa (top) and 100 Pa (bottom). All data points are included.

to consider the relation described by Versteeg and van Keulen (1986), using our parameters for the relation between photosynthesis and growth, and correcting for the loss of radiation due to the greenhouse cover and the imperfect absorption of radiation by the crop (Fig. 15.12). Although the relation of Versteeg and van Keulen was used with outdoor crops between 0° and 40° latitude only, and the fit with our calculations is not too bad, the structure of the data points suggests that the bi-phasic nature of their relation does not represent the real situation.

Great improvement in the description of the relation between AVRAD and GTW could be obtained by introducing a second factor, the proportion of diffuse radiation (FRDIF). A very good description for the whole range of daily light integrals occurring in the locations considered (Fig. 15.13) was:

$$\mathrm{GTW} = A + (B + C \cdot \mathrm{FRDIF}) \cdot \mathrm{AVRAD}. \qquad (15.3)$$

The coefficients A, B and C depend on the carbon dioxide concentration (Table 15.5). From Eq. (15.1) follows that A should be closely related to the daily rate

Table 15.5
Coefficients of Eq. (15.3)[1] at low and high CO_2 pressure

Parameter	CO_2 pressure	
	34.0 Pa	100.0 Pa
A	−1.20	−1.04
B	0.57	0.79
C	0.86	1.00

[1] $GTW = A + (B + C \cdot FRDIF) \cdot AVRAD$, with AVRAD in $MJ\,m^{-2}\,d^{-1}$ and GTW in g dry weight $m^{-2}\,d^{-1}$.

of maintenance respiration. Indeed, the values observed are close to the rate of maintenance respiration used in the present calculations ($2.3\,g\,m^{-2}\,d^{-1}$ glucose equivalents, or $2.3 \cdot 0.72 = 1.66$ dry weight equivalents, which is the basis for comparison with coefficient *A*). The underestimation of these values at both levels of carbon dioxide is a logical consequence of extrapolation of a simple equation covering a wide range of conditions. Our results suggest that the bi-phasic relation described by Versteeg and van Keulen (1986) could have been caused by an increase of the proportion of direct radiation at higher radiation levels.

A practical problem in using Eq. (15.3) is that quite often no data are available on FRDIF. In that case the procedure described above (p. 337) can be followed to derive FRDIF from average daily radiation at a given location and date.

DISCUSSION

A general method has been presented to predict potential production and yield of greenhouse crops, given the annual course of global radiation and the location. Obviously, this wide scope makes it necessary to accept a great number of assumptions and simplifications.

The basic assumptions are that potential production is limited only by photosynthesis and that at a given carbon dioxide pressure all variations in crop photosynthesis can be attributed to radiation conditions. These assumptions are a logical consequence of the definition of potential production used, implying that all other factors are of minor importance.

In many cases the actual production will be lower than potential production for reasons that have partly been discussed in the previous sections:

(1) The canopy does not intercept the radiation as efficiently as the model crop with a LAI of 3, or does so during only a part of the production cycle.
(2) Carbon dioxide levels may vary greatly, depending on enrichment, air exchange, and uptake by the crop.
(3) Temperatures inside the greenhouse may be above or below the optimum on account of high radiation or lack of heating.
(4) An unfavourable water status of the crop may negatively affect photosynthesis and growth of the crop.
(5) Problems may occur in the mineral nutrition of the crop.
(6) Pests and diseases may interfere.

In addition, when comparing actual with potential production the following points should be considered:

(1) Actual weather conditions may deviate considerably from the average conditions, often based on observations over many decades.
(2) The values presented for EFF, the efficiency of conversion of structural dry weight to harvestable product, may show considerable variation, due to variations in the dry matter content of the product and in the harvest index.
(3) Many other specific crop properties may differ from those used in the model crop and may have to be established more accurately in individual cases.
(4) Greenhouse transmission may differ considerably from that of the reference greenhouse, on account of construction, orientation and application of whitewash, screens, etc.
(5) In cases where artificial light is used as an additional light source, potential production may be exceeded.

These points make clear that caution is required in order to come to realistic interpretations of the predictions obtained. Nevertheless, this universal approach has a great value for horticulture and horticultural research, where, unfortunately, research efforts are distributed over many different crops and varieties.

ACKNOWLEDGEMENT

The many pieces of advice received from Mr. H. Gijzen, Research Institute for Agrobiology and Soil Fertility (AB-DLO), Wageningen, who also allowed us to use his FORTRAN version of the greenhouse light transmission model, are gratefully acknowledged.

LIST OF SYMBOLS AND ABBREVIATIONS USED

ASRQ	assimilate requirement (g CH_2O/g dry weight formed)
AVRAD	average daily global radiation
DMC	dry matter content of harvestable product (g dry/g fresh)
EFF	fresh weight of harvestable product formed per unit dry weight produced
FRDIF	diffuse radiation/(diffuse + direct radiation)
GPHOT	rate of gross CO_2 uptake by the crop
GTW	rate of growth of total dry weight of the crop
HI	harvest index (dry weight of harvestable product formed per g dry weight produced)
LAI	leaf area index, total leaf area (single sided) per unit greenhouse area
LAR	leaf area ratio, total leaf area of the crop/crop dry weight
MAINT	maintenance respiration per unit greenhouse area
PAR	photosynthetically active radiation (400–700 nm)

REFERENCES

Anonymous, 1989. *Kwantitatieve informatie voor de glastuinbouw 1989–1990.* Ministry of Agriculture and Fisheries, The Hague, Netherlands, Table G 176.

Bot, G.P.A., 1983. *Greenhouse climate: from physical processes to a dynamic model,* Dissertation. Wageningen Agricultural University, Wageningen, Netherlands, 240 pp.

Challa, H. and Schapendonk, A.H.C.M., 1986. Dynamic optimalization of CO_2 concentration in relation to climate control in greenhouses. In: H.Z. Enoch and B.A. Kimball (Editors), *Carbon Dioxide Enrichment of Greenhouse Crops.* CRC, Boca Raton, FL, pp. 147–160.

Critten, D.L., 1989. Light transmission into greenhouses. *Acta Hortic.,* 245: 377–383.

de Wit, C.T., 1965. *Photosynthesis of leaf canopies,* Agric. Res. Rep. 663. PUDOC, Wageningen, Netherlands, 57 pp.

de Wit, C.T. and van Heemst, H.D.J., 1976. Aspects of agricultural resources. In: W.T. Koetsier (Editor), *Chemical Engineering in a Changing World, Proc. Plenary Sessions, 1st World Congr. on Chemical Engineering,* Amsterdam, June 28–July 1, 1976. Elsevier, Amsterdam, pp. 125–145.

de Wit, C.T., van Laar, H.H. and van Keulen, H., 1979. Physiological potential of crop production. In: J. Sneep and A.J.T. Hendriksen (Editors), *Plant Breeding Perspectives.* PUDOC, Wageningen, Netherlands, pp. 47–82.

Gijzen, H. and Goudriaan, J., 1989. A flexible and explanatory model of light distribution and photosynthesis in row crops. *Agric. For. Meteorol.,* 48: 1–20.

Goudriaan, J., 1982. Potential production processes. In: F.W.T. Penning de Vries and H.H. van Laar (Editors), *Simulation of Plant Growth and Crop Production.* PUDOC, Wageningen, pp. 98–113.

Heuvelink, E. and Challa, H., 1989. Dynamic optimization of artificial lighting in greenhouses. *Acta Hortic.,* 260: 401–412.

Hurd, R.G. and Graves, C.J., 1985. Some effects of air and root temperatures on the yield and quality of glasshouse tomatoes. *J. Hort. Sci.,* 60(3): 359–371.

Iqbal, M., 1983. An Introduction to Solar Radiation. Academic Press, Toronto, Table 8.8.4, p. 230.

Kleijnen, J.P.C., 1987. *Statistical Tools for Simulation Practitioners. Statistics: Textbooks and Monographs,* Vol. 76. Marcel Dekker, New York, 429 pp.

Kozai, T. and Kimura, M., 1977. Direct solar light transmission into multi-span greenhouses. *Agric. Meteorol.,* 18: 339–349.

Kozai, T., Goudriaan, J. and Kimura, M., 1978. *Light Transmission and Photosynthesis in Greenhouses.* PUDOC, Wageningen, Netherlands, 99 pp.

Kropff, M.J. and Spitters, C.J.T., 1987. *Introduction to crop ecology,* Syllabus. Wageningen Agricultural University, Wageningen, Netherlands, 139 pp.

Kurata, K., 1990. Role of reflection in light transmissivity of greenhouses. *Agric. For. Meteorol.,* 52: 319–331.

Meek, D.W., Hatfield, J.L., Howell, T.A., Idso, S.B. and Reginato, R.J., 1984. A generalized relationship between photosynthetically active radiation and solar radiation. *Agron. J.,* 76: 939–945.

Moon, P., 1940. Proposed solar-radiation curves for engineering use. *J. Franklin Inst.,* 230: 583–618.

Nederhoff, E.M., 1990. Technical aspects, management and control of CO_2 enrichment in greenhouses. *Acta Hortic.,* 268: 127–138.

Penning de Vries, F.W.T., 1982. Systems analysis and models of crop growth. In: F.W.T. Penning de Vries and H.H. van Laar (Editors), *Simulation of Plant Growth and Crop Production.* PUDOC, Wageningen, Netherlands, pp. 9–19.

Penning de Vries, F.W.T. and van Laar, H.H. (Editors), 1982. *Simulation of Plant Growth and Crop Production.* PUDOC, Wageningen, 308 pp.

Spitters, C.J.T., Toussaint, H.A.J.M. and Goudriaan, J., 1986. Separating the diffuse and direct component of global radiation and its implications for modeling canopy photosynthesis, Part I. Components of incoming radiation. *Agric. For. Meteorol.,* 38: 217–229.

Spitters, C.J.T., van Keulen, H. and van Kraalingen, D.W.G., 1989. A simple and universal crop growth simulator: SUCROS87. In: R. Rabbinge, S.A. Ward and H.H. van Laar (Editors), *Simulation and Systems Management in Crop Protection.* PUDOC, Wageningen, Netherlands, pp. 147–181.

Stoffers, J.A., 1971. *Lichtdurchlässigkeit von Gewächshäuser in Blockbauweise,* IMAG Publication 39. IMAG, Wageningen, Netherlands, 29 pp. In German.

Szeicz, G., 1974. Solar radiation for plant growth. *J. Appl. Ecol.,* 11: 617–636.

van der Velden, N.J.A., Fonville, V.P. and Verhaegh, A.P., 1990. Energie-efficiency en CO_2 emissie in de glastuinbouw. LEI

Publications, Publication 4.126, Agricultural Economics Research Institute LEI, The Hague, The Netherlands, 52 pp. In Dutch.

Versteeg, M.N. and van Keulen, H., 1986. Potential crop production prediction by some simple calculation methods, as compared with computer simulations. *Agric. Sys.,* 19: 249–272.

Chapter 16

OPTIMAL DYNAMIC MANAGEMENT OF THE GREENHOUSE CLIMATE

Ido SEGINER

INTRODUCTION

The main difference between natural and agricultural ecosystems is that the latter are managed, at least to some degree. Certain agricultural systems, notably greenhouses, are managed (cultivated) more intensively than others (e.g., rain-fed field crops). As will be shown later, a useful measure of production intensity is the cost (of inputs) that a grower is willing to pay in order to produce the marginal unit of crop. Crop response is a concave-downward function of the inputs. Since the normal cultivation range corresponds to its ascending portion, a higher production intensity implies a higher yield. This higher yield, however, may not accomplish the best result from the point of view of the producer, if the additional cost involved in producing the extra yield is not covered by the additional income from its sale (due to diminishing returns). Hence, the production intensity has to be selected such that it maximizes a suitable measure of satisfaction of the producer. From the point of view of an individual grower, this measure (later referred to as performance criterion) is usually the net income over the life-span of the crop (season).

This approach is useful from a static, as well as from a dynamic point of view. The static point of view regards the whole season as a single decision unit, while the dynamic point of view considers in detail the sequence of operational decisions to be made through the season. Sometimes the individual decisions of the sequence may be regarded as quasi-static. The dynamic decisions should depend on the state of the evolving system (crop) and on its changing boundary conditions (weather and prices). In greenhouse production it is convenient to consider the system as encompassing not only the crop, but also the greenhouse structure and the control equipment (such as heating and irrigation systems). This places the upper physical boundary of the system just outside the cladding of the greenhouse, while the lower boundary is often placed just below the root zone.

The management of greenhouse production requires that decisions be made at different time scales: from selection of crop and planting time once every season, to practically continuous control of heating, ventilation and carbon dioxide enrichment. Some of the actions taken are due to active decisions, some others by default. Some are manual, while others are automatic. In this chapter only decisions related to the aerial environment of greenhouses will be considered. These decisions must be frequent, in order to adjust to the ever-changing meteorological conditions, and therefore are already automated in many greenhouses.

The *optimization* of greenhouse environmental control has a relatively short history. Traditionally, control engineers were content with maintaining the environment at the set-points (temperatures, humidities, carbon dioxide concentrations) specified by horticulturists (e.g., Davis and Hooper, 1991). These were based on experience and experimental results, and were usually aimed at maximizing the yield. Attempts to determine optimal set-points by using information contained in crop models started about twelve years ago. The emphasis was, and still is, on economic optimization and on dynamic adjustment to the changing weather.

One of the first optimization studies was presented by Challa and van de Vooren (1980). They attached economical value to earliness of cucumber production and could, as a result, calculate optimal greenhouse temperatures under a range of light conditions. The emphasis was on the *development* (rate of leaf initiation) of a young crop, and therefore did not require a close look at the effect of carbon dioxide. In another study, Challa and Schapendonk (1986) considered the stationary behavior of a mature cucumber crop. Focusing on *assimilation* as the dominant process, they related the value of additional cucumber dry matter to the cost of carbon dioxide enrichment. Heuvelink and Challa (1989) treated artificial lighting in a similar manner.

Each one of these studies treated a distinct phase in the life cycle of a crop, employing models with a single state variable (number of leaves in the first example and dry matter in the other two). Each phase was treated as quasi-steady in the short term, responding instantaneously to weather changes. Since current crop models may also contain information about the slow (long-term) evolution of a crop from one phase into another, this information could also be used, in principle, for control purposes. One would, in fact, expect the desirable crop environment to change with time, as the crop moves from the seedling phase through the complete-cover vegetative phase, to the reproductive phase.

An optimal solution for the transition from low to high leaf area index (LAI) was first presented by Gal et al. (1984). They specified a particular growth function, which led to a variable transformation (shown later, Eq. 16.35), that produced a simple result. Using tools of optimal control theory (Kirk, 1970), the treatment of single-variable models (Seginer, 1989) was later extended to two state variables (Seginer et al., 1991). Here the transition from low to high leaf area index in lettuce was reflected in shifting emphasis from production of leaf area to dry matter production. This shift (treated in some detail below, p. 370) is analogous to the above mentioned shift from development to photosynthesis in cucumber production. Further extension to models which cover the vegetative as well as the reproductive phase of a tomato crop proved to be difficult (Shina and Seginer, 1989), on account of the large number of state variables involved. In certain cases a solution is, nevertheless, possible (Seginer, 1991). Methods to overcome some of the computational problems associated with large crop models are suggested later in this chapter (pp. 376–378).

Before moving to a more precise description of the management problem, a mention should be made of the role of future events in management considerations. Since the future is only known to us in terms of forecasts, any decision through the season is taken under some level of uncertainty. Even if one ignores stresses incurred by future failure of other parts of the management system (water stress, diseases), there still are two important factors to be considered: future weather and future prices. The behavior of these boundary conditions may affect the choice of management strategy, as well as specific decisions. For instance, if prices for the produce peak at a certain date, a weather sequence which is cooler and duller than usual may be compensated for by more intensive management strategy (more heating, artificial light, carbon dioxide enrichment, etc.). Likewise, higher prices for the produce may also justify more intensive management, in order to increase the output for a given investment (greenhouse area). Advance knowledge of these future boundary conditions is helpful in selecting the correct management strategy.

Notwithstanding the statement just made, much of the material to be presented below is developed on the assumption that the relevant future is deterministically or statistically known to the decision maker at decision time. This assumption is a convenient starting point, which does not preclude adaptation to more realistic situations by means of either one, or a combination, of the following procedures.

(1) Computing expectations based on probability distributions for the future.
(2) Updating weather and price forecasts as soon as new information becomes available.

The bulk of this chapter is divided into two parts, which may be referred to as Methodology and Examples. In the first, several approaches to the optimal dynamic management of greenhouses are explored. Simple crop and greenhouse models are also presented. The second part, which relies on the material presented in the first part, provides examples of optimal control, including some numerical results. Extensions to the basic optimization approaches introduced in the first part are advanced when required. The chapter as a whole emphasizes methodology and qualitative results. The numerical results of the second part, although not unreasonable, are mostly based on very simple generic crop models, and should not be used to implement control in practice.

Since this problem is usually referred to as climate *control*, the terminology of optimal control theory will be used from here on. “Management” will be called “control”, “strategy” will be replaced by “policy” or “law” and “measure of satisfaction” will be termed “performance criterion”. Mathematical expressions will be used throughout the chapter, since these are the tools with which actual solutions are produced. They also serve as a concise language, which facilitates the qualitative analysis of a problem. It should be possible, however, to follow the main ideas and results without reference to mathematical manipulations, since verbal explanations are provided throughout the text.

SYSTEM MODELLING

The state of a system can be described by a set of state variables. Obviously, a complete description of the state of a biological system is not possible, because of its immense complexity. Useful *models* of the system contain only the minimum number of variables required to solve a *given problem*. It should be clear that a model is just a means for solving a problem, whether basic or applied. Hence, different problems may best be served by different models. This is particularly true when a model is complex, as some crop models tend to become.

Most crop modelers select state variables by intuition, sometimes leading to fairly large models (Jones et al., 1984, 1991). It is easy to see how the number of state variables can proliferate: consider a crop compartmentalized into 5 age groups for each of three organs (say, leaves, roots and fruits). If each compartment is represented by just two variables (say, weight and number), the total number of state variables becomes 30. Any further refinement, in terms of number of organs and age groups, or by inclusion of additional factors, such as soil-water distribution, may easily bring the total to well over a hundred.

Increasing the number of state variables of a model not only increases the effort involved in experimental support of its functional relationships and parameters, it usually also increases the effort required to apply it. With present-day computers the time required for a seasonal simulation of crop development is no longer prohibitive. If, however, a complete seasonal simulation serves as a single function evaluation in a multi-dimensional search, and must therefore be repeated a very large number of times, the problem of computer resources becomes pressing once again. The objective of a modeler should therefore be to select the minimal model which preserves the essential characteristics required for the problem under consideration.

In this chapter, mostly very simple models will be considered. While most of them contain only one or two state variables, their basic formulation is similar to that of the more complex models. The methods used to solve these simple problems are quite general, so that they can be adapted to more complex situations. The adaptation can be either by simplifying the models or by simplifying the decision-making process. An example of the former is presented on pp. 352–355, and an example of the latter is shown on pp. 374–376.

System equations

Let us consider a generic system with s state variables (such as leaf area or number of flowers), denoted by x_1 to x_s, and m exogenous (or boundary) variables (such as solar radiation), denoted by e_1 to e_m. The dynamics of the system (its development, its motion) can be expressed in terms of a set of differential equations (or difference equations), which show the rate of change with time of each state variable as a function of all state and exogenous variables

$$\begin{aligned} \frac{dx_1}{dt} &= f_1\{x_1, x_2, \ldots, x_s, e_1, e_2, \ldots e_m\}, \\ \frac{dx_2}{dt} &= f_2\{x_1, x_2, \ldots, x_s, e_1, e_2, \ldots e_m\}, \\ &\vdots \\ \frac{dx_s}{dt} &= f_s\{x_1, x_2, \ldots, x_s, e_1, e_2, \ldots e_m\}. \end{aligned} \quad (16.1)$$

The first of these equations may describe the rate of net gain in dry weight, the next the rate of change of leaf area, and so on. Using vector notation, the s system equations are condensed to

$$\frac{d\boldsymbol{x}}{dt} = \boldsymbol{f}\{\boldsymbol{x}\{t\}, \boldsymbol{e}\{t\}\}, \quad 16.2)$$

where bold letters represent vectors (for instance, $\boldsymbol{x} \equiv [x_1, x_2, \ldots, x_s]$). The system equation, or equation of motion, Eq. (16.2), with $\boldsymbol{e}\{t\}$ representing the ever-changing natural environment, may serve as a general description of an unmanaged ecosystem. As soon as management is introduced, a vector of control variables, $\boldsymbol{u}\{t\}$, representing irrigation, weed control or environmental control in general, must be included in the system equation, resulting in

$$\frac{d\boldsymbol{x}}{dt} = \boldsymbol{f}\{\boldsymbol{x}\{t\}, \boldsymbol{e}\{t\}, \boldsymbol{u}\{t\}\}, \quad (16.3)$$

where all variables may vary with time. To interpret this equation in the context of a greenhouse operation, let us assume that the development of the crop is a function of its state, $\boldsymbol{x}\{t\}$ (e.g., leaf area, number of flowers), and the environment inside the greenhouse, denoted by $\boldsymbol{e}_{\mathrm{i}}\{t\}$. Thus

$$\frac{\mathrm{d}\boldsymbol{x}}{\mathrm{d}t}=\boldsymbol{f}\{\boldsymbol{x}\{t\},\boldsymbol{e}_{\mathrm{i}}\{t\}\}. \tag{16.4}$$

In general, the rate of change of the greenhouse environment is itself a function of the state of the system (e.g., its heat content), the outside meteorological conditions, $\boldsymbol{e}_{\mathrm{o}}\{t\}$, and the control exerted by the air-conditioning equipment ("air-conditioning" being used here in the general sense):

$$\frac{\mathrm{d}\boldsymbol{e}_{\mathrm{i}}}{\mathrm{d}t}=\boldsymbol{e}\{\boldsymbol{x}\{t\},\boldsymbol{e}_{\mathrm{i}}\{t\},\boldsymbol{e}_{\mathrm{o}}\{t\},\boldsymbol{u}\{t\}\}. \tag{16.5}$$

Expressed in this way, the greenhouse environment may be considered a part of the system, with $\boldsymbol{x}$ and $\boldsymbol{e}_{\mathrm{i}}$ both serving as state variables.

In many cases the greenhouse is assumed to contain no heat capacity nor any other capacity. Under this assumption the greenhouse environment adjusts instantaneously to new meteorological conditions and to changes in the air-conditioning control. This is simulated by replacing the differential equation (16.5) with an algebraic equation:

$$\boldsymbol{e}_{\mathrm{i}}\{t\}=\boldsymbol{e}\{\boldsymbol{x}\{t\},\boldsymbol{e}_{\mathrm{o}}\{t\},\boldsymbol{u}\{t\}\}. \tag{16.6}$$

Note that in this case the $\boldsymbol{e}_{\mathrm{i}}$ are not state variables of the system.

Combining Eq. (16.4) with either Eq. (16.5) or Eq. (16.6) yields the general system equation (16.3), where $\boldsymbol{x}$ encompasses all state variables, and $\boldsymbol{e}$ represents only the outside environmental conditions. Note that prices, which represent the economic environment, do not affect directly the dynamic behavior of the system. They enter the problem, however, as soon as a performance criterion (p. 355) or a control-generating rule (control law, control policy) are defined, and tend to affect the evolution of the system indirectly, through the control decisions. They therefore play a different role to that of the physical exogenous (boundary) variables.

The system equation describes the development of the system, provided that the exogenous physical variables and the control variables are known. If that is the case, the system equations can be used to simulate this development. If a control-generating rule is available, which relates the control to the state of the crop, the outside weather and the market prices,

$$\boldsymbol{u}\{t\}=\boldsymbol{u}\{\boldsymbol{x}\{t\},\boldsymbol{e}\{t\}\}, \tag{16.7}$$

then the system equation may be reformulated to include the control rule. In that case the system equation (16.3) simplifies to

$$\frac{\mathrm{d}\boldsymbol{x}}{\mathrm{d}t}=\boldsymbol{f}_{\mathrm{c}}\{\boldsymbol{x}\{t\},\boldsymbol{e}\{t\}\}, \tag{16.8}$$

identical in appearance to Eq. (16.2), except that now it describes a closed-loop system. One may also consider both equations as including a control mechanism, except that in Eq. (16.2) it is a natural mechanism, while in Eq. (16.8) it is an artificial mechanism (of human origin). In addition, $\boldsymbol{e}\{t\}$ of Eq. (16.8) includes prices, which $\boldsymbol{e}\{t\}$ of Eq. (16.2) does not.

Simple crop and greenhouse models

The standard approach to crop growth simulation (Spitters et al., 1989) starts with the calculation of net photosynthesis rate (carbon dioxide assimilation rate) of the canopy, here written as

$$\frac{\mathrm{d}W}{\mathrm{d}t}=G\{P,T,X\}M\{L\}-R\{T\}N\{W\}. \tag{16.9}$$

The first term on the right-hand side of Eq. (16.9) represents the gross photosynthesis rate, and the second term represents the maintenance respiration rate. Both terms are corrected for light and growth respiration. G, the gross photosynthesis rate of a closed canopy, is assumed to be a function of the photosynthetically active radiation (PAR), P, reaching the top of the canopy, the ambient temperature, T, and the carbon dioxide concentration, X. If the crop does not completely cover the ground, G must be multiplied by a fraction, which is a function of the leaf area index, L. This fraction, $M\{L\}$, approaches 1 asymptotically as L increases. In the second term on the right-hand side of Eq. (16.9), R is the rate of maintenance respiration of a unit dry matter of tissue, and $N\{W\}$ represents the effective dry mass of the crop. Equation (16.9) is a specific form of Eq. (16.4), which uses as an argument the end product, $\boldsymbol{e}_{\mathrm{i}}\equiv[P_{\mathrm{i}},T_{\mathrm{i}},X_{\mathrm{i}}]$, of the interaction between the outside conditions, $\boldsymbol{e}_{\mathrm{o}}$, and the control, $\boldsymbol{u}$.

Simple forms of G, M, R and N, which seem to capture the essentials of the process, are as follows:

$$G\{P,T,X\}=\frac{\varepsilon PB\{T\}X}{\varepsilon P+B\{T\}X}, \tag{16.10}$$

$$M\{L\}=1-e^{-aL}, \tag{16.11}$$

$$R\{T\}=R_{\mathrm{R}}\exp\{b(T-T_{\mathrm{R}})\}, \tag{16.12}$$

$$N\{W\}=W. \tag{16.13}$$

In these equations, ε is the efficiency (at low radiation

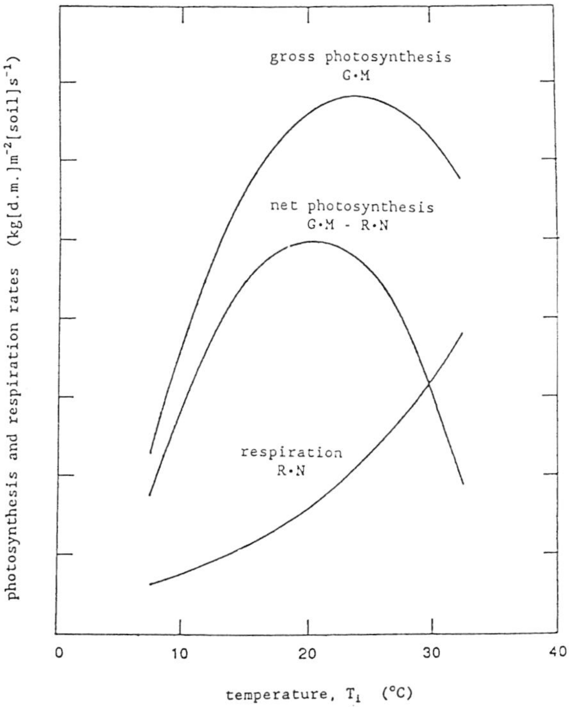

Fig. 16.1. The dependence of gross photosynthesis (GM), respiration (RN) and net photosynthesis (GM – RN) on temperature. The optimal temperature for net photosynthesis is lower than for gross photosynthesis and decreases as the crop develops. Schematic.

levels) of dry matter production from PAR; $B\{T\}$ is gross photosynthesis response to temperature; a, b and R_R are constant coefficients (the coefficient b is equal to $\ln(Q_{10})/10$ in common notation), and T_R is a reference temperature.

Since there is an optimal temperature for gross photosynthesis, a quadratic dependence of $B\{T\}$ on T is appropriate, namely

$$B\{T\} = B_R + \alpha(T - T_R) - \beta(T - T_R)^2, \quad (16.14)$$

where α and β are positive coefficients.

The dependence of the gross photosynthesis flux GM and the respiration flux RN (Eq. 16.9) on temperature is shown schematically, for a certain PAR level, in Fig. 16.1. The curve for gross photosynthesis is concave-downward, while the curve for respiration is convex. As a result, the optimal temperature for net photosynthesis (gross photosynthesis minus respiration) is lower than that for gross photosynthesis (in this example it is 20°C rather than 24°C). Furthermore, the optimal temperature for crop net photosynthesis

Table 16.1
Scientific names of crop plants mentioned in this chapter

Common name	Scientific name
Cucumber	*Cucumis sativus*
Lettuce	*Lactuca sativa*
Pepper	*Capsicum annuum*
Tomato	*Lycopersicon esculentum*

decreases as the crop develops, because $M\{L\}$ approaches an asymptote as L (namely W) increases, while N remains proportional to W.

Photosynthates produced by the crop are partitioned among its organs. A certain fraction is transformed into shoot structure. Denoting this fraction by r, one can write for a vegetative crop

$$W_s = rW \quad \text{and} \quad W_r = (1 - r)W, \quad (16.15)$$

where the subscripts s and r denote shoot and root, respectively.

The leaf area index, L, can now be obtained, if the mass of stems is negligible and a value is available for the characteristic leaf area ratio (LAR), λ, of the crop. The result,

$$L = \lambda W_s = \lambda r W, \quad (16.16)$$

shows that, under these assumptions, L can be eliminated altogether from the system equation, Eq. (16.9). This leads to a single-state-variable system, which can be represented by a modified Eq. (16.9):

$$\frac{dW}{dt} = G\{P,T,X\}M'\{W\} - R\{T\}N\{W\}. \quad (16.17)$$

Marsh and Albright (1991) modified this standard approach by considering a non-constant LAR. Data collected for lettuce (Marsh and Albright, 1991), tomato, cucumber and pepper (Bruggink and Heuvelink, 1987; Heuvelink, 1989), suggest that high temperatures and low light levels tend to produce thin leaves (high LAR) (see Table 16.1 for the scientific names of crops mentioned in this chapter). This can be used, in principle, by growers to control the LAR and therefore the leaf area of a crop. In particular, higher temperatures may be used to increase leaf area and hence light interception of young plants. In a mature crop, where light interception is already high, additional leaf area does not capture significantly more light, as indicated by Eq. (16.11).

Denoting the LAR of newly structured leaves by

$A\{T,P\}$, the rate of growth of leaf area can be related to the rate of shoot dry matter accumulation as

$$\frac{dL}{dW_s} = A\{T,P\}, \tag{16.18}$$

with A approximated by

$$A\{T,P\} = A_R + \gamma(T - T_R) - \delta(P - P_R), \tag{16.19}$$

where γ and δ are positive coefficients and A_R is the leaf area ratio of new leaves formed at reference temperature and reference PAR.

Combining Eqs. (16.9) and (16.18), and utilizing Eq. (16.15) to substitute for W_s, an equation of motion for the leaf area index, L, is obtained:

$$\frac{dL}{dt} = rA\{T,P\}\left[G\{P,T,X\}M\{L\} - R\{T\}N\{W\}\right]. \tag{16.20}$$

Equations (16.9) and (16.20) constitute the system equations for a crop model with the two state variables W and L.

One problem with simple (and even complex) crop models is that they may not represent processes which are important for optimization of environmental control. A common example is the control of night temperature in the greenhouse. Current models do not simulate chill damage. At night, therefore, the model provides no incentive for heating. On the contrary, low temperature produces an apparent gain by reducing respiration. To remedy this situation a lower constraint is imposed on the greenhouse temperature, say 10°C. The constraint, which reflects expert knowledge, completes the formulation of the crop model. In some cases all that is available to guide the control is a set of constraints (or expert rules) and the problem is then to satisfy these constraints at a minimum cost. Examples of the expert-knowledge approach can be found in the studies of Kozai (1985), Jacobson et al. (1988) and Gauthier and Guay (1990). An explicit use of constraint satisfaction was made by Cros and Martin-Clouaire (1991). The extraction of control rules from an expert source may, perhaps, be assisted by an automatic learning system (Kurata and Eguchi, 1990).

The system defined by Eqs. (16.9) and (16.20) includes just the crop, not its environment. The environment is treated as a boundary condition, represented by the exogenous variables P, T and X. For greenhouse crops it is more natural, as already mentioned, to treat the outside weather as the boundary and to use Eq. (16.5) or Eq. (16.6) to calculate the crop's ambient conditions, which in this way become part of the system. For simplicity it will be assumed here that the greenhouse has no heat or other capacity, so that Eq. (16.6) can be used to determine the inside environment. This assumption also means that no additional state variables (beyond W and L) are required to describe the crop–greenhouse system.

The simplest form of Eq. (16.6), which still retains the essential features of the greenhouse (except humidity), is as follows:

solar radiation:

$$P_i = \tau P_o, \tag{16.21}$$

heat balance:

$$qS_o + F - (U + Q)(T_i - T_o) = 0, \tag{16.22}$$

carbon dioxide balance:

$$\eta\left[C - \left(\frac{Q}{c_p}\right)(X_i - X_o)\right] - G = 0, \tag{16.23}$$

where τ is the transmissivity of the greenhouse to solar radiation, q is the heating efficiency of solar radiation, F is the heat flux delivered by the heater, U is the convective–radiative heat transfer coefficient of the greenhouse, Q is the transfer coefficient due to ventilation, η is the dry matter/carbon dioxide ratio (somewhat larger than the glucose/carbon dioxide ratio, 30/44), C is the carbon dioxide enrichment flux, and c_p is the specific heat of air at constant pressure (G is described in Eq. 16.10). The heat flux F, the carbon dioxide flux C, and the ventilation flux, to which Q is proportional, are the three control fluxes available to the grower in this case. If a shading screen is also available, τ could also serve as a control variable, but to model that case q and U of Eq. (16.22) would have to be formulated as functions of τ.

The energy balance (16.22) shows that T_i can be controlled by F and Q, and the carbon dioxide balance (16.23) shows that X_i can be controlled by means of C and Q. P_i cannot be dynamically controlled in the present case.

To complete the description of the system, the relation between PAR and global solar radiation must be specified. Here a constant ratio, ζ, will be assumed:

$$\frac{P_i}{S_i} = \frac{P_o}{S_o} = \zeta. \tag{16.24}$$

Proper substitutions of Eqs. (16.21) to (16.24) into Eqs. (16.10) and (16.12) and then into the system equations (16.9) and (16.20), yield the following implicit equations (to be compared with Eq. 16.3):

$$\frac{\mathrm{d}W}{\mathrm{d}t} = f_W\{W, L; S_o, T_o, X_o; F, C, Q\}, \tag{16.25}$$

$$\frac{\mathrm{d}L}{\mathrm{d}t} = f_L\{W, L; S_o, T_o, X_o; F, C, Q\}, \tag{16.26}$$

where

$$\boldsymbol{x} \equiv [W, L], \qquad \boldsymbol{e} \equiv [S_o, T_o, X_o] \qquad \boldsymbol{u} \equiv [F, C, Q]. \tag{16.27}$$

Many other dynamic crop and greenhouse models are available, but their general form is always that of Eq. (16.3).

OPTIMAL CONTROL

The system equations which were developed above, once having been supplied with the proper parameter values, can be used for crop simulation, provided the following additional information is available: initial and termination conditions, boundary conditions, $\boldsymbol{e}_o\{t\}$, and control sequence, $\boldsymbol{u}\{t\}$. The initial condition is often simply the state of the system (crop) at time zero. The termination condition is more likely to depend on the specific problem under consideration. For example, it may be specified as a termination time, t_f, or as a final state of the crop, $\boldsymbol{x}_f$, or as the point in time where the net returns peak. The boundary conditions, by definition, cannot be controlled by the operator of the system and must be provided, either as data or through on-line measurements. Finally, it is required to specify the control sequence, which is the subject of this chapter.

The first section below introduces the performance criterion, without which no optimization is possible, and the following sections describe two dynamic optimization methods, which will be used later in the examples. The first is a simplified search procedure which is often suggested for on-line decisions. The second is based on optimal control theory and solves, not just for the control sequence, but also for a control policy (a general rule which generates set-points or control fluxes for any given state and boundary conditions; control law). Other solution methods are also available, from enumeration to dynamic programming. In addition to the two methods to be described below in some detail, linear programming (Gutman et al., 1993) and non-linear programming (van Henten and Bontsema, 1991) have also been used for the optimization of greenhouse control. All these methods require a model of the crop, which, hitherto, has been a weak link in the optimization procedure.

Performance criterion

Different control sequences usually yield different outcomes, some more desirable than others. To select the best sequence, a suitable performance criterion must first be formulated. In accordance with the suggestion made earlier, the net income to the grower is selected here to serve as a performance criterion. Consequently, the higher the performance criterion the better.

Any performance criterion, J, can be expressed as the sum of a termination value and net running gain.

$$J = h\{\boldsymbol{x}\{t_f\}, \boldsymbol{e}\{t_f\}\} + \int_0^{t_f} g\{\boldsymbol{x}\{t\}, \boldsymbol{u}\{t\}, \boldsymbol{e}\{t\}\}\,\mathrm{d}t. \tag{16.28}$$

The first term on the right-hand side of Eq. (16.28) is the termination value of the system. It may be, for example, the price fetched for the final harvest of a crop, which, in turn, depends on the yield (an element of $\boldsymbol{x}\{t_f\}$) and on the market price (an element of $\boldsymbol{e}\{t_f\}$). The second term integrates, over the whole growing season, the increments of gain (sale of produce) and loss (operating costs) resulting from the greenhouse operation. For a crop harvested all at once (e.g. lettuce), the second term represents expenditure only and therefore is negative and decreases monotonically with time. The termination value, h, which represents the value of the crop at harvest time, is positive, and hopefully large enough to make a profit (positive J). If the crop is tomato or cucumber, h is normally zero, while g accumulates all gains and losses over the season.

The performance criterion can be used to compare the actual or simulated outcome of different control sequences (e.g., Jones et al., 1990). The different sequences may be the result of applying different control policies (such as Eq. 16.7). In that case the best policy for a particular season, or on the average, can be identified. However, finding the best policy among several intuitively devised policies goes only part of the way towards the goal of finding the best possible (optimal) policy and the resulting best possible control sequence.

The number of feasible control sequences for any given problem is enormous: every few minutes a

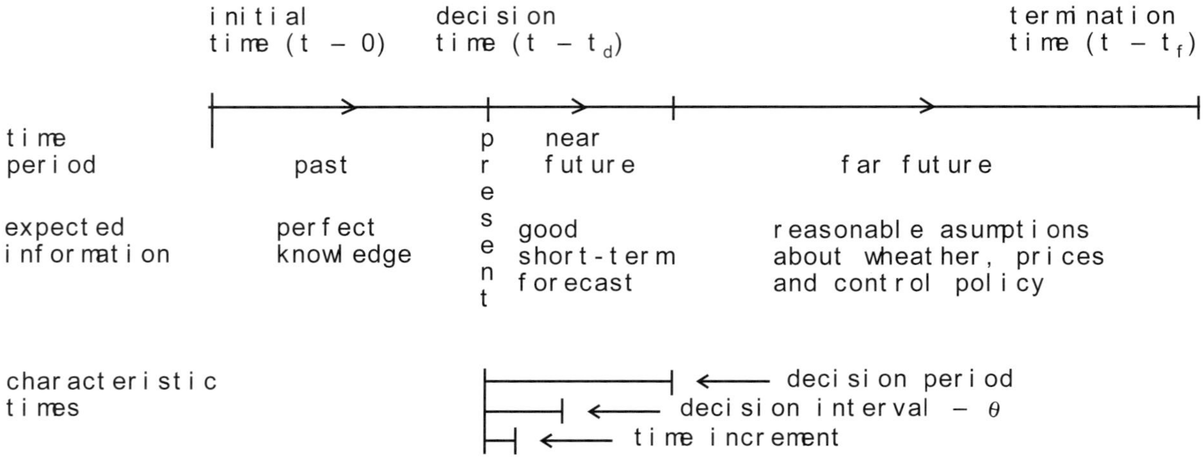

Fig. 16.2. A schematic representation of the various time periods involved in the sequential control search (SCS) method.

decision is made whether to pump more or less heat into the greenhouse, whether to draw the screen, or whether to increase or decrease carbon dioxide enrichment. Trying to find the best control sequence by enumeration, namely by trying all possible sequences, is of course out of the question.

In what follows, two approaches to the optimal control problem will be outlined. The first, which is perhaps intuitively simpler, searches for the best control sequence, without necessarily referring to a control policy. This approach, which may be regarded as a simple version of model predictive control (MPC: Garcia et al., 1989), will be called here sequential control search (SCS). It seems suitable for direct real-time applications. The second utilizes the relationships of Pontryagin's maximum principle (Kirk, 1970) to satisfy the necessary conditions for an optimal solution. Inherent in the latter approach is a control generating function, the Hamiltonian, which can also be interpreted as a control policy. To obtain this control policy, an off-line computation, covering at least one whole season, is used. Once the policy is available, the on-line (real-time) mode is invoked, utilizing the Hamiltonian to produce individual control decisions. This two-step decision structure is described more fully below (pp. 358–359). The Pontryagin maximum principle (PMP) can sometimes be used in a loose manner to obtain suboptimal solutions. The use of this technique will be demonstrated below (pp. 374–376).

Sequential control search (SCS)

The sequential control search approach may be described with reference to Fig. 16.2. Let us assume that, whenever a control decision has to be made (at $t = t_d$), the current state of the system is known, either from observation or from simulation. There is no need to know the history of the crop (before t_d), since all past information which is required for the current decision is contained in the present state of the system. The grower (or computer) now has to make a control decision, one of many sequentially made at time intervals θ (of the order of an hour or less for greenhouse environmental control and of the order of a day or more for field-crop irrigation). To make this decision, the grower has to have information about the future, and in particular information about the near future. The grower may need to know, for instance, what the solar radiation is likely to be in the time interval of length θ between the current and the next control decision. If solar radiation is expected to be high, it may be beneficial to move the required temperature (the temperature set-point) closer to the temperature which is optimal for net photosynthesis. A perhaps more straightforward example is related to irrigation of field crops: if the soil is dry, but the grower expects rain during the coming day, irrigation may be postponed. Otherwise, irrigation should be applied.

In principle, the boundary conditions (exogenous variables) for the rest of the season must be completely known before an optimal decision can be made. Fortunately, however, many problems can be solved in a satisfactory manner with only partial information about the future. In some cases this information is of a statistical nature (probability of occurrence), and methods such as stochastic optimal control or stochastic dynamic programming can be used. For instance, Swaney et al. (1983) applied the sequential control search in conjunction with a performance criterion, Eq. (16.28), modified to represent an expectation computed over several historical weather sequences. The examples presented later in this chapter will be treated differently: certain future information will be considered as perfectly known, while other facets of the future will be just rough estimates. Specifically, the near future may be given by a perfect short-term forecast, while the far future is only known in terms of past experience.

Figure 16.2 shows the near future as being longer than θ, the decision interval, and as coinciding with another interval of time, called here "decision period". The decision period is the interval of time for which a decision sequence is generated at decision time, t_d. Only the first decision of this sequence is implemented at t_d, while the rest may be modified when the next decision time is reached (at $t_d + \theta$). Nevertheless, a plan for the total decision period must be prepared, to make sure that the desired effect can be accomplished. The significance of the decision period is that it corresponds to the characteristic response time of the system, as will be illustrated first by a field-crop example.

Consider a light-soil region where the irrigation cycle (moisture capacity divided by maximum evapotranspiration rate) is 4 days long. The irrigated area is divided into 4 fields, to be irrigated consecutively as needed. Suppose that the irrigation system is in front of field 1 just after a rainfall, which brought the soil (of all fields) to field capacity. Since all fields will reach the moisture threshold which dictates irrigation on practically the same day, field 1 will have to be irrigated three days in advance, so that field 4 does not suffer water stress. However, if by the time that field 1 must be irrigated more rain is predicted to fall 3 days hence, the irrigation of field 1 may be postponed. In this example the decision interval is one day and the decision period is 4 days.

Another example, related to a solar greenhouse[1], is as follows: consider a solar greenhouse with a significant heat storage capability. Roughly speaking, the characteristic time of the storage is obtained by dividing the storage capacity by the net depleting flux. The larger the storage capacity, the larger the characteristic time and the farther into the future one should look for information relevant to one's control decisions. In other words, a good weather forecast for at least the characteristic time is required, otherwise the storage may be used up too fast or too slowly compared to its optimal usage.

Note that the two examples are concerned with constraints on a state variable (soil moisture content in the first example and storage heat content in the second). These constraints cause the state variable to fluctuate between an upper and a lower boundary, thus serving as an operational (temporary) storage, unlike the storage of, say, dry matter in the crop, which increases monotonically. The examples below (pp. 361–370) do not consider constraints on a state variable, but the system described on pp. 365–370 is similar to those mentioned here. It considers a somewhat abstract state variable: crop temperature exposure (temperature integral, commonly measured in degree days), but its basic behavior is just like that of systems with physical storage terms.

One can now proceed to describe the sequential control search technique: at decision time, t_d, the grower has to choose among a number, sometimes a large number, of control options. He may have to select among many possible set-points for temperature and carbon dioxide concentration, or he may only have to decide whether or not to irrigate. If the decision period is longer than the decision interval, *sequences* of control decisions which cover the decision period must be considered (Broner and Lambert, 1989). Feasible control decisions, or decision sequences, are now put to the test by running a simulation of the system to the end of the season (t_f), and evaluating the resulting performance criterion for the same assumed behavior of the far future. The decision (or decision sequence) which yields the best result is selected as the one to be implemented. If the set of feasible controls is large and/or the decision period is long, efficient search methods must be used. Once the proper control decision for t_d has been determined and implemented, the system advances to the next decision time, where the process is repeated.

Note that the assumptions about the future cover not only the boundary conditions (weather and prices) but also the control policy itself. Starting with some not-unreasonable policy for the future (say, constant set-points), the SCS selects at each decision time a decision

[1] A solar greenhouse is one deriving its heating energy from solar energy. The heat collected during the day is stored (in water, gravel, or other materials) to be used for heating at night.

which is as good as or better than the decision derived from the original control policy. This is so because the sequential control search inspects and compares all feasible control decisions, including the one obtained by the old policy. When the new, SCS-generated control decisions for the whole season are obtained, a new control policy can be formulated on this new basis.

Marsh and Albright (1991) used SCS to select the best daytime temperature set-points for greenhouse lettuce. They used a historic weather sequence and a constant set-point for the simulation into the far future. The resulting control sequence indicated that the recommended set-point decreases with time and increases with global radiation. One could, therefore, use a regression equation or a neural network to relate set-point decision to these two variables. The resulting relationship could then be used as a control policy for another season. Note that if one wished to conform to Eq. (16.7), time would have to be replaced by an appropriate state variable, possibly leaf area index.

As it turns out, a formulation of a more elaborate control policy for the far future is usually not required, because a crop model which does not contain operational storage elements (buffers) seems not to require far-future information.

Pontryagin's maximum principle (PMP)

The second approach to optimal control is based on Pontryagin's maximum principle. The principle is a set of necessary conditions which an optimal control sequence must satisfy. Only a very sketchy description of the method will be presented here (for proper exposition of this topic see Kirk (1970), or a similar textbook).

For convenience let us first copy the system equation, Eq. (16.3):

$$\frac{\mathrm{d}\boldsymbol{x}}{\mathrm{d}t} = \boldsymbol{f}\{\boldsymbol{x}\{t\}, \boldsymbol{e}\{t\}, \boldsymbol{u}\{t\}\}, \tag{16.29}$$

and the expression for the performance criterion, Eq. (16.28):

$$J = h\{\boldsymbol{x}\{t_\mathrm{f}\}, \boldsymbol{e}\{t_\mathrm{f}\}\} + \int_0^{t_\mathrm{f}} g\{\boldsymbol{x}\{t\}, \boldsymbol{u}\{t\}, \boldsymbol{e}\{t\}\}\,\mathrm{d}t. \tag{16.30}$$

Pontryagin et al. (1962) showed that the original problem of finding the optimal control sequence, $\boldsymbol{u}\{t\}$, by maximizing *globally* (over the growing season) the functional J, can be transformed to the problem of maximizing *locally* (for every t) over all feasible control, $\boldsymbol{u}$, the function called the Hamiltonian

$$H\{\boldsymbol{x}, \boldsymbol{e}, \boldsymbol{u}, \boldsymbol{p}\} = g\{\boldsymbol{x}, \boldsymbol{e}, \boldsymbol{u}\} + \boldsymbol{p}\{t\}\boldsymbol{f}\{\boldsymbol{x}, \boldsymbol{e}, \boldsymbol{u}\}. \tag{16.31}$$

The vector function $\boldsymbol{p}\{t\}$ is called the adjoint (co-state) function. It represents the value worth paying for the marginally produced unit of $\boldsymbol{x}$ (say, dollars per kg of dry matter). Intuitively, a higher value of $\boldsymbol{p}$ implies a higher production intensity, because the grower is willing to pay more to produce at a faster rate. Each of the state variables in $\boldsymbol{x}$ has an adjoint variable associated with it. If the value of a particular adjoint variable is high, the effort (expenditure) to increase the corresponding state variable should be high. If it is negative, an effort should be made to reduce the magnitude of the state variable. An adjoint variable which decreases with time through the season reflects the diminishing importance of the corresponding state variable (e.g., leaf area).

The right-hand side of Eq. (16.31) may be interpreted as follows: g from Eq. 16.30;, represents the *current* rate of net gain, while $\boldsymbol{pf}$ is the rate of investment in the *future*. In other words, $\boldsymbol{pf}$ represents the rate at which the potential for future gains is created. The correct balance between the two terms is maintained by the values of the adjoint variables. If one considers the vegetative phase of a greenhouse tomato crop, g is negative, because no fruit is available yet to be sold. That does not mean that effort should not be made to grow the leaves, which eventually will supply the carbohydrates to the saleable fruits. On the last day of the season, however, it is no longer justified to invest in any state variable that does not bear immediate fruits (e.g., leaves). As a result, certain adjoint variables approach zero towards the end of the season.

The rate of change of the adjoint variables with time must satisfy the adjoint equation

$$\frac{\mathrm{d}\boldsymbol{p}}{\mathrm{d}t} = -\frac{\partial H}{\partial \boldsymbol{x}}, \tag{16.32}$$

which shows that if the Hamiltonian, H, is not an explicit function of a particular state variable, then the corresponding adjoint variable remains constant with time.

The adjoint function, $\boldsymbol{p}\{t\}$, serves as a dynamic Lagrange multiplier and is initially unknown. In the process of solving for the optimal control, the adjoint function is also determined. Once it is known, the Hamiltonian becomes a control-generating function, since at a given time and for a given state, $\boldsymbol{x}$, and boundary conditions, $\boldsymbol{e}$, it is now possible to find a feasible control, $\boldsymbol{u}$, which maximizes the Hamiltonian. The Hamiltonian may be regarded, therefore, as a control policy in the sense that it is a rule which can produce a control decision. The adjoint variables may, therefore, be regarded as the parameters of that policy. If the parameter values (which may change with time) are incorrect, the control decision is not optimal.

A deterministic optimal control problem (when $\boldsymbol{e}\{t\}$ is assumed to be known in advance), may be solved numerically, utilizing the following computational sequence: assume that at some point along the solution path, a value for the vector $\boldsymbol{p}\{t\}$ is available. This value is used to maximize the Hamiltonian (16.31), yielding a value for the control vector $\boldsymbol{u}$. The system equation, Eq. (16.29), can now be used to advance the state vector, $\boldsymbol{x}$, utilizing the new value of $\boldsymbol{u}$, as well as the appropriate value of $\boldsymbol{e}$. Finally, Eq. (16.32) is used to modify $\boldsymbol{p}$, leading into the next point in time. This sequence is repeated until the termination condition is satisfied.

To start the computation, an initial guess $\boldsymbol{p}\{0\}$ must be available. Every initial value $\boldsymbol{p}\{0\}$ may yield a different value of the performance criterion, J, thus making J a function of $\boldsymbol{p}\{0\}$. The optimal value of $\boldsymbol{p}\{0\}$, namely that which maximizes J, must satisfy

$$\frac{\mathrm{d}J}{\mathrm{d}\boldsymbol{p}\{0\}} = 0, \tag{16.33}$$

and is obtained by searching over the s-dimensional $\boldsymbol{p}\{0\}$ space.

The optimal solution, besides maximizing the performance criterion (by satisfying Eq. 16.33), must also satisfy certain termination (transversality) conditions on $\boldsymbol{p}$ (like $p\{t_f\}=0$, as mentioned above). Since the maximum principle will later be used in a somewhat loose manner, the transversality conditions will no longer be useful and therefore will not be further discussed here.

When the maximum principle approach is applied to a problem in real time, future weather and prices are unknown and the initial values of the adjoint variables must be based on an off-line solution for a *likely* season. If these values are not the correct ones for the unfolding boundary conditions (weather and prices), the resulting control sequence will not be optimal. There is an indication (Seginer and Sher, 1992) that the initial values of the adjoint variables, $\boldsymbol{p}\{0\}$, are not sensitive to changes in the weather from one season to another. In any case, a fresh search for the correct adjoint-variable vector can be started at any time by disregarding its current value (obtained through Eq. 16.32), and resolving for the remainder of the season, while using Eq. (16.33) to point to the correct value. Since repeating a complete solution each time a decision is to be made may be computationally prohibitive, in practice a compromise approach may have be to be used. One possibility is to invoke a complete solution (for the rest of the season) only when the current season accumulates a considerable deviation from the climatic mean. Seginer and Sher (1992) demonstrated that, at least in certain cases, the adjoint variables can be normalized with respect to the expected market price, so that changes in price do not require a renewed seasonal optimization.

IMPLEMENTATION

Once the methodology for obtaining optimal dynamic control solutions for greenhouses is sorted out, the next step is to implement them. The methods on which this chapter focuses produce set-points, either directly (SCS) or by an explicit control policy (PMP). The determination of these set-points is constrained by guidelines produced by earlier decisions (such as choice of crop and greenhouse). At the same time, the set-points themselves serve as guidelines for a "lower" level of the control system, which is responsible for their implementation. This hierarchical description of the greenhouse control system is extended to several levels in the next section. The description is qualitative and is only intended to show the issue of set-point generation in the right perspective.

Complete on-line seasonal optimization for each set-point decision may require a prohibitively large computational effort, even with short-cut methods such as the SCS. Hence, the possibility is subsequently considered of using neural networks to mimic the optimal decision process.

Hierarchical control

The derivation of a control policy has been shown in previous sections to depend, among other factors, on the properties of the system. Specifically, the values of the adjoint variables, $\boldsymbol{p}\{t\}$ (serving to specify the production intensity), depend, among other factors, on the air-conditioning equipment available. For instance, one would expect the optimal production intensity for a greenhouse with a heating system and a carbon dioxide generator to be higher than for a greenhouse with just a simple ventilation system. The dependence of the control policy (intensity) on the available technology shows that when deciding on a strategy (plan), the tactical level (operation) must also be considered. This is true of any planning and design process, and results in a series of back-and-forth bouncing between decision levels during the planning stage.

Although the various decision levels cannot be uncoupled, the relations between the levels are not symmetrical. In *planning mode* (planning stage) the lower levels of the control hierarchy have more influence on the upper levels than in *operating mode*. Unless a serious mistake has been made in the planning stage, the lower levels become completely subordinate to the upper levels during operation. For instance, once a crop has been selected, it is not likely to be replaced before the end of the season. The heating system will have to do its best to provide the crop with the required temperature.

With this in mind, the various decision levels for a greenhouse system can now be described. The following scheme is similar to schemes suggested by others (e.g., Udink ten Cate et al., 1978; Challa and van Straten, 1991), but differs somewhat in its interpretation. The scheme starts at the highest managerial level still within the domain of the single farm, and ends at the purely mechanical operation of the actuators (motors, switches, etc.). The two top levels are outside the scope of this chapter and are sketched here just to give the lower levels some perspective. Each level represents in some sense the strategy for the next lower level. Starting with the top level, the general picture is as follows:

(Level 5) Decisions about type of crop (vegetables, flowers), size of holding, marketing channels, etc. Decisions made once every several years.

(Level 4) Decisions about area and time allocation for each crop during a given season. Decisions at this level are subject to available resources and economic environment provided by Level 5, and are made once every season.

(Level 3) Determination of control policy (e.g., initial values of adjoint variables) for crops selected in Level 4. This decision may be revised several times during a season, if conditions are found to deviate significantly from the assumptions made at a previous revision.

(Level 2) Generation of control decisions (fluxes or set-points) from the control policy determined in Level 3. Checked and revised several times an hour.

(Level 1) Maintenance of the set-points generated at Level 2. This operation is essentially continuous.

The present chapter deals mainly with Levels 3 and 2. The computations of Level 3 are meant to be performed off-line, possibly in a facility which serves several or many holdings, while the computations of Level 2 (as well as those of Level 1) are the responsibility of the greenhouse computer (microprocessor). As will be seen later, if a suitable control policy is available at all times, the generation of control decisions requires very little computing time. It may involve an evaluation of an explicit expression, such as Eq. (16.7) (perhaps even tabulated, as on p. 363) or the maximization of a Hamiltonian, such as Eq. (16.31), over all feasible control values.

An exception to this decision structure may be found if the sequential control search (SCS) method is adopted. That method does not refer to a control policy, thus making Level 3 unnecessary.

Neural networks[2] as a control policy

Independent of the method of optimization, each optimal control decision requires complete information on the current state of the system and on the future sequence of the exogenous variables. It follows that the control decision is a function of this information. Since in practice the state and the future can only be estimated, control decisions are based on these estimates.

Functional relationships can be approximated by various methods, regression analysis being one of them. There are two stages in fitting a functional relationship (curve) to available data. First, the relevant variables and the form of the function (e.g., linear) are selected, then follows parameter evaluation (identification), often with least-squares methods. Neural

[2] A neural network is a "black-box" input–output device, which utilizes combinations of s-shaped (logistic) functions to mimic functional relationships. Such a structure may be trained (fitted) to represent data-based information and the result may be used as a model for the phenomenon that the data represent.

networks (McClelland et al., 1986) are now routinely used (among other uses) to approximate complex multivariable, non-linear functions. The curve-fitting user still needs to select the input and output variables, but virtually all the rest (including the form of the function) is taken care of by the neural network (NN) training algorithm. The remarkable flexibility of NNs in capturing the details of complex relationships stems, to a large extent, from the large number of parameters (weights) embedded in the net.

Attempts to use NNs to mimic greenhouse-control algorithms have recently been made with some success (Seginer and McClendon, 1992; Seginer and Sher, 1992), and are reported below (pp. 377–378). The technique is simple: historical weather and price sequences are first used to generate optimal control sequences (using some optimization method). The input likely to be available in practice is then used in conjunction with these optimal control decisions to train a NN. It is important to have a training set which spans all likely future situations (patterns). Neural networks, like any other empirical model, should not be used to extrapolate beyond the available data.

The input to the NN should include important, yet easily measurable, elements of the state of the crop and of the current weather, as well as a weather forecast and a price forecast. Sometimes it is satisfactory to have substitutes for elements which are difficult to measure. As an example, time from planting is, under controlled conditions, well (but not necessarily linearly) correlated with the state of the crop, and may be used to substitute for it.

When the training of the NN is completed (its weights now having particular values), the NN may be embedded in a greenhouse controller, where it might serve as a control generating function, namely a control policy. On-line computing time with this arrangement is minimal.

CARBON DIOXIDE ENRICHMENT OF A CROP WITH A SINGLE STATE VARIABLE

The examples to be presented in this and the next three sections comprise a small and rather restricted sample out of the wide range of optimal dynamic greenhouse-control problems. The approach is basically deterministic, even though mention is made of the uncertainty of future developments. This approach seems justified when dealing with problems which are insensitive to assumptions about the future. In particular, systems with large operational storage terms (buffers) have been almost completely ignored. The single example dealing with such a situation (pp. 365–370) has been solved for a stationary season (where each day is a perfect repetition of the previous one), to avoid complications. Furthermore, problems with seasonal resource constraints, analogous to irrigation of field crops under a seasonal water quota, have not yet been considered in conjunction with greenhouse climate control. De-emphasizing future uncertainty is a characteristic of highly controllable systems, such as greenhouses. This is also the reason for touching upon risk-taking only very briefly below.

The main purpose of these examples is to provide a first impression of the possibilities that optimal control opens up, both in terms of better understanding of the greenhouse system and as a potential control tool.

This section deals with a model for a crop represented by a single state variable. Such a model is extremely simple compared to present-day crop models, not to mention the actual crop. It is surprising, therefore, that even this first approximation is capable of mimicking, and even refining, present-day control decisions. Generally speaking, the control policy for this case turns out to support the common view that environmental control is first and foremost a function of the outside weather (e.g., day versus night). It is also capable of producing *continuously varying* control decisions for a large number of control devices operating *simultaneously*. Whether these decisions are really optimal depends mostly on the validity of the crop model.

The next section describes a crop model which leads to a *constant* control policy, and the following section shows how the parameters of this constant policy depend on the economic constraints under which the system is operated. A more detailed treatment of the subject can be found in Seginer et al. (1986) and in Seginer (1989). A range of other attempts to optimize carbon dioxide enrichment can also be found in the literature (e.g., Challa and Schapendonk, 1986; Heissner et al., 1986; Critten, 1991; van Henten and Bontsema, 1991; Chalabi, 1992).

Constant control policy

Let us consider a simplified version of the crop model

with a single state variable described by Eq. (16.17), namely

$$\frac{dW}{dt} = G\{P,T,X\}M\{W\}. \tag{16.34}$$

The function G, which now represents *net* rather than gross photosynthesis rate, is a function of the meteorological conditions, $\boldsymbol{e}$, and the control decisions, $\boldsymbol{u}$, but not of the state of the crop, W. This description of the system, which is less general than Eq. (16.4), permits the transformation of W into a new state variable, Z, via

$$\frac{dW}{dZ} = M\{W\}. \tag{16.35}$$

As a result, Eq. (16.34) transforms into

$$\frac{dZ}{dt} = G\{P,T,X\}, \tag{16.36}$$

which is no longer a function of the state of the crop. Next, the general performance criterion, Eq. (16.28), is simplified (in both notation and substance) to

$$J = h\{t_f\} - \int_0^{t_f} \boldsymbol{cu}\,dt, \tag{16.37}$$

where $\boldsymbol{c}$ is the vector of unit prices of the control fluxes $\boldsymbol{u}$, and $\boldsymbol{cu}$ reflects the assumption that control cost is proportional to the fluxes employed (e.g., heat and carbon dioxide fluxes). The termination condition is $W\{t_f\} = W_f$, namely, a given final state of the crop (termination time is left free). This is a natural termination condition for crops which are harvested when ripe. It also seems to fit the problem of nursery-produced seedlings, which should be marketed when they reach a certain size. Incidentally, if the price is assumed to remain constant with time, the termination value may be completely removed from the performance criterion, leaving just the running costs, $C = \int \boldsymbol{cu}\,dt$.

Based on Eqs. (16.36) and (16.37), the Hamiltonian (16.31) becomes

$$H = -\boldsymbol{cu} + p_Z\{t\}\,G\{P,T,X\}, \tag{16.38}$$

which is equivalent to

$$H\{\boldsymbol{e}_o, \boldsymbol{u}\} = -\boldsymbol{cu} + p_Z\{t\}\,f\{\boldsymbol{e}_o, \boldsymbol{u}\}, \tag{16.39}$$

where p_Z is the adjoint variable associated with Z (Eq. 16.36). The last equation shows the Hamiltonian not to depend on the state variable (whether W or Z). In view of the adjoint equation, Eq. (16.32), the adjoint variable, p_Z, must, therefore, remain constant with time (this conclusion is independent of the termination condition). Since Eq. (16.39) is used to produce the optimal control in real time (by maximizing H over $\boldsymbol{u}$), it is clear that $\boldsymbol{u}$ is a function only of $\boldsymbol{e}$ and not of the evolving state. In other words, the set-points, in this particular case, are independent of the phase of development (here termed state) of the crop, unlike, for instance, the practice in poultry houses.

Note that a constant control policy does not imply constant control fluxes, nor even constant set-points. The set-points, although not responding to changes in the state of the system, do respond to changes in the meteorological conditions, not only between night and day (as commonly practiced), but also between bright and dull and between warm and cold periods.

The single-state-variable model (16.34) was used to simulate tomato seedlings grown in a greenhouse nursery (Seginer et al., 1986). Some results of that study are shown in Figs. 16.3, 16.4 and 16.5. Figure 16.3 shows the total running costs, C, as a function of the

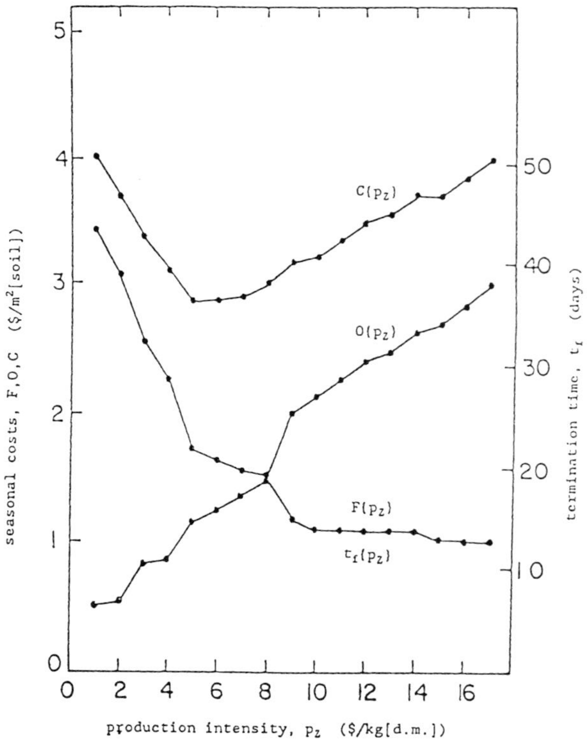

Fig. 16.3. Seasonal fixed costs, F (doubling as termination time, t_f), operational costs, O, and total cost, C, all as functions of production intensity (adjoint variable), p_Z, for production of tomato seedlings. Results for a greenhouse equipped with a carbon dioxide generator. After Seginer (1989).

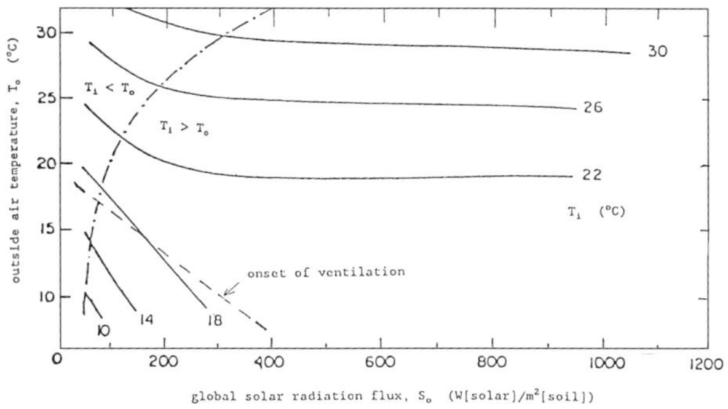

Fig. 16.4. Recommended temperature set-points (in °C) for a greenhouse without carbon dioxide enrichment system, as a function of outside meteorological conditions. The crop is tomato seedlings. The slanting dashed line is the boundary between a non-ventilated (bottom left) and ventilated greenhouse. After Seginer et al. (1986).

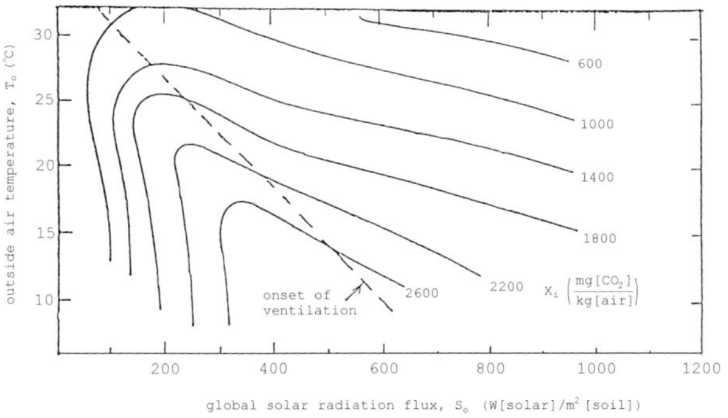

Fig. 16.5. Recommended set-points for carbon dioxide concentration for a tomato-seedling crop as a function of outside meteorological conditions. The slanting dashed line is the boundary between a non-ventilated (bottom left) and a ventilated greenhouse. To convert carbon dioxide concentration values from mg [CO_2] kg^{-1} [air] to µmol [CO_2] mol^{-1} [air], multiply values in the figure by 0.66 = 29/44. After Seginer et al. (1986).

value of the adjoint variable, p_Z, here conveniently equated with production intensity. $C\{p_Z\}$ is the sum of operating costs (such as carbon dioxide enrichment), $O\{p_Z\}$, and fixed costs (such as rent), $F\{p_Z\}$. The operating cost $O\{p_Z\}$ is seen to increase linearly with production intensity, p_Z, due to increased utilization of heat and carbon dioxide. This increased cost cannot be compensated, in the present case, by increased yield, because it was assumed that W_f is fixed. In addition it was assumed that the market price for the produce is also fixed, so that the termination value is independent of the termination time, t_f (this assumption will be relaxed in the next section). The remaining trade-off is, therefore, between the operating costs, $O\{p_Z\}$, and the fixed costs, $F\{p_Z\}$: the higher the production intensity, the faster the crop reaches its final state, reducing total rent or equivalent expenses in the process. This reduction in fixed costs is most pronounced at low cultivation intensities, leading to a fixed costs curve which is asymptotic. The total running costs, $C\{p_Z\}$ (which here is identical to the negative of the performance criterion) shows a minimum at about $p_Z = 6\,\$\,kg^{-1}$ [d.m.]. This is the price that the grower should be paying (at any instance along the growing season) for the production of the marginal unit of transformed dry matter. If conditions at some point in time are not conducive to photosynthesis, it is best not to force growth by increasing p_Z beyond $6\,\$\,kg^{-1}$ [d.m.], but rather to wait until conditions improve.

Sample results of recommended set-points for temperature and carbon dioxide concentration are shown in Figs. 16.4 and 16.5. Figure 16.4 shows the temperature set-point as a function of outside solar radiation and temperature. Roughly speaking, the figure can be divided into two regions: the lower left corner, where the set-point lines are diagonal, and the rest of the figure, where the lines are roughly horizontal. The lower left corner represents a non-ventilated greenhouse, whose equilibrium temperature increases as the heat load (due to both outside temperature and solar radiation) increases. Ventilation starts (dashed line) when the inside temperature increases above about 18°C (note that here the ventilation is *not* thermostatically controlled). From there on the inside temperature is maintained as close as possible to the outside temperature. The region in the upper left corner of the figure, where $T_i < T_o$, is the result of evaporative cooling of the crop, which is followed by cooling of the inside air.

Figure 16.4 is for a greenhouse without carbon dioxide enrichment. If a carbon dioxide generator is used, the onset of ventilation moves to a considerably higher heat load (dashed line in Fig. 16.5). In this case the onset of ventilation coincides approximately with a set-point of 28°C, about 8 K higher than the optimal temperature for photosynthesis. Apparently the high carbon dioxide concentration which can be maintained in the closed greenhouse compensates for the damaging effect of high temperature. It is common practice to increase the temperature set-point in greenhouses enriched with carbon dioxide (Hannan and Holley, 1978; their Table 4-1), but not as much as these results suggest.

Figure 16.5 shows the recommended carbon dioxide concentration set-points. The ridge in this figure is just below the line which indicates the onset of ventilation, the maximum recommended concentration being about 2000 μmol[CO_2] mol^{-1}[air] (to convert from kg[CO_2] kg^{-1}[air] to μmol[CO_2] mol^{-1}[air], multiply by 0.68×10^6). Note that when the greenhouse is not ventilated, carbon dioxide concentration is mainly a function of solar radiation. As the heat load and, as a result, the ventilation rate increase, the generated flux of carbon dioxide (not shown) increases too, but the resulting concentration declines. Figure 16.5 shows that there may be an advantage in simultaneous ventilation and carbon dioxide enrichment, provided that the ventilation rate is reduced to a level which produces the best trade-off between temperature and carbon dioxide concentration.

Although the quantitative results of Figs. 16.4 and 16.5 should not be taken at face value, the trends are most likely correct. Results of this kind may be produced at Control Level 3 (p. 360) and stored in the memory of a greenhouse computer as a series of tables (or as a neural network). Each time a control decision has to be made at Level 2, the computer must first interrogate the meteorological sensors, produce a forecast for the near future (which here may be only 15 min) and then search the tables for the best set-points for this expected future. Alternatively, Control Level 3 may be designed to transfer to Level 2 just the (single) value of p_Z (here about 6 \$ kg^{-1}) and let Level 2 maximize the Hamiltonian (16.39) over all feasible $\boldsymbol{u}$, once the expected weather ($\boldsymbol{e}$ in Eq. 16.39) is available. The first approach requires more memory and less computing time at Level 2 than the second approach.

Economic constraints

The tomato seedlings problem was solved in the preceding section under the implicit assumption that the grower is limited by a production quota. In that case the grower's policy should be to minimize production cost, which the preceding procedure has insured (Fig. 16.3). Other constraints on the grower's operation may require different performance criteria, hence resulting in different values for the adjoint variable (i.e., different production intensity). These other constraints will have no effect on the *constancy* of the adjoint variable, as long as the Hamiltonian (Eq. 16.39) is not a function of the state variable (Z; Eq. 16.36), and as long as the state does not hit a state constraint (pp. 365–370, below).

The effect of two economic constraints, namely limited production quota and limited greenhouse area, can be easily demonstrated for a climatically uniform year. In this artificial case the total production cost, $C\{p_Z\}$, and the termination time $t_f\{p_Z\}$, while being functions of the intensity of production, p_Z (Fig. 16.3), are independent of the starting time of a growing cycle. Under this assumption, and ignoring a possible effect of p_Z on produce quality, a rather general performance criterion per unit time, for a greenhouse of area A under continuous cropping, is:

$$J'\{p_Z\} = \frac{A(k\{t_f\{p_Z\}\} - C\{p_Z\})}{t_f\{p_Z\}}, \tag{16.40}$$

where k is the unit price of the produce (e.g., tomato seedlings). The numerator, which is the net income from the greenhouse over one crop cycle, allows market prices to change with time. For any given initial time (sowing date), the unit price for the produce, k, becomes a function of the termination time, which, in turn, is a function of the production intensity, p_Z.

Let us assume now that a production quota, Q (here expressed as seedling-tray area per unit time), is imposed on the grower. To meet the quota, the greenhouse area to be used, A, must satisfy

$$Q = \frac{A}{t_f\{p_Z\}}. \tag{16.41}$$

Hence, once a value of p_Z is selected, this determines the area to be planted (rented). Substituting A from Eq. (16.41) into Eq. (16.40) yields the relevant performance criterion for the quota-constrained case:

$$J'\{p_Z\} = Q(k\{t_f\{p_Z\}\} - C\{p_Z\}). \tag{16.42}$$

In order to maximize this expression, one requires that $\mathrm{d}J'\{p_Z\}/\mathrm{d}p_Z = 0$ (Eq. 16.33), namely:

$$\frac{\mathrm{d}k\{t_f\}}{\mathrm{d}t_f}\frac{\mathrm{d}t_f\{p_Z\}}{\mathrm{d}p_Z} - \frac{\mathrm{d}C\{p_Z\}}{\mathrm{d}p_Z} = 0. \tag{16.43}$$

In the simple case, where the price k is constant with time, Eq. (16.43) reduces to $\mathrm{d}C\{p_Z\}/\mathrm{d}p_Z = 0$, which places the optimal solution at the minimum production cost, *independent* of the value of k. This is the solution obtained in the previous section.

In general, since in this problem termination time decreases with increased production intensity (i.e., $\mathrm{d}t_f\{p_Z\}/\mathrm{d}p_Z$ non-negative; Gal et al., 1984), the sign of $\mathrm{d}C\{p_Z\}/\mathrm{d}p_Z$ must be opposite to the sign of $\mathrm{d}k\{t_f\}/\mathrm{d}t_f$. Specifically, if $\mathrm{d}k\{t_f\}/\mathrm{d}t_f > 0$ around t_f, $\mathrm{d}C\{p_Z\}/\mathrm{d}p_Z$ must be negative, corresponding to a p_Z value that is lower than appropriate for the original problem (namely, left of the minimum of $C\{p_Z\}$ in Fig. 16.3). In other words, if the market price of the produce increases in the neighborhood of marketing time, a certain delay in reaching the final (maturation) state W_f is desirable. Note again that k itself (in contrast with dk/dt) does not affect the optimal choice of p_Z. This means that a grower with a limiting production quota should not change the intensity of production, p_Z, in response to the *level* of market prices for the produce. A higher price level will obviously increase the profit, according to Eq. (16.40), but should not affect the greenhouse operation.

Let us suppose now that the greenhouse area, A, rather than the quota, Q, is the limiting factor. The optimal solution to this problem, obtained by maximizing Eq. (16.40) with respect to p_Z (A treated as constant), requires that:

$$t_f\{p_Z\}\left(\frac{dk\{t_f\}}{dt_f}\frac{dt_f\{p_Z\}}{dp_Z}-\frac{dC\{p_Z\}}{dp_Z}\right) - (k\{t_f\{p_Z\}\}-C\{p_Z\})\frac{dt_f\{p_Z\}}{dp_Z}=0, \qquad (16.44)$$

where both the price level, $k\{t_f\}$, and the rate at which it changes, $dk\{t_f\}/dt_f$, may affect the solution. Referring to the simple case where prices are constant with time ($dk/dt=0$), Eq. (16.44) reduces to

$$t_f\{p_Z\}\frac{dC\{p_Z\}}{dp_Z}+(k-C\{p_Z\})\frac{dt_f\{p_Z\}}{dp_Z}=0, \qquad (16.45)$$

Since $dt_f\{p_Z\}/dp_Z$ is never positive (as mentioned above), and $t_f\{p_Z\}$ is never negative, $k-C\{p_Z\}$ must have the same sign as $dC\{p_Z\}/dp_Z$. If the solution of Eq. (16.45) is such that $k>C\{p_Z\}$, then also $dC\{p_Z\}/dp_Z>0$, which means that the solution is to the right of the minimum total cost in Fig. 16.3. In other words, high prices for the produce justify, in the area-limiting case, higher intensities of cultivation. This is not surprising, since a higher production intensity reduces the residence time of the crop (t_f). This makes more cultivation cycles possible and results in marketing more produce.

The simple example of this section shows that the control policy is affected not only by physiological constraints (such as minimum permissible temperature) or engineering constraints (such as installed heater capacity), but also by economic constraints. One can think of several other constraints which may be classified as economic, for instance a shortage of labor. In particular, there may not be labor available on weekends. The latter situation should affect the control policy in a way that, for example, would minimize the loss due to not harvesting at the optimal time.

ENERGY SAVING THROUGH TEMPERATURE INTEGRATION

The previous section dealt with a situation where a crop model is available. Although often this is not the case for greenhouse crops (but see Leutscher and Vogelezang, 1990; Erwin and Heins, 1990; Jones et al., 1991; Lieth and Pasian, 1991), the grower has the next-best knowledge base, namely a recommended policy in terms of set-points. This policy, which is based on expert knowledge, may be crude, such as distinguishing only between day and night, or more refined, such as relating set-points to solar radiation. Recognizing the limitations of commercial greenhouses, these set-points are normally given as a range or as upper and lower limits (for instance: ventilate to maintain a temperature not higher than 28°C, heat to prevent the temperature from dropping below 16°C). Such recommendations are in fact better than rigid constant set-points, even if the latter could be maintained in practice. The reason is that they imply minimum energy expenditure as long as an action is not absolutely required.

This suggests that, even if a crop model is not available, systems with considerable operational storage (of, say, heat) offer an opportunity for optimization as long as an experience-based set-point *range* is provided. In that case the goal of the controller is simply to stay within the permitted set-point range (satisfy the constraints, p. 354) at the minimum cost. This is a limited goal, because the controller cannot consider effects on timing and level of yield, but, pending the availability of a sufficiently large storage capacity, the monetary savings may be considerable.

Usually the operational storage is conceived as a physical entity with capacity for storing mass or energy. The purpose of a storage is to match, as far as possible, the supply and demand curves. An obvious example is the storage of solar heat during daytime so that it can be used at night. The following example shows that an abstract storage term can also serve the same purpose.

A bang-bang control policy

Some experimental studies (Langhans et al., 1981; Hurd and Graves, 1984; Challa and Brouwer, 1985;

de Koning, 1990) have indicated that plants can tolerate a certain variation about the optimal temperature without ill effects, provided the deviations of the temperature and of the temperature exposure (temperature integral) I, given, for a greenhouse crop, by $I_{\rm i}$:

$$I_{\rm i} = \int T_{\rm i} dt, \tag{16.46}$$

are within bounds ($T_{\rm i}$ is the actual greenhouse air temperature). Thus, it is permissible for the greenhouse temperature to drop below the normal heating set-point, provided that there is a period when it is higher, so that the desired temperature integral (or mean temperature) is achieved over some prescribed time interval. As a consequence, it should be possible to shift some heating to times when the increased temperature is less expensive to produce (Ludolph and Hendriks, 1989), or benefits the growing process in some other way (e.g., by facilitating carbon dioxide enrichment, van Emmerik, 1990). The time interval over which the temperature has to be averaged, here a property of the crop, is the characteristic time considered on p. 357, but it will remain implicit throughout most of the following discussion.

A simple measure of deviation from the recommended temperature exposure is

$$I_{\rm d}\{t\} = \int (T_{\rm i} - T_{\rm ir}) dt, \tag{16.47}$$

where $T_{\rm ir}$ is the recommended temperature, and $I_{\rm d} = 0$ indicates complete adherence to its trajectory. The integrated deviation, $I_{\rm d}\{t\}$, is a state variable of the system, whose equation of motion is

$$\frac{dI_{\rm d}}{dt} = T_{\rm i} - T_{\rm ir}. \tag{16.48}$$

To stay within the limits of effective integration by the crop, $I_{\rm d}$ is constrained to an interval, such as

$$I_{{\rm d}n}\{t\} < I_{\rm d}\{t\} < I_{{\rm d}x}\{t\}, \tag{16.49}$$

where n and x denote minimum and maximum values (usually $I_{{\rm d}n}\{t\} < 0$). This interval can be used to normalize the state variable so that it spans the range 0 to 1:

$$i_{\rm d} = \frac{I_{\rm d} - I_{{\rm d}n}}{I_{{\rm d}x} - I_{{\rm d}n}}. \tag{16.50}$$

Since an explicit crop model is not involved, and the greenhouse model, Eq. (16.22), is algebraic, $I_{\rm d}$ (or $i_{\rm d}$) turns out to be the only state variable of the system.

The performance criterion for this problem is simply the negative of the seasonal cost of heating energy (ventilation cost is assumed here to be negligible compared to the cost of heating)

$$J = -c \int F dt, \tag{16.51}$$

where c is the unit price of energy. This leads to the Hamiltonian (Eq. (16.31):

$$H = -cF + p_{\rm I}(T_{\rm i} - T_{\rm ir}), \tag{16.52}$$

which, after substitution for $T_{\rm i}$ from Eq. (16.22), becomes

$$H = \left(\frac{p_{\rm I}}{U+Q} - c\right) F + \left(T_{\rm o} - T_{\rm ir} + \frac{qS_{\rm o}}{U+Q}\right) p_{\rm I}. \tag{16.53}$$

Since the Hamiltonian is not a function of the state variable, the value of $p_{\rm I}$ must be constant, at least while the state variable is not on its constraint. One also expects $p_{\rm I}$ to be positive in a heating problem, since it reflects a positive effort to increase the temperature exposure. The optimal control is found by maximizing the Hamiltonian over all feasible control fluxes, $\boldsymbol{u} \equiv [F, Q]$ (compare with Eq. 16.27), subject to the various constraints which the problem imposes. The minimum number of constraints compatible with the present problem are one on the temperature exposure, Eq. (16.49), one on the temperature itself

$$T_{{\rm i}n}\{t\} < T_{\rm i}\{t\} < T_{{\rm i}x}\{t\}, \tag{16.54}$$

and the obvious constraints on the equipment, $F \geqslant 0$ and $Q \geqslant 0$.

Let us now make the assumption that heating, F, and ventilation, Q, are mutually exclusive (no humidity and carbon dioxide control). Inspecting the Hamiltonian (16.53) for the ventilation mode, one finds that the Hamiltonian is maximized for the minimum Q compatible with the constraints on $I_{\rm d}$ and on $T_{\rm i}$. This means that during ventilation the system will operate at either $I_{\rm d} = I_{{\rm d}x}$ or $T_{\rm i} = T_{{\rm i}x}$, independent of the value of the adjoint variable $p_{\rm I}$.

More important is the behavior of the heater, since it is the only contributor to the performance criterion, Eq. (16.51). When the heating is on ($F > 0$, $Q = 0$), the Hamiltonian becomes

$$H = \left(\frac{p_{\rm I}}{U} - c\right) F + \left(T_{\rm o} - T_{\rm ir} + \frac{qS_{\rm o}}{U}\right) p_{\rm I}, \tag{16.55}$$

which, for a given $p_{\rm I}$, is linear in F. The coefficient of F, namely

$$s = \left(\frac{p_{\rm I}}{U} - c\right), \tag{16.56}$$

is termed the *switching* function, and may be positive or negative, depending on the values of U and p_I. If it is positive, the maximization of the Hamiltonian would require the maximization of F (subject to the constraints), while if it is negative, F must be minimized.

The preceding analysis shows that if the system is operated optimally, the control always lies on the boundary of its feasible domain. In particular, this so-called *bang-bang* control results in discontinuous (jumpy) control fluxes (such as the on–off operation of a greenhouse air-heater). A similar conclusion was reached by Aikman and Picken (1989), who considered the effect of variable wind on greenhouse temperature control.

Energy saving

The preceding simple formulation of the heating problem can handle a wide variety of energy sources and energy-conserving devices (such as screens), utilizing each at the most profitable time. Let us consider a few examples, using Eqs. (16.21) (slightly modified) and (16.22) to describe the greenhouse:

$$S_i = \tau S_o, \tag{16.57}$$

$$qS_o + F - (U + Q)(T_i - T_o) = 0, \tag{16.58}$$

The heating and ventilation control fluxes, F and Q, are considered here as continuous. In addition to these fluxes, some greenhouses are equipped with either a shade or a thermal screen or both. These have normally only two positions: either fully open or fully closed. The effect of these screens on the energy balance of a greenhouse is mainly through the parameters τ in Eq. (16.57) and U in Eq. (16.58): drawing a shade during daytime reduces the transmissivity to solar radiation, τ, and drawing a thermal screen during night-time reduces the heat-loss transfer coefficient, U. The shade is not a saver of heating energy, but it may be required in certain locations to prevent the inside temperature from climbing above its permitted range (Seginer, 1990).

The use of a thermal screen at night (or perhaps even on extremely dull and cold days) saves energy by reducing U, namely, by making the cost of an additional degree lower than during daytime. This suggests that, where daytime heating is required, and provided that the crop can tolerate a deviation I_d from the recommended temperature exposure over a period of one day, some heating can be shifted from day to night. Consequently, if a thermal screen is available, daytime temperature set-points would be lower and night-time set-points would be higher than without a screen. The best sequence of set-points can be obtained by first finding the appropriate value of the adjoint variable, p_I, and then using it to maximize the Hamiltonian (16.55). When the system is not on a state variable (I_d) constraint, the switching function, Eq. (16.56), is a function of U alone. Hence,

$$U_t = \frac{p_I}{c} \tag{16.59}$$

is a threshold value of U, which, when transgressed by lower values of U, triggers heating.

For example, suppose that wind speed becomes high enough so that $U > U_t$. This means that the greenhouse should not be heated to a temperature above the absolute minimum as required by the various constraints. Any further heating should be put off until the wind calms down to a level which satisfies $U \leqslant U_t$. If p_I is large, heat will be applied at higher wind speeds than otherwise. A similar argument holds for the effect of a thermal screen, provided that U_t is between the value of U for a screened house and that for the same house unscreened.

Consider now a situation where two sources of energy are available to the greenhouse operator. When a low-temperature heat source, such as reject heat or solar energy, is used, it is uneconomic to design the heating system for the maximum heat demand. Instead, it should be designed to meet a certain base heat load, while peak demands are covered by a second heater, for which fuel costs may be high but capital investment is low. To minimize heating costs, maximum use should be made of the base-load heater (de Koning, 1988). The solution of this case proceeds as follows: starting from a performance criterion with two heating terms (compare with Eq. 16.51)

$$J = -c_1 \int F_1 \mathrm{d}t - c_2 \int F_2 \mathrm{d}t;$$
$$c_2 > c_1, \quad 0 \leqslant F_1 \leqslant F_{1x}, \quad 0 \leqslant F_2 \leqslant F_{2x}, \tag{16.60}$$

two switching functions are obtained:

$$s_1 = \frac{p_I}{U} - c_1, \qquad s_2 = \frac{p_I}{U} - c_2 \tag{16.61}$$

which define two threshold values

$$U_{t1} = \frac{p_I}{c_1}, \quad U_{t2} = \frac{p_I}{c_2}; \quad U_{t2} < U_{t1}. \tag{16.62}$$

Whenever $U > U_{t1}$, no heat is applied (unless a

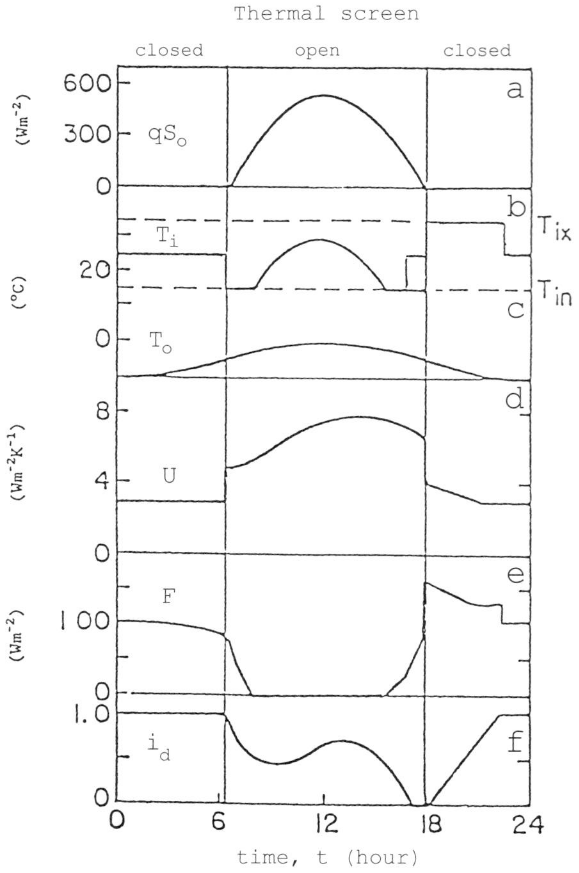

Fig. 16.6. Diurnal variation of outside meteorological conditions (S_o and T_o), control flux (F), and resulting conditions inside the greenhouse (T_i). Also shown are the transfer coefficient (U) as it is affected by wind and the thermal screen, and the normalized temperature exposure deviation (i_d). After Seginer (1988).

constraint requires it); when $U_{t2} < U \leqslant U_{t1}$, only the less expensive heat source is used (at $F_1 = F_{1x}$, if feasible); and when $U \leqslant U_{t2}$, both sources of heat are used simultaneously and at their peak capacity (again, so long as all constraints are satisfied). Note that a single value of p_I controls both heating systems (as well as the ventilation fan). A proper selection of p_I will make the heating operation as inexpensive as possible. Note that if a crop model with a single state variable was to replace the constraints on the temperature exposure, a single adjoint variable could be used to control *simultaneously* not only the two heaters and fan, but also a carbon dioxide generator, a shade, a thermal screen and other equipment.

Some results to illustrate the preceding analysis are shown in Figs. 16.6 to 16.9. First are shown the detailed decisions taken over a daily cycle for a synthetic

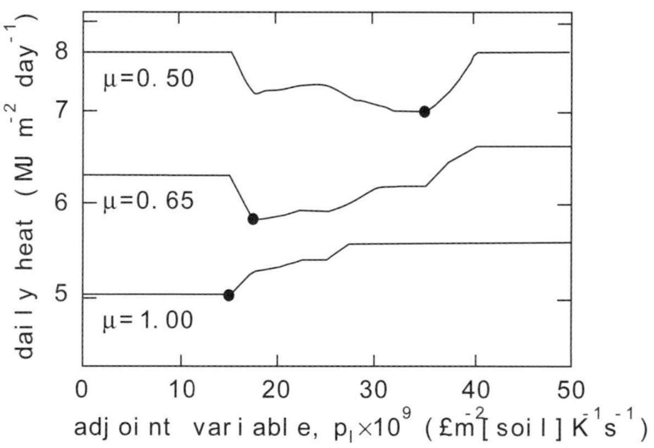

Fig. 16.7. Total heating energy required under three different solar radiation regimes (bottom curve is for a clear day; $\mu = 1$), for a range of the adjoint variable, p_I. Solid circles indicate the location of the optima. After Seginer (1988).

periodic weather (each day repeats the weather of the day before). A single heater with unlimited capacity is assumed (Seginer, 1988).

The heating efficacy of solar radiation and the outside temperature, shown in Figs. 16.6a and 16.6c, respectively, are symmetrical about noon. The temperature fluctuates between −10°C and 0°C (cold weather). The qualitative behavior of wind speed can be inferred from Fig. 16.6d, which shows the transfer coefficient, U. Wind speed reaches a maximum at 14:00 and is insignificant during most of the night. The value of U is also affected considerably by the thermal screen, which is opened at sunrise and closed at sunset (it, too, could have been controlled optimally, if a crop model was available; Bailey, 1988). The constraints on the inside temperature are shown by dashed lines in Fig. 16.6b.

The results for a not quite optimal constant value of p_I are shown in Figs. 16.6b, 16.6e and 16.6f. At midnight the thermal screen is in a closed position, the transfer coefficient, U, is low and the temperature exposure is at its upper constraint ($i_d = 1$; Fig. 16.6f). Heat (F; Fig. 16.6e) is supplied as necessary to maintain the temperature exposure at its upper constraint. From Eq. (16.48) this leads to a constant inside temperature, fixed at the recommended level (here $T_{ir} = 25$°C; Fig. 16.6b). At sunrise the thermal screen is opened and U suddenly increases. The heater operates now only to maintain the minimum permissible temperature ($T_{in} = 15$°C; Fig. 16.6b). As soon as solar heating is capable of maintaining the minimum temperature, the heater is turned completely off (about 08:00; Fig. 16.6e). Since even at noon T_i

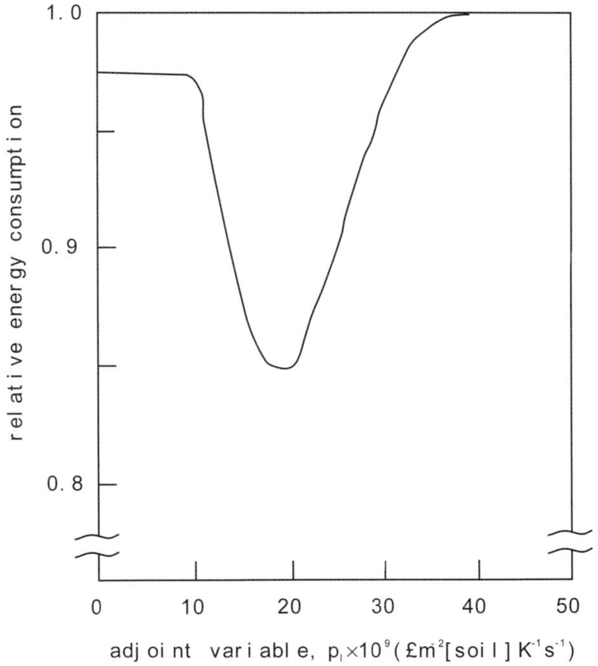

Fig. 16.8. Annual consumption of heating energy using a constant p_I, compared with conventional heating, both for a greenhouse with a thermal screen. For English conditions. After Bailey and Seginer (1989).

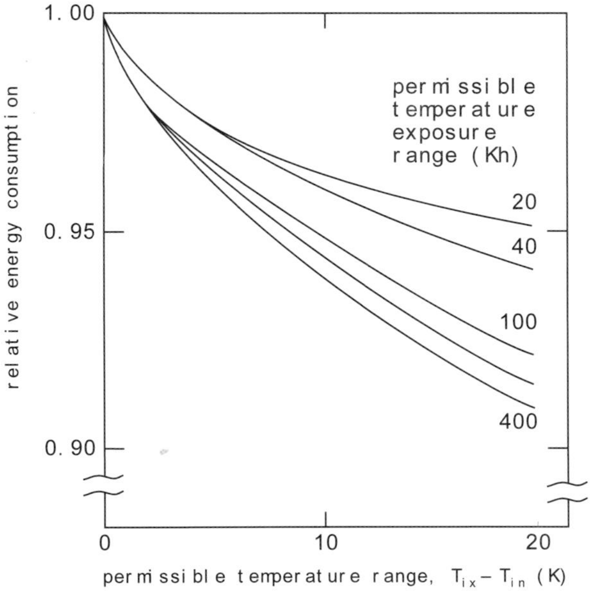

Fig. 16.9. Influence of the constraints on temperature and temperature exposure on energy consumption. English conditions. After Bailey and Seginer (1989).

does not reach the maximum temperature, T_{ix}, the ventilation fan is not turned on. Thus, during most of the daytime hours the heater and fan are both idle ($F=0$, $Q=0$). When the sun can no longer maintain the required minimum temperature, heating is turned on again (about 16:00; Fig. 16.6e). Shortly before the thermal screen is closed for the night, the temperature exposure reaches its lower constraint (about 17:00; Fig. 16.6f). As a result, more heat is delivered to the greenhouse to maintain the inside temperature at 25°C. As soon as the thermal screen is closed and U returns to its low night level, the controller seizes the opportunity to eliminate the deficiency of the temperature exposure. Heat is delivered at the maximum rate compatible with the maximum permissible temperature. This lasts for about 4 hours (until about 22:00), when the temperature exposure reaches its upper constraint – that is, the available "storage" is full again. Note that the system is at all times on at least one of the constraints (between about 8:00 and 16:00 the constraint is $F=0$). Note also the asymmetrical nature of i_d, which is depleted during daytime and replenished during the night.

It may seem counter-intuitive to recommend temperatures higher by night than by day (Fig. 16.6b). This increases night maintenance respiration without apparent gain. However, growth, namely transfer of carbohydrates from their temporary storage ("pool") to structural dry matter, requires elevated temperatures during at least part of the day. Hence, under the conditions of Fig. 16.6, this vital process will be shifted from day to night, when it is less expensive to provide higher temperatures.

The solution presented in Fig. 16.6 is not quite optimal, because it assumes, incorrectly, that the adjoint variable is constant throughout the entire daily cycle. The constancy of the adjoint variable (derived from Eq. 16.53) is valid, however, only as long as the state variable is not on one of its constraints. In other words, the optimal adjoint variable may have one value between 06:00 and 17:00 and a different value between 18:00 and 22:00.

Gutman et al. (1993) remedied the shortcomings of the preceding solution: first, by finding two different values for the adjoint variable, and second, by recognizing that the night-time solution is not unique, since filling-up the i_d "storage" may be postponed up to eight hours without loss. As a result, the correct solution is different in detail from Fig. 16.6.

Returning to the suboptimal solution with a constant adjoint variable, the effect of solar radiation level on the choice of p_I will now be illustrated. Solar radiation level was expressed in terms of the brightness factor, μ, which is the ratio between actual solar radiation level and its level on a clear (cloudless) day. Fig. 16.7 shows the effect that the choice of p_I has on the performance criterion (required heating, in energy units) for each of three solar radiation levels. When the sky is clear ($\mu = 1$, bottom curve) the best value of p_I is low ($p_I < 15 \times 10^{-9}$ £ m^{-2} K^{-1} s^{-1}). For duller days higher values of p_I are optimal (about 35×10^{-9} £ m^{-2} K^{-1} s^{-1}) when $\mu = 0.5$). In other words, duller seasons (less solar heating) require more marginal expenditure to keep the temperature exposure within its permissible range. Note that when $\mu = 0.5$, the level of the performance criterion resulting from too low and too high p_I values is the same. In the first case any advance replenishment of heat may be carelessly postponed by the risk-taking operator, until the constraint forces heating at any cost. In the second case (high p_I) the risk-aversive operator wishes to keep the "storage" full at all times, just "to be on the safe side".

The term "storage" will continue to be used in what follows for convenience; but note, again, that it does not represent an actual physical storage of heat. There are indications, however (Gary, 1988a,b), that temporary storage of carbohydrates, before they are converted to structural dry matter, may be related to the so-called temperature-integration capability of plants. In future studies, therefore, real carbohydrate storage may replace the fictitious temperature exposure as a state variable of the system.

Situations not covered by the example of this section are cases of dull (and/or cold) days followed or preceded by bright (and/or warm) days. In such a case the controller should take advantage of a bright day, so that the following dull day starts with a full heat "storage". Conversely, if it is clear that the current cold day will be followed by a warm day, the "storage" does not have to be full when the new day starts. A complete analysis of a real season may require the adjoint variable, p_I, to jump occasionally (perhaps twice a day) from one value to another. This has not been attempted yet, but can be accomplished with the approach of Gutman et al. (1993), provided that a short-term weather forecast is available. The forecast horizon should match the integration time constant of the crop.

At present only a suboptimal season-long solution is available (Bailey and Seginer, 1989). It employs a single (constant) value of p_I throughout the season. Figure 16.8 shows that about 15% of current energy usage of English greenhouses with thermal screens can be saved if the best single value of p_I is selected. More energy can be saved with a truly optimal p_I trajectory.

The ability to shift energy demands to times when less expensive energy is available depends on the size of the *effective* "storage" made available by the temperature and temperature-exposure constraints, Eqs. (16.54) and (16.49), respectively. Figure 16.9 shows how these constraints may affect the energy saving for conditions in England. Note again that, although an explicit crop model has not been used in specifying this heating problem, the effective "storage", as defined by the constraints, is in fact a statement about the "comfort zone" of the crop.

Before moving on to a new subject, a brief outline will be given of a control problem which is similar to the temperature-integral problem and could be treated in a similar manner. It has been shown that excessive humidity in the greenhouse may damage the crop by limiting transpiration and hence the transport of minerals (e.g., calcium) to locations where they are required (e.g., tomato fruit). As a result, it has been proposed to replace humidity set-points by transpiration "set-points" (Stanghellini and van Meurs, 1992). This approach can be made more flexible by requiring that the transpiration *integral* does not fall below a certain level over a characteristic time period (probably at least one day). A tolerable delay in calcium supply is analogous to such a delay in carbohydrate removal from its operational storage.

HEATING OF A CROP WITH TWO STATE VARIABLES

The simple crop with a single state variable (Eq. 16.36) led to a solution with a constant value for the adjoint variable, p. The temperature integral problem required a piece-wise constant p. Other models may require adjoint variables which vary continuously with time and whose evolution is controlled by the adjoint equation (Eq. 16.32). Changes in the values of the adjoint variables imply that the effort (expense) directed towards promoting the corresponding state variables changes with time. Specifically, the changing adjoint variables may prompt the grower to pay more attention to one state variable at the beginning of the season and to another near its end. This section deals with a vegetative crop with two state variables, where the emphasis shifts from leaf area to dry matter as the crop develops.

Problem characteristics

The crop under consideration here is greenhouse lettuce, for which there is some evidence that leaf area ratio, A, increases with temperature and decreases with light intensity, as in Eq. (16.19). The problem is to find the best daytime temperature set-points for a given weather sequence, the termination condition this time being a pre-specified termination time, t_f. The set-point is changed once every day, presumably when the next day's forecast becomes available. Marsh and Albright (1991) solved this problem by applying the sequential control search (SCS) method (pp. 356–358). In the next section results obtained with the SCS will be shown to coincide with the results obtained by the maximum principle (PMP, pp. 358–359). This means that the sequential control search method did, in fact, find the optimal solution, at least in this case. Before any numerical results are shown, however, it may be instructive to examine the qualitative behavior of the control policy, as reflected in the adjoint variables.

The basic crop model used to simulate the lettuce crop is given by Eqs. (16.9) and (16.20), which are copied here for convenience:

$$\frac{dW}{dt} = G\{P,T,X\}M\{L\} - R\{T\}N\{W\}. \tag{16.63}$$

$$\frac{dL}{dt} = rA\{T,P\}(G\{P,T,X\}M\{L\} - R\{T\}N\{W\}). \tag{16.64}$$

The energy balance equation is Eq. (16.22):

$$qS_o + F - (U + Q)(T_i - T_o) = 0, \tag{16.65}$$

with the proviso, as before, that the heat flux F and the ventilation flux Q are mutually exclusive. Carbon dioxide generating equipment is not available, and humidity control is not considered. Heat flux is charged at a unit price c, and ventilation flux is free. Note that the carbon dioxide balance equation, Eq. (16.23), has been omitted, for simplicity. It will be assumed, in what follows, that the standard carbon dioxide concentration is somehow maintained in the greenhouse. A more realistic formulation of the problem would include a carbon dioxide balance equation, but this may result in a simultaneous operation of ventilation and heating on cold bright days.

The performance criterion (compare with Eq. 16.28) contains three terms: price of produce, cost of rent and cost of heating

$$J = (k'\{t_f\}W_s\{t_f\} - \upsilon t_f) - \int_0^{t_f} cF\{t\}\,dt, \tag{16.66}$$

where $k'\{t_f\}$ is the market price of lettuce at harvest time, $W_s\{t_f\}$ is shoot dry matter at harvest time, and υ is the cost of rent (or all fixed costs). Since t_f is known in advance, υt_f is a constant and may be omitted from the optimization. The Hamiltonian (16.31) can now be obtained from Eqs. (16.63), (16.64) and (16.66):

$$H = p_W\frac{dW}{dt} + p_L\frac{dL}{dt} - cF, \tag{16.67}$$

or alternatively (see Eqs. 16.25–16.27):

$$H = (G\{F,Q\}M\{L\} - R\{F,Q\}N\{W\}) \times (p_W + p_L rA\{F,Q\}) - cF, \tag{16.68}$$

which shows explicitly the dependence of the Hamiltonian on the control fluxes F and Q.

The adjoint equations, Eq. (16.32), now read:

$$\frac{dp_W}{dt} = (p_W + p_L rA)R\left(\frac{\partial N}{\partial W}\right), \tag{16.69}$$

$$\frac{dp_L}{dt} = -(p_W + p_L rA)G\left(\frac{\partial M}{\partial L}\right). \tag{16.70}$$

It seems reasonable to assume that p_W and p_L are positive, because the grower is expected to make a positive effort (spend more) to achieve the growth of both W and L. With this assumption one can explore the expected changes of p_W and p_L with time. Equation (16.13) shows that $\partial N/\partial W = 1$, while Eq. (16.11) shows $\partial M/\partial L$ to decrease from a to 0 as the crop grows. Both derivatives being positive, Eqs. (16.69) and (16.70) now suggest that, to achieve an optimal control, p_W should increase with time, while p_L should decrease. In other words, the emphasis on producing dry matter should increase with time, while the emphasis on producing leaf area should decrease. Since leaf area is enhanced by high temperatures (γ positive in Eq. 16.19), a decreasing p_L is expected to result in a decreasing temperature set-point.

The value of p_L at termination time must satisfy (Kirk, 1970, p. 200) the transversality (termination) condition

$$p_L\{t_f\} = \frac{\partial h}{\partial L}\{t_f\}. \tag{16.71}$$

Since h (Eqs. 16.28, 16.66) is not a function of L (lettuce is sold by weight, not by leaf area), $p_L\{t_f\}$ must be zero. The other adjoint variable, p_W, must satisfy the condition

$$p_W\{t_f\} = rk'\{t_f\}. \tag{16.72}$$

When the optimal sequence of the adjoint variables is eventually computed (Seginer et al., 1991; Seginer

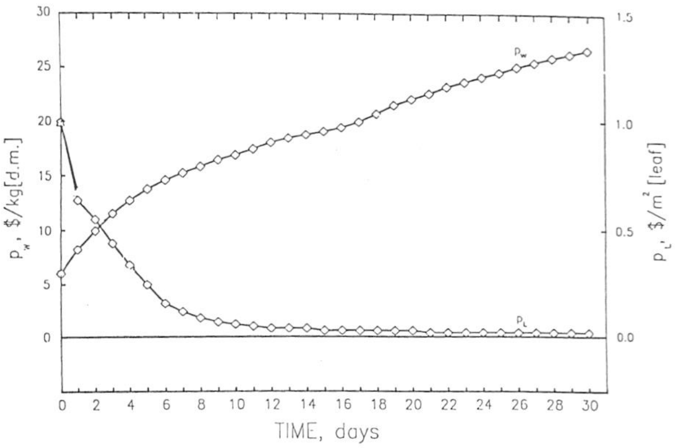

Fig. 16.10. Evolution of the adjoint variables p_W and p_L, over a period of one month, for the two-dimensional lettuce model. Results for free heating energy. After Seginer and McClendon (1992).

and McClendon, 1992), the results turn out to support the conclusions of the qualitative analysis. Figure 16.10 shows the results for a particular weather sequence (January 1986, in upstate New York): p_W increases monotonically with time, while p_L decreases to zero, as expected. Note that p_L becomes negligibly small within ten days. During that time the leaf area index, L, increases from 0.24 to 5.5, at which point the soil becomes completely covered, the canopy intercepting all available solar radiation.

An explanation for the shifting of emphasis from leaf area to dry matter production has been offered by Marsh and Albright (1991) and mentioned briefly in the paragraph following Eq. (16.17). If the leaves of the young plants are made as large as possible for the dry matter already available, more solar radiation can be captured and used for further carbon dioxide assimilation. As the crop completely covers the surface (soil), no additional gain can be had by larger leaf area and the emphasis shifts to maximum dry matter production. As a result, the temperature must change from the level which best serves leaf expansion to the (lower) temperature which is best for net photosynthesis. After all, it is shoot *dry matter*, W_s, that is sold on the market, as the performance criterion, Eq. (16.66), clearly asserts.

A different view of the behavior of the adjoint variables can also be offered. It turns out (Seginer and Sher, 1992), that for a crop with a single state variable, such as described by Eq. (16.17), the adjoint variable, p_W, first decreases rapidly, as does p_L in Fig. 16.10, and then increases slowly as does p_W in that figure. A possible explanation is as follows: since the young seedling grows exponentially, it follows that, under stationary environmental conditions, the cost of producing a unit of dry matter on an early day is much higher than it is several days later. This is immediately reflected in a decreasing adjoint variable. At a later stage of growth, when the cover is complete, more and more leaves are shaded and become a respirational burden. This reduces the rate of net production, increases the cost of producing an additional unit of dry matter, and therefore the adjoint variable. If the state variable W can be transformed into Z (Eq. 16.35), which represents the accumulation of *relative* growth, the initial decrease of p_W transforms into a constant value of p_Z. As a result, the control becomes a function of the meteorological conditions alone, not of the state of the crop.

The preceding qualitative analysis and discussion shows the ability of the maximum principle to provide a useful insight into a problem.

Solution by the sequential control search and by the maximum principle

Now that the qualitative solution is available, numerical solutions can be attempted. A solution obtained by the maximum principle will be, practically always, optimal. Since this is a well-behaved problem, a comparison between the results of the sequential control search and the maximum principle may be used to check on the optimality of the SCS solution.

As already mentioned, Marsh and Albright (1991), who solved the lettuce problem by SCS, formulated it originally so that the temperature set-point is changed only once a day. This made the number of decisions, and therefore the number of full-season simulations, only 1/24 of the number if hourly decisions had to be made. The required computing time for a seasonal solution is reduced, as a result, by a factor of 576 (24^2). The optimal solution to the problem with daily decisions is just an approximation to the more realistic problem with hourly or more frequent decisions. However, since the interest is mainly in *comparing* the two computation methods, the choice of decision interval should not make much difference. Note that in this problem the decision period coincides with the decision interval, because the weather for the next day must be known before the effect of one daily set-point can be distinguished from that of another.

The simulated on-line SCS solution can start with an arbitrary, but reasonable, initial policy for the future – for instance, $T_{ir} = 25°C$ (Marsh and Albright,

1991) – and a reasonable future weather. Each time a set-point decision is to be made (following the method described on pp. 356–358), a new 24-hour perfect weather forecast becomes available. When the decision is made, the simulation is advanced one day to the next decision time, and the process continues. When the computation reaches the end of the season, it can be repeated, using the now-available true weather. This new, presumably optimal, solution can be compared to the "on-line" solution. The difference between the performance criteria is the loss due to having incomplete information about the future when the on-line computation was carried out.

The SCS solution for the perfectly known future is obtained by successive iterations, each one using the former control sequence as the best estimate of the far-future control. If the performance-criterion surface has no local minima, the solution is likely to converge to its optimum.

Unlike the SCS solution, which can be applied directly on-line, the maximum principle solution requires first an off-line computation to produce suitable initial values $\boldsymbol{p}\{0\}$. These may be obtained by solving for several past seasons and finding the vector $\boldsymbol{p}\{0\}$ which maximizes the mean return. The new season is then handled on-line with that $\boldsymbol{p}\{0\}$, to produce a solution which is suboptimal (due to incomplete information about the future). Just as for the SCS solution, an optimal PMP solution can be obtained a posteriori for comparison with the suboptimal one.

Several solutions to the lettuce problem (for January 1986, as in Fig. 16.10) are shown in Fig. 16.11. The two methods, SCS and PMP, produced very close solution sequences (Seginer and McClendon, 1992). The difference between the two solutions in terms of the performance criterion was only about one in a thousand. The results show very clearly the expected reduction in set-point temperature as the importance of leaf area declines with time (p. 357). Another feature of the results (not shown in Fig. 16.11) is that set-points extending above the general trend are associated with higher solar radiation levels (Marsh and Albright, 1991). The sequence of set-points may, therefore, be viewed as the result of superimposing a solution solely dependent on weather (the up-and-down fluctuations), over a solution solely dependent on the state of the crop (the decreasing mean relationship). Note that the solution is constrained by $T_{ix}=30°C$. The unconstrained solution would recommend even higher temperatures over the first few days.

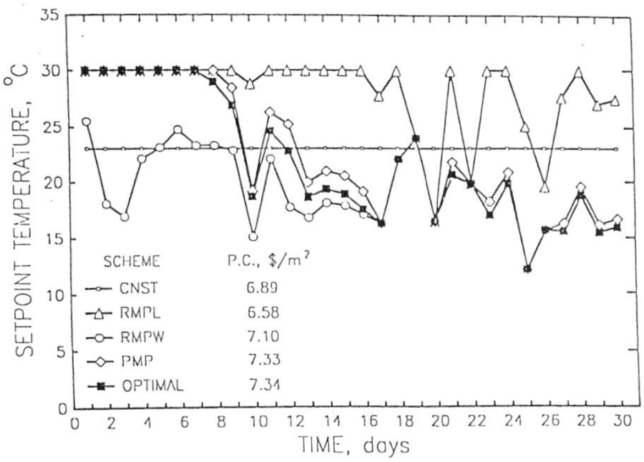

Fig. 16.11. Optimal and suboptimal sequences of daytime-temperature set-points for the two-dimensional lettuce model. Heating energy is free. Solid squares, sequential control search; open diamonds, maximum principle with p_W and p_L varying as in Fig. 16.10; open circles, RMP with $p_L=0$; triangles, RMP with $p_W=0$; horizontal line, best constant temperature. After Seginer and McClendon (1992).

The agreement between the two solutions suggests that SCS did in fact find the optimal solution. Moreover, it turns out that simulations of on-line computations (i.e., only once through), with considerably different future weather and future control sequences, produced almost identical results (Seginer, 1991). This leads to an important observation, namely that the solution of this problem is insensitive to assumptions about the future.

It should be mentioned that the solution shown in Fig. 16.11 has been obtained under the assumption that energy is free ($c=0$). This assumption simplifies the Hamiltonian (16.68) to

$$H=(G\{F,Q\}M\{L\}-R\{F,Q\}N\{W\}) \times (p_W+p_L rA\{F,Q\}). \tag{16.73}$$

One may recall that the optimal control (here $\boldsymbol{u}\equiv[F,Q]$) is found by maximizing $H\{\boldsymbol{u}\}$ over all feasible $\boldsymbol{u}$. Hence, multiplying p_W and p_L of Eq. (16.73) by the same factor would not change the optimal choice of $\boldsymbol{u}$. This means that the ratio $\rho\equiv p_W/p_L$, rather than the absolute values of each of the adjoint variables, is important in this particular case. The practical result is that, instead of searching over a two-dimensional space for the best initial pair $[p_W\{0\},p_L\{0\}]$, it is now sufficient to search over a one-dimensional space for the best ratio $\rho\{0\}$. The solution for real cases, where energy is not free, is more time-consuming, but presents no substantial difficulty.

It is evident from Fig. 16.11 that the assumption of

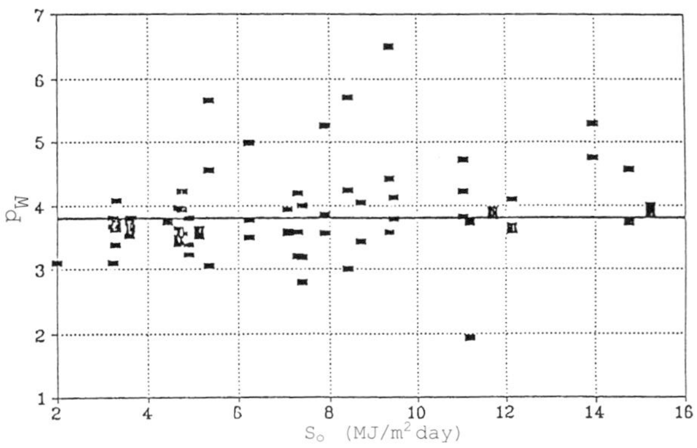

Fig. 16.12. Optimal monthly values of the adjoint variable p_W plotted against monthly mean solar radiation. After Seginer and Sher (1992).

free energy ($c = 0$) did not produce a trivial solution, as one might suspect. The "conflict of interests" between leaf area and dry matter, as well as between gross photosynthesis and respiration, produces the "checks and balances" that make the solution a non-trivial compromise.

As already mentioned, the off-line PMP computation mode provides the initial value $\boldsymbol{p}\{0\}$ for the on-line mode. This value contains (implicitly) all the assumptions about the future that the PMP method is making. The chosen $\boldsymbol{p}\{0\}$ may, of course, not be exactly correct for the new season, but there are indications that the problem may not be serious. At least in one case (Seginer and Sher, 1992), the loss of performance resulting from the use of a single (mean) value of $\boldsymbol{p}\{0\}$ for a wide range of weather conditions was found to be negligible. Figure 16.12 shows the optimal initial values $p_W\{0\}$ as a function of mean daily solar radiation for the month in question. The scatter is rather narrow and no correlation with solar radiation is apparent. Since the optimization is not very sensitive to $p_W\{0\}$, no perceptible loss of performance due to using the mean value of $p_W\{0\}$ could be detected (not shown here).

Approximate control policy

The main drawback of the maximum principle approach becomes apparent as the number of state variables increases. Even a problem with as few as three or four state variables may prove to be too large for the computers ordinarily used in agricultural applications (Shina and Seginer, 1989). This famous "curse of dimensionality" (Bellman and Dreyfus, 1962) makes the application of the PMP to large crop models virtually impossible. To overcome this difficulty, complex models could be replaced by simple ones, such as those developed on pp. 352–355. This, however, may be an oversimplification, leading to serious control errors. An alternative approach, to be pursued here, is as follows.

One may recall that $\boldsymbol{f}$ of Eq. (16.3) represents the complete (detailed) system model:

$$\frac{\mathrm{d}\boldsymbol{x}}{\mathrm{d}t} = \boldsymbol{f}\{\boldsymbol{x}\{t\}, \boldsymbol{e}\{t\}, \boldsymbol{u}\{t\}\}. \tag{16.74}$$

This equation can be used to simulate the evolution of the state, $\boldsymbol{x}$, provided that the exogenous (boundary) variables, $\boldsymbol{e}\{t\}$, and the control decisions, $\boldsymbol{u}\{t\}$, are known. When the simulation is carried to the end of the season, so that the final state, $\boldsymbol{x}_f$, becomes available, the performance criterion, J (Eq. 16.28) can be calculated. Assuming that the (complete) model is perfect and that the boundary conditions are known, the remaining issue is the control decisions, which normally are obtained by maximizing the complete Hamiltonian (16.31):

$$H\{\boldsymbol{u}\} = g\{\boldsymbol{x}, \boldsymbol{e}, \boldsymbol{u}\} + \boldsymbol{p}\{t\}\,\boldsymbol{f}\{\boldsymbol{x}, \boldsymbol{e}, \boldsymbol{u}\}, \tag{16.75}$$

which can be expanded to read:

$$H = g + \sum p_j f_j = g + p_1 f_1 + p_2 f_2 + \cdots + p_s f_s. \tag{16.76}$$

If some of the product terms, $p_j f_j$, are removed from Eq. (16.76), the remaining relationship, denoted by

$$H'\{\boldsymbol{u}\} = g\{\boldsymbol{x}, \boldsymbol{e}, \boldsymbol{u}\} + \boldsymbol{p}'\{t\}\,\boldsymbol{f}'\{\boldsymbol{x}, \boldsymbol{e}, \boldsymbol{u}\}, \tag{16.77}$$

defines a *reduced* Hamiltonian (the vectors $\boldsymbol{x}$, $\boldsymbol{e}$ and $\boldsymbol{u}$ are the complete original vectors). Using the reduced Hamiltonian instead of the complete one to produce control decisions, is equivalent to setting the omitted adjoint variables to zero for the entire length of the season.

It is now possible to advance the simulation, including control decisions, in the manner already described above (pp. 358–359), except with a reduced Hamiltonian. The corresponding adjoint equations are also reduced in number, so that Eq. (16.32) is now written as

$$\frac{\mathrm{d}\boldsymbol{p}'}{\mathrm{d}t} = -\frac{\partial H'}{\partial \boldsymbol{x}'}, \tag{16.78}$$

where $\boldsymbol{x}'$ is a reduced state-variable vector corresponding to the reduced adjoint-variable vector $\boldsymbol{p}'$.

The result of solving a control problem in this manner is at best suboptimal, but in some cases, as will be soon demonstrated, it may be satisfactory. This method may be referred to as the "reduced" maximum principle (RMP) method.

Before moving to the numerical results, a few comments concerning the proposed procedure should be made.

(1) The performance criteria calculated by this "approximate" procedure are correct for the decision sequence used, because the *simulation* is realized with the *complete system model*.
(2) The best solution out of those which were generated is selected using the *proper performance criterion* as a guide. This solution may not satisfy the transversality (termination) conditions (such as Eq. 16.71).
(3) One would want to remove from the Hamiltonian the "least important" adjoint variables, namely those whose values are close to zero, for example, p_L of Fig. 16.10. The selection of good candidates for omission may, however, require some subjective judgement, since information such as is shown in Fig. 16.10 becomes available only when the problem is solved properly.
(4) In the limit, when no variable is removed, the approximate procedure is identical with the original PMP procedure.
(5) The trajectories (sequences) of $\boldsymbol{p}'$ obtained by the approximate procedure are not the trajectories obtained (for the same variables) by the complete solution.

With these comments in mind let us consider the results of applying the RMP procedure to the lettuce model with two state variables. Let us suppose that p_L is set to zero (removed from the Hamiltonian). The Hamiltonian and the adjoint equation of the original problem (Eqs. 16.68–16.70) now reduce to

$$H = (G\{F,Q\}M\{L\} - R\{F,Q\}N\{W\})p_W - cF, \qquad (16.79)$$

and

$$\frac{dp_W}{dt} = p_W R \frac{\partial N}{\partial W}, \qquad (16.80)$$

respectively. The resulting temperature set-point sequence (for January 1986) is compared with the optimal solution in Fig. 16.11. The new results deviate from the old ones only during the first eight days, when originally p_L was significantly different from zero (Fig. 16.10). The new set-points are always lower than (or equal to) the old ones, supporting the view that it was the need for leaf expansion that pushed the temperature beyond its optimal value for photosynthesis. Figure 16.11 also shows the results when p_W, rather than p_L, is set to zero. Here the decision process is "not aware" of the needs of dry matter production, except indirectly through the seasonal performance criterion. The recommended set-points are now higher than optimal, most of the time even considerably so.

Table 16.2
Performance obtainable with various optimization schemes for the heating months January through April and November through December. Lettuce in upstate New York [1]

Optimization scheme	Mean PC [2] ($\$\,m^{-2}$)	Mean loss ($\$\,m^{-2}$)
Sequential Control Search (reference)	56.54	0.00
Maximum Principle policies		
Complete Hamiltonian (both p_W and p_L)		
Best hourly set-points*	56.93*	–0.39*
Individual monthly policy	56.53	0.01
Best monthly policy	56.45	0.09
Best seasonal policy	56.31	0.23
p_W only	55.46	1.08
p_L only	51.83	4.71
Constant set-point policies		
Individual monthly policy	54.19	2.35
Best monthly policy	54.14	2.40
Best seasonal policy	54.03	2.51

[1] Policies are derived from data of 1983–1986 applied to same period. The price of energy is assumed to be zero. All results, except those marked with an asterisk, are for daily set-points.
[2] PC is sum of performance criteria for the six heating months.

Table 16.2 is a brief summary of the results of Seginer and McClendon (1992), who compared several suboptimal policies in conjunction with the lettuce model with two state variables. Line (1) displays the reference values, obtained by SCS. The next four lines show results obtained with the PMP, employing the complete Hamiltonian. Lines (2) and (3) give the results obtained with the best possible initial value of $\rho\{0\}$ for each individual month (4 years$\times$12 months$=$48 individual $\rho\{0\}$ values). Line (3) is for the standard problem, and shows results practically identical with the reference values. Line (2) represents the only non-standard case, where set-points were adjusted every hour (and not once a day). The result is better, but by less than 1%.

Lines (4) and (5) give results of more feasible schemes. To obtain the results of Line (5), the single best value of $\rho\{0\}$ for all years and all heating months (which turned out to be $6\,m^2$ [leaf] kg^{-1} [d.m.], was used. Line (4) was obtained by finding the best value of $\rho\{0\}$ for each calendar month separately (12 separate values of $\rho\{0\}$). The results show the latter scheme to produce better results, but both Line (4) and Line (5) suggest only a marginal loss over the reference value.

More might be lost if only a single state variable is retained in the Hamiltonian. Retaining just W results in a rather small loss, but retaining only L leads to a mean loss of 4.71 $\$\,m^{-2}$ per season, which is significant.

The results of three constant set-point policies are shown next. As one would expect, moving from the best set-point for each individual month ($4\times12=48$ different values) to the best set-point for each calendar month (12 values), and then to the single best constant set-point for all the months (1 value), reduces the performance of the system. The differences between the three policies are, however, very small compared to their mean deviation from the optimal policy.

In summary, the reduced Hamiltonian with $p_L=0$ produced results which are considerably better than with $p_W=0$. The results are even better than can be obtained with a constant set-point policy. Although the use of a reduced Hamiltonian is not necessary when the system model is simple, it may produce an acceptable solution when the model is large and a complete optimal solution is not feasible.

The lettuce problem is in many respects unrealistic. One aspect, namely the length of the growing period (here, 30 days) deserves a few more words. Most crops are harvested when ripe. Ripeness is one aspect of produce quality, an issue which did not, until now, receive much attention, because it is difficult to model. Ripeness, however, can be estimated reasonably well by the temperature exposure of the fruit or flower. Too little exposure (in degree days) may be taken to mean unripe fruit, and vice versa. To make the control algorithm "see" this information, the temperature exposure of organs of interest must be added as state variables to the system. The price of the produce is then made a function of its state. For lettuce, which is harvested once only, the termination value (Eqs. 16.30, 16.66) becomes

$$h\{\boldsymbol{x}\{t_f\},\boldsymbol{e}\{t_f\}\} = k'\{W_s\{t_f\},I\{t_f\},t_f\}\,W_s\{t_f\}, \qquad (16.81)$$

where the unit price, k', is a function of size, W_s, (heads which are too small are not acceptable); of quality, as measured by the temperature exposure, I; and of the fluctuating market price, through the absolute termination time, t_f. For a continuously harvested crop, such as tomato, quality may enter the Hamiltonian (16.31) through the running net-gains term, g. TOMGRO (Jones et al., 1991), a multi-state-variable tomato model, uses temperature integration to drive the development of the various organs. Some results obtained with TOMGRO are discussed in the next section.

HIGHER-ORDER MODELS AND APPLICATION

This section considers two topics for which results are just beginning to emerge: solutions for multi-state-variable models and the use of neural networks for greenhouse control. The need for simplified application methods is greater where large models are involved, because complete on-line optimization may become unwieldy. The approach to these loosely related subjects is, at the moment, rather intuitive.

Tomato model with multiple state variables

Many current crop models have multiple state variables. Optimal control solutions for these models cannot be obtained with the maximum principle method due to the "curse of dimensionality". An attempt by Shina and Seginer (1989) to use a certain modification of the maximum principle approach in conjunction with a multi-state-variable tomato model (TOMGRO; Jones et al., 1991) was not very successful. In another attempt, the so-called "reduced" maximum principle approach (p. 374) produced acceptable results for a two-state-variable crop. These, however, were still inferior compared to the results of the SCS method (p. 372) for the same problem. In this section, application of the latter two methods to a multi-state-variable model will be described.

A daily temperature set-point problem, similar to the lettuce problem (p. 371), was formulated (Seginer, 1991) for a tomato crop represented by TOMGRO, which contains over 50 state variables. A typical season was about 8 months long. This combination makes it a large problem.

First, several SCS solutions were obtained under different a priori weather and control assumptions about the future. Practically identical solutions were obtained

Table 16.3
Performance obtainable with various optimization schemes for a tomato crop in Ohio weather. Season is January to August, 1963

Optimization scheme		Mean PC [1] ($\$ m^{-2}$)	Mean loss ($\$ m^{-2}$)
SCS			
(1)	Initial set-point 30°C (reference)	17.17	0.00
(2)	Second iteration	17.20	−0.02
Sensitivity to assumptions about the future			
(3)	Initial set-point 10°C	17.13	0.04
(4)	Initially 1962 weather	17.15	0.02
RPM			
(5)	Both p_N and p_W	16.74	0.43
(6)	p_N only	16.68	0.49
(7)	p_W only	9.51	7.66

[1] PC is performance criterion.

by solving just once-through (Table 16.3, Lines 3, 4). This gives support to the tentative conclusion that the performance criterion has no local maxima in a not-too-narrow neighborhood around the optimal solution. The reason for this, as already mentioned, is probably the lack of operational storage in TOMGRO.

Next, the RMP method was applied with two state variables in the reduced Hamiltonian (16.77). The two variables were total dry matter, as a measure of production (mainly a function of light; carbon dioxide enrichment was not considered), and number of nodes, as a measure of development (mainly a function of temperature). Three combinations of these two state variables were tried: together, and each one of them separately.

A summary of the RMP results is presented in Table 16.3, Lines (5) to (7). When both adjoint variables are retained in the Hamiltonian, the loss, compared to the SCS solution, is about 0.4 $\$ m^{-2}$ (out of about 17 $\$ m^{-2}$). About the same loss is maintained when only p_N is retained. When, however, only p_W is retained, the loss jumps to 7.66 $\$ m^{-2}$. This shows that, for the present problem, a proper development rate is considerably more important than a proper dry matter accumulation. It may also be that development (N) is better correlated with other important state variables than is dry weight.

It turns out that the value of p_N is almost constant throughout the season (Seginer, 1989), which may be explained as follows: during the first part of the season promotion of development is required in order to initiate production as early as possible, whereas toward the end of the season its main purpose is to ripen green fruit already formed.

The preliminary evidence of this section suggests that the SCS may be a useful optimization tool for greenhouse problems with multi-state-variable crop models, provided that operational storage is not a significant factor. It is also evident from the RMP results that only few of the 50 odd state variables are really independent of each other. This should not be surprising, since with control of air temperature alone, it is not possible to regulate *independently* the growth of individual organs (state variables). Hence they become well correlated among themselves.

Application with neural networks

A control policy is an instrument which transfers past experience into future operations. This past experience can be captured in expert rules, adjoint-variable values, neural networks or any other method of mapping available input (state, weather and prices) into control decisions. Examples of using adjoint-variable and constant set-point policies were given above (e.g., pp. 374–376). The use of neural networks (NNs) for this purpose has been outlined above (p. 360) and will now be illustrated. This particular example is based on a model with a single state variable and a single set-point, but the method can be applied in an almost identical manner to optimal solutions of multi-state-variable systems with many simultaneous decisions. The main issue in either case is the selection of a set of inputs which is both easy to obtain and relevant to the problem.

Seginer and Sher (1992) obtained optimal solutions for a single-state-variable version of the lettuce model (Eq. 16.17). These solutions were used to train several NNs, each one with a different set of inputs. The available inputs were as follows:

State	1	Day from transplanting
	2	Dry matter at end of previous day
Climate	3	Mean daytime temperature for season (month)
	4	Mean daily solar radiation for season (month)

Weather	5	Next day's mean outside temperature
	6	Standard deviation of next day's temperature
	7	Next day's solar radiation
Prices	8	Expected price at harvest time

Five neural-network configurations were tried, each one (except NNA) missing a different input category. Table 16.4 summarizes the amount of disagreement between the set-points generated by the neural network and the optimal set-points, in terms of the root mean square (rms) of differences between corresponding set-points for both training and verification sets of data.

Table 16.4
Root mean square (rms) of differences between optimal and NN-recommended temperature set-points for networks with different inputs [1]

NW	Input categories				Input nodes	Δ_T (K)	Δ_V (K)
NNA	state	climate	weather	prices	8	1.5	3.3
NNB		climate	weather	prices	6	2.1	4.3
NNC	state		weather	prices	6	1.4	2.5
NND	state	climate		prices	5	2.9	5.4
NNE	state	climate	weather		7	2.3	3.0

[1] Abbreviations: NW, Network; Δ_T, Δ_T: rms difference between generated and optimal setpoints for Training and Verification, respectively.

The table shows clearly that weather is the most critical input and climate (future weather) is the least important one, in agreement with the account on pp. 372–374. The importance of state and price information is somewhere in the middle.

NNC and the best constant set-point were used as control policies in simulation studies over a range of weather conditions. Compared to the optimal solutions, the loss of performance with the NNC was at most equivalent to 2% of the yield, while losses with the constant set-point policy reached up to 10%. The best policy in that study was the adjoint-variable policy (not shown here) for which losses were negligible.

CONCLUSION

From the material presented above it is clear that the application of dynamic optimization methods to the environmental control of greenhouses is in its infancy. It may develop into a practical tool or it may not, depending on its contribution to the profit or to less tangible benefits to the grower. The probability of success depends on several factors.

(1) A system model should be available which mimics faithfully those features of the real system which are essential to the problem under consideration. The central component of the system, namely the crop, is difficult to model, particularly if produce *quality* is to appear explicitly in the performance criterion. Optimal control with the wrong model may lead to results which are inferior to those obtained with perhaps less sophisticated, but more robust, control policies.

(2) The potential gain over present practices must be significant and consistent. An average net gain of the order of 5% may be enough to justify transition to dynamic optimal control. Some results suggesting an improvement of this order have been shown (Figs. 16.3, 16.7, 16.8, 16.10; Table 16.2), and much more cannot be expected in the neighborhood of an optimum, where changes of the performance criterion are small.

(3) Situations which are strongly influenced by unknown future events (such as fluctuating market prices) are difficult to handle under any control scheme. The small extra benefits of a more sophisticated control method may be overshadowed by the large effects of uncontrollable factors. It should be remembered, however, that factors which affect the income do not always affect the control policy (e.g. quota constraint, pp. 364–365).

(4) The computational effort by which a control policy is to be obtained must not be excessive. Preferably a control policy should only be computed (off-line at Level 3) once every season, and on-line computations should be reduced to the minimum.

This chapter treated the optimal control problem on the assumption that a perfect crop model (or its equivalent in terms of set-point constraints) is available. The solutions themselves were deterministic, but the effect of applying a policy based on past years to a future year was briefly discussed in conjunction with Figs. 16.11 and 16.12 and Tables 16.2 and 16.4. The loss of performance turned out to be minor when a policy previously obtained was applied directly (using a Hamiltonian or a constant set-point), and negligible when the previous policy was only applied to the far future (as with the sequential control search). It is not clear whether these encouraging results will be repeated for other greenhouse-control problems – for instance, problems with significant operational storage. It is quite clear that the latter problems would require good weather forecast for the near future (decision period, Fig. 16.2). However, once this is provided, the transferability of control policy from past seasons to a future season may not necessarily be more difficult than for problems without operational storage.

Note that a seasonal *resource* quota, such as a water quota, presents difficulties which are similar to those introduced by an operational storage. Here, however, the characteristic time is as long as the whole season. This makes weather forecasts only slightly more informative than climatological statistics. Fortunately, *greenhouse* production is seldom, if ever, constrained by limited physical inputs (except the natural ones, such as solar radiation), but one could imagine a situation where fuel would be rationed, or, more likely, time would be limiting. The latter constraint may be critical for crops whose market price peaks on certain dates (such as flowers for Christmas: Seginer, 1989, his Fig. 2). Lack of information about the weather during the few days just before harvest makes intelligent decisions at the beginning of the season difficult.

Systems with real operational storage, such as solar greenhouses (p. 357), have not been considered, although much can be gained by intelligent control of the available storage. A partial solution to a problem involving a significant temperature-exposure "storage" has, however, been worked out. The results (Fig. 16.9) show the improvement to be gained by proper manipulation (Fig. 16.6) of this storage. It should be remembered that these results were obtained with a constant value of p_I. Improved results are to be expected if p_I is changed to meet the conditions imposed by the changing weather.

Optimal control should be more beneficial at high cultivation intensities: greenhouses with carbon dioxide enrichment have more to gain from a sophisticated control system than greenhouses without carbon dioxide generators. There are two reasons for this: first, the cost of the additional resource (CO_2); and, second, the need to balance between more conflicting requirements. A clear example is the trade-off between the effects of ventilation on temperature and carbon dioxide. As Fig. 16.5 shows, this may lead to simultaneous ventilation and carbon dioxide enrichment, requiring a delicate control of the ventilation flux.

There seem now to be two possible approaches which will make suboptimal control of large systems attainable: the sequential control search (SCS) and the reduced maximum principle (RMP). Both methods retain the complete model for simulation purposes, resulting in correct evaluation of the performance criterion. SCS has the big advantage that its computing time does not increase exponentially with the number of state variables. It does not require an explicit policy statement to be computed off-line first, and it does not seem sensitive to assumptions about the far future (beyond the characteristic time of the operational storage, if one exists). The results presented in Fig. 16.11 and Table 16.3 suggest that the SCS is capable of finding the *exact* optimal solution, at least to a certain class of problems. If a large operational storage is present in the system, RMP may prove to have an advantage over SCS.

An explicit control policy which is a function of the state of the system may be used to guide crop modelers in their work. For example, Fig. 16.11 shows that the inclusion of leaf area in the Hamiltonian contributes very little to the performance of the system. This suggests that, following a careful sensitivity analysis, leaf-area expansion may receive less experimental attention than dry weight accumulation.

As indicated in the introduction, the material of this chapter stresses methodology and qualitative results, the former more than the latter. For the benefit of the reader who is interested in at least some practical guidelines, here is a summary of such results.

(1) An optimal controller requires a very small number of parameters to be set compared to the number of set-point options of modern greenhouse computers. The single parameter policy of pp. 361–364 was capable of producing an unending sequence of set-points for practically any number of control devices.
(2) The optimal policy may have to be adjusted for different economic constraints. The production intensity was shown on pp. 364–365 to depend on the expected market price of the produce under the area-limiting constraint, but not under the quota constraint.
(3) An operational storage, whether physical or abstract, may be used to decrease the gap between supply and demand of inputs: the temperature-exposure "storage" of pp. 365–370 was used to shift energy demand to periods when it could be supplied more cheaply.

The usefulness of dynamic optimization based on crop models is a test case for the modelling effort as a whole. These models are usually of minimal physiological or other scientific contents; the primary reason for their development is their eventual use in management decisions. Much effort has already been put into the application of crop models to field-crop management (e.g., Swaney et al., 1983; Hornbaker and Mapp, 1988; Wilkerson and Mishoe, 1990). There are certain differences in emphasis between field and greenhouse crops. For instance, the aerial environment of field crops cannot be economically manipulated. As a result, decisions in field-crop management are considerably less frequent, making the SCS more attractive than the PMP. Irrigation decisions, which are major resource-allocation decisions for outdoor extensive crops, are minor in the greenhouse. This allows modelers of greenhouse crops to disregard water-stress considerations (but not the cooling effect of transpiration).

A major problem of applying crop models to management decisions in greenhouse production is the large number of greenhouse crops. None of these crops can receive as much research attention as a major field crop like corn or cotton. Fortunately, the generic form of most crop models is similar and the effort of adding a new crop may not be as overwhelming as seems at a first sight. The first candidates, as the examples in this chapter suggest, are vegetable crops, where the quantities are large and problems of quality are not as delicate as with flowers.

Hopefully the material presented in this chapter will help to enhance the understanding of how control decisions should be made, and support crop modelers in their efforts to develop ever more useful models.

ACKNOWLEDGEMENT

The first draft of this paper was written when I was on sabbatic leave at the Department of Agricultural Engineering, University of Georgia, Athens, GA, USA, in 1990. The current version was completed in 1992, when the author was at Technion, Haifa, Israel.

NOTATION

Main symbols (SI units)[3]

A	leaf area ratio of leaf increment (m^2 [leaf] kg^{-1} [leaf d.m.])
A	greenhouse area (m^2 [soil])
a	coefficient (Eq. 16.11) (m^2 [soil] m^{-2} [leaf])
B	gross photosynthesis response to temperature (kg [d.m.] kg [air] kg^{-1} [CO_2] m^{-2} [soil] s^{-1})
b	coefficient (Eq. 16.12) (K^{-1})
C	carbon dioxide enrichment rate (kg [CO_2] m^{-2} [soil] s^{-1})
C	seasonal total running costs (\$ m^{-2} [soil])
$\mathbf{c}$	vector of unit prices (#)
c	unit price of energy (\$ J^{-1} [heat])
c_p	specific heat of air at constant pressure (J [heat] kg^{-1} [air] K^{-1})
$\mathbf{e}$	vector of exogenous variables (boundary conditions) (#)
F	heat flux from heater (W [heat] m^{-2} [soil])
F	seasonal fixed (rent) costs (\$ m^{-2} [soil])
$\mathbf{f}$	system function (equation of motion) (#)
G	closed-canopy gross or net photosynthesis rate (kg [d.m.] m^{-2} [soil] s^{-1})
g	instantaneous gain function (\$ m^{-2} [soil] s^{-1})
H	Hamiltonian function (\$ m^{-2} [soil] s^{-1})
h	termination value function (\$ m^{-2} [soil])
I	temperature exposure (temperature integral) (K s)
i	normalized temperature exposure
J	seasonal performance criterion (\$ m^{-2} [soil])
J'	performance criterion averaged over season (\$ m^{-2} [soil] s^{-1})
k	unit market price for crop (\$ m^{-2} [soil])
k'	unit market price for crop (\$ kg^{-1} [leaf d.m.])
L	leaf area index (m^2 [leaf] m^{-2} [soil])

[3] Some of the symbols are used for more than one entity. Absolute temperatures and temperature differences are indicated by K. Temperatures measured from the freezing point of water by °C. [d.m.] denotes dry matter. - denotes a dimensionless entity. # indicates that the units depend on the particular item under consideration.

M	gross photosynthesis correction for partial canopy (on leaf area basis)
M'	gross photosynthesis correction for partial canopy (on dry matter basis)
m	number of exogenous (boundary) variables
N	effective crop mass for respiration (kg [d.m.] m^{-2} [soil])
N	number of nodes
O	seasonal operating costs ($ m^{-2} [soil])
P	PAR (photosynthetically active radiation) flux (W [PAR] m^{-2} [soil])
$\boldsymbol{p}$	adjoint vector (#)
Q	transfer coefficient due to ventilation (W [heat] m^{-2} [soil] K^{-1})
Q	production quota (m^2 [soil] s^{-1})
q	heating efficiency of global solar radiation (J [heat] J^{-1} [solar])
R	respiration rate (kg [d.m.] kg^{-1} [d.m.] s^{-1})
r	fraction shoot to total dry matter (kg [d.m.] kg^{-1} [d.m.])
S	global solar radiation flux (W [solar] m^{-2} [soil])
s	number of state variables
s	switching function ($ J^{-1} [heat])
T	temperature (K)
t	time (s)
U	convective–radiative heat transfer coefficient (W [heat] m^{-2} [soil] K^{-1})
$\boldsymbol{u}$	vector of control variables (#)
υ	rate of fixed (rent) costs ($ m^{-2} [soil] s^{-1})
W	mass of dry matter per unit soil area (kg [d.m.] m^{-2} [soil])
X	carbon dioxide concentration (kg [CO_2] kg^{-1} [air])
$\boldsymbol{x}$	vector of state variables (#)
Z	transformed W (Eq. 16.35) (kg [d.m.] m^{-2} [soil])
α	coefficient (Eq. 16.14) (kg [d.m.] kg [air] kg^{-1} [CO_2] m^{-2} [soil] K^{-1} s^{-1})
β	coefficient (Eq. 16.14) (kg [d.m.] kg [air] kg^{-1} [CO_2] m^{-2} [soil] K^{-2} s^{-1})
γ	coefficient (Eq. 16.19) (m^2 [leaf] kg^{-1} [d.m.] K^{-1})
δ	coefficient (Eq. 16.19) ((m^2 [leaf] m^2 [soil])kg^{-1} [d.m.] W^{-1} [PAR])
ε	photosynthetic efficiency (kg [d.m.] J^{-1} [PAR])
ζ	P/S ratio (J [PAR] J^{-1} [solar])
η	dry matter to carbon dioxide ratio (approx. 30/44) (kg [d.m.] kg^{-1} [CO_2])
θ	length of decision interval (s)
λ	leaf area ratio of whole plant (m^2 [leaf] kg^{-1} [leaf d.m.])
μ	level of solar radiation (brightness)
ρ	$p_W\{0\}/p_L\{0\}$ratio (m^2 [leaf] kg^{-1} [d.m.])
τ	transmissivity of greenhouse to solar radiation

Subscripts

c	closed-loop
d	decision (when attached to t)
d	deviation (when attached to I and i)
f	termination (final) value
I	temperature exposure
i	inside greenhouse
j	an arbitrary term
L	leaf area
m	number of exogenous (boundary) variables
n	minimal value
o	outside greenhouse
R	reference value
r	root (when attached to W)
r	recommended (when attached to T)
s	number of state variables (when attached to x)
s	shoot (when attached to W)
t	threshold (switching) value
W	dry weight
x	maximal value
Z	transformed dry weight

Superscripts

′ modified

Other notation

{} enclose the arguments of functions
≡ identity

Acronyms

LAR leaf area ratio
NN neural network
PMP Pontryagin's maximum principle
PAR photosynthetically active radiation
RMP "reduced" maximum principle
SCS sequential control search

REFERENCES

Aikman, D.P. and Picken, A.J.F., 1989. Wind-related temperature setting in glasshouses. *J. Hortic. Sci.*, 64(6): 649–654.

Bailey, B.J., 1988. Control strategies to enhance the performance of greenhouse thermal screens. *J. Agric. Eng. Res.*, 40: 187–198.

Bailey, B.J. and Seginer, I., 1989. Optimum control of greenhouse heating. *Acta Hortic.*, 245: 512–518.

Bellman, R.E. and Dreyfus, S.E., 1962. *Applied Dynamic Programming.* Princeton University Press, Princeton, NJ, 363 pp.

Broner, I. and Lambert, J., 1989. Optimal scheduling of irrigation machines, I. Model development. *J. Irrig. Drain. Eng.*, 115(5): 862–879.

Bruggink, G.T. and Heuvelink, E., 1987. Influence of light on the growth of young tomato, cucumber and sweet pepper plants in the greenhouse: effect on relative growth rate, net assimilation rate and leaf area ratio. *Sci. Hortic.*, 31: 161–174.

Chalabi, Z.S., 1992. A generalized optimization strategy for dynamic CO_2 enrichment in a greenhouse. *Eur. J. Oper. Res.*, 59: 308–312.

Challa, H. and Brouwer, P., 1985. Growth of young cucumber plants under different diurnal temperature patterns. *Acta Hortic.*, 174: 211–217.

Challa, H. and Schapendonk, A.H.C.M., 1986. Dynamic optimalization of CO_2 concentration in relation to climate control in greenhouses. In: H.Z. Enoch and B.A. Kimball (Editors), *Carbon Dioxide Enrichment of Greenhouse Crops,* Vol. I. CRC, Boca Raton, FL, pp. 147–160.

Challa, H. and van de Vooren, J., 1980. A strategy for climate control in greenhouses in early winter production. *Acta Hortic.*, 106: 159–164.

Challa, H. and van Straten, G., 1991. Reflections about optimal climate control in greenhouse cultivation. In: *Mathematical and Control Applications in Agriculture and Horticulture,* IFAC Workshop Series 1991, No. 1, pp. 13–18.

Critten, D.L., 1991. Optimization of CO_2 concentration in greenhouses: a modelling analysis for the lettuce crop. *J. Agric. Eng. Res.*, 48: 261–271.

Cros, M.-J. and Martin-Clouaire, R., 1991. Determination of climatic setpoints by constraint satisfaction. In: *Mathematical and Control Applications in Agriculture and Horticulture,* IFAC Workshop Series 1991, No. 1, pp. 271–276.

Davis, P.F. and Hooper, A.W., 1991. Improvement of greenhouse heating control. *IEE Proc. D,* 138(3): 249–255.

de Koning, A.N.M., 1988. More efficient use of base load heating with a temperature integrating control program. Effect on development, growth and production of tomato. *Acta Hortic.*, 229: 233–237.

de Koning, A.N.M., 1990. Long-term temperature integration of tomato. Growth and development under alternating temperature regimes. *Sci. Hortic.*, 45: 117–127.

Erwin, J.E. and Heins, R.D., 1990. Temperature effects on lily development rate and morphology from the visible bud stage until anthesis. *J. Am. Soc. Hortic. Sci.*, 115(4): 644–646.

Gal, S., Angel, A. and Seginer, I., 1984. Optimal control of greenhouse climate: methodology. *Eur. J. Oper. Res.*, 17: 45–56.

Garcia, C.E., Prett, D.M. and Morari, M., 1989. Model predictive control: theory and practice – a survey. *Automatica,* 25(3): 335–348.

Gary, C., 1988a. Relation entre température, teneur en glucides et respiration de la plante entière chez la tomate en phase végétative. *Agronomie,* 8(5): 419–424.

Gary, C., 1988b. Un modèle simple de simulation des relations microclimat–bilan carboné chez la tomate en phase végétative. *Agronomie,* 8(8): 685–692.

Gutman, P.-O., Lindberg, P.-O., Ioslovich, I. and Seginer, I., 1993. A non-linear optimal greenhouse control problem solved by linear programming. *J. Agric. Eng. Res.*, 55: 335–351.

Hannan, J.J. and Holley, W.D., 1978. *Greenhouse Management.* Springer, Berlin, 530 pp.

Heissner, A., Augustin, P., Schmidt, M., Reinisch, K. and Puta, H., 1986. Optimierung von Energieaufwand und Ertrag bei der Produktion von Gewächshausgemüse auf der Grundlage von Ertragsmodellen. *Tag. Ber. Akad. Landwirtsch. Wiss. DDR,* 238: 55–67.

Heuvelink, E., 1989. Influence of day and night temperature on the growth of young tomato plants. *Sci. Hortic.*, 38: 11–22.

Heuvelink, E. and Challa, H., 1989. Dynamic optimization of artificial lighting in greenhouses. *Acta Hortic.*, 260: 401–412.

Hornbaker, R.H. and Mapp, H.P., 1988. A dynamic analysis of water savings from advanced irrigation technology. *West. J. Agric. Econ.*, 13(2): 307–315.

Hurd, R.G. and Graves, C.J., 1984. The influence of different temperature patterns having the same integral on the earliness and yield of tomatoes. *Acta Hortic.*, 148: 547–554.

Jacobson, B.K., Jones, P.H., Jones, J.W. and Paramore, J.A., 1988. Real-time greenhouse monitoring and control with an expert system. *Comp. Electron. Agric.*, 3: 273–285.

Jones, J.W., Mishoe, J.W. and Boote, K.J., 1984. SOYGRO: Soybean crop growth model. In: Vrinda Kumble (Editor) *Proc. of International Symposium on Minimum Data Sets for Agrotechnology Transfer.* ICRISAT, India, pp. 83–94.

Jones, J.W., Dayan, E., Allen, L.H., van Keulen, H. and Challa, H.,

1991. A dynamic tomato growth and yield model (TOMGRO). *Trans. ASAE,* 34(2): 663–672.

Jones, P., Jones, J.W. and Hwang, Y., 1990. Simulation for determining greenhouse temperature setpoints. *Trans. ASAE,* 33(5): 1722–1728.

Kirk, D.E., 1970. *Optimal Control Theory. An Introduction.* Prentice-Hall, Englewood Cliffs, NJ, 452 pp.

Kozai, T., 1985. Ideas of greenhouse climate control based on knowledge engineering techniques. *Acta Hortic.,* 174: 365–373.

Kurata, K. and Eguchi, N., 1990. Machine learning of fuzzy rules for crop management in protected cultivation. *Trans. ASAE,* 33(4): 1360–1368.

Langhans, R.W., Wolfe, M. and Albright, L.D., 1981. Use of average night temperatures for plant growth for potential energy savings. *Acta Hort.,* 115: 31–37.

Leutscher, K.J. and Vogelezang, J.V.M., 1990. A crop growth simulation model for operational management support in pot plant production. *Agric. Sys.,* 33: 101–114.

Lieth, J.H. and Pasian, C.C., 1991. A simulation model for the growth and development of flowering rose shoots. *Sci. Hortic.,* 46: 109–128.

Ludolph, D. and Hendriks, L., 1989. Dynamische Regelstrategien – Grenzen und Möglichkeiten. *Zierpflanzenbau,* 4/89: 150–154.

Marsh, L.S. and Albright, L.D., 1991. Economically optimum day temperatures for greenhouse hydroponic lettuce production, Part I. A computer model. *Trans. ASAE,* 34(2): 550–556; Part II. Results and simulations. *Trans. ASAE,* 34(2): 557–562.

McClelland, J.L., Rumelhart, D.E. and PDP Research Group, 1986. *Parallel Distributed Processing: Explorations in the Microstructure of Cognition,* Vol. 1. MIT Press, Cambridge, MA, 547 pp.

Pontryagin, L.S., Boltyanskii, V.G., Gamkrelidze, R.V. and Mishchenko, E.F., 1962. *The Mathematical Theory of Optimal Processes.* Wiley–Interscience, New York, 360 pp.

Seginer, I., 1988. Optimal control techniques for energy conservation in controlled environments. In: *FAO CNRE 3rd Technical Consultation on Energy Conservation,* Helsinki, Finland, 1988, *FAO CNRE Bull.,* 23: 160–169.

Seginer, I., 1989. Optimal greenhouse production under economic constraints. *Agric. Sys.,* 29: 67–80.

Seginer, I., 1990. Expected usefulness of CO_2 enrichment and shading in a desert climate. *Acta Hortic.,* 268: 149–156.

Seginer, I., 1991. Optimal greenhouse temperature trajectories for a multi-state-variable tomato model. In: *Mathematical and Control Applications in Agriculture and Horticulture,* IFAC Workshop Series 1991, No. 1, pp. 73–79.

Seginer, I. and McClendon, R.W., 1992. Methods for optimal control of the greenhouse environment. *Trans. ASAE,* 35(4): 1299–1307.

Seginer, I. and Sher, A., 1992. *Neural nets for greenhouse climate control,* ASAE Paper No. 92 7013, 18 pp.

Seginer, I., Angel, A., Gal, S. and Kantz, D., 1986. Optimal CO_2 enrichment strategy for greenhouses. A simulation study. *J. Agric. Eng. Res.,* 34: 285–304.

Seginer, I., Shina, G., Albright, L.A. and Marsh, L.S., 1991. Optimal temperature setpoints for greenhouse lettuce. *J. Agric. Eng. Res.,* 49: 209–226.

Shina, G. and Seginer, I., 1989. Optimal management of tomato growth in greenhouses. *Acta Hortic.,* 248: 307–313.

Spitters, C.J.T., van Keulen, H. and van Kraalingen, D.W.G., 1989. A simple and universal crop growth simulator: SUCROS87. In: R. Rabbinge, S.A. Ward and H.H. van Laar (Editors), *Simulation and Systems Management in Crop Protection.* PUDOC, Wageningen, Netherlands, Ch. 4.1, pp. 147–181.

Stanghellini, C. and van Meurs, W.Th.M., 1992. Environmental control of greenhouse transpiration. *J. Agric. Eng. Res.,* 51: 297–311.

Swaney, D.P., Mishoe, J.W., Jones, J.W. and Boggess, W.G., 1983. Using crop models for management: impact of weather characteristics on irrigation decisions in soybeans. *Trans. ASAE,* 26(6): 1808–1814.

Udink ten Cate, A.J., Bot, G.P.A. and van Dixhoorn, J.J., 1978. Computer control of greenhouse climates. *Acta Hortic.,* 87: 265–272.

van Emmerik, P., 1990. Temperatuur correctie programma biedt voordelen. *Weekblad Groenten en Fruit,* (23 March): 46–47.

van Henten, E.J. and Bontsema, J., 1991. Optimal control of greenhouse climate. In: *Mathematical and Control Applications in Agriculture and Horticulture,* IFAC Workshop Series 1991, No. 1, pp. 27–32.

Wilkerson, G.G. and Mishoe, J.W., 1990. PESTDEC: a simulation approach to insecticide application decision making for soybeans. *Agric. Sys.,* 32: 143–157.

Haifa, 1992

SYSTEMATIC LIST OF GENERA

All taxa mentioned in the text are here included in their taxonomic position. Division of genera is to the Family level in higher plants, to the Order level in arthropods and vertebrates, to the Phylum, Subphylum or Class elsewhere. Names of intermediate taxa mentioned are inserted immediately below the higher taxon to which they are ascribed, unless they are themselves divided. Within Class, Order or Family, lower taxa are listed in alphabetical order.

MONERA (PROKARYOTA)

BACTERIA (EUBACTERIA)

Agrobacterium
Bacillus
Clavibacter
Clavibacterium
Corynebacterium
Erwinia
Pseudomonas
Xanthomonas

FUNGI

ASCOMYCOTA

ERYSIPHALES

Alternaria
Aschersonia
Ascochyta
Botrytis
Cercospora
Cladosporium
Colletotrichum
Coniothyrium
Didymella
Diplocarpon
Erysiphe
Fusarium
Heteropatella
Heterosporium
Letosphaeria
Marssonina
Mycosphaerella
Oidium
Paecilomyces
Phialophora
Phoma
Phomopsis
Phyllosticta
Pleospora
Pyrenochaeta
Sclerotinia
Septoria
Sphaerotheca
Stemphylium
Trichoderma

BASIDIOMYCOTA

Armillaria
Calyptella
Phragmidium
Puccinia
Rhizoctonia
Uromyces
Ustilago

CHYTRIDIOMYCOTA

Olpidium

DEUTEROMYCOTA

Ampelomyces
Beauveria
Fulvia
Gliocladium
Metarhizium
Sclerotium
Thielaviopsis
Ulocladium
Verticillium

OOMYCOTA

Albugo
Bremia
Peronospora
Phytophthora
Pseudoperonospora
Pythium

PLASMODIOPHOROMYCOTA

Spongospora

ZYGOMYCOTA

Zoophthora

PLANTAE

BRYOPHYTA

MUSCI

Sphagnum

PTERIDOPHYTA

FILICOPSIDA

Adiantaceae
Adiantum

Polypodiaceae
Nephrolepis
Platycerium
Rumohra

PINOPHYTA (GYMNOSPERMAE)

Araucariaceae
Araucaria

Podocarpaceae
Podocarpus

MAGNOLIOPHYTA (ANGIOSPERMAE)

LILIOPSIDA (MONOCOTYLEDONES)

Agavaceae
Cordyline
Dracaena
Sansevieria
Yucca

Amaryllidaceae
Hippeastrum
Narcissus

Araceae
Aglaonema
Anthurium
Caladium
Dieffenbachia
Epipremnium
Monstera
Philodendron
Spathiphyllum
Syngonium
Zantedeschia

Arecaceae
Butia
Chamaedorea
Chrysaidocarpus
Howea

Bromeliaceae
Aechmea
Ananas
Cryptanthus

Iridaceae
Crocus
Freesia
Gladiolus

Liliaceae
Asparagus
Hyacinthus
Lilium
Muscari
Tulipa

Marantaceae
Calathea
Maranta

Orchidaceae
Cattleya
Cymbidium

Poaceae
Hordeum
Saccharum
Sorghum
Triticum
Zea

MAGNOLIOPSIDA (DICOTYLEDONES)

Acanthaceae
Aphelandra
Crossandra
Justicia
Pachystachis

Apiaceae
Apium
Petroselinum

Aquifoliaceae
Ilex

Araliaceae
Brassaia
Dizygotheca
Hedera
Schefflera
Tupidanthus

Asteraceae
Aster
Callistephus
Chrysanthemum
Cichorium
Cineraria
Cosmos
Dahlia
Dendranthema
Gerbera
Helianthus
Lactuca
Liatris

Senecio
Solidaster
Xanthium
Zinnia

Balsaminaceae
Impatiens

Begoniaceae
Begonia

Brassicaceae
Brassica
Matthiola
Raphanus

Cactaceae
Mammillaria
Schlumbergera

Campanulaceae
Campanula

Caryophyllaceae
Dianthus
Gypsophila

Chenopodiaceae
Beta

Cornaceae
Cornus

Crassulaceae
Kalanchoë

Cucurbitaceae
Benincasa
Citrullus
Cucumis
Cucurbita

Ericaceae
Rhododendron

Euphorbiaceae
Codiaeum
Euphorbia

Fabaceae
Glycine
Lathyrus
Medicago
Phaseolus
Pisum

Gentianaceae
Eustoma
Exacum

Geraniaceae
Pelargonium

Gesneriaceae
Achimenes
Episcia
Saintpaulia
Sinningia
Streptocarpus

Hydrangeaceae
Hydrangea

Malvaceae
Hibiscus

Moraceae
Ficus

Myrsinaceae
Ardisia

Myrtaceae
Chamaelaucium
Eucalyptus

Nyctaginaceae
Bougainvillea

Onagraceae
Fuchsia

Oxalidaceae
Oxalis

Piperaceae
Peperomia

Pittosporaceae
Pittosporum

Plumbaginaceae
Limonium

Polemoniaceae
Phlox

Polygonaceae
Rumex

Primulaceae
Cyclamen
Primula

Punicaceae
Punica

Ranunculaceae
Delphinium

Rosaceae
Fragaria
Prunus
Rosa

Rubiaceae
Coffea

Rutaceae
Citrus

Scrophulariaceae
Antirrhinum
Calceolaria

Solanaceae
Browallia
Capsicum
Lycopeersicon
Solanum

Tropaeolaceae
Tropaeolum

Verbenaceae
Clerodendron

Violaceae
Viola

Vitaceae
Vitis

ANIMALES

PROTOZOA

NEMATODA

Heterorhabditidae
Steinernematidae

Heterorhabditis
Meloidogyne
Steinernema

ARTHROPODA

ARACHNIDA

ACARI

Eriophyidae
Laelapidae
Phytoselidae
Pycmotidae
Tarsonemidae
Tenuipalpidae
Tetranychidae

Aculops
Amblyseius
Brevipalpus
Bryobia
Euseius
Hypoaspis
Iphiseius
Metaseiulus
Phytonemus
Phytoseiulus
Phytoseius
Polyphagotarsonemus
Pyemotes
Rhizoglyphus
Schizotetranychus
Seiulus
Septanychus
Steneotarsonemus
Tarsonemus
Tenuipalpus
Tetranychus
Typhlodromus
Tyrophagus

COLLEMBOLA

Onychiuridae
Sminthuridae

Archorutes
Bourletiella
Folsomia
Hypogastrura
Onychiurus
Orchosella
Tomocerus

CRUSTACEA

Malacostraca

ISOPODA

Armadillidium
Oniscus
Porcellio
Trichoniscus

DIPLOPODA

Diploiulus
Nopoiulus
Orthomorpha
Oxidis

INSECTA

COLEOPTERA

Coccinellidae
Curculionidae
Elateridae
Scolytidae

Agriotes
Anthonomus
Chilocorus
Cholus
Cryptolaemus
Delphastus
Diorymerellus
Hippodamia
Lindorus
Listroderes
Nephus
Otorhynchus
Pantomorus
Stethorus
Xyleborus

DERMAPTERA

Forficulidae
Forficula

DIPTERA

Anthomyiidae
Agromyzidae
Asilidae
Cecidomyiidae
Ephydridae
Mycetophilidae
Sciaridae
Syrphidae
Tephritidae
Tipulidae

Amauromyza
Aphidoletes
Braydesia
Chromatomyia
Citibaena
Clinodiplosis
Dasineura
Delia
Liriomyza
Merodon
Philophylla
Psilla
Rhopolmyia
Scatella
Sciara
Therodiplosis
Tipula

HEMIPTERA

Anthocoridae
Miridae

Calocoris
Dichphus
Lygcoris
Lygus
Macrolphus
Ortus
Poecilocapsus

HOMOPTERA
Coccoidea

Aleyrodidae
Aphididae
Cicadellidae
Coccidae
Diaspididae
Eriococcidae
Margorodidae
Pseudococcidae

Aleurotulus
Aleyrodes
Amorphophora
Aonidiella
Aphis
Aspidiotus
Aulacaspis
Aulacorthum
Bemisia
Brachycaudus
Brevicoryne
Ceratoophis
Chromphalus
Chrysomphalus
Coccus
Diaspis
Dysaphis
Eriocuccus
Eucalymnatus
Florinia
Furcadiaspis
Geococcus
Hauptidia
Hemiberlesia
Howardia
Icerya
Idiopterus
Lepidosaphes
Macrosiphoniella
Macrosiphium
Micromyzus
Myzophis
Myzus
Nasonovia
Orthezia
Parlatoria
Parthenolecanium
Phenocuccus
Pinnaspis
Planococcus
Pseudococcus
Pulvinaria
Rhizoecus
Rhopalosiphonius
Rhopalosiphum
Saisettia
Toxoptera
Temnaspidotus
Trialeuroides
Trionymus
Zygina

HYMENOPTERA
Aphelinidae
Braconidae
Encyrtidae
Eulophidae
Eurytomidae
Formicidae
Myrmaridae
Pteromalidae
Tenthredinidae

Ametastegia
Anagrus
Aphelinus
Aphidius
Aphytis
Bombus
Ceraniscus
Chrysocharis
Cystogaster
Dacnusa
Diaeretiella
Diglyphus
Diversineervus
Encarsia
Encyrtus
Ephedrus
Eretomocerus
Eurytoma
Formica
Hartigia
Lepomatidea
Leptomastix
Linepithema
Lysiphlebus
Metaphycus
Moncrotonus
Opius
Pgnalio
Praon
Prenolepis
Psudoaphycus
Scutellista
Solenopsis
Tetracnemoidea
Tetramorium
Thripobius
Trioxys

ISOPTERA
Termitidae
Reticulotermes

LEPIDOPTERA
Gelechiidae
Geometridae
Gracillaridae
Noctuidae
Pterophoridae
Tortricidae

Adoxophyes
Agrotis
Argyrotaenia
Autographica
Cacoecimorpha
Carpitarsia
Choristoneura
Chrysodeixis
Clepsis
Epichoristodes
Euplexia
Gracillaria
Heliothis
Lacanobia
Mamesta
Naenia
Noctua
Peridroma
Phlogophora
Phlyctaenia
Phthorimaea
Platynota
Platyptilia
Plusia
Sabulodes
Spodoptera

NEUROPTERA
Chrysopidae

Chrysoperla

THYSANOPTERA
Thripidae

Anaphothrips
Echinothrips
Frankliniella
Heliothrips
Hercinothrips
Liothrips
Microcephalothrips
Parthenothrips
Scirtothrips
Taeniothrips
Thrips

SYMPHYLA
Scutigerellidae

Scutigerella

MOLLUSCA
GASTROPODA
STYLOMMATOPHORA

Arionidae
Helicidae
Limacidae
Zonitidae

Agriolimax
Arion
Helix
Limax
Milax
Oxychilus
Zonitoides

AUTHOR INDEX[1]

[1] Page references to text are in roman type, to bibliographical entries in italics.

SYSTEMATIC INDEX

GENERAL INDEX

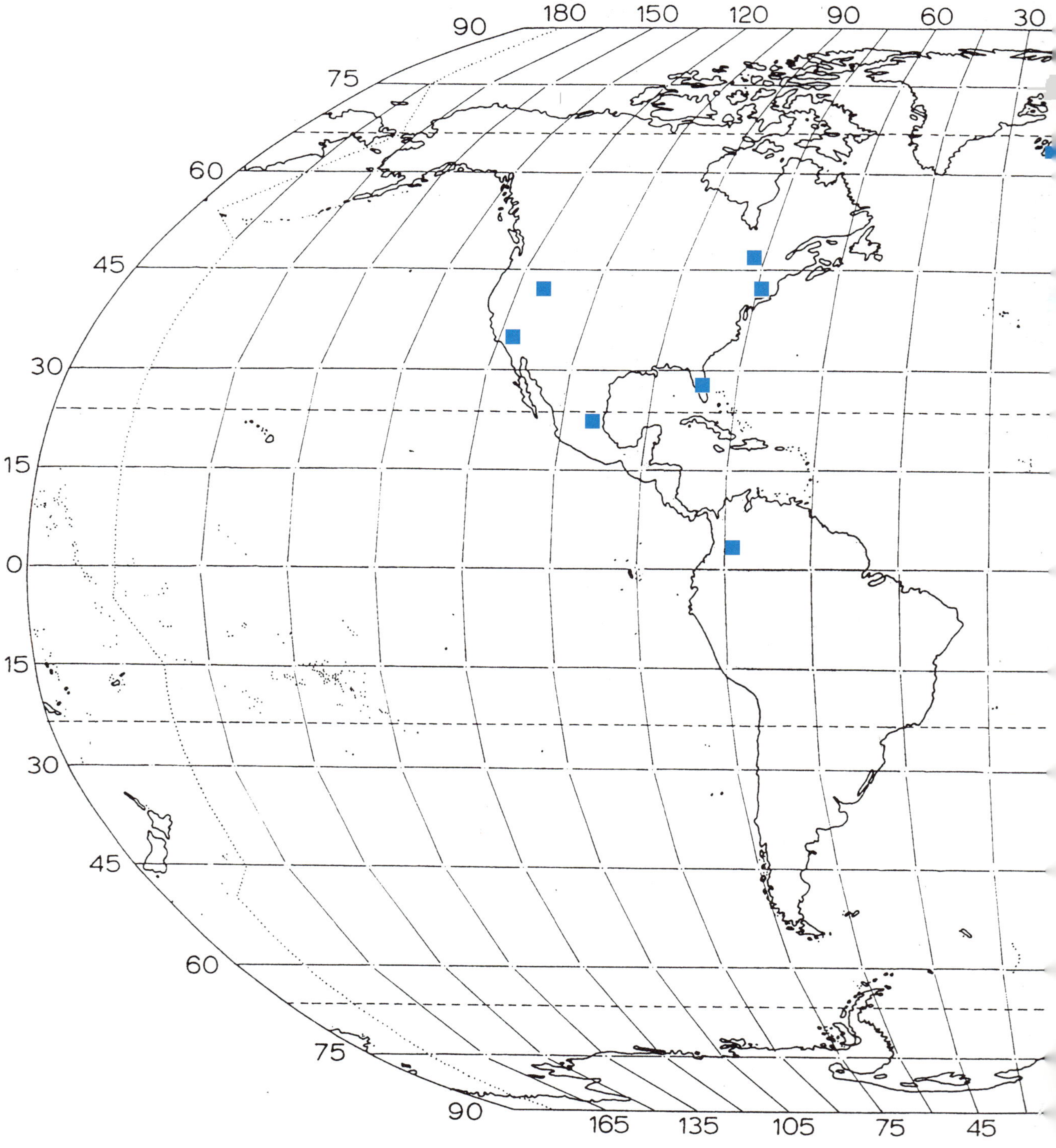

REGIONS IN WHICH THERE A
GREENHOUSES: DETAILS OF N
ARE GIVEN IN TABLE 1.1

Historically the major greenho
wealthy population centers in
for year-round crop productio
vation has spread to regions w
climatic conditions where skill